Lecture Notes in Civil Engineering 526

Lecture Notes in Civil Engineering (LNCE) publishes the latest developments in Civil Engineering—quickly, informally and in top quality. Though original research reported in proceedings and post-proceedings represents the core of LNCE, edited volumes of exceptionally high quality and interest may also be considered for publication. Volumes published in LNCE embrace all aspects and subfields of, as well as new challenges in, Civil Engineering. Topics in the series include:

- Construction and Structural Mechanics
- Building Materials
- Concrete, Steel and Timber Structures
- Geotechnical Engineering
- Earthquake Engineering
- Coastal Engineering
- Ocean and Offshore Engineering; Ships and Floating Structures
- Hydraulics, Hydrology and Water Resources Engineering
- Environmental Engineering and Sustainability
- Structural Health and Monitoring
- Surveying and Geographical Information Systems
- Indoor Environments
- Transportation and Traffic
- Risk Analysis
- Safety and Security

To submit a proposal or request further information, please contact the appropriate Springer Editor:

- Pierpaolo Riva at pierpaolo.riva@springer.com (Europe and Americas);
- Swati Meherishi at swati.meherishi@springer.com (Asia—except China, Australia, and New Zealand);
- Wayne Hu at wayne.hu@springer.com (China).

All books in the series now indexed by Scopus and EI Compendex database!

Lecture Notes in Civil Engineering 526

Lecture Notes in Civil Engineering (LNCE) publishes the latest developments in Civil Engineering—quickly, informally and in top quality. Though original research reported in proceedings and post-proceedings represents the core of LNCE, edited volumes of exceptionally high quality and interest may also be considered for publication. Volumes published in LNCE embrace all aspects and subfields of, as well as new challenges in, Civil Engineering. Topics in the series include:

- Construction and Structural Mechanics
- Building Materials
- Concrete, Steel and Timber Structures
- Geotechnical Engineering
- Earthquake Engineering
- Coastal Engineering
- Ocean and Offshore Engineering; Ships and Floating Structures
- Hydraulics, Hydrology and Water Resources Engineering
- Environmental Engineering and Sustainability
- Structural Health and Monitoring
- Surveying and Geographical Information Systems
- Indoor Environments
- Transportation and Traffic
- Risk Analysis
- Safety and Security

To submit a proposal or request further information, please contact the appropriate Springer Editor:

- Pierpaolo Riva at pierpaolo.riva@springer.com (Europe and Americas);
- Swati Meherishi at swati.meherishi@springer.com (Asia—except China, Australia, and New Zealand);
- Wayne Hu at wayne.hu@springer.com (China).

All books in the series now indexed by Scopus and EI Compendex database!

Guangliang Feng

Editor

Proceedings of the 10th International Conference on Civil Engineering

 Springer

Editor
Guangliang Feng
Wuhan Institute of Rock and Soil Mechanics
Chinese Academy of Sciences
Wuhan, China

ISSN 2366-2557   ISSN 2366-2565 (electronic)
Lecture Notes in Civil Engineering
ISBN 978-981-97-4354-4   ISBN 978-981-97-4355-1 (eBook)
https://doi.org/10.1007/978-981-97-4355-1

This Springer imprint is published by the registered company Springer Nature Singapore Pte Ltd.
The registered company address is: 152 Beach Road, #21-01/04 Gateway East, Singapore 189721, Singapore

If disposing of this product, please recycle the paper.

Preface

Civil engineering is closely related to people's life, and the quality of civil engineering directly affects people's life and personal safety and affects the development of society to a great extent. Therefore, it is very necessary to analyze the development status of civil engineering and to improve the problems existing in the development of civil engineering.

This book contains the proceedings of the 10th International Conference on Civil Engineering (ICCE 2023) which was held on December 23, 2023, as a hybrid conference (both physically and online via Zoom) at Nanchang Institute of Technology in Nanchang, China. The conference is hosted by Nanchang Institute of Technology and Civil Engineering Academy of Jiangxi Province, co-organized by Journal of Rock and Soil Mechanics, Beijing Engineering Management Science Institute, Key Laboratory for Safety of Water Conservancy and Civil Engineering Infrastructure in Jiangxi Province, Key Laboratory of Sichuan Province for Road Engineering, Southwest Jiaotong University. More than 180 participants were able to exchange knowledge and discuss the latest developments at the conference. The book contains 83 peer-reviewed papers, selected from more than 320 submissions and ranging from the theoretical and conceptual to strongly pragmatic and addressing industrial best practice.

The book shares practical experiences and enlightening ideas from civil engineering and will be of interest to researchers and practitioners of civil engineering everywhere.

Guangliang Feng

Preface

Civil engineering is closely related to people's life, and the quality of civil engineering directly affects people's life and personal safety and affects the development of society to a great extent. Therefore, it is very necessary to analyze the development status of civil engineering and to improve the problems existing in the development of civil engineering.

This book contains the proceedings of the 10th International Conference on Civil Engineering (ICCE 2023) which was held on December 23, 2023, as a hybrid conference (both physically and online via Zoom) at Nanchang Institute of Technology in Nanchang, China. The conference is hosted by Nanchang Institute of Technology and Civil Engineering Academy of Jiangxi Province, co-organized by Journal of Rock and Soil Mechanics, Beijing Engineering Management Science Institute, Key Laboratory for Safety of Water Conservancy and Civil Engineering Infrastructure in Jiangxi Province, Key Laboratory of Sichuan Province for Road Engineering, Southwest Jiaotong University. More than 180 participants were able to exchange knowledge and discuss the latest developments at the conference. The book contains 83 peer-reviewed papers, selected from more than 320 submissions and ranging from the theoretical and conceptual to strongly pragmatic and addressing industrial best practice.

The book shares practical experiences and enlightening ideas from civil engineering and will be of interest to researchers and practitioners of civil engineering everywhere.

Guangliang Feng

Contents

About the Editor

Guang-Liang Feng PhD, Professor, is working at Wuhan Institute of Rock and Soil Mechanics, Chinese Academy of Sciences. His main research is deep rockmass mechanics and underground engineering safety. He has been selected in several provincial and ministerial level talent programs, such as Outstanding Youth Program of Hubei Province, Top Young Talent Program of Hubei Province, and Youth Talent Support Project of China Association for Science and Technology.

He served as editorial board member of about ten mainstream SCI/EI journals. He presided three projects of National Natural Science Foundation of China and one sub-project of National Key Research and Development. Authored/co-authored more than 100 SCI/EI papers which have been cited more than 3000 times and six of them are ESI highly cited and hot papers as the first/corresponding author. Authorized more than 40 international and national invention patents. As the main draftsman, he compiled one International Society for Rock Mechanics Suggested Method and two energy industry standards of the People's Republic of China codes. Five provincial-level construction methods were compiled. Meanwhile, two first prizes of technological inventions awards (ranked 1st and 3rd), one first prize of natural science award (ranked 2nd), and two first prizes of technological progress award (ranked 3rd and 5th) in provincial and ministerial level were won.

Aiming at the scientific problem of rockburst mechanism, warning, and mitigation in deep tunnel, a technology and device for rockmass fracture high-precision perception was invented, the development evolution and mechanism of rockburst were revealed, and a quantitative warning theory and software platform of rockburst were established. The theory and technology have been listed into six international and industry standards and applied to major deep projects, such as Sichuan-Tibet railway and hydropower station in Pakistan.

Experimental Study on Parameters of Hardening Soil Model with Small Strain Stiffness for Muddy Silty Clay and Silty Sandy Soil in Yangtze Floodplain Area

Yaning Wang$^{(\boxtimes)}$, Desheng Wei, Lei Zhai, Huilai Qin, Zheng Chen, and Yuxuan Zhu

China Construction Second Engineering Bureau Co. Ltd., Beijing 100160, China
`yaningwang11@gmail.com`

Abstract. To obtain test parameters for the hardening soil model with small strain stiffness (HSS) of the muddy silty clay and silty sandy soil in the Nanjing Yangtze River floodplain, the following tests were conducted: consolidation test, consolidated-drained triaxial test, triaxial loading and unloading test, resonant column test and other common geotechnical tests. These test parameters included reference tangent modulus Eoedref, reference secant modulus E50ref, reference unloading and reloading modulus Eurref, failure ratio Rf, etc. A standardized process for determining HSS model parameters of certain types of soils, including relevant laboratory tests procedures, calculations, and analyses, has also been discussed. The test findings conclude with a summary of the model parameters for the typical strata in the Yangtze River floodplain region, providing engineering reference value for the deep foundation excavation and other relevant projects in the Yangtze River floodplain area.

Keywords: Hardening Soil Model With Small Stain Stiffness · Triaxial Test · Oedometer Test · Resonant Column Test

1 Introduction

The soil layers in the Yangtze River floodplain area mainly consist of muddy silty clay in the upper floodplain and silty sand in the lower floodplain, which are widely distributed. These soils exhibit typical engineering properties such as low foundation bearing capacity, easy deformation after disturbance, and time-consuming for stability. In order to better understand the response behavior of these soils in engineering activities, samples of the muddy silty clay and silty sand in this area are collected and tested in the following three aspects:

- Physical property tests: natural gravity, liquid limit, plastic limit, plastic index;
- Mechanical property tests: consolidation test, consolidated-drained triaxial test, triaxial loading and unloading test;
- Dynamic property test: resonant column test.

G. Feng (Ed.): ICCE 2023, LNCE 526, pp. 1–15, 2024.
https://doi.org/10.1007/978-981-97-4355-1_1

2 Hardening Soil Model with Small Strain Stiffness (HSS)

The HSS model could reflect the nonlinear behavior of soil stiffness in the small strain range, which is one of the constitutive models of soft soil widely utilized in the numerical analysis of geotechnical engineering. The HSS model contains 11 standard Hardening Soil (HS) model parameters and 2 additional small strain parameters [1], which are specified as follows: $E_{50}{}^{ref}$ is the reference secant modulus, which is the secant slope corresponding to 1/2 destructive strength under the reference confining pressure; $E_{oed}{}^{ref}$ is the compression modulus of consolidation test under reference confining pressure; $E_{ur}{}^{ref}$ is the unloading and reloading modulus; m is the power index for stress-level dependency of stiffness; c' is the effective cohesion of soil mass; φ' is the effective internal friction angle of soil mass; Ψ is the dilatancy angle of soil mass; R_f is the failure ratio; v_{ur} is the Poisson's ratio for unloading and reloading; $G_0{}^{ref}$ is the initial shear modulus under the reference confining pressure; $\gamma_{0.7}$ is the shear strain threshold; P^{ref} is the reference stress; K_0 is the coefficient of lateral earth pressure at rest.

3 Test Content and Results

3.1 Soil Samples

The sampling site is located at No. 201, Yanjiang Road, Gulou District, Nanjing City, Jiangsu Province. The samples, which consist of muddy silty clay and silty sandy soil, were collected using thin-wall sampling method and preserved by wax-sealing, as shown in Fig. 1. The muddy silty clay on site has a thickness of 11.5 to 17.6 m. It is grayish black in color, flow plastic, sensitive to touch, with a smooth breaking section and relatively uniform particle size, and contains a large amount of organic matter and some shellfish fragments. The thickness of silty sandy soil on site ranges from 18 to 23.4 m. These two corresponding soil layers are typical strata in Nanjing Yangtze River floodplain area.

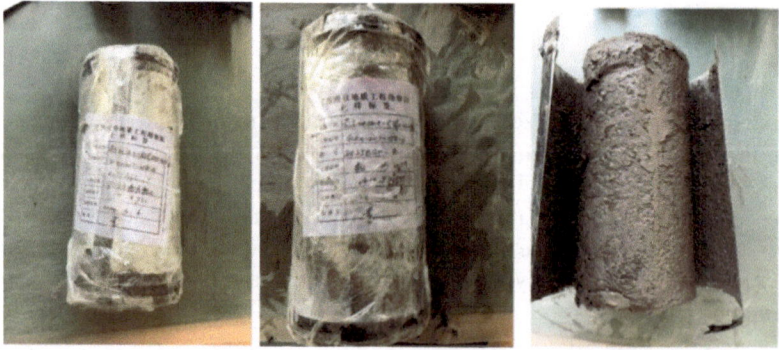

Fig. 1. Thin-wall samples from Nanjing Yangtze River floodplain area.

3.2 Physical Property Tests of Undisturbed Soil Samples

The basic physical properties of undisturbed soil include natural water content (w), density (ρ), specific gravity (G_S), Atterberg limits, etc. The test methods shall be in accordance with the Standard for Soil Test Methods (GB/T50123–2019) [2].

Water Content Test. The natural water contents of soil samples were measured using the drying method. 15–30 g representative samples were dried in oven for 48 h with the temperature controlled at 60–70 °C during the test. The samples were weighed before and after the drying process, respectively. The calculation formula of natural water content (w) is as follows:

$$w = (m/m_d - 1) \times 100 \tag{1}$$

where w is water content of the soil sample (%), m is the mass of wet soil (g), m_d is mass of dry soil (g).

Density Test. The densities of soil samples were obtained using the cutting ring method. The soil sample was trimmed into a soil column which is slightly larger than the diameter of the ring sampler using a wire saw. Then the ring sampler lubricated with Vaseline was pressed vertically to cut the soil until the soil sample extended out of the ring sampler. The remaining soils at both ends were cut off and leveled. The ring sampler were weighed before and after, respectively. The density calculation formula is as follows:

$$\rho = (m_1 - m_2)/V \tag{2}$$

where ρ is the density of the soil sample (g/cm³), m_1 is the total mass of wet soil and ring sampler (g), m_2 is mass of ring sampler (g), V is the volume of ring sampler (cm³), which is 60 cm³.

Specific Gravity Test. The specific gravities of soil samples were measured using the pycnometer method. Dry the pycnometer before filling it with a 100 ml pycnometer with 15 g of dried soil. Half-fill the bottle with pure water, then shake the pycnometer. The following day, place the bottle on the sand bath and boil it for at least an hour. Fill the pycnometer with pure water, and wait until the suspension on the upper part of the bottle is clear and the temperature of the suspension in the bottle is stable. Plug the bottle stopper to make excess water overflow from the capillary tube of the bottle stopper. Weigh the combined mass of the bottle, water, and soil after drying the water outside the bottle. Immediately after weighing, take a temperature reading of the water in the bottle. Check the total mass of the bottle and the water based on the relationship that has been drawn between the temperature and the total mass. The calculation formula is as follows:

$$G_S = \left(\frac{m_d}{m_1 + m_d - m_2} \right) G_{wt} \tag{3}$$

where G_S is the specific gravity of the soil particles, m_d is the mass of the dry soil (g), m_1 is the total mass of the bottle and water (g), m_2 is the total mass of the bottle, water and soil (g), G_{wt} is specific gravity of pure water at t °C, accurate to 0.001.

This test requires two parallel measurements and the allowable parallel difference is 0.02. The arithmetic mean should be used.

Atterberg Limits Test. The Atterberg limits tests were conducted with the FG-III liquid plastic limit test equipment, as shown in Fig. 2. The cone sinking depths in the soil at different water contents were measured, and the relationship diagram between the water content and the cone penetration depth is drawn, as shown in Fig. 3. The water content in the diagram corresponding to the cone penetration depth of 2 mm is the plastic limit water content, and the water content in the diagram corresponding to the cone penetration depth of 17 mm is the liquid limit water content. According to the plastic limit and liquid limit obtained from the test, the plasticity index of soil is defined as:

$$I_P = w_L - w_P \tag{4}$$

where I_P is the plasticity index, w_L is the liquid limits (%), w_P is the plastic limits (%).
The liquidity index of soil is calculated according to the following formula:

$$I_L = \frac{w - w_P}{I_P} \tag{5}$$

where I_L is the liquidity index, w is natural moisture content of soil (%).

Liquid limit, plastic limit and plasticity index comprehensively reflect clay content in soil. The plasticity index is higher when more clay is present. The liquidity index reflects the natural state of the soil. When the liquid index is greater than 1, the soil is in flowing state.

Fig. 2. Atterberg limits test.

The physical properties test results of muddy silty clay and silty sandy soil samples are shown in Table 1.

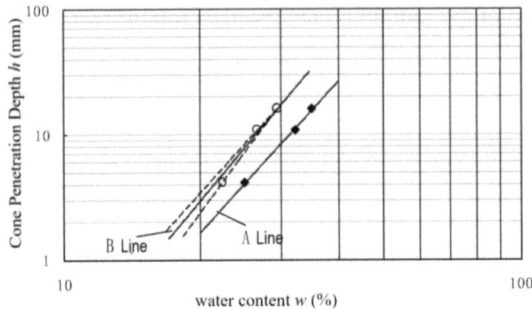

Fig. 3. Relationship between cone penetration depth and water content.

Table 1. Physical properties test results of undisturbed soil samples.

Soil sample	General parameters						Atterberg limits		
	Water content (%)	Density (g/cm³)		Specific gravity	Void ratio	Plastic limit (%)	Liquid limit (%)	Liquidity index	Plasticity index
		Wet	Dry						
	w	ρ	ρ_d	G_S	e_0	w_p	w_L	I_L	I_P
Muddy silty clay	40.1	1.81	1.29	2.71	1.094	20.7	41.4	/	21
Silty sandy soil	31.2	1.81	1.38	2.69	0.952	/	/	/	/

3.3   Mechanical Property Tests of Undisturbed Soil Samples

Consolidation Test. The consolidation and compression characteristics of undisturbed soil mainly include the compressibility coefficient, compressibility modulus, consolidation coefficient and other characteristic parameters. The test method to determine these parameters should follow the steps specified in the Standard for Soil Test Methods (GB/T50123–2019) [2]. The consolidation pressure grades are 12.5, 25, 50, 100, 200, 400, 800 and 1600 kPa, respectively. The relationship between deformation and time under each consolidation pressure shall be measured, based on which the compression and consolidation characteristic parameters are determined.

The consolidation test was carried out on an oedometer with a sample area of 30 cm² and an initial height of 2 cm. The applied loads were 12.5, 25, 50, 100, 200, 400, 800, 1600 kPa, each of which lasted for 24 h. The compression deformation of the sample under each level of load was measured, the void ratio of the sample under each level of load was then calculated and the compression curve ($e - p$ curve) was drawn in Fig. 4.

The initial void ratio of the sample is calculated as follows:

$$e_0 = \frac{\rho_w G_S (1 + 0.01 w_0)}{\rho_0} - 1 \tag{6}$$

where G_S is the specific gravity of soil particles, ρ_w is the density of water (g/cm³), ρ_0 is initial density of the sample (g/cm³), w_0 is the initial moisture content of the sample (%).

According to the compression curve ($e - p$ curve) obtained from the test, the compression coefficient and compression modulus were calculated. Generally, the compressibility of soil is evaluated by the compressibility coefficient within the pressure range from 100 to 200 kPa. The formula for calculating the compressibility coefficient and modulus is as follows:

$$a_v = \frac{e_i - e_{i+1}}{p_{i+1} - p_i} \tag{7}$$

where e_i is the void ratio under certain pressure p_i (kPa).

The compression modulus within the pressure range from 100 to 200 kPa is calculated using the following formula:

$$E_s = \frac{1 + e_0}{a_v} \tag{8}$$

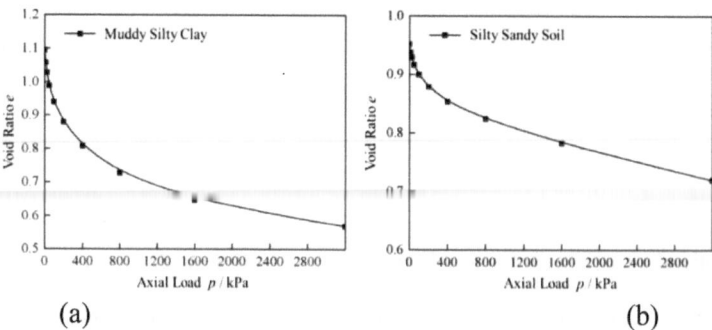

(a)   (b)

Fig. 4. Relationships between void ratio and load in consolidation test of: (a) Muddy silty clay; (b) Silty sandy soil.

The relationship between the axial load and the axial strain in the standard consolidation test of each sample is shown in Fig. 5. The axial deformation of each sample increases with the increment of the axial load applied. The trend of ε_d is consistent with the axial load p: the initial curve is relatively flat and the slope of the curve increases with the increment of the axial load applied. The muddy silty clay sample generated larger axial deformation than the silty sandy soil sample. The tangent slope of each curve at the load (p) of 100 kPa was calculated, which is the reference tangent modulus E_{oed}^{ref} when the reference stress P^{ref} is 100 kPa. The consolidation test results are shown in Table 2.

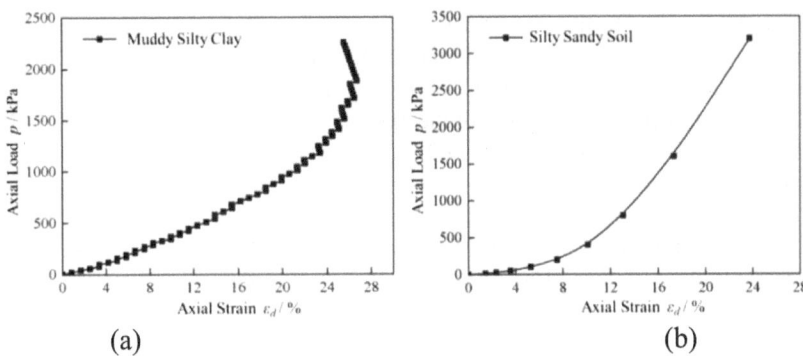

Fig. 5. Relationships between load and strain in consolidation test of: (a) Muddy silty clay; (b) Silty sandy soil.

Table 2. Consolidation test results of undisturbed soil samples.

Soil sample	Compressibility coefficient $a_{v1-2}\left(\text{MPa}^{-1}\right)$	E_{oed}^{ref} (Mpa)	E_{s1-2} (Mpa)
Muddy silty clay	0.60	3.80	3.50
Silty sandy soil	0.22	8.93	9.09

Consolidated-Drained Triaxial Test. The reference secant modulus $E_{50}{}^{ref}$, failure ratio R_f and strength parameters of soil mass (c' and φ') were obtained from the consolidated-drained triaxial test, which is conducted with triaxial shear apparatus. The sample is 39.1 mm in diameter and 80 mm in height.

The consolidated-drained triaxial test is mainly carried out in four steps: (1) back pressure saturation: In this study, the back pressure saturation of soil samples was directly conducted under the condition that the confining pressure was always 5 kPa higher than the back pressure; (2) B value check: The back pressure remained unchanged while the confining pressure was increased by 30 kPa, and then the pore water pressure coefficient B was measured. If B value is greater than 0.95, the sample is considered to be saturated. Otherwise, the back pressure saturation should be continued until B value is greater than 0.95; (3) Consolidation: The samples were consolidated under effective isotropic confining pressure σ_3; (4) Shear: The shear rate was set to be 0.008 mm/min. The test was terminated when the strain value of the sample reached 20%. During the test, the drainage volume during consolidation, axial load, shear displacement, volume change and other parameters were measured. Relevant calculations were conducted per specification requirements.

As shown in Fig. 6, the consolidated-drained triaxial tests were conducted with confining pressure of 100 kPa to obtain the relationship between the deviatoric stress q and axial strain ε_d of the samples. At the beginning of the test, the deviatoric stress of each sample increased with the increment of the axial strain. When the axial strain reached a certain range, the deviatoric stress tended to be unchanged or slightly decreasing. The deviatoric stress value corresponding to the axial strain of 15% is taken as the failure value q_f and the deviatoric stress value corresponding to the stable section or peak of the curve is taken as the asymptotic value q_a. Then the failure ratio of each sample R_f can be calculated $(R_f = q_f/q_a)$. The slope of the line connecting the origin and the point corresponding to $0.5q_f$ is the reference secant modulus E_{50}^{ref} of the sample.

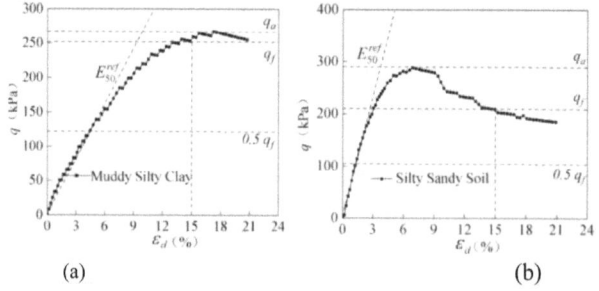

Fig. 6. Stress strain relationship curves from consolidated-drained triaxial tests for: (a) Muddy silty clay; (b) Silty sandy soil.

According to the obtained stress strain relationships under four different confining pressures of 100, 200, 300 and 400 kPa, four Mohr circles and corresponding failure envelope were drawn in Figs. 7 and 8, for muddy silty clay and silty sandy soil, respectively. The intercept and slope of each failure envelope are taken as the effective cohesion c' and effective friction angle φ'. The consolidated drained triaxial test results are shown in Table 3.

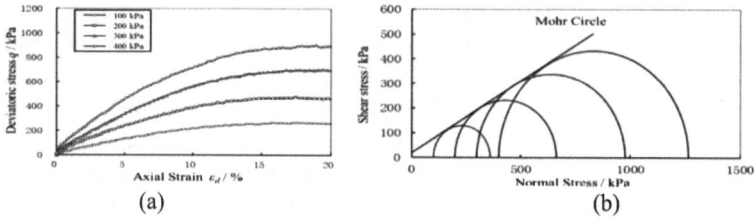

Fig. 7. (a) Stress strain relationship curves and (b) Mohr circles for muddy silty clay.

Triaxial Loading and Unloading Test. The consolidated-drained triaxial loading and unloading test is mainly carried out in four steps, the first three of which are the same as the consolidated-drained triaxial test mentioned in the previous chapter. The fourth step

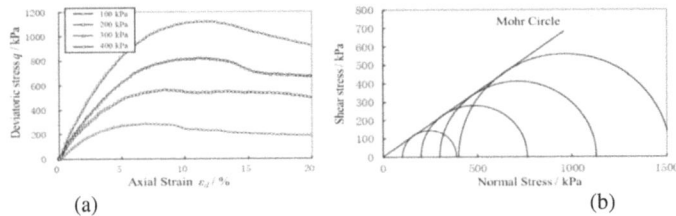

Fig. 8. (a) Stress strain relationship curves and (b) Mohr circles for silty sandy soil.

Table 3. Consolidated-drained triaxial test results of undisturbed soil samples.

Soil sample	E_{50}^{ref} (MPa)	R_f	Effective cohesion c' (kPa)	Effective friction angle φ' (degree)
Muddy silty clay	3.53	0.96	18.3	30.2
Silty sandy soil	6.9	0.73	1.6	35.4

is to apply axial loading, unloading and reloading to the sample. The load applied for the first time is about 40% of the expected failure deviatoric stress of the sample. When the load is added to the target value, the axial unloading shall be carried out immediately until the load is zero. Then the axial reloading shall be conducted until the load applied is about 60% of the expected failure deviatoric stress of the sample. The axial loading, unloading and reloading processes are all stress-controlled.

The consolidated-drained triaxial loading and unloading test can determine the reference unloading and reloading modulus E_{ur}^{ref} of soil. Figure 9 shows the stress strain relationship curves during consolidated-drained triaxial loading and unloading tests under the confining pressure of 100 kPa for muddy silty clay and silty sandy soil, respectively.

The deviatoric stress increased with the increment of axial strain at the initial stage of loading, and the curve is relatively flat; During the initial unloading stage, the axial strain slightly increased before decreasing, showing unloading rebound; During reloading process, the initial stress-strain curve is very steep, indicating that the axial strain changed little with the increment of deviatoric stress until the deviatoric stress reached the stress level before unloading. Then the stress-strain curve became relatively flat again and increased along the trend of the initial loading curve. In the process of unloading and reloading, the stress-strain curve showed a hysteresis loop. The average slope of the unloading and reloading curves, which connects the two endpoints of the hysteresis loop, serves as the reference unloading and reloading modulus E_{ur}^{ref} of the soil mass, as shown in Table 4.

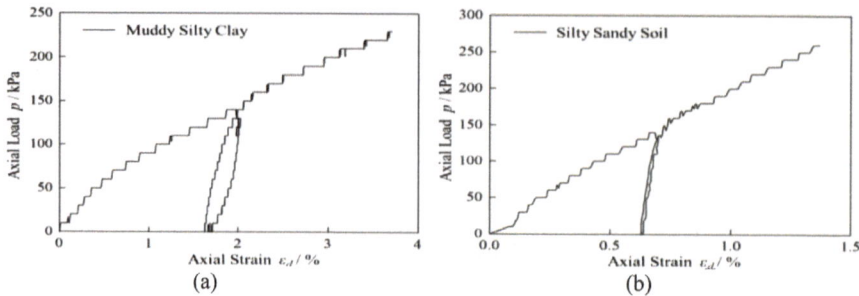

Fig. 9. Stress strain relationship curves during triaxial loading and unloading test for: (a) Muddy silty clay; (b) Silty sandy soil.

Table 4. Triaxial loading and unloading test results of undisturbed soil samples.

Soil sample	E_{ur}^{ref} (MPa)
Muddy silty clay	23.5
Silty sandy soil	44.8

3.4 Dynamic Property Tests of Undisturbed Soil Samples

Resonant Column Test and Data Analysis. GZZ-50 resonant column test equipment, which is shown in Fig. 10, has incorporated necessary data treatment in the instrument program based on the basic principle of resonant column test, and can output the test results directly. The relationships between confining pressure and maximum dynamic shear modulus, shear strain and shear modulus, shear strain and material damping ratio of remolded soil samples are obtained using resonant column tests.

(a)   (b)

Fig. 10. Resonant column test instrument.

It should be noted that the analysis results using different fitting methods for the dynamic stress-strain relationships of soil are not exactly the same. In this study, values of shear modulus and material damping ratio are obtained using resonant column test within shear strain (γ) less than 10^{-4}, and fitted using Wang and Stokoe (2022) model [3] for shear strain (γ) greater than 10^{-4}.

Based on Menq (2003) [4], the relationship between the maximum dynamic shear modulus and the effective average consolidation stress can be described using following empirical formula:

$$G_{max} = C \cdot P_a \cdot \left(\frac{\sigma_m}{P_a}\right)^n \tag{9}$$

where G_{max} is the maximum dynamic shear modulus, C is G_{max}/P_a value when $\sigma_m/P_a = 1$, n is the slope of the fitting line in log scale, P_a is the atmospheric pressure.

The vertical effective self-weight stress of muddy silty clay and silty sandy soil at the depth of sampling is 300 kPa and 400 kPa, respectively. According to Wang and Stokoe (2022) [3], the relationships between shear strain and shear modulus, shear strain and material damping ratio of uncemented soils can be described using Eqs. 10 and 11, respectively.

$$\frac{G}{G_{max}} = \frac{1}{\left(1 + (\gamma/\gamma_{mr})^a\right)^b} \tag{10}$$

where both "a" and "b" are the curvature parameters which control the shape of the shear modulus reduction curve, and γ_{mr} is the modified reference strain at which $G/G_{max} = 0.5^b$.

$$D = \frac{d \cdot (\gamma/\gamma_D)^c + D_{min}}{(\gamma/\gamma_D)^c + 1} \tag{11}$$

where "d" is the practical upper boundary limit of the material damping ratio, "c" is the curvature parameter which controls the shape of the material damping increment curve, D_{min} is the material damping ratio of soil in the small-strain range and γ_D is the reference shear strain at which $D = (d + D_{min})/2$.

According to the PLAXIS manual [1], the initial shear modulus G_0^{ref} of each layer under the reference confining pressure can be obtained using following formula:

$$G_0 = G_0^{ref} \left(\frac{c' \cot \varphi' - \sigma_3}{c' \cot \varphi' + P^{ref}}\right)^m \tag{12}$$

Dynamic Properties of Muddy Silty Clay and Silty Sandy Soil. Based on the data obtained from resonant column tests, the relationships between confining pressure and maximum dynamic shear modulus for muddy silty clay and silty sandy soil are shown in Table 5, and plotted in Fig. 11, respectively. The relationships between shear strain and shear modulus for muddy silty clay and silty sandy soil are plotted in Fig. 12, and the normalized shear modulus reduction curves are shown in Fig. 13. The relationships between shear strain and material damping ratio for muddy silty clay and silty sandy soil are plotted in Fig. 14. The small-strain stiffness input parameters are calculated and shown in Tables 5 and 6.

Table 5. Relationships between confining pressure and maximum dynamic shear modulus for undisturbed soil samples

Soil sample	Consolidation ratio K_c	Confining pressure (kPa)	G_{max} (MPa)	n	C
Muddy silty clay	1.0	100	36.01	0.739	354.9
		200	57.85		
		300	78.95		
		400	101.08		
Silty sandy soil	1.0	100	79.38	0.573	788.4
		200	115.72		
		300	147.78		
		400	175.64		

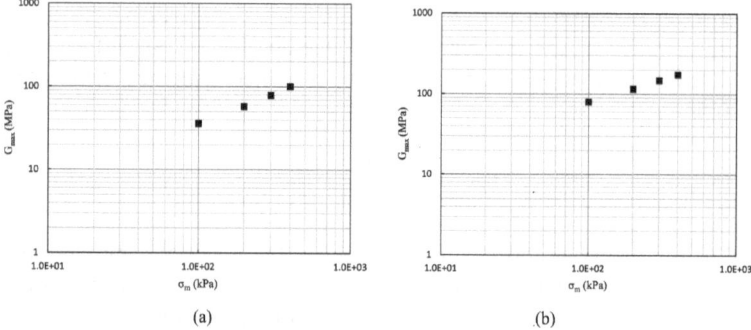

(a)   (b)

Fig. 11. Relationships between confining pressure and maximum dynamic shear modulus for: (a) Muddy silty clay; (b) Silty sandy soil.

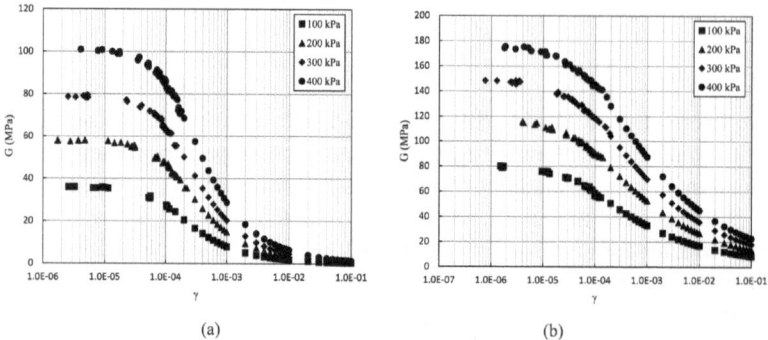

(a)   (b)

Fig. 12. Relationships between shear strain and shear modulus for: (a) Muddy silty clay; (b) Silty sandy soil.

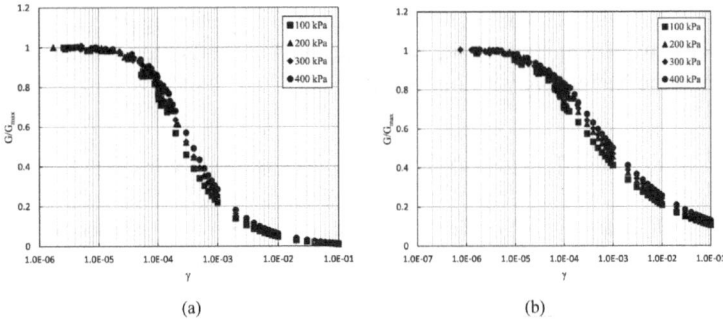

Fig. 13. Relationships between shear strain and normalized shear modulus for: (a) Muddy silty clay; (b) Silty sandy soil.

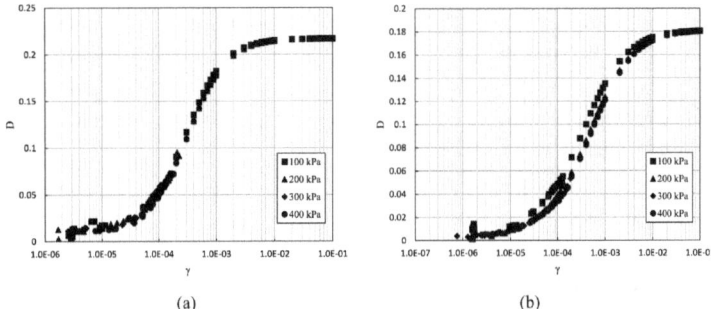

Fig. 14. Relationships between shear strain and material damping ratio for: (a) Muddy silty clay; (b) Silty sandy soil

Table 6. Small-strain stiffness additional input parameters for undisturbed soil samples.

Soil layer	Confining pressure σ_m (kPa)	G_0 (MPa)	G_0^{ref} (MPa) ($\sigma_m = 100$ kpa)	$\gamma_{0.7}$ (10^{-4}) ($\sigma_m = 100$ kpa)
Muddy silty clay	300	37.04	47.44	2.9
Silty sandy soil	400	76.92	90.45	3.5

4 Summary

In this study, based on the data obtained from physical property tests, mechanical property tests, and dynamic property test, all input parameters for the hardening soil model with small strain stiffness (HSS) of the muddy silty clay and silty sandy soil in the Nanjing Yangtze River floodplain area have been determined. The summary of all input parameters for HSS model is shown in the Table 7. A standardized process for determining HSS model parameters of certain types of soils, including relevant laboratory tests procedures, calculations, and analyses, has also been discussed. These research results provide a basis for the numerical simulation of engineering practice, such as deep foundation excavation modeling, site response analysis, and other relevant engineering calculation in the Yangtze River floodplain area.

Table 7. Summary of input parameters for HSS model.

Parameter	Physical Significance	Unit	Muddy Silty Clay	Silty Sandy Soil	Methods of Acquiring
E_{50}^{ref}	Reference secant modulus	kPa	3.53	6.9	Consolidated-drained triaxial test
E_{oed}^{ref}	Compression modulus	kPa	3.80	8.93	Consolidation test
E_{ur}^{ref}	Unloading and reloading modulus	kPa	23.5	44.8	Triaxial loading and unloading test
M	Power for stress-level dependency	/	0.61	0.56	Consolidated-drained triaxial test
c'	Effective cohesion of soil mass	kPa	18.3	1.6	Consolidated-drained triaxial test
φ'	Effective internal friction angle of soil mass	degree	30.2	35.4	Consolidated-drained triaxial test
Ψ	Dilatancy angle of soil mass	degree	0	5.4	Consolidated-drained triaxial test
R_f	Failure ratio	/	0.96	0.73	Consolidated-drained triaxial test
ν_{ur}	Poisson's ratio for unloading and reloading	/	0.33	0.30	Consolidated-drained triaxial test
G_0^{ref}	Initial reference shear modulus	MPa	47.44	90.45	Resonant column test
$\gamma_{0.7}$	Shear strain threshold	10^{-4}	2.9	3.5	Resonant column test
p^{ref}	Reference stress	kPa	100	100	/
K_0	Coefficient of lateral earth pressure at rest	/	0.5	0.42	$K_0^{nc} = 1 - sin\,\varphi'$

Acknowledgments. This research was supported by R&D Program of China Construction Second Engineering Bureau Co. Ltd (2021ZX000001, 2021ZX050001).

References

1. PLAXIS 2D, Material Models Manual, pp. 1–188 (2010)
2. Ministry of Water Resources of the People's Republic of China. Standard for geotechnical testing method, GB/T 50123-2019, Beijing, China (2019)
3. Wang, Y., Stokoe, K.H.: Development of constitutive models for linear and nonlinear shear modulus and material damping ratio of uncemented soils. ASCE-JGGE **148**(3), 04021192 (2022). https://doi.org/10.1061/(ASCE)GT.1943-5606.0002736
4. Menq, F.Y.: Dynamic properties of sandy and gravelly soils. Ph.D. Dissertation, The University of Texas at Austin, Austin (2003)

Technical Degradation Prediction of Bridge Components Based on Semi-Markov Degradation Model

Zhiqiang Shang[1,2], Yerong Hu[3(✉)], Xiangyin Chen[1,2], Shiyu Liu[3], and Zejun Zhang[1,2]

[1] Shandong High-Speed Group Co., LTD., Jinan 250101, China
[2] Shandong Provincial Key Laboratory of Highway Technology and Safety Assessment, Jinan 250101, China
[3] College of Highway Engineering, Chang'an University, Xi'an 710064, China
1106362734@qq.com

Abstract. To overcome the limitations of traditional Markov bridge degradation prediction models, which fail to consider the interactions between different component degradation mechanisms and struggle to accurately capture the true degradation conditions of bridge components, introducing an improved version of the traditional Markov model by incorporating the Weibull distribution. This enhancement results in a semi-Markov model that offers a probability distribution for predicting the technical condition of bridge components. Taking advantage of periodic inspection data from a highway section in Shandong Province, China. With this data, the states of bridge components are defined in the semi-Markov degradation model. The improved semi-Markov model integrates a two-parameter Weibull distribution and involves determining the parameters of the Weibull distribution, transition probability matrix, and state distribution vector. The semi-Markov degradation model, in contrast to commonly used Markov degradation models, accounts for both the state and duration of each state, resulting in significantly more accurate predictions of the degradation process of bridge components, achieving a prediction accuracy of 96%. The developed semi-Markov bridge degradation model facilitates the timely detection of changes in the technical condition of bridge components by updating the transition probability matrix according to variations in the duration of each state, thereby improving the efficiency of subsequent bridge maintenance decision-making.

Keywords: Bridge Component Degradation Model · Weibull Distribution · Semi-Markov · Monte Carlo · Bridge Component Condition Prediction

1 Introduction

The failure of bridge structures can result in substantial economic, social, and environmental costs [1–3]. Therefore, it is crucial to proactively understand the degradation patterns of bridge structural performance [4]. Long-term degradation models for bridges play a vital role in optimizing whole-life maintenance strategies. Existing models can generally be classified into mechanism-based models and statistical-based models.

G. Feng (Ed.): ICCE 2023, LNCE 526, pp. 16–27, 2024.
https://doi.org/10.1007/978-981-97-4355-1_2

Mechanism-based degradation models begin by analyzing the different factors and principles that impact the degradation of bridge structures. These models focus on the specific deterioration mechanisms of specific bridge components. Currently, there is significant research emphasis on predicting concrete carbonation, steel reinforcement corrosion, chloride ion intrusion, and other related aspects. For example, Dai [5] investigated the deterioration process of steel corrosion in concrete members to predict the life of the structure.

Statistical-based degradation models analyze historical data on structural degradation to capture the degradation patterns of bridges.By considering both the historical and current condition of the bridge, these models can predict its future state. Currently, the main prediction methods can be broadly categorized into deterministic models and stochastic models. Deterministic models establish relationships between factors influencing the degradation process and bridge condition levels using mathematical and statistical methods. The output of these models is represented by deterministic values and does not involve probabilities. These models assume that the relationship between future bridge condition levels is deterministic as time progresses [6]. Deterministic models can be classified into methods such as linear extrapolation, regression, and curve fitting [7–9]. For example, Zhou [10] used regression analysis to predict the technical condition of bridges. One limitation of deterministic models is that they do not consider the interaction between degradation mechanisms of different bridge components, such as the interaction between deck panels and joints [11]. On the other hand, stochastic models account for the randomness of the bridge degradation process. They are capable of capturing the stochastic nature and uncertainty associated with the degradation process. In stochastic models, the prediction of bridge condition levels is influenced by inherent uncertainties, such as traffic loads, weather conditions, material properties, exposure to deteriorating agents, etc. Stochastic models create probabilistic prediction models. The main types of stochastic models include Markov chain methods and reliability theory-based methods. These methods represent the state levels of bridge components or parts at a specific moment in the form of probability distributions rather than deterministic values. Statistical-based degradation models are more suitable for simulating the current degradation of bridges.

Stochastic models, which effectively account for the randomness of various factors, are currently the most widely used predictive models for structural performance. Prominent bridge management systems in the United States, such as PONTIS, BRIDGIT, BLCCA, and Bridge LCC, rely on Markov chains as the theoretical foundation for bridge degradation prediction models. Jiang [12], utilizing Markov chains, developed a performance prediction model for bridges based on status-level data of bridge degradation from the Indiana Department of Transportation. Pontis [13], as part of Bridge Management System (BMS) sponsored by the Federal Highway Administration, employed Markov chains to develop core component degradation models. In Markov chain degradation models, the performance level of a bridge structure is defined by a set of discrete states, and the bridge transitions from one state to another based on constant transition probabilities. Thompson [14] analyzed the California bridge dataset to quantify observed Markov transition probabilities. Similar work can be found in Puz [15], where they proposed a probabilistic model for the full lifecycle performance of structures based on homogeneous Markov processes to calculate degradation over time. Ng [16] introduced relaxation time homogeneity of Markov processes in a more general semi-Markov process stochastic model. At present, most of the bridge management systems in the United States still use Markov chain as the theoretical basis of bridge degradation prediction model to predict the change of bridge NBI rating [17].

However, Markov models have certain limitations. Firstly, they do not consider the age of the structure or components. Secondly, they assume that future states depend solely on the current state and do not account for past events such as bridge maintenance. Lastly, they may exhibit an initial rapid degradation rate. To overcome these limitations, additional model are required to describe the time distribution of each state.

To accurately understand the degradation patterns of bridge performance and enable scientific and precise bridge maintenance, this study relies on regular inspection data from a highway bridge in Shandong province. Considering the time and condition of bridge components, an improved Markov model based on the Weibull distribution is introduced to develop a semi-Markov model that closely reflects reality. This model is utilized to predict the distribution of technical condition levels for highway bridge components, providing necessary support for subsequent intelligent bridge maintenance decision-making.

2 Construction Process of Semi-Markov Bridge Component Degradation Model

Based on the theoretical research conducted on bridge performance degradation prediction in Shandong Province, the Markov model has been selected as the approach for predicting bridge performance degradation. The semi-Markov model, which is an extension of the Markov chain model, takes into account the state of the bridge and introduces the Weibull distribution to model the duration of each state. Using existing detection data and considering both time and state variables, the Markov model is enhanced by incorporating the Weibull distribution.

2.1 State Division of Semi-Markov Bridge Degradation Model

After preprocessing the technical condition data of bridge components, it is necessary to define the states of the components in the semi-Markov bridge component degradation model. Since Standards for Technical Condition Evaluation of Highway Bridges (JTGT H21-2011) does not provide a specific definition for the technical condition level of components, this chapter refers to the classification threshold table for the technical condition of bridge components in the Standards for Technical Condition Evaluation of Highway Bridges (JTGT H21-2011) and defines five semi-Markov states as follows:

- State 1: Technical condition score [95,100]
- State 2: Technical condition score [80,95)
- State 3: Technical condition score [60,80)
- State 4: Technical condition score [40,60)
- State 5: Technical condition score [0,40)

These defined states serve as a basis for modeling the degradation process of bridge components using the semi-Markov framework.

2.2 The Weibull Distribution Parameters of the Duration of Each State of the Bridge are Determined

The Standards for Technical Condition Evaluation of Highway Bridges (JTGT H21-2011) specifies that the duration of a bridge component in a particular state follows a random variable. $T_1, T_2, \ldots, T_{n-1}$ respectively represent the random variable of time of duration which bridge components is in each state i, the duration T_i in state i is specified to obey a two-parameter Weber probability distribution, meaning T_i Weibull$(b_i, \frac{1}{a_i})$. It is further defined that the duration of a component in any given state follows a two-parameter Weibull probability distribution.

$$F_i(t) = \Pr[T_i \le t] = 1 - e^{-(a_i t)^{b_i}} \tag{1}$$

$$S_i(t) = \Pr[T_i > t] = 1 - F_i(t) = e^{-(a_i t)^{b_i}} \tag{2}$$

$$f_i(t) = \frac{\delta F_i(t)}{\delta t} = a_i b_i (a_i t)^{b_i - 1} e^{-(a_i t)^{b_i}} \tag{3}$$

In the Eq. (1), $F_i(t)$ represents the Cumulative Density Function(CDF) corresponding to the duration of time t; $S_i(t)$ represents the Survival Function (SF), also known as the cumulative survival rate, which refers to the probability that the component remains in state i for a duration longer than t; $f_i(t)$ represents the Probability Density Function (PDF) corresponding to the duration of time t; $T_{i \to j}$ represents the time required for the process from state i to state j; $f_{i \to j}(T_{i \to j})$, $F_{i \to j}(T_{i \to j})$, $S_{i \to j}(T_{i \to j})$ represent the PDF, CDF, and SF of $T_{i \to j}$, respectively.

If the proportion of components in each state is known for each year, the parameters a_i and b_i can be calculated by fitting the survival function curve that corresponds to the duration. This fitting process allows for the estimation of the Weibull distribution parameters. Once the parameters are determined, a transition probability matrix can be established based on the historical data. This matrix represents the probabilities of transitioning from one state to another in each time step. By utilizing this transition probability matrix, it becomes possible to predict the future annual state distribution of bridge components. This approach provides a useful framework for forecasting the deterioration and performance of bridge components based on the available historical data and the estimated Weibull distribution parameters.

2.3 Modeling of Semi-Markov Bridge Component Degradation Process

Determination of State Transition Matrix. Under the assumption that the degradation process of bridge components is unidirectional and does not allow direct transitions from one state to another without passing through an intermediate state, a relatively simple Markov transition probability matrix can be derived. Therefore, the degradation process cannot jump from state 1 to state 3 without going through state 2. This leads to the relatively simple Markov transition probability matrix:

$$P^{t,t+1} = \begin{bmatrix} p_{11}^{t,t+1} & p_{12}^{t,t+1} & 0 & 0 & 0 \\ 0 & p_{22}^{t,t+1} & p_{23}^{t,t+1} & 0 & 0 \\ 0 & 0 & p_{33}^{t,t+1} & p_{34}^{t,t+1} & 0 \\ 0 & 0 & 0 & p_{44}^{t,t+1} & p_{45}^{t,t+1} \\ 0 & 0 & 0 & 0 & p_{55}^{t,t+1} \end{bmatrix} \tag{4}$$

In the Eq. (4), $P_{ij}^{t,t+1}$ represents the transition probability from state i at time t to state j at time $t+1$, with i and j ranging from 1 to 5. The matrix P captures the probabilities of transitioning between different states in the degradation process, reflecting the underlying dynamics of the bridge component degradation. In a semi-Markov process, the system is in a specific state for a random duration, and the distribution of states depends on the current state and the next state. To exclude two-state degradation, the time step (Δt) should be sufficiently small. In this chapter, we assume Δt to be 1 year. If a bridge component is in state 1 at time t, the conditional probability of transitioning to the next state in the next time step can be represented as:

$$P[X(t+1) = 2|X(t) = 1] = P^{1,2}(t) = \frac{f_1(t)}{S_1(t)} \tag{5}$$

In the Eq. (5), $f_1(t)$ represents the probability density function (PDF) of the duration of the component in state 1; $S_1(t)$ represents the survival function (SF), which is the probability that the duration exceeds t. If the process is in state 2 at time t, the conditional probability of transitioning to the next state in the next time step is:

$$P[X(t+1) = 3|X(t) = 2] = P^{2,3}(t) = \frac{f_{1\to2}(t)}{S_{1\to2}(t) - S_1(t)} \tag{6}$$

In the Eq. (6), $f_{1\to2}(t)$ represents the PDF which representing the sum of the duration time $T_{1\to2}$ in states 1 and 2; $S_{1\to2}(t) - S_1(t)$ represents that probability which $T_{1\to2} < t$ minus the probability which $T_1 < t$, which is equivalent to the condition $X(t) = 2$.

$$P[X(t+\Delta t) = i+1|X(t) = i] = P^{i,i+1}(t) = \frac{f_{1\to i}(t)\Delta t}{S_{1\to i}(t) - S_{1\to(i-1)}(t)} \quad (i = 1, 2, \ldots, n-1) \tag{7}$$

All conditional probabilities $P^{i,i+1}(t)$ can be calculated by the above Eqs. (4)–(7) to generate the transition probability matrix of the semi-Markov process.

The State Distribution Vector is Determined. In a semi-Markov chain, the probability of being in state i at time t is described by the state distribution vector $D(t)$, which represents the probabilities of being in each state at time t:

$$D(t) = \{d_1^t, d_2^t, \ldots d_n^t\} \quad (d_i^t \geq 0, \sum_{i=1}^{n} d_i^t = 1) \tag{8}$$

In the Eq. (8), d_i^t is the probability of a bridge component being in state i at time t. The state distribution vector in time (t + k) is

$$D(t + k) = D(t)P^{t,t+1}P^{t+1,t+2} \ldots P^{t+k-1,t+k} \tag{9}$$

By utilizing the transition probability matrix established in Sect. 3.4.1, the state distribution vector for the semi-Markov degradation process can be obtained at any given time t. By solving for the state distribution vector, the probabilities of each state at any specific time can be determined. This information allows for further calculations such as average duration in a particular state, state transition probabilities, and distribution of state transition times. These calculations provide valuable insights and serve as reference and basis for subsequent bridge maintenance decision-making.

3 Case Study

This paper takes the North Line of Shandong Expressway as a case study. There are a total of 8 interchanges along the route, and out of which 222 bridges have been selected Considering the completeness of historical inspection data, 40 prestressed concrete continuous girder bridges were chosen from these 222 bridges, comprising a total of 285 prestressed concrete box girder components. Inspection data from the regular inspection reports for these 40 bridges from 2008 to 2019 have been collected, processed, and analyzed. The semi-Markov degradation model has been utilized for prediction and analysis based on this data.

3.1 Semi-Markov State Division and Statistics

Based on the five semi-Markov states defined in Sect. 3.2, the component states are divided and aggregated. The percentages of components in each state for each year are calculated and summarized, as shown in Table 1.

Table 1. The percentage of each state of the component of the prestressed reinforced concrete bridges

Year	Percentage by State							
	1	2	3	4	5	1–2	1–3	1–4
2008	100.00%	0.00%	0.00%	0.00%	0.00%	100.00%	100.00%	100.00%
2009	66.67%	33.33%	0.00%	0.00%	0.00%	100.00%	100.00%	100.00%
2010	36.84%	28.77%	34.39%	0.00%	0.00%	65.61%	100.00%	100.00%
2011	29.82%	22.11%	16.84%	31.23%	0.00%	51.93%	68.77%	100.00%
2012	20.00%	7.37%	31.93%	35.79%	4.91%	27.37%	59.30%	95.09%
2013	19.30%	10.18%	28.42%	36.49%	5.61%	29.47%	57.89%	94.39%
2014	18.60%	4.91%	30.18%	39.30%	7.02%	23.51%	53.68%	92.98%
2015	18.60%	10.18%	20.00%	43.51%	7.72%	28.77%	48.77%	92.28%
2016	18.25%	2.81%	24.56%	46.32%	8.07%	21.05%	45.61%	91.93%
2017	17.54%	4.21%	20.35%	47.72%	10.18%	21.75%	42.11%	89.82%
2018	14.39%	2.81%	18.25%	52.28%	12.28%	17.19%	35.44%	87.72%
2019	13.68%	1.75%	19.65%	50.18%	14.74%	15.44%	35.09%	85.26%

3.2 Modeling of Semi-Markov Degradation Processes

According to Sect. 2.3, the Matlab program for constructing the semi-Markov degradation model is written, and the age of the main beam components of the prestressed reinforced concrete bridges in Table 1 and each state percentage are substituted and calculated, so that the Weibull distribution parameters of the duration of each state in Table 2 can be easily obtained.

Table 2. Weibull distribution parameters and fitting degree of each state duration

States	Weibill distribution parameters					
	a_i	b_i	SSE_i	R_i	AR_i	$RMSE_i$
States 1	0.3870	0.5745	0.0291	0.9615	0.9577	0.0539
States 1–2	0.2091	0.9990	0.0802	0.9232	0.9155	0.0896
States 1–3	0.1040	1.1127	0.0496	0.9264	0.9190	0.0704
States 1–4	0.0261	1.5009	0.0013	0.9562	0.9518	0.0112

In the Table 2, SSE_i represents the sum of the squared differences between the actual and predicted values, indicating the effect of random errors; R_i represents the multiple determination coefficient, measuring the success of the fit in explaining the variation in the data; $RMSE_i$ represents the square root of the mean square error, which is the expected value of the squared difference between the estimated and true parameter values.

Comparing the goodness-of-fit values from Table 2, including SSE_i, R_i, and $RMSE_i$, it can be concluded that the obtained goodness-of-fit. Overall, the fit is good, with small errors and high effectiveness. Considering $RMSE_i$ as the final measure of model accuracy, it can be inferred that the accuracy has reached above 96%. Based on these fitting results, the Weibull distribution parameters can be further used to predict the cumulative survival rate and corresponding probability density function distribution for the duration of different component states of bridge elements, as shown in Figs. 1 and 2.

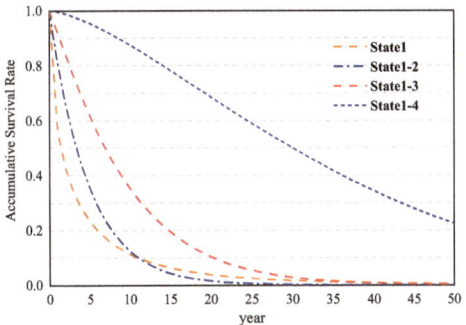

Fig. 1. Prediction of the cumulative survival rate of prestressed reinforced concrete components in each state duration

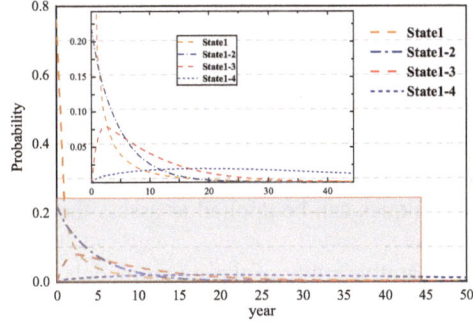

Fig. 2. Prediction of the cumulative survival rate of prestressed reinforced concrete components in each state duration

Based on the survival function of each state duration of the bridge components and the corresponding probability density function distribution, the conditional probability matrix of transformation from t to $(t + 1)$ year to the next state can be easily obtained from Eqs. (5), (6), (7) and (8), for example, the conditional probability matrix of transformation from 4th to 5th year is as follows

$$
p^{4,5} = \begin{bmatrix} 0.8154 & 0.1846 & 0 & 0 & 0 \\ 0 & 0.4219 & 0.5781 & 0 & 0 \\ 0 & 0 & 0.7155 & 0.2845 & 0 \\ 0 & 0 & 0 & 0.9565 & 0.0435 \\ 0 & 0 & 0 & 0 & 1 \end{bmatrix} \tag{10}
$$

From the probability matrix, we can obtain the state distribution vector for any given year of prestressed concrete components along the route. Let's assume the state distribution vector for the first year of prestressed concrete components within the route is:

$$
D_1 = \begin{bmatrix} 0.6667 & 0.3333 & 0 & 0 & 0 \end{bmatrix} \tag{11}
$$

The state distribution vector of the box girder components in the fifth year can be obtained

$$
\begin{aligned}
D_5 &= D_1 \times p^{1,2} \times p^{2,3} \times p^{3,4} \times p^{4,5} \\
&= \begin{bmatrix} 0.2158 & 0.1074 & 0.2940 & 0.2799 & 0.0138 \end{bmatrix}
\end{aligned} \tag{12}
$$

3.3 Comparison of Semi-Markov Model and Traditional Markov Model

In this section, we will compare the accuracy of the traditional Markov model and the proposed semi-Markov model, which have different mathematical forms and parameter estimation methods. The accuracy of the models can be evaluated by comparing their predictions with actual data. This can be done using various performance measures such as mean squared error, root mean squared error, or correlation coefficients. Additionally, cross-validation techniques can be applied to assess the out-of-sample predictive performance of the models.

By conducting a comparative study of these two models, we can determine which model provides better accuracy and prediction capabilities for the given system. The Markov state transition probability matrix based on the data in Sect. 3.2:

$$
p^{4,5} = \begin{bmatrix} 0.4737 & 0.3860 & 0.1404 & 0 & 0 \\ 0 & 0.4286 & 0.5238 & 0.0476 & 0 \\ 0 & 0 & 0.7473 & 0.2527 & 0 \\ 0 & 0 & 0 & 0.9804 & 0.0196 \\ 0 & 0 & 0 & 0 & 1 \end{bmatrix} \tag{13}
$$

Assume that the state distribution vector of a prestressed reinforced concrete components in the route is

$$
D_1 = \begin{bmatrix} 0.6667 & 0.3333 & 0 & 0 & 0 \end{bmatrix} \tag{14}
$$

The state distribution vector of the components in the fifth year can be obtained

$$D_5 = D_1 \times p^{1,2} \times p^{2,3} \times p^{3,4} \times p^{4,5} = [0.1559\ 0.2187 0.4368\ 0.1844 0.0043]$$

Based on Fig. 3, it can be observed that the proposed semi-Markov model, enhanced by introducing the Weibull distribution, exhibits significant improvement in accuracy compared to the traditional Markov model. It shows a closer fit to the actual measured data and can more accurately predict the degradation trend of bridges. This provides a scientific basis for subsequent bridge maintenance activities, enabling the rational planning of maintenance schedules, optimization of repair processes and materials selection. Consequently, it helps extend the service life of bridges while reducing maintenance costs.

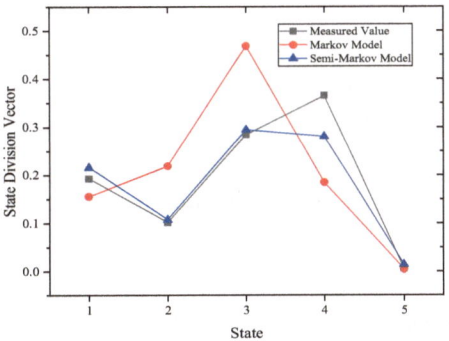

Fig. 3. Comparison of semi-markov model and traditional Markov model

4   Conclusion

This study addresses the insufficient consideration of time and limited research on component-level degradation models in current bridge condition prediction methods. Based on the assessment scores of highway bridge component conditions and considering the impact of maintenance on these scores, a semi-Markov model is constructed to predict the probability distribution of the condition levels of highway bridge components over time. The main conclusions are as follows:

(1) The study divides the assessment scores of bridge component conditions into multiple discrete states, referring to the boundary values for assessing degradation states in specifications. The Weibull distribution parameters for the duration of each state are fitted based on the realistic data. This establishes a Weibull-based semi-Markov model for predicting the condition of bridge components.

(2) The proposed model can update the transition probability matrix based on changes in state durations and predict the probability distribution of bridge component condition levels for any future year. It can timely identify changes in component conditions, provide references for subsequent bridge maintenance decisions, and improve maintenance efficiency.

(3) Compared to the commonly used Markov models, the proposed approach is more realistic and comprehensive in considering the durations of different states. However, since the model requires fitting parameters for the Weibull distribution of state durations, a larger amount of regular bridge inspection data is needed to further improve the model's accuracy.

Funding. Fund program: Science and Technology Project of Shandong Department of Transportation (2021B51).

References

1. Decò, A., Frangopol, D.M.: Risk assessment of highway bridges under multiple hazards. J. Risk Res. **14**(9), 1057–1089 (2011)
2. Decò, A., Frangopol, D.M.: Life-cycle risk assessment of spatially distributed aging bridges under seismic and traffic hazards. Earthq. Spectra **29**(1), 127–153 (2013)
3. Zhu, B., Frangopol, D.M.: Risk-based approach for optimum maintenance of bridges under traffic and earthquake loads. J. Struct. Eng. **139**(3), 422–434 (2013)
4. Chen, S., Hu, M., Fang, S., et al.: Research on degradation prediction model of bridge technical condition based on degradation factor mechanism. Municipal Technol. **38**(3), 54–60 (2020)
5. Dai, Y., Liu, X.: Life prediction of concrete structure based on corrosion and deterioration model of steel bar. Water Conservancy Hydropower Technol. **51**(S2), 412–415 (2020)
6. Ranjith, S., Setunge, S., Gravina, R., et al.: Deterioration prediction of timber bridge elements using the Markov chain. J. Perform. Constr. Facil. **27**(3), 319–325 (2013)
7. Morcous, G., Rivard, H., Hanna, A.: Modeling bridge deterioration using case-based reasoning. J. Infrastruct. Syst. **8**(3), 86–95 (2002)
8. Chen, F., Wei, D., Tang, Y.: Wavelet analysis based sparse LS-SVR for time series data. In: 2010 IEEE Fifth International Conference on Bio-Inspired Computing: Theories and Applications (BIC-TA), pp. 750–755 (2010)
9. Wei, D., Liu, H.: An adaptive-margin support vector regression for short-term traffic flow forecast. J. Intell. Transport. Syst. **17**(1), 317–327 (2013)
10. Fang, Z.: Research on technical condition evaluation and prediction of highway bridge-taking prestressed concrete girder bridge as an example. Dalian University of Technology (2014)
11. Setunge, S., Hasan, M.S.: Concrete bridge deterioration prediction using Markov chain approach (2011)
12. Jiang, Y., Sinha, K.: Bridge service life prediction model using the markov chain. Transportation Research Record, pp. 24–30. National Research Council, Washington, D.C. (1990)
13. Pontis. Version 4.0 Technical Manual Report. Department of Transportation, FHWA, U.S. (2001)
14. Thompson, P.D., Johnson, M.B.: Markovian bridge deterioration: developing models from historical data. Struct. Infrastruct. Eng. **1**(1), 85–91 (2005)
15. Puz, G., Radic, J.: Life-cycle performance model based on homogeneous Markov processes. Struct. Infrastruct. Eng. **7**(4), 285–296 (2011)
16. Ng, S.-K., Moses, F., Balkema, A.A., et al.: Bridge deterioration modeling using semi-Markov theory. Struct. Safety Reliab. **1**, 113–120 (1998)
17. -68A N P. Successful Approaches to Utilizing Bridge Management Systems for Strategic Decision Making in Asset Management Plans (2021)

Experimental Study on Back Fill Material of a High Fill Slope

Song Liu[✉]

Shanghai East China Civil Aviation Airport Construction Supervision Co. Ltd., Shanghai, China
691600251@qq.com

Abstract. Aiming at the problem of high fill slope engineering in China's airports, a high fill slope project of a new airport was selected as the research object. Through on-site engineering investigation, it was found that the backfill material was mainly moderately weathered stone, with high strength and not easy to break, accompanied by silty clay. The particle shape was mainly high angular and the surface was rough. In order to explore the mechanical properties of backfill materials for high fill slope of airport, the material parameters were calibrated by indoor screening test, stone wear and liquid-plastic limit test, natural water content test and saturated water absorption test.

Keywords: Airport Engineering · High Fill Slope · Indoor Test

1 Introduction

With the increasing number of new and expanded civil airports in China, the available land resources are becoming increasingly tense [1]. More airports choose to be built in mountainous valleys, crisscrossed areas and other harsh geographical environments, and a large number of high fill slope projects have appeared [2]. The settlement and uneven settlement of high fill slope engineering will occur after the completion of the project, which will seriously affect the safety of people's lives and property [3, 4]. Therefore, it is necessary to explore the material properties of different regions in depth. Based on the actual project, this paper determines the engineering geological conditions of the study area through engineering investigation, and calibrates the material parameters in the indoor screening test, stone wear and liquid-plastic limit test, and saturated water absorption test, laying the material parameter foundation for subsequent research.

2 Overview of Engineering Geological Conditions

A proposed airport is intended to be a dual-use facility for both military and civilian purposes, with a total area exceeding over 4,000 acres. The construction standards for the airport are based on the design of a feeder airport with an annual passenger capacity of one million and an annual cargo capacity of 4,000 tons. The airport plans to construct a runway measuring 2800 m × 45 m, with a designed shoulder width of 1.5 m. Additionally,

© The Author(s) 2024
G. Feng (Ed.): ICCE 2023, LNCE 526, pp. 28–33, 2024.
https://doi.org/10.1007/978-981-97-4355-1_3

blast pads measuring 60 m × 48 m will be situated at each end of the runway to prevent damage from jet blasts. The airport's flight area will be constructed to meet Grade 4C standards, with dimensions for the clearway set at 2920 m × 300 m and dimensions for the safety areas at the ends of the runway set at 240 m × 90 m. The site elevation, based on the 1985 National Elevation Datum, is preliminarily established at 165 m. Finite Element Model Establishment.

3 Construction Site Sampling

The samples are taken from the field project, and the selection results of different positions are quite different. In order to make the collected soil samples represent the real soil as much as possible, the soil samples are collected as much as possible at each position to avoid the influence of uneven selection on the test results. After the soil samples are transported back to the laboratory, they are uniformly mixed and tested. This sampling 1t. The acquisition process is shown in Fig. 1.

Fig. 1. The acquisition process

4 Tests and Results

4.1 Indoor Screening Test

Referring to the 'Highway Soil Test Procedure' screening test procedure, the pore size (mm) of the coarse sieve (round hole) is selected as: 60, 40, 20, 10, 5, 2; the pore size (mm) of fine sieve was 2.0, 1.0, 0.5, 0.25, 0.075. The total soil mass before screening was 10,000 g, and the soil mass less than 2 mm was 710 g, and the soil less than 2 mm accounted for 7.1% of the total soil mass. The remaining records are shown in Table 1.

Table 1. The remaining records

Rough screening analysis				Fine screening analysis				
Aperture (mm)	Cumulative sieve soil (g)	Soil mass less than the aperture(g)	Less than the pore diameter of the soil mass percentage (%)	Aperture (mm)	Cumulative sieve soil mass (g)	Soil mass less than the aperture (g)	Less than the pore diameter of the soil mass percentage (%)	Percentage of total soil mass (%)
				2	0	710	100	2.6
60	3210	6790	67.9	1	149	561	79.0	2.0
40	4385	5615	56.15	0.5	353	357	50.28	1.4
20	5509	4491	44.91	0.25	494	216	30.42	0.8
10	7574	2426	24.26	0.075	655	55	7.75	0.3
5	8590	1410	14.1					
2	9290	710	7.1					

When the content of particles exceeding the particle size is greater than 5%, the equal mass substitution method is adopted. The method is to replace the content of super-diameter particles with equal mass according to all the coarse materials allowed by the instrument (from coarse materials with a particle size of 5 mm to the maximum particle size). The new gradation composition can be calculated according to the following formula. The calculation results are shown in Table 2:

$$p_i = \frac{100 - p_m}{p_m - p_5}(p_{0i} - p_5) + p_{0i} \tag{1}$$

Table 2. The calculation results

Aperture (mm)	60	40	20	10	5	2
Less than the pore diameter of the soil mass percentage (%)	100	81.24	63.30	30.32	14.10	7.10

According to the coarse particle screening record (conversion algorithm), the filler particle gradation curve is drawn, as shown in Fig. 2.

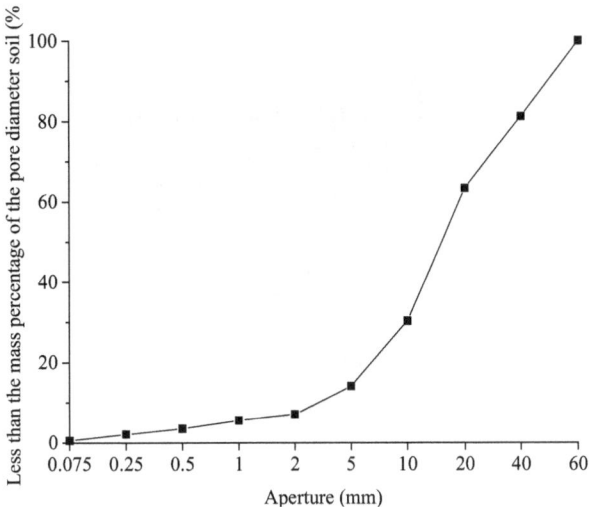

Fig. 2. The acquisition process

From the above Fig., the 0–5 mm gradation line is smooth, the particle size is continuous, and the slope is slow. The curve of 5–20 mm began to steep, and the particle content was the most, about from 14% to 63%. The steepness of the 20–60 mm line segment is slightly reduced. After calculation, the non-uniformity coefficient Cu is 6.172; the curvature coefficient Cc is 1.51, which meets the requirements of Cu ≥ 5 and 1 ≤ Cc ≤ 3, and the gradation is good.

5 Stone Abrasion and Liquid Limit, Plastic Limit

Referring to the 'Highway Geotechnical Test Procedures' joint determination method of liquid limit and plastic limit, representative samples were taken from coarse materials (particle size greater than 5 mm) and fine materials (particle size less than 0.5 mm) for stone wear and physical properties tests such as liquid limit and plastic limit. The experimental process is shown in Fig. 3, and the experimental results are shown in Table 3.

Fig. 3. Stone wear and liquid plastic limit test

Table 3. Test results of stone wear and liquid plastic limit

Coarse material (particle size greater than 5 mm)	Fine material (particle size less than 0.5 mm)
Stone wear value Q: 27.4%	Liquid limit w_L: 23% Plastic limit w_p: 14%

6   Saturated Water Absorption Test

Refer to the 'Highway Geotechnical Test Procedures' standard hygroscopic moisture content, the saturated water absorption rate of granular materials with particle sizes of 0–0.5 mm, 0.5–5 mm, and 5–40 mm is determined by grading. HT-30 saturated surface dry mold test instrument is selected for fine materials of 0–0.5 mm, and 101A-3 electric hot blast constant temperature drying box is selected for coarse materials of 0.5–40 mm. After drying, soak until the water absorption is saturated, then wipe the surface dry, and then weigh the wet mass. The saturated water absorption is:

$$w_a = \frac{(m - m_s)}{m_s} \times 100 \tag{2}$$

The test results are shown in Table 4.

Table 4. Saturated water absorption test results

coarse stuff (5–40 mm)	coarse stuff (0.5–5mm)	fine stuff (0–0.5 mm
w_a: 2.2%	w_a: 3.0%	w_a: 5.7%

7 Conclusion

Through the construction site sampling and indoor screening test, the calculated non-uniformity coefficient Cu is 6.172, and the curvature coefficient Cc is 1.51, which meets the good grading requirements of Cu ≥ 5 and 1 ≤ Cc ≤ 3, and the grading is good. Through the soil sample parameter test, the stone wear value of coarse material (particle size greater than 5 mm) is 27.4%, the liquid limit of fine material (particle size less than 0.5 mm) is 23%, and the plastic limit is 14%. The saturated water absorption rate of coarse material (5–60 mm) is 2.2%, the saturated water absorption rate of coarse material (0.5–5 mm) is 3.0%, and the saturated water absorption rate of fine material (0–0.5 mm) is 5.7%.

References

1. Jie, Y., Wei, Y., Wang, D., et al.: Numerical study on the settlement of high-fill airports in collapsible loess geomaterials: a case study of Lvliang Airport in Shanxi Province, China. J. Central South Univ. **28**(3), 939–953 (2021)
2. Zhang, H.J., Song, Y., Xu, S.L., et al.: Combining a class-weighted algorithm and machine learning models in landslide susceptibility mapping: a case study of Wanzhou section of the Three Gorges Reservoir, China. Comput. Geosci. **158**, 104966 (2022)
3. Pham, Q.D., Ngo, V.L., Tran, T.V., et al.: Evaluation of asaoka and hyperbolic methods for settlement prediction of vacuum preloading combined with prefabricated vertical drains in soft ground treatment. J. Endineering Technol. Sci. **54**(5), 859–872 (2022)
4. Yin, L., Sun, X., Ping, Y., et al.: Stability analysis for subgrade settlement prediction by curve fitting methods. IOP Conf. Ser.: Earth Env. Sci. **170**(3), 032050 (2018)

Vibration and Stability Analysis of Functionally Gradient Flow Pipe with Axial Motion

Jie Zhou[1,2(✉)], Haowen Mou[1], Junchao Gao[1], and Xueping Chang[2]

[1] Tianjin Aerospace Relia Technology Co., Ltd., Tianjin 300000, China
17713653269@163.com
[2] College of Mechanical and Electrical Engineering, Southwest Petroleum University, Chengdu 610065, China

Abstract. The axially moving functionally graded pipe is widely used in many fields of industry, and its vibration and stability analysis is the key to its design. In this paper, the vibration characteristics and stability of functionally graded pipes with axial motion are analyzed. Based on the extended Hamilton variational principle, the dynamic equation of the pipe with internal flow velocity, material volume fraction index and axial velocity is established. The modified Galerkin method is used to solve the dynamic equation. The influence of the internal flow velocity, material volume fraction index, axial velocity and acceleration on the dynamic characteristics and stability of the system is analyzed. The characteristic curves of volume fraction index, axial velocity, acceleration and natural frequency are given. The results show that the natural frequency and critical velocity of the system increase with the increase of volume fraction index, and the designed volume fraction index can adjust the natural frequency of FGM pipeline system. When the system has axial acceleration, the greater the acceleration, the system will reach the critical value of axial instability in a short time.

Keywords: Functionally Graded Materials · Volume Fraction Index n · Axial Acceleration · Critical Velocity · Natural Frequency

1 Introduction

Functionally gradient materials (FGM) [1] are composed of two or more different materials. In functionally gradient materials, power series, S-shaped function and exponential function are mainly used to describe the changes of mechanical properties of functionally gradient materials in functionally gradient structures [2]. The mechanical properties and research of functionally gradient materials are reviewed in References [2]. Its volume fraction index changes smoothly and continuously along the preferred direction [3, 4]. In the study reported in References [5], the linear vibration of FGM pipe conveying fluid is studied.

The results show that the stability of the pipe is significantly improved when the conventional isotropic material is replaced by a functional gradient material (FGM). In fact, the improved stability of FGM pipes is mainly due to their increased stiffness. Deng et al. [6] analyzed the instability of multispan viscoelastic pipes made of

© The Author(s) 2024
G. Feng (Ed.): ICCE 2023, LNCE 526, pp. 34–46, 2024.
https://doi.org/10.1007/978-981-97-4355-1_4

functional gradient materials using the wave propagation method and the reverbera-tion ray matrix algorithm. Tang and Yang [7] investigated the dynamics and stability of a bi-directional FGM infused nanotube. The governing equations and corresponding boundary conditions were obtained using Hamilton's principle and solved by differen-tial product method. The results showed that the bi-directional material distribution can significantly change the critical flow rate, fundamental frequency and stability. Wang and Liu [8] evaluated the effect of power-law exponents on the deflection and stability of a clamped FGM pipeline by using the Sin method. An and Su [9] analyzed the linear vibration and amplitude of a functionally graded pipeline for transporting fluids by using the generalized integral transform method.

The dynamic response of the axial motion system under the action of moving mass has important practical research value. Li Weiming et al. studied the dynamic response of the simply supported beam under the continuous velocity change of the moving mass. [10], Wang Yingze et al. studied the vibration of the barrel under the action of multiple moving masses. [11], Liu Ning et al. studied the vibration characteristics of the axially moving cantilever beam under the action of moving mass. [12], Wu et al. studied the dynamic response of the axially moving beam under the moving mass based on the finite element method. [13], CHEN et al. studied the vibration analysis of the axially moving beam and the chord [14]. DING et al. derived several different forms of the control equations of the axially moving beam, and analyzed the differences between the control systems by numerical analysis. [15], Qi Yafeng, Liu Ning et al. studied the vibration response analysis of the axially moving simply supported beam [16]. In the above research, the influence of the axial velocity on the vibration characteristics of the fluid-conveying pipe is rarely considered.

The article takes the axial motion functional gradient flow tube as the research object, and the description of the mechanical properties of its material is described by a power series, and the FGM linear Euler-Bernoulli model containing the inward flow velocity with axial motion is established. The governing equations of the system were derived using Hamilton's principle. Then the Galerkin method is utilized to solve its dynamical system. Finally, the effects of several parameters on the vibration characteristics and stability of the axial motion functional gradient flow tube are discussed.

2   Establishment of Mathematical Models

Considering the mathematical model of the axial velocity of the FMG pipe, an analysis model is established as shown in Fig. 1. A functionally graded axially moving pipe with an average radius of r and a length of l, r_i and r_o represent the inner and outer diameters of the pipe, U represents the velocity of the fluid, and v represents the axial velocity of the pipe. The pipeline adopts the Euler-Bernoulli beam model. In this paper, the power series is used to describe the effective performance change of functionally graded materials along the thickness direction. When the mechanical properties change in the form of power function, the effective mechanical properties of the pipeline can be expressed as [17].

$$P_f = P_m V_m + P_c V_c \tag{1}$$

in which P_m, P_c is the material properties of metals and ceramics(such as Young's modulus and mass density), V_m and V_c are the volume fractions of metals and ceramics, respectively, and the composition is expressed as

$$V_c = \left(\frac{R_0 - r}{R_0 - R_i} \right)^n, \quad V_c + V_m = 1 \tag{2}$$

Among them, n is the volume fraction power-law index; the labels i and o represent the inner and outer layers, respectively. Since the mechanical properties of the functional gradient flow transfer pipe are in the radial direction along the pipe, the effective bending stiffness and mass per unit length of the functional gradient material can be written as [17, 18]

$$(EI)^* = \int Ez^2 dA = \int_0^{2\pi} \int_{Ri}^{Ro} E(r)r^2 \sin^2 \theta r dr d\theta \tag{3}$$

$$m^* = \int \rho dA = \int_0^{2\pi} \int_{Ri}^{Ro} \rho(r) r dr d\theta \tag{4}$$

where: $(EI)*$ is the effective bending stiffness of the functionally graded tube; $m*$ is the mass per unit length of the functionally graded tube.

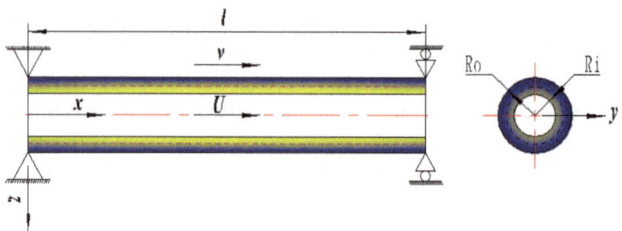

Fig. 1. Model of axial motion system

The dynamic modeling of the model is carried out by Hamilton principle.

$$\delta \int_{t_1}^{t_2} (T - V) dt + \int_{t_1}^{t_2} \delta W dt = 0 \tag{5}$$

In this study, the tube has an axial velocity, so the tube has an axial displacement. The change rate of the deflection $w(x, t)$ with time, that is, the calculation of the lateral velocity and acceleration should be based on the field velocity. The change rate of the field coordinate x with time is to calculate the derivative of the deflection to time, and the change rate of the x coordinate must be considered. Therefore:

$$\frac{dw}{dt} = \frac{\partial w}{\partial t} + v \frac{\partial w}{\partial x} \tag{6}$$

$$\frac{d^2w}{dt^2} = \frac{\partial^2 w}{\partial t^2} + 2\frac{\partial^2 w}{\partial x \partial t}v + \frac{\partial w}{\partial x}\dot{v} + \frac{\partial^2 w}{\partial x^2}v^2 \tag{7}$$

The fluid kinetic energy can be expressed as:

$$T_f = \frac{1}{2}M_f \int \left[\left(\frac{\partial w}{\partial t} + U\frac{\partial w}{\partial x} \right)^2 + U^2 \right] dx \tag{8}$$

The potential energy of the functionally graded tube without considering the effect of gravity can be expressed as:

$$V_p = \frac{1}{2} \int (EI)^* \left(\frac{\partial^2 w}{\partial x^2} \right)^2 dx \tag{9}$$

The kinetic energy of the functionally graded pipe can be expressed as:

$$T_p = \frac{1}{2}m^* \int \left(\frac{dw}{dt} \right)^2 dx \tag{10}$$

By substituting Eq. (6), (8), (9), (10) into Eq. (5), the dynamic governing equations of the system can be obtained:

$$(EI)^* \frac{\partial^4 w}{\partial x^4} + \left(M_f U^2 + m^* v^2 \right) \frac{\partial^2 w}{\partial x^2} + m^* \dot{v}\frac{\partial w}{\partial x} + 2(M_f U + m^* v)\frac{\partial^2 w}{\partial x \partial t} + (M_f + m^*)\frac{\partial^2 w}{\partial t^2} = 0 \tag{11}$$

3 Numerical Calculation

The vibration model of the axial power tube is a complex time-varying system, and the natural frequency and vibration shape of the structure are constantly changing with time, so the Galerkin method is used to approximate the solution, so the bending displacement of the beam can be expressed as Fig. 2:

$$w(\xi, \tau) = \sum_{i=1}^{n} \varphi_i(\xi)q_i(\tau) \tag{12}$$

Where $q_i(\tau)(\tau=1, 2, 3...n)$ is the reduced generalized displacement, and $\varphi_i(\xi)$ represents the the modal function of the i th order of the system and satisfies the boundary condition. In this paper, we assume that:

$$\varphi_i(\xi) = \sin(n\pi x/l) \tag{13}$$

Substituting Eqs. (12) and (13) into Eq. (11), multiplying both sides by $\varphi_j(\xi)$ at the same time, and then integrating on the interval [0, 1], we get:

$$\mathbf{M}\ddot{q} + \mathbf{C}\dot{q} + \mathbf{K}q = 0 \tag{14}$$

where, $\mathbf{q} = (q_1, q_2, q_3....q_n)$, $\mathbf{M, C, K}$ are mass matrix, damping matrix and stiffness matrix, respectively.

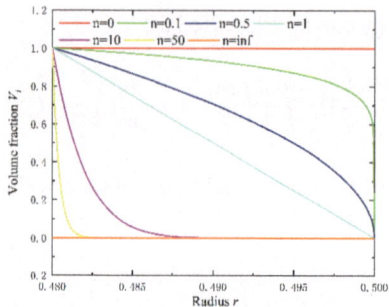

Fig. 2. Volume fraction Vi of inner material with different volume fraction index n change along the radial direction

Where:

$$\mathbf{M} = \begin{bmatrix} \mathbf{I} & \mathbf{0} \\ \mathbf{0} & \mathbf{I} \end{bmatrix} \qquad \mathbf{C} = \begin{bmatrix} 2(M_f U + m^* v)B & 0 \\ 0 & 2(M_f U + m^* v)B \end{bmatrix} \qquad (15)$$

$$\mathbf{K} = \begin{bmatrix} (EI)^* \Lambda + (M_f U^2 + m^* v^2)C^* + m^* \dot{v}B & 0 \\ 0 & (EI)^* \Lambda + (M_f U^2 + m^* v^2)C^* + m^* \dot{v}B \end{bmatrix} \qquad (16)$$

$$C_{ij}^* = \begin{cases} -(i\pi/l)^2 & i = j \\ 0 & i \neq j \end{cases} \quad \Lambda_{ij} = \begin{cases} (i\pi/l)^4 & i = j \\ 0 & i \neq j \end{cases} \quad B_{ij} = \begin{cases} \frac{4ij}{i^2 - j^2} & i + j \text{ is odd} \\ 0 & i + j \text{ is even} \end{cases} \qquad (17)$$

where, C_{ij}^*, Λ_{ij}, B_{ij} are the matrix elements of row i and column j. The solution of Eq. (14) can be assumed as

$$\mathbf{q} = \mathbf{S} * \exp(i\varpi t) \qquad (18)$$

where, $\mathbf{S}^\cdot = (S_1, S_2, S_3 \dots S_n)$, Substituting Eq. (17) into Eq. (14), we can get:

$$\left(-\varpi^2 \mathbf{M} + i\varpi \mathbf{C} + \mathbf{K}\right)\mathbf{S}^* = 0 \qquad (19)$$

Equation (19) has an asymmetric solution, so the determinant of the matrix coefficient must be 0.

$$\left| -\varpi^2 \mathbf{M} + i\varpi \mathbf{C} + \mathbf{K} \right| = 0 \qquad (20)$$

4   Numerical Results Calculation and Discussion

4.1   Example Verification

The equation is dimensionless

$$\frac{\partial^4 \eta}{\partial \xi^4} + \left(u^2 + V^2\right)\frac{\partial^2 \eta}{\partial \xi^2} + \frac{\partial^2 \eta}{\partial \tau^2} + \gamma \frac{\partial \eta}{\partial \xi} + 2\left(u\sqrt{\beta} + V\sqrt{\delta}\right)\frac{\partial^2 \eta}{\partial \tau \partial \xi} = 0 \qquad (21)$$

where:

$$u = \sqrt{\frac{M}{(EI)^*}}UL \quad V = \sqrt{\frac{m^*}{(EI)^*}}vL \quad \beta = \frac{M}{M + m^*} \quad \delta = \frac{m^*}{M + m^*} \quad \gamma = \dot{v}\frac{m^*}{(EI)^*}L^3$$

(22)

When $\sigma = 0$, $V = 0$, $\delta = 0$, $\gamma = 0$, the model is a pipe conveying model without considering gravity and internal and external damping. In this state, the natural frequency and crit0ical flow velocity of different flow velocity U are calculated (Tables 1 and 2).

Table 1. Comparison of dimensionless natural frequencies [19]

U	ω	numerical results	Literature Results
$U = 0$	ω_1	9.8696	π^2
	ω_2	39.4784	$4\pi^2$

Table 2. Comparison of dimensionless critical velocity [19]

U_{cr}	Present results	Literature Results
The first modal frequency	3.14	π
The second-order modal frequency	6.28	2π

Table 3. Study material parameters

Material	$\rho_P\left(Kg/m^3\right)$	E(GPa)	G(GPa)
SiC	3210	440	188
Ti-6Al-4V	4515	115	44.57

It can be seen from the results of the above table that when $U = 0$, the dimensionless natural frequencies of the first two orders are 9.8698 and 39.4784, which are consistent with the results of Reference [19].Therefore, the program has certain accuracy in solving the complex frequency of the system. When $\beta = 0.1$, the dimensionless critical flow velocities corresponding to the first-order modal frequency and the second-order modal frequency are 3.14 and 6.28, which are consistent with the results of Reference [19], and the correctness of the algorithm is verified.

4.2 Example Analysis

Here, Deng [11] wave propagation method and reverberation ray matrix algorithm are selected to solve the computational parameters of the multi-span viscous multifunctional gradient flow tube, and the materials of the functional gradient are selected to be composed of SiC and Ti-6Al-4V. Since the material of the functional gradient changes into a gradient in the direction of the thickness, so in the functional gradient tube, the volume fraction of the SiC decreases from the outer surface gradually until it is reduced to zero. The volume fraction of the Ti-6Al-4V gradually decreases from the inner surface until it is reduced to zero.

Figure 3 depicts the variation of the natural frequency with the volume fraction index n. The system frequency increases with the increase of the volume fraction index, and the frequency is strongly affected by the volume fraction index when the volume fraction index n is between 0 and 10, and it is gradually reduced by the volume fraction index after n is greater than 10, and the frequency is almost unaffected by the volume fraction index when $n = 50$. From Table 3, it can be known that the Young's modulus of SiC is much larger than that of Ti-6Al-4V, and it can also be seen from Fig. 1 that when the volume fraction index is less than 10, the content of SiC changes faster with the increase of the index n, which has a larger effect on the dynamic properties of the system. When the volume fraction index n is greater than 10, V_i is gradually reduced by the influence of the index n. When the index n tends to infinity, Vi is almost unaffected by the index n. Therefore, we can conclude that the natural frequency of the functional gradient conveying pipeline is more sensitive to the volume fraction index n in the stage of lower flow velocity. As the volume fraction index n increases, its effect on the intrinsic frequency and stability of the pipe decreases. After the index $n > 50$, it has almost no effect on the dynamics of the functional gradient pipeline.

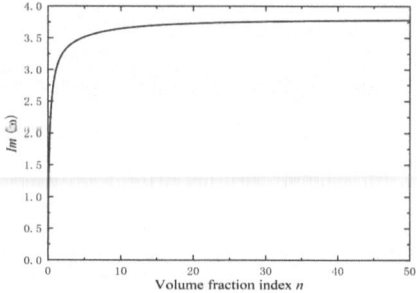

Fig. 3. Variation of natural frequency with functional gradient fractional index n

Figure 4 shows the change of the first three natural frequencies of the functionally graded material tube with the flow rate under the action of different volume fraction index n. In the form obtained in this paper, the imaginary part Im(ω) of the frequency is the vibration frequency of the pipeline, and the real part Re(ω) of the frequency represents the growth or attenuation of the vibration response. The stability of the system can be determined by the real and imaginary parts of the natural frequency. When Im(ω) > 0, Re(ω) = 0, the system is in a stable state; When Im(ω)=0, Re(ω) < 0, the system is in static instability. When Im(ω)<0, Re(ω) < 0, , the system is in dynamic instability. It can be seen from the fig.(a) that when $n = 0$, the first-order modal instability occurs at $U = 29.3\,m/s$ When the flow rate continues to increase to $U = 58.6\,m/s$, the first-order

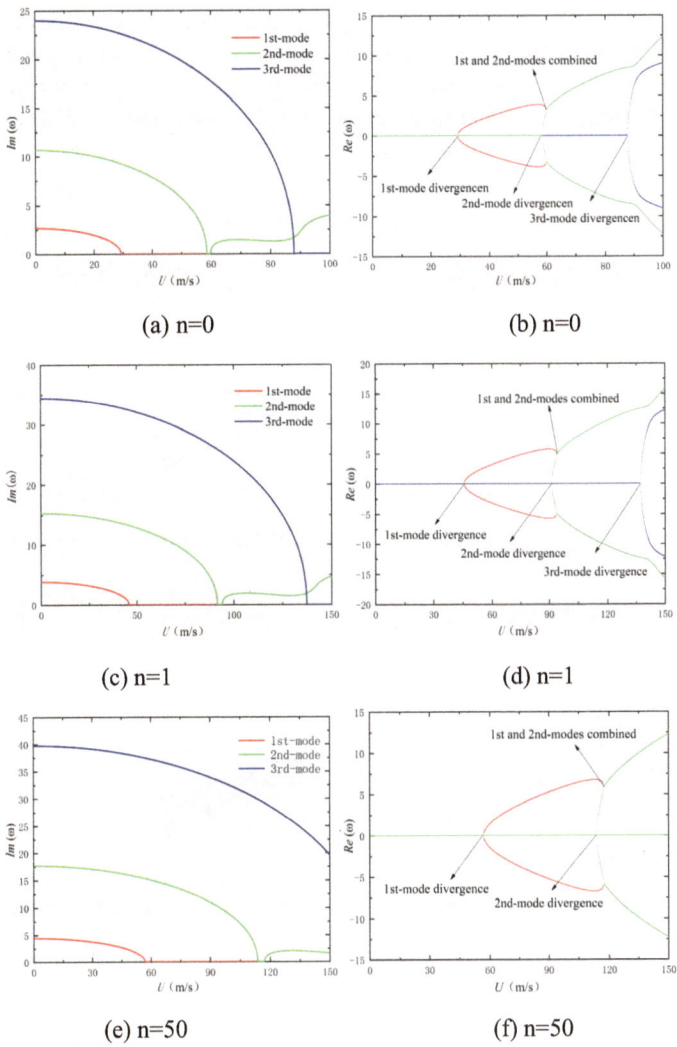

Fig. 4. Relationship between natural frequency and flow velocity under different volume fraction index

modal frequency and the second-order modal frequency overlap. The imaginary part of the frequency is positive and the real part of the frequency is negative, indicating that the system has a coupled dynamic instability. It can be seen from Fig. 4 (c) that when $n = 1$, the first-order modal instability occurs at. It can be seen from Fig. 4 (e) that when $n = 10$, the first-order modal instability velocity $U = 55.5\,m/s$. It can be seen from Fig. 4 (g) that when $n = 50$, the first-order modal instability velocity occurs at $U = 56.9\,m/s$. In different volume fraction index n, when $U < 59.8\,m/s$, the system does not occur dynamic instability. The above analysis shows that the volume fraction index n of the functionally graded material has a great influence on the natural frequency and critical flow velocity of the system. The natural frequency and critical flow velocity of the system increase with the increase of the volume fraction index n. In practical engineering, the natural frequency of the system can be changed by adjusting the volume fraction index n, which has important guiding significance for the application of FGM in practical engineering.

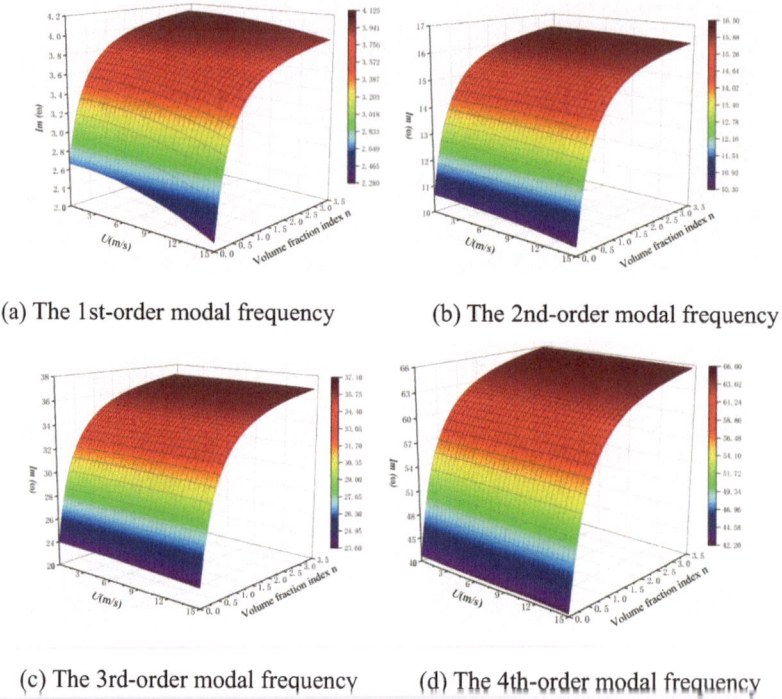

(a) The 1st-order modal frequency   (b) The 2nd-order modal frequency

(c) The 3rd-order modal frequency   (d) The 4th-order modal frequency

Fig. 5. Variation of the first four modal frequencies of the system with velocity and n

Figure 5 is the first four natural frequencies of the system with the change of the fluid flow rate and the volume fraction index n. It can be seen from the Fig. 5(a) that when. When $U = 15$ m/s, $n = 0$, the frequency of the first-order mode ω=2.6643, when $U = 0$ m/s, $n = 3.4$, the frequency of the first-order mode ω=4.115, that is, the point with the largest frequency appears in the place where the flow velocity is low and the volume fraction index is small, while the point with the smallest frequency appears in the place where the flow velocity is high and the volume fraction is small.

Figure 6 depicts the change of the critical velocity of the system with the volume fraction index n. It can be seen from Fig. 6 that when the volume fraction $n < 10$, the volume fraction index n has a great influence on the critical velocity. When $n > 10$, the volume fraction index n is smaller for the critical flow velocity and gradually tends to a fixed value.

It can also be concluded from Fig. 6 that the critical flow rate of the system increases from $U = 29.3$ m/s when $n = 0$ to $U = 56.9$ m/s when $n = 50$, and the critical flow rate increases by 51.5%. In practical engineering applications, the critical flow rate of the system can be improved by adjusting the volume fraction index n.

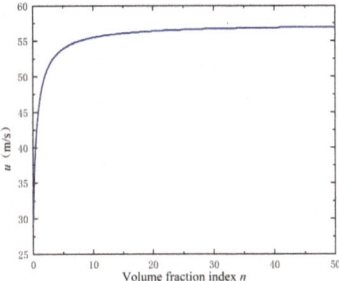

Fig. 6. Variation of critical velocity with functional gradient fraction exponent n

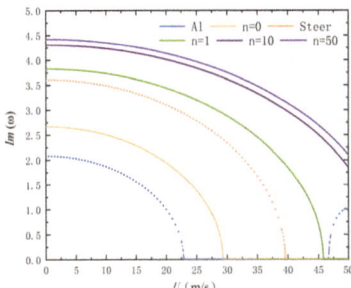

Fig. 7. First frequency comparison between functionally Graded materials and conventional materials

Figure 7 is a comparison of the mechanical properties of functionally graded materials and materials commonly used in engineering. From Fig. 7, we can see that functionally graded materials can improve their mechanical properties compared to conventional engineering materials. Compared with aluminum, the mass per unit length of the functionally graded material increases, and its critical flow rate also increases. Compared with steel, the mass per unit length of functionally graded materials is reduced, but the critical flow rate is increased. Therefore, functionally graded materials can improve the stability of the system, which has important guiding significance for the selection of materials in engineering.

Figure 8 shows the modal frequencies with time when the system has an axial velocity of $v = at$. In Fig. 8, it can be seen that when the acceleration is large, the critical value of the axial motion of the system is reached faster for the same time with a large acceleration. The stability of the system can be determined by the real and imaginary parts of the intrinsic frequency. When $Im(\omega) > 0$ and $Re(\omega) = 0$, the system is in a stable state; while when the acceleration is larger, the time for the intrinsic frequency of the system to converge to 0 is shorter, i.e., the larger the acceleration is, the system is more prone to destabilization with the change of time. By calculating different volume fraction indices n, it is observed that the larger the volume fraction index, the slower the time for the intrinsic frequency of the system to converge to 0 under the same conditions, i.e., the time for the occurrence of instability is suppressed.

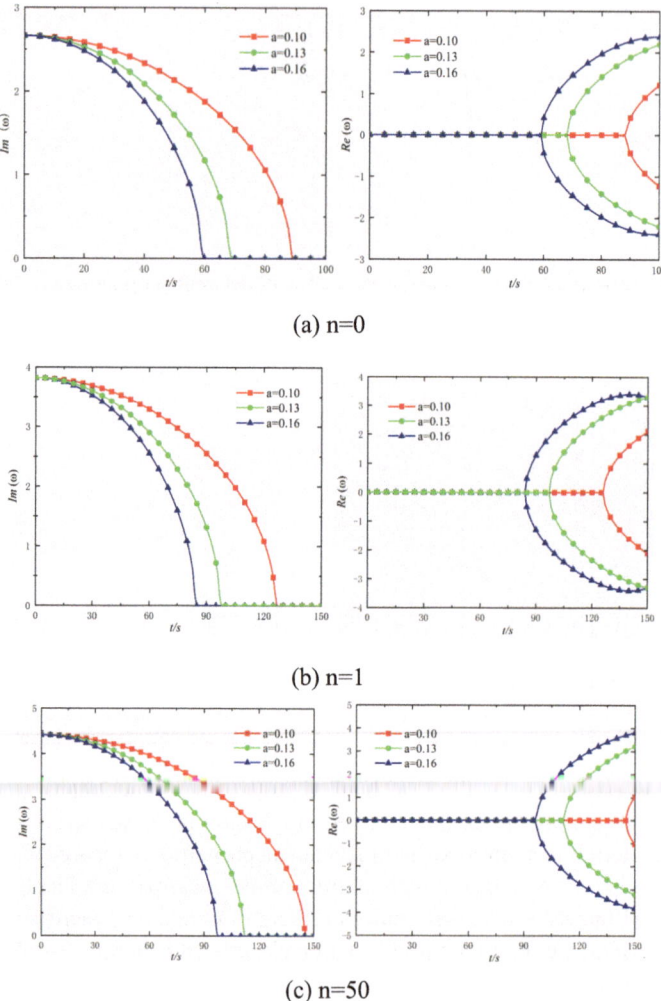

Fig. 8. Variation of modal frequency with time under different accelerations

5 Conclusion

In this paper, the vibration characteristics of a functional gradient axial kinematic tube are investigated. The critical flow rate and intrinsic frequency of the system are discussed with respect to the volume fraction index n. Both the critical flow rate and intrinsic frequency of the system increase with the increase of the volume fraction index n, and the critical flow rate and intrinsic frequency of the system are more affected when n < 10. That is, in practical engineering applications, under the condition of knowing the external excitation frequency, the excitation frequency can be avoided by adjusting the volume fraction index n, so as to reduce the vibration and improve the stability of the system. And through the comparison of mechanical properties of conventional materials and functional gradient materials, it can be found that the functional gradient materials can improve the stability of the system than conventional engineering materials, which is an important guidance for the selection of materials in engineering. When the system exists an acceleration of axial motion, the larger the acceleration, the system will reach the critical value of axial motion tends to be destabilized in the shorter time, and the increase of the volume fraction index n will inhibit the time for the system to reach the critical velocity, i.e., the stability of the system can be changed by changing the volume fraction index n.

References

1. Koizumi, M.: FGM activities in Japan. Compos. Part B Eng. **28**(1–2), 1–4 (1997)
2. Birman, V., Byrd, L.W.: Modeling and analysis of functionally graded materials and structures. Appl. Mech. Rev. **60**, 195–216 (2007)
3. Tang, Y., Lv, X., Yang, T.: Bi-directional functionally graded beams: asymmetric modes and nonlinear free vibration. Compos. B Eng. **156**, 319–331 (2019)
4. Yang, T., Tang, Y., Li, Q., Yang, X.D.: Nonlinear bending, buckling and vibration of bi-directional functionally graded nanobeams. Compos. Struct. **204**, 313–319 (2018)
5. Eftekhari, M., Hosseini, M.: On the stability of spinning functionally graded can-tilevered pipes subjected to fluid thermomechanical loading. Int. J. Struct. Stab. Dyn. **16**, 1550062 (2016). Selmi, A., Hassis, H.: Composites Part C: Open Access **4**, 100117 (2021)
6. Deng, J., Liu, Y., Zhang, Z., Liu, W.: Stability analysis of multi-span viscoelastic functionally graded material pipes conveying fluid using a hybrid method. Eur. J. Mech. A-Solid **65**, 257–270 (2017)
7. Tang, T., Yang, T.: Bi-directional functionally graded nanotubes: fluid conveying dynamics. Int. J. Appl. Mech. **10**(4), 1850041 (2018)
8. Wang, Z.M., Liu, Y.Z.: Transverse vibration of pipe conveying fluid made of functionally graded materials using a symplectic method. Nucl. Eng. Des. **298**, 149–159 (2016)
9. An, C., Su, J.: Dynamic behavior of axially functionally graded pipes conveyingfluid. Math. Probl. Eng. **2017**, 1–11 (2017)
10. Weiming, L., Hanbin, L., et al.: Influence of moving mass velocity on dynamic response of simply supported beam. J. Huazhong Univ. Sci. Technol,: Nat. Sci. Ed. **36**(9), 117–120 (2008)
11. Yinze, W., Xiaobing, Z.: Vibration response analysis of expansion wave launcher with moving mass. J. Aerodyn. **24**(8), 1714–1719 (2009)
12. Ning, L., Guolai, Y.: Analysis of vibration characteristics of axially moving cantilever beam with moving mass. J. Vib. Shock **31**(3), 106–109 (2012)
13. Wu, J.J., Whittaker, A.R., Cartmell, M.P.: Dynamic responses of structures to moving bodies using combined finite element and analytical methods. Int. J. Mech. Sci. **43**, 2555–2579 (2001)
14. Chen, L.-Q., Yang, X.-D.: Nonlinear free transverse vibration of an axially moving beam: comparison of two models. J. Sound Vib. **299**(1–2), 348–354 (2007)
15. Ding, H., Chen, L.: On two transverse nonlinear models of axially moving beams. Sci. China Ser. E: Technol. Sci. **52**(3), 743–751 (2009). https://doi.org/10.1007/s11431-009-0060-1

16. Qi, Y., Liu, N., Yang, G.: Vibration analysis of axially moving simply-supported beam.J, Ordnance Equipment Eng. (12), 126–129 (2016)
17. Setoodeh, A.R., Afrahim, S.: Nonlinear dynamic analysis of FG micro-pipes conveying fluid based on strain gradient theory. Compos. Struct. **116**, 128–135 (2014)
18. Tang, Y., Yang, T.Z.: Post-buckling behavior and nonlinear vibration analysis of a fluid-conveying pipe composed of functionall graded material. Compos. Struct. **185**, 393–400 (2018)
19. Ni, Q., Zhang, Z.L., Wang, L.: Application of the differential transformation method to vibration analysis of pipes conveying fluid. Appl. Math. Comput. **217**, 7028–7038 (2011)

Analysis of Temperature Field Characteristics in Seasonal Frost Region Airport Pavement Subgrade

Yonghua Ma[1]([⊠]), Zhimin Zhang[1], and Guoliang Yang[2]

[1] Hohhot Urban Transportation Investment and Construction Group Co., Ltd., Hohhot, China
2426901445@qq.com
[2] Inner Mongolia Autonomous Region Civil Aviation Airport Group Co., Ltd., Hohhot, China

Abstract. To study the variation patterns of the temperature field in seasonal frost region airport subgrades, this research establishes a heat conduction equation with phase change latent heat as an internal heat source. During the freezing and thawing processes in seasonal frost regions, a numerical simulation model of airport subgrades is developed, with thermal characteristic parameters of the soil as research variables. The study focuses on investigating the characteristic changes in the temperature field of seasonal frost region airport subgrades, ultimately obtaining the variation patterns of shallow temperature within the natural subgrade and surface temperature.

Keywords: Airport Engineering · Temperature Field · Numerical Simulation

1 Introduction

Seasonal frost in temperature fluctuations can pose safety hazards to the stability of airport subgrades, making the composition, structure, and mechanical properties of soil in seasonal frost regions more complex than ordinary soils [1]. Cheng [2] analysed the embankment temperature field using both steady-state and transient thermal analysis methods, explaining the solution process of the embankment temperature field and proposing research ideas for frozen soil embankments. Feng [3] conducted numerical simulations of the highway embankment temperature field using Finite Element Method (FEM), and the results showed that the temperature change in the embankment lags behind the air temperature change. The aforementioned studies have proposed various methods for analysing the temperature field characteristics of frozen soil embankments, but research on modelling methods based on thermal characteristic parameters is relatively limited. Building upon existing theoretical research, this paper utilizes COMSOL finite element software to analyse the characteristic changes in the temperature field of a natural embankment.

© The Author(s) 2024
G. Feng (Ed.): ICCE 2023, LNCE 526, pp. 47–54, 2024.
https://doi.org/10.1007/978-981-97-4355-1_5

2 Theoretical Analysis

2.1 Fundamentals of Temperature Field Description

Temperature Field.The temperature field is a collection of temperatures at various points in space that describe the thermal distribution of the studied object at a specific moment. It is typically represented as a function of both space and time, and its expression is as follows:

$$T = f(x, y, z, t) \tag{1}$$

where T represents temperature in degrees Celsius (°C); x, y, z are spatial coordinates in a Cartesian coordinate system; t stands for time.

Temperature Gradient.For the studied object, within the same temperature field at a specific moment, temperatures are equal on the same isothermal surface. Heat transfer occurs between different isothermal surfaces, with the maximum heat transfer and temperature change occurring in the normal direction. The temperature change along the normal direction is commonly referred to as the temperature gradient (grad T), as shown in the following equation:

$$gradT = \frac{\partial T}{\partial n} n \tag{2}$$

where $\partial T/\partial n$ represents the temperature directional derivative in the normal direction; n stands for the unit vector in the normal direction, with x, y, z components.

Fourier's Law. According to Fourier's theorem, the heat flux vector and temperature gradient can be combined to establish the following equation:

$$q_n = -\lambda \frac{\partial T}{\partial n} \tag{3}$$

where λ is the thermal conductivity coefficient (W·m − 1·K − 1).

2.2 Governing Temperature Field Equation

Based on Fourier's Law, to calculate the heat flux vector, it is necessary to determine the temperature gradient, as illustrated in Fig. 1. In this study, by employing the first law of thermodynamics (the law of energy conservation and transformation), a connection between temperatures at various points within an object is established. The heat generated by frozen soil phase change is treated as a heat source, ultimately leading to the formulation of the soil thermal conduction equation.

Due to the fact that the temperature and moisture of the airport pavement subgrade change primarily in the vertical direction due to freeze-thaw cycles and settlement, it can be treated as a two-dimensional problem. Neglecting the influence of factors such as moisture evaporation, the non-steady-state heat conduction equation can be expressed using Eq. (4):

$$\rho C(\theta) \frac{\partial T}{\partial t} = \lambda(\theta) \nabla^2 T + L \cdot \rho_I \frac{\partial \theta_I}{\partial t} \tag{4}$$

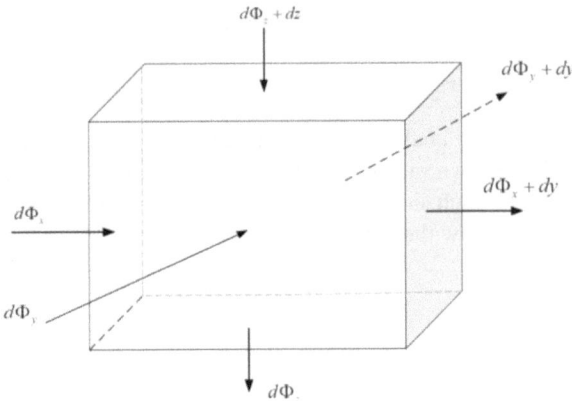

Fig. 1. Thermal Conduction Within the Unit Cell

where C represents the specific heat capacity of the soil ($J.kg^{-1}.K^{-1}$); ρ represents the density of the soil ($kg.m^{-3}$); λ represents the thermal conductivity coefficient ($W.m^{-1}.K^{-1}$); L represents the latent heat of water phase change, usually taken as 334.5 J/m^3; ρ_I represents the density of ice ($kg.m^{-3}$); θ_I represents the volumetric content of ice (%).

Since the subgrade soil within the pavement is in a negative temperature state, the water inside the soil consists of both ice and pore water. Due to the different densities of unfrozen water and ice, the unfrozen water content is defined as follows:

$$\theta_I = (\theta - \theta_u) \cdot \frac{\rho_1}{\rho_w} \tag{5}$$

where θ represents the volumetric water content (%); θ_u represents the volumetric unfrozen water content (%); ρ_w represents the density of water ($kg\ m^{-3}$).

The migration and redistribution of moisture in the soil, as well as the distribution of heat and temperature, are interrelated. Considering the water-heat coupling issue during soil freeze-thaw processes [4], the latent heat of phase change is treated as an internal heat source in the established heat conduction equation:

$$\rho C_{vs}\frac{\partial T}{\partial t} = \lambda \nabla T^2 + L\rho_I \frac{\partial \omega_I}{\partial t} \tag{6}$$

∇ represents the spatial differential operator; ρ is the density of soil in $kg\ m^{-3}$; ρ_I is the density of ice in $kg\ m^{-3}$; C_{vs} is the specific heat capacity of soil in $J \cdot (kg\ °C)^{-1}$; λ is the thermal conductivity of soil in $W \cdot (m\ °C)^{-1}$; ω_I is the ice content in soil, expressed as a percentage; L denotes the latent heat of phase change, taken as 335,000 $J\ kg^{-1}$; t stands for time in seconds; and T represents the temperature of the soil in degrees Celsius.

3 Finite Element Model Establishment

3.1 Geometric Parameters

The numerical simulation analysis is performed using a full cross-sectional two-dimensional model as shown in Fig. 2. The airport runway has a width of 45 m, with 15 m-wide shoulders on each side. Therefore, the model length is set to 75 am, and the depth is taken as 30 m below the natural ground surface.

Fig. 2. Numerical Model Geometry Dimensions.

3.2 Basic Parameters

This study adopts soil thermal characteristic parameters as the representation parameters for the temperature field model. The soil thermal characteristic parameters include the specific heat capacity and thermal conductivity of the soil. From a perspective of material composition, Xu [5] proposed a method for determining thermal characteristic parameters with mass-weighted average properties. Applying the step function to the additive model, we obtain the following expression:

$$\begin{cases} C = \rho_d C_s + \rho_w C_w \theta_u + \rho_i C_i \theta_i \\ C_s = C_{sf} + (C_{su} - C_{sf})H(T) \end{cases} \tag{7}$$

$$\begin{cases} \lambda = \rho_d \lambda_s + \rho_w \lambda_w \theta_u + \rho_i \lambda_i \theta_i \\ \lambda_s = \lambda_{sf} + (\lambda_{su} - \lambda_{sf})H(T) \end{cases} \tag{8}$$

where Cu and Cf represent the specific heat capacity of unfrozen and frozen soil, respectively, measured in J kg^{-1} ·K^{-1}; Csu and Csf denote the specific heat capacity of the unfrozen and frozen soil skeleton, respectively, measured in J kg^{-1} K^{-1}; λ_u and λ_f stand for the thermal conductivity of unfrozen and frozen soil, respectively, measured in W m^{-1} K^{-1}; λ_{su} and λ_{sf} represent the thermal conductivity of the unfrozen and frozen soil skeleton, respectively, measured in W m^{-1} K^{-1}; θ, θ_u, and θ_i refer to the total water content, unfrozen water content, and ice content, respectively, expressed as a percentage; Cw and Ci indicate the specific heat capacity of water and ice, respectively, measured in J kg^{-1} K^{-1}; ρd, ρw and ρi represent the density of soil, water, and ice, respectively, measured in kg m^{-3}.

In the additive model, the relevant parameters of water and ice vary at a relatively small rate with changes in the external environment and can be treated as constants. The values are given in the Table 1 below:

Table 1. Values of Parameters in the Additive Model.

ρ_w	ρ_i	C_w	C_i	λ_w	λ_i
1000 kg·m^{-3}	918 kg·m^{-3}	4200 J·kg^{-1}·K^{-1}	2100 J·kg^{-1}·K^{-1}	0.63 W·m^{-1}·K^{-1}	2.31 W·m^{-1}·K^{-1}

The initial moisture content is 18.7%. However, the specific heat capacity and thermal conductivity of the soil skeleton primarily depend on the mineral composition and organic matter content, and are temperature-dependent [6]. The measured values of similar soils may also vary.

3.3 Boundary Conditions

Zhu [7] proposed the "attachment layer principle," which replaces the complex subgrade surface with a stable "attachment layer" as the upper temperature boundary condition. The expression is shown as follows.

$$T(t) = 9.22 + 16.37\sin(\frac{2\pi t}{365} + \frac{\pi}{2}) + \frac{0.032t}{365} \tag{9}$$

4 Analysis of the Natural Subgrade Temperature Field

In this section, a natural subgrade with a depth of 30 m and a width of 75 m is taken as an example. The temperature boundary conditions, soil material parameters, and initial temperature conditions determined in the previous section are applied to establish the temperature field model of the natural subgrade. Figures 3 and 4 show the temperature field contour maps of the subgrade.

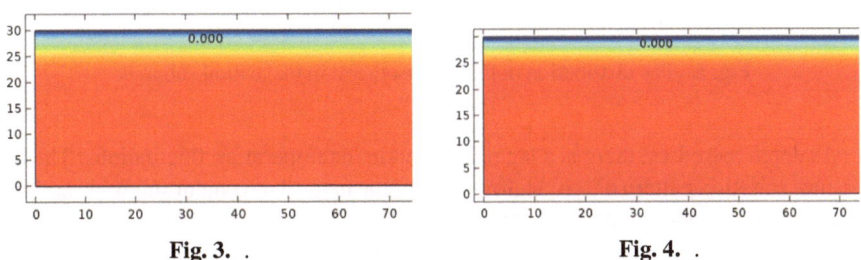

Fig. 3. .   **Fig. 4.** .

It can be observed that the significant variation of the natural subgrade temperature field throughout the entire freeze-thaw cycle mainly occurs within the depth range of −

5 m. Beyond the −5 m depth range, the subgrade soil temperature remains around 10–12 °C, and the 0 °C isotherm extends to approximately −2 m depth below the ground. Next, we focus on studying the temperature field variation within the depth range of −5 m of the natural subgrade, as shown in Table 2.

Table 2. Pattern of Temperature Variation.

Location	Surface	Composition	−0.5 m	$^{-1}$ m	$^{-1}$.5 m	−2 m	−3 m
Maximum Temperature (°C)	26.46	24.13	22.23	20.39	19.12	16.94	14.10
Minimum Temperature (°C)	−6.26	−3.83	$^{-1}$.70	−0.40	0.47	2.46	5.96
Temperature Amplitude (°C)	16.36	13.98	11.97	10.40	9.33	7.24	4.07

The variation pattern of the 0 °C isotherm with respect to time is obtained, as shown in Fig. 5.

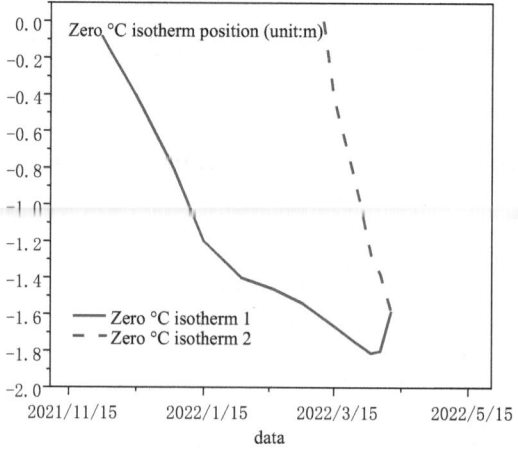

Fig. 5. The distribution of the 0 °C isotherm in the natural subgrade.

As depth increases, there is a lag phenomenon in temperature fluctuations. The temperature variation pattern of the shallow layer of the natural pavement subgrade is similar to that of the surface temperature. With increasing depth, the influence of atmospheric temperature gradually diminishes, and the geothermal curve becomes smoother. The maximum surface temperature is 26.46 °C, and the maximum temperature at a depth of −5 m is 14.10 °C, decreasing by 12.36 °C. The minimum surface temperature is

−6.26 °C, and the minimum temperature at a depth of −5 m is 5.96 °C, increasing by 12.22 °C. The amplitude of surface temperature and the temperature at a position 5 m below the surface over an annual cycle are approximately 16.36 °C and 4.07 °C, respectively, with a decrease of 12.29 °C in temperature amplitude.

By the end of November 2021, the surface temperature started to become negative. Around December 15th, the 0 °C isotherm line was at a depth of about −0.4 m underground. As the external temperature continued to drop, by January 15th, 2022, the 0 °C isotherm line had reached a depth of about $^{-1}$.2 m in the natural pavement subgrade. With the ongoing migration of the freezing front into the subgrade, the 0 °C isotherm line continued to descend. By February 15th, 2022, the 0 °C isotherm line had descended to a depth of about $^{-1}$.46 m, and the frozen part within the subgrade continued to increase. By March 10th, 2022, the external temperature had risen to positive values. A second 0 °C isotherm line appeared within the subgrade, with the first 0 °C isotherm line at a depth of about $^{-1}$.6 m underground. At this point, the frozen soil was sandwiched between two warm soil bodies, resulting in bidirectional melting with a relatively rapid rate. By April 1st, 2022, this was the moment of maximum freezing depth for the year, with the freezing depth being approximately $^{-1}$.81 m. By April 10th, 2022, as the temperature increased, the internal soil continued to warm. With the external environmental temperature continuously rising, the two 0 °C isotherm lines coincided, and the temperature inside the frozen soil was projected to transition entirely to a positive temperature state, resulting in the disappearance of the 0 °C isotherm line within the soil.

5   Conclusion

With increasing depth, there is a lag phenomenon in temperature fluctuations. The temperature variation pattern in the shallow layer of the natural subgrade is similar to that at the surface, but as the depth increases, the temperature variation curve tends to be smoother. During the cold season, when the atmospheric temperature is lower than the temperature of the natural subgrade, the surface temperature of the subgrade starts to decrease. When it reaches the freezing point, freezing begins at the surface and gradually extends to the interior of the subgrade, causing the 0 °C isotherm to gradually descend with freezing time. During the warming period, the atmospheric environment returns to positive temperatures, leading to a gradual recovery of the surface temperature of the subgrade. The thawing of the surface frozen soil begins, and due to the fact that the lower part of the frozen soil remains at positive temperatures, the thawed soil during the warming period will be sandwiched between two positive-temperature soil masses, leading to bidirectional thawing.

References

1. Chen, L.Y., Liu, X., Zeng, C., He, X.Z., Chen, F.G., Zhu, B.S.: Temperature prediction of seasonal frozen subgrades based on CEEMDAN-LSTM hybrid model. Sensors **22**(15), 5742 (2022). https://doi.org/10.3390/s22155742
2. Cheng, Y., Shi, Z., Zu, F.: Temperature field distribution and thermal stability of roadbed in permafrost regions. Int. J. Heat Technol. **39**(1), 241–250 (2021). https://doi.org/10.18280/ijht.390127

3. Feng, R., Lijian, Wu., Wang, Bin: Numerical simulation for temperature field and salt heave influential depth estimation in sulfate saline soil highway foundations. Int. J. Geomech. **20**(10), 04020196 (2020)
4. Wang, F.Y., Pang, W.V., Li, Z.Q., Wei, H.B., Han, L.L.: Experimental study on consolidation-creep behavior of subgrade modified soil in seasonally frozen areas. Materials **14**(18), 5138 (2021). https://doi.org/10.3390/ma14185138
5. Xu, L., Wang, J., Zhang, L.: Frozen Soil Physics. Science Press, Beijing (2001)
6. Meng, S., Sun, Y., Wang, M.: Fiber Bragg grating sensors for subgrade deformation monitoring in seasonally frozen regions. Struct. Control Health Monit. **27**(1), e2472 (2020). https://doi.org/10.1002/stc.2472
7. Zhu, L.: Study on the attached layer of different underlying surfaces in the plateau permafrost region. J J. Glaciol. Geocryol **10**, 8–14 (1988)

Use of Temporal Convolutional Network with an Attention Mechanism and a Bidirectional Gated Recurrent Unit to Capture and Predict Slope Debris Flow Risk

Kai Wei, Qing Li$^{(\boxtimes)}$, Yi Yao, and Yeqing Sun

National and Local Joint Engineering Laboratories for Disaster Monitoring Technologies and Instruments, China Jiliang University, Hangzhou 31008, China
lq13306532957@163.com

Abstract. A novel approach for predicting slope debris flow risk is proposed to address the issue of single-factor data modeling in current slope debris flow risk prediction. The DA-TCN-BiGRU approach combines the dual attention mechanism, temporal convolutional network, and bidirectional gated recurrent unit. Based on the slope debris flow simulation platform, rainfall, soil shear wave velocity, surface displacement, soil pressure and soil moisture data are collected. The data warning features of debris flow risk are captured using the TOSIS entropy method, and the risk level of the slope debris flow is represented based on this. Compared to similar models, this model achieves better slope debris flow risk prediction results.

Keywords: Slope Debris · Risk Prediction · Attention Fusion · Temporal Convolution Network · Bidirectional Gated Recurrent

1 Introduction

Slope debris flow is a common natural disaster in mountainous and hilly areas. They are typically formed by the movement of a large amount of sediment and rock on steep slopes. Slope debris flow is characterized by their rapid flow, high velocity, high volume, high concentration, and destructive nature. The complex generation process of slope debris flows involves the interaction of a number of variables, including rainfall, terrain, soil properties, vegetation cover, and human activity [1]. The main triggering factor for debris flows is usually water, with rainfall being the most significant. Currently, researchers both domestically and internationally mainly predict the possibility of debris flow occurrence through rainfall forecasts, often using a critical rainfall threshold for a specific area to achieve prediction.

Through analyzing historical data of the Huanren Reservoir watershed, Chusheng Xing and others proposed a method of multi-model integrated rainfall forecast for the future 1–3 days in the Huanren Reservoir watershed using artificial neural network (ANN), extreme learning machine (ELM), and support vector machine (SVM) prediction

G. Feng (Ed.): ICCE 2023, LNCE 526, pp. 55–67, 2024.
https://doi.org/10.1007/978-981-97-4355-1_6

models [2]. This study demonstrates the feasibility of using machine learning models for multi-model integrated rainfall forecast and shows that it can improve the accuracy of short-term rainfall prediction. Tang et al. gathered 254 debris flow data and daily cumulative rainfall data in the study area and used the long short-term memory (LSTM) method to anticipate short-term rainfall [3]. They used a statistical classification approach to define the rainfall warning threshold for debris flows. By comparing the predicted values with the threshold, they were able to determine the warning level and the likelihood of debris flow occurrence, thus creating an integrated warning method. Pradeep [4] have proposed a new lightweight weather prediction model based on the structure of Time Convolutional Neural Networks (TCN) and LSTM. This model can be used to forecast the weather for a selected fine-grained geographical location for up to 9 h. Hirschberg J [5] and others used 17 years of rainfall records in the Swiss Alps region and 67 instances of mudslides to determine the critical rainfall threshold. They employed a random forest model (RF) for prediction, which improved the extraction of mining and development information from the data and enhanced the accuracy of the warning performance.

Traditional methods of studying slope debris flows rely heavily on the collection and analysis of geological, hydrological, and meteorological data, as well as the construction of empirical and statistical models [6]. For instance, empirical and statistical methods typically use factors such as rainfall and terrain as predictive indicators, and use empirical formulas or statistical models to make predictions and issue warnings.

The integration of various sensors with suitable prediction models is a successful strategy to obtain more accurate slope debris flow prediction, overcoming the drawbacks of conventional techniques. In this study, a new approach for slope debris flow prediction is proposed, focusing on three main aspects. First, tests simulating the occurrence of slope debris flows were carried out, gathering five different types of sensor data to examine the precursor warning characteristics of the data, and utilizing the TOPSIS entropy approach to determine the danger level of slope debris flows. Second, a brand-new technique for forecasting slope debris flow is presented. It is dubbed DA-TCN-BiGRU and integrates the dual attention mechanism, temporal convolutional neural network, and bidirectional gated recurrent unit. Last but not least, the model is tested using data from a slope debris flow, and experimental findings show that the suggested model outperforms comparable models in terms of accuracy in warning detection and prediction.

2   Slope Debris Flow Simulation Platform

Slope debris flows are fluids with a mixed phase composition of water and solids, and the production process is highly intricate. Heavy rainfall is an external trigger for the onset of slope debris flows, but steep topography and the availability of solid materials are inherent elements that also contribute to their occurrence [7].

The platform for simulating slope debris flows was built with the intention of simulating actual slope debris flows. Figure 1 shows a physical representation of the slope debris flow simulation platform, which consists of a rainfall simulation device, sensor measurement device, and soil loading test box. The system for simulating rainfall creates the necessary amount of strong rainfall for the occurrence of slope debris flows. The soil loading test box replicates mountainous terrain, and its hydraulic support and lifting rods can be adjusted at different angles to represent the range of possible mountain slopes seen in nature.

Fig. 1. Slope debris flow simulation experimental platform

Six sensors, including a tipping-bucket rain gauge, a ground displacement sensor, a soil pressure sensor, a shear wave velocity sensor, and two soil moisture sensors, have been mounted in the experimental platform that we are using to simulate debris flow on slopes. Figure 2 depicts the sensors' mounting locations.

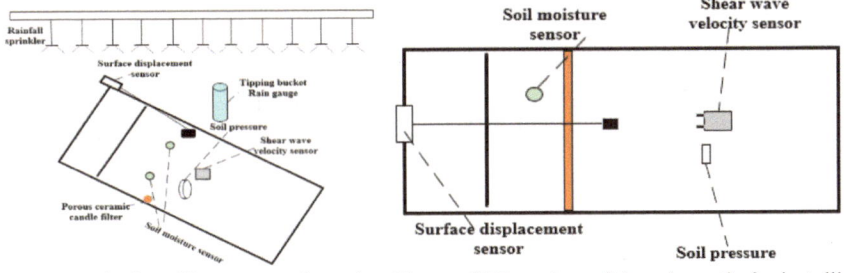

(a)Diagram for installing sensors from the side   (b)Top view of the schematic for installing sensors

Fig. 2. Schematic of sensor installation on slope debris flow simulation platform

3   Methodology

3.1   Topsis-Entry Method

Sensors are used to continuously track a number of factors, such as rainfall, shallow soil moisture content, deep soil moisture content, surface displacement, and soil shear wave velocity, during the slope debris flow simulation process.

There are many methods for risk assessment of debris flow sensor data on slope. In this study, TOPSIS entropy method [8] was adopted to obtain the risk degree of debris flow on slope.

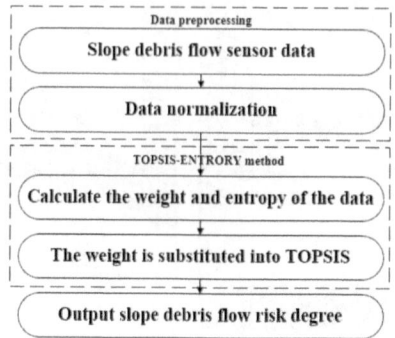

Fig. 3. TOPSIS-ENTRORY calculation flow chart

Figure 3 depicts the flowchart demonstrates how to handle sensor data using the TOPSIS entropy approach. First, the sensor data of slope debris flow is subjected to a simple preprocessing. Then, the normalized data is used to calculate the weight p_{ij} and entropy value e_j using the following formulas:

$$p_{ij} = x_{ij} / \sum_{i=1}^{N} x_{ij} \tag{1}$$

$$e_j = -\frac{1}{\ln N} \sum_{i=1}^{N} p_{ij} \ln p_{ij}, \, e_j \in [0, 1] \tag{2}$$

Calculate the information entropy for each data and compute the information utility value, as shown in Eq. (3).

$$d_j = 1 - e_j \tag{3}$$

To determine the magnitude of the weights for the sensor data, the following steps can be followed:

$$\omega_j = d_j / \sum_{j=1}^{N} d_j \tag{4}$$

Normalizing and standardizing the data, and constructing a weighted matrix as follows:

$$z_{ij} = x_{ij} / \sqrt{\sum_{i=1}^{N} x_{ij}^2} \tag{5}$$

$$z_{ij}^* = z_{ij} \cdot w_j \tag{6}$$

Find the optimal solution z_{ij}^{*+} and the worst solution z_{ij}^{*-} and determine the optimal distance D_i^+ and the worst distance D_i^-. Construct the similarity C_i as follows:

$$\begin{cases} z_{ij}^{*+} = \max(z_1^+, z_2^+, z_3^+, \cdots, z_i^+) \\ z_{ij}^{*-} = \max(z_1^-, z_2^-, z_3^-, \cdots, z_i^-) \end{cases} \tag{7}$$

$$\begin{cases} D_i^+ = \sqrt{\sum_j (z_{ij}^* - z_{ij}^{*+})^2} \\ D_i^- = \sqrt{\sum_j (z_{ij}^* - z_{ij}^{*-})^2} \end{cases} \tag{8}$$

$$C_i = D_i^- / (D_i^+ + D_i^-) \tag{9}$$

3.2 Models

Temporal Convolutional Network. Convolutional Neural Networks (CNNs) are commonly used in image processing, while CNNs used for time-series prediction are called Time Convolutional Neural Networks (TCNs). Due to the size of their convolutional kernels, traditional CNNs are unable to successfully extract features before and after temporal information from sequential input. In 2018, Shaojie Bai [9] et al. used CNNs with dilated convolutions in sequence prediction modeling, allowing CNNs to have a causal convolutional temporal constraint model that captures longer dependency relationships. TCNs have a bigger receptive field as a result. TCNs have a more straightforward and effective model structure. TCNs have also been expanded by numerous academics to include multivariate time-series prediction.

The TCN is composed of multiple residual blocks [9]. In each residual block, the output of the convolutional layer is added to the input of the residual block and fed into the next residual block. To adjust the width of the residual tensor, a *1x1* convolution is added to perform this operation. As a result, the receptive field width of TCN is twice the size of the original causal layer. Therefore, the size of the receptive field can be obtained using Eq. (10).

$$r = 1 + \sum_{i=0}^{n-1} 2(k-1)b^i = 1 + 2(k-1)\frac{b^n - 1}{b - 1} \tag{10}$$

$$n = \left\lceil \log_b\left(\frac{(l-1)(b-1)}{2(k-1)} + 1\right)\right\rceil \tag{11}$$

In the above equation, the variables k and b stand for the size of the convolutional kernel and the dilation base, respectively, and both satisfy the constraint k, b. According to Eq. (11), where n denotes the quantity of residual blocks and l denotes the input tensor's relationship to that quantity. The residual block's input and output are kept equal in length by the 1×1 convolutional process, and the output is protected from future information by the dilated causal convolution.

Bidirectional Gated Recurrent Unit. For processing and making predictions with regard to sequential data, the Gated Recurrent Unit (GRU) is a condensed version of the LSTM neural network [10]. Figure 4 illustrates the GRU organizational structure.

$$r_t = sigmoid\left(W_r[h_{t-1}, x_t]\right) \tag{12}$$

$$z_t = sigmoid\left(W_z[h_{t-1}, x_t]\right) \tag{13}$$

$$\hat{h}_t = \tanh\left(W[r_t \odot h_{t-1}, x_t]\right) \tag{14}$$

$$h_t = (1 - z_t) \odot h_{t-1} + z_t \odot \hat{h}_t \tag{15}$$

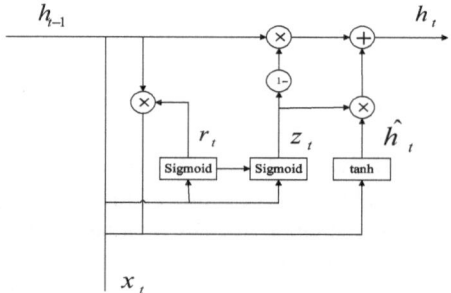

Fig. 4. The structure of GRU

The specific equations are shown in Eqs. (12) and (15), The GRU's update gate is represented by z_t, while its reset gate is represented by r_t. The activation function is the sigmoid function, while the hyperbolic tangent activation function is represented by *tanh*. The equivalent weight matrices are *Wr*, *Wz*, and *W*.

The output in conventional GRU simply depends on the previous input and hidden state at each time step. However, the Bidirectional Gated Recurrent Unit (BiGRU) considers both the previous and future inputs and hidden states of each time step in the input sequence. In BiGRU, the forward GRU determines the time intervals from the start to the finish while the backward GRU determines the time intervals from the finish to the start. The output of the complete sequence is created by concatenating the outputs from the two directions [11].

Figure 5 depicts the formula for BiGRU is similar to GRU, but it calculates the forward and backward GRU outputs separately. The updated BiGRU equation is as follows:

$$\vec{h}_t = GRU\left(x_t, \vec{h}_{t-1}\right) \tag{16}$$

$$\overleftarrow{h}_t = GRU\left(x_t, \overleftarrow{h}_{t+1}\right) \tag{17}$$

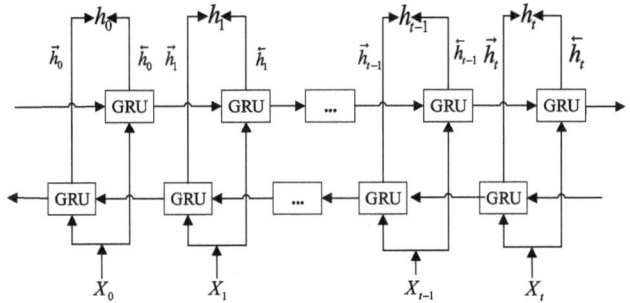

Fig. 5. The structure of BiGRU

$$h_t = [\vec{h}_t, \overleftarrow{h}_t] \qquad (18)$$

Attention mechanism. Attention mechanism was initially used in image tasks to weight important features of the image[12]. The computation of attention weights can be thought of as a query in a key-value pair. The attention mechanism essentially obtains a weight matrix through this procedure. The first step is to determine how comparable the query (Q) and the keys (K) are. This can be done by taking the vectors' dot products. The weights must then be normalized in order to produce weights that may be used immediately. The weighted summation of the weights and values is done in the third step to get the attention values.

$$\alpha_t = \text{softmax}(Q^T K) = \frac{\exp(Q^T K)}{\sum_j \exp(Q^T K)} \qquad (19)$$

$$a = \sum_t \alpha_t V_t \qquad (20)$$

$$Q = W^{q_i} X_t \qquad (21)$$

$$K = W^{k_i} X_t \qquad (22)$$

$$V = W^{v_i} X_t \qquad (23)$$

In Eqs. (19) and (20), α_t represents the attention weight at time t, Softmax is the activation function, α is the weighted sum of weights and variables, Q, K, and V respectively represent the query, key, and value of the attention mechanism. By using Eqs. (21) and (23), Wqi, Wki, and Wvi are the corresponding weights.

DA-TCN-BiGRU Model. Figure 6 shows the model architecture. A time series data set made up of slope debris flow serves as the model's input. The hidden layer's data at time t-1 and the data from n sensors at time t are inputs to the input stage attention mechanism (I-Attn), which outputs the attention weights at time t. I-Attn is sent through TCN after going through a residual block framework. The attention mechanism following TCN (T-Attn) is then created by multiplying the weight vector produced by the attention

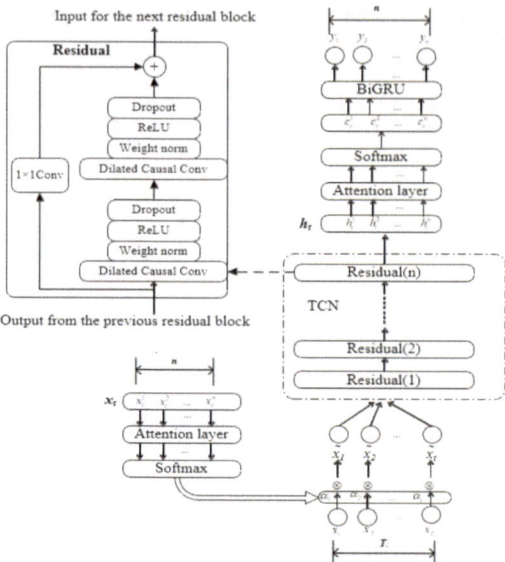

Fig. 6. The structure of DA-TCN-BiGRU

mechanism by the output of TCN. The final prediction value is then output after being passed via a BiGRU layer.

A sliding window is used to implement this model's dynamic sliding prediction in order to accommodate dynamic data., as illustrated in Fig. 7, since sensor data in the real experiments of slope debris flow is returned to the host computer in the form of a continuous array. The result of Fig. 7 is the slope debris flow danger degree for upcoming time steps T_o, using an input of 6-dimensional sensor data of length T_i. Each time step causes the sliding window to advance, producing the predicted value.

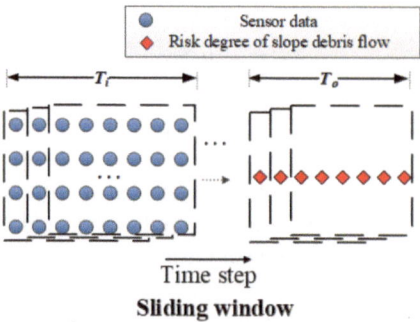

Sliding window

Fig. 7. Sliding window of slope debris flow sensor data

4 Simulation Experimental Data and Model Validation of Slope Debris Flow

4.1 Simulation Experimental Data and Analysis

During the entire simulation process of slope debris flow slope, sensors are used to monitor the process. For experimental purposes, the debris flow simulation platform configures the rainfall circumstances as pre-rainfall and strong rainfall.

Early rainfall: The rainfall intensity is set at 10 mm/h, and it will last for a total of 60 min, divided into 2 stages of raining for 60 min and then stopping for 60 min.

Heavy rainfall: The rainfall intensity is set at 100 mm/h, and it will last for a total of 30 min, divided into 2 stages of raining for 30 min and then stopping for 60 min.

During simulated rainfall, the sediment box was maintained at a 30 angle using a hydraulic lifting rod. The monitoring system collected data every 1 s, resulting in approximately 20,000 data points. These data will be used for modeling slope debris flows.

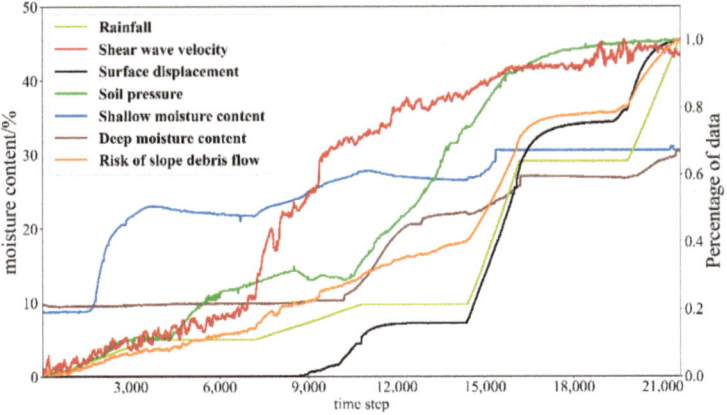

Fig. 8. Experimental data and risk degree of slope debris flow

Figure 8 depicts the curve that was plotted based on the dataset of debris flow slope that was acquired. The technique of debris flow analysis and its link to various sensor data led to the following conclusions:

- Rainfall is a triggering factor for the occurrence of slope debris flows. Its variation directly affects the change in soil moisture content.
- During the first stage of rainfall, the initial infiltration capacity of the soil is greater than or equal to the rainfall intensity. This causes the rainfall water to gradually penetrate into the ground, increasing the moisture content in the shallow soil layer. Eventually, the shallow soil layer becomes partially saturated, and the moisture content does not change further. However, in the deeper soil layer, the initial moisture content is higher than that of the shallow soil layer due to the presence of groundwater. As the rainfall continues and the intensity increases, the infiltration capacity of the surface

soil gradually approaches or becomes lower than the rainfall intensity. This leads to the formation of surface water pooling, resulting in surface runoff and triggering soil erosion and movement.

- As the moisture content of the entire soil layer changes, the shear strength of the soil also changes. The shear wave velocity represents the magnitude of soil shear strength. When the soil moisture content does not reach the critical moisture content, there is a positive correlation between soil shear strength and shear wave velocity. In contrast, there is a negative correlation between soil shear strength and shear wave velocity when the soil moisture content reaches the critical value. Before surface displacement occurs, the shear wave velocity shows a clear upward trend and the soil pressure also increases, both of which serve as precursory warning features for slope debris flow.

- Before the slope enters the stage of sliding flow, the soil moisture content reaches saturation and no longer shows a significant increasing trend. With the occurrence of heavy rainfall, the soil enters the stage of sliding and becomes in a flowing state. As the surface soil is eroded and washed away by surface runoff on the slope, the soil becomes loose, forming channels, and further accelerating the flow velocity of water, promoting the rapid movement and transportation of soil until the soil reaches stability and forms a deposition area at the bottom.

- The risk level of slope debris flow is determined using data from slope debris flow sensors, using the TOPSIS entropy weight method. This method takes into account a combination of surface displacement, soil pressure, shear wave velocity, and rainfall intensity, providing a comprehensive assessment of the hazard level. It highlights the presence of early warning signs and can provide advanced indication of the risk situation that slope debris flow may encounter.

4.2 Model Test

The DA-TCN-BiGRU model has been extensively described in Sect. 3 of this paper. The DA-TCN-BiGRU model was compared against the LSTM, GRU, TCN, BiGRU, and Bidirectional Long Short-Term Memory (BiLSTM) models in order to assess its performance. The performance was evaluated using metrics such as Root Mean Square Error (RMSE) [13], Mean Absolute Error (MAE) [13], and Mean Absolute Percentage Error (MAPE) [13], and the specific equations are shown in Eqs. (24) and (25).

$$MAE = \frac{1}{N} \sum_{i=1}^{N} |\hat{y}_i - y_i| \tag{24}$$

$$RMSE = \sqrt{\frac{1}{N} \sum_{i=1}^{N} (\hat{y}_i - y_i)^2} \tag{25}$$

$$MAPE = \frac{100\%}{N} \sum_{i=1}^{N} \left| \frac{\hat{y}_i - y_i}{y_i} \right| \tag{26}$$

Model running environment: R5-5600G CPU, Windows 11, NVIDIA GeForce GTX 1070 GPU, 16GB RAM, Python 3.6, Keras 2.6.0, TensorFlow 2.6.0. Model testing with sliding windows comes in two flavors. One kind predicts a data length of 10 given an input data length of 100. The other method type predicts a data length of 50 given an input data length of 100. These two categories reflect different input data lengths and prediction lengths. The following settings are made for the DA-TCN-BiGRU model: filters = 32, batch size = 128, kernel size = 8, and gru_units = 16. It's vital to remember that Softmax serves as the attention mechanism's activation function. The TCN model's parameters are as follows: filters = 32, batch size = 128 and kernel size = 8. The depth is set to 32 layers for the LSTM, GRU, BiLSTM, and BiGRU models, while the number of units is set to 16. Adam is the optimization method, ReLU is the activation function, and 0.001 is the starting learning rate. The loss function will determine how to change the learning rate. To prevent overfitting, a regularization dropout of 0.2 is applied. Each model is tested 20 times, and the average performance metrics for each model are calculated and summarized in the following table:

Table 1. Study of different prediction lengths for deep learning models

Models	Metric	Input-Prediction	
		100–10	100–50
LSTM	RMSE	0.03545	0.07657
	MAE	0.03237	0.07141
	MAPE	3.83105	8.41730
GRU	RMSE	0.04912	0.06035
	MAE	0.04493	0.05730
	MAPE	5.31827	6.82432
BiLSTM	RMSE	0.04647	0.05624
	MAE	0.04129	0.05256
	MAPE	4.85380	6.23408
BiGRU	RMSE	0.03661	0.04909
	MAE	0.03360	0.04649
	MAPE	3.97548	5.53723
TCN	RMSE	0.02569	0.03653
	MAE	0.02106	0.03274
	MAPE	2.47684	3.90941
DA-TCN-BiGRU	RMSE	0.01364	0.02739
	MAE	0.01067	0.02502
	MAPE	1.23066	2.98312

Table 1 shows that the DA-TCN-BiGRU model outperforms other prediction models in terms of both "100–10" and "100–50" sliding windows, and it exhibits steady performance.

5   Conclusion and Discussion

The prediction of debris flow risk on slopes is a cross-disciplinary research field involving geotechnical engineering, computer science, and other disciplines. In this study, we conducted simulated experiments on slope debris flows and obtained four types of sensor data. Through the TOPSIS entropy method, we obtained an objective measure of the risk of debris flow on slopes, which characterizes the risk of debris flow occurrence by effectively combining factors from both the atmosphere and the ground. To address the nonlinearity and high complexity of data modeling in debris flow prediction methods, we propose the DA-TCN-BiGRU prediction model, which considers the impact of important information on the prediction and effectively extracts features from the sensor data. Through comparative experiments, it can be concluded that the DA-TCN-BiGRU model has certain effectiveness and feasibility in predicting debris flow risk on slopes, and it has certain engineering practical significance.

Acknowledgments. The authors would like to acknowledge the financial support given to this work, with the support of National Key Research and Development Program Project 2022YFC3003403, Zhejiang Key Research and Development Program Project 2021C03016.

References

1. He, K., Liu, B., Hu, X., et al.: Rapid characterization of landslide-debris flow chains of geologic hazards using multi-method investigation: case study of the Tiejiangwan LDC. Rock Mechanics And Rock Engineering **55**(8), 5183–5208 (2022)
2. Huang, J., Hales, T.C., Huang, R., et al.: A hybrid machine-learning model to estimate potential debris-flow volumes. Geomorphology **367**, 107333 (2020)
3. Oh, C.-H., Choo, K.-S., Go, C.-M., et al.: Forecasting of debris flow using machine learning-based adjusted rainfall information and RAMMS model. Water **13**(17), 2360 (2021)
4. Hewage, P., Behera, A., Trovati, M., et al.: Temporal convolutional neural (TCN) network for an effective weather forecasting using time-series data from the local weather station. Soft Comput. **24**(21), 16453–16482 (2020). https://doi.org/10.1007/s00500-020-04954-0
5. Hirschberg, J., Badoux, A., Mcardell, B.W., et al.: Limitations of Rainfall Thresholds For Debris-Flow Prediction In: An Alpine Catchment . Copernicus Gmbh (2021)
6. Cheung, D.J., Giardino, J.R.: Debris Flow Occurrence Under Changing Climate And Wildfire Regimes: A Southern California Perspective. Geomorphology **422**, 108538 (2023)
7. Yang, Z., Zhao, X., Chen, M., et al.: Characteristics, dynamic analyses and hazard assessment of debris flows in Niumiangou valley of Wenchuan county. Appl. Sci. **13**(2), 1161 (2023)
8. Zhang, D., Yang, J., Li, F., et al.: Landslide risk prediction model using an attention-based temporal convolutional network connected to a recurrent neural network. IEEE Access **10**, 37635–37645 (2022)
9. Bai, S., Kolter, J.Z., Koltun, V.: An empirical evaluation of generic convolutional and recurrent networks for sequence modeling. Arxiv Preprint Arxiv: 180301271 (2018)
10. Ma, C., Xu, X., Yang, J., et al.: Safety monitoring and management of reservoir and dams. Water **15**(6), 1078 (2023). https://doi.org/10.3390/w15061078
11. Dogani, J., Khunjush, F., Seydali, M.: Host load prediction in cloud computing with discrete wavelet transformation (DWT) and bidirectional gated recurrent unit (Bigru) network. Comput. Commun. **198**, 157–174 (2023)

12. Vaswani, A., Shazeer, N., Parmar, N., et al.: Attention is all you need. In: Advances in Neural Information Processing Systems, vol. 30 (2017)
13. Wang, W., Lu, Y.: Analysis of the mean absolute error (MAE) and the root mean square error (RMSE) in assessing rounding model. OP Conf. Ser.: Mater. Sci. Eng. **324**, 012049 (2018). https://doi.org/10.1088/1757-899X/324/1/012049

Deformation Analysis of Deep Buried Soft Rock Tunnel and Adaptive Control Measure of TBM Construction

Maochu Zhang[1,2,3](✉), Tianyou Yan[2,3], Guoqiang Zhang[2,3], and Jianhe Li[2,3]

[1] Changjiang Design Group Co., Ltd., Wuhan 430010, Hubei, China
mczhangucas@163.com
[2] Changjiang Survey, Planning, Design and Research Co., Ltd., Wuhan 430010, Hubei, China
[3] Key Laboratory of Water Grid Project and Regulation of Ministry of Water Resources,
Wuhan 430010, Hubei, China

Abstract. This study focuses on the deformation mechanism of deep buried soft rock tunnel and adaptive prevention and control measures for TBM construction. For the case of large deformation in deep buried soft rock tunnel, the maximum resistance of support measures such as steel arches, anchor rods, and shotcrete is determined based on the support plan. Based on the determination of surrounding rock types, deformation monitoring data, and support resistance, the convergence constraint method is used to theoretically invert the mechanical parameters of deep buried soft rock, and combined with relevant standard parameter values for verification, reasonable mechanical parameters for deep buried soft rock are obtained. Based on the finite element method, open TBM excavation simulation is conducted for tunnel, and the mechanical response of surrounding rock is studied. The reasons for support structure failure are analyzed, and adaptive prevention and control measures for soft rock deformation are studied.

Keywords: Deep Buried Tunnel · Soft Rock · Deformation Analysis · TBM · Control Measure

1 Introduction

With the advancement of national water network and transportation power construction, China has become the country with the highest demand and difficulty in tunnel construction in the world [1]. A large number of water diversion tunnels with burial depths of more than 1,000 m and lengths of tens of kilometres have been planned and constructed in China. In the central Yunnan water diversion project, the length of the tunnel reaches 611.99 km, and the maximum burial depth reaches 1450 m [2]. The total length of the Hanjiang to Weihe river water diversion tunnel is 98.26 km, with a maximum burial depth of 2012 m [3]. In the construction of the water diversion project from Three Gorges reservoir to Hanjiang river, the total length of the water diversion tunnel is 194.8 km, and the maximum depth of the tunnel is 1,182 m [4]. Deep buried long tunnels are playing an increasingly important role in major infrastructure projects in China.

G. Feng (Ed.): ICCE 2023, LNCE 526, pp. 68–79, 2024.
https://doi.org/10.1007/978-981-97-4355-1_7

Deep buried long tunnels are subject to alignment constraints, and most of them have to pass through mountainous areas with complex geological and tectonic backgrounds, and face a lot of engineering geological problems such as high ground stress, rock burst, and large deformation of the soft rock, etc., which are extremely risky for engineering construction. Among them, there are more and more engineering examples of large deformation tunnels with soft surrounding rockmass extruded under high ground stress, which seriously restrict the progress of the tunnel construction. During the construction process of the soft rock tunnel section, due to the weak self bearing capacity of the surrounding rock and the short self stabilization time, the surrounding rock during the stress adjustment stage after excavation exhibits characteristics such as fast deformation speed, large deformation amount, and long duration, resulting in a large total deformation amount and a high force on the support structure. In severe cases, it can lead to support failure [5].

In order to study the deformation mechanism of deep buried soft rock, predict the magnitude of soft rock tunnel deformation, and study the adaptive prevention and control measures of TBM construction, a theoretical inversion of mechanical parameters of deep buried soft rock is conducted based on the convergence-confinement method. Then reasonable mechanical parameter values of deep buried soft rock are obtained. Based on the finite element method, the simulation of the tunnel excavation with open TBM is carried out, and the mechanical response of surrounding rock and the magnitude of deep buried soft rock deformation are studied. Through analysis of the causes of support structure failure, study on adaptive prevention and control measures for soft rock tunnel deformation is carried out.

2   Theoretical Inversion of Mechanical Parameters for Deep Buried Soft Rock Tunnel

2.1   Theoretical Inversion Principles Based on the Convergence-Confinement Method

The Convergence-Confinement method is a theory and method that applies elastic-plastic theory and rock mechanics to underground engineering to explain the interaction between surrounding rock and support. Its principle is shown in Fig. 1. The pressure of surrounding rock decreases with the increase of deformation of the tunnel, and the surrounding rock has been deformed to a certain extent when support measures are taken.

The elastic-plastic analysis method for the deformation and pressure of surrounding rocks in circular tunnels is first proposed by Fenner [6], and then improved by Kastner [7]. The classical pressure theory and bulk pressure theory cannot effectively describe the mechanical behavior of deep buried tunnel after excavation. Currently, elastic-plastic pressure theory is commonly used in engineering. This paper adopts the modified Fenner formula based on elastic-plastic theory to study the interaction between surrounding rock and support [8].

$$u_p = \frac{Mr_0}{4G} \left(\frac{P_i + c \cot \varphi}{(P + c \cot \varphi)(1 - \sin \varphi)} \right)^{\frac{\sin \varphi}{1 - \sin \varphi}} \tag{1}$$

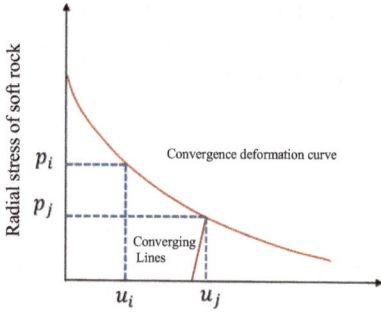

Fig. 1. The principle of Convergence Constraint Theory.

where: u_p is the plastic deformation of the tunnel (m), P_i is the support resistance (MPa), r_0 is the excavation radius of the tunnel (m), c is the cohesion of the rock mass (MPa), φ is the angle of internal friction of the rock mass (°), G is the shear modulus of the rock mass (GPa), and P is the average stress (MPa).

2.2 Calculation of Support Resistance

Open TBM excavation tunnels generally use steel arches, shotcrete and anchor rods as support measures. According to Eq. (1), it can be seen that the plastic deformation of the tunnel is closely related to the support resistance, therefore, it is necessary to calculate the support resistance that can be provided by the various support measures, and on this basis, the relevant rock mechanics parameters can be inverted based on the monitoring deformation of the surrounding rock. In soft rock large deformation tunnel, the maximum support force that can be provided by each support measure is calculated by the following formula.

The support force of the anchor:

$$P_{1,max} = \frac{T_b}{S_c S_l} \tag{2}$$

where: $P_{1,max}$ is the maximum support force of the anchor (MPa), T_b is the average pullout force of the pullout test (MN), S_c is the annular spacing of the anchor (m), and S_l is the longitudinal spacing of the anchor (m).

The support force of the steel arches:

$$P_{2,max} = \frac{\sigma_{ys} A_s}{s r_0} \tag{3}$$

where: $P_{2,max}$ is the maximum support force of steel arch, σ_{ys} is the yield strength of steel arch, A_s is the cross sectional area of steel arch, s is the spacing of steel arch, and r_0 is the radius of tunnel.

Shotcrete support force calculation based on backwall cylinder theory:

$$P_{3,max} = \frac{1}{2} S_c \left[1 - \frac{r^2}{(r+t)^2} \right] \tag{4}$$

where: $P_{3,max}$ is the maximum support force of shotcrete, S_c is the compressive strength of shotcrete (MPa), t is the thickness of shotcrete, and r is the inner diameter of shotcrete.

Therefore, the total support resistance of anchor, steel arch and shotcrete is calculated as:

$$P_{max} = P_{1,max} + P_{2,max} + P_{3,max} \tag{5}$$

2.3   On-Site Monitoring Information

A deep buried hydraulic tunnel in Southwest China experienced serious soft rock deformation during TBM construction, and the supporting structure was severely damaged due to the tunnel deformation, as shown in Fig. 2. When the tunnel undergoes large deformation of the soft rock, monitoring is carried out on the cross-section and arch deformation, and the monitoring results are shown in Figs. 3 and 4.

Fig. 2. Soft rock tunnel deformation causing distortion of steel arch frames.

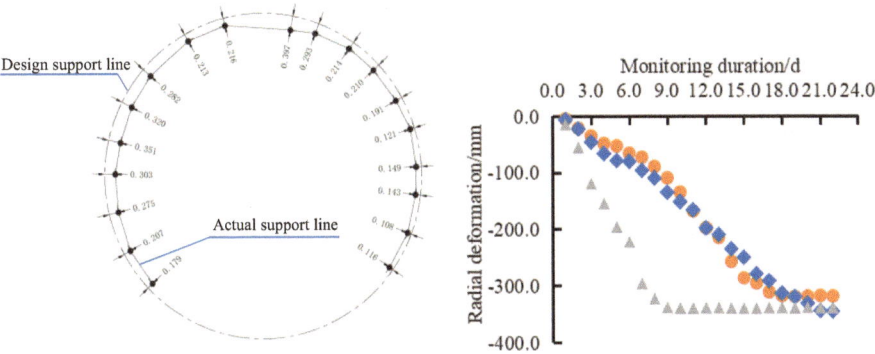

Fig. 3. Monitoring deformation of soft rock tunnel section.

Fig. 4. The monitoring deformation of arch displacement.

According to the on-site monitoring data, within a short period of time (3 days) of TBM excavation, the maximum deformation of soft rock increased by 12 cm. After

22 days of adopting support measures for the soft rock tunnel, the surrounding rock deformation became stable, reaching a maximum deformation of 37 cm. The deformation amount and degree of soft rock are relatively high.

2.4 Parameter Theory Inversion and Validation

The initial support parameters of the tunnel are listed in Table 1. According to each support parameter, the maximum support force that can be provided by each support measure is calculated by Eqs. (2), (3) and (4), and the total support resistance P_{max} is obtained by Eq. (5), $P_{max} = 1.35$ MPa.

Table 1. Support parameter table.

System anchor spacing(m)	Steel arch frame		Sprayed concrete layer	
	Type	Spacing(m)	Type	Thickness
1.25	I20b	0.4	C25	20

After determining the relationship between support force and deformation, with monitoring deformation as the target value, the soft rock strength index is inverted. The φ value is determined according to class V surrounding rock, and then the c value is back calculated. Finally, the mechanical parameters of soft rock are obtained as shown in Table 2. The short-term strength parameters are the inverse parameters of TBM excavation during shield tail support. As the tunnel deformation increases, the surrounding rock around the tunnel becomes relaxed, and the mechanical parameters of the rock mass gradually deteriorate. The long-term strength parameters are soft rock parameters inverted based on monitoring data.

Table 2. The inversion of rock mass mechanical parameters table.

Inversion stage	Buried depth(m)	Severe(kN/m^3)	$\varphi(°)$	c(MPa)	E(GPa)
short-term	1168	22	35.0	1.3	4.0
long-term	1168	22	28.0	0.54	1.4

In order to verify the reasonableness of the mechanical parameters of the soft rock, according to the M-C criterion rock equivalent strength can be obtained from Eq. (6), brought into the inversion of c, φ value, the equivalent rock strength of the soft rock is calculated to be 1.8 MPa.

$$\sigma_{cm} = \frac{2c\cos\varphi}{1 - \sin\varphi} \tag{6}$$

According to the relevant Chinese regulation [9], the internal friction angle of class V surrounding rock: $21.8 \leq \varphi \leq 28.8$, $0.05 \leq c \leq 0.30$, the integrity coefficient of fracture rock mass $K_v \leq 0.15$, the saturated uniaxial compressive strength of the tested rock: 10 MPa $\leq R_b \leq$ 20 MPa. Taking $K_v = 0.15$, then the rock mass equivalent strength is calculated: 1.5 MPa $\leq R_c \leq$ 3 MPa.

$$R_c = K_v R_b \tag{7}$$

where: K_v is the integrity coefficient of the rock mass and R_b is the saturated uniaxial compressive strength of the rock.

The equivalent compressive strength of the rock mass obtained according to the strength index of the rock mass determined by the inversion is similar to the equivalent rock mass strength obtained by the method suggested by the code, so the mechanical parameters of the theoretical inversion have a high reliability.

3 Adaptive Prevention and Control Measures for Open TBM Excavation of Deep Buried Soft Rock Tunnel

3.1 The Simulation of the TBM Excavation Based on Finite Element Method

When predicting large deformation of soft rock and determining support measures, it is necessary to compare and analyze multiple schemes. Due to the large calculation amount of three-dimensional models, it is not convenient to the design and comparison of multiple schemes. RS2 is a powerful elastic-plastic finite element analysis software, which is suitable for underground rock excavation calculation. Therefore, this paper relies on deep buried soft rock tunnel and uses two-dimensional finite element software RS2 for modeling and calculation. The load release rate is used as the basic indicator to simulate tunnel TBM excavation, and the deformation response and adaptive prevention and control measures of soft rock tunnel are studied.

When the tunnel is excavated, the surface of the tunnel will not immediately undergo complete deformation. Due to the unloading effect generated by nearby excavation, the rock in front of the excavation face has already deformed before excavation. As the excavation progresses, the face of the tunnel continues to move forward, and the stress in the surrounding rock changes continuously, causing the tunnel wall to continue to deform. Usually, the tunnel does not reach its "two-dimensional" deformation state until the excavation face is moved a few diameters in front of it. As shown in the schematic diagram of Fig. 5.

For tunnels with poor surrounding rock conditions, the influence of the plastic characteristics of the surrounding rock on the deformation of the tunnels should not be ignored, and the Vlachopoulos-Diederichs tunnel deformation curves [10] can be considered in the plastic zone of the surrounding rock (see Fig. 6), therefore, this curve is selected as the reference curve for the longitudinal deformation characteristics of the tunnel in this paper.

The spatial effect of three-dimensional excavation is achieved by controlling the stress release coefficient of the surrounding rock, and different support reactions are provided to the excavation surface at different excavation steps to achieve stress release of the surrounding rock. The displacement ratio method built-in in RS2 software is used to calculate the support reaction coefficient of the surrounding rock when supporting

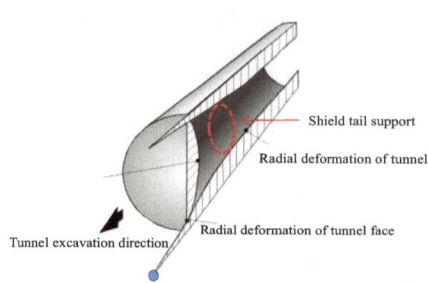

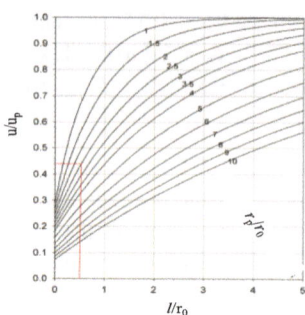

Fig. 5. Tunnel excavation deformation diagram.

Fig. 6. Vlachopoulos-Diederichs characteristic curve of longitudinal deformation of tunnel.

measures are taken. As shown in Fig. 6, the ratio of the distance from the face to the excavation radius of the tunnel during support is taken as the horizontal coordinate ($s = l/r_0$), and the ratio of the radius of the unsupported plastic zone to the radius of the tunnel ($t = r_p/r_0$) is taken as the value of the displacement-dependent curve.

3.2 Soft Rock Excavation Unloading Response

The 2D numerical model built using 2D finite element software is shown in Fig. 7. The tunnel deformation before the installation of support measures is calculated using the Vlachopoulos-Diederichs method. The maximum tunnel wall displacement u_{max} away from the tunnel face and the radius of the plastic zone away from the tunnel face are obtained based on finite element analysis.

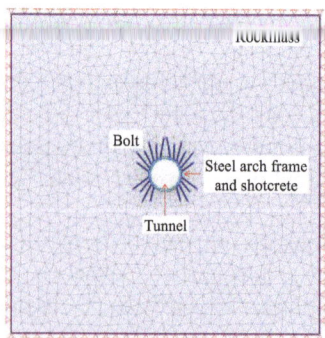

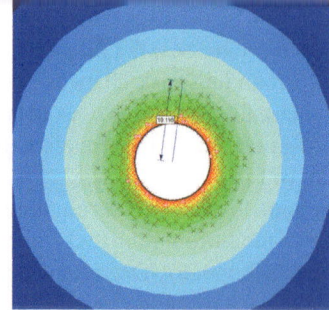

Fig. 7. Two-dimensional analysis model of tunnel cross-section.

Fig. 8. The depth of plastic zone after tunnel excavation.

According to the results of numerical calculation, when TBM excavation and boring, using short-term strength parameters. When the load is completely released, the maximum deformation of the surrounding rock is 13 cm, and the length of the open TBM shield body is taken as 6 m, so the distance from the excavation face to the shield tail installing steel arch and other support measures will be 6 m. The excavation radius of the tunnel is 4.85 m. Thus, the ratio of the distance from the excavation face to the excavation radius of the tunnel during support is calculated as 1.24 (s). The distribution of the plastic zone in the surrounding rock after excavation is shown in Fig. 8. The radius of the unsupported plastic zone is 10.12 m, and the ratio of the radius of the unsupported plastic zone to the radius of the tunnel is calculated to be 2.08 (t).

According to the value of s and t, with reference to the longitudinal deformation characteristic curve of the tunnel (Fig. 6), the ratio of the shield tail rock deformation to the maximum deformation of the surrounding rock is 0.70, from which the deformation of the shield tail rock is calculated to be 9.1 cm. Based on the curve of the relationship between the load release coefficient and the deformation of the arch top in Fig. 9, it is determined that the load release coefficient at the support of the shield tail is 0.04.

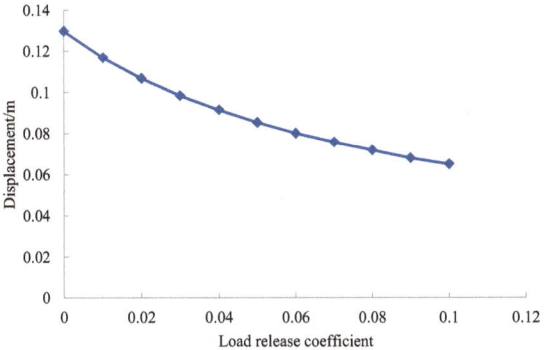

Fig. 9. The curve of the relationship between load release coefficient and arch deformation after tunnel excavation.

Table 3. Surrounding rock response after tunnel excavation.

Tunnel location	Vault	Floor	Left wall	Right wall
Deformation(cm)	79.1	81.1	77.1	79.4

During the excavation and support process of the tunnel, the deformation vectors of the surrounding rock around the tunnel all point towards the inside of the tunnel. As shown in Table 3. The deformation displacements of the top arch and bottom plate are 79.1 cm and 81.1 cm respectively. The convergence displacement of the side wall is 77.1–79.4 cm, and the depth of the plastic zone reaches 9.58 m.

Figure 10 shows the stress diagram and yield state of the support structure. It can be seen that the overall axial force of the steel arch and shotcrete combination support is 4545 kN, and yield phenomenon occurs. The system anchor rods have all undergone tensile failure, with a maximum axial tensile force of 10 kN after yielding.

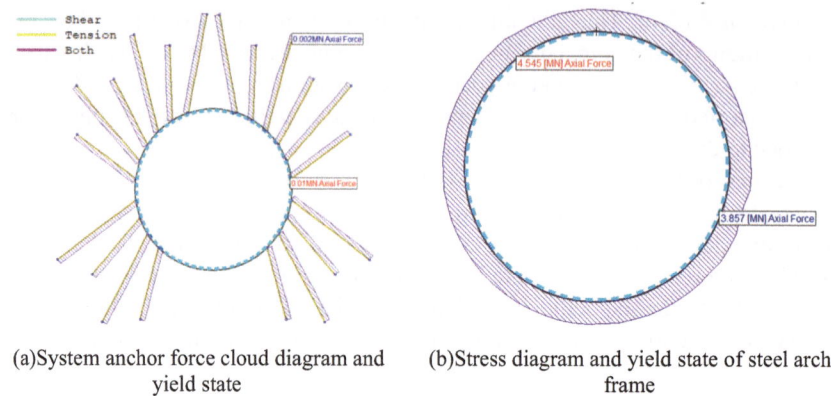

(a)System anchor force cloud diagram and yield state

(b)Stress diagram and yield state of steel arch frame

Fig. 10. Stress cloud diagram and yield state of support structure.

3.3 Principle of Support and Effect of Retractable Steel Arches

The retractable steel arch support system permits the installation of steel arches with sliding devices near the tunnel surface with minimal deformation of the tunnel. The sliding joints will allow further deformation of the tunnel before the arches are subjected to axial loads. There will be no axial forces in the steel arch until the locking strain is reached, but the steel arch will resist bending moments until the locking strain is reached. This system will prevent extreme deformation in the tunnel, but will also prevent failure of the steel arch by ensuring that the supports are not subjected to very high stresses.

A retractable arch is used, with two retractable locking joints are set around the steel arch frame. The sliding gaps of the steel arch within each retractable joint are d_1 and d_2, with a total retractable space of $d_1 + d_2$. The schematic diagram of the principle of retractable arch frame is shown in Fig. 11.

For a steel arch with two sliding gaps, the strain calculation during locking is as follows:

$$\frac{\Delta L}{L} = \frac{d_1 + d_2}{2\pi r} \tag{8}$$

where: ΔL is the sliding gap of the steel arch, which is composed of sliding gaps d_1 and d_2, m; L is the circumference of the steel arch, m. The sliding gap of the steel arch is the sliding gap of the steel arch, which is composed of sliding gaps d_1 and d_2.

The finite element analysis software is used to carry out numerical simulation study on the support of retractable steel arch for the water transfer tunnel through the deep buried soft rock cave section. The calculation model and conditions are the same as those in Fig. 7.

The calculated values of the surrounding rock response of the tunnel excavated with retractable steel arch support are listed in Table 4, and the influence law of the retractable arch frame structure is shown in Figs. 12 and 13.

From Figs. 12 and 13, it can be seen that the tunnel excavation is supported by a retractable steel arch, the deformation displacement of the top and bottom of the tunnel

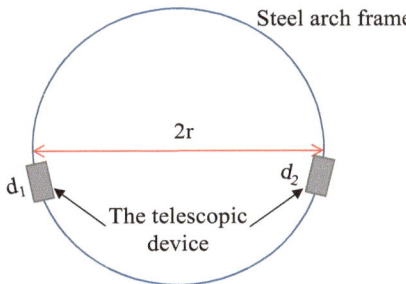

Fig. 11. Schematic diagram of the principle of retractable arch frame

Table 4. The response of surrounding rock supported by retractable steel arch frame after tunnel excavation.

Tunnel location	Vault	Floor	Left wall	Right wall
Deformation(cm)	115.0	117.0	112.0	115.0

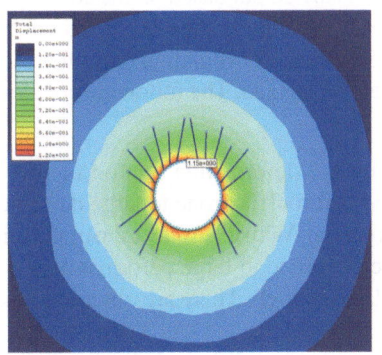

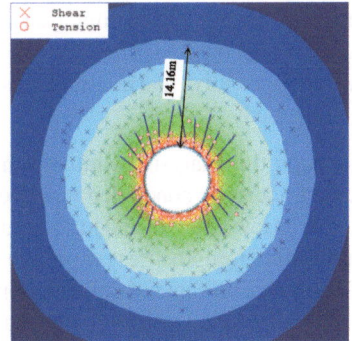

(a)The cloud map of tunnel surrounding rock deformation	(b)The cloud diagram for plastic differentiation of tunnel surrounding rock

Fig. 12. The response of soft rock excavation using scalable steel arch support.

is 115 cm and 117 cm respectively, and the convergence displacement of the side wall is 112–115 cm, and the depth of the plastic zone reaches 14.16 m.

In the short-term parameter working condition, the force of steel arch is 0, which indicates that the reserved deformation gap has not been used up, and there is still space for expansion and contraction of steel arch. After the use of retractable steel arch support, the maximum displacement of soft rock tunnel arch in the long-term parameter working condition is 115 cm, which is higher than that observed in the case of no sliding gap displacement of 79.1 cm, but much smaller than that in the case of no liner displacement of 163 cm. Compared with the maximum axial force of 4545 kN for the steel arch without sliding gap, the maximum axial force of the retractable steel arch with sliding gap is only

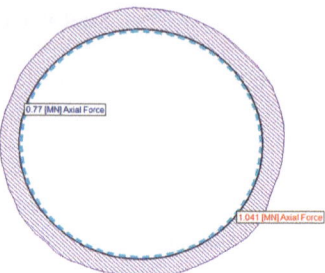

Fig. 13. Support structure stress.

1041 kN, and the maximum stress is 202.45 MPa, which is less than the yield strength of the steel arch, so the structural safety can be ensured.

4   Conclusion

For the case of large deformation in deep buried soft rock tunnel, theoretical inversion of mechanical parameters of deep buried soft rock is carried out based on the convergence-confinement method. The simulation of the open TBM excavation is carried out based on finite element method. The mechanical response of the surrounding rock and research of adaptive preventive and control measures are investigated. The following conclusions can be obtained:

(1) Through the analysis of monitoring data on the deformation section of soft rock tunnel, it can be concluded that deep buried soft rock is prone to large deformation due to its low strength and high stress environment. With the excavation of the tunnel, at a lower strength stress ratio, the soft rock is prone to large deformation, and the magnitude and scale of deformation are relatively high. There are obvious temporal and spatial effects on the deformation of soft rock tunnel. As TBM is excavated forward, the stress on the support continues to increase, ultimately leading to failure.

(2) The Convergence-Confinement method based on elastic-plastic theory can quickly deduce the mechanical parameters of deep buried soft rock based on the maximum resistance provided by the support structure, combined with the classification of surrounding rock and deformation monitoring data. The rationality of parameter values has been verified by relevant specifications.

(3) In the soft rock tunnel section excavated with open TBM, when the support structures are used at the shield tail, the deformation of the surrounding rock is relatively small, and the depth of the plastic zone is not large. As TBM continues to excavate, the support time and the distance from the excavation face increase, the deformation of the soft rock increases, the depth of the plastic zone deepens, and the mechanical properties of the rock mass continue to deteriorate. The stress on the support structure continues to increase until it is destroyed.

(4) The use of retractable steel arch can effectively ensure the safety of the support structure by dissipating the plastic energy of the surrounding rock and releasing the deformation of the surrounding rock through sliding gaps, thereby reducing the

surrounding rock pressure acting on the support structure. This measure has reliable connection performance and good energy consumption ability, and can be applied in open TBM excavation of soft rock tunnels.

Acknowledgments. This work is supported by the fellowship of China National Postdoctoral Program for Innovative Talents (BX20230303), and the post doctoral innovation practice positions in Hubei province (Grant no.: 2023CXGW03). This work is also an autonomous project of Changjiang Design Group Co., Ltd. (Grant no.: CX2021Z01).

References

1. Niu, X.Q., Zhang, C.J.: Some key technical issues on construction of ultra-long deep-buried water conveyance tunnel under complex geological conditions. Tunn. Construct. **39**(4), 523–536 (2019)
2. Yang, Q.G., Zhang, C.J., Yan, T.Y., Liu, Q., Li, J.H.: Integrated research and application of construction and safe operation of long-distance water transfer projects. Chin. J. Geotech. Eng. **44**(7), 1188–1210 (2022)
3. Zhang, J., Li, W., Li, L.M., Wan, J.W., Ding, W.H., Jia, C.: Influence of tectonic characteristics of surrounding rocks on engineering geology of water conveyance tunnel in the Hanjiang-to-Weihe river valley water diversion project. Mod. Tunn. Technol. **58**(3), 23–32 (2021)
4. Wang, J.L., Xiang, J.B., Yan, H.M., Deng, Z.R., Jia, J.H., Xu, Q.: Route selection for water diversion project from Three Gorges reservoir to Hanjiang river based on engineering geology. J. Changjiang River Sci. Res. Inst. **40**(5), 100–105 (2023)
5. Yang, Y., Bai, J.J., Wang, X.Y., Wang, J.D.: High-resistance controlled yielding supporting technique in deep-well oil shale roadways. Int. J. Min. Sci. Technol. **24**(2), 229–236 (2014)
6. Fenner R. (1938) A study of ground pressure. Gluckauf, 74: 681–695 & 705–715
7. Kastner, H.: Statik Des Tunnel-und Stollenbaues. Translated by Tongji University. Shanghai Science and Technology Press, Shanghai (1980)
8. Xu, G.C., Zheng, Y.R., Qiao, C.S., Liu, B.G.: Underground Engineering Support Structure and Design, p. P171. China Water Resources and Hydropower Press, Beijing (2013)
9. Mohurd: Code for engineering geological investigation of water resources and hydropower (GB/T 50487-2008). China Planning Press, Beijing (2008)
10. Vlachopoulos, N., Diederichs, M.S.: Improved longitudinal displacement profiles for convergence confinement analysis of deep tunnels. Rock Mech. Rock Eng. **42**(2), 131–146 (2009)

Effect of Dry-Wet Cycle on Slope Stability of Laterite Subgrade

Rong Xie[1], Shengxia Hu[1(✉)], Li Yong[2], Sen Lin[1], Kuang Bo[1], and Wang Haojie[1]

[1] School of Civil and Architectural Engineering, East China University of Technology, Nanchang, China
201960120@ecut.edu.cn

[2] CRRC Intelligent Transportation Engineering Technology Co., Ltd., Beijing, China

Abstract. In order to explore influencing factors of stability of laterite subgrade slope under the action of dry and wet cycle, the indoor direct shear tests of compacted laterite under different dry and wet cycles were carried out, and the strength parameters were obtained. The stress correction method was introduced to analyze the influence of the number of dry and wet cycles on the strength of laterite and the influence of stress correction on the analysis results. Based on the results of direct shear tests, Geostudio finite element software was used to simulate the stability and influencing factors of laterite soil subgrade slopes of different heights under different dry and wet cycle times. The results show: (1) The shear strength of laterite decreased with the increasing of the number of dry and wet cycle times, and the shear strength index of laterite after stress correction was added compared with the before correction. (2) With the increasing of the times of dry and wet cycles, the slope stability coefficient gradually decreases as an exponential function. The stability coefficient after correction is increased compared with before correction, and it have great significance to improve the reliability of direct shear test data by using stress correction method in landslide treatment and filling engineering. (3) The stability coefficient of the subgrade slope decreases monotonically with the increase of slope height, and the smaller the slope height of the slope, the better the overall structural stability.

Keywords: Compacting Laterite · Dry-Wet Cycle · Direct Shear Test · Stress Correction · Slope Stability

1 Introduction

With the continuous improvement of China's traffic network, highway engineering as an important facility to promote the development of local economy, its construction scale is expanding day by day, and the stability of roadbed slope directly affects the smooth passage and safety of highway [1]. Periodic dry-wet alternation will cause the weakening of the strength parameters of the subgrade soil mass, resulting in the decline of the slope anti-sliding ability. When the sliding force of the slope soil mass is greater than its anti-sliding force, the slope will slide and lose stability, thus inducing landslide accidents [2–4]. In recent years, many scholars have conducted a series of studies on the

© The Author(s) 2024
G. Feng (Ed.): ICCE 2023, LNCE 526, pp. 80–90, 2024.
https://doi.org/10.1007/978-981-97-4355-1_8

influence of dry and wet cycling on the stability of soil slopes through laboratory tests and finite element simulation software. Lian [5] et al. studied the creep characteristics and long-term strength parameters of loess after dry-wet cycle by triaxial creep tests, which the correlation between the failure mechanism of loess landslide and the creep behavior and structure of loess after dry-wet cycle (creep tests) and SEM tests was discussed in detail. Ma Jiangping [6] analyzed the influence of wet and dry cycles on soil mechanical properties of roadbed slope by dry and wet cycle indoor tests and direct shear tests. Wang Deyong [7] studied the stability characteristics of cut slope considering the strength variation of granite residual soil under the condition of dry-wet cycle using numerical software.

In light of the above study, the rainfall evaporation cycle causes the erosion of fine particles of the roadbed, which will reduce the stability of the roadbed slope and affect its long-term service performance. Therefore, it is necessary to study the influence of dry and wet cycling on the stability of laterite roadbed slope. In this paper, the direct shear tests of laterite under different dry and wet cycles are carried out, and the stress correction method is introduced to obtain the change law of laterite shear strength under dry and wet cycles before and after stress correction. Then, the stability of laterite roadbed slope is analyzed by using the direct shear test results before and after correction, and the rule of slope stability safety factor varying with the number of dry-wet cycles is discussed. The research results of this paper have certain theoretical reference significance for landslide control and subgrade filling projects.

2 Test Scheme

2.1 Test Soil Sample

The test soil was taken from a roadside site of a highway in Nanchang City. According to the indoor soil tests, the basic physical properties of red soil were determined as shown in Table 1.

Table 1. Basic physical properties of red soil.

Natural moisture content(%)	Natural dry density(g·cm^{-3})	Plastic limit(%)	Liquid limit(%)	Plastic limit index Ip	Specific gravity(g·cm^{-3})
21	1.7	21	49	28	2.7

2.2 Experimental Methods and Equipment

The red soil was crushed, dried and screened, and the ring knife sample with 21% water content and a dry density of 1.7 g/cm^3 was prepared by static pressing the sample. The dry and wet cycle times were controlled as 0,1,3,5, and the dry and wet cycle of the sample was realized by combining wetting and drying. Dry and wet cycle method: the prepared ring knife sample was soaked in water for 4 h (this is "wet"); The humidified sample is then dried to initial mass in the oven and then stopped drying (this is "dry"). After completion, seal curing for 24 h to make the moisture inside and outside the sample evenly distributed. The ZJ strain controlled direct shear instrument was used to perform fast shear tests on the samples that completed the controlled dry and wet cycles. The test results are listed in Table 2.

Table 2. Shear strength of laterite under different dry-wet cycles.

Vertical pressure (kPa)	Shear strength (kPa)			
	0 times	1 times	3 times	5 times
100	120.21	63.33	36.21	32.23
200	151.68	102.38	57.26	43.99
300	183.91	132.72	76.22	62.57
400	213.3	151.3	87.22	78.12

3 Analysis of Shear Strength Index of Red Soil Sample Under Dry and Wet Cycle

In the process of direct shear test, with the increase of shear displacement, the effective shear area gradually decreases (see Fig. 1), and the distribution of shear stress and normal stress on the shear plane also gradually changes. The whole direct shear test is a dynamic process, so there will be certain errors in the index of soil shear strength obtained by direct shear test [8, 9].

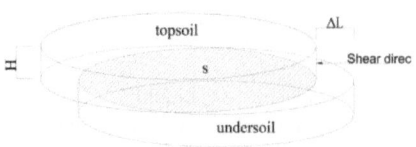

Fig. 1. Shear process diagram

According to the single-point area stress correction theory of direct shear test [10], stress correction expressions (1) and (2) were used to correct the shear strength data in Table 2 to obtain more accurate shear strength indicators. The results are reported in Table 3.

$$\tau^* = \alpha\tau$$

$$\alpha = \left[\frac{1}{90}\arccos\left(\frac{x}{2r}\right) - \frac{x}{2r}\sqrt{4 - \left(\frac{x}{2r}\right)^2}\right]^{-1} \tag{1}$$

$$\sigma^* = \sigma + CR\beta$$
$$\beta = \frac{h(1-\alpha)}{\Delta L}n \qquad (2)$$

where, α is the shear stress correction coefficient; x is the staggered distance of the upper and lower shear boxes; r is the radius of the sample in the shear box; β is the correction factor of normal stress; C is the dynamometer coefficient; R is the dynamometer meter reading; h is the height of the sample on the upper part of the shear box (10mm); ΔL is shear displacement.

Table 3. Modified shear strength of laterite under different dry and wet cycles.

Number of D-W cycles	Corrected stress (kPa)	Normal stress (kPa)			
		100	200	300	400
0	σ^*	72.95	165.87	258.62	352.01
	τ^*	131.02	165.33	200.46	232.50
1	σ^*	85.75	176.96	270.14	365.96
	τ^*	69.03	111.60	144.66	164.92
3	σ^*	91.85	187.12	282.85	380.38
	τ^*	39.47	62.41	83.08	95.07
5	σ^*	92.75	190.10	285.92	382.42
	τ^*	35.13	47.95	68.20	85.15

By collating the direct shear test data of the samples corresponding to the vertical stresses at all levels before and after stress correction under different dry and wet cycles, the relationship between the shear strength of red soil samples under different stress conditions before and after correction can be obtained, as presented in Fig. 2.

Table 3 and Fig. 2 show that the normal stress on the corrected effective shear plane is less than before the correction, and gradually increases with the increasing of the times of dry and wet cycles, while the shear stress on the corrected effective shear plane is greater than that before the correction. It can be seen from the changes of the shear strength with the number of dry and wet cycles under different normal stresses that the shear strength under different stress conditions decreases as a whole with the increasing of the number of dry and wet cycles, that is, the shear strength is negatively correlated with the number of dry and wet cycles, which is consistent with the existing test results[11].

As shown in Fig. 3, the dry-wet cycles led to a continuous decrease in the shear strength parameters. After five dry-wet cycles, the cohesion decreased by 74.23 kPa and the internal friction Angle decreased by 8.42°. Among them, the first dry-wet cycle cohesion decays the most, from 89.4 kPa in the initial state to 38.87 kPa, and then the variation decreases. The reason is that under the action of dry and wet circulation, damage cracks occur in the soil, and the cracks destroy the integrity and continuity of

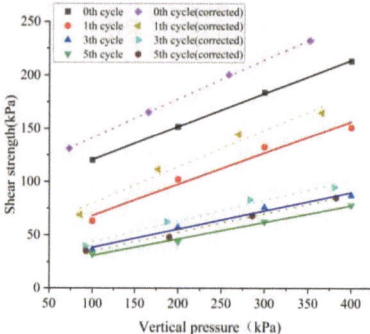

Fig. 2. Shear strength of laterite before and after correction.

the red clay, greatly weaken the bond strength of soil particles, and cause the cohesion of the soil to drop sharply.

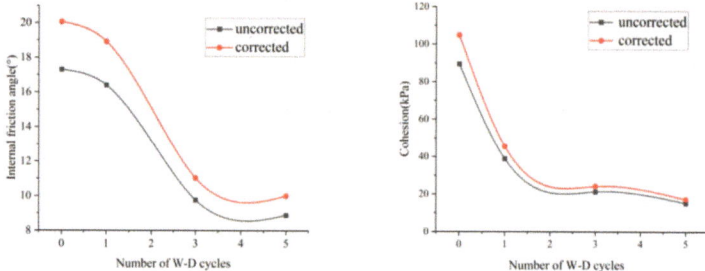

Fig. 3. The cohesiveness of red soil changed with the number of dry and wet cycles before and after correction.

In addition, the shear strength index of soil after the single-point area stress correction is improved compared with that before the correction. Taking the dry-wet cycle as an example, the cohesion and friction Angle of soil are increased by 17% and 16% respectively compared with that before the correction, indicating that there is a large error in the data of the shear strength index in the direct shear test without correction. The application of stress correction can improve the accuracy of the soil shear strength index obtained by the direct shear test. The reliability of stability analysis of laterite subgrade slope is improved.

4   Influence of Wet and Dry Cycles on the Stability of Laterite Roadbed Slope

4.1   Numerical Analysis Model

The dry and wet cycle has a significant effect on the shear strength of laterite, and also affects the stability of laterite roadbed slope. In order to further study the influence rule of laterite roadbed slope stability under the condition of dry and wet cycling, the test results were applied to Geostudio finite element analysis software, which used simplified Bishop method for slope stability analysis, to calculate and analyze the simplified roadbed slope based on an example.

The top width of the roadbed is 12m, and the height of the roadbed is H. According to the requirements of the Code for Design of Roadbed (JTG D30-2015) that the height of the lateritic roadbed slope should not exceed 10m, the slope platform is set as 2 m, the slope rate of the first grade roadbed side is 1:1.5, and the rest is 1: 1.75 (as shown in Fig. 4), respectively simulate the slope stability of the common slope height of 8 m, 9 m, 10 m three horizontal standards.

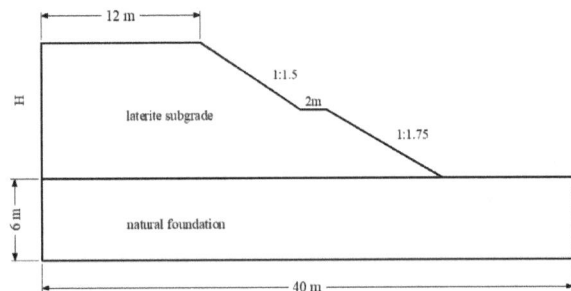

Fig. 4. Slope calculation model.

The bottom of the model is fixed and the soil layers on both sides are subject to horizontal constraints. The constitutive model adopts the Moore-Coulomb model. The model is divided into two layers, the upper layer is filled with compacted red soil, the lower layer is natural foundation, and the material of each layer is assumed to be homogeneous.

4.2 Calculation Parameters

According to the data obtained from the indoor dry and wet cycle tests and the relevant literature consulted, the mechanical parameters of each layer of the slope are shown as Table 4, where c_n and ϕ_n are the cohesion-force and internal friction Angle of the soil during the NTH cycle.

Table 4. Mechanical parameters of each layer.

	Cohesion (kPa)	Internal friction Angle (°)	Unit weight (KN·m$^{-3)}$
Laterite subgrade	C_n	ϕ_n	20
Natural foundation	200	28.4	23.5

4.3 Analysis of Calculation Results

The strength parameters of red soil under the action of dry and wet cycles are substituted into the above compacted red soil roadbed slope model for stability analysis. The most dangerous sliding surface of red soil roadbed slope with a height of 10m under different dry and wet cycles is illustrated in Fig. 5. From the Figure, as the number of dry and wet cycles increases, the position of the most dangerous sliding surface moves to the edge slope surface.

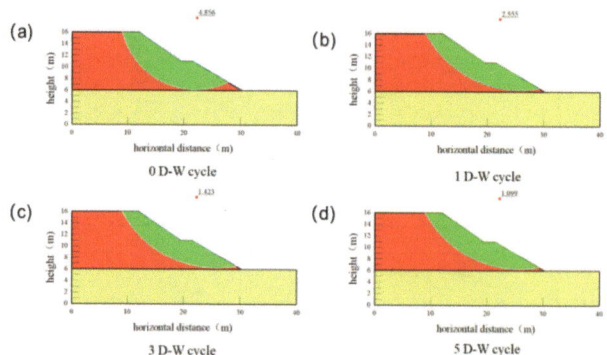

Fig. 5. The most dangerous slip plane with a height of 10 m in the dry and wet cycle downhill slope.

From the test data, it is obtained that the stability coefficient of laterite roadbed slope at different slopes varies with the number of dry and wet cycles, as illustrated in Fig. 6. It can be seen from the calculation results in the Figure, before the dry-wet cycle, the slope is in a stable state as a whole, and the stability coefficient is greater than 4, so the slope is relatively safe at this time, while the stability coefficient begins to decline after the dry-wet cycle. Taking the slope height of 8m as an example, the safety factor of the slope becomes 3.010, 1.674 and 1.283, respectively, after the first, third and fifth times of wetting and drying. The slope coefficients decreased by 49.6%, 72% and 78.5% respectively after one, three and five times of wetting and drying. It can be seen that the damage effect of dry and wet cycles on soil stability increases with the increase of cycles.

The combination of rainfall and evaporation is one of the important reasons for slope instability in laterite roadbed. The cyclic action of rainfall and evaporation not only destroys the structural bond between soil particles, but also causes the erosion of laterite structure, which leads to the gradual reduction of the strength of soil in the roadbed slope. During the evaporation process, water escapes, solid particles keep the same size, move and rearrange, which reduces the void space and hardens the soil to produce shrinkage cracks (see Fig. 7). The formation of cracks exposes the deep soil to the atmosphere and expands the channel of water infiltration and evaporation, which is an important prerequisite for slope instability. In the rainfall stage, continuous rainfall forms runoff, which flows into the inclined body along the crack, has the function of scouring, carrying fine particles, softening the soil, and making the crack cut deeper

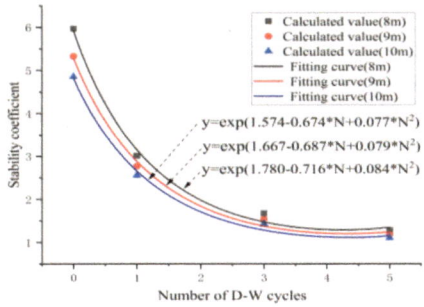

Fig. 6. The stability coefficient of laterite subgrade slope changes with the dry-wet cycle.

into the soil. In addition, under the action of continuous rainfall, the soil body weight increases, the sliding force increases, and the shear strength decreases, which ultimately leads to the landslide and instability failure of the roadbed slope[12] Therefore, in order to ensure the stability of the roadbed slope during the operation period, it is necessary to do a good job of slope waterproofing and regular site survey to find cracks and seal them in time to prevent rainwater from pouring into the roadbed slope and reducing the soil strength, thus causing landslides.

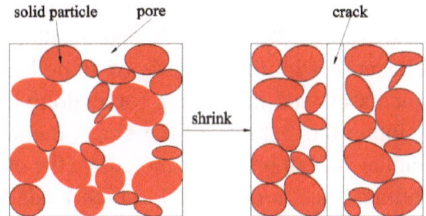

Fig. 7. Diagram of soil shrinkage and cracking.

Figure 6. Also shows that with the increase of the number of dry and wet cycles at different slope heights, the stability coefficient of laterite roadbed slope presents a trend of first rapid and then slow decline, and the change law of this coefficient with the dry and wet cycles can be described by exponential function (3), and the fitted correlation coefficient R^2 is greater than 0.99.

$$Fs = e^{A+BN+CN^2} \tag{3}$$

where: Fs is the stability coefficient; N is the number of dry and wet cycles; A, B and C are the fitting parameters of the exponential function.

The slope model with slope height of 10m was used to calculate and analyze the change of stability coefficient of red soil subgrade slope under different dry-wet cycles before and after stress correction, and the results are shown in Fig. 8. According to the Fig. 8, the stability coefficient of homogeneous laterite subgrade slope after correction changes with dry-wet cycles similar to that before correction. The stability coefficient decreases rapidly under the first three dry-wet cycles, and then the decreasing speed slows down and gradually tends to be flat or slightly decreases. In addition, the stability coefficients obtained after modification are increased by 17.2%, 16.9%, 13.5% and 12.9% respectively. Therefore, using stress correction method to improve the reliability of direct shear test data is of great significance in landslide control and fill engineering.

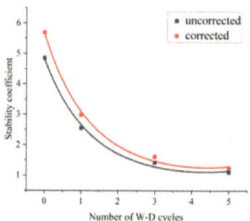

Fig. 8. The stability coefficient of laterite subgrade slope changes with the dry-wet cycle before and after correction.

The variation of the stability coefficient of laterite roadbed slope with different slope heights is shown in Fig. 9. From the Fig., the stability coefficient of roadbed slope decreases monotonically with the increase of slope height, that is, it decreases with the increase of slope height. The smaller the slope height of the slope, the better the overall structural stability. With the increase of slope height, the stability coefficient of the roadbed slope decreases from 1.283 to 1.099, and the slope will become unstable and fail. This is mainly because with the increase of slope height, the corresponding slope soil body weight increases, and the sliding force along the slope side gradually increases. In order to maintain the stability of the slope, the anti-sliding force is also required. Therefore, when the side slope height increases, the original balance between sliding force and anti-sliding force is destroyed, which leads to the reduction of the safety factor of slope stability.

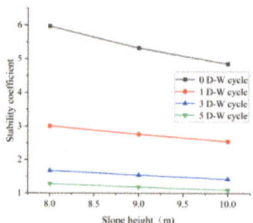

Fig. 9. Variation of stability coefficient of laterite subgrade slope with different slope heights.

5   Conclusions

In this paper, direct shear tests are carried out on the laterite samples that are made by soil taken from the site in Nanchang city, and the accuracy of the strength parameters obtained by tests is improved by stress correction method. Then, the test results are applied to the numerical analysis model for the stability analysis of laterite roadbed slope, and the influence of dry and wet cycling on strength parameters and slope stability is discussed and analyzed. The following conclusions are obtained:

- Under different stress conditions, the shear strength of laterite decreases as a whole with the increase of the times of dry and wet cycles, and the shear strength index of laterite after stress correction is improved compared with that before correction. The stability coefficient calculated by the stress correction method is increased compared with that before correction. Therefore, using stress correction method to improve the reliability of direct shear test data is of great significance in landslide control and fill engineering.
- With the gradual increasing of the number of dry and wet cycles, the stability coefficient of the slope decreased significantly, and the overall stability of the laterite roadbed slope showed a downward trend, along with the slope tended to become unstable after 5 dry and wet cycles.Therefore, in the remediation and treatment of laterite subgrade slope and other projects, the relevant parameters should adopt the value of attenuation after multiple dry and wet cycles, otherwise the safety and quality of the project cannot be guaranteed enough.
- The stability coefficient of roadbed slope decreases monotonically with the increase of slope height, and the smaller the slope height of the slope, the better the overall structural stability.

Acknowledgments. The authors thank the financial support from the doctoral research start-up fund project of the East China University of Technology (Grant Nos.DHBK2019239).

References

1. Nong, M.: Stability analysis and protection measures of roadbed high slope. J Western Transport. Sci. Technol. **1**(10), 8–11+33 (2021)
2. Du, J., Tong, F.: Effects of dry-wet cycle and rainfall on the stability of loess slope. J. Guangxi Univ. (Nat. Sci. Ed.). **45**(04), 783–791 (2019)
3. Yang, X., Chen, D., Liu, Y.: Fracture evolution and slope stability of granite residual soil under dry-wet cycle. J. Xiamen Univ. (Nat. Sci. Ed.) **61**(04), 591–599 (2022)
4. Liu, G.: Influence of dry and wet cycle mechanism on stability of weak expansion geotechnical slope. J. China Water Transp. **20**(04), 247–248 (2020)
5. Lian, B., Wang, X., Zhan, H., et al.: Creep mechanical and microstructural insights into the failure mechanism of loess landslides induced by dry-wet cycles in the Heifangtai platform. China. J Eng. Geol. **300**, 106589 (2022)
6. Ma, J.: Influence of dry and wet cycling on soil mechanical properties of roadbed slope. J. Highw. Road China (02), 104–105+108 (2023)
7. Wang, D., Zhou, S., Han, Y., et al.: Research on the stability of granite residual soil slope under dry and wet cycle. J. Highw. **66**(01), 1–6 (2021)
8. Kai, Y., Xin, Y., Yongshuang, Z., et al.: Analysis of direct shear test data based on area and stress correction. J. Chin. J. Rock Mech. Eng. **33**(01), 118–124 (2014)
9. Cheng, L., Song, Q., Li, L., et al.: Study on shear strength index of wheat stalk soil under direct shear test considering stress correction. J. Sci. Technol. Eng. **17**(33), 319–327 (2017)

10. Lei, S., Hui, H.: Analysis of shear stress-shear displacement and strength of soil considering shear area correction. J. Mech. Pract. **44**(03), 640–645 (2022)
11. Chen, J.S.: Crack evolution of red clay under dry and wet cycle and its influence on shear strength. J. Hydrogeol. Eng. Geol. **45**(01), 89–95 (2018)
12. Gao, Q.F., Zeng, L., Shi, Z.N.: Effects of desiccation cracks and vegetation on the shallow stability of a red clay cut slope under rainfall infiltration. J. Comput. Geotech. **140**(1), 104436 (2021)

Formation and Evolution Mechanism of the ZhengGang Giant Ancient Landslide

Wei Cheng[1,2], Junyao Luo[1(✉)], Taiqiang Yang[1], Xiaolong Jiang[1], Xuefeng Fan[1], Yang Yang[1], Yelin Feng[1], and Qingfu Huang[1]

[1] Power China Kunming Engineering Corporation Limited, Kunming 65000, China
luojunyao_kmy@powerchina.cn
[2] Faculty of Architecture, Civil and Transportation Engineering, Beijing University of Technology, Beijing 100124, China

Abstract. During the continuous uplift of the Qinghai-Tibet Plateau, the genetic mechanism of giant ancient landslides in the deep-cutting rivers has attracted widespread attention. The formation of giant ancient landslides is closely related to the evolution history of rivers. Using the methods of field investigation and numerical simulation, the formation and evolution process of the Zhenggang giant ancient landslide of the Lancang river upstream is studied. The correlation between the landslide formation and the valley evolution is analyzed. The toppling deformation body evolution of a giant ancient landslide and its genetic mechanism is studied. The results show that the bank slope unloading caused by the down-cutting of the river is the main effect factor of the landslides occurrence. The structure of soft and hard inter-bedded, and rainfall infiltration are all important influencing factors. The bank slope unloading caused by the down-cutting of the river promoted the bending and toppling of the steep rock mass. The giant ancient landslide takes the toppling fracture zone as the slip zone slipped.

Keywords: Qinghai-Tibet Plateau · Zhenggang Ancient Landslide · Toppling Deformation Body · Genetic Mechanism · Bank Slope Unloading

1 Introduction

The Qinghai-Tibet Plateau is uplifted by the collision and compression of the Indian and Eurasian plates. Many large-scale landslides have occurred in the deep-cutting rivers, which genetic mechanism has attracted more and more attention [1–3]. The formation of the landslide is closely related to the evolutionary history of the river. The formation mechanism of giant rock ancient landslides is studied by combining the method of field investigation and numerical simulation. The relationship between the occurrence of landslides and the evolution of the valley is analyzed. The formation process and mechanism of the toppling deformation bodies transformed into landslides are studied. The results show that the weak interlayer parallel to the slope direction is the main cause of the landslide. It is also affected by the factors such as the strength of the rock mass, the degree of the bank slope unloading, and rainfall infiltration [4–6]. The down-cutting

G. Feng (Ed.): ICCE 2023, LNCE 526, pp. 91–106, 2024.
https://doi.org/10.1007/978-981-97-4355-1_9

of the river leads to the slope unloading toward the empty direction. The front edge of the toppling deformation body is eroded and loses part of the anti-sliding force, which causes the steeply dipping rock layer to gradually topple. The slip zone evolved from the toppling fracture zone is the decisive factor for the formation of giant ancient landslides [7, 8]. Construction excavation and continuous rainfall will further induce landslides [9, 10].

During the continuous uplift of the Qinghai-Tibet Plateau, many rivers cutting deeply, such as the Lancang river, Nujiang River, and Jinsha River [8, 11, 12]. Under the interaction of internal and external dynamic geological processes, the topography of the study area has changed significantly. Climate, lithology, valley deep-cutting, and other factors are all important factors influencing the occurrence of large-scale landslides. Existing studies have shown that the shape of the river and the profile of the river play a key role in the process of large-scale landslide formation. During the continuous and rapid uplift of the Qinghai-Tibet Plateau, the complex geological environment and climate change have caused large-scale geological disasters along the river [13–15].

With global climate change, the frequency of extreme weather events has gradually increased, leading to frequent occurrences of geological disasters. Due to the rainfall infiltration effect, the landslides induced by rainstorms occur frequently. With the increase of pore water pressure during rainfall, the effective stress of rock and soil mass decreases significantly. The shear strength of the slope is weakened [16–18]. Most of the landslides occur that attributed to rainfall. The cracks caused by strong rainfall and bank slope unloading are considered to be important interlayer surfaces of bedding landslides [2, 19, 20]. Slope ridges are affected by hydrostatic pressure, and the rise of groundwater level on the sliding surface is a favorable trigger for the occurrence of landslides. It has been found that most of the landslides are closely related to human engineering activities such as engineering excavation and mining [21–23]. Large-scale landslides along the Lancang river and other rivers have received widespread attention.

In the Lancang river, there are also several large-scale landslides of different scales, such as Gendakan landslide, front toppling deformation body, Meilishi 4# landslide, Meilishi 5# landslide, etc. [24, 25]. It is very effective to make full use of river profile and river topography to infer the influence of internal and external dynamic geological action on a landslide. Existing studies on the causes of large-scale landslides along the river show that the landslide process has an important relationship with geological environmental conditions, geological age, and valley evolution. However, there are few studies on the relationship between the giant ancient landslide that evolved from the toppling deformation and the evolution of the river [3, 7, 26, 27].

The purpose of this study is aim to investigate the formation process of landslides, river cutting process and discuss the relationship between landslides and valley evolution. Through field investigation and numerical analysis, the boundaries of landslides, the causes of landslides, and the formation mechanism of giant ancient landslides are determined. At the same time, the controlling factors of landslides are studied. The research results are helpful to describe the landslide process related to the evolution of rivers. The flow chart of the research method is as follows: The "Regional Geological Background" section describes the geological background of the landslide area. The

"Zhenggang Landslide" section describes the basic characteristics and deformation characteristics of the landslide. "Numerical analysis of the formation and evolution process of Zhenggang landslide" describes the toppling deformation process of the slope in the process of valley down-cutting. In the "Discussion" section, the incubation process of the toppling deformation of steeply inclined layered rock mass and the evolution process and genetic mechanism of giant ancient landslides are discussed in depth. "Conclusion" Partially concluded (Fig. 1).

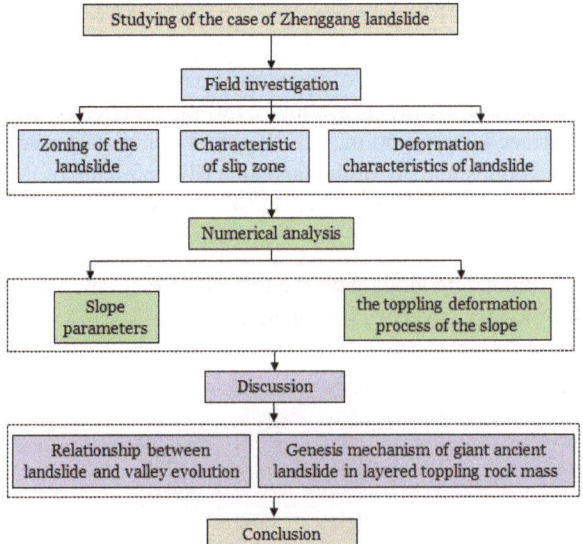

Fig. 1. Research methodology flowchart.

2   Regional Geological Background

The study area is located on the left bank of the Lancang river upstream, about 3–4 km long and 1–2 km wide. The strong unloading of rock masses occur during the rapid down-cutting process of rivers. The valley is in the shape of a "V" (Fig. 2). The elevation of the mountain top is about 4000–6000 m. The maximum height difference is about 1500–3200 m. The terrain is high in the north and low in the south. The mountains spread out in a north-south direction as a whole. The slopes on both sides of the Lancang river are relatively steep, generally ranging from 20° to 45°. Some sections of the river are relatively wide, with well-developed terraces, and generally ranging from 10–25°. The three major rock types, sedimentary rock, magmatic rock, and metamorphic rock, are exist in the study area. The strata are mainly Devonian (D), Carboniferous (C), and Triassic (T). The lithology is mainly basalt sandwich slate, andesite, quartz sandstone, metamorphic sandstone, mudstone, and a small amount of thin limestone.

The study area is located in the Hengduan Mountains in the southeast of the Qinghai-Tibet Plateau. The dam site of the Gushui Hydropower Station is located in the earthquake intensity area of VII. The annual average temperature in the study area is 4.7 °C. The annual average rainfall is 633.7 mm. According to on-site investigations, the following four-level terraces have developed in the Lancang river section in the area:

1) First-level terrace: It is the modern riverbed of the Lancang river (the annual water level is 2070–2078 m above sea level), mainly composed of sand and gravel deposits on the floodplain.
2) Second-level terrace: The elevation difference from the current river surface is about 15 m, developed on the left bank of the upstream of the landslide. The terrace elevation is 2080 m (front edge)–2100 m (rear edge). The exposed width is about 40 m. It is composed of sand gravel layer and silt fine soil. A base terrace formed by a dual structure, the middle and rear part of the terrace passes through 214 National Road.
3) Third-level terrace: The elevation is about 2130 m. The height difference from the current river surface is about 60 m. It is the most complete terrace, and it is more developed on the opposite bank of the landslide (the left bank of the river).
4) Fourth-level terrace: only remains in this landslide area, with a development elevation of 2250–2280 m.

The bank slope unloading along with the terrace formation during valley cutting is a crucial factor for the formation of Zhenggang landslide.

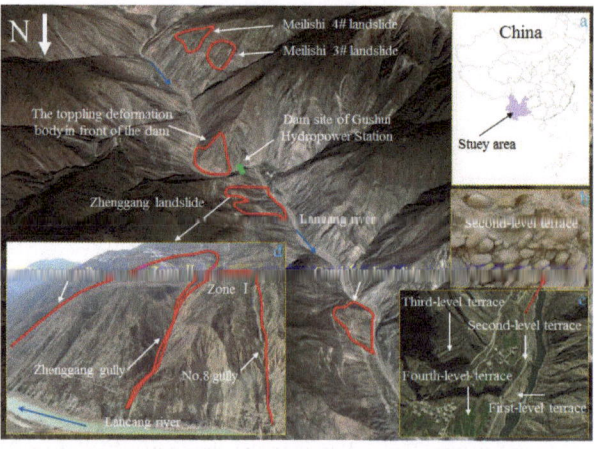

Fig. 2. The typical landslide in the study area (a. the location of Zhenggang landslide; b. Second-level terrace Pebble layer; c. First to fourth-level terrace; d. Full view of Zhenggang landslide).

3   Zhenggang Landslide

3.1   Basic Characteristics and Zone

The Zhenggang landslide deposits are located in the east of the Zhenggang mountain beam, about 900 m away from the Gushui Hydropower Station dam site upper. The maximum vertical length of the landslide is about 1010 m. The maximum horizontal width is about 1100 m. The top elevation is about 2720 m. The bottom elevation is about 2170 m. The thickness of the landslide accumulation body is about 15–60 m. The average thickness is about 26.9 m. The volume is about 4750×10^4 m^3. It is a giant rocky ancient landslide. The landslide accumulation body is affected by the erosion and down-cutting of Zhenggang gully. The old landslide accumulation body is divided into Zone I and Zone II (Fig. 3). The size of zone II is larger than that of zone I, all of which are long tongue-like with a wide bottom and a narrow top. The overall morphology is "M" shaped.

The landslide deposits in Zone I have an elevation of 2180 m–2650 m. The left side is bounded by the No. 8 gully. The right side is bounded by Zhenggang gully. The front edge is relatively steep, with a slope of about 40°, and an elevation of about 2000 m at the front edge. The middle part tends to be gentle, with a length of about 430 m, a width of about 390 m, and an overall slope of about 10°. The rear edge has a steep wall of landslide with a slope angle of about 42°.

The distribution elevation of the landslide deposits in Zone II is 2180 m–2730 m. The left side is bounded by the Zhenggang gully. The right side is bounded by Yagong gully. The front edge is steep, with a slope of about 40°. The middle part tends to be gentle, with a length of about 190 m, a width of about 450 m, and an overall slope of about 10°. The back scarp at the rear edge is obvious, with a slope of about 30°.

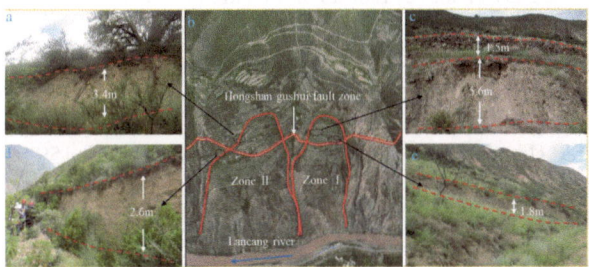

Fig. 3. Deformation characteristics and zoning of Zhenggang landslide (c, e. Tension cracks in the landslide accumulation zone I; a, d. Tension cracks in the landslide accumulation zone II).

3.2 Material Composition

The material composition of Zhenggang landslide can be divided into residual slope accumulation layer, ice-water accumulation layer, collapse slope accumulation layer, bottom slip accumulation layer, and slip zone soil.

1) The residual slope layer is mainly crushed sandy soil and silt, with a small amount of rock. The content of crushed rock and rock is about 30%–40%. The diameter is about 1 cm–10 cm.

2) The ice-water accumulation layer can be divided into two layers, mainly composed of blocks, gravel, sand, and silt. The diameter of the blocks in the block and crushed rock layer is about 5 cm–25 cm. The diameter of the crushed rock is less than 5 cm. The gaps are filled with sandy silt and a small amount of clay. The overhead phenomenon

is obvious and loose. The diameter of the gravel in the gravel sand and silt layer is about 2 cm–3 cm. The gaps are mainly filled and cemented by sand, silt, and clay. The degree of compaction is good.

3) The collapsing layer is mainly composed of gravel soil mixed with sandy gravel soil. The diameter of gravel is about 0.5 cm–2 cm. The content is about 20%–30%. The diameter of block stone and crushed stone soil are about 6 cm–30 cm. The content of block stones is about 45%–55%. The content of crushed stone is about 20%.

4) The bottom-slip accumulation layer includes blocks, gravel soil, and broken rock mass. The diameter of the block and gravel soil is about 10–40 cm, and the maximum can reach 100 cm–300 cm. Broken rock masses are mostly blue-gray metamorphic sandstone, gray-brown slate, and light gray limestone. Locally it has a layered structure, and the fractured rock mass at the trailing bed attitude is N25°–30°W, SW$\angle$20°–40°.

3.3 The Landslide Structural Characteristics

According to the PD1704 survey results of the landslide accumulation body in Zone I, it can be known that 0 m–4.5 m is the residual slope accumulation layer; 4.5 m–9.0 m is the collapse slope accumulation layer; 9.0 m–32.6 m is the bottom slip accumulation layer; and 32.6 m–68 m is the bottom slip accumulation layer.

According to the PD144 survey results of the landslide accumulation body in Zone II, it can be known that: 0 m–4.3 m is the residual slope accumulation layer; 4.3 m–16.0 m is the collapse slope accumulation layer; 16.0 m–107.5 m is the bottom slip accumulation layer; and 107.5 m–133.3 m is the bedrock: the lithology is limestone, which is broken, and the bed attitude is S29°–35°E, SW$\angle$0°–20°; 133.3 m–159.0 m is the bedrock: the lithology is layered slate, and the bed attitude is N32°W, SW$\angle$5°; 159.0 m–176 m is the bedrock: the lithology is gray-green basalt, SN, W$\angle$55°, which is relatively broken and in the shape of fragments.

3.4 Characteristics of the Slip Zone

1) The slip zone of the landslide accumulation body in Zone I is exposed at 73.7 m on the right wall of PD1704. The slip zone soil is gray and brown clay, which is plastic. The fine particles in the slip zone soil are sub-circular. Larger particles are sub-circular to sub-angular. The diameter of gravel is about 0.5 cm–1 cm. The composition is slate. The thickness is about 10 cm–20 cm, it is a dark gray clay interlayer. The bed attitude of the slip surface is N25°W/NE$\angle$20°–25° (Fig. 4a).

2) The slip zone of the landslide accumulation body in Zone II is exposed in the PD144 right wall 70–80 m. The thickness is about 7 cm. The clay content in the zone is relatively high. There are some white calcite bands in the zone. The bed attitude is 140°$\angle$14°. The upper part of the sliding zone is mainly breccia gravel soil with a particle size of about 1 cm–10 cm. The lower part of the sliding zone is crushed stone soil, which is gray and brown overall, with a coarse-grain content of about 20%–30%, and a particle size of about 6 cm–10 cm. The rock composition is sandstone and slate (Fig. 4b).

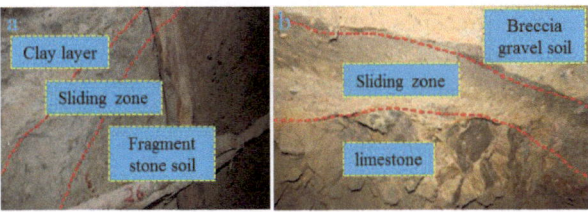

Fig. 4. Characteristics of the sliding belt (a. PD1702 at 26 m on the right wall; b. PD144 at 90 m at the adit).

3.5 Zoning Deformation Characteristics

According to the results of on-site survey, combined with the structural and deformation characteristics of the landslide, the Zhenggang landslide I zone is divided into a strong deformation area at the front edge, a strong deformation area on the right side, a strong deformation area in the middle, and a strong deformation area at the rear edge. The Zhenggang landslide II zone is divided into the front edge and its left strongly deformed area, the rear edge and its right strongly deformed area, and the middle strongly deformed area.

Zone I deformation characteristics

a) The strong degeneration area at the front edge of Zone I is dominated by the deformation and destruction of the overburden on the slope surface. The material composition of the overburden is broken stone soil. The lithology of the broken stone parent rock is mainly slate, limestone, and sandstone. The deformation zone is eroded by the Zhenggang gully and the Lancang river.

b) In the strong deformation zone on the right side of Zone I, four tension fractures and one shear fracture develop at the right boundary of the middle part. Three tension fractures I-1, I-2 and I-3 are inclined to Zhenggang gully. The strikes are approximately 160°. The 1-4 tension cracks intersect with the Zhenggang gully. The I-5 shear cracks develop along the boundary line of the basement in the I area.

c) Neither of the two tensile fractures I-6 and I-11 and two shear fractures is open which developed in the strong deformation zone in the middle of Zone I.

d) The strongly deformed area collapsed down along with the gully at the rear edge of Zone I, and the elevation is 2510 m. The collapsed area is mainly the overburden of the slope surface. The material composition is broken stone soil. The collapsed section is steeply. It can be seen that the upper overburden and the lower sliding bed are obviously staggered. Compared with the recent strong deformation area, this area has been strongly deformed since the last time. Now it has entered into a relatively stable period (Fig. 3c, e).

Zone II Deformation Characteristics

a) The front edge of zone II and the strongly deformed area on the left side are dominated by the overburden on the slope surface. The material composition is mainly crushed rock soil. With tensile cracks, II-1 developed and local collapsed II-9. Multiple nearly parallel secondary tension cracks.

b) There are three shear fractures II-2, II-4, II-6, and pull trough II-3 in the back edge of zone II and the strongly deformed zone on the right side. II-2 is located at the rear edge of zone II near the right boundary bedrock ridge, with an elevation of 2500 m, extending in an arc shape, intersecting with the main sliding direction of the landslide, and an angle less than 30°. II-3 elevation is 2560 m, and the main sliding direction of the landslide is 45°. II-4 and II-6 are nearly parallel and perpendicular to the slope direction.

c) There are many nearly parallel shear fractures II-5 in the middle of the strong deformation zone of Zone II, with an elevation of 2480 m, obvious signs of shear deformation, step-like, extending direction perpendicular to the main sliding direction of the landslide. The material composition is mainly gravel soil and cultivated soil which is loosely structured (Fig. 3a, d).

4 Numerical Analysis of Zhenggang Landslide Formation and Evolution Process

4.1 Model and Parameters

For the toppling deformation process of steeply inclined layered rock mass study, this paper uses UEDC4.0 version for modeling analysis. UDEC software adopts the discrete element method, which is often used in the research of jointed slopes and tunnel excavation. The modeling process is completed with the aid of Auto CAD software etc. The length of this model is about 1600 m. The height of this model is about 850 m. The boundary displacement is constrained. There is no horizontal displacement on both sides and no vertical displacement at the bottom. The influence of weak joints on the slope is considered in this model. Because the slope rock layer toppling deformation under gravity, the model stress field only considers gravity (Fig. 5).

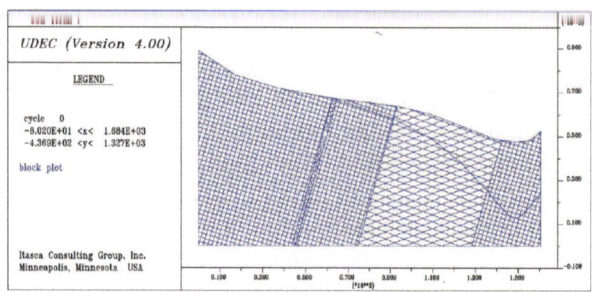

Fig. 5. Two-dimensional discrete element model of Zhenggang landslide.

According to the results of on-site investigation and indoor test, referring to similar projects, and comprehensive local experience values, the physical and mechanical parameters of the rock and soil mass of the calculation model are determined as follows (Table 1).

Table 1. Physical and mechanical parameters of Zhenggang landslide.

Stratigraphic code	r (kN/m³)	C (Pa)	φ (°)	Bulk modulus (Pa)	Shear modulus (Pa)	tensile strength (Pa)
	natural	natural	natural			
Slip belt	21.0	2.6E+4	29	5.6E+07	1.2E+07	4.5E+04
Sliding body	22.0	5.0E+4	31.0	8.3E+07	1.8E+07	5.0E+04
P$_1$j	26.8	1.5E+6	54.8	1.2E+10	7.6E+09	2.0E+07
T$_3$hn	26.7	1.7E+6	54.5	9.6E+09	6.0E+09	1.0E+06
F1 Fault	19.0	1.0E+5	21.3	9.0E+08	2.4E+08	1.5E+05
Structural plane	C (KPa)	φ (°)	normal stiffness (MPa/m)		tangential stiffness (MPa/m)	
Joints oft plane	200	24	9500		19000	
	350	22	8000		17000	

4.2 Numerical Calculation Results

(1) When the model calculates to 20000 steps, the layered rock mass in the middle of the slope and the front edge basalt show no obvious deformation. The anti-dipping rock layer at the foot of the slope is slightly deformed. The shallow anti-dip layered rock mass at the rear edge develops micro-cracks that are nearly parallel to the plane. The entire rear edge is in a weak toppling creep stage (Fig. 6).

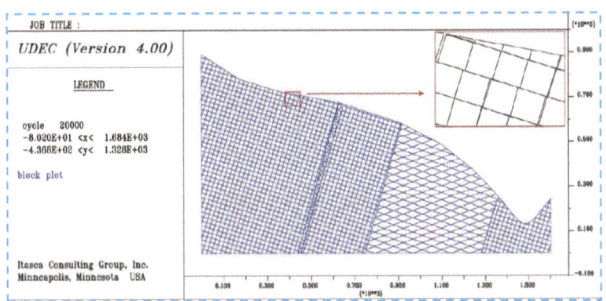

Fig. 6. The degree of toppling deformation at 20000 steps.

(2) When the model calculates to 36000 steps, the tensile cracks at the rear edge of the slope begin to develop. And then the tensile-shear fracture surface formed. The toppling and creep of the anti-dipping rock mass further intensely. The middle layered rock mass begins weakly stretched. The cutting layer also begins to develop. The anti-dipping rock mass at the foot of the slope bends and topples strongly in the direction of the river. The massive basalt on the front edge produces a shear-slip along the existing structural inside planes (Fig. 7).

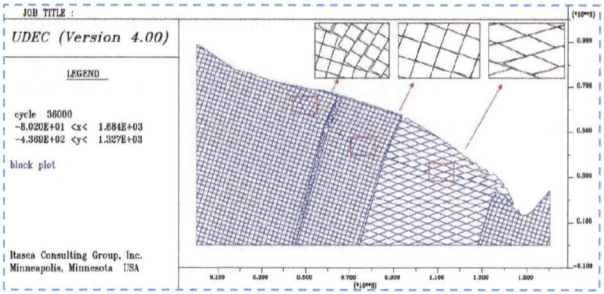

Fig. 7. The degree of toppling deformation at 36000 steps.

(3) When the model calculates to 82000 steps, there is an obvious relaxation phenomenon between the front basalt blocks. The shear-slip surface tends to be partially connected. The front slope surface develops local slumping. The middle and rear edges rock mass is violently toppled and broken. The fracture surface has tended to penetrate. The scale of the surface cracks has increased significantly. The overburdened rock mass is gradually formed, which belongs to extremely strong toppling (Fig. 8).

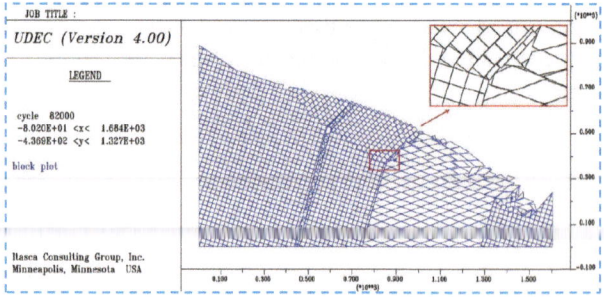

Fig. 8. The degree of toppling deformation at 82000 steps.

(4) When the model calculates to 280000 steps, an integrated slip surface has been formed in the bank slope. The whole overburdened rock mass downward slides gradually, and finally forms a landslide. The internal slip surface of the front-edge massive basalt is provided by the self-developed along-slope structural surface and gentle toppling crack. The internal slip surface of the middle-rear edge anti-dipping rock mass is developed from its toppling fracture surface. When the partial slip surface is penetrated, the entire slip surface of the landslide is formed. In addition, the middle and rear edge of the falling overburden developed partially connected fracture surfaces, which provided the possibility for the secondary sliding of the landslide in the later stage (Fig. 9).

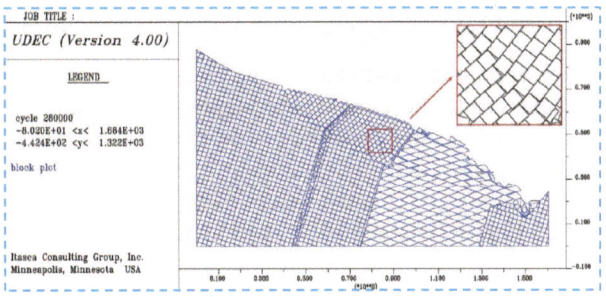

Fig. 9. The degree of toppling deformation at 280000 steps.

5 Discussion

5.1 The Steeply Dipping Layered Rock Mass Toppling Formation Process

1) Slight toppling deformation stage: it was in the early wide valley stage of the Lancang river, with relatively low and gentle valley slopes on both sides of the river.
2) Toppling-bending creep stage: as the river further undercuts, the stress release increases. The rock mass begins to topple and deform toward the empty direction. The layered rock mass that has undergone toppling deformation produces relative displacement along the horizontal direction of the potential shear-slip structure. The bending-tension deformation is intensified. The bending of the rock mass gradually increases and accumulates the tensile strain generated in the layer. Under the combined action of the gravitational bending moment and lateral slip, when the accumulated tensile stress exceeds the tensile strength of the rock mass, it is accompanied by the further dislocation of the interlayer rock mass. The shear fracture surface appears which intensifies the toppling-bending deformation. At this stage, the slope body appears layer-wise tensile failure, and apparently. There are multi-level toppling crack grooves and anti-slope sills.
3) The stage of strong bending and near-through fracture: after valley formation, the unloading effect of the rock formation is enhanced. The shear slip of the cut layer continues to develop. The self-weight bending moment continues to increase. The slope gradually undergoes gravitational deformation toward the horizontal and empty direction. Obvious tensile cracks began to appear in the stress concentration area at the middle and rear of the slope. The toppling deformation has developed to the stage of toppling and toppling fracture stage. The rock strata-undergoes strong shear-slip dislocation along with the layer. At the same time it combines with the dominant structural surface to form a shear-slip surface. Macroscopically, the tensile crack tends to penetrate, and resulting in shear-slip.
4) Slipping instability stage: when the rock strata bend at a large angle in the root, the slope will undergo creep bending and tearing under the action of gravity. The bending and fracture surface will penetrate under the effect of gravity. A penetrating slip surface outside the gently inclined slope will be formed. The slope will slip and lose stability, then forming a giant landslide.

5.2 The Giant Ancient Landslide Formation and Evolution Process

1) In the early stage of toppling deformation, the river is wide and the valley slope is small. The rock mass has almost no obvious toppling deformation.
2) Early toppling deformation, the acceleration of the crustal uplift speed. The intensified river downward cutting, forming high and steep valley slopes. The rock mass began to cause obvious toppling deformation toward the river.
3) The river cutting further accelerates and the rock mass toppling deformation is intensified, forming a series of toppling fracture planes in the rock mass. Strong toppling deformation and toppling overburden area are formed in a certain depth of the slope. The collapsed section of the site is further expanded and penetrated. Eventually, a continuous failure surface is formed, and a landslide occurs.

5.3 The Genetic Mechanism of the Toppling Deformation Body Evolving into a Giant Ancient Landslide

For the Zhenggang landslide, the underlying basalt with a thickness of about 400m should constitute the foundation of the entire slope, which is the resistance body of the overlying and deformed rock mass. However, the shear outlet of the zhenggang landslide is located inside the basalt. The investigation found that the basalt is many shear dislocation zones formed in the early structural process (Fig. 10).

1) The weak layer zone 1 outside the inclined slope is bluish-gray, about 0.5–1 m thick, composed of clay breccia, with a particle size of 1–3 cm, sub-angular, and its bed attitude is 20–30°∠25–30°.
2) The weak layer zone 2 outside the inclined slope is blue-gray, yellow-brown, about 30–40 cm thick, composed of clay breccia, with a particle size of about 0.5–2 cm, and sub-angular.
3) The weak layer zone 3 outside the inclined slope, the bed attitude is 20–30°∠25–30°. The thickness is about 20–30 cm. The upper layer is yellow-brown. The lower layer is blue-grey, composed of clay breccia. The particle size is 0.5–2 cm, second Angular.
4) The weak layer zone 4 outside the inclined slope is about 40–50 cm thick, composed of yellow-brown clay breccia, with a grain size of 0.5–3 cm, sub-angular, and the rock mass between the two layers is extremely broken and loose, with development in between 2–5 cm thick soft layer belt.

The existence of the weak zone makes it possible for the overlying toppling and fractured rock mass to appear shear dislocation along the weak zone. However, because these weak layers are not exposed on the surface, their extension length is not completely connected. Therefore, only limited to this, the overturning rock mass may still be insufficient to remove the basalt to form a landslide. Further investigation found that the basalt unloading zone. In addition, two groups of gently-dipping structural surfaces are also developed in the rock body (Fig. 11). (1) The gentle slope outside of the unloading crack (bed attitude 45°∠10°–20°). (2) The gentle slope inside of the unloading crack (bed attitude 240°∠10°–30°). The two groups of unloading cracks all have a gentle dip angle. The tendency of the first group unloading cracks direction is the same as the main sliding direction of the landslide, which provides a good channel for the upper overburdened rock mass to be cut out from the front edge.

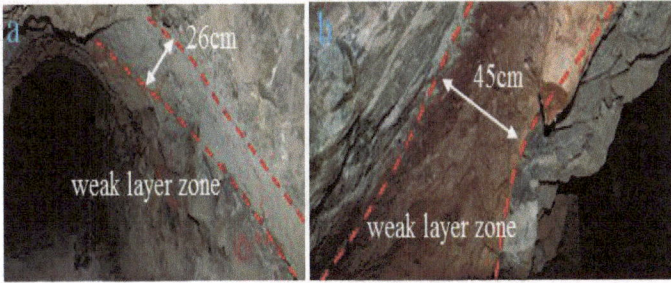

Fig. 10. Shearing dislocation zone of adit (a. The weak layer zone 3; b. The weak layer zone 4).

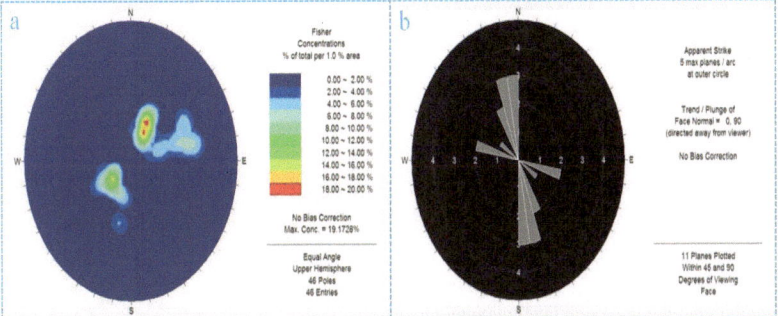

Fig. 11. a. The isometric map of the poles of the gently-dipping structural plane in the basalt; b. Rose diagram of basalt gently dipping structural surface strike.

The formation mode of the giant ancient landslide in Zhenggang is as follows, the upper part of the slope is along the gradually penetrating toppled rock mass. The lower part is along with the soft layer in the basalt along the slope. The front edge is cut out along the gently-dipping unloading fissure. The formation mechanism of the giant ancient landslide in Zhenggang is as follows, the Lancang river is eroding rapidly and formation many high-steep bank slopes. The steep-dip layered rock mass of the bank slope bends and topples towards the valley under the action of gravity and the pressure of the overlying rock mass. With the continuous erosion of the valley, the layered rock mass bending and toppling phenomenon intensified. The rock mass at the rear edge of the bank slope falls strongly. When the toppling deformation develops to a through slip surface in the layered rock mass, the overburdened rock mass will slide along the through slip surface, and finally forms a giant landslide (Fig. 12).

6   Conclusion

Based on the above analysis, the following understanding can be drawn:

1) The giant rocky ancient landslide in the Lancang river is jointly affected by regional structure and valley down-cutting. The weak layer structure is produced by complex tectonic geology. During the valley down-cutting, the structural surface of the slope

rock mass is produced by time-dependent creep effects. The front edge erosion leads to the collapse of the steeply inclined layered rock mass.

2) The multi-level down-cutting of the river is closely related to the evolution process of the river. The cutting force along the river makes the slopes on both sides steep, the slope stress redistributes, and the slope surface deformation. The deformation process of the toppling deformation body is divided into, the micro-pumping deformation stage, the toppling-bending creep stage, the strong bending, the cracknear-through stage, and the slipping instability stage.

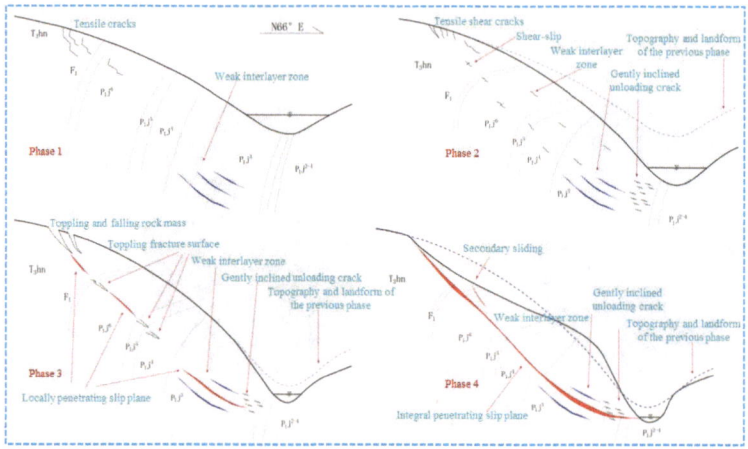

Fig. 12. Schematic diagram of formation mechanism of Zhenggang landslide.

3) The steeply inclined layered rock mass structure of the bank slope is the main internal factor for the toppling deformation body evolution into a giant ancient landslide. The down-cutting of the river is the main external factor. The through surface formed by the connection of the toppling fracture zone is the decisive factor for the occurrence of giant ancient landslides. The intensely valley erosion promotes the process of toppling deformation, which leads to the concentration of shear stress on the rock layer. Rainfall infiltration reduces the strength of the weak layer and accelerates the occurrence of sliding. The Zhenggang giant ancient landslide belongs to the slip-shear type.

References

1. Fan, X.M., Scaringi, G., Korup, O., et al.: Earthquake-induced chains of geologic hazards: patterns, mechanisms, and impacts. Rev. Geophys. **57**, 421–503 (2019). https://doi.org/10.1029/2018RG000626

2. Tu, G.X., Huang, D., Huang, R.Q., et al.: Effect of locally accumulated crushed stone soil on the infiltration of intense rainfall: a case study on the reactivation of an old deep landslide deposit. Bull. Eng. Geol. Environ. **78**, 4833–4849 (2019). https://doi.org/10.1007/s10064-019-01460-y

3. Tu, G.X., Deng, H.: Characteristics of a deep-seated flexural toppling fracture and its relations with down-cutting by the Lancang river: a case study on a steeply dipping layered rock slope. Southwest China. Eng. Geol. **275**, 105574 (2020). https://doi.org/10.1016/j.enggeo

4. Logan, J.M.: Friction in rocks. Rev. Geophys. **13**(3), 358–361 (1975). https://doi.org/10.1029/RG013i003p00358

5. AghaKouchak, A., Mirchi, A., Madani, K., et al.: Anthropogenic drought: definition, challenges, and opportunities. Rev. Geophys. **59**, e2019RG000683 (2021). https://doi.org/10.1029/2019RG000683

6. Cohen-Waeber, J., Bürgmann, R., Chaussard, E., et al.: Spatiotemporal patterns of precipitation-modulated landslide deformation from independent component analysis of InSAR time series. Geophys. Res. Lett. **45**, 1878–1887 (2018). https://doi.org/10.1002/2017GL075950

7. Tu, G.X., Deng, H., Shang, Q., et al.: Deep-seated large-scale toppling failure: a case study of the Lancang slope in southwest China. Rock Mech. Rock Eng. **53**, 3417–3432 (2020). https://doi.org/10.1007/s00603-020-02132-0

8. Deng, H., Zhong, C.Y., Wu, L.Z., et al.: Process analysis of causes of Luanshigang landslide in the Dadu River. China. Environ Earth Sci. **80**, 737 (2021). https://doi.org/10.1007/s12665-021-10069-y

9. Orland, E., Roering, J.J., Thomas, M.A., et al.: Deep learning as a tool to forecast hydrologic response for landslide-prone hillslopes. Geophys. Res. Lett. **47**(16), e2020GL088731 (2020). https://doi.org/10.1029/2020GL088731

10. Hong, Y., Adler, R., Huffman, G.: Evaluation of the potential of NASA multi-satellite precipitation analysis in global landslide hazard assessment. Geophys. Res. Lett. **33**, L22402 (2006). https://doi.org/10.1029/2006GL028010

11. Wang, Y.S., Wu, L.Z., Gu, J.: Process analysis of the Moxi earthquake-induced Lantianwan landslide in the Dadu River. China Bull. Eng. Geol. Env. **78**(7), 4731–4742 (2019). https://doi.org/10.1007/s10064-018-01438-2

12. Wu, L.Z., Deng, H., Huang, R.Q., et al.: Evolution of lakes created by landslide dams and the role of dam erosion: a case study of the Jiajun landslide on the Dadu River. China. Quat Int. **503A**, 41–50 (2019). https://doi.org/10.1016/j.quaint.2018.08.001

13. Deng, H., Wu, L.Z., Huang, R.Q., et al.: Formation of the Siwanli ancient landslide in the Dadu River China. Landslides **14**(1), 385–394 (2017). https://doi.org/10.1007/s10346-016-0756-9

14. Li, S.H., Luo, X.H., Wu, L.Z.: A new method for calculating failure probability of landslide based on ANN and a convex set model. Landslides **18**, 2855–2867 (2021). https://doi.org/10.1007/s10346-021-01652-2

15. Lin, F., Wu, L.Z., Huang, R.Q., et al.: Formation and characteristics of the Xiaoba landslide in Fuquan, Guizhou. China. Landslides. **15**(4), 669–681 (2018). https://doi.org/10.1007/s10346-017-0897-5

16. Wu, L.Z., Huang, J., Fan, W., et al.: Hydro-mechanical coupling in unsaturated soils covering a non-deformable structure. Comput. Geotech. **117**, 103287 (2020). https://doi.org/10.1016/j.compgeo.2019.103287

17. Wu, L.Z., Huang, R.Q., Li, H.L., et al.: The model tests of rainfall infiltration in two-layer unsaturated soil slopes. Europ. J. Environ. Civ. Eng. **25**(9), 1555–1569 (2021). https://doi.org/10.1080/19648189.2019.1585961

18. Zhu, S.R., Wu, L.Z., Peng, J.B.: An improved Chebyshev semi-iterative method for simulating rainfall infiltration in unsaturated soils and its application to shallow landslides. J. Hydrol. **590**, 125157 (2020). https://doi.org/10.1016/j.jhydrol.2020.125157

19. Li, A.R., Deng, H., Zhang, H.J., et al.: The shear-creep behavior of the weak interlayer mudstone in a red-bed soft rock in acidic environments and its modeling with an improved

Burgers model. Mech. Time-Depend. Mater. **27**, 1–18 (2021). https://doi.org/10.1007/s11 043-021-09523-y

20. Li, A.R., Deng, H., Zhang, H.J., et al.: Developing a two-step improved damage creep constitutive model based on soft rock saturation-loss cycle triaxial creep test. Nat. Hazards **108**, 2265–2281 (2021). https://doi.org/10.1007/s11069-021-04779-6

21. Shi, J.S., Wu, L.Z., Wu, S.R., et al.: Analysis of the causes of large-scale loess landslides in Baoji, China. Geomorphology **264**, 109–117 (2016). https://doi.org/10.1016/j.geomorph. 2016.04.013

22. Sun, P., Wang, G., Wu, L.Z., et al.: Physical model experiments for shallow failure in rainfall-triggered loess slope, northwest China. Bull. Eng. Geol. Env. **78**(6), 4363–4382 (2019). https:// doi.org/10.1007/s10064-018-1420-5

23. Wu, L.Z., Zhou, Y., Sun, P., et al.: Laboratory characterization of rainfall-induced loess slope failure. CATENA **150**, 1–8 (2017). https://doi.org/10.1016/j.catena.2016.11.002

24. Tu, G.X., Deng, H.: Formation and evolution of a successive landslide dam by the erosion of river: a case study of the Gendakan landslide dam on the Lancang river. China. Bull Eng Geol Environ. **79**, 2747–2761 (2020). https://doi.org/10.1007/s10064-020-01743-9

25. Wang, F., Tang, H.M.: The mechanism and evolution of toppling deformation and failure of the inter-bedded slope in the upstream of the Yalong River. J. Eng. Geo. **25**(06), 1501–1508 (2017). https://doi.org/10.13544/j.cnki.jeg.2017.06.013. (in Chinese)

26. Ning, Y.B., Tang, H.M., Zhang, B.C., et al.: Study on the evolution process and instability mechanism of the Lancang river deep toppling body. Chin. J. Rock Mech. Eng. **40**(11), 2199–2213 (2021). https://doi.org/10.13722/j.cnki.jrme.2020.1071. (in Chinese)

27. Wang, F., Tang, H.M., Zhang, G.C., et al.: Development characteristics and formation and evolution mechanism of deep toppling bodies in the upstream of Yalong River. J. Mt. Res. **36**(03), 411–421 (2018). https://doi.org/10.16089/j.cnki.1008-2786.000337. (in Chinese)

Potential Instability Modes and Support Stability Analysis of Soil Accumulation Slopes Along the Lexi Expressway

L. S. Wu[1], C. F. Li[2], T. B. Li[2(✉)], Y. Ren[2], Y. Wen[1], J. S. Zhang[2], and J. F. Yang[1]

[1] Sichuan Lexi Expressway Co., Ltd., Chengdu 610041, China
[2] State Key Laboratory of Geohazard Prevention and Geoenvironment Protection, Chengdu 610059, China
ltb@cdut.edu.cn

Abstract. Geological conditions along the Mabian-Zhaojue section of Leshan -Xichang Expressway are complicated, and during the construction process, landslides and geologic hazards are frequent under the influence of engineering excavation disturbance. In view of the deformation, damage and development and evolution of soil slope under excavation disturbance widely existed along Leshan-Xichang Expressway, taking the soil accumulation body slope on the left side of the section from ZK125 + 654 to ZK125 + 775 along the route as the research object, the potential deformation and damage mechanism and the stability of excavation support were analyzed by using FLAC 3D software. The results show that under the unsupported excavation condition, the slope is mainly dominated by creep-slip deformation of the slope surface; under the unsupported rainstorm condition, the slope undergoes large-scale destabilization damage with excavation. In the excavation and support process, as well as in the rainstorm condition after the completion of support, the stability of the slope as a whole and the support structure are better. The slope is a creep-slip-pulling earth slide in which the rock and soil bodies on the surface of the slope body are sheared off at the potential shear sliding surface in the slope body under the multiple effects of engineering disturbance, rainfall and gravity, and the slope body as a whole is subjected to unloading and traction at the leading edge and squeezing and pushing at the trailing edge, and then it slides along the circular arc shaped shear sliding surface in the accumulation body.

Keywords: Soil Accumulation Slope · Flac 3D · Deformation Damage Mechanism · Stability Analysis

1 Introduction

The Western Sichuan Plateau Gradient Zone is located on the eastern edge of the Tibetan Plateau and is an area with extremely complex topographic and geological conditions in China. The region has steep topography, high seismic intensity, fragile geological environment and frequent geological disasters. Influenced by the complex

© The Author(s) 2024
G. Feng (Ed.): ICCE 2023, LNCE 526, pp. 107–123, 2024.
https://doi.org/10.1007/978-981-97-4355-1_10

topography and geological structure of the Western Sichuan Plateau Gradient Zone, landslides, avalanches and mudslide disasters have developed along the highways in the region, which have serious impacts on their normal operation and traffic safety. The Lexi Expressway is located at the edge of the Western Sichuan Plateau Gradient Zone. During the construction process, due to the influence of excavation disturbance, landslide disasters occur frequently along the route, and slope management has become an urgent problem to be solved in the construction of the expressway. Therefore, it is very important to study the whole process of deformation-sliding evolution of slopes under engineering disturbances so as to take timely and effective management measures for slopes.

The deformation and failure mechanism and stability analysis of slopes is a classical research direction in engineering geology [1–3]. Domestic and foreign scholars have done more research on this issue and have achieved rich research results. On the basis of analyzing the geological environmental conditions, structural characteristics, and deformation and fracture characteristics of slopes, Li et al. analyzed the instability mechanism and stability of soil-like slopes [4]. Li W G proposed a mechanical model for the instability and failure of soil slopes based on the theory of elasticity, and obtained the ultimate length of soil slopes in the ultimate equilibrium state [5]. Li A H classified bedding slope according to slope lithology, rock combination characteristics, rock dip angle, and rock thickness, and summarized 8 deformation and failure modes of bedding slope [6]. Dong used similarity experiments to study the deformation, development, and failure process of soft and hard interlayer anti tilting slopes, and analyzed the influence of different soft and hard lithology on the tilting deformation process and failure law through 3DEC [7]. Zhang K Y used the finite element strength reduction method to calculate soil slopes, analyzed the variation law of unit stress states at typical positions on the sliding surface, established the relationship between unit stress states and the overall stability of the slope, proposed unit instability criteria, and applied the proposed criteria and program to conduct finite element numerical simulation of the progressive failure process of the slope [8]. Wang G S studied the impact of the progressive failure process of slopes on stability, proposed a new contact element model to simulate the contact friction state on the sliding surface, and conducted numerical simulation and stability analysis of the progressive failure process of slope [9]. Wu and Hsieh simulated the debris movement and deposition process of slope in the Taiwan Chi-Chi earthquake, and better realized the damage pattern of the actual slope after the earthquake [10]. Zhao B Q Used the SLOPE/W module in Geostudio software to analyze the evolution law of bedding rock slopes controlled by weak interlayers from local failure to overall failure sliding [11]. Zeng Y W combined the finite element method with the limit equilibrium method to study the stability analysis of slope and analyzed the relationship between stability and deformation of slope [12]. Liu X R combined shaking table test and numerical simulation method to analyze the deformation destabilization mechanism and stability of down-gradient slope in the Three Gorges reservoir area [13]. Zhang J X used FLAC3D to simulate the deformation and damage process of rocky slope with weak interlayers under different working conditions, analyzed the deformation and failure mechanism of the slope, and put forward the corresponding management countermeasures accordingly [14].

Existing research has focused more on the deformation characteristics and instability mechanisms of slopes, and there is less research on the deformation, failure, and development evolution of slopes disturbed by engineering along highways. This article selects the high slope of soil accumulation along the ZK125 + 654–ZK125 + 775 section of the Lexi Expressway. Based on on-site geological investigation and relevant rock and soil experimental data collection, a calculation model for excavation stability of soil accumulation slope is established. The three-dimensional numerical simulation analysis method is used to simulate the evolution characteristics and stability of slope deformation and failure under various working conditions such as unsupported excavation and supported excavation. This can provide a theoretical basis for the layout of subsequent monitoring points and the evaluation of support schemes for the slope.

2 Lexi Expressway ZK125 + 654–ZK125 + 775 Section Slope Introduction

2.1 Slope Engineering Geological Conditions

The proposed Layimu Interchange of Mabian to Zhaojue section of Leshan to Xichang Expressway is located in Layimu Village, Layimu Township, Zhaojue County, Liangshan Prefecture, Sichuan Province. The slope is located in the site of Layimu Interchange, near the exit end of Layimu Tunnel, on the left side of the section with pile number ZK125 + 654–ZK125 + 775, as shown in Fig. 1.

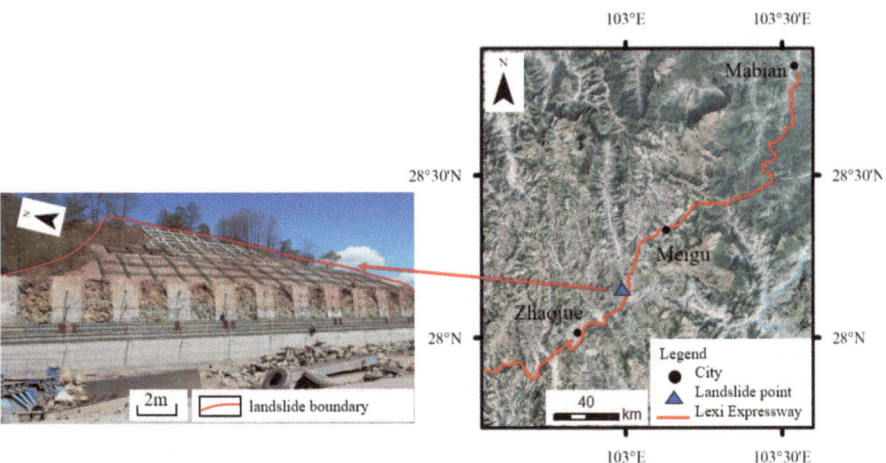

Fig. 1. Accumulation Slope of ZK125 + 654–ZK125 + 775 Section of Lexi Expressway

The slope is located in southwest Sichuan Hengduan mountain system northeast edge of Daliangshan high mountain, Sichuan basin to the southwest mountain transition zone. The mountains in the area are mostly oriented north-south, between the ridge and valley, the site is a middle mountain area in the erosion tectonics, the valley is "V" type, the highest peak near the Longtoushan, the elevation of up to 3500 m.

The site is located in the middle and lower part of the left slope of Xigou (a tributary of Zhuhe River), with a gentle transverse slope and a slope angle of 10–25°, with local areas reaching 40–60°. The slope surface is mainly composed of colluvial gravel and silty clay containing gravel, with locally exposed bedrock and underdeveloped vegetation. There are a large number of residential buildings distributed at the lower part of the slope.

The stratigraphy of the site mainly consists of Cenozoic Quaternary new avalanche slope accumulation (Q_4^{c+dl})gravelly soil, pulverized clay and Mesozoic Triassic Lower Feixianguan Formation (T_1^f) siltstone, the section is shown in Fig. 2.

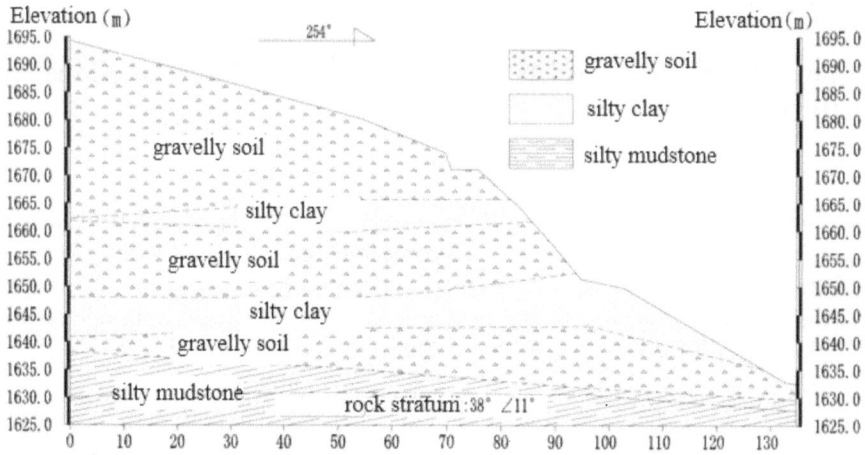

Fig. 2. Profile of Accumulation Slope (Modified by Sichuan Provincial Highway Planning, Survey, Design and Research Institute)

2.2 Slope Disposal Measures

This slope is a fifth grade slope, with a slope ratio of 1:1.25 for the second to fifth grade slopes. The width of the first grade slope platform is 5.5 m, the width of the second grade slope platform is 10 m, and the width of the third and fourth slope platforms is 5 m. According to the design documents, a comprehensive treatment method of setting anchor cable piles on the outer side of the cutting ditch platform, setting pressure grouting steel anchor pipe frame beams after the slope grading slows down, and setting steel pipe piles on the slope platform for support and protection is adopted.

(1) Steel anchor pipe frame beam

The slope is protected by 4 × 3 m pressure grouting steel anchor pipe frame, and the frame girder is cast-in-place C30 concrete, with a beam width of 40 cm and a thickness of 50 cm, and the beam is set up with an expansion and contraction joint every 15–20 m, with a joint width of 2 cm, and filled with asphalt sisal wadding. The steel anchor pipe is made of φ70 × 5 mm steel pipe with a length of 15 m.

(2) Steel Pipe Piles

Three rows of steel pipe piles are set up in the first and third level slope platforms respectively, with an outer diameter of 140 mm and a length of 23.9 m, which are laid out in plum blossom type. The steel pipe is made of No.3 steel with yield strength not less than 240 MPa and tensile strength 380–470 MPa.

(3) Pile slab wall

Anti-slip piles are laid at 15.25 m from the left side of the highway center line, the designed pile diameter is: pile width × pile height = 2.5 m × 3.5 m, pile length is 32–34 m, anchoring section is 21–23 m, pile top elevation is 1660.79–1661.79 m, pile spacing is 5.5 m, there are 16 piles in total, which are poured with C30 concrete, and the retaining plate is hung outside between the piles. The prestressing anchor cable in the anti-slip pile adopts 6 bundles of φ15.2 mm low relaxation strand (1860 MPa).

3   Model Establishment and Parameter Selection

3.1   Establishment of Calculation Model

By using Rhino6.0 software and Griddle built-in plugins, a three-dimensional mesh model of the slope was established using three cross-sectional views, one plan view, and one elevation view of the left accumulation slope of ZK125 + 654–ZK125 + 775 as references. The model was imported into FLAC 3D 6.0 software for calculation and analysis. The model has a length of 302 m, a width of 291 m and a height of 116 m, with a total of 946,921 meshes. There are four types of rock and soil mass in total: crushed stone soil, clay, strongly weathered silty mudstone, and moderately weathered silty mudstone, calculation using the Mohr-Coulomb elastoplasticity criterion. The left and right sides of the X-axis and the front and back sides of the Y-axis of the model are set as normal displacement constraints, the upper surface is set as a free interface, and the ground surface is set as a fixed constraint. The specific model diagrams are shown in Figs. 3 and 4.

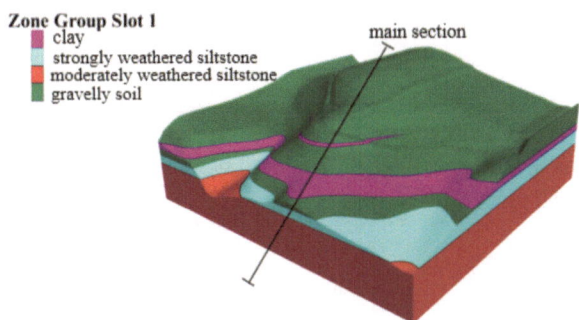

Fig. 3. Slope model in initial state

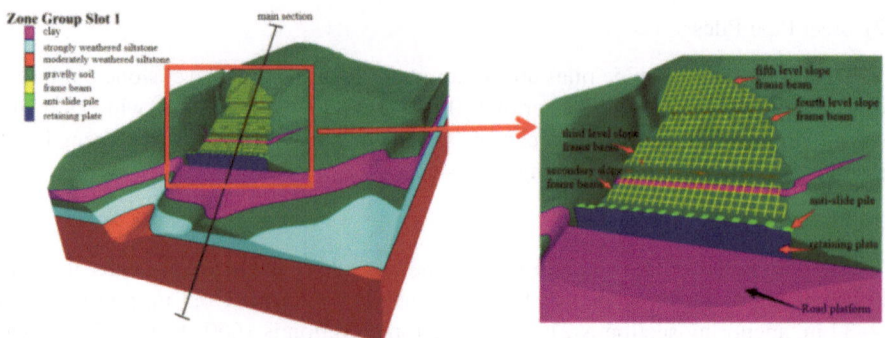

Fig. 4. Slope model after excavation and support

3.2   Calculation Parameter Selection

Based on the preliminary investigation data of the slope, the geotechnical test data of on-site drilling and sampling, and then through the experience of analogy and reference to the neighboring areas in the same region and other comprehensive considerations [15, 16], the basic physical and mechanical parameters of the geotechnical body involved in the calculation model are shown in Tables 1 and 2.

Table 1. Physical and mechanical parameters of the foundation of the geotechnical body

Rock formation	Operating condition	Weight (KN/m^3)	Cohesive force (KPa)	Friction angle (°)
Gravelly soil	Natural	20	50	35
	Ssaturated	25	30	21
Silty clay	Natural	20	55	30
	Saturated	23	33	18
Strongly weathered silty mudstone	Natural	24	150	30
	Saturated	26	90	18
moderately weathered silty mudstone	Natural	25	220	35
	Saturated	27	176	28

Table 2. Physical and mechanical parameters of geotechnical bodies

Rock formation	Operating condition	Elastic modulus (MPa)	Poisson's ratio	Shear modulus (MPa)	Bulk modulus (MPa)	Tensile strength (MPa)
Gravelly soil	Natural	100	0.31	38.17	87.71	-
	Rainstorm	80	0.33	30.08	78.43	-
Silty clay	Natural	120	0.30	46.15	100.00	-

(continued)

Table 2. (*continued*)

Rock formation	Operating condition	Elastic modulus (MPa)	Poisson's ratio	Shear modulus (MPa)	Bulk modulus (MPa)	Tensile strength (MPa)
	Rainstorm	96	0.32	36.36	88.89	-
Strongly weathered silty mudstone	Natural	400	0.28	156.25	303.03	0.48
	Rainstorm	400	0.28	156.25	303.03	0.489
moderately weathered silty mudstone	Natural	550	0.26	218.25	381.94	0.82
	Rainstorm	550	0.26	218.25	381.94	0.82

The relevant parameters of the support structure are mainly obtained through the design documents, as shown in Tables 3, 4 and 5. The rainstorm condition adopts the calculation method of parameter reduction, and the parameter values are selected based on the investigation report, the data of neighboring areas in the same region [16], and "the Technical Specification for Building Slope Engineering" GB50330-2013 combined with the site conditions.

Table 3. Mechanical parameters of steel anchor pipe and anchor cable

Density (kg/m^3)	Cross-sectional area (m^2)	Elastic modulus (MPa)	Yield strength (MPa)	Mortar bond strength (MPa)	Mortar stiffness (MPa/m)
7.81E+03	6.53E−03	2.05E+05	500	20	1000

Table 4. Mechanical parameters of steel pipe pile

Density (kg/m^3)	Cross-sectional area (m^2)	Elastic modulus (MPa)	Yield strength (MPa)	Mortar bond strength (MPa)	Internal friction angle of mortar (°)
7.85E+03	2.54E−02	2.50E+05	240	20	50

Table 5. Mechanical parameters of frame beams and pile-slab walls

Supporting structure	Density (kg/m^3)	Elastic modulus (MPa)	Poisson's ratio	Bulk modulus (MPa)	Shear modulus (MPa)
Frame beam	2.5E+03	3.0E+04	0.3	2.50E+04	1.15E+04
Anti-slide pile	3.0E+03	5.0E+04	0.2	2.78E+04	2.08E+04
Retaining plate	2.5E+03	3.5E+04	0.2	1.94E+04	1.46E+04

4 Analysis of Deformation and Failure Process of Unsupported Excavation State of Slope

4.1 Natural Slope Characterization

The pre-excavation morphology of the slope was selected as the first stage of deformation characterization, and the slope model was subjected to initial ground stress equilibrium and its displacement and shear strain increments were obtained as shown in Fig. 5.

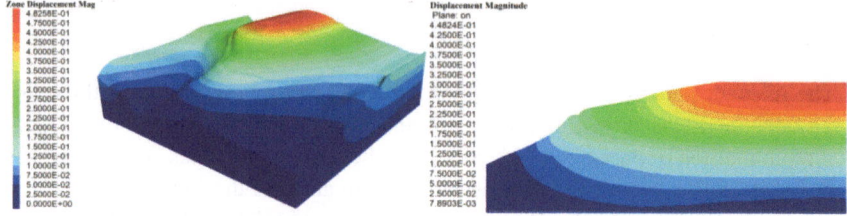

(a) Initial state slope displacement cloud maps (unit: m)

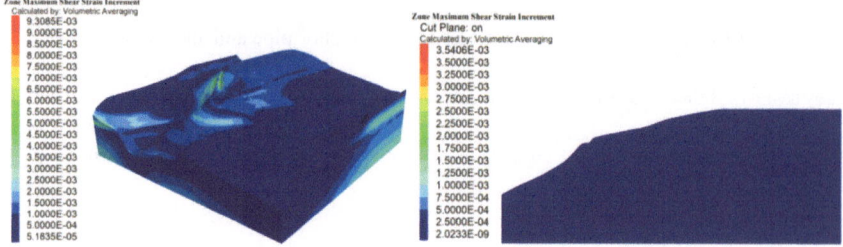

(b) Cloud maps of initial state slope shear strain increment

Fig. 5. Cloud maps of slope simulation in initial state

It can be seen that the overall stability of the accumulation slope in the initial state is very high, and the safety reserve is good, Fs = 1.61. There is no obvious shear strain increment zone or plastic zone within the slope. If only subjected to gravity, the slope will not experience large-scale overall instability and failure in the future for a long period of time.

4.2 Characteristics of Deformation and Failure of Unsupported Excavation State of Slope

In order to investigate the potential deformation damage mechanism and mode of this slope, an unsupported excavation model was set up. The simulation cloud maps under each step of excavation are shown in Fig. 6.

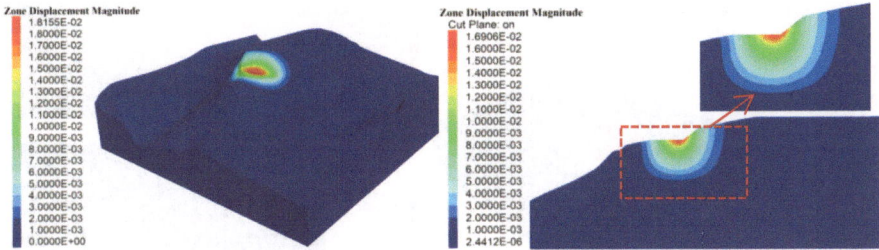

(a) The first step of unsupported excavation

(b) The second step of unsupported excavation

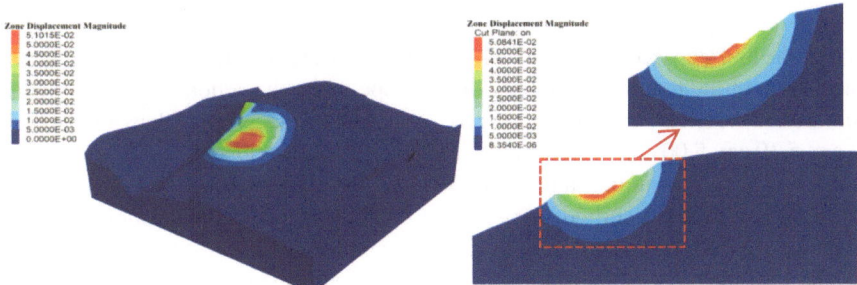

(c) The third step of unsupported excavation

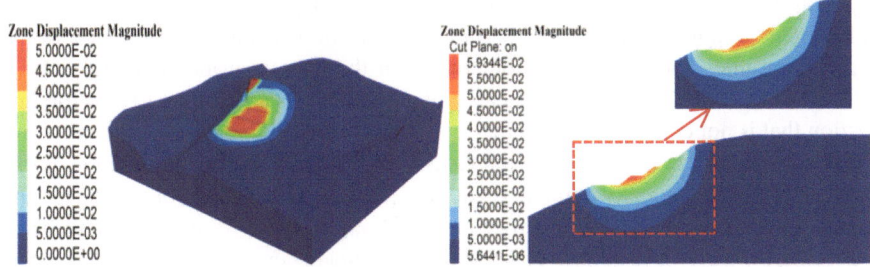

(d) The fourth step of unsupported excavation

Fig. 6. Cloud maps of slope displacement after unsupported excavation in the first–fifth steps and under rainstorm condition (unit: m)

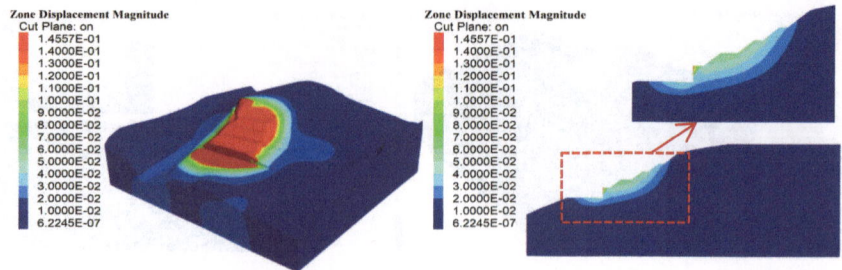

(e) The fifth step of unsupported excavation

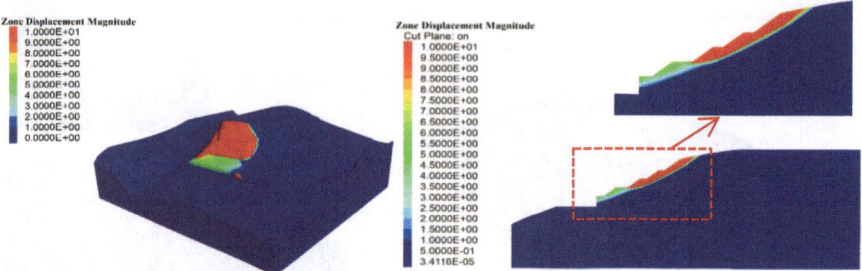

(f) The fifth step of unsupported excavation rainstorm condition

Fig. 6. (*continued*)

After the first step, second step, third step and fourth step excavation of the slope, the displacement of the rock and soil body of the slope under the action of gravity mainly occurs around the excavation surface, and the maximum displacement values are 1.82 cm, 3.64 cm, 5.10 cm, 5.00 cm, respectively. Comparing with the cloud diagram of the slope displacement under the initial state, the maximum displacement of the slope body of the slope after excavation is shifted from the top of the slope to the excavation platform and with the excavation proceeding, its distribution pattern is almost the same, and the range has increased. After the fifth step of excavation of the slope, the increase of its maximum displacement value becomes larger than before, increasing nearly three times, with the maximum value of 14.56 cm, and the distribution range of the slope displacement cloud map increases slightly further. The distribution range of the maximum value of displacement is significantly increased to include the entire shallow surface layer of the excavation profile. The stability coefficients Fs of the five excavations are 1.53, 1.47, 1.32, 1.11, and 1.03, respectively, indicating that the slope is continuing to develop in a direction that is not conducive to stability.

After the slope excavation under the rainstorm condition, the value of slope deformation changed dramatically compared with the previous excavation condition, and the maximum displacement value of the rock and soil body changed from 14.56 cm to 10 m, an increase of nearly 100 times. The stability coefficient Fs = 0.97 and the slope is unstable. From the contour map, the displacement distribution range is more concentrated, in the three-dimensional map is concentrated in the whole excavation surface, in the topography of the circle closed, the contour in the profile map is concentrated in the back of the excavation surface is distributed in the form of a circular arc, presenting the characteristics of the landslide, from the map there are two slip surfaces formed, one of which will be included in the whole excavation surface, and the other is distributed in the back of the third, fourth, and fifth level slopes, and the amount of the sliding body displacement is greater on this slip surface.

4.3   Analysis of Slope Deformation and Failure Mechanism

Through on-site investigation, data collection and analysis of the slope, and comprehensive numerical simulation calculation results, a deep understanding of the left accumulation slope of ZK125 + 654–ZK125 + 775 has been obtained, and its potential instability failure mode and deformation mechanism have been determined:

(1) Natural slope period

The slope is mainly composed of three types of lithologies: gravelly soil, silty clay, and silty mudstone. The silty mudstone is divided into two types: strongly weathered and moderately weathered. Gravel soil and silty clay are interbedded, with a thickness of about 60 m. The underlying silty mudstone is a typical soil accumulation slope. According to the excavation exposure, the accumulation on the slope is relatively loose, and rainfall and engineering disturbance are unfavorable factors for the stability of the slope.

(2) Excavation period from first step to fourth step

In actual construction excavation, support and excavation are carried out simultaneously, and even support is carried out before excavation. In order to explore the potential deformation and failure mechanism of the slope and identify the locations with high potential risks, an unsupported excavation mode is set up. After the first to fourth steps of excavation, due to the unloading effect, small-scale tensile cracks and plastic zones are formed locally behind the excavation surface. As the slope toe is excavated, the plastic zone and cracks expand, but no large through cracks or plastic zones are formed. The slope is still in a creep period, and the slope soil continues to shear and creep towards the excavation direction under gravity. At this time, there is no controlled sliding surface in the slope body, and the slope is still stable.

(3) Fifth step excavation period

After the fifth step of unsupported excavation, the displacement and shear strain increment of the slope have significantly increased, and the plastic zone has also expanded significantly. The plastic zone of the first level slope has developed to a certain depth, and at the same time, the plastic zones of all levels of slopes are also tending to connect. Under the action of excavation unloading and gravity, the creep of the surface layer of the slope tends to accelerate, and the surface of the slope sinks. At the same time, the tensile cracks at the rear edge also accelerate to develop deeper and have a trend of continuity.

Based on the above analysis, the potential deformation and failure mechanism and mode of the left side accumulation slope of ZK125 + 654–ZK125 + 775 are defined as a creep tension type soil landslide where the geotechnical body on the surface of the slope are sheared off at the potential shear sliding surface under the multiple effects of human engineering disturbance, rainfall, and gravity. The slope is pulled by the front edge unloading and pushed by the rear edge, sliding along the circular arc shear sliding surface in the accumulation, The deformation and failure process is shown in Fig. 7.

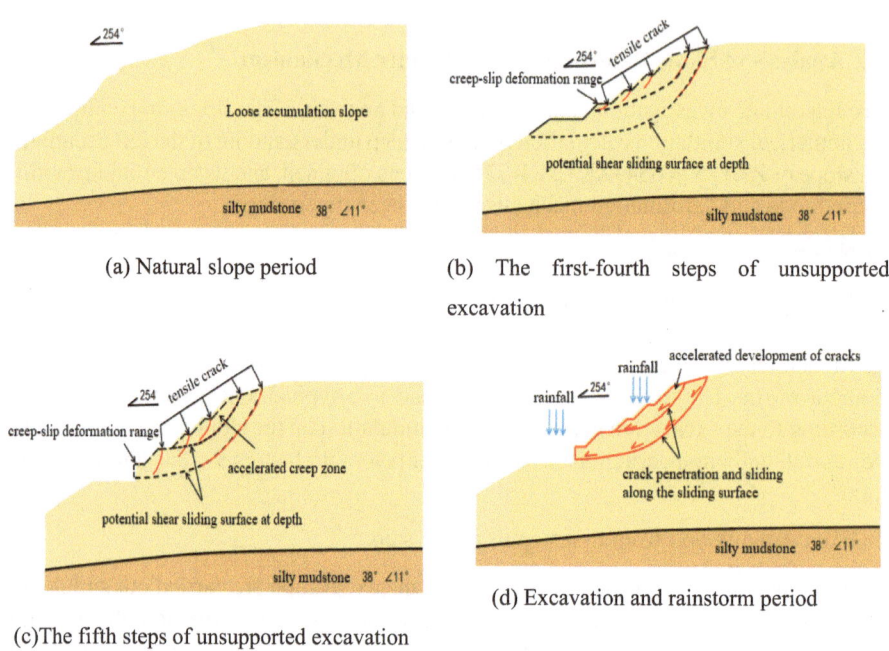

(a) Natural slope period

(b) The first-fourth steps of unsupported excavation

(c)The fifth steps of unsupported excavation

(d) Excavation and rainstorm period

Fig. 7. Schematic diagram of slope deformation and failure mechanism

5 Analysis of Deformation and Failure Process of Slope Support Excavation State

5.1 Characteristics of Deformation and Failure of Supported Excavation State of Slope

According to the design documents, the fifth step of excavation requires excavation and pouring of anti-skid piles. After the anti-skid piles are formed, the soil is excavated and retaining plates are applied step by step during the excavation project. This process is complex and difficult to replicate using simulation software. Therefore, this process is directly set as the fifth step of excavation and support for calculation and analysis. The simulated cloud maps is shown in Fig. 8.

(a) The first step of supported excavation

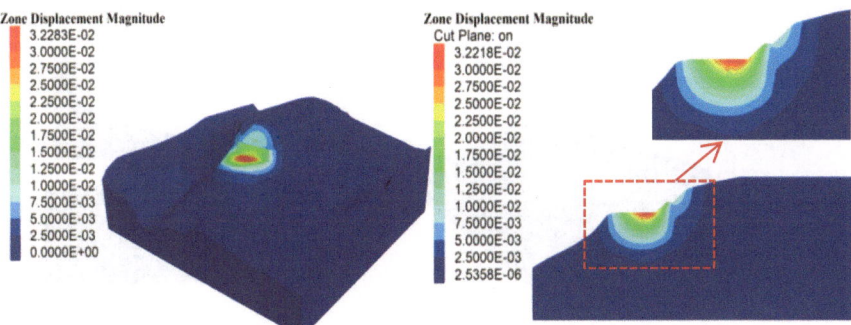

(b) The second step of supported excavation

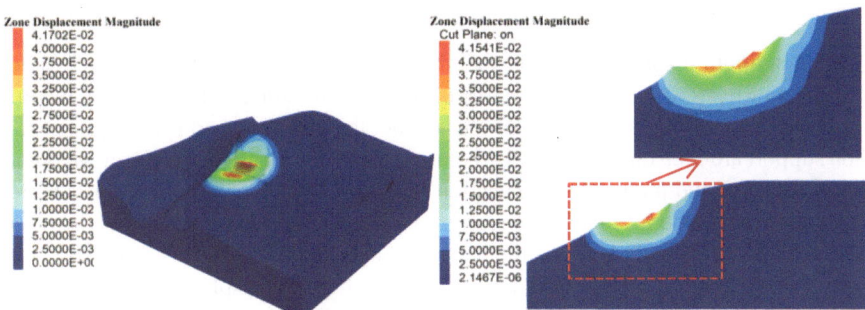

(c) The third step of supported excavation

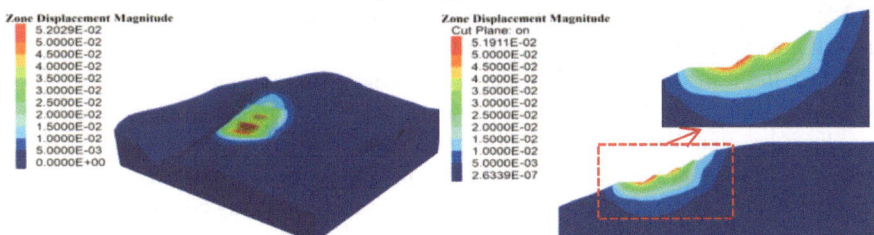

(d) The fourth step of supported excavation

Fig. 8. Cloud maps of slope displacement after supported excavation in the first–fifth steps and under rainstorm condition (unit: m)

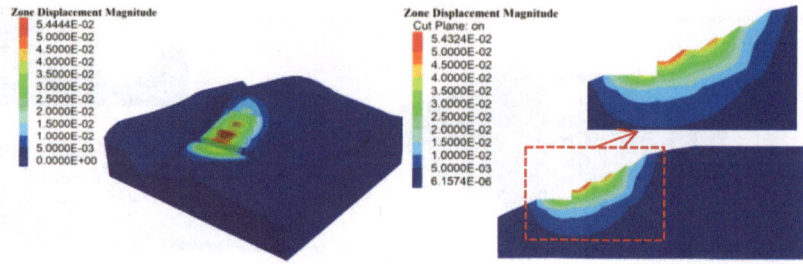

(e) The fifth step of supported excavation

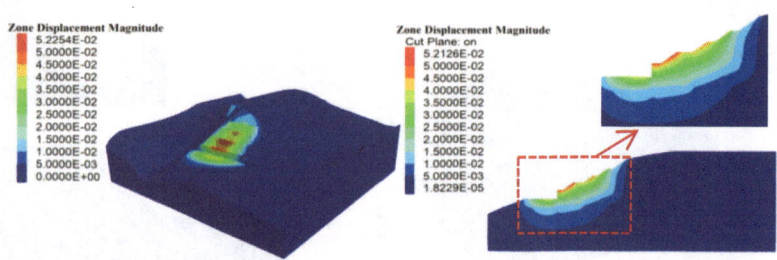

(f) The fifth step of supported excavation and rainstorm condition

Fig. 8. (*continued*)

The maximum displacement after the first step of support is 0.92 cm, which is distributed below the fourth level slope platform and in the middle and lower part of the fifth level slope. The maximum values after the second, third, fourth, and fifth steps of excavation support are 3.23 cm, 4.17 cm, 5.19 cm, and 5.44 cm, respectively. Compared with the pre support cloud maps, after excavation and support, the displacement distribution range and maximum value distribution range in the 3D cloud map under each excavation step are significantly reduced, and they are no longer connected like when there is no support; This feature is also present in the profile, where the displacement contour line is divided from a continuous patch distribution into shallow surfaces concentrated on all levels of slopes and slope platforms. The stability coefficients Fs of the five excavations are 1.59, 1.57, 1.56, and 1.54, respectively, indicating that the slope has high stability and will not experience large-scale instability failure.

Under the rainstorm condition, the displacement distribution range in the three-dimensional map after the fifth step of excavation is similar to that without support, but the value is less than 1% of that without support. The distribution pattern of displacement contour lines in the profile map is also significantly different from that without support, still maintaining the distribution characteristics of natural working conditions, and there is no obvious convergence of displacement contour lines into strips. This indicates that the surface displacement and deep displacement of the slope have significantly decreased compared to before support. The stability coefficient Fs = 1.42, and the slope stability is good, no different from that under the natural condition.

5.2 Analysis of Deformation Characteristics of Support Structures

The displacement cloud maps of the support structure under the multi-step excavation steps of the slope is shown in Fig. 9.

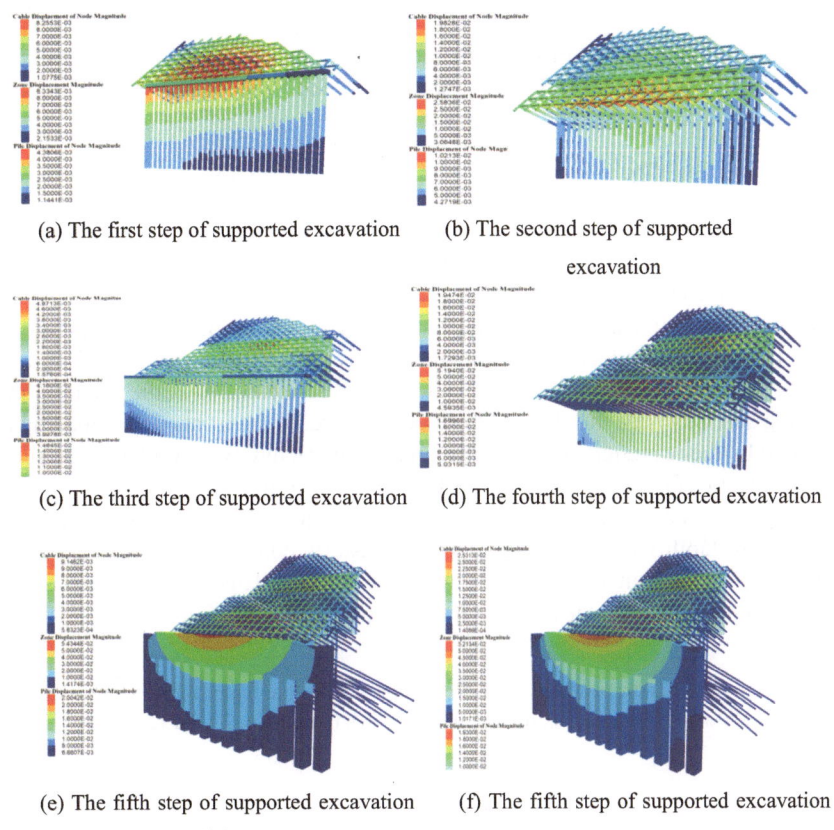

(a) The first step of supported excavation   (b) The second step of supported

excavation

(c) The third step of supported excavation   (d) The fourth step of supported excavation

(e) The fifth step of supported excavation   (f) The fifth step of supported excavation

and rainstorm condition

Fig. 9. Cloud maps of supporting structure displacement after supported excavation in the first–fifth steps and under rainstorm condition (unit: m)

During the excavation and support process, the maximum displacement of steel anchor pipes and prestressed anchor cables is about 3.6 cm, with a very small distribution range; The maximum displacement of the frame beam is about 5 cm, and the distribution range of the maximum value is small, and the overall displacement distribution is relatively uniform; The maximum displacement of the steel pipe pile is 2 cm, which is mainly distributed in the shallow surface area of the fourth level slope platform and the middle of the second level slope platform, and gradually decreases around them, with the overall distribution still relatively uniform; The overall maximum displacement of the pile sheet wall is 5.4 cm, with the maximum value mainly distributed in the upper part of the retaining plate between piles A7 # and A11 #. The overall displacement distribution is centered around this and gradually decreases towards the surrounding areas.

The deformation of the entire support structure is relatively small and evenly distributed, resulting in good support effect.

6 Conclusions

The deformation and failure mechanism and support stability of the left accumulation slope of ZK125 + 654–ZK125 + 775 were analyzed through numerical simulation methods, and the following conclusions were obtained:

(1) Under natural conditions, unsupported excavation of slopes will not cause large-scale instability and failure, mainly due to creep deformation on the surface of the slope. The overall impact of excavation from the first to fourth steps on the slope is relatively small. After excavation in the fifth step, the creep range of the slope increases, the creep speed accelerates, and the slope is in an unstable state.

(2) The potential deformation and failure mechanism and mode of the slope are that under the multiple influences of human engineering disturbance, rainfall, and gravity, the geotechnical body on the surface of the slope are sheared off at the potential shear sliding surface inside the slope. The slope is pulled by the front edge unloading and pushed by the rear edge, resulting in a creep pull fracture type soil landslide sliding along the arc shaped shear sliding surface in the accumulation body.

(3) During the excavation and support process, the slope will not experience overall or local instability and failure; Under the rainstorm condition after the support, the slope still has a high safety reserve and high stability. This support structure has a good support effect on the slope of the accumulation body.

Funding. This study is supported by the Research on Key Technologies for the Construction and Operation of DaLiangshan and Xiaoliangshan Highways (Grant No. 2021-ZL-15).

References

1. Zhang, Y.F., Fan, J.W., Yuan, K.: Disaster-induced mechanisms and prevention and control new technologies of major landslides. Chin. J. Rock Mech. Eng. **42**(8), 1910–1927 (2023)
2. Zeng, L., Fu, Y.H., Li, T., et al.: The analysis of seepage characteristics and stability of carbonaceous mudstone embankment slope in rainfall condition. Adv. Mater. Res. **1615**, 446–449 (2012)
3. Song, Z., Wang, K., Cun, J.H., et al.: Genetic mechanism and treatment measures of landslide for cutting slope. J. Railway Eng. Soc. **32**(4), 27–31+53 (2015)
4. Manyi, L., Hongyan, Z., Yanzhen, W., et al.: Instability mechanism and stability analysis of soil-like slope. J. Yangtze River Sci. Res. Inst. **33**(5), 111–115 (2016)
5. Li, W.G., Zhang, J.C.A.: Mechanical model for an inclined bedding slope based on elastic theory. J. Eng. Geol. **2**, 218–221 (2005)
6. Li, A.H., Zhou, D.P., Feng, J.: Failure modes of bedding rock cutting slope and design countermeasures. Chin. J. Rock Mech. Geotech. Eng. **28**, 2915–2921 (2009)
7. Dong, M.L., Zhang, F.M., Lv, J.Q., et al.: Study on deformation and failure law of soft-hard rock inter-bedding toppling slope base on similar test. Bull. Eng. Geol. Env. **79**(9), 4625–4637 (2020)

8. Zhang, K.Y., Li, G.S., Du, W., et al.: Simulation of progressive failure process of soil slope. J. Tianjin Univ. (Sci. Technol.) **52**, 99–105 (2019)
9. Wang, G.S.: The progressive failure of slope and the stability analyses. Chin. J. Rock Mech. Eng. **19**(1), 29–33 (2000)
10. Wu, J.H., Hsieh, P.H.: Simulating the postfailure behavior of the seismically-triggered Chiu-fen-erh-shan landslide using 3DEC. Eng. Geol. **287**, 106113 (2021)
11. Zhao, B.Q., Long, J.H., Fang, J.: Stability and failure mode of bedding rock slope under control of weak layer. Coal Technol. **35**(no11), 216–218 (2016)
12. Zeng, Y.W., Tian, W.M.: Slope stability analysis by combining fem with limit equilibrium method. Chin. J. Rock Mech. Eng. **24**, 5355–5359 (2005)
13. Liu, X.R., He, C.M., Liu, S.L., et al.: Dynamic stability of slopes with inter beddings of soft and hard layers under high-frequency microseism. Chin. J. Geo-tech. Eng. **41**(3), 430–438 (2019)
14. Zhang, J.X., Pei, X.J., Huang, R.Q., et al.: Study of deformation mechanism and treatment measures of bedding rock deformation body. Res. Soil Water Conserv. **18**(3), 177–181 (2011)
15. Zeng, X.S., Feng, W.K., Yi, X.Y.: Relationship between grain size distribution characteristics and kinetic process of rock avalanche in Baisha Village, Leibo County. Sci. Technol. Eng. **19**(6), 38–43 (2019)
16. Yin, Z.Q., Pang, M.F., Ding, Y., et al.: Research progresses on landslide disaster patterns and their response to geomorphic evolution in Meigu river basin. J. Eng. Geol. **31**(3), 905–915 (2023)

Research on Salt Corrosion Resistance Design
of Highway Concrete Structures

Xiong Tang[1], Guoqiang Xiang[1], and Jie Zou[2(✉)]

[1] Sichuan Transportation Construction Group Co., Ltd, Chengdu, China
[2] China Highway(Beijing) Engineering Materials Techology Co., Ltd, Beijing, China
2541353789@qq.com

Abstract. In response to the problem of salt corrosion on highway concrete structures in high-altitude areas, this article studies the effects of water cement ratio, admixtures, and admixtures on the salt corrosion resistance of concrete through salt freezing cycle tests and salt freezing erosion resistance tests. The results indicate that under the conditions of meeting the workability of concrete, when the water cement ratio is below 0.4, the resistance to salt freezing and erosion is excellent. When the air content of concrete is between 0 and 3.8%, the salt freezing resistance and corrosion resistance of concrete increase with the increase of air content. When the air content exceeds 3.8%, the impact of the increase of air content on its performance decreases. It is advisable to use silica fume as the active mineral admixture for concrete in the admixture. If fly ash or mineral powder is used as the mineral admixture, the curing period must be extended.

Keywords: Road Engineering · Salt Corrosion Resistance · Mineral Incorporation

1 Introduction

In the infrastructure construction of China in recent decades, cement concrete has become the most important building material in the world due to its low cost, easy to obtain materials locally, and good integrity and modelability [1, 2]. Small concrete components of road ancillary facilities such as curbs and anti-collision piers also use this material. This type of component does not withstand large loads such as vehicles, and therefore does not generate significant internal forces. Therefore, high-strength concrete is not required. However, due to the damage caused by freeze-thaw, salt corrosion, and other conditions, the surface peeling and other diseases of hydraulic concrete occur prematurely, resulting in a short service life and poor durability, resulting in an astonishing frequency of replacement of curbstones [3, 4]. Especially in high-altitude areas, during winter snowfall or freezing weather, deicing salt is usually sprayed for fast and open traffic. Coupled with factors such as freeze-thaw cycles and ultraviolet radiation, roadside structures are particularly corroded, and their appearance and functionality are far from reaching their lifespan [5].

© The Author(s) 2024
G. Feng (Ed.): ICCE 2023, LNCE 526, pp. 124–133, 2024.
https://doi.org/10.1007/978-981-97-4355-1_11

This article relies on the Sichuan Jiuma Expressway project, which is located in a high altitude area (over 3000 m) with a minimum temperature of −26.6 °C, an average annual temperature difference of 52.1 °C, and an average annual freeze-thaw cycle of 118 times. The concrete structure undergoes significant seasonal freeze-thaw cycles, with frequent alternations of positive and negative temperatures not only accelerating the freeze-thaw failure of concrete, but also leading to the superposition of freeze-thaw and shrinkage cracking damage, Further leading to difficulty in ensuring strength and durability. The article studies the effects of concrete water cement ratio, admixtures, and admixtures on the interface structure and pore structure of concrete, studies better corrosion resistance design of concrete, enhances its service life, and reduces later maintenance costs.

2 Raw Materials and Testing

2.1 Raw Materials

PO 42.5 cement is selected, and its technical performance is shown in Table 1.

Table 1. Physical and Mechanical Properties of Cement

project	Specific surface area m²/kg	Standard consistency %	stability	setting time (min)		flexural tensile strength (MPa)		compressive strength (MPa)	
				Initial setting	Final set	3d	28d	3d	28d
Specification requirements	-	-		≥90	≤600	≥4.5	≥7.5	≥17	≥42.5
cement	353	27.6	合格	209	267	6.5	9.4	36.1	54.8

The fine aggregate is river sand, with a fineness modulus of 2.87 and a silt content of 0.9%. Which meets the requirements of Zone II grading in the national standard "Building Sand" (GB/T 14684-2001).

The coarse aggregate is 4.75–26.5 mm continuously graded limestone crushed ston.

Water reducing agent JG-2 is a high-efficiency water reducing agent, and the air entraining agent is a triterpenoid saponin air entraining agent.

The Class I fly ash used in the fly ash test meets the requirements of Class I fly ash. The physical properties and chemical composition of Class I fly ash are shown in Tables 2 and 3. The physical performance indicators of silicon powder are shown in Table 4.

2.2 Test Methods

According to the requirements of GB/T50080-2002 "Standard for Testing the Performance of Ordinary Concrete Mixtures" for air content testing, a direct reading air content tester is used for measurement.

This article considers the dual effects of plateau climate and deicing salt environment, and designs an indoor accelerated test (salt freezing test) for the coupling effect of

Table 2. Physical properties of fly ash

Density/(g*cm-3)	45um sieve residue/%	Water demand ratio/%	Loss on ignition/%	Moisture content/%	Mass fraction of sulfur trioxide/%
	≤12	≤1.0	≤5.0	≤95	≤3.0
2.8	10.8	0.1	2	94	0.3

Table 3. Chemical Composition of Fly Ash (Unit:%)

w(silica)	w(alumina)	w(iron oxide)	w(calcium oxide)	w(Sulfur Trioxide)	w(Potassium oxide)	w(Sodium oxide)	w(magnesium oxide)
52.54	33.62	7.05	3.56	0.68	0.5	0.31	0.36

Table 4. Physical Properties of Silicon Powder

45umsieve residue/%	Specific surface area/(m^2/kg)	activity index/%	loss on ignition/%	moisture content/%	silicon dioxide mass fraction/%
1	18000	121	1.6	0.7	92

corrosive salt and freeze-thaw cycles. At present, there are significant differences in the preparation of specimens, the contact method of salt solution, the selection of salt solution concentration, the setting of freeze-thaw system, and the evaluation parameters of salt freezing and erosion damage between domestic and foreign concrete salt freezing test methods. By comparing and studying domestic and international standards, it is summarized that there are several main ways of contact between specimens and salt solutions: (1) specimens are completely immersed in salt solutions for salt freezing cycles, such as the test specifications for harbor concrete; (2) The test piece was immersed in a salt solution for 4–6 mm on one side for salt freezing cycle testing. For example, the CDF test method was proposed by the TC117-FDC Professional Committee of the European International Federation of Materials Testing Laboratories (RILEM); (3) The surface of the test piece is covered with a 4-6mm thick salt solution for salt freezing cycle testing, such as the Swedish SS137244 (Boras) method and the American ASTM C672 method.

Considering the objective and reasonable simulation of the destructive effect of deicing salt on concrete structures, as well as the simple, accurate, and easy operation method, the salt freezing cycle test was carried out by immersing the specimen in a salt solution for 4–6 mm on one side. The specific schematic diagram and equipment are shown in Fig. 1. The solution is a 4% NaCl salt solution, and the computer control system can control the limit temperature (+20 °C ~ −40 °C) and the rate of temperature rise and fall (cooling rate greater than 10 °C/h) inside the chamber. At the same time, corresponding software is used to automatically collect the temperature of the test chamber and the surface of the specimen. Each freeze-thaw cycle of the specimen is 6 h, and the frozen specimen is 3.5 h. The melting time is 2.5 h, which means 4 cycles per day. The time required for the specimen to decrease from 15 °C to −20 °C shall not exceed 2 h, and the time required for −20 °C to rise to 15 °C shall not exceed 1.5 h. After freeze-thaw of the specimen, ultrasonic cleaning is used.

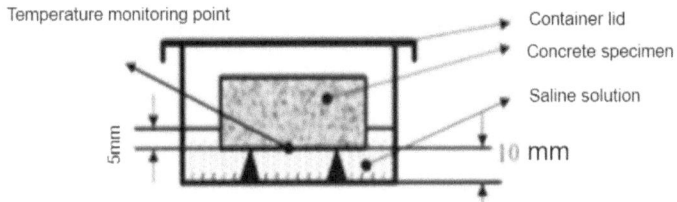

Fig. 1. Schematic diagram and equipment of salt freezing test method

The formed surface of highway concrete structures determines the key factor in their corrosion resistance. In order to better simulate and reflect the salt frost resistance characteristics of the road surface, this article chooses the formed surface as the salt frost test surface. At the same time, in order to reduce the boundary effect of the specimen during the salt freezing process, the test area of the specimen was appropriately enlarged during the testing process. The forming method and method of the test specimen are shown in Fig. 2. Place the newly mixed concrete into a trial mold (with a wooden bottom mold) made of PVC pipes with an inner diameter of 250 mm and a length of 75 mm, vibrate and compact it on a concrete vibration table, and treat the formed surface of the concrete with a wooden trowel. After 24 h, remove the wooden bottom mold and place the PVC material test mold and concrete together in the standard curing room for standard curing.

Fig. 2. Forming Method of Salt Frozen Specimens

3   Research on the Mix Design of Salt Resistant Cement Concrete

The water cement ratio is the most important parameter in concrete mix design. It directly affects the porosity and pore structure inside the concrete, and is an important parameter for measuring the compactness and permeability of concrete. It largely determines the strength and long-term durability of concrete. Five different water cement ratios were compared. Due to the fact that the main cause of concrete salt freezing and peeling is cement slurry, the mix design was based on a fixed volume of cement slurry. Five levels of water cement ratios, namely 0.36, 0.40, 0.44, 0.48, and 0.52, were used. After 28 days of curing, the salt freezing cycle test was conducted in a salt freezing testing machine, and the results are shown in Fig. 3.

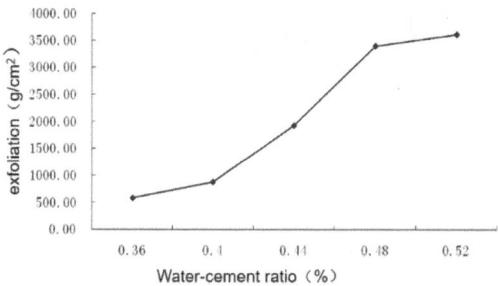

Fig. 3. Effect of water cement ratio on the resistance of concrete to salt freezing and erosion

From Fig. 3, it can be seen that as the water cement ratio gradually increases, the amount of concrete peeling gradually increases, and the salt frost resistance gradually decreases. When the water cement ratio is greater than 0.4, the resistance of concrete to salt freezing and erosion exceeds 1000 g/m^2, which exceeds the recognized and acceptable requirements for salt freezing damage. Therefore, in environments with salt freezing damage, such as non aerated concrete, it is recommended to control the water cement ratio of concrete below 0.4 in order to prevent salt freezing and erosion damage.

The slurry to aggregate ratio mainly affects the workability of concrete, which in turn affects its durability, and to a certain extent also affects its strength, elastic modulus, and dry shrinkage. Usually, in order to ensure the workability of concrete, a large amount of cementitious material is required. But as the slurry to aggregate ratio increases, the elastic modulus of concrete will decrease and the shrinkage of concrete will also increase. At the same time, due to the fact that the salt freezing damage of concrete is mainly caused by surface erosion, and the slurry content directly determines the amount (or degree) of concrete erosion, the slurry to aggregate ratio of concrete also greatly affects the salt freezing resistance of concrete. In order to ensure that concrete has good workability and excellent resistance to salt freezing and erosion, a reasonable range of slurry to aggregate ratio should be selected during mix design. The test results are shown in Fig. 4.

From Fig. 4, it can be seen that when the slurry to aggregate ratio fluctuates within the range of 258.6:741.4–287.4–712.6, there is an optimal range for the concrete's salt frost resistance performance as the slurry to aggregate ratio changes. That is, when the slurry to aggregate ratio is within the range of 265.8:734.2–280.2:719.8, the concrete's

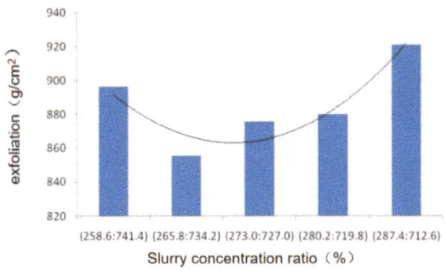

Fig. 4. Effect of Slurry to Aggregate Ratio on the Resistance of Concrete to Salt Freezing and Erosion

salt frost resistance performance is better. When the slurry to aggregate ratio exceeds this range, the amount of concrete peeling under salt freezing conditions increases due to the increase in slurry content in the concrete. When it falls below this range, it may be due to the low content of slurry in the concrete, resulting in a decrease in the compactness of the concrete and a slight increase in the amount of salt freezing and erosion of the concrete. However, the fluctuation of the slurry to aggregate ratio leads to a much smaller amplitude of change in the salt freezing and erosion damage of concrete compared to the fluctuation caused by changes in the water cement ratio. Therefore, it can be seen from the law of the influence of changes in the slurry to aggregate ratio on salt freezing and peeling damage that in order to improve the salt freezing and peeling resistance of concrete, the optimal range of slurry to aggregate ratio for concrete should be selected between 265:745 and 280:720.

4   Research on the Influence of Admixtures on the Salt Corrosion Resistance of Concrete Surface

With the development of cement concrete technology, the addition of mineral admixtures in cement concrete is receiving increasing attention. The application of mineral admixtures can improve many properties of concrete and is also an important component material of green concrete in the 21st century.

4.1   Fly Ash

With the increasingly widespread application of fly ash in concrete, especially in the harsh climate and environment areas of northern China, the salt freezing durability of fly ash concrete has received widespread attention as an important indicator of concrete. The test results are shown in Fig. 5.

From Fig. 5, it can be seen that under the standard curing condition of fly ash concrete for 28 days, the salt freezing resistance of the concrete gradually decreases with the increase of fly ash content. The main reason may be that the activity of fly ash was not fully utilized in the early stage. When the replacement amount of fly ash is large, the strength of fly ash concrete is lower, the porosity is higher, and the salt freezing resistance of fly ash concrete is slightly worse. But when the concrete is cured

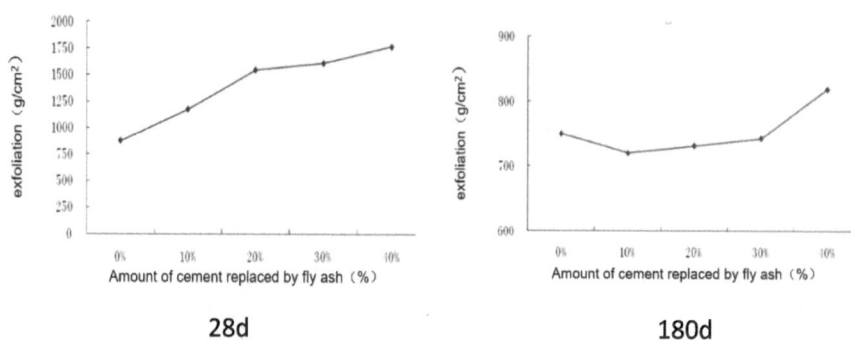

28d   180d

Fig. 5. Effect of fly ash content on the salt freezing resistance of concrete

to the age of 180 days, due to the sufficient secondary hydration of fly ash, the hardening structure of fly ash concrete is more dense, and the salt freezing resistance of fly ash concrete is significantly improved. Therefore, in the presence of salt freezing conditions, it is advisable to avoid using fly ash as a mineral admixture. If fly ash is added to the concrete, maintenance should be strengthened to ensure that the fly ash is fully hydrated.

4.2   Granulated Blast Furnace Slag Powder

Granulated blast furnace slag powder is a glassy substance formed by rapid cooling of blast furnace molten material. Its main components are calcium oxide, silicon oxide, and aluminum oxide, with a total content of about 95% or more. It has high activity and can generate hydraulic cementitious substances in the presence of an activator. The test results are shown in Fig. 6.

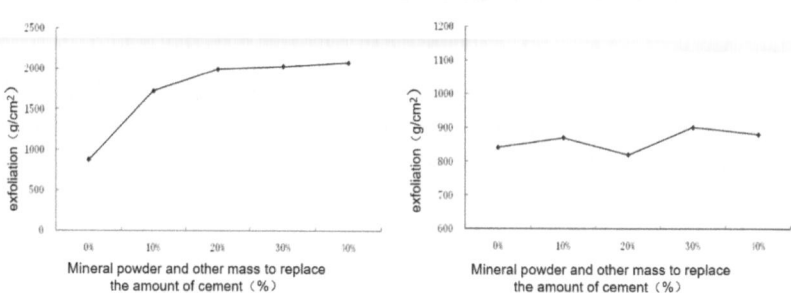

Fig. 6. Effect of Mineral Powder Content on the Salt Freezing Resistance of Concrete

From Fig. 6, it can be seen that the salt freezing resistance of mineral powder concrete is basically the same as that of fly ash. Under the 28 day standard curing test of mineral powder concrete, the salt freezing resistance of concrete gradually decreases with the increase of mineral powder content. When the concrete is cured to 180 days, the salt freezing resistance of mineral powder concrete is basically the same as that of the reference concrete.

4.3 Silica Ash

Silica fume, also known as silica micro powder, also known as micro silica powder or silica ultrafine powder, is generally referred to as silica fume. Silica ash is a ultrafine siliceous powder material formed by the rapid oxidation and condensation of SiO_2 and Si gases generated during the smelting of ferrosilicon alloys and industrial silicon with oxygen in the air. The experimental results are shown in Fig. 7.

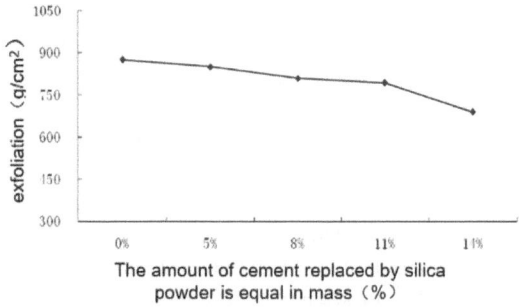

Fig. 7. Effect of silica fume on the salt freezing resistance of concrete

From Fig. 7, it can be seen that as the amount of silica fume gradually increases, the amount of concrete peeling gradually decreases and the salt frost resistance gradually increases. The main reason is that silica fume has a large specific surface area, high activity, and can quickly undergo secondary hydration reactions, improving the compactness of concrete and improving its salt freezing resistance. Therefore, when the road surface concrete requires salt freezing, if mineral admixtures are selected, silica fume should be selected as the active mineral admixture. If fly ash or mineral powder is selected, it is necessary to strengthen maintenance to avoid early damage to the concrete due to salt freezing.

5 Research on the Effect of Air Entraining Agent on the Salt Corrosion Resistance of Concrete Surface

Since the 1950s, foreign countries have generally required the addition of air entraining agents in the preparation of concrete in freeze-thaw environments, thus effectively solving the general problem of frost damage to concrete. In cold regions and concrete pavements that use deicing salts for deicing, air entrained concrete has been widely used to improve the durability of concrete against freezing and thawing. Adding a large number of uniform bubbles to concrete can cut off the pore channels inside the concrete, greatly reducing capillary action and improving impermeability. When water is filled into the capillary pores leading to the surface, the pores in the capillary pore pathway locally expand the capillary pores, which can buffer the ice crystal pressure when water freezes and significantly improve the frost resistance of concrete. The test results are shown in Fig. 8.

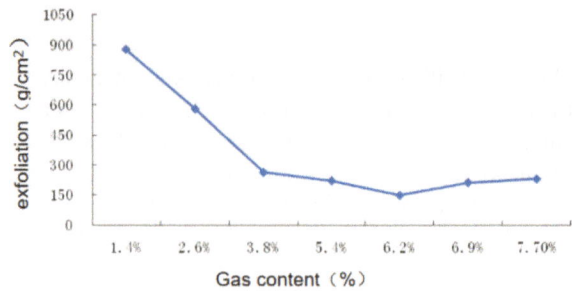

Fig. 8. Effect of Air Content in Concrete on Salt Freezing and Erosion Resistance

From Fig. 8, it can be seen that the introduction of bubbles in concrete can significantly reduce the amount of salt freezing and erosion of concrete, and improve its salt freezing resistance. And with the increase of air content in concrete, the salt frost resistance of concrete further improves. However, when the air content in concrete reaches 3.8%, further increase in air content no longer significantly improves the salt frost erosion resistance of concrete. Therefore, it is recommended to add an air entraining agent and control the air content above 4% when preparing salt frost resistant concrete.

6 Conclusion

(1) When preparing salt frost resistant road surface machine made sand concrete, it is necessary to strictly control the water cement ratio of the concrete, reasonably control the air content of the concrete, and try to choose a lower slurry aggregate ratio under the condition of meeting the workability of the concrete.

(2) When preparing salt frost resistant concrete, it is advisable to use silica fume as the active mineral admixture of the concrete. If fly ash or mineral powder is used as the mineral admixture, the curing period must be extended to prevent the concrete from bearing salt frost erosion damage too early.

(3) When preparing salt frost resistant concrete, when the water cement ratio is below 0.4, the salt frost resistance and peeling performance are excellent; When the air content in concrete is between 0 and 3.8%, the salt frost resistance increases with the increase of air content. When the air content exceeds 3.8%, the continuous increase in air content will no longer significantly improve the salt freezing and peeling resistance of concrete.

Funding. Project: Research on corrosion resistance technology of highway concrete structure in high altitude area.

References

1. Ye, M.: Research status and development application of concrete materials. Build. Mater. Décor. **35**, 49 (2018)
2. Zhoa, R., et al.: Research progress on concrete bridges and their high-performance materials in 2019. J. Civil Environ. Eng. (Chin. Eng.) **42**(05), 37–55 (2020)
3. Li, X.: Research on Concrete Frost Resistance Design and Preventive Measures in the Qinghai Tibet Plateau Region. Southeast University, Nanjing (2015)
4. Tan, Y., Chen, H., Wang, Z., Xue, C., He, R.: Performances of cement mortar incorporating Superabsorbent Polymer (SAP) using different dosing methods. Materials **12**(10), 16–19 (2019)
5. Kovler, K., Roussel, N.: Properties of fresh and hardened concrete. Cem. Concr. Res. **41**(7), 775–792 (2011)

Effects of Coal Metakaolin on Compressive Strength and Microstructure of Cemented Soil

Jianping Liu[1,2(✉)], Weimin Chen[1], Jixia Zhang[1], Xiangqian Xie[1], and Guoshuai Xie[1]

[1] Engineering Test Center, Power China Huadong Engineering Corporation Limited, Hangzhou 310014, China
liujianpingtyut@126.com
[2] College of Civil Engineering and Architecture, Zhejiang University, Hangzhou 310058, China

Abstract. In this paper, compressive strength tests are conducted on the cemented soil with five coal metakaolin (CMK) contents to study the effects of CMK on the cemented soil compressive strength and microstructure. It is found that CMK can enhance the strength of cemented soil. Especially, when the CMK content is 3%, its compressive strength is obviously higher than that of unmixed CMK cemented soil. The effects of CMK on the strength of cemented soil is analyzed and discussed by means of X-ray diffraction (XRD) and scanning electron microscopy image (SEM) analysis. The strength mechanism of cemented soil with CMK is explained by the analysis of hydration products and microstructures of cemented soil. The reason for the macroscopic mechanical property change of cemented soil is proved by the microscopic mechanism. In addition, the relationship of cemented soil compressive strength and age is statistically analyzed. The statistical formula is set up to predict the long-term compressive strength of cemented soil by age. The predicted strength agrees well with the measured strength, indicating that the statistical formula has good predictability.

Keywords: Coal Metakaolin · Cemented Soil · Phase Components · Microstructure · Strength Prediction

1 Introduction

Cemented soil is an artificial mixture of materials by mechanical mixing or jet punching the foundation of the natural soft soil and cement slurry (or powder) stirred together, mainly a mixture of cement and soil [1]. Due to low cost, wide range of materials, simple construction technology, good performance and so on, cemented soil is widely used in foundation treatment and road engineering [2]. In recent years, cement production has caused a lot of energy problems and environmental problems. To reduce the demand for cement consumption and further improve its mechanical properties, many scholars used a certain amount of additives to replace the cement in the cemented soil [3]. Lang et al. used cement and steel slag powder to treat mucky soil and found that steel slag powder can effectively alleviate the strength loss caused by humic acid in mucky soil [4]. Furlan et al. studied the effect of fly ash on the mechanical properties and microstructure of

G. Feng (Ed.): ICCE 2023, LNCE 526, pp. 134–146, 2024.
https://doi.org/10.1007/978-981-97-4355-1_12

cemented soil, and found that fly ash can increase its compressive strength and improve its pore structure [5]. Avci et al. studied the influence of sodium silicate on the mechanical properties and permeability of cemented sandy soil, and found that sodium silicate can significantly improve its compressive strength and permeability [6]. Wang et al. studied the effect of nano-SiO_2 on the mechanical modification coastal of cemented soil in the early age, and found that nano-SiO2 can improve its splitting tensile strength and elastic modulus, but it would aggravate its brittleness [7].

Coal metakaolin (referred to as CMK) is the coal kaolin ($Al_2O_3 \cdot 2SiO_2 \cdot 2H_2O$, referred to as AS_2H_2) that calcined dehydration at a certain temperature (500 °C–900 °C) to form anhydrous calcium aluminate ($Al_2O_3 \cdot 2SiO_2$, referred to as AS_2) [8]. Coal kaolin, also known as coal gangue, is a by-product of coal. This mineral is mainly distributed in the north China region, which is a unique and precious resource in China [9]. CMK is a highly active mineral admixture with high pozzolanic activity. It is mainly used as a concrete admixture and can also be used to make high-performance geopolymer, but it is rarely studied in cemented soil. Therefore, the research on cemented soil with metakaolin has some positive social and economic value.

In this paper, the compressive strength of cemented soil with different CMK contents and ages was performed by the indoor simulation experiment. Combined with the phase composition and microstructure of the hydration product, the strength mechanism of cemented soil was revealed by means of X-ray diffraction (XRD) and scanning electron microscopy (SEM) image analysis. The relationship between the compressive strength and the age of cemented soil was analyzed. The compressive strength prediction formula of cemented soil by age was set up.

2   Materials and Testing Methods

2.1   Experimental Materials

The soil used in the experiment was sandy soil, sourced from a construction site in Taiyuan, Shanxi Province, China. The soil was dried, crushed, and passed through by a 2 mm sieve. The particle composition of the soil is shown in Table 1, according to Chinese standard GB50123-2019. The cement is P.O 42.5 ordinary Portland cement produced by Taiyuan Lionhead Cement Co., Ltd.. The CMK was HP-90B coal metakaolin produced by Shanxi Jinyang Co., Ltd.. It was white powders, the particles of less than 2 μm (6250 mesh) accounted for (90 ± 3)% of the total mass. The chemical composition of cement and CMK is shown in Table 2.

Table 1. Particle composition of soil.

Particle Size/mm	2.00 –0.50	0.50–0.25	0.25–0.075	0.075–0.005	<0.005
Content/%	18.0	25.0	21.0	24.0	12.0

Table 2. Chemical composition of cement and CMK.

Oxides	SiO_2	Al_2O_3	TiO_2	Fe_2O_3	Na_2O	CaO	K_2O	MgO	SO_3	other
Cement/%	18.81	5.86	/	3.34	0.31	66.35	0.41	1.04	2.53	1.35
CMK/%	52.62	45.42	0.85	0.45	0.25	0.166	0.13	0.11	/	0.004

2.2 Sample Preparation

The specific mix proportions of cemented soil were shown in Table 3. The cement and CMK were cementitious materials. The amount of cementitious material added was 15% of the dry soil mass. In the mix proportions, CMK was used to replace part of the cement to prepare cemented soils. The content of water required for mixing was 20% of the dry soil mass according to Chinese standard JGJ 79-2012.

Table 3. Mix proportions.

Number	Material mass content/%			
	Soil	CMK	Cement	Water
M0C15	100	0	15	20
M2C13	100	2	13	20
M3C12	100	3	12	20
M4C11	100	4	11	20
M5C10	100	5	10	20

The dried soil, CMK and cement were weighed into the mortar mixer and stirred for 2 min. After the solid particles were fully and evenly stirred, the corresponding quality tap water was added to stir for another 2 min to make the solid-liquid mixture well mixed. The mixture was divided into a 70.7 mm × 70.7 mm × 70.7 mm steel mold. The whole steel mold was put on vibration shaking for 1 min to ensure bubbles discharged. Then the surface of the samples was flattened, and the steel mold was wrapped with plastic wrap. After 24 h, the steel mold was dismantled. All cemented soil samples were numbered and placed in a water tank for curing.

2.3 Sample Preparation

The compressive strength of cemented soil samples was tested by a microcomputer controlled electronic universal testing machine (WDW-100) at the age of 3 days, 7 days, 14 days, 28 days, 60 days and 90 days. The maximum load of this testing machine was 100 kN, the error of load and displacement was better than ± 1%, and the loading rate was 0.1 kN/s. In the analysis of experimental data, the compressive strength was calculated by six parallel samples according to Chinese standard JGJ/T 233–2011.

After 28 days of compressive strength testing, selected about 10 g specimen from each sample, placed them in an ethanol solution for dehydration, and ground them through a 0.075 mm sieve. The specimens were measured by an XRD-6000 X-ray diffractometer produced by Shimadzu Co., Ltd.. The scanning angle was 4°, the ending angle was 60° and the speed was 6°/min. The phases were analyzed by JADE6.5 software.

The samples were cut into 20 mm × 20 mm × 5 mm specimens, then polished and cleaned. An SBC-12 ion sputtering instrument was used to plate gold film on the surface of the specimens. Then the specimens were put into TM-3000 scanning electron microscope produced by HITACHI Co., Ltd.. The acceleration voltage of the scanning electron microscope is 15 kV, and the magnification is 2000 times.

3   Results and Discussion

3.1   Analysis of Compressive Strength Test Results

The relationship between the compressive strength of cemented soil and the CMK content at all ages is shown in Fig. 1. In general, the compressive strength of cemented soil mixed with CMK was significantly improved compared with that of unmixed cemented soil. When the CMK content was less than 3%, the compressive strength of cemented soil increased with the increase of the CMK content. When the CMK content was more than 3%, the compressive strength of cemented soil decreased with the increase of the CMK content. When the CMK content was 3%, the compressive strength reached the maximum. This indicates that the use of CMK to replace a certain amount of cement contributes to improving the compressive strength of cemented soil.

This may be because the cement content decreases with the increase of the CMK content. When the CMK content is small, the cement can not only ensure the strength produced by its early hydration, but also can produce an appropriate amount of $Ca(OH)_2$ to react with the active ingredients (Al_2O_3, SiO_2) in the CMK [10]. This makes further improving in the compressive strength of cemented soil. When the CMK content increases to a certain level, the $Ca(OH)_2$ produced by the early cement hydration has completely reacted with the CMK. At this time, the CMK is not involved in the reaction and production of hydration products to provide strength [11]. However, the role of cement hydration to provide strength is weakened due to the reduction of the cement content. Finally, the compressive strength of cemented soil begins to decrease.

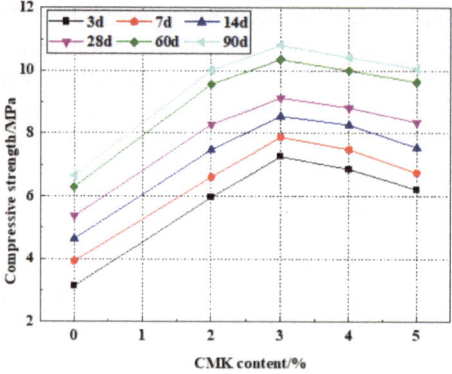

Fig. 1. Compressive strength of cemented soil at all ages.

3.2 Analysis of XRD Phase Results

Figure 2 shows the XRD phase analysis of cemented soil with five contents of CMK at the age of 28 d. It can be seen from the Fig. That the main phase components of cemented soil were quartz, calcium aluminate hydrate (CAH), calcium hydroxide (CH), calcium aluminosilicate hydrate (CASH), calcium silicate hydrate (CSH), ettringite (AFt), anorthite (CAS) and other compounds. According to the hydration reaction equation of cement and CMK, it can be basically determined that CAH, CH, CASH, CSH and Aft were the main hydration products. Specifically, during the cement hydration process, tricalcium silicate (C_3S), dicalcium silicate (C_2S), tricalcium aluminate (C_3A), and tetracalcium aluminoferrite (C_4AF) reacted to form CSH, CH, tricalcium aluminate hydrate (C_3AH_6), calcium ferric aluminate hydrate (CFH), etc. The main reaction formulas are as follows [12]:

$$C_3S + 3H \rightarrow CSH + CH \tag{1}$$

$$C_2S + 2H \rightarrow CSH + CH \tag{2}$$

$$C_3A + 6H \rightarrow C_3AH_6 \tag{3}$$

$$C_4AF + 7H \rightarrow C_3AH_6 + CFH \tag{4}$$

$$3CS + C_3A + 32H \rightarrow C_6AS_3H_{32} \tag{5}$$

When the CMK was added, it can accelerate the hydration effect and pozzolanic effect of cement. It is mainly that the active substances (SiO_2 and Al_2O_3) in the CMK were activated by the alkaline environment where CH was generated by the cement hydration. At this time, the interface between the cemented soil slurry and the aggregate would accumulate a large amount of CH and become a weak link. CMK had a large number of disconnected chemical bonds, which would rapidly undergo a secondary hydration reaction with CH to generate calcium silicate hydrate. It makes the content of CH reduced and the orientation of CH changed. According to the content ratio of CMK to CH in the hydration process of cement, the hydration products were mainly CSH, tetracalcium aluminate hydrate (C_4AH_{13}), tricalcium aluminate hydrate (C_3AH_6) and anorthite aluminate hydrate (C_2ASH_8). The specific reactions are as follows [13]:

$$AS_2/CH = 0.5, AS_2 + 6CH + 9H \rightarrow C_4AH_{13} + 2CSH \qquad (6)$$

$$AS_2/CH = 0.6, AS_2 + 5CH + 3H \rightarrow C_3AH_6 + 2CSH \qquad (7)$$

$$AS_2/CH = 1.0, AS_2 + 3CH + 6H \rightarrow C_2ASH_8 + CSH \qquad (8)$$

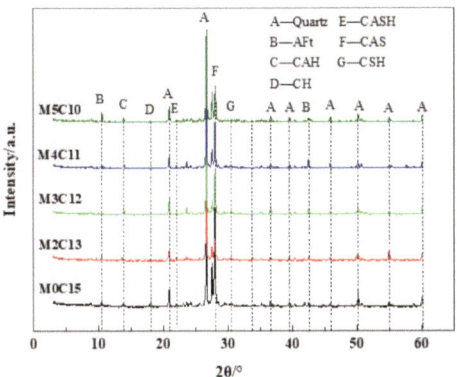

Fig. 2. XRD phase analysis of 28d cemented soil.

It can be seen from Fig. 4 that the diffraction peak intensities of CH and AFt are relatively higher in M0C15, and the diffraction peaks of CAH, CASH and CSH are relatively higher in M3C12. This is due to the high cement content in the cemented soil without mixing CMK, and the hydration reaction is concentrated in the cement reaction. When CMK is used to replace cement, the hydration reaction of CMK will consume CH to generate more CAH, CASH and CSH [14]. However, it does not mean that the more cement is replaced by coal measure metakaolin, the more CAH, CASH and CSH are generated. In this study, when the mass ratio of CMK to cement was 1:4, the hydration reaction was complete and the hydration products were the most in the cemented soil. A reasonable amount of cement and CMK can help promote to generate hydration products in the cemented soil. This is the reason why CMK can effectively improve the compressive strength of cemented soil.

3.3 Analysis of XRD Phase Results

Figure 3 shows the microstructure of cemented soil with five contents of CMK at the age of 28 d. As can be seen from the Fig., the main hydration products of cemented soil were a large number of flocculent CSH/CASH/CAH gels, acicular AFt and platy CH crystals. In Fig. 3 (a), the amount of CSH/CASH/CAH in cemented soil was relatively small, the gap between soil particles was large, the overall structure is loose. This is the reason why the strength of cemented soil without CMK is relatively low. In Fig. 3(b)–(c), the CSH/CASH/CAH gels were cross-connected with Aft crystals after CMK was added, which filled the pores with each other and wrapped the soil particles more densely. Especially, when the CMK content was 3%, the number of cemented soil pores was significantly lower than that of cemented soil without CMK. In Fig. 3(d)–(e), when the CMK content exceeded 3%, the CSH/CASH/CAH gels and Aft crystals began to decrease obviously, and the pores between soil particles increased. This indicates that the overall structural strength of cemented soil is reduced.

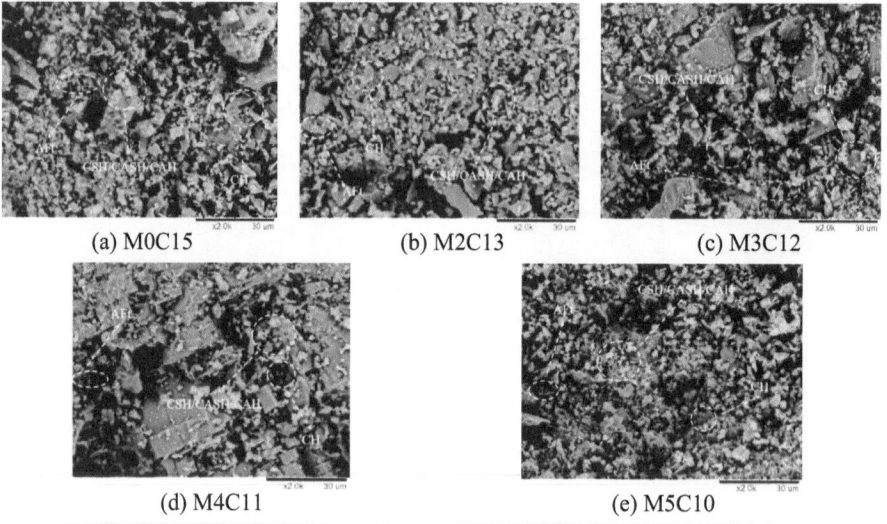

(a) M0C15   (b) M2C13   (c) M3C12

(d) M4C11   (e) M5C10

Fig. 3. Microstructure of cemented soil magnified 2000 times at 28d.

According to the microstructure analysis, it can be explained that CMK has a filling effect in the cemented soil, mainly in two aspects. On the one hand, the particle size of CMK is about an order of magnitude smaller than that of cement (the CMK selected is 6250 mesh). The space between cement particles and soil particles is filled by the CMK particles. On the other hand, the pores between cemented soil frameworks are filled by the main hydration product of CMK and cement (CSH/CASH/CAH gels, AFt crystals) [15]. The pore structure is optimized to reduce the internal porosity. The compactness of the structure is increased to promote strength enhancement.

3.4 Prediction of Long-Term Compressive Strength

The relationship between the compressive strength of cemented soil and the age is shown in Fig. 4. When other conditions were the same, the compressive strength increased with age, but the rate of increase decreased with age.

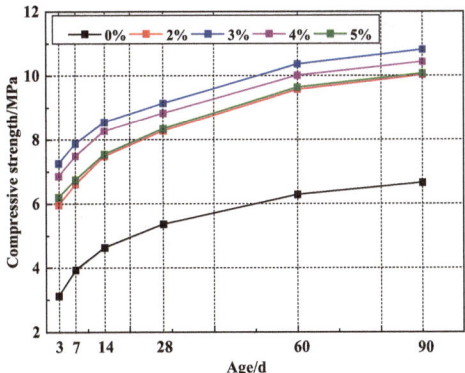

Fig. 4. Relationship between compressive strength and age.

The results show a general linear relationship between the compressive strength of cemented soil and age under the same conditions [16]. For the different CMK contents, the prediction formula of the cemented soil compressive strength and age is established as follows:

$$q_{u,t} = A_t \cdot q_{u,t0} \tag{9}$$

where $q_{u,t0}$ is the known compressive strength of cemented soil at age t_0, $q_{u,t}$ is the prediction compressive strength of cemented soil at age t, A_t is the prediction coefficient.

Based on the data results, the compressive strength of different ages was regarded as the known compressive strength $q_{u,t0}$. The relationship between the compressive strength of each age and the known compressive strength was analyzed. Figure 5 shows the relationship between the compressive strength at different ages and the compressive strength at 7d ($t_0 = 7$d), 14d ($t_0 = 14$d) respectively. The linear relationship between the compressive strength of each age was linearly fitted and the linear slope of each age was obtained. The corresponding slope was the prediction coefficient At, and the fitting correlation coefficient was R_t. The prediction coefficients A_t and the fitting correlation coefficient R_t is listed in Table 4.

It can be seen from Table 4 that the relationship between $q_{u,t}$ and $q_{u,t0}$ was determined mainly by the prediction coefficient A_t. Figure 6 shows that the variation of the prediction coefficient A_t and the age t can be fitted by the power function.

$$A_t = at^b \tag{10}$$

where a, b is the fitting parameters, R is the corresponding fitting correlation coefficient. For different ages t_0, the fitting parameters are shown in Table 5, and b is a fixed value ($b = 0.395$).

The relationship between the fitted parameter a and the age t_0 is shown in Fig. 7, and the obtained equation is:

$$a = t_0^{0.135} \tag{11}$$

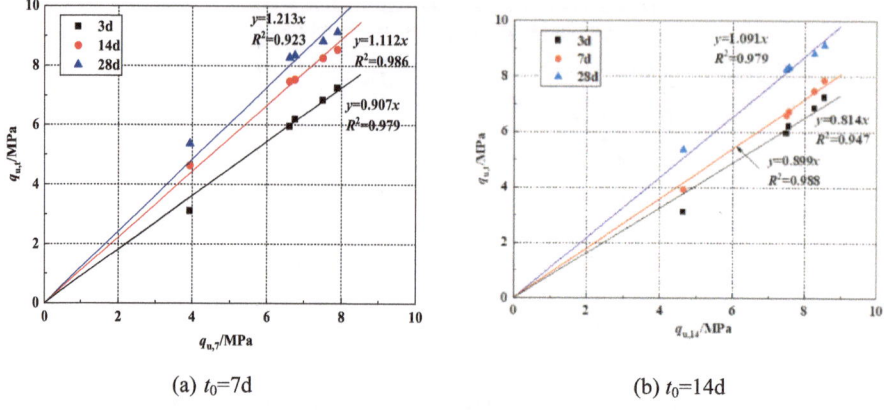

(a) t_0=7d   (b) t_0=14d

Fig. 5. Relationship of cemented soil between qu, t and qu, $t0$.

Table 4. Prediction coefficient A_t and fitting correlation coefficient R_t.

t/d	$t_0 = 3$		$t_0 = 7$		$t_0 = 14$		$t_0 = 28$	
	A_3	R_3	A_7	R_7	A_{14}	R_{14}	A_{28}	R_{28}
3	1	1	0.907	0.979	0.814	0.947	0.745	0.896
7	1.101	0.972	1	1	0.899	0.988	0.823	0.951
14	1.224	0.913	1.112	0.986	1	1	0.916	0.983
28	1.334	0.784	1.213	0.923	1.091	0.979	1	1

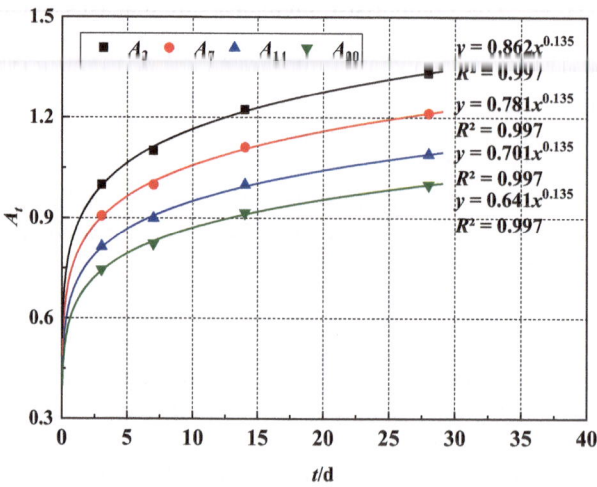

Fig. 6. Relationship between A_t and t.

Table 5. Fitting parameters a, b of different t_0.

t_0/d	a	B	R^2
3	0.641	0.135	0.997
7	0.701	0.135	0.997
14	0.781	0.135	0.997
28	0.862	0.135	0.997

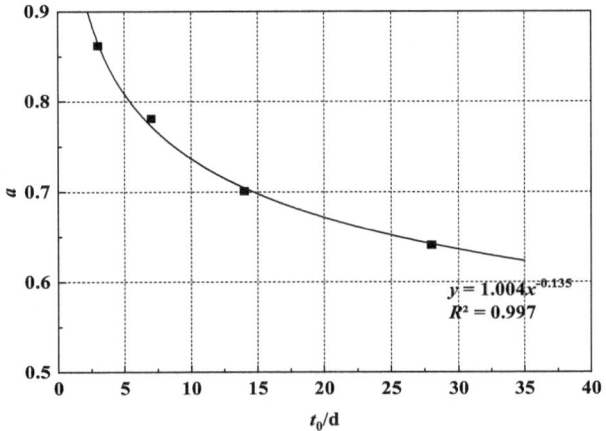

Fig. 7. Relationship between a and t_0.

Equation 11 is put into Eq. 10:

$$A_t = t_0^{0.135} \cdot t^{-0.135} \tag{12}$$

When the specific age t_0 is known, the cemented soil compressive strength $q_{u,\,t}$ can be estimated by the compressive strength $q_{u,\,t0}$, the equation is as follows:

$$q_{u,t} = t_0^{0.135} \cdot t^{-0.135} \cdot q_{u,t_0} \tag{13}$$

In actual engineering, the compressive strength of cemented soil is generally selected at the age of 28d to predict the compressive strength of 60d and 90d. In this paper, $q_{u,28}$ was used as the known compressive strength, and $q_{u,60}$, $q_{u,90}$ were calculated by Eq. 14. The predicted values of 60d and 90d were compared with the measured values, the specific results are shown in Table 6. It is found that the predicted values fall within 6% of the deviation of the measured values. This proved that the formula proposed to predict the later compressive strength of cemented soil is reasonable in this paper.

$$q_{u,t} = 1.568t^{-0.135} \cdot q_{u,t_0} \tag{14}$$

Table 6. Predicted value and measured value for cemented soil $q_{u,60}$, $q_{u,90}$.

CMK content /%	Measured value $q_{u,60}$/MPa	Predicted value $q_{u,60}$/MPa	Deviation /%	Measured value $q_{u,90}$ /MPa	Predicted value $q_{u,90}$ /MPa	Deviation /%
0	6.29	5.95	−5.37	6.66	6.29	−5.60
2	9.57	9.18	−4.10	10.02	9.69	-3.26
3	10.36	10.13	−2.22	10.81	10.70	−1.01
4	10.01	9.79	−2.23	10.43	10.34	-0.89
5	9.64	9.25	−4.00	10.07	9.78	−2.92

Note: $Deviation = \frac{Predicted\ value - Measured\ value}{Predicted\ value} \times 100\%$

4 Conclusions

In this paper, the influence of the CMK content and age on the compressive strength of cemented soil was investigated. The solidification mechanism of the cemented soil was explored by microscopic testing technology. The main conclusions can be drawn.

(1) The compressive strength of cemented soil can be significantly improved by using part of CMK instead of cement. When the total amount of cementitious materials was 15% and the CMK content was 3%, the compressive strength of the cemented soil reached the maximum.

(2) CMK played a role in accelerating the cement hydration effect, pozzolanic effect and filling effect, which reacted with CH generated by cement hydration to generate CAH, CASH and CSH gels. These hydration products and cement hydration products enhanced the compactness of soil particles through bonding, filling, and compaction, forming a dense whole, thereby improving the strength of cemented soil.

(3) The CMK and sandy soil used to study were in Shanxi Province, China. The two raw materials had obvious regional characteristics. The engineering properties and material compositions of raw materials in different regions were different. Based on this, a statistical formula between compressive strength and age of cemented soil ($q_{u,t} = t_0^{0.135} \cdot t^{-0.135} \cdot q_{u,t_0}$) was proposed in this paper.

Acknowledgment. The authors are grateful to Prof. Xiaohong Bai of Taiyuan University of Technology in this paper. This study was funded by the National Natural Science Foundation of China (No.51879180, No.41807256), PowerChina Huadong Engineering Corporation Limited (No.KY2023-SY-02-01, No.KY2019-SY-01).

References

1. Wei, L.T., Xu, Q., Wang, S.Y., Wang, C.L., Chen, J.F.: Development of transparent cemented soil for geotechnical laboratory modelling. Eng. Geol. **262**, 105354 (2019)

2. Devarangadi, M., Uma Shankar, M.: Effect on engineering properties of ground granulated blast furnace slag admixed with laterite soil, cement and bentonite mixtures as a liner in landfill. J Cleaner Prod **329**, 129757 (2021)
3. Kulkarni, P.P., Mandal, J.N.: Strength evaluation of soil stabilized with nano silica-cement mixes as road construction material. Constr. Build. Mater. **314**, 125363 (2022)
4. Lang, L., Song, C.Y., Xue, L., Chen, B.: Effectiveness of waste steel slag powder on the strength development and associated micro-mechanisms of cement-stabilized dredged sludge. Constr. Build. Mater. **240**, 117975 (2020)
5. Furlan, A.P., Razakamanantsoa, A., Ranaivomanana, H., Amiri, O., Levacher, D., Deneele, D.: Effect of fly ash on microstructural and resistance characteristics of dredged sediment stabilized with lime and cement. Constr. Build. Mater. **272**, 121637 (2021)
6. Avci, E., Deveci, E., Gokce, A.: Effect of sodium silicate on the strength and permeability properties of ultrafine cement grouted sands. J. Mater. Civ. Eng. **33**(8), 04021203 (2021)
7. Wang, W., Liu, J.J., Li, N., Ma, L.: Mechanical properties and micro mechanism of nano-SiO2 modified coastal cement soil at short age. Acta Mater Compositae Sin **39**(4), 1701–1714 (2022)
8. Papa, E., Medri, V., Landi, E., Ballarin, B., Miccio, F.: Production and characterization of geopolymers based on mixed compositions of metakaolin and coal ashes. Mater. Des. **56**, 409–415 (2014)
9. Cao, R.C., Fang, Z., Jin, M., Shang, Y.: Study on the activity of metakaolin produced by traditional rotary kiln in China. Miner **12**(3), 365 (2022)
10. Kolovos, K.G., Asteris, P.G., Cotsovos, D.M., Badogiannis, E., Tsivilis, S.: Mechanical properties of soilcrete mixtures modified with metakaolin. Constr. Build. Mater. **47**, 1026–1036 (2013)
11. Zhang, T.W., Yue, X.B., Deng, Y.F., Zhang, D.W., Liu, S.Y.: Mechanical behaviour and microstructure of cement-stabilised marine clay with a metakaolin agent. Constr. Build. Mater. **73**, 51–57 (2014)
12. Maskell, D., Heath, A., Walker, P.: Use of metakaolin with stabilised extruded earth masonry units. Constr. Build. Mater. **78**, 172–180 (2015)
13. Deng, Y.F., Yue, X.B., Liu, S.Y., Chen, Y.G., Zhang, D.W.: Hydraulic conductivity of cement-stabilized marine clay with metakaolin and its correlation with pore size distribution. Eng. Geol. **193**, 146–152 (2015)
14. Navarro-Blasco, I., Pérez-Nicolás, M., Fernández, J.M., Duran, A., Sirera, R., Alvarez, J.I.: Assessment of the interaction of polycarboxylate superplasticizers in hydrated lime pastes modified with nanosilica or metakaolin as pozzolanic reactives. Constr. Build. Mater. **73**, 1–12 (2014)
15. Wu, Z.L., Deng, Y.F., Liu, S.Y., Liu, Q.W., Chen, Y.G., Zha, F.S.: Strength and micro-structure evolution of compacted soils modified by admixtures of cement and metakaolin. Appl. Clay Sci. **127–128**, 44–51 (2016)
16. Consoli, N.C., Cruz, R.C., Floss, M.F.: Variables controlling strength of artificially cemented sand: influence of curing time. J. Mater. Civ. Eng. **23**(5), 692–696 (2011)

Cause Analysis and Research of Influence Factors of Cracks in Railway Tunnel Lining Concrete

Yuhui Zhao[1] and Chaoyang Yu[2(✉)]

[1] China Highway Engineering Consulting Corporation, Hangzhou 311122, China
[2] Sichuan-Tibet Railway Co., Ltd, Chengdu 610041, China
wtu77_001@163.com

Abstract. Aiming at the cracking of railway tunnel lining concrete, the cracking reasons were analyzed, the non-load factor was determined as the main reason, and the main factors which affecting the shrinkage cracking of lining concrete were discussed in this paper. On this basis, the influences of binder content, unilateral water content, sand ratio and limestone powder content on cracking sensitivity of lining concrete were studied, by using dry-shrinkage, flexural strength and splitting tensile strength as the evaluation indexes of cracking sensitivity of concrete. The results show that the binder content has a significant effect on the dry-shrinkage of lining concrete, and the 56d dry-shrinkage value of concrete increases by about 8% with the increase of 20kg binder. The flexural strength and spitting tensile strength of lining concrete firstly increase and the decrease with the decrease of unilateral water content, and the unilateral water content should be controlled at about 140 kg/m^3–145 kg/m^3. The increase of sand ratio is unfavorable to the cracking sensitivity of lining concrete. The flexural strength and spitting tensile strength of lining concrete tend to decrease with the increase of limestone powder content, and the limestone powder content should be controlled within 20%.

Keywords: Railway Tunnel · Lining Concrete · Cracking Sensitivity · Influence Factors

1 Foreword

With the completion of the "eight vertical and eight horizontal" main network of China's railway, the railway network between major cities in central and eastern regions is becoming more and more perfect. At present, in order to complement the short board of railway network construction in China, the state is actively promoting the pace of railway network construction in the western and southwest regions. The difficulties of railway construction in western and southwest China include complicated geological conditions, lack of high-quality raw materials, and great difficulty in construction. In Yunguichuan area, more than half of the railway lines are tunnel structures, and the proportion of tunnels on individual lines can reach more than 70%, which is significantly different from the central and eastern regions where the railways are mainly bridge structures.

© The Author(s) 2024
G. Feng (Ed.): ICCE 2023, LNCE 526, pp. 147–156, 2024.
https://doi.org/10.1007/978-981-97-4355-1_13

The most important structure in railway tunnel is lining concrete structure, which has the characteristics of dense reinforcement, narrow construction space, concrete is not easy to vibrate and compact, vault part is easy to empty, lining concrete is not easy to cure, so the construction of railway tunnel lining concrete has always been a difficult point in railway engineering. Due to the "high complexity and strong concealment" of tunnel structure, once the quality disease occurs in railway operation, the maintenance and replacement difficulty is far greater than that of bridge structure and subgrade structure. Therefore, how to ensure the construction quality of lining concrete has become the primary concern of engineering and technical personnel.

This paper analyzes the causes and influencing factors of railway tunnel lining concrete cracking, and focuses on the influence of mix ratio parameters on the cracking sensitivity of lining concrete.

2 Causes and Influencing Factors of Railway Tunnel Lining Concrete Cracking

2.1 Causes of Lining Concrete Cracking

Among the quality defects of railway tunnel lining concrete, cracking is the most common quality defect that affects the safety and durability of the concrete structure, second only to cavitation and uncompaction. In the statistical classification of railway tunnel operation risks[1], the quality diseases of lining concrete cracks accounted for 20.0% of the 5 types and 15 risks of the whole railway tunnel operation safety risks, second only to the quality diseases of lining concrete caverns 28.8%.

Concrete is a kind of brittle material. It is easy to crack under load and non-load, which seriously affects the durability of concrete structure. Data [2–4] show that cracking of tunnel lining concrete mainly occurs during construction and within 1–2 years of tunnel through. Many railway tunnel projects show that the tunnel lining concrete cracking is mainly divided into three forms: circumferential cracking, longitudinal cracking and oblique cracking, among which the circumferential cracking is the main. According to the statistical results of Wang Jiahe [5], among the cracks of 696 tunnels at home and abroad, the average proportion of circumferential cracks is 64.5%, and the highest proportion is 71%. Research [6] shows that the circumferential cracking of tunnel lining concrete is mainly caused by non-load action, that is, the cracking is caused by shrinkage and deformation of lining concrete when it is not under load. The longitudinal cracks and oblique cracks are mainly caused by loading action. Therefore, the cracking of tunnel lining concrete is mainly caused by non-load factors, that is, it is closely related to the shrinkage and deformation of concrete. For lining concrete, there are four main reasons for its greater shrinkage. First, temperature stress shrinkage is greater. The general thickness of lining concrete is about 0.5m, and the thickness of special sections can reach 1m, which is similar to mass concrete. The temperature inside the tunnel is relatively stable before the tunnel passes through, and its internal hydration temperature rises higher. In addition, it is difficult to carry out warming and moisturizing maintenance after the mold removal of the lining concrete, so it is easy to crack due to temperature stress shrinkage. Second, the lining concrete itself shrinks and deforms greatly. Because it is required to smoothly pour in a close space with dense reinforcement, the amount of cementing material in the lining concrete is high, and the slump is large, especially in the vault part, to ensure that the pouring is dense and not empty, the concrete slump is larger, usually more than 220mm, so the plastic shrinkage, autogenous shrinkage and dry shrinkage of the concrete are relatively large. Once the concrete is not cured well, it is easy to crack due to greater shrinkage. Third, the waterproof board on the back of the lining and the initial spray concrete will constrain the shrinkage deformation of the lining concrete, resulting in tensile stress on the concrete surface, thus causing cracking. Fourth, the concrete construction and maintenance is poor. Holes or non-compaction caused by the inadequacy of vibration, and segregation or bleeding caused by overvibration, will increase the risk of concrete cracking. The lack of timely maintenance after mold removal of lining concrete also aggravates the water loss of concrete surface, resulting in shrinkage cracking.

2.2 Influencing Factors on Shrinkage Cracking of Lining Concrete

Amount of Cementing Material. The amount of cementing material is the main factor affecting the shrinkage and deformation of concrete. The greater the amount of cementing material, especially the higher the amount of cement in the cementing material, the greater the self-shrinkage of the hydration product and the easier the concrete cracking. In addition, the amount of cementing material determines the amount of aggregate, which is an important factor affecting the shrinkage and deformation of concrete [7, 8]. The ratio of shrinkage S_c of concrete to shrinkage S_p of cement paste depends on the aggregate content α, that is, $S_c = S_p(1-\alpha)n$, where n is the empirical coefficient. The higher the aggregate content, the smaller the shrinkage deformation of concrete [8]. When the water-binder ratio is fixed, the amount of cementing material reflects the proportion of cement slurry and aggregate, that is, the slurry aggregate ratio[8]. The higher the amount of cementing material, the larger the slurry aggregate ratio, and the larger the shrinkage deformation of concrete. In order to ensure the strength and working performance of tunnel lining concrete, a higher amount of cementite material is usually used.

Water Consumption Per Cubic Meter of Concrete. The drying shrinkage of concrete is mainly caused by the continuous evaporation of excess water in concrete. Generally, the amount of water required for cement hydration is only about 25% of the cement mass, that is, the water-cement ratio is 0.25, but the water-cement ratio of concrete applied in actual engineering is much larger than this. It is beneficial to reduce the drying shrinkage of concrete by using concrete with less water consumption.

Aggregate Varieties. Aggregate varieties also have a great influence on the dry shrinkage of concrete[9]. General low shrinkage aggregates are quartzite, limestone, granite and basalt. In addition, the drying shrinkage of concrete is reduced when the aggregate size is large, and the drying shrinkage of concrete is increased when the aggregate moisture content is high.

Mineral Admixtures. Railway tunnel lining concrete is mainly mixed with fly ash. An Mingzhe [10] showed that the mixture of fly ash reduced the self-shrinkage of concrete, especially in the early stage. However, the reduction effect of fly ash on concrete shrinkage is mainly reflected in the later stage of concrete hardening. In the early stage of hydration, due to the slow hydration rate of fly ash, more free water can be evaporated in concrete, so it is easier lose water because of dry, which aggravates the early shrinkage of concrete. Therefore, it is very necessary for the early moisture curing of the lining concrete mixed with fly ash. The addition of slag powder will increase the shrinkage deformation of concrete, which is unfavorable to the crack prevention of the lining concrete.

Moisture Curing. Early moisture curing is the most effective measure to reduce the shrinkage of concrete, but because of the structural characteristics of lining concrete, it is difficult to achieve effective moisture curing in actual construction, which is unfavorable to the crack prevention of lining concrete.

3 Experiment

3.1 Raw Material and Experiment Mix Ratio

In this paper, Beijing Jinyu P·O42.5 grade ordinary Portland cement was used, with specific surface area of 347 m^2/kg and 3d and 28d compressive strength of 23.0 MPa and 47.0 MPa, respectively. Fly ash was produced by Shenyang Guodian Kangping New Materials Co., LTD.,which is Class F grade I fly ash, with fineness of 11%, firing loss of 1.42%, water requirement ratio of 94%. Limestone powder was used with MB value of 0.9 and mobility ratio of 105%. The water reducing agent was produced by Jiangsu Subote New Materials Co., LTD., which is PCA-I retarding water reducing agent. The fine aggregate is made of machine-made sand with fineness modulus of 2.8, stone powder content of 6% and apparent density of 2650 kg/m^3. The coarse aggregate is 5 mm–31.5 mm continuously graded three-graded gravel, with an apparent density of 2700 kg/m^3 and a mud content of 0.5%.

The test adopts the mix ratio of lining concrete with strength grade C40 commonly used in railway tunnel engineering, the gas content is between 4.0% and 5.0%, and the slump is 200 mm. The base mix ratio is shown in Table 1.

Table 1. Mix proportion(kg/m^3)

Cement	Fly ash	Fine aggregate	Coarse aggregate	Water	Water reducing agent
336	84	716	1074	155	4.2

3.2 Experiment Method

Drying shrinkage can directly reflect the shrinkage of concrete, flexural strength and splitting tensile strength can also effectively characterize the cracking sensitivity of concrete. Dry shrinkage, flexural strength and splitting tensile strength were used to evaluate the influence of key parameters on the cracking sensitivity of lining concrete. The flexural strength and splitting tensile strength of concrete were tested according to the "Standard for test methods of concrete physical and mechanical properties" (GB/T 50081-2019), and the drying shrinkage of concrete was tested according to the "Standard for test methods of long-term performance and durability of ordinary concrete" (GB/T 50082-2009).

4 Influence of Key Parameters on Cracking Sensitivity of Lining Concrete

4.1 Content of Cementitious Materials

Based on the base mix proportion in Table 1, the influence of different cementitious materials (total content of cement and fly ash) on the dry-shrinkage of lining concrete were studied when they were 360 kg/m^3, 380 kg/m^3, 400 kg/m^3 and 420 kg/m^3, respectively, while the water-binder ratio and other parameters remained unchanged. The results were shown in Fig. 1.

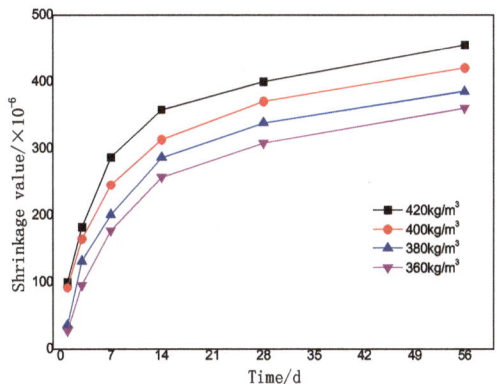

Fig. 1. Influence of content of cementitious materials on dry-shrinkage of concrete

The results of Fig. 1 show that the content of cementitious materials has a significant effect on the dry-shrinkage of lining concrete, and the shrinkage value of concrete increases by about 8% with the increase of 20 kg/m^3 cementitious materials. The content of cementitious materials increases, and the content of aggregate decreases accordingly when the water-binder ratio is fixed, therefore, the change in the content of cementitious materials actually reflects the change of bone pulp ratio. The greater the content of cementitious materials, the smaller the bone pulp ratio, and the greater the dry-shrinkage value of concrete.

4.2 Unilateral Water Consumption

Based on the base mix ratio in Table 1, the influence of water consumption of 155 kg/m^3, 150 kg/m^3, 145 kg/m^3, 140 kg/m^3 and 135 kg/m^3 on flexural strength and splitting tensile strength of concrete were studied while keeping other parameters constant. The results were shown in Figs. 2 and 3.

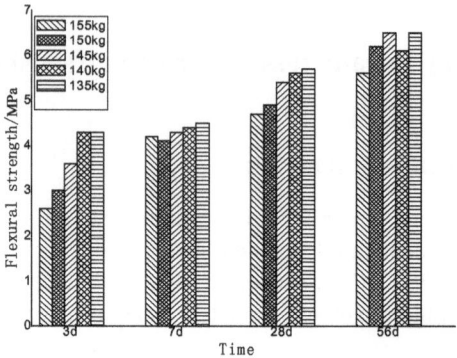

Fig. 2. Influence of water consumption on flexural strength of concrete

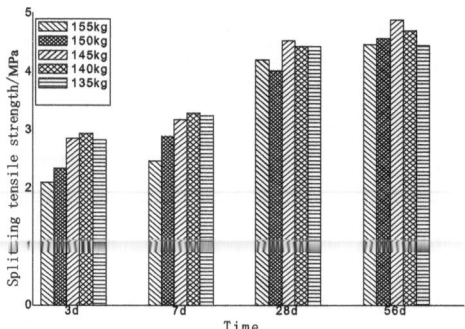

Fig. 3. Influence of water consumption on splitting tensile strength of concrete

From Figs. 2 and 3, it can be seen that with the decrease of unilateral water consumption, both the flexural strength and splitting tensile strength of concrete tend to increase significantly, especially before 28d. The splitting tensile strength is more sensitive to the change of unilateral water consumption, when the unilateral water consumption is lower than 145 kg, the increase rate of the splitting tensile strength decreases significantly and shows a decreasing trend at 56d. When the unilateral water consumption is lower than 140kg, the increase rate of the flexural strength also decreases significantly. The flexural strength test is a test method to measure the tensile stress of the reaction material under the action of transverse load, which is usually tested by three-point bending method. The splitting tensile strength is another commonly used test method to characterize the tensile strength of concrete materials, which is usually tested by the two-point concentrated load method. Both methods are used to replace the direct tensile strength test which is difficult to operate to test the tensile strength of the material. The tensile strength is a key index to determine their cracking properties for concrete material [11]. Therefore, flexural strength and splitting tensile strength are often used to characterize the cracking sensitivity of concrete material. The results show that the cracking sensitivity of lining concrete firstly increase and the decrease with the decrease of unilateral water consumption, and there is an optimal unilateral water consumption. At the same time, the unilateral water consumption should not be too low in order to ensure the working performance and hydration characteristics of concrete, it should be controlled at about 140 kg/m^3–145 kg/m^3.

4.3 Sand Ratio

Generally speaking, the best sand ratio of the concrete mix in good condition ranges from 36% to 45%. Based on the base mix ratio in Table 1, the influence of different sand ratio (36%, 39% and 42%) on the flexural strength, splitting tensile strength and dry-shrinkage of concrete under the same amount of cementitious materials were studied. The results were shown in Figs. 4, 5 and 6 respectively.

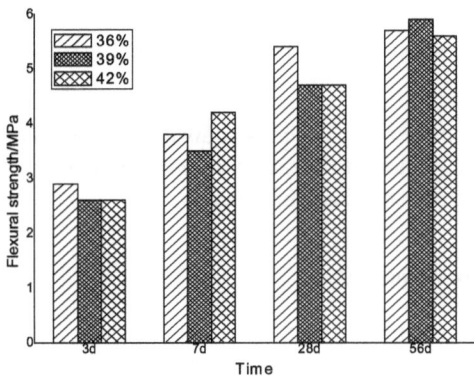

Fig. 4. Influence of sand ratio on flexural strength of concrete

From Figs. 4 and 5, it can be seen that the flexural strength and splitting tensile strength of the lining concrete before 28d showed a decreasing trend in general and basically the same after 56d with the increase of sand ratio. The results show that the increase of sand ratio is unfavorable to the early cracking sensitivity of lining concrete, because the limiting ability of microcrack propagation in concrete is reduced with the reduction of coarse aggregate content.

From Fig. 6, it can be seen that the dry-shrinkage of lining concrete presents a linear growth trend with the increase of sand ratio, and the 56d dry-shrinkage value increases by 10% with the increase from 36% to 39% sand ratio. The 56d dry-shrinkage value increase by 6% compared with the sand rate of 39% When sand ratio continues to increase to 42%. The increase of sand ratio means the decrease of coarse aggregate content and

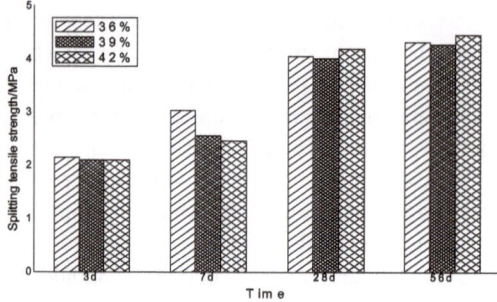

Fig. 5. Influence of sand ratio on splitting tensile strength of concrete

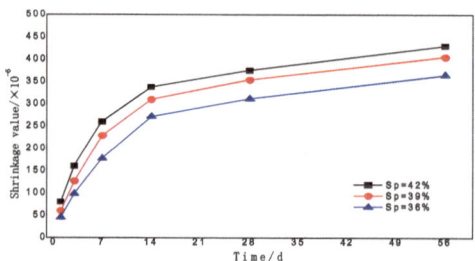

Fig. 6. Influence of sand ratio on dry-shrinkage of concrete

the increase of pulp to bone ratio, which is equivalent to the increase of the amount of cementitious materials. Although river sand dose not participate in hydration reaction, it will also increase the dry-shrinkage of concrete.

4.4 Limestone Powder Content

Machine-made sand is used to produce concrete on nearly half of railway lines with the increasing shortage of natural river sand and high-quality fly ash resources in the country, and as a by-product of machine-made sand production, limestone powder is used to prepare concrete has become an inevitable trend as a new mineral admixture. Based on the base mix ratio in Table 1, when limestone powder is used to replace fly ash, the influence of limestone powder content of 10%, 20% and 30% respectively (including 6% limestone powder in machine-made sand) on flexural strength and splitting tensile strength of concrete were studied under the total amount of cementing material remains unchanged at 420kg/m^3. The results were shown in Fig. 7 and Fig. 8.

From Figs. 7 and 8, it can be seen that the flexural strength and splitting tensile strength of different aged lining concrete decreases significantly with the increase of limestone powder content, and the splitting tensile strength being more evident. In addition, the flexural strength and splitting tensile strength of concrete has little difference when the stone powder content is 10% and 20%, but the reduction is significant when the stone powder content reaches 30%.As an inert material, limestone powder does not participate in cement hydration, but mainly accelerates cement hydration by forming

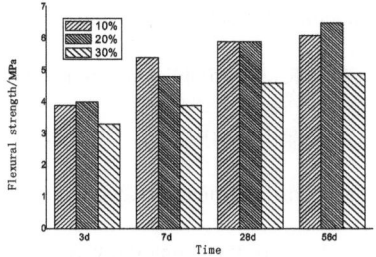

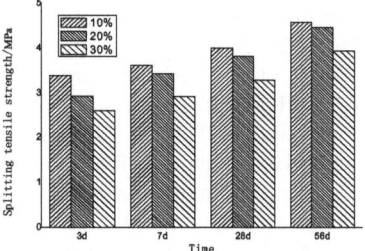

Fig. 7. Influence of limestone powder content on flexural strength of concrete

Fig. 8. Influence of limestone powder content on splitting tensile strength of concrete

crystalline nucleation points and plays a filling role in concrete. Due to the extremely weak hydration, limestone powder will reduce the hydration rate of cement, reduce the generation of hydration gel products, and reduce the flexural strength and splitting tensile strength of concrete after exceeding a certain dosage. At the same time, due to the delay of cement hydration rate, the amount of free water that can be evaporated in concrete will be more, and the dry shrinkage of concrete will be increased to a certain extent.

5   Conclusion

(1) Among the mix ratio parameters that affect the cracking sensitivity of lining concrete, the amount of cementitious materials and unilateral water consumption have obvious effects on the shrinkage cracking of concrete. The content of cementitious materials should be as low as possible, and the water content should be controlled at about $140 \text{ kg/m}^3 - 145 \text{ kg/m}^3$. The increase of sand ratio is unfavorable to the shrinkage of lining concrete, so it is advisable to choose as low sand ratio as possible to prepare lining concrete.

(2) As a new mineral admixture, the excessive addition of limestone powder will increase the crack sensitivity of the lining concrete, and the limestone powder content should be controlled within 20%.

References

1. Minqing, Z., Jingyu, H., Shuishan, W., et al.: Research on the control standard for the end of concrete pouring of tunnel arch lining. J. Railway Eng. Soc. **9**(9), 29–33 (2021)
2. Zhang, Y.: The research on temperature crack control of mass concrete base on a lake tunnel. Nanchang University, Nanchang (2015)
3. Zhu, B.: Temperature Stress and Temperature Control of Large Volume Concrete. China Electric Power Press, Beijing (1999)
4. Yuan, Y.: Control for Early Cracks of Concrete Structures[M]. Science Press, Beijing (2004)
5. Wang, J., Huang, F., Li, H., et al.: Cause analysis and prevention measures of temperature cracks in railway tunnel lining concrete. Railway Eng. **60**(9), 73–77 (2020)
6. Xie, L., Yang, Q.: Cause analysis of and preventive advices for circumferential cracks occurring in the lining of a tunnel. Highway **65**(5), 347–352 (2020)

7. The European Project Group: The European guidelines for self-compacting. Concrete **5**, P6–P7 (2005)
8. Wu, Z., Lian, H.: High Performance Concrete. China Railway Publishing House, Beijing, P269, P126–P127 (1998)
9. Vinogradov, B.H.: Influence of Aggregate on Concrete Properties, pp. P105-107. China Architecture & Building Press, Beijing (1985)
10. Mingzhe, A.: Study on Autogenous Shrinkage of High-Performance Concrete. Tsinghua University, Beijing (1999)
11. Chengju, G.: Physics and Chemistry of Concrete, pp. P191-198. China Railway Publishing House, Beijing (2004)

Evaluation and Analysis of the Effect of Anti-slide Piles Around Pile Foundations of Super Large and Large Span Projects

Nan Zhang[✉], Ya Dai, Bo Wang, Boyi Li, Weizhou Xu, and Chao Ye

Economic Research Institute, State Grid Jiangsu Electric Power Co. LTD., Nanjing, China
zhangnan_ERI@163.com

Abstract. In this study, the three-dimensional finite element numerical analysis was carried out on the anti-sliding characteristics of cement mixing piles around the foundation of the north collapse tower, and the influence of the length and number of rows of anti-slide piles around the pile foundation on the deformation of the pile foundation was systematically studied. When the anti-slide pile length increases from 0 m to 20 m, the pile foundation deformation decreases with the increase of the anti-slide pile length, but the decrease rate is less than 5%. When the length of the anti-slide piles is 10 m and 20 m, after the anti-slide piles on both sides of the large-span foundation increase from 3 rows to 5 rows, the horizontal and vertical deformation of the pile foundation decreases, but the deformation of the pile foundation decreases. Less than 2%. By installing tie beams, the four pile groups of the large-span foundation form an integral structure. The overall horizontal resistance is already strong, but the horizontal resistance provided by cement mixing piles is limited.

Keywords: Anti-Sliding Characteristic · Pile Foundation Deformation · Large-Span Foundation · Horizontal Resistance · Numerical Analysis

1 Introduction

Under the action of wind load, large span transmission project of crossing tower inevitably produce additional stress and deformation. If the additional stress or deformation of pile foundation is too large, it will affect the normal use of large span project. Therefore, domestic scholars put forward new construction technology and construction methods to enhance the long-term service safety of large span projects.

Construction techniques such as high-strength prestressed pipe pile and post-grouting of foundation can not only reduce the construction period, but also improve the ability of resisting external deformation of large-span foundation [1–3]. In order to determine the bearing capacity of large-span pile foundation, Zhang Tianguang et al. [4, 5] found that the self-balancing test method of pile foundation can quickly and accurately determine the bearing capacity of large-span pile foundation, which provides guidance for engineering design optimization.

© The Author(s) 2024
G. Feng (Ed.): ICCE 2023, LNCE 526, pp. 157–166, 2024.
https://doi.org/10.1007/978-981-97-4355-1_14

In order to optimize the design of pile foundation for large-span projects, many scholars used three-dimensional numerical simulation to analyze the stress deformation law of pile foundation [6–9]. Based on the quadratic programming method, Fang Xiaowu et al. [6] optimized the layout of large-span pile foundations. Through three-dimensional numerical analysis, the development depth and width of soil cracks on the side of the pile were obtained [7].

For the large span pile foundation project, the current research focuses on the construction method and bearing capacity characteristics of the foundation pile foundation. In order to limit the deformation of large span pile foundation, anti-slide pile is often constructed on the side of pile foundation, but the reinforcement effect of anti-slide pile is not clear. Through the systematic three-dimensional finite element numerical analysis, the reinforcement effect of anti-slide pile is studied, which provides guidance for the optimal design of large span pile foundation.

2   Three-Dimensional Finite Element Simulation

2.1   Engineering Situation

The Yangtze River Crossing Project of Fengcheng ~ Meili 500 kV line in Jiangsu adopts the "with-direct-direct-with-resistance" crossing mode. The length of the tensioning section is 4.055 km and the spacing distribution is 755 m-2550 m-750 m. The total height of the span tower is 385 m. The concrete filled steel tube tower is made of Q420C, and the interior is filled with C50 self-compacting concrete. The foundation adopts the cast-in-place pile scheme of the cap. The four foundations are independent and the concrete connecting beams are used between the caps. The upper layer of the foundation soil is about 10 m thick of muddy silty clay, while the lower layer is composed of fine sand, silty clay, and medium sand, without any bedrock. Three rows of anti-slide piles are arranged in front of and behind each pile foundation. In the initial design scheme, the length of anti-slide piles of the north span tower is 10.0 m and the number of piles is 360.

2.2   Calculated Operating Condition

Based on the previous study, the lateral deformation of pile foundation can be significantly reduced by installing straining beams. In this numerical calculation, the height of the straining beam is 2.0 m, and the number of straining beam supporting piles is 2. When the front and back of the anti-slide pile are 3 rows, the pile length is 0.0 m, 10.0 m, 15 m and 20 m, respectively. When the length of anti-slide pile is 10 m and 20 m, the number of anti-slide pile rows is 5, as shown in Table 1.

Table 1. Working condition of finite element calculation

Working condition	Number of anti-slide pile rows	Length of anti-slide pile (m)
Numerical value	0	0
	3、5	10
	3	15
	3、5	20

2.3 Wind Load Condition

The loads on the foundation under four kinds of strong winds are shown in Table 2. There are 6 sets of three-dimensional finite element numerical calculation grid in the evaluation and analysis of anti-slide pile effect. A total of 24 three-dimensional finite element numerical calculations are carried out considering the deformation of pile foundation under 4 kinds of wind conditions.

2.4 Finite Element Mesh and Boundary Condition

The software ABAQUS is used to analyze the deformation and stress of oversized large-span engineering pile foundation. Three-dimensional finite element grid partition USES the actual distribution of soil layer, corresponding to the stratigraphic boundary according to the actual survey line access, will be merged into thin soil property close ten layers.

The finite element mesh is divided into 519457 elements and 509692 nodes. Normal displacement constraints are applied to the sides of the formation, and three-direction displacement constraints are applied to the bottom. The coordinate system Z axis used in the finite element calculation model is vertical and upward is positive.

Table 2. Numerical calculation of wind load condition

Working condition	node	F_X	F_Y	F_Z
90° wind	3000	−10143	−7379	77413
	3001	−4376	1520	−17314
	3002	−9920	7270	75630
	3003	−4367	−1615	−17789
60°wind	3000	−12672	−11471	103914
	3001	−2834	−1341	−2028

(continued)

Table 2. (*continued*)

Working condition	node	F_X	F_Y	F_Z
	3002	−6277	2239	39090
	3003	−4642	−3466	−23034
45°wind	3000	−12717	−12662	109579
	3001	−818	−3576	14074
	3002	−3837	−502	16866
	3003	−4218	−4217	−22577
0°wind	3000	−6857	−9514	71521
	3001	6856	−9514	71516
	3002	1076	−3815	−12547
	3003	−1077	−3815	−12548

Note: F_X and F_Y are horizontal forces; F_Z is the vertical force, and the pressure is positive and the pull is negative

2.5 Constitutive Model and Soil Layer Parameters

The Mohr-Coulomb model is used to simulate the mechanical properties of the foundation soil layer and the linear elastic model is used to simulate the pile foundation and cap. The formation material parameters are set based on the geological exploration data of the north span tower foundation. The soil layer names and soil layer parameters are shown in Table 3.

Table 3. Name and parameter of foundation soil layer

Soil layer	Weight density γ (kN/m³)	Thickness m	Modulus of compression E_s (Mpa)	Cohesion c' (kPa)	Friction Angle φ' (°)
①Silty clay	17.8	2	3.5	8	24
②Mucky silty clay	17.4	12	3.0	5	27
③Silty clay mixed with silt	18.0	5	4.0	7	28

(*continued*)

Table 3. (*continued*)

Soil layer	Weight density γ (kN/m^3)	Thickness m	Modulus of compression E_s (Mpa)	Cohesion c' (kPa)	Friction Angle φ' (°)
④Silty sand mixed with silt	17.8	4	5.5	4	32
⑤Silt	18.6	8	11.5	4	33.5
⑥Silty clay	18.6	5	6.1	9	26
⑦Silty clay	18.4	4	4.5	9	25
⑧Silty clay mixed with silt	19.2	4	7.3	10	26
⑨Silty fine sand	19.8	12	13.5	4	36
⑩Medium sand	20.0	24	15.5	3	36

3   Influence of the Length of Anti-Slide Pile

3.1   Effect of Anti-Slide Pile Length on Three-Dimensional Deformation of Foundation

This section focuses on the analysis of the effect of the length of anti-slide pile and wind direction on the three-dimensional deformation of foundation. Four wind conditions of 0°, 45°, 60° and 90° are considered in the numerical calculation. Limited to space, only the three-dimensional deformation cloud image of the foundation under the wind condition of 90° is given.

Figure 1 shows the cloud image of foundation deformation caused by 90° wind load. With the increase of the length of cement mixing pile, the deformation of foundation decreases, but the decreasing range is limited. Under four kinds of wind loads, the deformation of foundation decreases, but the reduction range is not more than 3%.

By setting up straining beams, the four pile foundations of the large span form an integral structure, with strong overall horizontal resistance. The length of the cement mixing pile is limited, which does not form a whole with the pile foundation, and the mechanical properties of the soil within the range of the cement mixing pile are poor. Therefore, the horizontal resistance provided by cement mixing piles is limited.

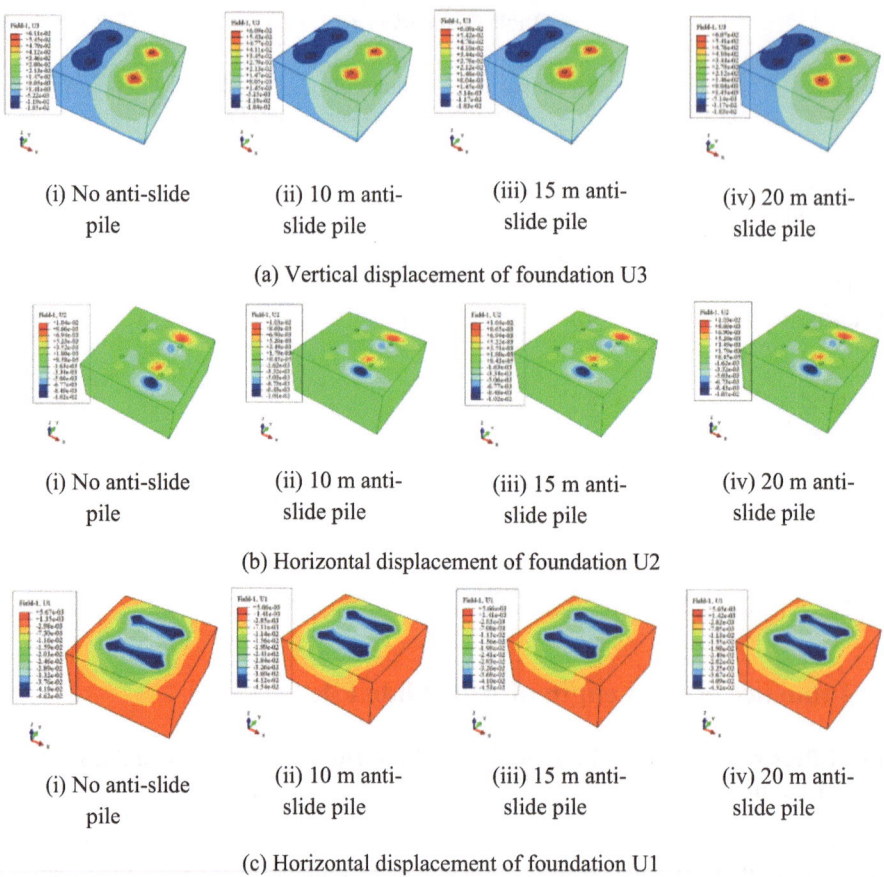

(i) No anti-slide pile   (ii) 10 m anti-slide pile   (iii) 15 m anti-slide pile   (iv) 20 m anti-slide pile

(a) Vertical displacement of foundation U3

(i) No anti-slide pile   (ii) 10 m anti-slide pile   (iii) 15 m anti-slide pile   (iv) 20 m anti-slide pile

(b) Horizontal displacement of foundation U2

(i) No anti-slide pile   (ii) 10 m anti-slide pile   (iii) 15 m anti-slide pile   (iv) 20 m anti-slide pile

(c) Horizontal displacement of foundation U1

Fig. 1. Three-dimensional deformation cloud image of foundation under 90° wind load

3.2 Effect of Anti-Slide Pile Length on Three-Dimensional Deformation of Pile Foundation

Figure 2 shows the cloud map of pile foundation deformation caused by 90° wind load. When the length of anti-slide pile is 0 m, 10 m, 15 m and 20 m, the maximum displacement of pile foundation along the two horizontal directions (U1, U2) and vertical directions (U3) are (46.0, 45.3, 45.2, 45.1 mm), (4.84, 4.67, 4.65, 4.63 mm) and (61.1, 60.9, 60.8, 60.7 mm), respectively.

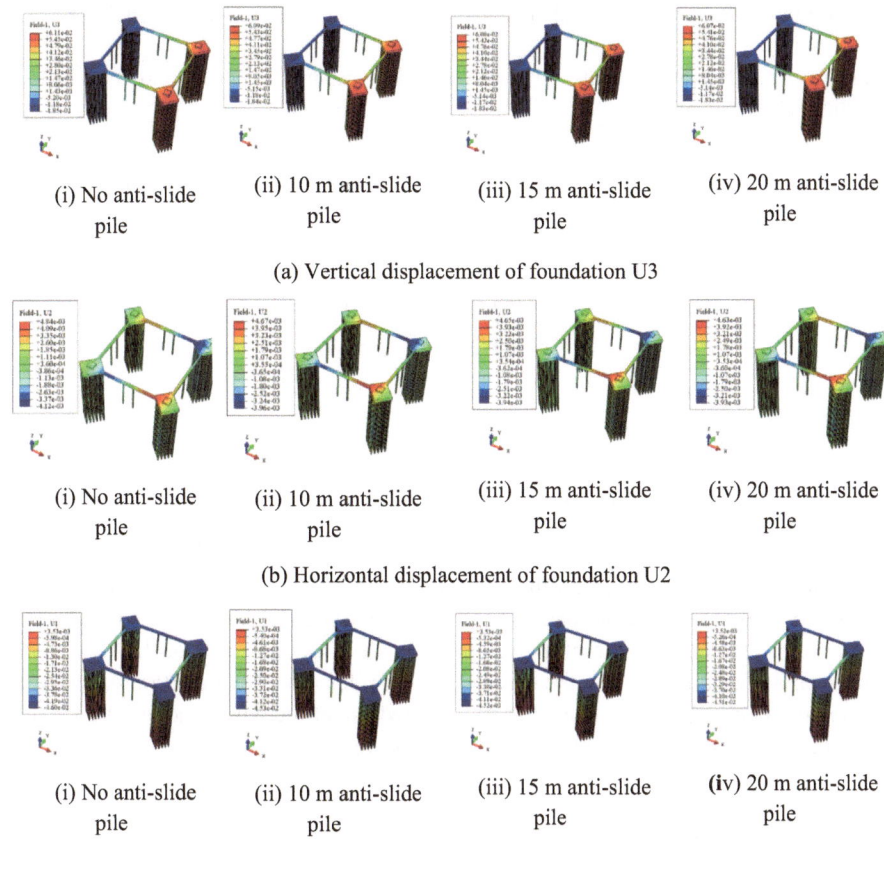

(i) No anti-slide pile | (ii) 10 m anti-slide pile | (iii) 15 m anti-slide pile | (iv) 20 m anti-slide pile

(a) Vertical displacement of foundation U3

(i) No anti-slide pile | (ii) 10 m anti-slide pile | (iii) 15 m anti-slide pile | (iv) 20 m anti-slide pile

(b) Horizontal displacement of foundation U2

(i) No anti-slide pile | (ii) 10 m anti-slide pile | (iii) 15 m anti-slide pile | (iv) 20 m anti-slide pile

(c) Horizontal displacement of foundation U1

Fig. 2. Three-dimensional deformation cloud image of pile foundation under 90° wind load

3.3 Effect of Anti-Slide Pile Length on Maximum Deformation of Pile Foundation

Table 4 summarizes the maximum displacement of pile foundation in three directions caused by 90° wind load under each working condition. With the increase of the length of cement mixing pile, the deformation of pile foundation is reduced, but the reduction range is limited and the reduction range is less than 2%. Cement mixing piles provide limited horizontal resistance. Under 60°, 45° and 0° wind loads, the deformation of pile foundation decreases with the increase of anti-slide pile length. The results show that the horizontal and vertical deformation of pile foundation can be reduced to a certain extent, but the reduction is less than 5%.

Table 4. Maximum deformation of pile foundation under each working condition

Working condition (90°)	No anti-slide pile	10.0 m Anti-slide pile	15.0 m Anti-slide pile length	20.0 m Anti-slide pile
U1(mm)	3.53	3.53	3.53	3.52
	−46.0	−45.3	-45.2	−45.1
U2(mm)	4.84	4.67	4.65	4.63
	−4.12	−3.96	−3.94	−3.93
U3(mm)	61.1	60.9	60.8	60.7
	−18.5	−18.4	−18.3	−18.3

4   Influence of Anti-Slide Pile Row Number

4.1   Effect of Different Anti-Slide Pile Rows on Three-Dimensional Deformation of Foundation

Table 5 summarizes the extreme values of foundation deformation under different rows of stirred piles under 90° wind load. When the length of cement mixing pile is 10 m and 20m respectively, the increase of mixing pile from 3 rows to 5 rows has little effect on foundation deformation. When the cement mixing piles are 10m and 20m respectively and the mixing piles increase from 3 rows to 5 rows under the wind load of 60°, 45° and 0°, the deformation reduction of pile foundation is very small, not more than 2%.

Table 5. Maximum foundation deformation under different working conditions

Working condition (90°)	10 m anti-slide pile (3 rows)	10 m anti-slide pile (5 rows)	20 m anti-slide pile (3 rows)	20 m anti-slide pile (5 rows)
U1 (mm)	5.66	5.66	5.65	5.65
	−45.4	−45.1	−45.2	−44.7
U2 (mm)	10.3	10.3	10.3	10.2
	−10.1	−10.2	−10.1	−10.0
U3 (mm)	60.9	60.8	60.7	60.6
	−18.4	−18.3	−18.3	−18.2

4.2   Effect of Different Anti-Slide Pile Rows on Three-Dimensional Deformation of Pile Foundation

Table 6 summarizes the extreme deformation values of pile foundation under different rows of stirred piles under 90° wind load. When the length of cement mixing pile is 10 m and 20m respectively and the mixing pile is increased from 3 rows to 5 rows, the deformation reduction of pile foundation is not more than 3%. When the cement mixing piles are 10m and 20m respectively and the mixing piles increase from 3 rows to 5 rows under the wind load of 60°, 45° and 0°, the deformation reduction of pile foundation is very small, not more than 3%.

Table 6. Maximum deformation of pile foundation under each working condition

Working condition (90°)	10 m anti-slide pile (3 rows)	10 m anti-slide pile (5 rows)	20 m anti-slide pile (3 rows)	20 m anti-slide pile (5 rows)
U1 (mm)	3.53	3.53	3.52	3.52
	−45.3	−45.0	−45.1	−44.7
U2 (mm)	4.67	4.60	4.63	4.55
	−3.96	−3.90	−3.93	−3.85
U3 (mm)	60.9	60.8	60.7	60.6
	−18.4	−18.3	−18.3	−18.2

5 Conclusion

The influence of anti-slide pile length and number of rows around pile foundation on the deformation of foundation and pile foundation is systematically studied with three-dimensional finite element numerical analysis aiming at the anti-slip characteristics of cement mixing piles around the north span tower foundation. Based on the numerical results, the main conclusions are as follows:

(1) Under four kinds of wind loads, the deformation of pile foundation decreases with the increase of anti-slide pile length. It is shown that the installation of anti-slide pile can reduce the horizontal and vertical deformation of pile foundation to a certain extent, but the reduction is less than 5%.

(2) When the length of anti-slide piles is 10 m and 20 m and the rows of anti-slide piles on both sides of the large-span foundation are increased from 3 rows to 5 rows, the horizontal and vertical deformation of pile foundation caused by wind load in four directions is reduced somewhat, but the deformation reduction of pile foundation is less than 3%.

(3) By setting the straining beam, the four pile groups of the large span foundation form an integral structure, and its overall horizontal resistance has been strong. The length of cement mixing pile is limited. It does not form a whole with pile foundation and the mechanical properties of soil in the range of cement mixing pile are poor. Therefore, cement mixing pile provides limited horizontal resistance.

References

1. Ding, Z., Zhu, Z.: Application of high strength prestressed pipe pile (PHC) in large span foundation J. Electric Power Constr. **30**(3), 98-99 (2009)

2. He, C., Zhu, Y., Luo, J., Hou, Y., Jiang, H.: Application of post-grouting pile in the transmission project of the Yellow River Crossing. J. Water Resour. Arch. Eng. **13**(5), 60–64, 116 (2015)
3. Dong, K., Wang, C.: Design and construction of 500kV long-span and large-diameter deep-hole cast-in-place piles in the Yellow River J. Henan Electric Power (01), 22–27 (1994)
4. Zhang, T., Sun, D., Fu, M., Ao, Q.: Self-balancing test of pile foundation over the Yellow River Crossing of 500kV line J. Electr. Power Constr. **28**(01), 20–22 (2007)
5. Yang, T., Zhang, Z., Jin, Y.: Experimental study on self-balancing method of foundation cast-in pile in Zhoushan Great Crossing sea. J. Shanxi Arch. **36**(35), 84–85 (2010)
6. Fang, X., Zeng, L., Huang, W.: Calculation and optimization of large span tower infrastructure at sea. J. Electric Power Surv. Des. **01**, 65–68 (2010)
7. Wei, Y., Xiao, L., Wei, B., Wang, S.: Study on pile-soil-cap interaction characteristics of high voltage transmission tower based on three-dimensional nonlinear finite element method. J. Smart Power **42**(2), 32–35 (2014)
8. Gan, F., Huang, J., Xu, J., Ma, T.: Research on pile distance optimization of large span foundation considering pile-soil interaction J. Electric Power **24**(2), 94-100 (2009)
9. Lv, X., He, Z.: ANSYS analysis of pile foundation of a power transmission line long span project. J. Hongshui River **29**(3), 124–126, 130 (2010)

Research on Geological Information Updating and Prediction of Tunnel Surrounding Rocks Based on Machine Learning

Yang Ren[1,2], Cunbin Yang[1,2(✉)], Yuehua Wu[1,2], and Tianbin Li[1,2]

[1] State Key Laboratory of Geohazard Prevention and Geoenvironment Protection, Chengdu 610059, China
2630822993@qq.com

[2] College of Environment and Civil Engineering, Chengdu University of Technology, Chengdu 610059, China

Abstract. Surrounding rock geological information is the basic index for tunnel disaster prediction and forecasting, so dynamic updating and prediction of surrounding rock geological information in an efficient and intelligent manner can provide an important support for disaster prediction and forecasting. In this paper, a surrounding rock geological information updating and prediction model based on RNN (RNN) is constructed to carry out research on the prediction of the four geological parameter indexes of rock mass integrity, rock hardness, rock weathering degree, and water abundance of tunnel surrounding rocks. The various index data of excavated tunnel sections are collected, the original data is normalized, a time series prediction model for tunnel surrounding rock geological information is established based on the training set samples, and finally the updated prediction of surrounding rock information is achieved. Preliminary application shows that the updating prediction accuracy of surrounding rock geological information of three indexes including rock mass integrity is up to 87.5%, and the prediction accuracy of rock weathering degree reaches 75%. This is an approach with relatively high engineering practical value.

Keywords: Tunnel Engineering · Surrounding Rock Geological Information · Recurrent Neural Network (Rnn) · Time Series

1 Introduction

The rapid development of underground tunnel engineering has brought about an increasingly complex environment of tunnel surrounding rocks and frequent geological disasters such as rock bursts, water inrush, and mud inrush. How to safely and efficiently carry out tunnel construction operations under complex geological conditions has become a major challenge, and higher requirements have been put forward for tunnel disaster prediction and forecasting. Traditional geological analysis and geophysical exploration methods are helpful to grasp the rock structure and its changing law, but due to the relative independence of different methods, barriers in data extraction, strong subjective

© The Author(s) 2024
G. Feng (Ed.): ICCE 2023, LNCE 526, pp. 167–176, 2024.
https://doi.org/10.1007/978-981-97-4355-1_15

judgment and poor timeliness, they cannot meet the needs of modern tunnel construction and disaster prediction. In the updating prediction of the surrounding rock geological information of tunnels, the geological parameters of unexcavated tunnels are usually predicted based on the geological information of existing excavated tunnels through the time series relationship of the tunnel geological information. However, as the tunnel develops towards longer, larger, and deeper directions, the geological information predicted based on subjective experience has low accuracy and cannot provide guidance for practical engineering. Therefore, it is particularly important to conduct research on the updating prediction of surrounding rock geological information during tunnel construction to improve the accuracy and efficiency of prediction results.

Today, there are a variety of methods for predicting surrounding rock geological information, but most of them are based on advanced geological information prediction using geophysical means. Mathematical model methods based on computer technology have also been applied and developed rapidly. Wang et al. and Li et al. have established artificial neural network theories for identifying the surrounding rock classification of tunnels [1, 2]. Yao et al. proposed to predict the change of tunnel surrounding rocks by artificial BP neural network based on machine learning. They also compared the accuracy between artificial neural networks and support vector machines to predict tunnel surrounding rocks [3, 4]. After studying a variety of advanced prediction technologies and typical cases of unfavourable geology, Jiang et al. proposed to determine the prediction parameters for predicting the geological conditions in front of a tunnel face, and make intelligent prediction of the indexes [5]. More and more scholars have applied various machine learning algorithms to tunnel information prediction, and Liu et al., Zhang et al. and Tian et al. even applied them to rock burst prediction. They used neural network models to predict the risk and intensity of rock bursts, which has further improved the importance of information prediction [6–8]. Many scholars such as Li et al., Xu et al., Xue et al. and Guan et al. have proposed the use of advanced algorithms for advanced geological prediction, and some of them have achieved high accuracy, with an accuracy of up to 95% [9–12]. This indicates that artificial intelligence will gradually replace traditional empirical methods in tunnel engineering in the future [13, 14]. Machine learning has been used by some scholars to predict tunnel surrounding rock geological information, but due to low prediction efficiency and accuracy, it cannot provide more detailed and accurate geological information of surrounding rocks for tunnel disaster prediction and forecasting.

In this paper, a surrounding rock geological information updating and prediction model based on RNN is constructed to carry out research on the prediction of the four geological parameter indexes of rock mass integrity, rock hardness, rock weathering degree, and water abundance. After the various index data of excavated tunnel sections of are collected as a training set, the data are normalized, and the discrete data are eliminated, the training data set is used as the input set of the recurrent neural network, and a hidden layer is constructed for data training to obtain the surrounding rock geological information updating and prediction model based on RNN, so that the prediction results of each index can be obtained quickly. A preliminary test of this method has been conducted in a tunnel in Kangding. The results show that the updating prediction accuracy of indexes such as rock mass integrity is high, which can meet the practical application of engineering.

2 Tunnel Surrounding Rock Geological Information Updating and Prediction Model Based on RNN

2.1 Recurrent Neural Network Theory

RNN originated from the Hopfield Network proposed by Saratha Sathasivam. Fig. 1 shows the flowchart of a recurrent neural network. It adds the relationship between the previous and subsequent time series on the basis of fully connected neural network, which is very effective for data with sequence characteristics. Compared with the neural network used in the early stage, recurrent neural network can mine the time series information and semantic information in the data. It stores past information by introducing state variables, and it can determine, together with previous input information, the current information output. In the traditional neural network models, from the input layer to the hidden layer and then to the output layer, the layers are fully connected, while the nodes in each layer are unconnected. However, this ordinary neural network is helpless in processing time series data [15].

Raw data should be processed before the establishment of a prediction model to avoid large errors as much as possible. Since the algorithm system is nonlinear, and its initial value has a great influence on the convergence of the system and the accuracy of the algorithm, the data is normalized before input, so that relatively large inputs are assigned at positions where the function gradient is relatively large. For data already within [0, 1], normalization is not required. The normalization formula used is shown in Eq. (1):

$$X_1 = \alpha \frac{X - \min(X)}{\max(X) - \min(X)} + \beta \tag{1}$$

where, α and β are normalization coefficients, X_1 is the normalized value, and $\max(X)$ and $\min(X)$ are respectively the maximum and minimum values of the data when predicting geological information during the normalization process.

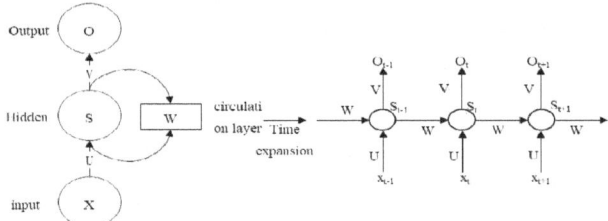

Fig. 1. RNN flowchart

As shown in Fig. 1, after receiving the input X(t) at time t, the value of the hidden layer is S(t) and the output value is O(t), where the value of S(t) depends not only on

X(t), but also on S(t-1).

$$\text{Hidden layer}: S_t = f(U \cdot X_t + W \cdot S_{t-1}) \tag{2}$$

$$\text{Output layer}: O_t = \text{softmax}(V \cdot S_t) \tag{3}$$

where, x_0 is the index parameter. After inputting the initial parameters, U, V, and W are the shared parameters of each layer respectively, which reduces the learning parameters required in the network and improves the learning efficiency. Output the input indexes. For example, after inputting the index x_0, input the other mileage data of the index parameters $\{x_0, \ldots, x_{t-1}, x_t, \ldots\}$ in sequence, and after transmission to the hidden layer, S_t is obtained by modifying the indicator parameters through S_{t-1}. During the output process, the output S_t is normalized.

2.2 Prediction and Updating Process of Surrounding Rock Geological Information

A tunnel surrounding rock geological information updating and prediction model based on RNN will be established in this paper. The general idea and process are shown in Fig. 2, and the main implementation steps are as follows.

(1) Data collection and screening: Collecting the surrounding rock geological information of excavated tunnel sections, selecting prediction indexes and performing quantitative processing on them, substituting the training set data into Eq. (1) for normalization, and eliminating the discrete data.
(2) Prediction model construction: Substituting the cleaned data into Eq. (2) to construct the hidden layer and get the output layer, comparing and analysing the output layer with the measured data, and changing the parameters U, V and W to optimize the geological information prediction model and obtain an ideal prediction model.
(3) Test set checking: Using the established prediction model to update and predict the surrounding rock geological parameter indexes of unexcavated tunnel sections, and comparing the prediction results with the actual excavation results to verify the test set.
(4) Error analysis: Comparing the obtained index parameter prediction data with the excavation results, evaluating the accuracy of the prediction model, and making a brief comparison with other prediction methods.

3 Engineering Application

A total of 33 groups of data were selected from some mileage sections of a tunnel in Kangding for engineering application, of which 25 groups of data were used as training sets and 8 groups of data were used as verification sets. A surrounding rock geological information updating and prediction model based on RNN was constructed using the 25 groups of training set data, and then the surrounding rock geological information of the 8 groups of tunnel faces ahead was updated and predicted. The predicted data results were compared with the actual surrounding rock information after excavation, and the quality evaluation of the model was obtained.

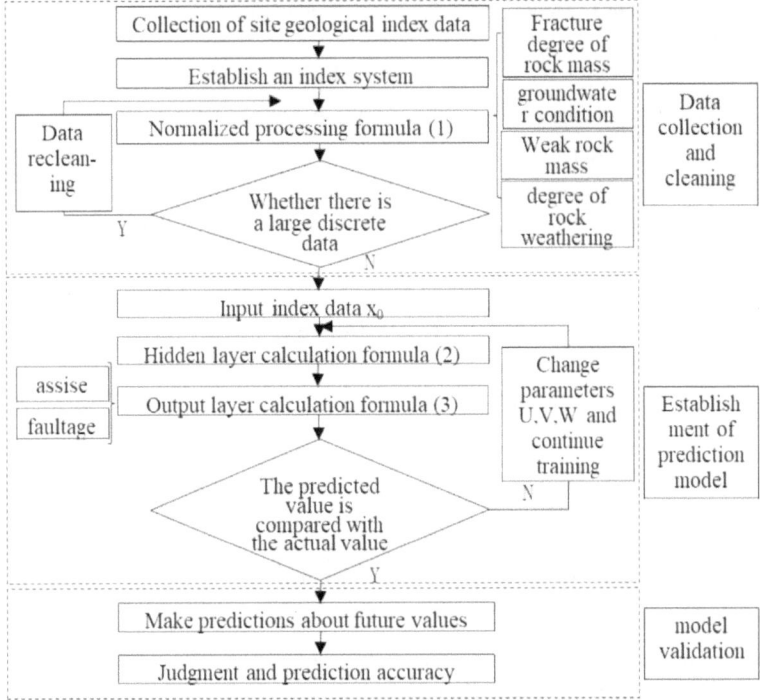

Fig. 2. Prediction process of tunnel index parameters based on RNN

3.1 Data Processing

Li et al. and Zhang et al. proposed the parameter indexes of rock mass integrity, rock hardness, and water abundance, and applied them to engineering [16, 17]. Nie et al. and Fu et al. used such indexes as rock strength, surrounding rock integrity, groundwater, and rock weathering capacity to predict abrupt large deformations in tunnels [18, 19]. In this paper, rock mass integrity, rock hardness, water abundance and rock weathering degree are taken as prediction indexes. The geological indexes of the surrounding rocks in the excavated sections of a tunnel in Kangding were collected, and the indexes were classified and quantified according to the *Code for Rock and Soil Classification of Railway Engineering (TB-10077-2019)*, as shown in Table 1. After substituting the quantified results in Table 1 into Formula 1 for normalization, and the normalized data can be used for training of the surrounding rock geological information updating and prediction model based on RNN.

3.2 Model Construction

The optimal training mode for the RNN model is a 3-layer RNN structure, with 5 and 20 layers of hidden layer neurons. The learning default value, the maximum number of iterations and the training error target are set to 0.1, 2000 and 0.001, respectively.

Table 1. Geological parameters of tunnel surrounding rock

Index	GI	QI	Index	GI	QI
Fracture degree of rock mass	Integral	1	Weak rock mass	Hard	1
	Relatively integral	2		Relatively hard	2
	Relatively fragmented	3		Relatively weak	3
	Fragmented	4		Weak	4
Groundwater condition	Drenched	1	Degree of rock weathering	Intense weathering	1
	Linear water	2		Intermediary weathering	2
	Humid	3		Moderate weathering	3
	Dry anhydrous	4			

Remark: GI, geological information; QI, quantization index.

The obtained prediction model of the four indexes of rock mass integrity, rock hardness, water abundance, and rock weathering degree is shown in Fig. 3.

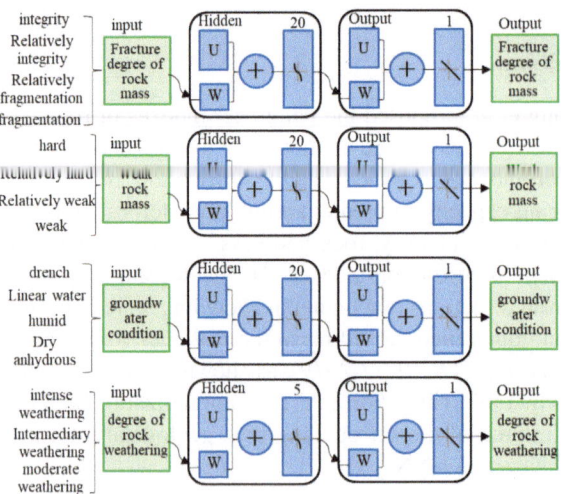

Fig. 3. Construction of a RNN model

3.3 Model Validation

The constructed model has been used to predict the geological information of the surrounding rock in the unexcavated section ahead of the tunnel face. Table 2 shows the comparison between the predicted data of four index parameters: rock mass integrity, water abundance, rock hardness, and rock weathering degree, and the actual excavation data.

Table 2. Comparison of actual data and predicted data.

	N Data	1	2	3	4	5	6	7	8
F	Real data	Integral	Fragmented	Fragmented	Integral	Relatively integral	Relatively integral	Relatively fragmented	Fragmented
	Predicted data	Integral	Fragmented	Fragmented	Integral	Relatively fragmented	Relatively integral	Relatively fragmented	Fragmented
	A /%	87.5%							
G	Real data	Linear water	Drenched	Humid	Humid	Humid	Drenched	Linear water	Humid
	Predicted data	Linear water	Drenched	Humid	Humid	Dry anhydrous	Drenched	Linear water	Humid
	A /%	87.5%							
W	Real data	Hard	Relatively weak	Relatively weak	Hard	Relatively hard	Hard	Relatively weak	Relatively weak
	Predicted data	Hard	Relatively weak	Relatively weak	Hard	Relatively hard	Hard	Relatively hard	Relatively weak
	A /%	87.5%							
D	Real data	MW	MW	MW	MW	IDW	MW	IW	IDW
	Predicted data	MW	IDW	MW	MW	IDW	MW	MW	IDW
	A /%	75.0%							

Remark: F, fracture degree of rock mass; G, groundwater condition; W, weak rock mass; D, degree of rock weathering; A, Accurate rate; MW, Moderate weathering; IDW, Intermediary weathering; IW, Intense weathering.

Figures 4, 5, 6 and 7 show the comparison between the predicted data results based on the recurrent neural network algorithm and the actual data. According to the analysis of the prediction results of the four surrounding rock parameter indexes, although there were some prediction deviations in individual data in the prediction of the three indexes of rock mass integrity, water abundance, and rock hardness, there are no overstep prediction results. For example, the prediction results of rock mass integrity were relatively broken, while the actual excavation results were relatively complete, but the difference was small, in 8 prediction sets, the predicted result is the same as the actual data in 7 sets, and the prediction accuracy is 87.5% according to Eq. (4). In the prediction of rock weathering degree, 7 groups of data are the same as the actual data, the prediction accuracy was 75%, and there was leapfrog misjudgement in individual data. Preliminary analysis showed that the data was affected by fault fracture zone. In addition, the small number of data samples used in this study was also a factor affecting the deviation, so more research data samples will be supplemented in subsequent studies. However, when compared with other prediction methods [20, 21], the recurrent neural network model established in this paper also has higher accuracy in predicting the four surrounding rock geological parameter indexes.

$$\text{prediction accuracy} = \frac{\text{predicted data}}{\text{real data}} \times 100\% \tag{4}$$

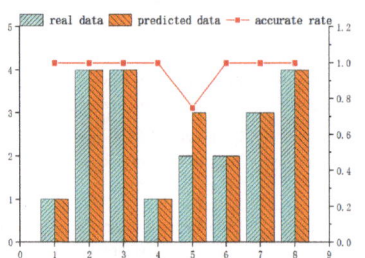

Fig. 4. Comparison of fracture degree of rock mass

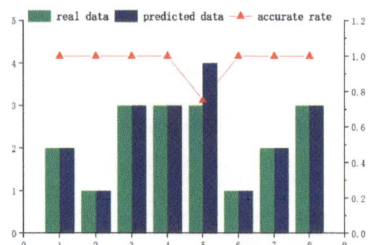

Fig. 5. Comparison of water abundance

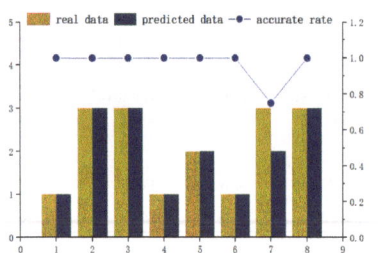

Fig. 6. Correlation map of weak rock mass

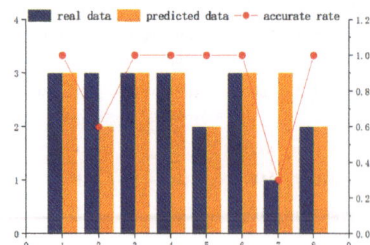

Fig. 7. Comparison of rock weathering

4   Conclusions

In order to meet the requirements for fast updating of surrounding rock geological information and high accuracy of prediction results during tunnel construction, a surrounding rock geological information updating and prediction model based on RNN has been constructed. After engineering application, the following conclusions are drawn:

(1) After full consideration of the surrounding rock indexes commonly used in underground tunnel engineering disaster prediction and forecasting, this paper selects four parameter indexes: rock mass integrity, rock hardness, water abundance, and rock weathering degree, and uses machine learning to conduct research on the dynamic updating and prediction of these four indexes.

(2) A surrounding rock geological information updating and prediction model based on RNN has been constructed. The process and characteristics include: adding the relationship between the previous and subsequent time series on the basis of fully

connected neural networks, establishing a three-layer neural network model, and continuously optimizing the geological information prediction model through comparative analysis of the output layer and measured data, ultimately obtaining an ideal geological information prediction model for surrounding rocks.

(3) The application of this method in engineering practice shows that the prediction accuracy for the three indexes of rock mass integrity, rock hardness and water abundance is 87.5%, and the prediction accuracy for rock weathering degree is 75%. The overall prediction accuracy is high.

Acknowledgement. The research project is financially supported by the Natural Science Foundation of Sichuan Province (No. 2022NSFSC0411), the National Natural Science Foundation of China (Nos. 41772329 and U19A20111), the State Key Laboratory of Geohazard Prevention and Geoenvironment Protection Independent Research Project (No. SKLGP2021Z011).

References

1. Wang, J.X., Zhou, Z.H., Zhao, T., et al.: Research on surrounding rock stability classification based on Alpha stable distributed probability neural network. Rock Soil Mech. **37**(S2), 649–657, 664 (2016)
2. Li, Y.D., Ruan, H.N., Zhu, Z.D., et al.: Classification of underground engineering surrounding rock based on improved BP neural network. People's Yellow River **36**(1), 130–133 (2014)
3. Sun, J., Yao, B.-Z., Yang, C.-Y., Yao, J.-B.: Tunnel surrounding rock displacement prediction using support vector machine. Int. J. Comput. Intell. Syst. **3**(6) (2010)
4. Yao, B.-Z., Yang, C.-Y., Yu, B., Jia, F.-F., Yu, B., et al.: Applying support vector machines to predict tunnel surrounding rock displacement. Appl. Mech. Mater. **29–32**, 1717–1721 (2010)
5. Jiang, C.X.: Classification of underground engineering surrounding rock based on improved BP neural network. People's Yellow River **36**(1), 130–133 (2014)
6. Liu, H.M., Xu, F.Y., Liu, B.J., et al.: Classification of underground engineering surrounding rock based on improved BP neural network. People's Yellow River **36**(1), 130–133 (2014)
7. Zhang, K., Zhang, K., Li, K., et al.: Principal component analytics-neural network rockburst grade prediction model.China Saf. Sci. J. **31**(3), 96–104 (2021)
8. Tian, R., Meng, H.D., Chen, S.J., et al.: Classification prediction of rockburst intensity based on deep neural network. Acta Coal Sinica **45**(S1), 191–201 (2019)
9. Li, B., Jia, X., Gao, W., Chen, G.: Source 2022 Lecture Notes on Data Engineering and Communications Technologies, vol. 136, pp. 783-789 (2022)
10. Xu, B.: Research of advanced geological prediction in tunnel excavation with ultra-long broken zone. IOP Conf. Ser.: Earth Environ. Sci. **330**(2) (2019)
11. Xue, Y.-G., et al.: Study of geological prediction implementation method in tunnel construction. Yantu Lixue/Rock Soil Mech. **32**(8), 2416–2422 (2011)
12. Guan, Z., Deng, T., Du, S., Li, B., Jiang, Y.: Classification prediction of rockburst intensity based on deep neural network. Acta Coal Sinica **45**(S1), 191–201 (2019)
13. Wang, K., Zhang, L., Fu, X.: Time series prediction of tunnel boring machine (TBM) performance during excavation using causal explainable artificial intelligence (CX-AI). Autom. Constr. **147** (2023)
14. Ma, K., Shen, Q., Sun, X., Ma, T., Hu, J., Tang, C.: Rockburst prediction model using machine learning based on microseismic parameters of Qinling water conveyance tunnel. J. Central South Univ. **30**(1), 289–305 (2023)

15. Li, L.B., Gong, X.N., Gan, X.L., et al.: Classification prediction of rockburst intensity based on deep neural network. Acta Coal Sinica **45**(S1), 191–201 (2019)
16. Chen, Y.: Research on dynamic risk assessment and support decision of tunnel large deformation disaster. Chengdu University of Technology (2017)
17. Ou, G.Z.: Research on dynamic risk assessment and support decision of tunnel large deformation disaster. Chengdu University of Technology (2022)
18. Nie, Y.: Application status and technical analysis of highway tunnel advanced geological prediction. J. Underg. Space Eng. (04), 766-771 (2018)
19. Fu, X.: Research on application of rockburst and large deformation risk assessment in high ground stress tunnel construction stage. Chengdu University of Technology (2022)
20. Li, T.B., Meng, L.B., Zhu, J., et al.: Comprehensive analysis method of tunnel advance geological prediction. Chin. J. Rock Mech. Eng. **28**(12), 2429–2436 (2009)
21. Zhang, Q.S., Li, S.C., Sun, K.G., et al.: Application status and technical analysis of highway tunnel advanced geological prediction. J. Underg. Space Eng. (04), 766–771 (2008)

Planosol Soil Condition Improvement Effect of a New Plow of Three-Stage Subsoil Interval Mixing

Chunfeng Zhang[1,2], Baoguo Zhu[1,2(✉)], Qingying Meng[1,2], Nannan Wang[1,2], Haoyuan Feng[1,2], and Ken Araya[3]

[1] Jiamusi Branch, Academy of Agricultural Sciences of Heilongjiang, Jiamusi 154007, Heilongjiang, China
zhubaoguo82@163.com
[2] Key Laboratory of Breeding and Cultivation of Main Crops in Sanjiang Plain, Jiamusi 154007, China
[3] Environmental Science Laboratory, Senshu University, Bibai 079-0197, Hokkaido, Japan

Abstract. In this paper, a three-stage subsoil interval mixing four-gang plough (TSIMF) was manufactured to enhance the property of planosol soil in 2016. The working width is set to 2 m to obtain a low operation cost. The total draught of the TSIMF was about 100 kN, the running resistance of the tractor included. The traction force (draught) of TSIMF was determined and the soil condition improved with the TSIMF was discussed. A field test of soybean with TSIMF showed that the soil moisture was about 2% volume greater than that in the CK (subsoiler) field. The soil temperature with the TSIMF was about 1 ℃ greater than the CK in the plowed layer, and the yield was 120% of the treatment of the CK. Sanjiang Plain is one of the most important commodity grain bases in China, and the albic soil accounts for about 25% of the total cultivated land. Improving the property of the albic farming land attaches great significance to the stability and increase of national grain production.

Keywords: Planosol · Soil Improvement · Interval Mixing · Improvement Effect · Traction Force

1 Introduction

Planosol, diffusely distributed in Heilongjiang and Jilin provinces, People's Republic of China [1, 2], is a particular soil variety featuring white dense clayey subsoil. This subsoil structure is hard enough that the plant roots can hardly get through the layer to deeper profiles seeking nutrients and water when suffering a continuous adverse situation. Furthermore, the clayey subsoil is so tight that even the water molecules are incapable of moving up and down easily, which leads to drought in high-temperature weather and waterlogging in rainy conditions for dry crops. Hence, Planosol improvement attaches

---

C. Zhang and B. Zhu—These authors contributed equally to this work

© The Author(s) 2024
G. Feng (Ed.): ICCE 2023, LNCE 526, pp. 177–185, 2024.
https://doi.org/10.1007/978-981-97-4355-1_16

great importance. However, the albic soil usually recovers to its previous structure after a year's planting with the normal improving method. Farmers have to implement deep tillage yearly to maintain the soil quality. In this paper, a three-stage subsoil interval mixing four-gang plough (TSIMF) was manufactured to improve the property of planosol soil. The improvement effect can last five years once used, saving labor, material, and financial costs.

2   Materials and Methods

2.1   Albic Soil Structure and TSIMF

A typical original planosol profile in a forest in State Farm 854 in Heilongjiang province, P. R. of China is exhibited in Fig. 1. The first horizon (Ap) with a layer of about 0.2 m is a plant-growth beneficial soil that is of colossal humic acids. The second horizon (Aw) which takes a similar thickness to the first layer comes to an impermeable dense lessivage structure. And there is a diluvial heavy clay in the third horizon (B) beneath 0.4 m depth [3, 4].

Fig. 1. Typical original planosol profile of a forest in State Farm 854, Heilongjiang, China. (Each gradation on the scale represents 0.1 m).

Plants suffer from both drought and flood in different seasons separately owing to the impermeable Aw horizon. The general plant's root can hardly penetrate the horizon of which the hardness is over 5.0 MPa and the soil microorganisms also disappear under impermeable layers [5]. We determined the soil hardness with a cone penetrometer, whose cone angle is 30° and base diameter is 16 mm.

A three-stage subsoil interval mixing four-gang plough (TSIMF, Fig. 2) was created in 2016 [6] to make a mixture of the Aw and the B layers in an equal amount with the first (Ap) horizon undisturbed meanwhile. With this mechanism, 2-m working width was obtained, and soft and hard subsoil rows (tilled and untilled subsoil, whose width is 0.5 m each) are alternately produced, so a certain trafficability can be maintained after tilling operation with the TSIMF.

The principle of subsoil mixing of the horizon Aw and B with an equal amount via TSIMF has already been reported in previous researches [7–11]. In this paper, the traction force (draught) of the TSIMF on an actual planosol field was determined. Then the improved soil conditions and the yield harvested from the field tilled with the TSIMF were studied.

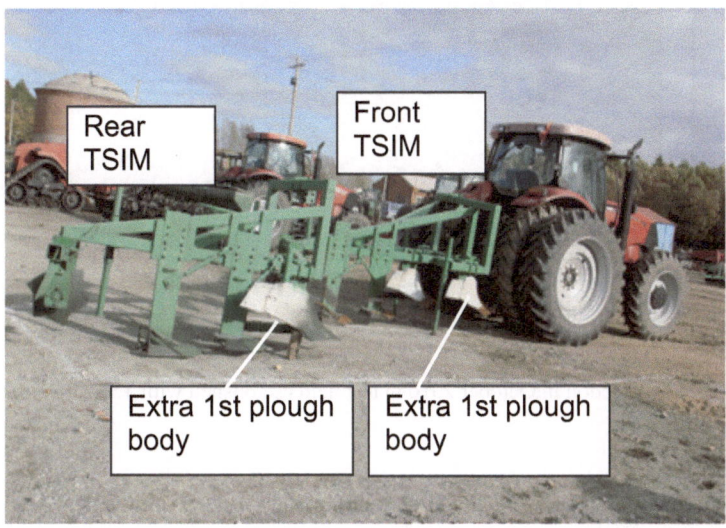

Fig. 2. Three-stage subsoil interval mixing four-gang plough (TSIMF) created in 2016.

2.2 Soil Section

Soil sections of 1 m length × 0.7 m depth (Figs. 1 and 3) were prepared at the test fields. Then the soil horizon, soil texture, and root penetration depth were measured.

Triplicate soil samples (total = 18 samples) were taken by the soil sampler (105 mm^3) at each soil depth layer (0–0.1, 0.1–0.2, 0.2–0.3, 0.3–0.4, 0.4–0.5 and 0.5–0.6 m) to determine the physical soil properties.

2.3 Traction Force (Draught) Test

A TSIMF set was mounted on an off-powered wheel tractor (the first tractor) for supporting (Fig. 2). It was dragged by the second tractor (not shown) on an actual planosol field with a traction dynamometer which a 150 kN-capacity strain gauge was installed [11, 12]. Therefore, only the TSIMF's horizontal force (draught) was gauged once the running resistance of the supporting tractor (first tractor) was reduced.

Working depth, ≈

Untilled subsoil layer,   Tilled and mixed subsoil layer,
≈ 0.5 m wide   ≈ 0.5 m wide

Fig. 3. Recent soil section in October 2016 in test field established with TSIMF in May 2016. P and Q: see [6].

The running resistance of the first (supporting) wheel tractor with the TSIMF mounted was first determined (Fig. 2). The first tractor was drawn by the second tractor without any plough operation of the TSIMF, that is, by lifting the lever of the hydraulic device of the first tractor.

The weight transfer caused by the plough bodies of the TSIMF (Fig. 2) was not included in the running resistance of the first (supporting) tractor.

The draught of three elements of the TSIMF was measured separately as follows. The four 1st-plough bodies were initially operated alone without the 2nd and 3rd plough joined (Fig. 2). Then the two 2nd-plough bodies were attached, and the draught of four 1st-plough bodies and two 2nd-plough bodies was determined. The two 3rd-plough bodies were added subsequently, and the total draught of all eight elements was acquired. Three to five measurements were carried out for each configuration.

The working depth of the 1st plough was set to 0–0.2 m, that of the 2nd plough was 0.2–0.4 m and that of the 3rd was 0.4–0.6 m beneath the soil surface (Fig. 2).

2.4   Test Fields

A field test of 20 hectares was implemented applying the TSIMF in May 2016 on the State Farm 854 in the Three-river Plain of Heilongjiang province, P. R. of China, and all the investigations of soil and crops were implemented in October 2016.

The land prepared by a conventional rotary tiller (0–0.15 m deep) and a subsoiler (0–0.3 m deep) [6] was treated as a control (CK).

2.5 Soil Properties

The basic fertility indicators of the albic soil are: organic matter 26.07 g/kg, pH 6.3, alkali hydrolysable nitrogen 82.46 mg/kg, available phosphorus 8.25 mg/kg, and available potassium 57.23 mg/kg. Soil water (moisture) content in volume %, EC value in dS m-1, and soil temperature in °C were determined in the actual field by a commercial tester. They were also measured at each soil depth layer (0–0.1, 0.1–0.2, 0.2–0.3, 0.3–0.4, 0.4–0.5, and 0.5–0.6 m).

2.6 Yield

Soybean-corn rotation is treated the same as State Farm 854 does. At the test fields, soybean Kenfeng 16 were seeded in May 2016 and harvested in October 2016.

Pods per Plant and Seeds per Plant. The average values of pods per plant, which were picked up from thirty other cultivated plants selected randomly in each field, were calculated. And an average of seeds per plant are treated with the same method.

Predicted Yield. The whole soybean per 3 m^2 was harvested at each test field and dehydrated. Three points were selected randomly in each experiment treatment land. The average yield was calculated according to the mass of the dry soybean with a moisture content of about 13% w.b. measured with a pan balance.

3 Results and Discussions

3.1 Soil Profile

Figure 3 shows a recent soil profile in October 2016 in the test field established with the TSIMF in May 2016. An interval blending of the subsoil layers was distinctly observed, in which the uncultured subsoil layer (P) width was around 0.5 m and crop roots could be hardly found beneath 0.2 m (see [6] about points P and Q).

The thickness of the tilled and mixed subsoil layer (Q) also turned to about 0.5 m, in which the Aw horizon (Fig. 1) was already invisible. A few crop roots were even discovered 0.6 m deep. The topsoil of 0–0.2 m in both P and Q included plenty of crop roots.

3.2 Measured Traction Force (Draught)

Figure 4 shows the results of the traction force (draught) test. The running resistance of the wheel tractor mounted (Fig. 2) was about 20 kN (standard deviation was 2.22 kN) on the planosol field, whose value was two times larger than prediction [6].

The total traction force of four 1st- the horizon Ap was about 40 kN. Therefore, the traction force of a single 1st-plough body occupied a quarter of the force of the

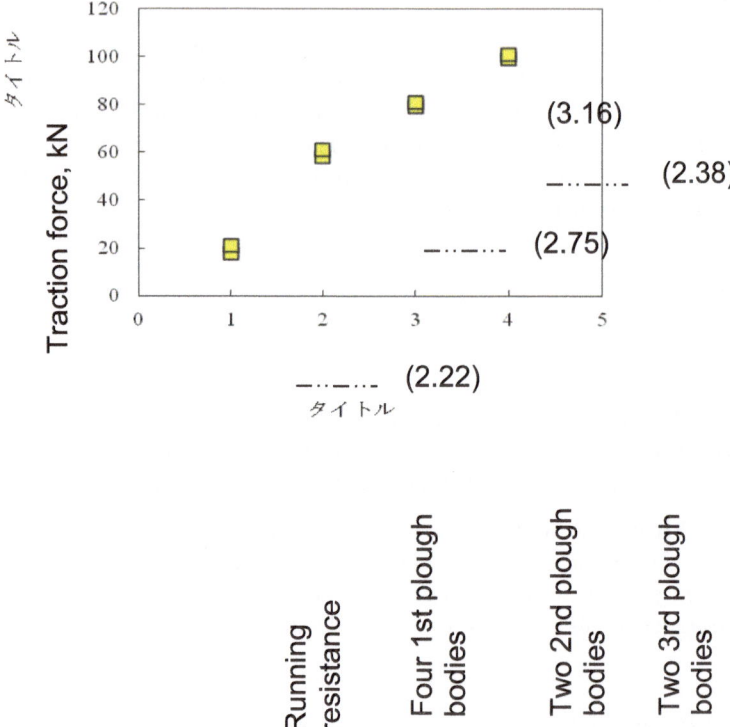

Fig. 4. Results of traction force (draught) test on a planosol field. Notes: The brackets in the Fig. shows the standard deviation of the traction force, kN.

total, which fairly coincided with the predicted value [6]. The traction of the two 2nd-plough bodies tilling the horizon Aw came to be about 20 kN, namely 10 kN each, which conformed to the prediction [6]. And the 3rd-plough bodies which were tilling the horizon B took the same situation as the 2nd-plough bodies [6]. For this combination, the total draught of the TSIMF, which was about 100 kN including the running resistance of the tractor (standard deviation was 3.16 kN), appeared to be a little greater than the previous prediction [6].

3.3 Soil Water Content

The soil water (moisture) content of the planosol soybean field is shown in Tables 1 and 2. The soil moisture in the TSIMF soybean field (Table 2) was an average of 2 volume % greater than that in the CK field (Table 1). This is due to the increase of air and water-space (porous media) of the soil with deep tillage of the TSIMF.

Table 1. Soil properties of planosol in CK (subsoiler) soybean field in October 2016.

Soil depth, mm	0–100	100–200	200–300	300–400	400–500	500–600
Soil moisture, volume %	24.17	24.03	23.90	29.43	31.27	32.18
EC, dS m^{-1}	0.777	0.804	0.830	1.095	1.198	1.201
Soil temperature, °C	4.70	5.05	6.23	7.38	7.93	8.10

Table 2. Soil properties of planosol in TSIMF soybean field in October 2016.

Soil depth, mm	0–100	100–200	200–300	300–400	400–500	500–600
Soil moisture, volume %	22.23	25.67	30.73		32.33	32.56
EC, dS m^{-1}	0.793	0.797	0.807	0.913	1.040	1.054
Soil temperature, °C	5.87	6.57	7.53	7.76	7.80	8.10

3.4 EC Value

The EC values of the planosol in the TSIMF soybean field are also shown in Table 2. There was no big difference between those in the CK field (Table 1) and those in the TSIMF field (Table 2). The EC value at any layer in any field was about 1.0 dS m^{-1}.

3.5 Soil Temperature

The soil temperature in the planosol soybean fields is also shown in Tables 1 and 2. The temperature of the TSIMF field (Table 2) was about 1 °C higher than that of the CK (Table 1) in most soil layers. Especially at the topsoil layer (0–0.3 m deep), there was about a 1.5 °C difference. This is due to the increase in the air phase of the soil with deep tillage of the TSIMF.

3.6 Yield

Table 3 shows the average number of pods per plant. In the TSIMF field, it was greater than that in CK. The average number of soybean per plant in the TSIMF field was also greater than that in CK (subsoiler) field. The average yield (kg m-2) is in Table 3. The yield with the usage of the TSIMF was 120% of the CK field.

Table 3. Average yield of soybeans in State Farm 854 in October 2016.

	Subsoiler	TSIMF
Plant height, m	0.726	0.851
No. of pods per plant	23.1	34.8
No. of seeds per plant	58.7	86.1
Yield, kg m^{-2}	23.62	28.47
Relative yield	1.0	1.20

4   Summary and Conclusions

A new three-stage subsoil interval mixing four-gang plough (TSIMF) was designed, produced, and determined for promoting planosol conditions. The traction force (draught) of the TSIMF on an actual field was determined. Then, soil conditions improved and crop yield from the field tilled with the TSIMF was discussed.

The soft and hard subsoil rows (tilled and untilled subsoil, whose width is 0.5 m each) are alternately produced, so a certain trafficability can be maintained after tilling operation with the TSIMF. The TSIMF's total draught was around 100 kN including the running resistance of the tractor. The soil moisture in the TSIMF soybean field was about 2 volume % greater than that in the CK field (subsoiler). The soil temperature in the TSIMF field was about 1 °C greater than those in the CK field (subsoiler). The yield in the TSIMF field was 120% of the CK field (subsoiler). The running cost of TSIMF could reduce by 75% compared with conventional subsoiler operation in a 5-year benefit period.

Acknowledgments. This work was supported by the National Key Research and Development Program of China (No. 2022YFD150080301, 2022YFD150080501) and Strategic Priority Research Program of the Chinese Academy of Sciences (Grant No. XDA28100202, XDA28010403).

References

1. Tseng, C., Chuang, C., Li, N.: On the genesis and classification of Paichiang soils. Acta Pedol. Sin. **11**(2), 111–129 (1963)
2. Gong, Z.: Chinese Soil Taxonomy, pp. 544–552. Science Press, Beijing (1999)
3. Akazawa, T.: Soil and pasture in three-river plain (I). J. Hokkaido-Black Dragon Sci. Cooper Inst. **17**, 13–25 (1986)
4. Akazawa, T.: Soil and pasture in three-river plain (II). J. Hokkaido-Black Dragon Sci. Cooper Inst. **26**, 11–30 (1987)
5. Zhao, D., Hong, F.: Soil solums in three-river plain and soil improvement. J. Chin. Sci. Agric. Sin. **1**, 54–60 (1983)
6. Zhu, B., et al.: Planosol soil conditions improved with four-kinds of ploughs. Part 2: Soybean field in State Farm 854. J. JSAM-Hokkaido **57**, 11–16 (2017)

7. Jia, H., Liu, F., Zhang, H., Zhang, C., Araya, K., Kawabe, H.: Improvement of planosol solum: part 8: analysis of draught of a three-stage subsoil mixing plough. J. Agric. Eng. Res. **70**, 185–193 (1998b)

8. Araya, K., Zhao, D., Liu, F., Jia, H.: Improvement of planosol solum: part 1: experimental equipment and methods and preliminary soil bin experiments. J. Agric. Eng. Res. **63**, 251–260 (1996a)

9. Araya, K., Zhao, D., Liu, F., Jia, H.: Improvement of planosol solum: part 3: Optimization of design of drop-down ploughs in soil bin experiments. J. Agric. Eng. Res. **63**, 269–274 (1996c)

10. Araya, K., Zhao, D., Liu, F., Jia, H.: Improvement of planosol solum: part 5: soil bin experiments with a three-stage subsoil mixing plough. J. Agric. Eng. Res. **65**, 143–149 (1996e)

11. Liu, F., Jia, H., Zhang, C., Zhang, H., Araya, K., Kawabe, H.: Improvement of planosol solum: part 10: mixing of wheat straw and corn stalk into subsoil. J. Agric. Eng. Res. **71**, 221–226 (1998b)

12. Araya, K.: Soil failure caused by subsoilers with pressurized water injection. J. Agric. Eng. Res. **58**, 279–287 (1994)

Simulation Analysis of the Effects of Liquefaction on Pile Foundations Under Seismic Response

Yaping Zhang[1], Yong Zhang[1], Dongdong Xue[1], Shangyou Zhou[1], Mingshan Zhang[1], and Hao Wen[2(✉)]

[1] CCCC Second Harbor Engineering Company Ltd., Wuhan 430040, China
[2] Faculty of Engineering, China University of Geosciences, Wuhan 430074, China
whenhowow@163.com

Abstract. As a widely used structural element in construction projects, pile foundations have been subject to varying degrees of damage during major earthquakes, especially in liquefiable sites. In order to study the effects of liquefaction on pile foundations under seismic conditions, this paper relies on an actual engineering case of the Xiong'an New Area to Beijing Daxing International Airport Express Line. A numerical model of the dynamic response of pile foundations in liquefied sites under seismic actions is established using the finite difference software FLAC3D. A comparative analysis is conducted between liquefied and non-liquefied sites. The results indicate that compared to non-liquefied sites, liquefaction of the site increases the bending moment on the pile shaft, with increased values for both the maximum bending moment and axial force. The main locations where pile damage may occur are determined. Although the influence of liquefaction on the axial force of pile foundations is relatively small, overall, liquefaction has a significant impact on pile foundations, reducing their bearing capacity.

Keywords: Pile Foundation · Liquefied Site · Numerical Simulation

1 Introduction

As a structure widely used in bridge engineering, pile foundation has the advantages of high bearing capacity, small settlement and not easy to damage. In order to avoid or reduce the loss of life and property caused by earthquake, researchers have carried out a lot of research on the damage or failure of pile foundation in liquefied site due to seismic response. Zheng et al. [1] summarized the earthquake damage of bridge pile foundations in liquefaction sites in several major earthquakes and pointed out that sand liquefaction is one of the important factors leading to the serious damage or even collapse of bridge structures. Ishihara, Yasuda et al. [2] studied the shaking table test of pile-soil group in liquefaction site, and found that soil deformation caused by liquefaction would lead to large bending moments in the middle of pile foundation and pile foundation at the surface. Tokimatsu [3] and Tamura [4] obtained that the bending moment of pile

© The Author(s) 2024
G. Feng (Ed.): ICCE 2023, LNCE 526, pp. 186–192, 2024.
https://doi.org/10.1007/978-981-97-4355-1_17

foundation is closely related to the liquefaction of sand by shaking table test. When the sand is liquefied, the bending moment of pile foundation has a significant trend of increase. In addition to carrying out model experiments, the numerical simulation method is also an important means to study the seismic response of pile-soil structures. Based on the two-dimensional finite difference program, Haldar and Sivakumar [5] selected the nonlinear soil constitutive model, and used the strength reduction method to study the failure mechanism of pile foundation in liquefied soil. Uzuoka et al. [6] established a three-dimensional numerical model by referring to the relevant earthquake damage survey data and taking the collapsed buildings in the Hanshin earthquake in Japan as the background. Jiang et al. [7] used FLAC3D finite difference software to discuss the changes of pile bending moment and pile-soil interaction force under earthquake action. Dai et al. [8] analyzed the variation of total stress and effective stress of sand foundation by using finite element numerical method for the seismic problem of pile group foundation in liquefied site.

In general, domestic and foreign scholars have conducted a lot of research on the mechanical response and deformation and failure characteristics of pile foundations in liquefied sites under seismic conditions by using shaking table experiments and numerical simulation methods, and have achieved some guiding results. In this paper, the project from Xiong'an New Area to Beijing Daxing International Airport Express Line is selected as the engineering background. The numerical calculation model of seismic response of bridge pile foundation in liquefied site is established by FLAC3D finite difference program. According to whether the site is liquefied, the dynamic response of soil and pile foundation, the bending moment, shear force and deformation of pile body in dynamic system are compared and analyzed.

2 Project Overview

The express line from Xiong'an New Area to Beijing Daxing International Airport is an important part of the high-speed railway transportation network in the 'four vertical and two horizontal' area of Xiong'an New Area. Among them, Xiong'an Station to Bazhou Economic and Technological Development Zone Station crosses the existing Tianjin-Baoding Railway in the form of a bridge between pier C143 # and C146 #, and the angle between the line and the existing railway is 59°. The on-site photos of the cross-Tianjin-Baoding Railway and the elevation map of the cross-Tianjin-Baoding Railway are shown in Fig. 1 and Fig. 2.

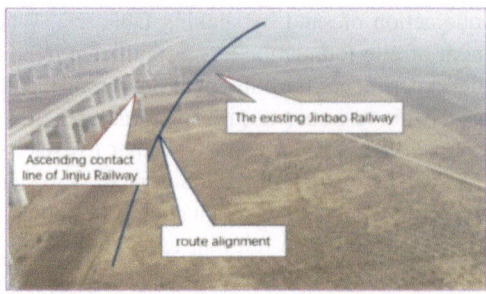

Fig. 1. On-site photos of the Cross-Tianjin-Baoding Railway

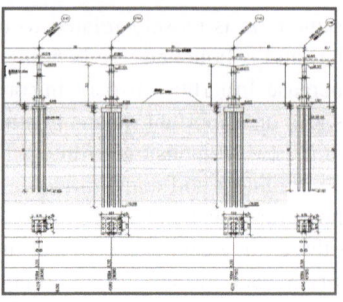

Fig. 2. Elevation drawing of the Cross-Tianjin-Baoding Railway

3   Numerical Simulation

According to the geological engineering investigation data, the typical geological profile under a borehole can be obtained and the strata under the site are simplified in FLAC3D as shown in Fig. 3, that is, the soil layer is divided into four layers. The size of the calculation model is 40 m × 40 m × 60 m. The soil constitutive model is Mohr-Coulomb model, and the relevant physical and mechanical parameters of the soil used are shown in Table 1.

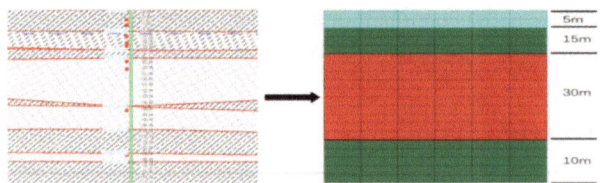

Fig. 3. Initial soil layer model

Table 1. Relevant soil physical and mechanical parameters

	thickness (m)	dry density (kg/m³)	Porosity	cohesion (kPa)	friction angle (°)	bulk modulus (MPa)	shear modulus (MPa)
silty sand	5	1630	0.655	12.3	24	14.71	5.640
clay	10	1650	0.652	27	16	14.71	5.640
fine sand	30	2000	0.46	0	25	29.41	11.28
clay	30	1680	0.612	28	17	14.71	5.640
pile	12	2400	/	/	/	1.6E+4	7.69E+3

After the seismic load is applied, considering that the seismic waves will be reflected on the surrounding boundary, in order to reduce the impact of reflection, the free field boundary is used in the dynamic calculation process of numerical simulation. At this time, the free field generated grids are coupled with the main part of the grid, which can better simulate the free field vibration. Figure 4 shows the grid diagram generated by selecting the free field boundaries used.

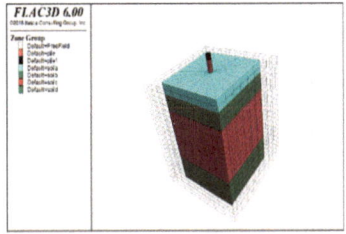
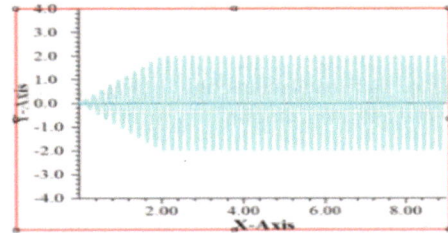

Fig. 4. Setting up free-field boundaries   **Fig. 5.** Acceleration time history curve for vibratory load

4   Simulation Analysis of Seismic Response of Pile Foundation in Liquefaction Site

In this paper, in order to compare and analyze, the site conditions are treated differently in the calculation, and the site is divided into two conditions: liquefaction and non-liquefaction. One is the PL-Finn model [9], which uses the pore water pressure accumulation of the soil under dynamic conditions to cause liquefaction. The other uses the Mohr-Coulomb model without considering the pore pressure accumulation. The input vibration waveform at the bottom of the model is consistent, and the peak acceleration reaches 0.2 g within 2 s. The acceleration time history curve of the vibration load is shown in Fig. 5.

4.1   Dynamic Analysis

In order to facilitate the analysis of the numerical simulation results, the monitoring points p1, p2 and p3 are set from near to far away from the pile foundation, as shown in Fig. 6. Under the action of vibration load, the time history curves of pore pressure ratio of sand layer under three measuring points are shown in Fig. 7. It can be seen that the pore pressure is constantly changing under the influence of the peak acceleration of the earthquake, and the overall trend of the calculated value at each monitoring point is roughly the same. In the initial stage, the pore pressure ratio increases in an uneven and irregular trend, reaching a peak around 0.5 s, and then maintaining a certain range for a period of time. From the change frequency and the change peak, it can be analyzed that the farther the measuring point from the pile foundation, the smaller the change range of the excess pore pressure ratio and the more stable it is.

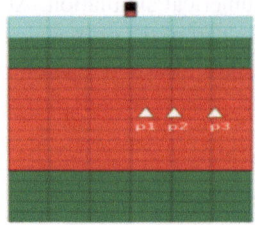
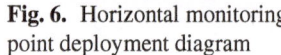

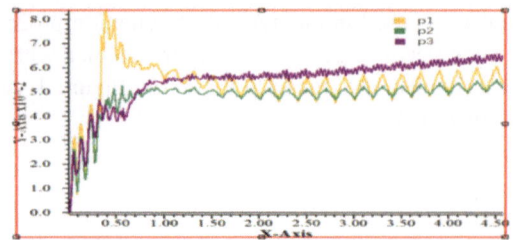

Fig. 6. Horizontal monitoring point deployment diagram

Fig. 7. Time history curve graph for excess pore pressure ratio

4.2 Comparative Analysis of Liquefaction and Non-liquefaction Site Conditions

Figure 8 shows the bending moment envelope diagram of pile foundation under liquefied and non-liquefied conditions. It can be seen from the Fig that the bending moment of pier body increases first and then decreases, and the changing trend is the same. However, for the non-liquefied site, the bending moment of the pile reaches an extreme value of 335 kN·m at a distance of 40 m from the pile top, while for the liquefied site, the bending moment reaches an extreme value of 883 kN·m at a distance of 43 pm from the pile top. It can be seen that the bending moment effect of the pile increases due to liquefaction. In the liquefied soil layer, due to the lack of effective lateral support of the pile body caused by liquefaction, the bending moment of the pile body increases faster than that of the non-liquefied soil layer. As can be seen from the slope of the corresponding curve in the figure, the pile body is also most likely to withstand bending moments beyond the ultimate bending resistance of the pile itself in this liquefaction area, resulting in bending failure. This is consistent with the macro-seismic damage of the liquefied lateral expansion pile foundation obtained by other scholars [10].

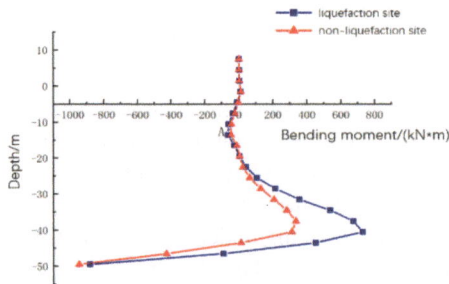

Fig. 8. Enveloping diagram of the bending moment for pile foundation

Figure 9 shows the enveloping diagram of axial force for pile foundation in liquefied and non-liquefied state. It can be seen from the Fig that the axial force of pile body increases gradually. The variation trend of the axial force envelope curve is consistent, but for the non-liquefied site, due to the influence of the friction resistance of the soil layer on the pile side, the axial force reaches a peak of 6612 kN at 45 m from the pile top, and continues to increase after a short decrease. For the liquefiable site, after the liquefaction of the liquefied soil layer, as the pore water pressure dissipates, the foundation consolidation settlement is caused, and the pile body is affected by the lateral negative friction resistance of the soil, resulting in the increase of the axial force of the liquefied pile in the site compared with the extreme axial force of the non-liquefied site. It can be seen from the Fig that the axial force of the pile foundation in the liquefied site reaches an extreme value of 7780 kN. However, in general, the influence of soil liquefaction on the axial force of pile foundation is relatively small.

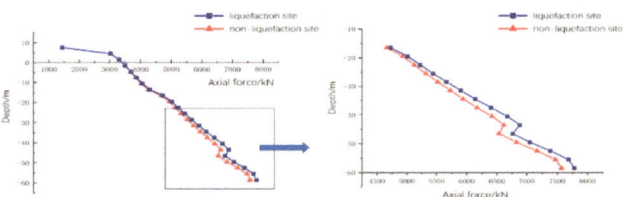

Fig. 9. Enveloping diagram of axial force for pile foundation

5   Conclusion

In this paper, through the numerical simulation analysis of pile foundation in liquefied and non-liquefied sites under earthquake action, the following results can be obtained:

Under site liquefaction condition, the bending moment of pile body increases first and then decreases, and the extreme value of bending moment is also greater than that of non-liquefied conditions. The position of the peak bending moment of the pile is near the 45 m below the surface (the junction of liquefied layer and non-liquefied layer), so it can be concluded that this position is the main location for pile body failure and the main location for seismic design of pile foundation. The variation trend of axial force envelope curve of pile foundation in liquefied site and non-liquefied site is consistent, but the extreme value of axial force in liquefied site is larger, and the influence of soil liquefaction on axial force of pile foundation is relatively small. On the whole, the liquefaction of the site has a great impact on the pile foundation, which will reduce the bearing capacity of the pile, and the impact should be considered in the actual design.

References

1. Zheng, X.: Research on seismic response performance of bridge pile foundation in liquefaction site, p. 54. Dalian Maritime University, Dalian (2008)
2. Yasuda, S., Ishihara, K., Morimot, I., Orense, R., Ikeda, M., Tamura, S.: In: Proceedings of the 12th World Conference on Earthquake Engineering (Auckland) (2000)
3. Tokimatsu, K., Suzuki, H., Sato, M.: Soil Dyn. Earthq. Eng. **10**, 753–762 (2005)
4. Tamura, S., Tokimatsu, K.: Seismic Performance and Simulation of Pile Foundations in Liquefied and Laterally Spreading Ground, pp. 83–96 (2006)

5. Haldar, S., Babu, G.L.S.: Int. J. Geomech. **10**, 74–84 (2010)
6. Uzuoka, R., Sento, N., Kazama, M., Zhang, F., Yashima, A., Oka, F.: Soil Dyn. Earthq. Eng. **27**, 395–413 (2007)
7. Jiang, K.D., Qian, D.L., Dai, Q.: J. Hefei Univ. Technol. **39**, 1372–1375 (2016)
8. Yan, D., Guoxing, C., Zhihua, W.: J. Disaster Prev. Mitig. Eng. **37**, 785–801 (2017)
9. Chen, Y., Xu, D.: Foundation and Engineering Examples of FLAC/FLAC3D, p. 122. China Water Resources and Hydropower Press, Beijing (2009)
10. Motamed, R., Towhata, I.: J. Geotech. Geoenviron. **136**, 477–489 (2010)

Evaluation of Treatment Effect of Highway Subgrade Reconstruction Damaged by Large Landslide

Shihong Liu[1], Qing Lei[1], Bo Jiang[2(✉)], and Yao Zeng[2]

[1] Tongren Highway Administration Bureau of Guizhou Province, 182 Jinli Avenue, Wanshan District, Tongren, Guizhou, China

[2] Guizhou Transportation Planning Survey and Design Academe Co. Ltd., No. 100 Yangguan Avenue, National High Tech Industrial Development Zone, Guiyang, Guizhou, China

479634279@qq.com

Abstract. Highway in mountainous area is easily threatened by landslide and other disasters, so it is necessary to comprehensively evaluate the prevention and control effect of the treatment when the subgrade is restored and rebuilt after landslide. In order to deeply study the landslide treatment effect of subgrade reconstruction, taking the landslide damage reconstruction of G326 highway in Guizhou Province as the object, the deformation and failure process of landslide induced by heavy rainfall is simulated, and the prevention effect of different treatment plans such as "cutting + anti-slide pile + anchor cable" is evaluated. The results show that with the increase of rainfall, the landslide deformation gradually expands from the vicinity of the front national G326 highway to the back edge; Under the combined treatment of "cutting + anti-slide pile + anchor cable", the maximum displacement of the main sliding area of landslide is 3.1 mm, and the maximum displacement of the trailing edge is 1.7 mm, which makes the overall deformation in a controllable range. As a result, it has certain reference value for the reconstruction of highway subgrade engineering damaged by landslide.

Keywords: Landslide · Numerical Simulation · Rainfall · Impact Assessment

1 Introduction

Highway in mountainous area may not only be threatened by landslide in the construction process, but also be affected by landslide in normal operation. Once the landslide occurs during the operation period, it not only damages the highway subgrade or bridge, but also threatens the safety of vehicles and pedestrians. Not only the highway landslide can cause the damage of subgrade, but also the landslide treated by anchor pile has signs of revival and deformation of anti-slide pile. For example, the landslide of Trevor-Froncysyllte highway in England slips again after one year of treatment, resulting in the inclination of anti-slide pile at most about 15° [1]; Diezma landslide of A-92 highway in Spain caused the deformation of anchorle again after 8 years of restoration and reconstruction [2]. Anti-slide pile is a common protective measure in landslide treatment engineering [3].

© The Author(s) 2024

G. Feng (Ed.): ICCE 2023, LNCE 526, pp. 193–199, 2024.

https://doi.org/10.1007/978-981-97-4355-1_18

Anchor pile is mainly used to solve the problems of large thrust and deformation control of landslide. The deformation and failure mechanism of anchor pile [4], the internal mechanism of landslide reinforcement [5], and the design and calculation method have been studied extensively at home and abroad. The above research mainly focuses on the design and mechanical mechanism of anchor pile supporting structure, but there is little research on the supporting effect of composite structure mainly based on anchor pile in damaged highway reconstruction.

Based on the G326 highway landslide damage reconstruction as a basis, this paper studies different treatment support plan, and evaluates different support plan prevention effect.

2   General Situation of Highway Subgrade Damaged by Landslide

At about 7:00 on July 8th, 2020, a landslide occurred in G326 K41 + 300 ~ K43 + 000 section of national highway, Tianbao Formation, Shiban Village, Ganlong Town, Songtao County, Guizhou Province. The landslide is about 1100 m in length, 60 ~ 450 m in width, 330 m in height difference between front and rear edges, 2 ~ 10 m in thickness and 136×10^4 m^3 in volume.After sliding, a clear sliding surface can be seen in the middle and back of the landslide, which is a light gray-green nodular marl layer with an occurrence of $260°\angle 26°$. The sliding bed is medium-thick nodular marl of Dawan Formation of Ordovician with a small amount of thin limestone and shale in Fig. 1.

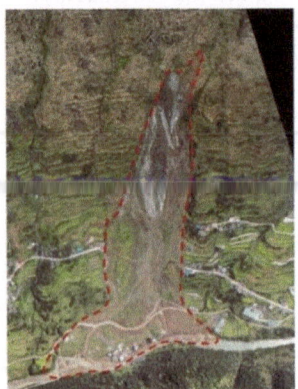

Fig. 1. Analysis diagram of sliding direction of landslide

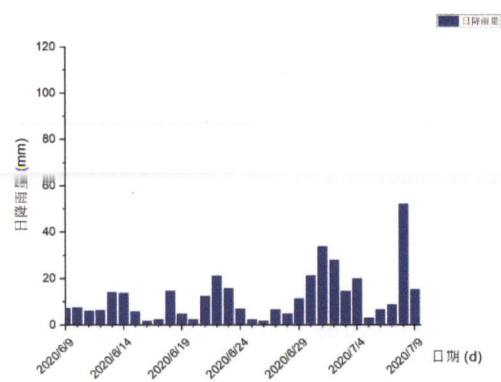

Fig. 2. Precipitation before landslide

3   Simulation Study on Deformation and Failure of Landslide

According to the engineering geological conditions and investigation results of landslide, the deformation and failure process of landslide is studied by numerical simulation method. The calculation parameters are selected based on the test of indoor rock compressive strength and shear strength, and comprehensively considering regional engineering experience, as shown in Table 1.

Table 1. Mechanical parameters of landslide rock

Materials	Severe weight /kN·m^{-3}	Elastic modulus /MPa	Cohesion /kPa	Angle of internal friction /°	Poisson's ratio
Gravel soil	20.3	30	15	25	0.35
Sandstone-mudstone interbed	26.1	1900	36	16	0.30

The deformation and failure of landslide is closely related to heavy rainfall, so the deformation process of landslide under actual rainfall conditions (Fig. 2) is simulated.

Figure 3 is a partial process diagram of landslide deformation and failure development under the actual continuous rainfall condition. At the initial stage of rainfall, the landslide deformation is small, and the main deformation is concentrated near the national G326 highway in front of the landslide (Fig. 3a); On July 8, 2020, the landslide showed obvious deformation as a whole, and the deformation amount and deformation range continued to increase (Fig. 3b).

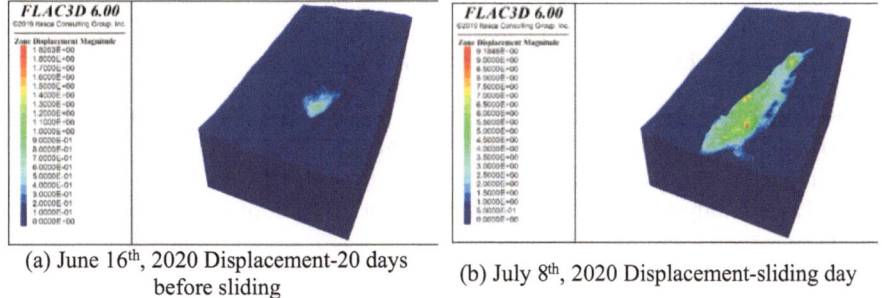

(a) June 16th, 2020 Displacement-20 days before sliding

(b) July 8th, 2020 Displacement-sliding day

Fig. 3. Cloud diagram of landslide displacement and plastic zone with rainfall process

4 Analysis of Prevention and Control Effect of Subgrade Reconstruction in Landslide

4.1 Subgrade Reconstruction Prevention Plan

In order to restore the normal operation of National G326 Highway, the area with the most serious impact on the highway is treated by anti-slide pile + anchor cable + clearing slope + facing wall + subgrade restoration + foot wall.

4.2   Calculation Model and Parameters

According to the layout of the design scheme, a numerical calculation model was established. The physical and mechanical parameters of the rock and soil are shown in Table 1, and the parameters of the anchor cable and anti slip pile are shown in Table 2.

Table 2. Physical and mechanical parameters of supporting structure

Materials	Modulus of elasticity /GPa	Tensile strength /GPa	Adhesion kN·m^{-1}	Poisson's ratio
Anti-slide pile	30.0	/	/	0.20
Anchor cable	200	13.2	0.56	0.15

In FLAC 3D, cable element is used to simulate anchor cable, and solid structure element is used to simulate anti-slide pile. Mohr-Coulomb criterion is used to describe the stress-strain relationship of rock and soil.

According to the design plan, the simulation calculation of two working conditions is mainly carried out: ① Cutting + anti-slide pile plan; ② Cutting + anti-slide pile + anchor cable plan.

4.3   Governance Effect of Different Control Plans

Influence of the Combination of Cutting and Anti-slide Pile on Landslide Deformation. On the basis of the above-mentioned cutting, a row of anti-slide piles with a spacing of 5 m is set at the landslide subgrade, totaling 20 piles. The main section is selected as the study section, and the deformation of landslide after treatment is shown in Fig. 4.

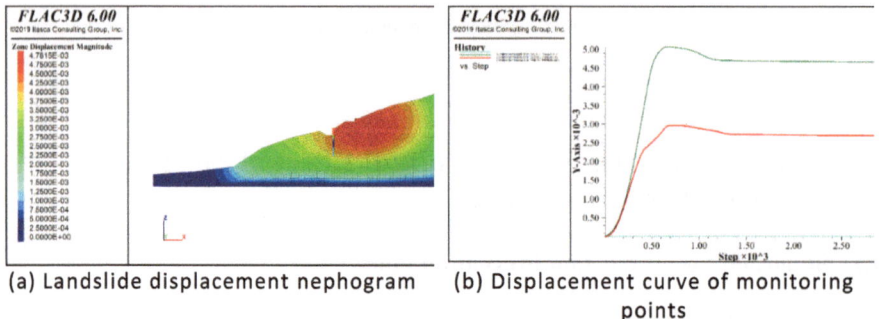

(a) Landslide displacement nephogram   (b) Displacement curve of monitoring points

Fig. 4. Landslide deformation under cutting + anti-slide pile treatment

As can be seen from Fig. 4, the displacement of landslide subgrade is reduced from 9.8 mm to 4.8 mm after the combination treatment of cutting and anti-slide pile.

The displacement of front and rear edges of landslide is basically unchanged, and the anti-slide pile has a good supporting effect on landslide deformation in the treatment area.

Influence of the Combination of Cutting, Anti-slide Pile and Anchor Cable on Landslide Deformation. The combination of cutting, anti-slide pile and anchor cable is used to support the landslide, and the stress produced at the subgrade after cutting is concentrated and dispersed to the sliding body through the anti-slide pile and anchor cable, thus improving the stress state at the subgrade and achieving the purpose of controlling the deformation of the landslide. The simulation results are shown in Fig. 5.

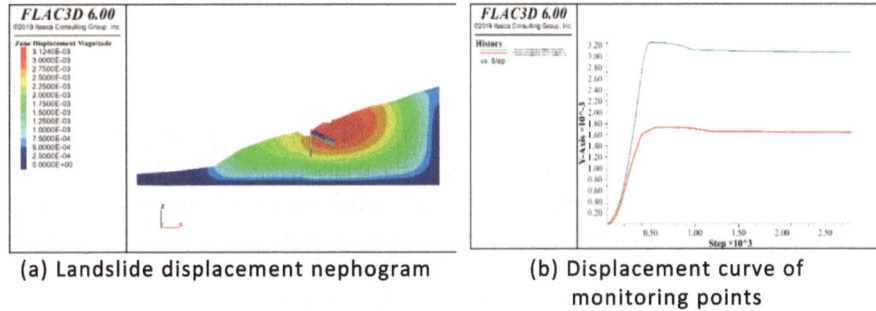

(a) Landslide displacement nephogram   (b) Displacement curve of monitoring points

Fig. 5. Landslide deformation under cutting + anti-slide pile + anchor cable treatment

On the basis of the above support, after adding anchor cables to the anti-slide piles at the subgrade, the displacement of the middle and rear edge of the landslide is reduced. The displacement at the subgrade is reduced from 4.8 mm to 3.1 mm under the support of cutting and anti-slide pile, and the displacement at the trailing edge of landslide is also reduced from 3.0 mm to 1.7 mm. It can be seen that the protection of anti-slide pile to landslide deformation is mainly reflected in the retaining effect of pile foundation on landslide soil, while anchor cable improves the stress state of landslide soil at the layout, so that the stress that should be concentrated in subgrade is evenly distributed on the landslide, thus reducing the deformation of landslide.

Figure 6 shows the change of anchor cable in support. It can be seen that the maximum axial force of anchor is 0.9 kN, which is mainly concentrated in the part below the elevation of the free surface, while the node displacement of anchor is evenly distributed as a whole, with the maximum node displacement of 3.0 mm. It is also verified that the anchor cable mainly improves the stress state of soil, and the stress is evenly distributed on the landslide soil.

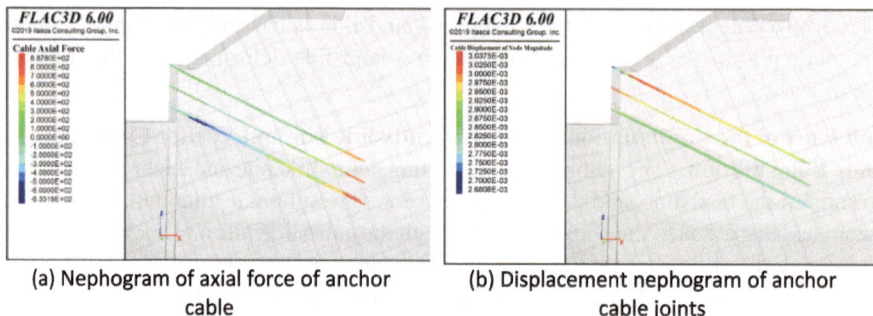

(a) Nephogram of axial force of anchor cable   (b) Displacement nephogram of anchor cable joints

Fig. 6. Stress and displacement of anchor cable supporting structure

5 Conclusion

(1) The landslide deformation is small at the initial stage of rainfall, mainly concentrated near the national G326 highway in front of the landslide; With the increase of rainfall, the deformation of the front part of the landslide increases continuously, and the deformation concentration area gradually appears near the right front part and the trailing edge of the landslide; On July 8th, 2020, the landslide showed obvious deformation as a whole, and the deformation amount and deformation range continued to increase, and the whole landslide was in a plastic state.

(2) The maximum displacement of landslide under different supporting conditions is decreased, among which the maximum displacement of main sliding area and trailing edge is 9.8 mm and 3.0 mm respectively; Under the combined treatment of cutting and anti-slide pile, the maximum displacement of the main sliding area of landslide is 4.8 mm, and the maximum displacement of the trailing edge is 3.0 mm; Under the combined treatment of cutting, anti-slide pile and anchor cable, the maximum displacement of the main sliding area of landslide is 3.1 mm, and the maximum displacement of the trailing edge is 1.7 mm.

(3) The protection of anti-slide pile to landslide deformation is mainly reflected in the retaining effect of pile foundation on landslide soil, while anchor cable improves the stress state of landslide soil at the layout, so that the stress that should be concentrated in subgrade is evenly distributed on the landslide, thus reducing the deformation of landslide. The combined treatment plan of "cutting + anti-slide pile + anchor cable" is adopted in subgrade restoration and reconstruction project, so that the overall deformation of landslide is in controllable range.

Acknowledgments. This study belongs to the technology project of Guizhou Provincial Highway Bureau, project number: 2022QLK04; Key technology project in the transportation industry, project number: 2021-ZD1-021; Key technology project in the transportation industry, project number: 2022-MS1-058; Guizhou Provincial Department of Transportation Science and Technology Project, Project Number: 2022-122-003.

References

1. Aderhold, J., et al.: J. Cryst. Growth **222**, 701 (2001)
2. Nichol, D., Graham, J.R.: Remediation and monitoring of a highway across an active landslide at Trevor, North Wales. Eng. Geol. **59**(3–4), 337–348 (2001)
3. Rodriguez-Peces, M.J., Azanon, J.M., Garcia-Mayordomo, J., et al.: The Diezma landslide (A-92 motorway, Southern Spain): history and potential for future reactivation. Bull. Eng. Geol. Env. **70**(4), 681–689 (2011)
4. Ming, L., Hua, H., et al.: Analysis on double-row anti-slide piles of slope in Yanan. J. Chang'an Univ. (Nat. Sci. Ed.) **31**(2), 63–67 (2011)
5. Kanagasabai, S., Smethurst, J.A.: Three-dimensional numerical modelling of discrete piles used to stabilize landslides. Can. Geotech. J. **48**(9), 1393–1411 (2011)

Seismic Responses Analysis of Suspended Ceiling Structure Attached to Large LNG Storage Tank

Yu Fu[(✉)], Jiuhuan Cheng, and Juan Su

Offshore Oil Engineering Co., Ltd, Tianjin 300461, NJ, China
fuyu1_cumtb@163.com

Abstract. The suspended ceiling (SC) of LNG storage tank on land is mainly composed of hanger rods, reinforcing rings, and aluminum panels. The hanger rods are prone to damage during earthquakes, which will affect the safe operation of the whole storage tank. In order to facilitate the study of the response of the SC under horizontal seismic shaking, the finite element models of the hanger rods with four different sections, including rectangular, circular, thin-walled circular tube, and groove shape, were established using the ABAQUS software. By inputting EL-Centro horizontal seismic waves, the dynamic time-history analysis is carried out for four hanger rod models with different sections. The response characteristics of acceleration, displacement and internal force with time are obtained, and the comparative analysis results of the maximum peak acceleration (MPA), maximum peak displacement (MPD), and maximum peak internal force (MPIF) at each measuring point are obtained. The results show that the shape of the section has obvious influence on the seismic response of the hanger rods, and the amplitudes of the acceleration-time and displacement-time curves of the hanger rods with different sections are different, and the MPA, MPD and MPIF of the hanger rods are closely related to the shape of the section.

Keywords: LNG · Storage Tank · Suspended Ceiling · Finite Element · Seismic Response

1 Introduction

As a clean energy source, LNG will account for a significant increase in the future energy consumption structure. In the entire industrial chain, LNG storage and transportation are very important link. With the development of technology, the storage tank volume is increasing, and the safe operation and maintenance of LNG storage tanks have become the focus and difficulty. As important energy infrastructure, large-scale LNG storage tanks also have profound significance for national economic construction and energy security. The SC is an important part of the insulation structure of onshore LNG storage tank, which affects the evaporation rate of the storage tank during operation. At the same time, the SC, a kind of suspended structure, is located in the internal space of the storage tank. Under the horizontal earthquake shaking, the hanger rods need to bear various

© The Author(s) 2024
G. Feng (Ed.): ICCE 2023, LNCE 526, pp. 200–212, 2024.
https://doi.org/10.1007/978-981-97-4355-1_19

forces, resulting in large deformation of the structure. Once the structural unsafe factors occur during operation, the tank should be shut down for maintenance, which will cause great economic losses and affects the safety of other structures, such as the inner tank [1].

Scholars have studied the seismic performance and dynamic response of the SC through experimental researches and numerical simulations. Soroushian et al. [2] summarized the damage phenomena of suspended ceilings in three shaking table tests conducted by the University at Buffalo, University of Nevada, Reno, and the Earthquake Engineering Research Center of Japan. Yao et al. [3] studied the impact of the installation of lateral supports on the seismic performance and seismic response of suspension structures by using multidimensional shaking table test method. Takhirov et al. [4] proposed a method for reinforcing the connection between the SC system boundary and a new seismic clamping structure based on the boundary damage characteristics of the SC system. Zaghi et al. [5–7] established five common damage models of suspended ceilings based on OpenSEES, and analyzed their damage evolution process. Kou Miaomiao [8] and Zhang Peng et al. [9] established a simplified analytical model of suspended ceilings based on ANSYS, and took the response spectrum of the floor of the three-story frame structure obtained from the time-history analysis as an incentive to study the law of the influence of the installation of the cable and the floor height on the seismic response of the continuous SC.

It can be seen that there are more studies on the SC structure of traditional buildings at present, but there are fewer studies on the SC of special building structures, such as LNG storage tanks. In this paper, the SC structure of a 200,000 m³ LNG storage tank is taken as the research object. Based on finite element simulation analysis, the dynamic response characteristics of the acceleration, displacement and internal force of the hanger rods with four kinds of cross-sections under horizontal earthquake are studied, and the differences in the seismic response results of hanger rods with different sections are compared and analyzed. The research in this paper can provide help for the optimization of SC design and the improvement of seismic performance in storage tank engineering.

2 Composition of SC System

The top of the inner tank of the onshore LNG storage tank is open, and the SC structure is located at the top of the inner tank, which is composed of hanger rods, reinforcing rings, and aluminum panels. The detailed structure is shown in Fig. 1. The LNG storage tank is insulated mainly by covering the SC with elastic felt, and the SC function is to provide support for the insulation layer. This design can reduce the heat exchange between the internal low-temperature medium and the external environment, so as to achieve thermal insulation. In order to make the cold protection effect better, a thicker cold insulation layer needs to be laid, which requires the SC to bear more weight. At the same time, the SC also needs to bear its own weight and other live load.

Figure 2 shows the relevant design parameters of the 200,000 m³ LNG storage tank SC. The radius of the SC reaches 40.4 m, and the aluminum panels of the SC are all made of 5 mm thick slabs in a certain sequence, and a sealed overall structure is formed by welding at the overlapping place. As a connection between the dome and the SC, the current design of the hanger rod is a 304L flat steel rod with a cross-section size of 50 mm × 8 mm. The number and length of the steel rod in each ring are different, but in order to balance the force, the length of the same ring of steel rod is the same. There are a total of 12 rings of the hanger rods, and the length of the hanger rods from the center ring to the outer ring gradually becomes shorter. The connector at the bottom of each flat steel rod is connected to the reinforcing ring of the SC by bolts. The reinforcing ring on the aluminum plate of the SC is vertically arranged (i.e. the long side direction of the section is along the vertical direction). All the reinforcing rings are concentric circles with the radius from R1 to R12. There are 12 rings of reinforcing ring in total, which are vertically welded on the aluminum plate by flat material. The thickness of the middle reinforcing ring (vertical) is 18 mm, and the most outer reinforcing ring is welded into a T-shape by flat aluminum with a thickness of 30 mm.

In this paper, four kinds of hanger rods with the same cross-section area and different cross-section shapes are designed. The cross-sectional areas are all 400 mm^2, and the sections are rectangular, thin-walled circular tube and groove shape respectively. All the materials and lengths are the same. In order to determine the difference between the same area and different sectional shapes of the hanger rods, the moments of resistance and section modulus of the hanger rods are calculated according to the theoretical formula. The calculation results show that the moment of inertia of thin-walled circular tube section is the largest, the section modulu of groove section is the largest, as shown in Table 1.

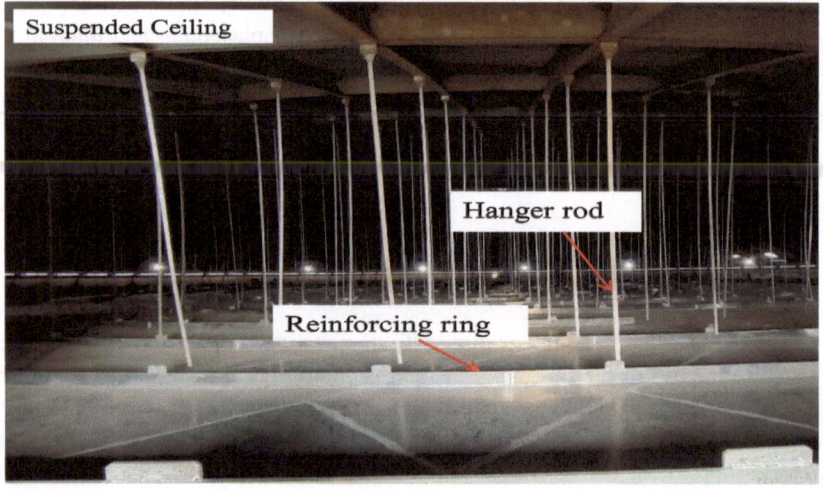

Fig. 1. Suspended ceiling

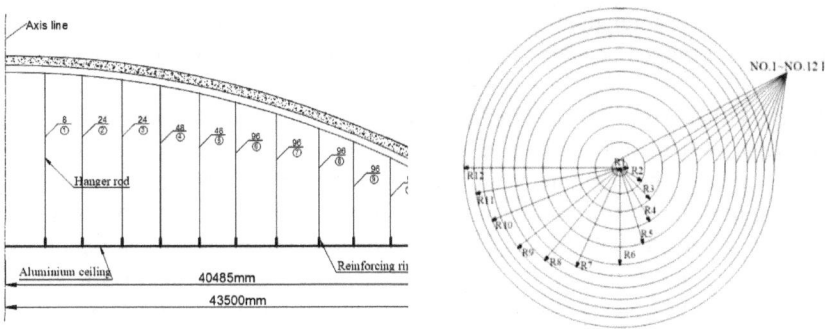

(a) Section view of the ceiling   (b) Reinforcing ring layout of the SC

Fig. 2. Schematic diagram of SC

Table 1. Parameters of cross-section

Number	Sectional shape	Sectional area (mm^2)	Moment of inertia ($\times$ 104 mm^4)	Moment of resistance ($\times$ 103 m^3)
Section1	Rectangle	400	0.213	0.533
Section2	Circle	400	1.272	1.127
Section3	Thin-walled circular tube	400	3.867	2.417
Section4	Grooved	400	3.739	2.460

3   Analysis Model of SC

3.1   Finite Element Model

Three-dimensional finite element analysis models of hanger rod with four kinds of sections were established by finite element software ABAQUS, as shown in Fig. 3. The actual shape of the structure was taken into account to ensure the accuracy of the calculation results, and the model was simplified as far as possible to improve the calculation efficiency. In the model, the beam element model is used for the hanger rod and reinforcing ring, and the shell element is used for the aluminum plate. Then the section properties are set according to their respective characteristics to complete the setting of various material parameters, and the section direction of the beam element is set by rotating method. In the model, the aluminum plate and the reinforcing ring of the SC adopt a common node, so that the beam and the shell are overlapped. In the process of modeling, the hanger rod model is established separately, and then it is bound to the reinforced ring model as a whole by the combination method.

The present test results show that reinforcing ring (tee) and hanger rod basically maintain elastic state under earthquake shaking, and elastic beam element is used in the model [10]. In the simulation, the top of the hanger rod is simplified as a fixed. A rigid connection is arranged between the bottom of the hanger rod and the reinforcing ring [11].

When setting the boundary conditions required for the model, the main considerations are displacement boundaries and loads. The main permanent load in the SC is the self-weight of the structure, which mainly includes the self-weight of the hanger rod, aluminum panel, and reinforcement ring, as well as some connecting components. However, compared to the self-weight of the structure, this part of the structural weight can be ignored. In addition to its own weight, another important load is the insulation load laid on the SC. Due to the uniform distribution of the insulation layer, the total weight of the insulation structure can be calculated based on the actual thickness and density of the insulation load. Then, the average load is obtained by dividing the weight by the area, and then applied to the upper surface of the SC in a uniform load manner.

When setting the boundary conditions of the model, the load should be considered in addition to the displacement boundary. The permanent load on the SC is the self-weight of the structure, which mainly includes the weight of the hanger rods, aluminum panels and reinforced rings, and also includes some other connecting screw. But compared with the structural self-weight, the weight of the connecting screw can be ignored. In addition to its self-weight, another main load is the weight of the insulation material laid on the SC. Because the insulation layer is uniform distribution, the total weight of the insulation layer can be calculated according to the actual thickness and density, and then the value of the average load is calculated. Finally, it is applied as a uniformly distributed load on the upper surface of the SC.

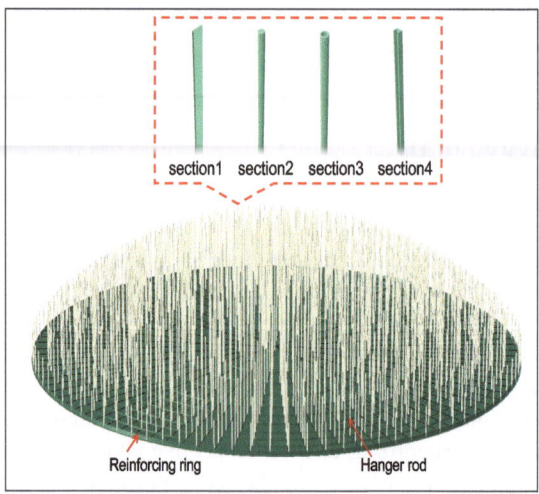

Fig. 3. Finite element model of SC

3.2 Measuring Points

According to the symmetry characteristics of the model, the measuring points are arranged in the X direction parallel to the seismic action and the Y direction perpendicular to the seismic action. A total of five measuring points from A1 to A5 are arranged in the X direction, and the same number of measuring points from B1 to B5 are also arranged in the Y direction. Measuring point A1 is located on the ring of hanger rod nearest to the center, measuring point A5 is located on the ring of hanger rod farthest from the center, and A2, A3 and A4 are located the 2nd, 4th and 6th rings where the number of hanger rod changes, respectively. There are a total of 10 measuring points, and their specific layout is shown in Fig. 4(a). The corresponding hanger rods and its position are shown in Fig. 4(b).These measuring points were used to record the dynamic response of SC structure under seismic loading and obtain the corresponding dynamic characteristics.

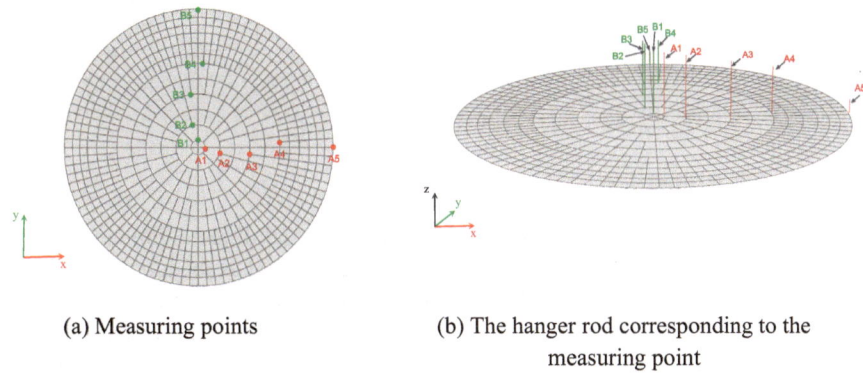

(a) Measuring points

(b) The hanger rod corresponding to the measuring point

Fig. 4. Measuring points layout

3.3 Damping

In general, structural damping should be considered in the analysis of structural dynamic characteristics. In this paper, Rayleigh damping system is adopted for the time-history analysis of the SC. The two proportional coefficients of damping are calculated according to Formula (1)–(3) [12], and the damping ratio of structural vibration mode is 5% [13]. Damping can be calculated by the following formula.

$$C = \alpha[M] + \beta[N] \tag{1}$$

$$\alpha = \xi \frac{2\omega_1\omega_2}{\omega_1 + \omega_2} \tag{2}$$

$$\beta = \frac{2\xi}{\omega_1 + \omega_2} \tag{3}$$

where α and β are two proportional coefficients, with dimensions of s^{-1} and s, respectively; ξ is damping ratio; and ω_1, ω_2 are the fundamental natural frequencies of the SC in X direction and Y direction respectively.

Based on the principle that the selected two natural frequencies should include the frequency that has a great influence on the structural response, the 1st natural frequency and the 10th natural frequency are selected to calculate Rayleigh damping. The calculation results are shown in Table 2.

Table 2. Rayleigh damping coefficient

Number	ω_1	ω_{10}	α	β
Section1	0.2445	0.2530	0.0781	0.0320
Section2	0.2281	0.5974	0.1037	0.0193
Section3	0.3806	1.0158	0.1739	0.0114
Section4	0.3470	0.6845	0.1446	0.0154

3.4 Materials Properties

According to ASTM B209M specification [14], the material of the aluminum panels is B209 5083-O, and the reinforcing rings are the same material as the aluminum panels. The material of the hanger rods is A276Gr 304L stainless steel, which bears the main load transfer function. The constitutive relation of hanger rod, reinforcing ring and aluminum plate is only considered elasticity. At the same time, it is necessary to consider the large deformation of the aluminum ceiling, so the large deformation switch should be set to open when determining the material properties. The material parameters of the finite element model of the SC structure are shown in Table 3.

Table 3. Material information

Ceiling component	Material	Density (kg·m^{-3})	Elastic Modulus (MPa)	Poisson ratio
Hanger rod	A276Gr 304L	7850	2×105	0.3
Reinforcing ring	B209 5083-O	2710	78000	0.34
Aluminum panels	B209 5083-O	2710	78000	0.34

4   The Selection of Seismic Waves

In this paper, the time-history analysis method is selected to analyze dynamic response of SC structure. According to the regulations, when the dynamic time-history analysis of the structure is carried out, acceleration time-history curves of the actual strong earthquake records and artificial simulated should be selected according to the classification of building site and the design earthquake groups, and the number of actual strong earthquake records should not be less than 2/3 of the total [15]. This paper only takes a real seismic wave as an example to analyze and illustrate, and some representative nodes' displacement and rods' dynamic inner force are given, as the amount of data of results of the dynamic analysis is so great. The seismic wave used for calculation and analysis in this paper is EL-Centro wave, and the time-history curve of the seismic wave is shown in Fig. 5. The peak acceleration of EL-Centro wave is 342 cm/s^2, and the duration is taken as 30 s.

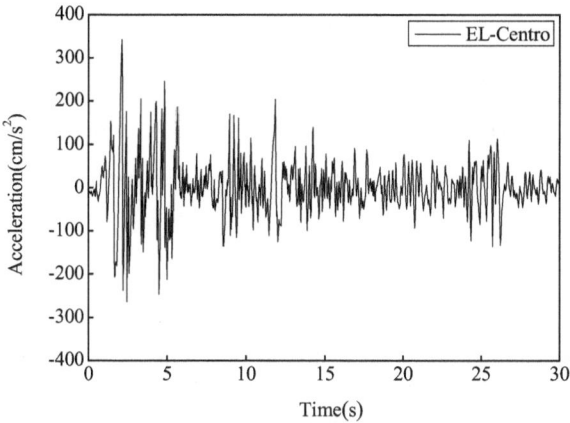

Fig. 5. Acceleration time history of ground motions

5   Response Analysis and Discussion

5.1   Acceleration Response

The relative acceleration response of each measuring point can be obtained through the acceleration time-history data. The MPA of each measuring point can be calculated using by Formula 4:

$$A_j = max\{|a_i(t)|\} \tag{4}$$

$a_i(t)$ is the acceleration time-history of different measuring points of the model, A_j is the MPA corresponding to each measurement point.

Figure 6(a) and (b) compare the MPA of four hanger rod models with different sections. It can be seen that the magnitude relation of the MPA of the hanger rod models are as follows: Section 4 > Section 3 > Section 1 > Section 2, indicating that the section shape of the hanger rod affects the MPA of the structure under earthquake load. In Fig. 6(a), the acceleration response of the A5 measuring point of the hanger rod at the most outer ring is more obvious than that at other positions in the parallel seismic input direction. Compared with the initial input acceleration, the MPA of the models corresponding to Section 1, Section 2, Section 3 and Section 4 increase by 141%, 46.5%, 154% and 189% respectively.

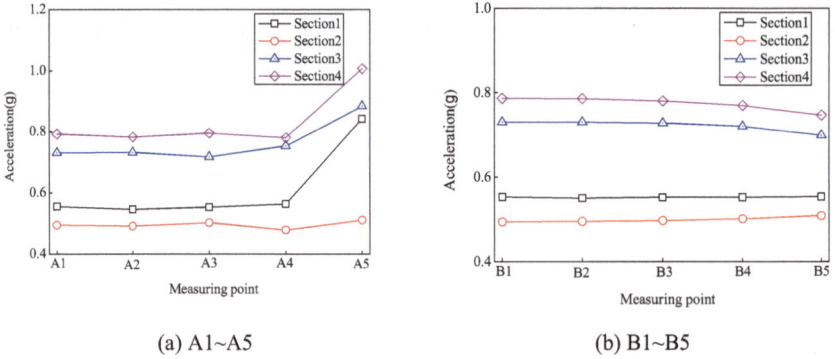

(a) A1~A5   (b) B1~B5

Fig. 6. Comparison of maximum peak acceleration (MPA)

5.2 Displacement Response

The MPD of each measuring point of the model is also calculated according to the method of formula (4). Figure 7(a)–7(b) shows the comparison of the MPD of four types of cross-sectional hanger rod models at different measuring points. It can be seen that the displacement response of Section 2 is obviously smaller than the other three sections. The relationship between the MPD of each measuring points of the four sectional hanger rod models is: Section 3 > Section 4 > Section 1 > Section 2. However, the peak displacement response of the same section at each measuring point is basically the same, indicating that the maximum displacement difference between each ring of the hanger rods is very small.

5.3 Internal Force Response

The failure of the hanger rod connection point is an important feature of the SC failure, and the magnitude of the hanger rod axial force is an important cause of the failure of the SC connection point. Figure 8 shows the maximum peak axial force (MPAF) of the hanger rods corresponding to each measuring point. It can be seen that in measuring points A1—A4 and B1–B3, the MPAF of the hanger rod corresponding to Section 1 is greater than that of other sections. The MPAF increases significantly at measuring points A5 and B5, indicating that the axial force on the most outer ring hanger rod is subjected to the largest axial force and the axial force response under seismic action

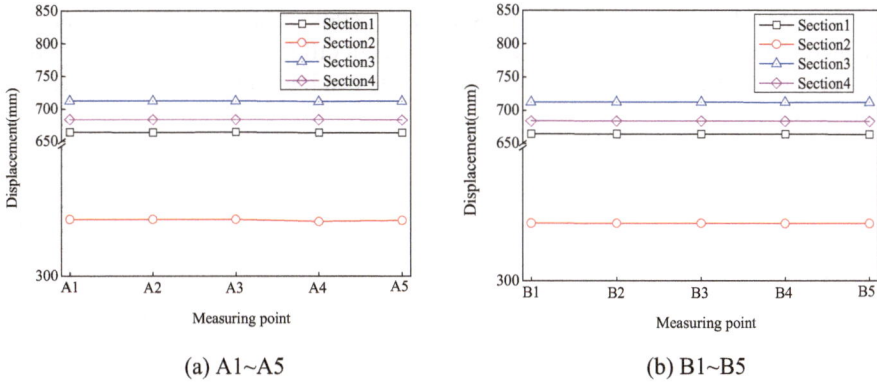

(a) A1~A5

(b) B1~B5

Fig. 7. Comparison of maximum peak displacement (MPD)

is more obvious. The difference in the MPAF of the hanger rod with different section forms indicates that although the section area is the same, the combined action of the axial force and bending moment on each hanger rod will affect the final axial force.

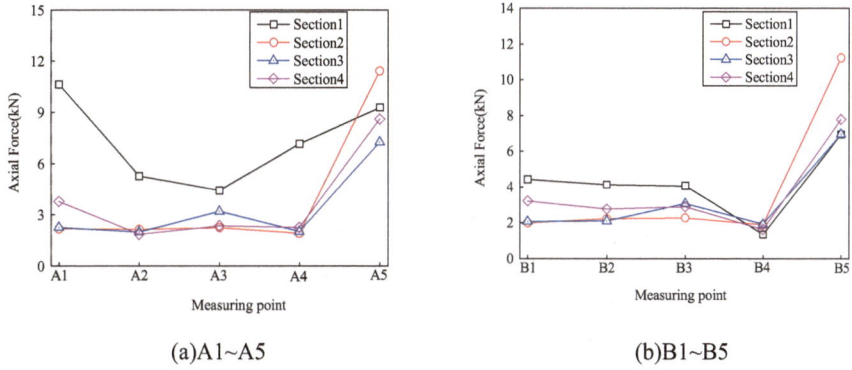

(a)A1~A5

(b)B1~B5

Fig. 8. Comparison of maximum peak axial force (MPAF)

Figure 9 compares the maximum peak bending moments (MPBM) of the hanger rod models with different cross-section at each measuring point. It can be seen from Fig. 9(a) that the MPBM corresponding to different cross-sections at A1 –A5 measuring points are: Section 3 > Section 4 > Section 2 > Section 1. The maximum bending moment appears at A5 measuring point, and the MPBM of the hanger rods with four sections increases significantly, indicating that the most outer ring hanger rod bears greater bending moment in the direction of seismic wave input. Figure 9(b) shows the comparison of the MPBM of the hanger rods from B1 to B5. It can be seen that the maximum value of the hanger rod with Section 1 occurs at B1 measurement point, while the maximum values of Section 2, Section 3, and Section 4 occur at B2 measurement point. It can be seen from Table 2 that the moment of inertia and moment of resistance of the hanger rod in different sections are different, resulting in the difference in the bending moment of the hanger rod with the corresponding section. In addition, the magnitudes of the inertia moments of asymmetric cross-sections in different directions are not equal, so when the direction of force action changes, the bending moment of the hanger will exhibit differences.

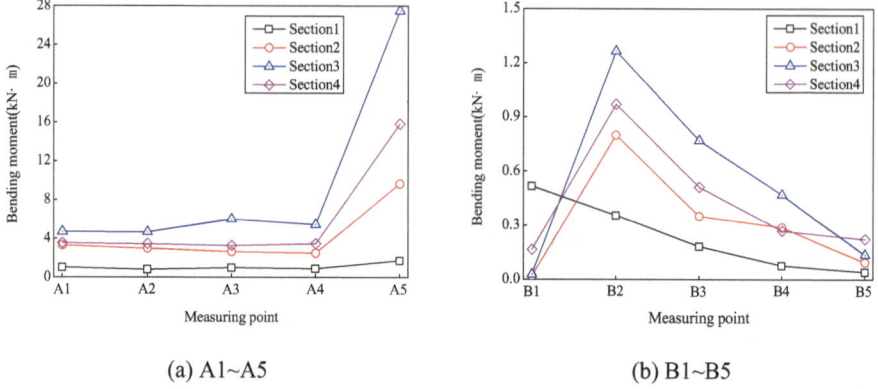

(a) A1~A5   (b) B1~B5

Fig. 9. Comparison of maximum peak moment

6   Conclusions

The safety of the suspender structure under earthquake action is very important for the overall safety of the LNG storage tank structure. Through the simulation calculation of four different cross-section suspender models including rectangular, circular, thin-walled circular tube, and groove shape under horizontal earthquake action, and the comparative analysis of the displacement, acceleration, and internal force of the SC structure, the following conclusions are obtained.

(1) Through the comparative analysis of displacement-time curve and peak accelera-tion, it is found that the peak acceleration responses of four kinds of hanger rods with different cross-sections are different. The acceleration response of the hanger rod is affected by the cross-sectional form of the hanger rod. Compared with the initial input acceleration ratio, the MPA of the models corresponding to rectangular cross-Section 1, circular cross-Section 2, thin-walled circular tube cross-Section 3, and groove-shaped cross-Section 4 increased by 141%, 46.5%, 154%, and 189%, respectively.

(2) The MPD of the hanger rod under seismic action is significantly affected by the sectional form of the hanger rod. The MPD of the hanger rod in the same section at different measuring points are very close. The relationship between the MPD of the four sections is: thin-walled circular tube > groove shape > rectangular > circular.

(3) The internal force response of the hanger rod under earthquake action is an important factor affecting the structural safety. The moment of inertia and resistance moment of different sections of the hanger rod are different. Under horizontal earthquake

action, the MPIF of the hanger rod with different sections is significantly different, and the MFIF of the most outer ring of the hanger rod is greater in the direction parallel to the earthquake action.

References

1. Zhai, X., Zhao, X., Wang, Y.: Numerical modeling and dynamic response of 160,000–m^3 liquefied natural gas outer tank under aircraft impact. J. Perform. Constr. Facil. **33**(4), 04019039 (2019)
2. Soroushian, S., Rahmanishamsi, E., Ryu, K.P., Maragakis, E.M., Reinhorn, A.M.: A comparative study of sub-system and system level experiments of suspension ceiling systems. In: Proceedings of the 10th US National Conference on Earthquake Engineering Frontiers of Earthquake Engineering, July 2014
3. Yao, G.C., Chen, W.C.: Vertical motion effects on suspended ceilings. In: Proceedings of the 16th World Conference on Earthquake Engineering, January 2017
4. Takhirov, S.M., Gilani, A.S., Straight, Y.: Seismic evaluation of lay-in panel suspended ceilings using static and dynamic and an assessment of the US building code requirements. In: Improving the Seismic Performance of Existing Buildings and Other Structures 2015, pp. 483–496 (2015)
5. Soroushian, S., Maragakis, E.M., Jenkins, C.: Capacity evaluation of suspended ceiling components, part 1: experimental studies. J. Earthq. Eng. **19**(5), 784–804 (2015)
6. Soroushian, S., Maragakis, M., Jenkins, C.: Axial capacity evaluation for typical suspended ceiling joints. Earthq. Spectra **32**(1), 547–565 (2016)
7. Zaghi, A., Soroushian, S., Echevarria Heiser, A., Maragakis, M., Bagtzoglou, A.: Development and validation of a numerical model for suspended-ceiling systems with acoustic tiles. J. Archit. Eng. **22**(3), 04016008 (2016)
8. Miaomaio, K.: Research on Seismic Performance of Nonstructural Components. Tianjin University, Tianjin (2014). (in Chinese)
9. Zhang, P., Lu, Y.: Seismic response analysis of suspended ceiling in frame structure. In: Tianjin University and Tianjin Steel Structure Association. Proceedings of the 14th National Symposium on Modern Structural Engineering, pp. 1467–1473 (2014). (in Chinese)
10. Jiang, H., Wang, Y., He, L.: Study of seismic performance of Chinese-style single-layer suspended ceiling system by shaking table tests. Adv. Civ. Eng. **2021**, 1–14 (2021)
11. Jun, S.C., Lee, C.H., Bae, C.J., Lee, K.J.: Shake-table seismic performance evaluation of direct-and indirect-hung suspended ceiling systems. J. Earthq. Eng. **26**(9), 4833–4851 (2022)
12. Jiang, H., Huang, Y., He, L., Wang, Y., Wang, H.: Numerical modeling and experimental validation for suspended ceiling system with free boundary condition. J. Build. Eng. **61**, 105285 (2022)
13. Anajafi, H., Medina, R.A., Santini-Bell, E.: Inelastic floor spectra for designing anchored acceleration-sensitive nonstructural components. Bull. Earthq. Eng. **18**(5), 2115–2147 (2020)
14. ASTM B Standard: Standard Specification for Aluminum and Aluminum-Alloy Sheet and Plate (2014)
15. Ministry of Housing and Urban-Rural Development of the People's Republic of China: Code for Seismic Design of Buildings: GB 50011-2010. 2016 ed. China Architecture & Building Press, Beijing (2016). (in Chinese)

Development of Pre Drilling In-Situ Rock Mass Shear Measurement System

Zicheng Zhong[✉]

Xi'an Research Institute (Group) Co. Ltd, China Coal Technology and Engineering Group
Corp., Xi'an 710077, China
zhongzicheng@cctegxian.com

Abstract. In engineering construction, it is of great significance to calculate the
shear strength and internal friction Angle of rock mass. Compared with laboratory test and theoretical analysis, in-situ test can maintain the essential state of
rock mass to the maximum extent, and obtain multiple groups of test data conve-
niently and quickly. The pre drilling in situ rock mass shear measurement system
is independently developed. The system is described from the aspects of shear
principle, overall design, hardware design and software platform. The system can
get the shear strength parameters of rock mass quickly and efficiently, and the
linear correlation of test data is high.

Keywords: In-Situ Test · Shear Apparatus · Data Collection · Mechanical
Parameters

1 Introduction

The shear strength of rock mass is an important parameter in the design of large-scale
rock mass engineering, which is directly related to the above-ground high-rise civil
construction and the excavation of the bank slope of the surface high dam [1]. At the same
time, with the use of underground space, especially the rapid mining of coal, roadway
and stope surrounding rock stability control, rock burst, coal and gas outburst and other
disasters become more prominent. However, it is difficult to accurately grasp the strength
parameters of rock mass because the stress environment of rock mass is complex and
contains randomly developed spatial discrete structural planes. The laboratory test is
limited by size, mining disturbance and rock separation, and the test conditions are
different from the actual environment. The traditional large-scale field test has a long
period, complex operation, a lot of manpower and material resources, and there are
constraints of soil and rock mass properties. In view of the limitations of the existing
methods, it has become one of the important research topics in the field of geotechnical
engineering to find a fast, simple and reliable means for rock mass in-situ testing [1, 2].

In the 1960s, Handy et al. from the University of Iowa proposed an in-situ test method
using Borehole shear test (BST) to measure rock and soil shear strength parameters, and
developed related equipment. After more than 40 years of development, borehole shear
test has been applied in situ to test the shear strength index of fine grained sedimentary

G. Feng (Ed.): ICCE 2023, LNCE 526, pp. 213–219, 2024.
https://doi.org/10.1007/978-981-97-4355-1_20

soils such as loess, hard clay, glacial moraine, Marine soft clay and residual soil [3]. Borehole shear test has been used in slope stability analysis and pile bearing capacity design abroad. In 2009, the Survey and Design Institute of Machinery Industry introduced a set of soil borehole shear instrument [4, 5] from the United States to test the shear strength indexes of typical Malan loess and leishite loess in Xi'an area. The China Institute of Water Resources and Hydropower Research has introduced a set of rock borehole shear instrument from the United States, and applied it to the hollow cylinder of concrete in the laboratory and the rock foundation of Xiangjiaba Hydropower Station.

At present, the equipment used in the borehole shear test in China is mainly focused on the BST borehole shear meter, Iowa borehole shear meter, and Phicometre geotechnical dual-purpose borehole meter in France [6]. The shear modulus test is mostly focused on soil, and the rock mass test is less. Aiming at the shortcomings of the existing test, developed in situ rock mechanics properties was designed to measure system, the system is compared with the United States and France similar equipment, in addition has the advantages of simple structure, light weight small size, can quickly measure the shear strength of rock mass advantage for many times, magnetostrictive sensor installation, can be in situ detection intrusion rock mass displacement size, a data acquisition module, Test data can be collected and stored in real time [7]. In addition, the end of the rock shear instrument is designed with a conversion joint, which can be used in coal mine when combined with the power head drill. The shear modulus of coal and rock is tested at 360 in situ, which expands the range of rock detection.

2 Shear Principle

Rock borehole shear instrument similar to the principle of direct shear test indoor and field, rock borehole shear apparatus [8] is in a certain depth in the borehole drilled in the two parallel wedge pressed into the drill hole wall, the shear apparatus between two parallel cutting tool to form a thin layer of rock, and then through the tyra shear apparatus tail, variable diameter joint implementation into dentate convex surface and hole direct shear failure of surrounding rocks. The experiment can be regarded as a forced direct shear failure test along this thin rock slice. In the rock borehole shear test, if the area of the rock slice embedded between dentate bulges is A, the normal force acting on the rock slice and the lifting force of the connecting rod are P and T, then the normal stress and shear stress acting on the rock slice are respectively.

$$\sigma = \frac{P}{A} \tag{1}$$

$$\tau = \frac{T}{2A} \tag{2}$$

The in-situ rock mass mechanical property measurement system is composed of rock mass intrusion module, shear strength measurement module, and data acquisition and processing module. It can detect different types of rock mass, generate test reports of mechanical properties, and analyze and process test data in the later stage. In the rock intrusion module, the manual hydraulic pump compressively invades the rock mass by the tool of the rock shear apparatus, and the magnetostrictive sensor measures the intrusion displacement. Shear strength measurement module by drilling or hollow plunger jack to provide axial tension, by shear instrument to shear rock axial; The data acquisition and processing module is mainly responsible for the acquisition of data in the shear instrument, and the realization of data reception, display, graphics, storage, analysis and other functions. The overall block diagram of module design is shown in Fig. 1.

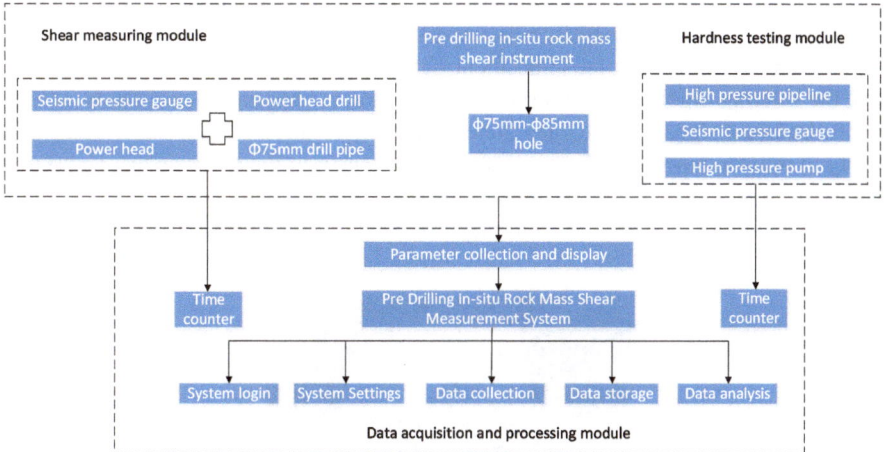

Fig. 1. Overall module design block diagram.

3   The Design of System Hardware

3.1   Pre Drilling In-Situ Rock Mass Shear Instrument

The predrilling in situ rock mass shear instrumentv [9] is designed as a lumen cylindrical structure, equipped with an elastic plate and a wedge tool, with an outer diameter of 75 mm (with a wedge tool and a metal positioning sleeve), and connected with a sensor adapter and a drill pipe adapter at the upper end. The sensor adapter is provided with a tubing interface, which is used for pressurizing oil injection or pressure relief of the pressure chamber, so that the elastic sheet and wedge cutter can be stretched out and the rock mass can be tested.

The structure of the shear instrument is mainly composed of shear device, pressure device, displacement device, diameter converter and other parts, as shown in the Fig. 2. The pressure device mainly converts the hydraulic pressure in the tube provided by the manual hydraulic pump into the axial forward force of the slider, and the displacement device uses the axial movement of the slider to make the tool on the elastic shear sheet move radially into the rock mass. The diameter changer head is mainly used to facilitate the shear meter from the drilling position, extraction, and subsequent rock shear axial tension, hollow plunger jack to provide shear force, shear meter through the jack hollow, using casing or caliper fixed upward pull; When the drill provides shear force, the reducer is connected to the drill pipe and shear in the direction of power head tension.

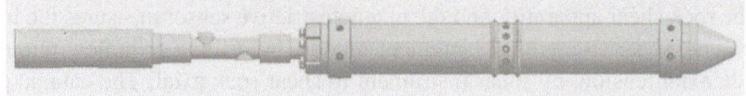

Fig. 2. Pre drilling in-situ rock mass shear instrument.

3.2 Parameter Collection and Display

The digital display table is used for accurate display of displacement value, power supply of sensor and transmission of displacement signal. In order to improve the display accuracy, the filtering algorithm and discount calculation are applied to accurately process the displacement signal. The display meter is directly connected to the 220 V AC power supply to provide power for the magnetostrictive sensor in the shear meter. The sensor displacement signal is accurately processed by the display table and transferred to the computer serial port to enter the data acquisition and management system (Fig. 3).

Fig. 3. Parameter collector.

The instrument is equipped with filtering algorithm, first smooth filtering and then inertial filtering. Smooth filtering is a common preprocessing method in signal analysis, which has good suppression effect on periodic interference and high smoothness.

The displacement obtained through the measurement process may have errors due to various reasons such as sensors, transmitters, leads or instruments. The correction function provided by instruments can effectively reduce the errors and improve the measurement and control accuracy of the system. In order to further improve the accuracy of the collected displacement data, the display table was modified by conversion.

3.3 Design of Auxiliary Accessories

Auxiliary accessories include manual hydraulic pump, anti-vibration pressure gauge, leveling base, hydraulic pipeline, data transmission line, etc. Depending on the mode of shear force provision, a drill, drill pipe connected to the shear meter, or a plunger jack and hollow steel pipe are used for axial shear.

4 System Software Platform Design

The design idea of rock shear data acquisition and processing system software is to build a software system with strong expansibility, high reusability, alarm and flexible control. Running the Windows operating platform, the software system based on intrusion displacement parameters, formed by the sensor is converted signal synchronization of real-time data acquisition and processing, the system of data acquisition and processing, curve fitting, report output and multifarious work can be done in an instant, can not only intuitive and clearly display and print processing results, Moreover, the fitting curve characterizing the change law of physical quantity can be drawn, so that the whole software system integrates the advanced nature and practicability.

Specific design, in order to improve the operability and convenience of the system. The computer acquisition interface adopts the display mode of combining data and curve, which is intuitive and clear. In the center of the screen, the displacement curve is displayed in real time with time as the abscissa and displacement as the ordinate. The displacement data on the right side of the curve is in millimeters and accurate to thousandths. In order to improve the data scalability, the interface can display a maximum of 16 channels of curves and data values simultaneously. Above the data report, print, save picture button, real-time record generated data; The lower part of the screen is equipped with various function keys, and there are operation hints, you can check the reference at any time; The function key includes the automatic curve refresh time, which can be adjusted according to the difference of intrusion time of different lithologies. Set up/down displacement alarm values to detect rock intrusion outliers in time to improve the accuracy of testing (Fig. 4).

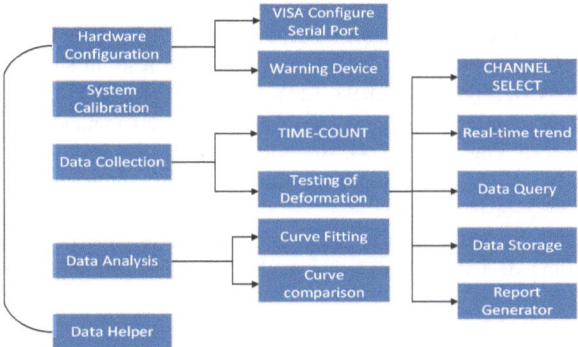

Fig. 4. System software platform.

5 Conclusion and Prospect

Through the test of the predrilling in situ rock shear system, the rock intrusion and data acquisition performance of the system is significantly improved compared with similar foreign equipment, which is reflected in the following points:

1) The system has the advantages of simple structure, light weight and small volume, and little disturbance to rock mass. It has the advantages of in situ testing and can quickly measure the shear strength of rock mass for many times.
2) The selected magnetostrictive sensor has the characteristics of high accuracy, good stability, high sensitivity and fast response time. Compared with similar equipment, it can accurately measure the intrusive rock mass displacement and improve the shear effect.
3) The software system has the characteristics of strong expansibility, high reuse and flexible control. It has the function of real-time data recording and overcomes the shortcoming of large error of handwriting by eye observation.

Acknowledgments. This research was financially supported by Key R&D Project in Shaanxi Province (Grant No.2023-YBGY-340).

References

1. Jin, C., He, C., He, L.: On rock-soil in-situ test and on-site monitoring technology. J. Shenyang Jianzhu Univ. (Nat. Sci.) **26**(04), 690–694 (2010)
2. Kang, H., Lin, J.: New development in geomechanics measurement and test technology of mine roadway surrounding rock. Coal Sci. Technol. **29**(7), 27–30 (2001)
3. Yang, Y., Pan, B.: Micromechanics-based numerical simulation of rock fragmentation process induced by static indentation. China Sci. Paper **12**(09), 1024–1029+1043 (2017)
4. Li, S., Li, D., Yu, S.: Meso-simulation for fracturing process of rock specimen under action of indenter. J. Xi'an Univ. Sci. Technol. **36**(06), 769–774 (2016)
5. Li, D., Li, S., Yu, S., et al.: Fractal characteristics of rock fragmentation process induced by indenters. Chin. J. Geotech. Eng. **36**(06), 769–774 (2016)
6. Shen, X.K., Cai, Z.Y., Cai, G.J.: Applications of in-situ tests in site characterization and evaluation. Chin. Civ. Eng. J. **49**(02), 98–120 (2016)
7. Yu, Y., Gao, Y.: Preliminary study on cobalt hole shear test method of soil shear strength parameters. Geotech. Eng. Tech. **29**(4), 169–172, 208 (2015)
8. Yu, Y., Zheng, J., Liu, Z., et al.: Borehole shear test and its application in Loess. Rock Soil Mech. **37**(12), 3635–3641 (2016)
9. Zhong, Z., Zhang ,Y., Shao, J., et al.: Preboring formula original position rock association formula measuring device and measuring method. Shaanxi Province: CN109187226B, 28 May 2021

Modeling Resilient Self-centering Concrete Walls with Repairable Structural Fuses to Predict Earthquake Performance

Nouraldaim F. A. Yagoub[1,2(✉)], Aqdas Shehzad[1], Hikma Ally[4], and Xiuxin Wang[1,3(✉)]

[1] School of Civil Engineering, Southeast University, Nanjing 210096, China
{233179921,gsdean2}@seu.edu.cn
[2] Department of Civil Engineering, Faculty of Engineering Science, University of Nyala, Nyala, Sudan
[3] Full Key Laboratory of Concrete and Prestressed Concrete Structures of Ministry of Education, Southeast University, Nanjing 210096, China
[4] School of Transportation and Civil Engineering, Hohai University, Nanjing, China

Abstract. In the last decades, earthquake-resilient structural systems have become popular in rocking structures and are considered a viable option for buildings in seismic regions. Self-centering concrete shear wall systems offer numerous benefits, including reduced seismic damage. Designing buildings, especially in areas with weak earthquakes, needs a simple damper for energy dissipation in terms of design, execution, and ease of removal after the seismic. Extensive experimental studies have demonstrated excellent seismic performance of the self-centering shear walls. However, the analytical models currently used still have some limitations for modeling the gap rocking behavior. This study presents a self-centering concrete wall with energy dissipation (ED) steel angle devices and evaluates it to achieve seismic-resilient building structures. The angle devices are externally installed on the wall corners to achieve controllable energy dissipation and are easily replaceable. The numerical study was performed using displacement control cyclic loading, and verification of the self-centering (SC) reinforced concrete RC wall was first introduced. Subsequently, five different configurations with different thicknesses of ED steel angles were investigated. The outcome demonstrates that the proposed system structure has excellent load-bearing capability, energy absorption, lower damage, and self-centering capability. In addition to improving the self-centering wall's lateral stiffness, strength, and energy dissipation, increasing the angle damper thickness can also increase residual drift if it surpasses a certain threshold. Compared to rocking RC walls, the proposed RC walls offer a promising solution for low-performance structural systems required by resilient and sustainable civil infrastructure.

Keywords: Numerical Modeling · Self-Centering Concrete Shear Wall · Rocking · Earthquake Resilience · Repairable Energy Dissipation Fuse

G. Feng (Ed.): ICCE 2023, LNCE 526, pp. 220–231, 2024.
https://doi.org/10.1007/978-981-97-4355-1_21

1   Introduction

Reinforced concrete wall structures can bear vertical and lateral loads caused by wind or earthquakes. Concrete buildings do not fall easily and stay standing after an earthquake, but the high residual drift ratio makes it much more likely that the buildings will be destroyed [1]. The installation should be demolished after the earthquake, or it might require expensive structural repairs that are neither practical nor cost-effective. In the past ten years, low-damage structural systems have attracted the attention of both academics and practicing engineers. To increase the performance of repairable shear walls, some researchers have suggested using high-strength materials like FRP reinforcement [2, 3], PC strands [4], and SMA (shape memory alloy) [5].

The ability of a concrete structure, subsequent to being unloaded, to revert back to its initial position is known as Self-centeredness aptitude. This capability can be attained through three methods: (a) using unbonded post-tensioning strands[6–9], (b) memory shape alloy steels[10, 11], and (c) pre-pressed disc springs[12, 13]. Various systems have been developed using these methods, such as a self-centering link beam that is reinforced by post-tensioned Shape Memory Alloy (SMA) rods [14], Unbonded post-tensioned prefabricated concrete moment framework [15], Shear wall made of unbonded post-tensioned concrete [16, 17], The wall is designed with self-centering capabilities and incorporates disc spring devices or SMA bars [13], as well as an unbonded post-tensioned linked wall system [18]. Shen et al. [19, 20] examined the performance of a post-tensioned concrete linked wall system that includes a steel coupling beam. Additionally, these strands contribute to the system's ability to automatically align itself. Consequently, following the earthquake, the structure would go back to its initial undisturbed position and exhibit either no remaining or minimal remaining displacement. The benefits of unbonded post-tensioned coupling beams, as compared to monolithic cast-in-place RC coupling beams and embedded steel coupling beams, include (a) enhanced aesthetics due to less visible beam and wall details, (b) ability to endure substantial nonlinear displacements without substantial structural harm, (c) self-centering ability that minimizes residual capacity of the structure following a major earthquake, and (d) expedited and simplified post-earthquake repair of the system. Additional ED devices are necessary for this system because of the potential insufficiency of energy dissipation by self-centering concrete walls. The structural deformation behavior can be simplified as "bilinear elastic," as depicted in Fig. 1(a). To enhance the ability of the self-centering shear wall to dissipate energy, researchers explored the possibility of incorporating a damper element to induce a "flag-shaped" deformation pattern in the structure, as depicted in Fig. 1(b).

To enhance the energy absorption capability of the wall, the hybrid wall, which adds energy absorption fuses to the self-centering wall, is proposed. Restrepo and Rahman[21] first suggested steel rebars embedded in the wall-to-foundation to provide sufficient energy dissipation capacity. Metal devices that presented energy absorption were presented to easily repair hybrid walls after earthquakes. Marriott et al. [22] Suggested an innovative wall design that combines tension-compression elements yielding steel fuses exterior of the wall. Li et al. [23] devised a novel wall-to-foundation connection utilizing buckling-restrained steel sheets through testing methods. For the vertical joints of walls, a unique U-shaped flexural plate (UFP) [24] was designed. Wall rankings and tests on the wall with the O-shaped plates revealed its substantial seismic capacity [25], so Henry et al. [26] proposed welding an O-shaped plate on the wall and using columns to distribute energy.

The current study presents earthquake-resilient RC walls installed with replaceable ED steel angles damper (SC-SAD) to improve earthquake performance. This wall system can produce the following effects: SC capability provided by the unbonded strands, enhanced energy dissipation by external steel angles, limited damage in RC walls with major inelastic deformations concentrated in steel angles, and an earthquake-resilient design requiring little to no repair even after earthquakes. A numerical model of the RC wall was subsequently created and validated using data from representative tests. Last but not least, the computational (finite element) investigation and evaluation of the seismic performance of the RC walls was done in terms of hysteresis curves, skeleton curves, stiffness degradation, residual displacement, and self-centering and energy dissipation capabilities.

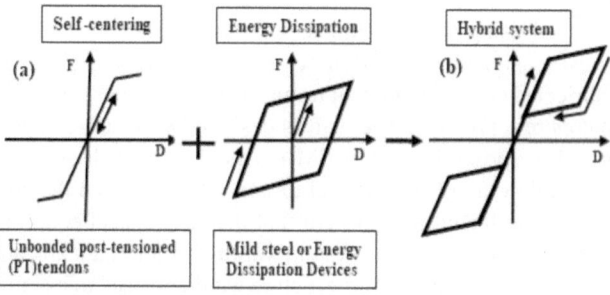

Fig. 1. Post-tensioned SC wall system with energy dissipaters a) post-tensioning re-centering and energy dissipation b) Hybrid system response.

2 Material Properties

The ABAQUS concrete damaged plasticity model (CDP) was used for the concrete materials in the FEM. This model was mainly made for reinforced concrete structures that are loaded cyclically or dynamically. The concrete damage plasticity model is generally built on two primary uniaxial concrete data sets and five plasticity parameters. The yield surface function, the potential flow, and the material's viscosity are all determined by the five parameters φ, e, $fb0/fc0$, Kc, and λ, Multiple calculations were performed to increase the analysis's precision and convergence, and the results show that the model's plasticity parameter values are 38,0.1,1.16, 0.6667, and 0.0005, respectively. The elastic modulus and Poisson's ratio are two more variables that must be defined to define the concrete material. The classical elastic-perfectly plastic stress-strain material model is adopted for the steel reinforcements and pre-stress tendons, and tests obtain the essential properties. In addition, the ED angle devices are made mainly of steel. The Q345 strength grade describes the angle made of steel. This investigation uses a model that is elastically and exhibits excellent plasticity to mimic the behavior of steel angles accurately. Table 1 and Fig. 2 outlines the particular characteristics of all of the materials.

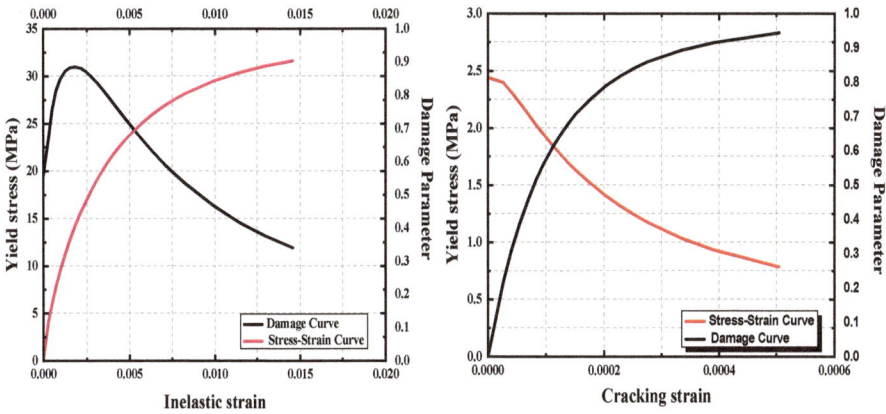

Fig. 2. Mechanical properties of concrete material.

Table 1. Properties of concrete, reinforced bars, ED steel angle, and strands.

The material	Variable	The value
The concrete	Strength	40 (MPa)
	Elastic modulus	29.915 (GPa)
	Poisson's ratio	0.2
The Reinforcement	Tensile strength	430 (MPa)
	Highest level of strength	570 (MPa)
	Modulus elasticity	210 (GPa)
	The Poisson proportion	0.3
Strands	Area (mm2)	100
	Highest level of strength	1836(MPa)
	Tensile strength	1746(MPa)
	Modulus elasticity	205(GPa)
	The Poisson proportion	0.3
Steel sheets	Tensile strength	375(MPa)
	Highest level of strength	425(MPa)
	Modulus elasticity	210 (GPa)
	The Poisson proportion	0.3

3 Test Specimen

Restrepo et al. [21] tested three self-centering shear walls that had been pre-stressed to look into how well self-centering concrete wall structures handle earthquakes. The prototype for a 4-story building structure was used to design the test specimens. The

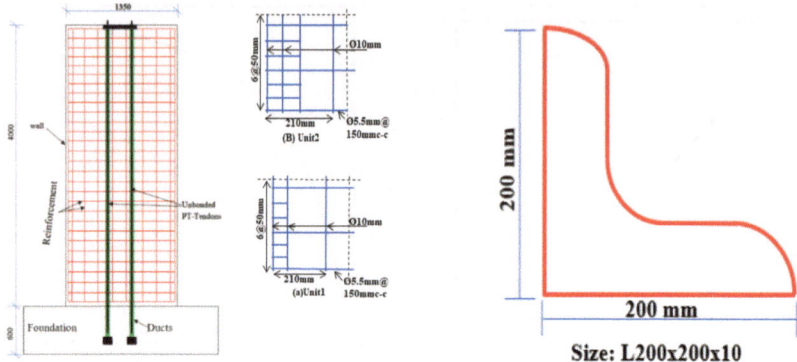

Fig. 3. Specifications of specimens' construction.   **Fig. 4.** Design details of the ED steel angle.

tested samples were called "Unit1, Unit2, and Unit3." This research paper uses Unit 1 as an example, and Fig. 3 displays the wall's design details. The wall is 3700 mm tall and 125 mm thick. The shear wall has two sets of pre-stressed tendons (PTs). They are set up 175 mm from the centerline of the wall. Each pre-stressed tendon comprises a steel strand that is 12.7 mm wide and has a cross-sectional area of 100 mm2. Their starting level of pre-stress is 0.5. The two corners of the test sample have additional enhancements to strengthen the local concrete and prevent the concrete's premature failure at the wall toe while the wall is moving back and forth.

First, Unit 1 has no ED bars or other energy-dissipating devices at the joints. It means it is a typical pre-stressed rocking wall structure written as SC in this study. Second, this research shows a brand-new precast self-centering concrete wall structure made of ED steel angles device. It is called SC-SAD. In Fig. 4, you can see more information about the design parameters. The angle damper made of steel used in this study has a cross-section that is L75 mm × 75 mm × 4 mm. It was done with an angle width of b = 10 mm. The ED steel angle dampers' legs were attached to the corners of the walls and the foundation. It can create stable hysteretic loops that can handle more ED loads.

4 Finite Element Analysis

This study presents a numerical simulation analysis using the ABAQUS software to investigate the energy dissipation of the proposed system. Consequently, the proposed method parameters conduct a finite element analysis (FEA), as shown in Fig. 7. The strong portion of the sample utilizes the C3D8R part, while the pre-stressed and stressed reinforcements adjust the truss element. A small pad simulates the anchor at the top of the wall and the bottom end of the foundation, corresponding to the pre-stressed tendons. Regular rebar and concrete are embedded in regions. Concrete is hardened into a damaged flexibility in concrete. The touch interfaces of further elements are formed with hard contact in the desired direction and the circumferential contact characteristic direction. The coefficient of specific friction is established based on the physical characteristics of the touching layer. The experimental parameters, including loading approaches and environment at borders, are identical to those of the subsequent studies. The bond slip occurs among steel. Bars and concrete are ignored when the precast concrete wall contacts reinforcing steel bars and profiled steel. The tie partially simulates wall contact with ED steel angle damper legs. A "surface-to-surface" connection models the wall to the foundation. The contact surface is "contact hard," meaning it can be Split off from the outer layer. The perpendicular attribute is "penalty," and the percentage of interaction is 0.5. This method can mimic the self-centering rocking wall's corner lifting characteristics while avoiding concrete tension to make the structure more logical. We investigate the seismic response of a self-centering rocking wall subjected to cyclic horizontal forces on the upper wall. Figure 5 depicts the horizontal cyclic load-displacement loading system. The SC and SC-steel angle damper (SC-SAD) models are analyzed numerically. Figure 6 shows the first step in validating the precision of the mathematical models by comparing the simulation findings of SC with the test results. This diagram depicts the nonlinear elastic response typical of a rocking body. Almost no residual lateral displacements were detected during the reaction, even after applying drift ratios of 3%. According to test results, [21]: As shown in Fig. 1(a), the hysteresis curve of the SC specimen without any energy absorption device is of the typical "rocking wall" type, indicating that its energy dissipation capacity is poor. Figure 6 demonstrates that the overall patterns of the two hysteresis curves have a high degree of unity with one another. Numerical simulation and experimental results agree well and show similar trends, so they can accurately reflect the proposed system's mechanical properties like bearing and energy dissipation capacities in Fig. 7.

5 Result Analysis

The loading process was modeled through the force system, which consists of the pre-stress tendons and axial force to the wall body, resisting the bending moment caused by the horizontal load before the self-centering rocking wall begins to rotate. When the wall starts to turning, the right and left devices of energy absorption, which have more significant force, begin to yield tension in the left. As the turning continues to increase, the two instruments of energy absorption begin to yield in. To evaluate the FEM's accuracy even further, regional variables for reaction like the unbonded tendon stress are looked into. To make it possible to compare the FEM calculations and test results. The FEM results showed a good correlation with the experimental results. The inaccurate estimation of strand stress during lateral force, as depicted in Fig. 8, may have been caused by minor seating losses at the post-tensioning anchor and the test-related deformation of the loading beam, which were not considered in the Finite Element Model (FEM).

5.1 Analysis of Hysteresis Curve

The hysteresis loop of the load-movement is shown in Fig. 9. In Fig. 9a, the starting stage of loading, the structure is in an entirely elastic step, the hysteresis loop is almost a straight path, and the area of energy absorption is tiny; with the increase of the load, the uplifting of the wall increases, the energy dissipation capacity little increases.

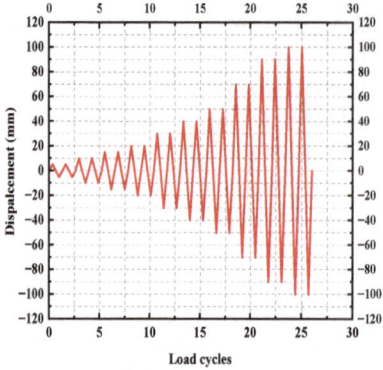

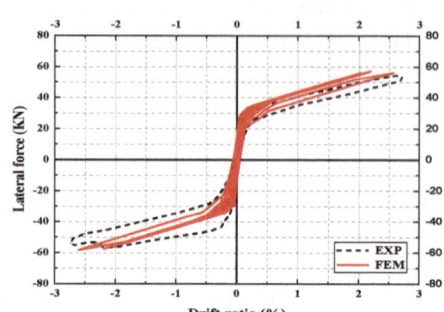

Fig. 5. Displacement control process of numerical model.

Fig. 6. Differences between the predicted hysteretic curves and the experimental findings.

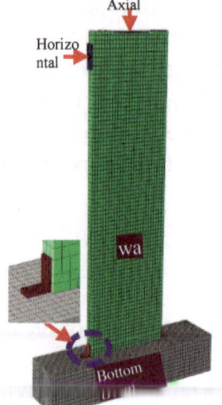

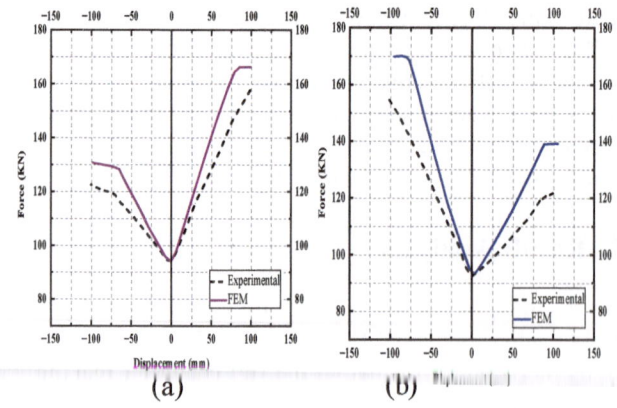

Fig. 7. Finite element model loading

Fig. 8. Relationship between modelling and test outcomes: (a) The stresses of the PT1 bars; (b) The stresses of the PT2 rods.

5.2 Influence of Thickness of Steel Angle Damper on Energy Dissipation Capacity

Figure 9 (a, b, c, d, and e) shows the hysteresis curve of the load-sway. With the augment in the load, the energy dissipation capacity of the steel angle damper starts to play; The structure's rotation angle is increasing, causing the steel angle devices in the corners of the wall to enter the plastics phase; the energy absorption capacity augment until it attended its peak. Based on the above analysis, the energy absorption capacity of the steel angle device is dominant for the SC-SAD under cyclic loading. So, it is essential to look into how SAD's thickness affects SC-SAD's mechanical properties when it is loaded and unloaded many times. To accomplish this goal, five cases have been chosen to investigate the effect that varying thicknesses of SAD have on the energy dissipation of SC-SAD. These results can be found in Fig. 9. For a small angle thickness of 10mm, the energy dissipation is about 2.5 KN-m. For the case of an angle thickness of 12.5mm, the energy dissipation is about 8 KN-m. In the case of an angle with a thickness of 15mm, the energy dissipation is approximately 11 KN-m. For an angle thickness of 17.5mm, the energy dissipation is about 15 KN-m. In the case of an angle with a thickness of 20mm, the energy dissipation is approximately 19 KN-m. For all cases, the energy absorption gradually increases with the displacement increase δ. For the same shape and strength as the angle damper, the selection of the thickness of the angle damper will help determine energy dissipation, which is quite crucial to the design of the angle damper. The deformation for the angle thickness of 10 mm to 20 mm in the FEA simulation is shown in Fig. 9 (a, b, c, d, and e) and Fig. 10, respectively.

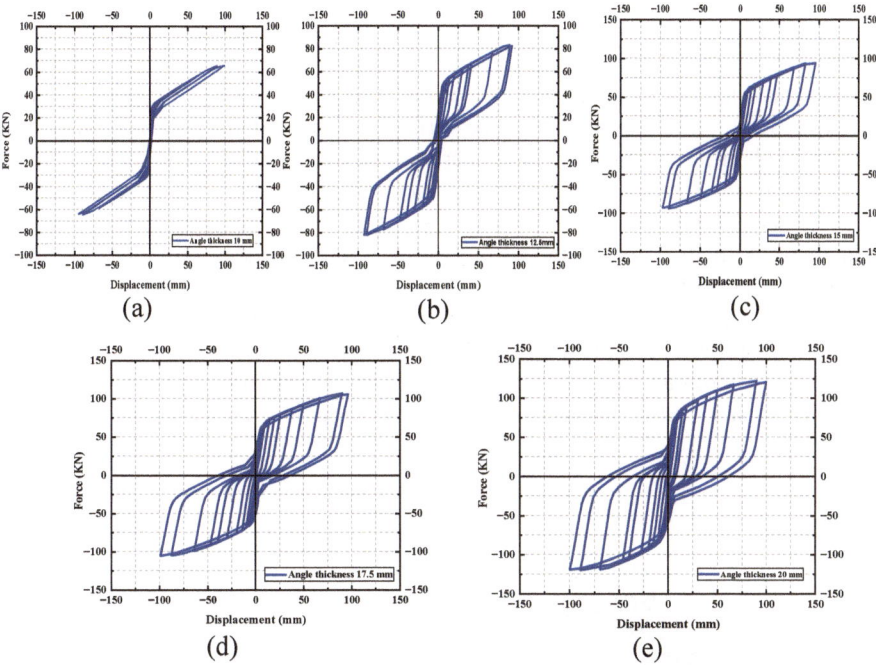

Fig. 9. Lateral force-top displacement hysteretic curves of the (a) angle plate thickness 10 mm (b) angle plate thickness 12.5 mm (c) angle plate thickness 15 mm (d) angle plate thickness 17.5 mm(e) angle plate thickness 20 mm.

5.3   Analysis of Skeleton Curve

The skeleton loops and cumulative energy dissipators coefficient ξ loops eshedtablis on the hysteresis loops for walls with the ED angle thickness of 10 mm, 12.5 mm, 15 mm, 17.5 mm, and 20 mm are illustrated in Figs. 11 and 12, indicating that the early stiffness and force, as well as the energy absorption ability of SC-SAD, are substantially improved with the augment of SAD's thickness under the premise of other conditions being the same. However, the self-centering capability was decreased. It is because the same recovery forces presented by the pre-stressed tendons should overcome the different SAD's resistances due to varying thicknesses of SADs, resulting in higher starting stiffness, larger bearing capacity, and energy absorption capacity but more obvious residual malformation with the increased thickness of SADs. The envelope curves extracted from the cycle peaks are shown in Fig. 11.

The equivalent viscous dampening proportion assesses the structure's ability to dissipate energy. Figure 13 depicts how the same amount of damping viscous alters depending on the displacement that is being measured. Overall, the same viscous damping proportion increases with the increase in displacement of the SC-SAD system. For example, if the angle thickness is 20 mm, the comparable viscous pressure damping proportion of the sample reaches a peak value of 0.01274 importance the drift is 0.0025%. When the drift is 0.0075%, the same viscous damping proportion of the sample gets a maximum importance of 0.02196. When the drift is 0.01575%, the comparable viscous damping ratio of the model reaches a maximum importance of 0.02612. When the drift is 0.025%, the same viscous damping proportion of the model gets a maximum importance of 0.0259. To better understand the variation of the system cyclic response of an SC-SAD with different angle thickness parameters, residual drift is presented here, as shown in Fig. 14. The variation in selected thickness parameters of the SAD (10, 12.5, 15, 17.5, and 20 mm) It is observed that the residual drift increases noticeably with an increase in the thickness of the SAD. However, this comes at a penalty of reduced re-centering capabilities. The reduction in re-centering is more apparent in more thickness angles, but for a low and medium range of thickness of the SAD, the SC-SAD can re-center after the complete cycle. Increasing the thickness of the angle damper is beneficial to improving the self-centering wall's lateral stiffness, strength, and energy absorption capacity. Still, it simultaneously causes a significant residual drift.

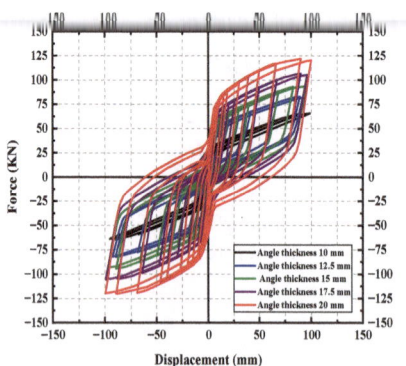

Fig. 10. Load-displacement curves of specimens with different angle thickness ratios.

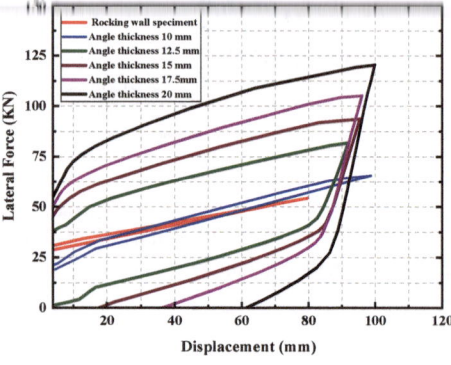

Fig. 11. Backbone curves.

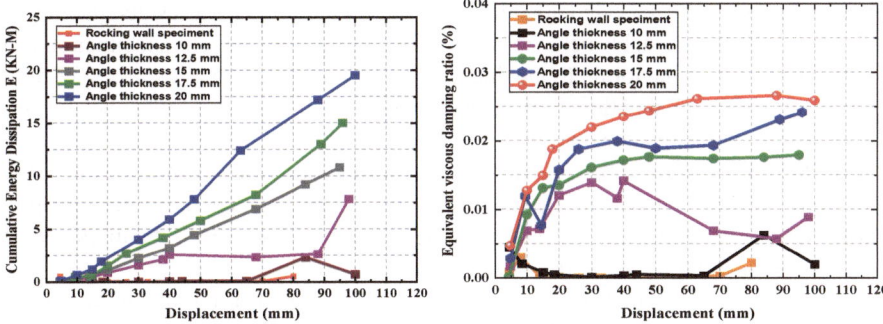

Fig. 12. Energy dissipation results.   **Fig. 13.** Equivalent viscous damping ratio.

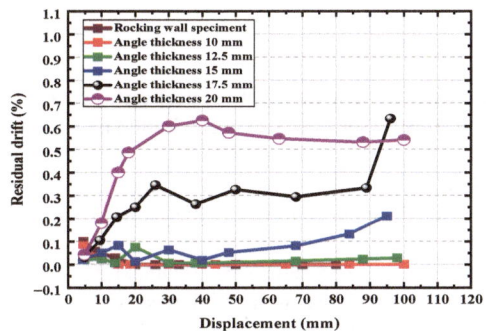

Fig. 14. Residual Drift ratio.

6   Summary and Conclusions

This study suggests using replaceable ED devices to create an earthquake-resistant RC wall. The structure and design of this innovative precast RC wall were presented. The post-tensioned wall was investigated. Steel-angle dampers were used on the wall to dissipate the energy. The numerical modeling samples included a control specimen without an energy dissipation damper and five specimens with steel angle dampers with different damper thicknesses were tested under cyclic loading. The main findings of this study are summarized as follows:

The control sample with no energy absorption damper under cyclic loading showed bilinear-elastic load-movement behavior. At the same time, the models equipped with steel angle dampers presented flag-shaped hysteresis loops. In these models, PT strands played the role of self-centering, and steel angle devices played the role of energy absorption in the system. The angle damper self-centering wall suggested in the paper is straightforward and clearly defined. It is possible to achieve the following: no damage to the primary wall structure, replaceable members, simple installation, and quick restoration of structure-function. The force-displacement curve for the suggested system showed an ordinary "flag-shaped" hysteretic response, and it has excellent energy dissipation and self-centering capabilities. Based on the numerical results, the damper provided a completely stable behavior without increasing damage. The specimen showed regular hysteresis behavior up to the thickness of 15 mm, but with thicknesses 17.5 mm and 20 mm, the residual of the sample increased significantly. At the same time, the models had excellent energy dissipation, load-carrying capacities, and lateral stiffness under cyclic loads. They could withstand large nonlinearities without additional damage to wall and angle dampers. Increasing the ED device thickness is beneficial to improving the energy absorption and the self-centering wall's lateral stiffness. Still, it simultaneously causes a significant residual drift.

The plastic deformation of the structure is concentrated in the energy absorption device. The proposed precast RC walls with ED steel angle will provide a promising solution for performance seismic-resisting structural systems suitable for resilient and sustainable civil infrastructure.

References

1. Ruiz-García, J., Miranda, E.: Evaluation of residual drift demands in regular multi-storey frames for performance-based seismic assessment. Earthq Eng \& Struct Dyn. **35**(13), 1609–1629 (2006)
2. Zhao, J., Shen, F., Si, C., Sun, Y., Yin, L.: Experimental investigation on seismic resistance of RC shear walls with CFRP bars in boundary elements. Int. J. Concr. Struct. Mater. **14**, 1–20 (2020)
3. Hassanein, A., Mohamed, N., Farghaly, A.S., Benmokrane, B.: Deformability and stiffness characteristics of concrete shear walls reinforced with glass fiber-reinforced polymer reinforcing bars. ACI Struct. J. **117**(1), 183–196 (2020)
4. Yuan, W., Zhao, J., Sun, Y., Zeng, L.: Experimental study on seismic behavior of concrete walls reinforced by PC strands. Eng. Struct. **175**, 577–590 (2018)
5. Kian, M.J.T., Cruz-Noguez, C.A.: Seismic design of three damage-resistant reinforced concrete shear walls detailed with self-centering reinforcement. Eng. Struct. **211**, 110277 (2020)
6. Nouraldaim, F.A., xiuxin, Y.W.: Finite Element Analysis of Reinforced Concrete Shear Walls [Internet]. Springer Nature Singapore, pp. 481–496 (2023). http://hdl.handle.net/2142/14203
7. Kurama, Y., Sause, R., Pessiki, S., Lu, L.W.: Lateral load behavior and seismic design of unbonded post-tensioned precast concrete walls. Struct. J. **96**(4), 622–633
8. Smith, B.J., Kurama, Y.C., McGinnis, M.J.: Behavior of precast concrete shear walls for seismic regions: comparison of hybrid and emulative specimens. J. Struct. Eng. **139**(11), 1917–1927 (2013)
9. Ricles, J.M., Sause, R., Peng, S.W., Lu, L.W.: Experimental evaluation of earthquake resistant posttensioned steel connections. J. Struct. Eng. **128**(7), 850–859 (2002)
10. Wang, B., Zhu, S.: Seismic behavior of self-centering reinforced concrete wall enabled by superelastic shape memory alloy bars. Bull. Earthq. Eng. **16**, 479–502 (2018)
11. Wang, B., Zhu, S., Zhao, J., Jiang, H.: Earthquake resilient RC walls using shape memory alloy bars and replaceable energy dissipating devices. Smart Mater. Struct. **28**(6), 65021 (2019)
12. Xu, L.H., Fan, X.W., Li, Z.X.: Development and experimental verification of a pre-pressed spring self-centering energy dissipation brace. Eng. Struct. **127**, 49–61 (2016)
13. Xu, L., Xiao, S., Li, Z.: Hysteretic behavior and parametric studies of a self-centering RC wall with disc spring devices. Soil. Dyn. Earthq. Eng. **115**, 476–488 (2018)

14. Xu, X., Tu, J., Cheng, G., Zheng, J., Luo, Y.: Experimental study on self-centering link beams using post-tensioned steel-SMA composite tendons. J. Constr. Steel Res. **155**, 121–128 (2019)
15. Cheok, G.S., Stone, W.C., Kunnath, S.K.: Seismic response of precast concrete frames with hybrid connections. Struct. J. **95**(5), 527–539 (1998)
16. Yagoub, N.F.A., Xuxin W.: Predicting the Performance of Shear Wall Structures Using the Confidence Nets Model [Internet]. Springer Nature Singapore, pp. 257–266 (2024). https://doi.org/10.1007/978-981-99-4045-5_22
17. Guo, T., Zhang, G., Chen, C.: Experimental study on self-centering concrete wall with distributed friction devices. J. Earthq. Eng. **18**(2), 214–230 (2014)
18. Weldon, B.D., Kurama, Y.C.: Experimental evaluation of posttensioned precast concrete coupling beams. J. Struct. Eng. **136**(9), 1066–1077 (2010)
19. Kurama, Y.C., Weldon, B.D., Shen, Q.: Experimental evaluation of posttensioned hybrid coupled wall subassemblages. J. Struct. Eng. **132**(7), 1017–1029 (2006)
20. Shen, Q., Kurama, Y.C.: Nonlinear behavior of posttensioned hybrid coupled wall subassemblages. J. Struct. Eng. **128**(10), 1290–1300 (2002)
21. Restrepo, J.I., Rahman, A.: Seismic performance of self-centering structural walls incorporating energy dissipators. J. Struct. Eng. **133**(November), 1560–1570 (2007)
22. Marriott, D., Pampanin, S., Bull, D., Palermo, A.: Dynamic testing of precast, post-tensioned rocking wall systems with alternative dissipating solutions. Bull. New Zeal Soc. Earthq. Eng. **41**(39), 90–103 (2008)
23. Li, X., Kurama, Y.C., Wu, G.: Experimental and numerical study of precast posttensioned walls with yielding-based and friction-based energy dissipation Eng. Struct. [Internet] **212**(March), 110391 (2020)
24. Chen, Z., Popovski, M.: Material-based models for post-tensioned shear wall system with energy dissipators. Eng. Struct. [Internet] **213**(September 2019), 110543 (2020). https://doi.org/10.1016/j.engstruct.2020.110543
25. Nazari, M., Sritharan, S.: Influence of different damping components on dynamic response of concrete rocking walls. Eng. Struct. [Internet] **212**(June 2019), 110468 (2020). https://doi.org/10.1016/j.engstruct.2020.110468
26. Sritharan, S.: JI and RSH 2011 Self-centering precast concrete walls for buildings in regions with low to high seismicity. Auckland

Shear Strength Check of the Stud in Lightweight Composite Bridge Deck Accounting for Overloaded Vehicles

Ning fei Huo(✉), Jiamu Zhang, Wei Yin, Jun Cheng, and Peishuo Liu

Beijing Institute of Structure and Environment Engineering, Beijing 100076, China
ahuoningfei@163.com

Abstract. This paper evaluates the shear strength of the steel stud connectors in the steel-UHPC lightweight composite bridge deck composed of the orthotropic steel deck with closed U-ribs. The interlaminar shear stress between the UHPC pavement and the steel plate of orthotropic steel deck in the new steel-UHPC lightweight composite bridge deck is computed using the finite element method, in which the influence of wheel loading conditions and overloaded vehicles are studied. Results show that the largest interlaminar transverse shear stress between UHPC pavement and steel plate is induced by the action of triple-axle loading. The interlaminar shear stress between the UHPC pavement and the steel plate will increase significantly under the action of the overloaded vehicles.

1 Introduction

The orthotropic steel bridge deck is the preferred bridge deck of large and medium span bridges at domestic and foreign countries [1]. Most of orthotropic steel bridge decks are covered with asphalt pavement. The integral stiffness of orthotropic steel bridge deck cannot be effectively improved due to the low stiffness and poor high temperature stability of asphalt pavement. Under the cyclic load of vehicles, this type of pavement is prone to pavement damage and fatigue cracks at the welds of the orthotropic steel bridge deck [2]. The traditional solution is mainly to improve the weld details and increase the thickness of the roof for the orthotropic steel bridge deck. However, the traditional method does not fundamentally solve the problem. The application of Ultra-high Performance Concrete (UHPC) provides a new direction for solving above problems.

Shao's research team [3] proposed a new steel-UHPC lightweight composite bridge deck structure that connects UHPC pavement with steel deck through stud connectors. The steel stud connectors are the key to realize the joint work of the orthotropic steel bridge deck and the UHPC pavement [4]. Previous studies focused on the asphalt pavement. In recent years, the interlaminar shear stress of the new steel-UHPC lightweight composite bridge deck structure have been widely concerned [5]. Taking the light composite deck of two bridges on Dongting Lake with opening ribs as the engineering background, Zhang et al. (2017) analyzed the influence of local wheel load on the shear stress between the UHPC layer and steel roof interface layer [3]. The results show that,

G. Feng (Ed.): ICCE 2023, LNCE 526, pp. 232–239, 2024.
https://doi.org/10.1007/978-981-97-4355-1_22

the shear strength of steel studs under the action of standard load vehicles could meet the requirements of static bearing capacity [5].

The interlaminar shear stress distribution of steel-UHPC light composite deck composed of orthotropic steel deck with closed ribs is more complex than that of composite deck with open ribs. This paper calculates the interlaminar shear stress distribution between the UHPC pavement and the steel roof in the steel-UHPC lightweight composite deck with closed ribs of a bridge in North China, and checks the shear strength of the steel stud connectors. Selecting the standard five axle overload vehicles specified in the "General code for design of highway bridges and culverts", we analyze the influence of triple axle wheel's load and overloaded vehicles' conditions on the calculation of interlaminar shear stress between UHPC pavement and steel roof.

2   Finite Element Model of Interlaminar Shear Stress Analysis for Light Composite Bridge Deck

2.1   The Establishment of Local Finite Element Model

This study adopts the local finite element model established by Deng et al. (2017) [6]. The model is based on a bridge in North China. In order to improve the disease resistance of the bridge deck pavement and increase the fatigue life of the steel bridge welds, the new pavement adopts a new steel-UHPC lightweight composite bridge deck as shown in Fig. 1.

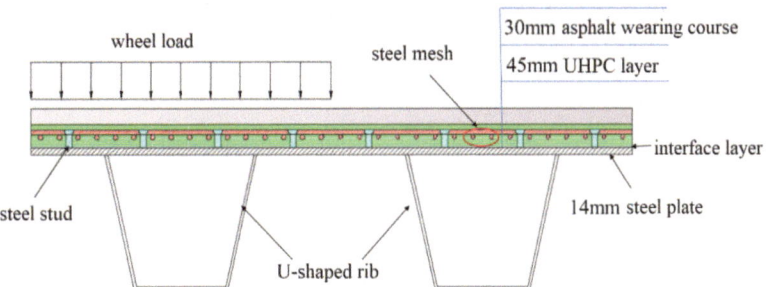

Fig. 1. The structure of new steel-UHPC lightweight composite bridge deck.

As shown in Fig. 2, the local finite element model of light composite bridge deck contains four diaphragms and five U-shaped ribs. The following assumptions are used in the calculation. The pavement layer is equivalent homogeneous isotropic elastomer. The bonding between pavement layers as well as between the pavement layer and the top plate of orthotropic steel deck are ideal. The displacement boundary conditions of the model are: restraining the displacement of x direction at the end of bridge deck, the displacement of y direction at the end of outer beam, and the displacement of z direction at the bottom of the beam.

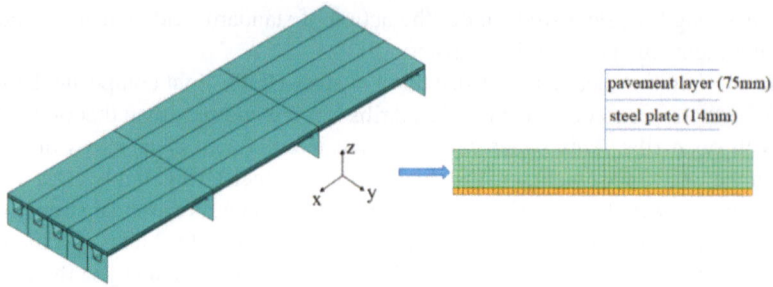

Fig. 2. Local finite element model of lightweight composite bridge deck.

2.2 Selection of Vehicle Load

Working condition 1: the calculation load of Deng et al. (2017) selects the standard five-axis vehicle specified in the "General Code for Design of Highway Bridges and Culverts". As shown in Fig. 3, the wheel load of each rear axle is 70 kN. The wheel load area is 0.2 m 0.6 m, and the vehicle wheel load is applied to the mid-span in Fig. 2. Deng et al. (2017) adopts three common transverse loading positions of the wheel: in-between-rib loading, riding-rib wall loading and over-rib loading. The loading methods are shown in Fig. 4. Because six-axis vehicles are increasing and the distance between the rear three axles in a six-axis vehicle is smaller than the distance between the two adjacent diaphragms, this study considers the effect of triple-axle wheel load on the interlayer shear stress of the new steel-UHPC light composite bridge deck structure. The wheel load of each rear axle is taken as 70 kN.

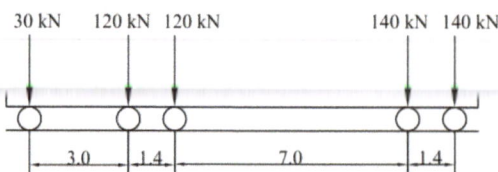

Fig. 3. The axle loads of standard five-axis vehicle (the length unit is in meter).

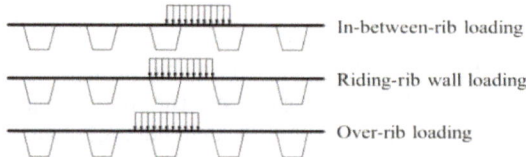

Fig. 4. Three transverse positions of wheel loading.

Working condition 2: performing a statistical analysis of vehicle load based on the WIM (Weigh-In-Motion) system. As shown in Fig. 5, the single axle load of the rear axle of five-axis overload vehicle is 252 kN, and the wheelbase is 1.4 m. According to the relative position between the wheel moving load centerline and the reference diaphragm, each wheel load condition moves forward with 100 mm per step.

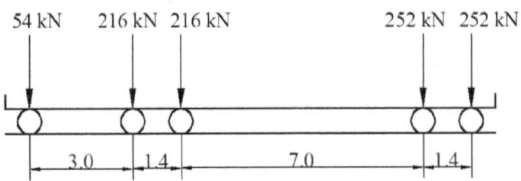

Fig. 5. The axle loads of overloaded five-axis vehicle (the length unit is in meter).

3 The Interlaminar Shear Stress of Steel-UHPC Light Composite Bridge Deck

Based on different wheel load forms and overloaded vehicles, we calculate the maximum interlayer shear stress between UHPC pavement and steel roof in the new steel UHPC light composite deck structure.

3.1 The Influence of Triple-Axle Wheel Load on Interlaminar Shear Stress

The results of Deng et al. (2017) show that the riding-rib wall loading generates the largest interlaminar shear stress, which is the most unfavorable transverse position of wheel loading. After analyzing the most unfavorable transverse position of wheel loading, we further study the influence of triple-axle wheel loads on the interlayer shear stress of the new steel-UHPC light composite bridge deck structure. Figure 6 shows the interlayer shear stress under triple axle load at the selected most unfavorable transverse position of wheel loading. As shown in Fig. 6, when the standard load vehicle is under triple axle wheel load, the interlayer shear stress continued to increase after the wheel load position 1.4 m away from the diaphragm and reached the maximum at the mid-span of the two diaphragms.

Table 1 shows the peak values of interlaminar shear stress of steel-UHPC lightweight composite bridge deck structure under different wheel loading methods. As shown in Table 1, the wheel loading method has a great influence on the maximum interlaminar shear stress. The maximum interlaminar shear stress generates under the action of triple axle load. Taking the maximum transverse shear stress between layers as an example, the maximum transverse shear stress between layers under the action of triple axle load is 1.2 times that of double axle load and 1.6 times that of single axle load, which indicates the necessity of considering triple axle load.

3.2 The Interlaminar Shear Stress of Steel-UHPC Light Composite Bridge Deck Under the Action of Overloaded Vehicles

We conduct the statistical analysis of vehicle load based on the WIM system. At the selected most unfavorable transverse loading position, the maximum interlaminar shear stress of different positions under biaxial load of overloaded vehicle are shown in Fig. 7.

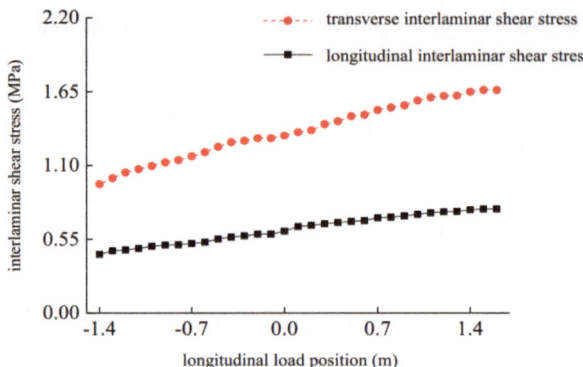

Fig. 6. The maximum interlaminar shear stress of different positions under triple axle load of standard vehicle.

Table 1. Peak values of interlaminar shear stress of Steel-UHPC lightweight composite bridge deck structure under different wheel loading methods.

	Single axle load	Double axle load	Triple axle load
The maximum transverse shear stress between layers/MPa	1.040	1.352	1.663
The maximum longitudinal shear stress between layers/MPa	0.522	0.658	0.776

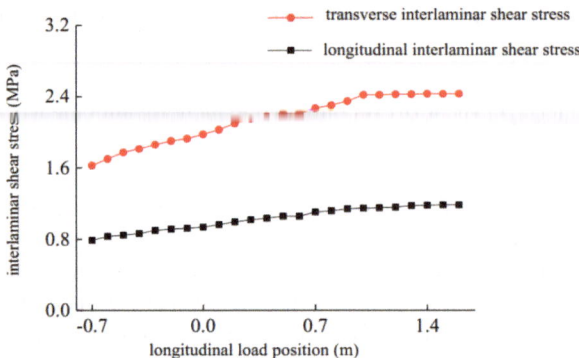

Fig. 7. The maximum interlaminar shear stress of different positions under biaxial load of overloaded vehicle.

According to results of Fig. 7, under the action of double axle load of overloaded vehicle, the change of interlaminar shear stress of steel-UHPC light composite bridge deck structure is the same as that under the action of double axle load of standard vehicle. After the wheel load position of 0.7 m away from the diaphragm, the value of the interlayer shear stress continues to increase due to the wheel load changing from single axle load to biaxial axle load, and reaches a maximum value at the mid-span of the two diaphragms.

Table 2 shows peak values of interlaminar shear stress of steel-UHPC lightweight composite bridge deck under the action of overloaded vehicles. The maximum shear stress between layers increases significantly under the action of overloaded vehicles. Taking the maximum transverse shear stress between layers as an example, compared with the double axle load of standard vehicle, the maximum transverse shear stress between layers under the action of the double axle load of overloaded vehicle increases by 80%. The shear stress on the studs increases significantly under the action of overloaded vehicles. Therefore, in order to ensure the effective connection between the UHPC layer and the steel roof, it is important to check whether the studs meet the interlayer shear requirements under the action of overloaded vehicles.

Table 2. Peak values of interlaminar shear stress of steel-UHPC lightweight composite bridge deck under the action of overloaded vehicles.

	Double axle load	Triple axle load	Double axle load
The maximum transverse shear stress between layers/MPa	1.352	1.633	2.435
The maximum longitudinal shear stress between layers/MPa	0.658	0.776	1.185

4   The Shear Strength Check of Steel Stud Connectors

According to the calculation results of Deng et al. (2017) [6], the shear strength of the studs corresponding to its shear force capacity is 1.84 MPa resulting from the current "Code for Design of Steel-Concrete Composite Bridges" of China.

It can be seen from Table 2 that the studs meet the strength requirements under standard vehicle load. However, under double-axle load of the overloaded vehicle, the maximum transverse shear stress between the UHPC pavement and the steel roof is greater than the maximum shear stress corresponding to the shear bearing capacity of studs. Then studs have the potential to be sheared. Therefore, it is necessary to restrict the passage of overloaded vehicles to meet the strength conditions of the stud connectors, and to ensure that the steel-UHPC light composite bridge deck structure can effectively work.

5   Conclusion

By considering the triple axle wheel load and overload vehicles, this study calculates the interlaminar shear stress of steel-UHPC light composite bridge deck that composed of orthotropic steel deck with U-shaped ribs. The following conclusions can be drawn from the calculation results in this paper.

1) The loading mode has a great influence on the interlayer shear stress calculation of the light composite bridge deck structure. The maximum interlayer shear stress is generated under the action of the triple axle load. Taking the maximum transverse shear stress between layers as an example, the maximum transverse shear stress between layers under the action of triple axle load is 1.2 times that of double axle load and 1.6 times that of single axle load. Therefore, the interlaminar shear stress in the steel-UHPC lightweight composite bridge deck structure generated by the vehicle load will be greatly underestimated when considering only the single-axle load.

2) The interlayer shear stress of the light composite bridge deck structure increases greatly under the action of overloaded vehicles. Taking the maximum transverse shear stress between layers as an example, compared with the double axle load of standard vehicle, the maximum transverse shear stress between layers under the action of the double axle load of overloaded vehicle increases by 80%.

3) According to the calculation results of the maximum interlaminar shear stress of the new steel-UHPC light composite deck structure under different load conditions, the maximum interlaminar transverse shear stress under the action of overloaded vehicles is higher than the shear strength limit corresponding to the shear capacity of the studs in the steel-UHPC light composite deck. Therefore, it is necessary to restrict the passage of overloaded vehicles.

Acknowledgments. This work is supported by the National Natural Science Foundation of China (Grant Nos. 12173062, 52305565), the Stably Supports Scientific Research Projects of the Technology and Industry for National Defense of China (WDZC702B2023303).

References

1. Tong, L.W., Shen, Z.Y.: Fatigue assessment of orthotropic steel bridge decks. Chin. Civil Eng. J. 03, 16–21 (2000). (In Chinese)
2. Huang, W.: Design of deck pavement for long-span steel bridges. Chin. Civil Eng. J. **40**(9), 65–77 (2007). (In Chinese)
3. Zhang, C.C., Zhang, S.H., Guo, X.G., et al.: Calculation of interlaminar shear stress of lightweight composite bridge and preliminary study on layout design of stud connectors. Steel Constr. **34**(3), 62–69 (2017). (In Chinese)
4. Yang, B.: Shear Performance of Studs of Composite Deck System Composed of Steel and CRRPC Layer. Hunan University, Changsha (2016). (In Chinese)
5. Zhang, S.H., Shao, X.D., Huang, X.J., et al.: Static and fatigue behavior of small stud shear connector for lightweight composite bridge deck. J. Highw. Transp. Res. Develop. **33**(11), 41–49 (2016). (In Chinese)
6. Deng, M., Huo, N., Shi, G., et al.: Shear strength analysis of the stud in steel-UHPC composite bridge deck **100**(1), 012176 (2017)

Optimized Fertilization's Beneficial Impact on Soil Nutrient Levels and Its Influence on the Principal Agronomic Traits of Maize

Enjun Kuang[1], Baoguo Zhu[2], Jiuming Zhang[1(✉)], Yingxue Zhu[1], Jiahui Yuan[1], Xiaoyu Hao[1], and Lei Sun[1]

[1] Heilongjiang Academy of Black Soil Conservation and Utilization, Key Laboratory of Black Soil Protection and Utilization, Ministry of Agriculture and Rural Areas, Harbin 150086, China
zjm_8049@163.com

[2] Jiamusi Branch of Heilongjiang Academy of Agricultural Sciences, Jiamusi 154007, Heilongjiang, China

Abstract. The soil nutrients, main agronomic indexes, and key yield factors during the maize growth period were studied by using a nutrient expert system to recommend optimal fertilization. The results showed that the fertilizer application had a significant impact on the growth and yield formation of maize, especially the N fertilizer application. N deficiency in maize seriously affected the yield and its component factors, while P and K deficiency had no significant effect on decreasing maize yield. The optimized fertilization treatment (NE) significantly reduced the bald tip length and increased the ear length and the yield of maize by 10.2% compared with the conventional fertilization treatment (FP). NE significantly increased the utilization rate of N, P, and K in maize, which was 20.1%, 12.4%, and 45.4% higher than that of FP. The trend of fertilizer N deficiency was opposite to NE, but P and K deficiency were not obvious. Compared with FP, soil organic matter of NE did not change significantly, and the pH value was increased. Nitrate nitrogen and ammonium nitrogen were decreased by 51.9% and 3.9%, respectively. A significant correlation between maize yield and alkali hydrolyzed nitrogen, organic matter. In conclusion, the optimized fertilization treatment had obvious effects on the growth indicators and yield components of maize, improving the fertilizer utilization rate, and providing technical support for rational fertilization of maize.

Keywords: Maize; Fertilizer Deficiency · Yield · Fertilizer Application Efficiency

1 Introduction

N, P, and K are the major elements restricting the growth and yield formation of maize [1]. The amount absorbed by maize affects the utilization rate of N, P, and K fertilizers. At present, the excessive application of chemical fertilizer not only causes a waste of

---

E. Kuang and B. Zhu—contributed equally to this work.

© The Author(s) 2024
G. Feng (Ed.): ICCE 2023, LNCE 526, pp. 240–249, 2024.
https://doi.org/10.1007/978-981-97-4355-1_23

resources but also poses a potential threat to the environment. There is a certain space for saving fertilizer when applying N, P, and K fertilizers [2]. The Sanjiang Plain is the main production area of spring corn in Heilongjiang province [3]. Maintaining the maize yield is of great significance for food security and stability in China [4]. However, to pursue high yield, the Sanjiang Plain has a high amount of fertilization and a serious phenomenon of blind fertilization, resulting in an imbalance in the proportion of soil N, P, and K, and limiting corn yield [5]. In the northeast black soil area, the yield will not be reduced by reducing the application of P fertilizer by 20%, and the partial productivity of P fertilizer will increase by 17.0%-21.6% [6]. There is no significant difference in corn yield when the straw is used to replace 30% and 60% K fertilizer [7]. Optimizing fertilization can reduce fertilizer input, stabilize production and increase production, reduce environmental pollution, and improve the utilization rate of N, P, and K fertilizers. Studies have shown that reasonable fertilization is conducive to improving the photosynthetic rate of plant leaves [8], prolonging the photosynthetic time, and significantly improving the yield and its indicators [9]. Under the wheat-maize rotation system, the maize yield of optimized fertilization increased by 5.3% [10]. Winter wheat of optimal fertilization in Weibei dryland increased wheat yield, reduced nitrate nitrogen accumulation in the soil profile, and improved nitrogen utilization rate [11].

How to effectively maintain the yield and control the amount of fertilizer is an urgent problem to be solved in agricultural production. This study recommended fertilization according to the nutrient expert system, and on this basis, weight loss, and clear the impact of different fertilizer elements on soil and corn yield, to provide basic data for the gradual realization of cost reduction and efficiency increase in agriculture.

2　Materials and Methods

2.1　Overview of the Test Site

The test site is located in the Agricultural Extension Center of Friendship Farm in Sanjiang Plain, (46.7548° N, 131.8498° E). The soil type is typical black soil, belonging to the continental monsoon climate of the cold temperate zone. The annual average temperature is 2.8 °C, the effective accumulated temperature is 2 170-2 700 °C, the annual precipitation is 512.4 mm, and the frost-free period is 120–130 days. The soil type is black soil, with organic matter content of 20.8 g kg^{-1}, an available phosphorus content of 23.86 mg kg^{-1}, an alkali hydrolyzed nitrogen content of 64.0 mg kg^{-1}, a pH of 6.16, an ammonium nitrogen content of 11.86 mg kg^{-1}, the nitrate nitrogen content of 8.42 mg kg^{-1}, and available potassium content of 169.60 mg kg^{-1}.

2.2　Experimental Design

The optimization fertilization experiment in 2021 and 2022 adopted the random block arrangement design, and a total of five treatments are set: nutrient expert-recommended fertilization system (NE), N deficiency treatment (NE-N), P deficiency treatment (NE-P), K deficiency treatment (NE-K) and conventional fertilization (FP). The fertilizer amount were shown in Table 1. The maize variety was Tianhe 2, and the sowing date was in the middle of May. During the whole growth period of maize, natural precipitation was the main irrigation practice.

Table 1. Effect of different fertilization treatments on nitrogen fertilizer utilization rate

Treatments	Fertilization time	N amount (kg ha^{-1})	P amount (kg ha^{-1})	K amount (kg ha^{-1})
FP	Base fertilizer (before sowing)	58.3	69.0	50
	Top dressing (jointing stage)	116.8		
NE	Base fertilizer (before sowing)	75.0	50.6	41.5
	Top dressing (jointing stage)	65.8		
	Top dressing (tasselling stage)	46.9		
NE-N	Base fertilizer (before sowing)	—	50.6	41.5
NE-P	Base fertilizer (before sowing)	75.0	—	
	Top dressing (jointing stage)	65.8		
	Top dressing (tasselling stage)	46.9		
NE-K	Base fertilizer (before sowing)	75.0	50.6	—
	Top dressing (jointing stage)	65.8		
	Top dressing (tasselling stage)	46.9		

2.3 Sample Collection and Determination Method

Soil sample collection: Five points were selected in different treatments according to the "S" shape before corn harvest, using a soil drill to collect 0–20 cm and 20–40 cm soil layers and mixed totally into a sample respectively. Picked out straw residues and small stones. The air-dried soil samples were extracted with 1 mol L^{-1} potassium chloride solution (the ratio of soil to liquid is 1:10) by shaking for 1 h. After filtering, use a continuous flow analyzer to determine nitrate nitrogen and ammonium nitrogen. Other soil nutrients were determined by conventional methods.

Yield measurement: in the mature period, the sampling area of each plot was 13 m^2, and another 10 plants were taken for indoor planting to calculate the yield.

2.4 Data Analysis

N utilization rate (NUE,%) = (N accumulation amount of plants in N application area at maturity - N accumulation amount in non N application area)/N application amount × 100;

The utilization rate of P fertilizer (PUE,%) = (P accumulation amount of plants in P application area at maturity – P accumulation amount in non-P application area)/P application amount × 100;

Utilization rate of K fertilizer (KUE,%) = (K accumulation of plants in K application area at maturity - K accumulation in non K application area)/K application amount × 100;

The SPSS 17.0 (SPSS, Inc., Chicago, IL, USA) analytical software package was used for all the statistical analyses. Single-factor analysis of variance (ANOVA) and least significant range (LSD) was used to test the different significance at the 5% level. Pearson correlation coefficients of maize yield and other indexes.

3 Results

3.1 Maize Yield and Its Components

The yield was more sensitive to any deficit in N, P, and K fertilizer, especially the fertilizer N. As shown in Table 2, the ear length of maize in the NE-N treatment was the shortest and the FP treatment was the longest, the ear length of NE-N was 41.7% lower than that in the NE treatment ($P < 0.05$), while NE-P and NE-K were slightly lower than NE treatment. The bald tip value of NE-N was 2.15 cm at the highest level and NE treatment was 0.2 cm at the lowest level, while the value of NE-P, NE-K, and FP treatments did not reach a significant difference higher than NE. In terms of yield, the yield of the NE-N treatment was the lowest one which was significantly different from other treatments, only 3 156 kg ha^{-1}, and the FP, NE-P, and NE-K treatments were lower than NE at 9.3%, 78.0%, and 6.4%, respectively. The NE-N treatment showed the lowest yield, the highest bald tip length, and the ear length the lowest which indicated that the yield and its indicators were seriously affected by fertilizer N deficiency, and the yield reduction of fertilizer P deficiency was the minimum, following by fertilizer K deficiency.

Table 2. Effect of different fertilization treatments on nitrogen fertilizer utilization rate

Treatment	Ear length (cm)	Bare tip of the ear (cm)	Row number per ear	Hundred seeds weight (g)	Yield (kg hm^{-2})	NUE (%)	PUE (%)	KUE (%)
NE	18.85 ± 1.03ab	0.2 ± 0.41b	16 ± 1.48a	50.02 ± 3.18b	14314 ± 622.8a	27.28 ± 1.36a	30.99 ± 2.96a	66.50 ± 4.14a
NE-N	13.3 ± 0.75d	2.15 ± 0.71a	15.1 ± 3.84a	32.89 ± 1.19c	3156 ± 193.6c	—	—	—
NE-P	17.2 ± 0.94c	0.35 ± 0.35b	15.8 ± 1.33a	51.31 ± 2.15ab	13882 ± 535.4ab	—	—	—
NE-K	18.35 ± 1.53ab	0.6 ± 0.46b	15.2 ± 1.4a	49.43 ± 3.19b	13401 ± 839.7ab	—	—	—
FP	19.00 ± 0.91a	0.3 ± 0.48b	15.6 ± 0.84a	52.66 ± 4.45ab	12987 ± 681.4b	22.70 ± 0.65b	27.58 ± 0.74b	45.70 ± 1.58b

a The values are mean standard deviation.
b Different lowercase letters represent a very significant difference ($P < 0.05$).

3.2 Effect of Different Treatments on Soil Nutrients, Nitrate Nitrogen, and Ammonium Nitrogen

It was shown in Fig. 1 that the nitrate nitrogen content of NE-N was the lowest at 0–20 cm and 20–40 cm soil layers, only 1.57 mg L^{-1} and 1.79 mg L^{-1}, and FP treatment was the highest at 6.56 and 5.67 mg L^{-1}, which had significant differences from other treatments. However, it was not different among NE-P, NE-K, and NE treatments. Nitrate nitrogen content was closely related to the amount of nitrogen fertilizer applied, for which N deficiency had a lower nitrate nitrogen content. Ammonium nitrogen is mostly adsorbed and fixed by soil particles, with less leaching in soil. Ammonium nitrogen content in NE was the highest at 0–20 cm and 20–40 cm layers. With the deeper of the soil layer, ammonium nitrogen content decreased slightly. At 0–20 cm, the content of ammonium nitrogen of NE-K was the lowest, 28.6% lower than NE while 18.6% of NE-N was lower than NE. At 20–40 cm, the content of ammonium nitrogen in NE was 24.5% higher than that in FP, with no significance.

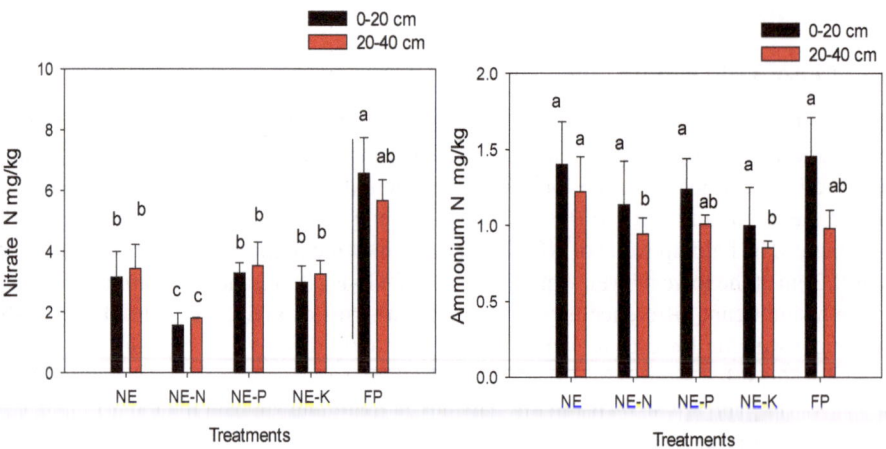

Fig. 1. Ammonium nitrogen and nitrate nitrogen under different treatments (Different lowercase letters indicate the significance analysis at the level of 0.05 between different treatments at the same level)

Fertilizer deficiency could decrease organic matter content at 0–20 cm soil layer, except for K deficiency which showed in Fig. 2. All the treatments had no obvious effect at the 20–40 cm soil layer. The content of organic matter of NE-N was the lowest and had a significant difference with NE ($P < 0.05$). The content pH did not fluctuate heavily, and NE treatment increased the pH value of both 0–20 cm and 20–40 cm soil layer, followed by NE-P as the lowest. There was no significant difference among NE-P, NE-K, and FP treatments.

The content of available nutrients in all treatments was higher at 0-20cm than in the 20–40 cm soil layer (Fig. 2). AN of NE-K and FP treatments were the highest at 0–20 cm soil layer, and the NE-N was the lowest one. The content of AP in FP was the highest at

0–20 cm and 20–40 cm, and the NE treatment was the lowest. The NE-N, NE-P, NE-K, and FP were 3.3%, 10.6%, 26.1%, and 46.6% higher than the NE treatment, respectively.

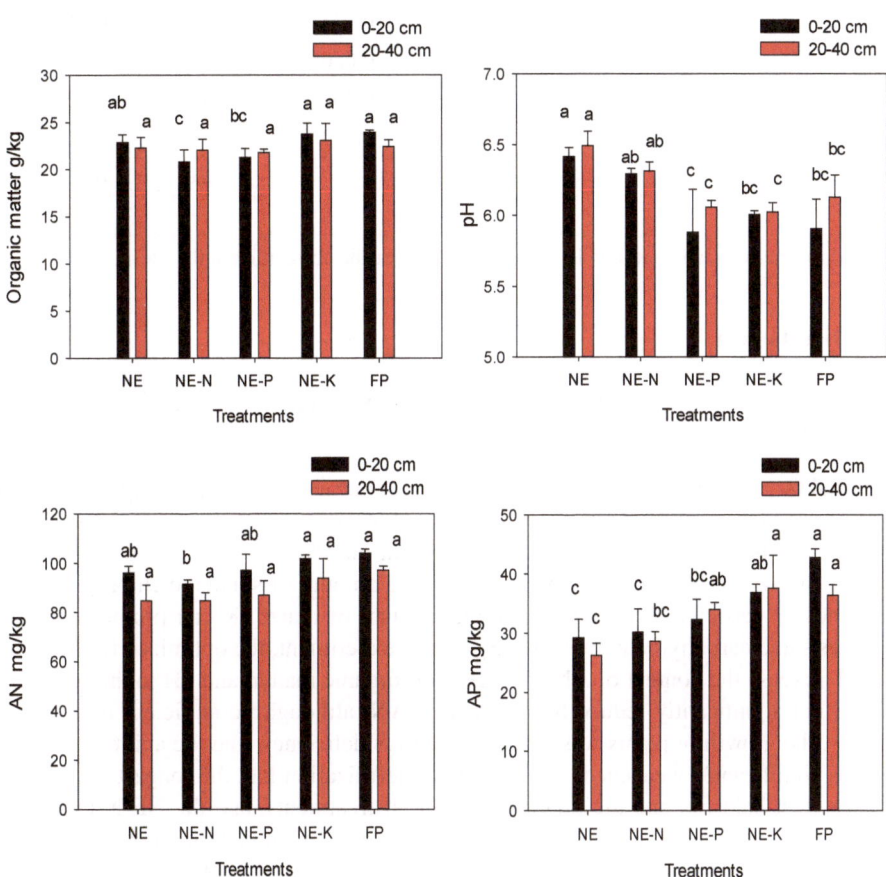

Fig. 2. Soil nutrient under different treatments (Different lowercase letters indicate the significance analysis at the level of 0.05 between different treatments at the same level)

3.3   The Relationship Between Yield and Other Indicators

It could be seen from Table 3 that the yield had a positive correlation with AN and organic matter. Nitrate nitrogen had a very significant positive correlation with AP, and a significant positive correlation with AN.

Table 3. Person correlation among yields and other indicators

	A-N	O-P	NO$_3$$^-$-N	NH$_4$$^+$-N	O.M	Yield
A-N	1.000	0.551*	0.632*	0.124	0.757**	0.514*
O-P		1.000	0.664**	0.086	0.526*	0.284
NO$_3$$^-$-N			1.000	0.462	0.566*	0.509
NH$_4$$^+$-N				1.000	0.004	0.163
O.M					1.000	0.517*
Yield						1.000

*Stand for the correlation at 0.05 level, and **stands for the correlation at 0.01 level.

4 Discussion

4.1 Effects of Optimized Fertilization on Soil Nutrients, Yield, and Fertilizer Utilization

Optimal fertilization can promote the matching of nutrient demand and supply of crops to promote high yield. N, P, and K are all essential nutrients for crop growth [12]. N is the material basis for plant growth and physiological metabolism. The net photosynthetic rate of maize can be improved by increasing the amount of synthetic chlorophyll and enzymes. P participates in the energy metabolism of maize. K can promote protein synthesis and carbohydrate transfer [13]. In this experiment, the optimized fertilization could increase the content of AN, AP, AK, soil organic matter, and pH in the soil. NE-N did not significantly reduce the content of AN, although no sufficient N fertilizer supply, the growth of plants was reduced due to N deficiency, and the ability to absorb N was significantly weakened. The optimized fertilization had the longest ear length, the shortest bald tip, and the highest yield in all treatments, and the NE-N treatment was the lowest. One of the reasons for the P deficiency without reducing yield was the amount of phosphorus applied in the local custom was higher than that in the optimized fertilization, and a large amount of phosphorus was fixed in the soil.

The fertilizer utilization rate is a representation of the situation in that the fertilizer input in crop production is absorbed and utilized by crops. In 1998, the utilization rate of N, P, and K fertilizers for major grain crops in China ranged from 30% to 35%, from 15% to 20%, and 35% to 50%, respectively [14]. Twenty years later, the utilization rate of N, P, and K fertilizers for major grain crops in China has gradually declined, and the utilization rates of N, P, and K fertilizers range from 10.8% to 40.5%, 7.3% to 20.1% and 21.2% to 35.9%, respectively. In recent years, it has become a common understanding to reduce the amount of fertilizer and improve the practice of fertilization. According to the survey of corn fields in many cities and counties in Heilongjiang Province in recent 20 years, the utilization rates of N, P, and K fertilizers have been improved to varying degrees, 22.3%-50.7%, 5.1%-37.6%, and 26.3%-76.4%, respectively [15]. In this experiment, the utilization rate of N, P, and K in NE was higher than that of FP. Similarly, in Heilongjiang Province, the optimized fertilization of maize carried out in Shuangcheng was 41.8%, 15.5%, and 48.8% of N, P, and K application [9], and the utilization rate of N is significantly higher than the results of this experiment, which may be because the AN content in Shuangcheng was three times that of this experiment. The optimized fertilization treatment with light simplified one-time fertilization can significantly increase the yield of maize, and at the same time improve the utilization rate of nitrogen, phosphorus, and potassium fertilizer, which was 12.8%, 1.8%, and 2.9% higher than the farmers' conventional fertilization. More importantly, it was a reasonable and effective practice to reduce nitrogen according to the local climate, soil, and production conditions, and to conFig.feasible fertilization technology so that the applied nitrogen fertilizer can be fully absorbed and utilized by crops [16].

4.2 Effects of Optimized Fertilization on Nitrate Nitrogen and Ammonium Nitrogen

Nitrate nitrogen was a former of nitrogen loss in soil moved down with water which tends to increase with the increasing amount of nitrogen [17]. Here, we described the NE treatment and applied a small amount of N fertilizer repeatedly to avoid the loss of water and fertilizer caused by too much N fertilizer. On the other hand, it could also supply N continuously, effectively avoiding the leaching loss of N caused by the one-time mass application and high rainfall intensity [18]. Zheng et al. studied how to reduce N and apply different proportions of organic fertilizer at the same time to ensure maize yield in Shanxi Loess and found that the content of nitrate nitrogen was lower after applying slow-release fertilizer, reducing the amount of potential leaching nitrogen [19]. Under the same amount of nitrogen fertilizer application, coated urea could slowly and continuously release nitrogen. Nitrate nitrogen tends to be low first, then high, and then low throughout the growth period, and remains at an environmentally friendly level. In this experiment, the nitrate nitrogen content of the local farmers used to apply fertilizer was the highest, and the NE-N content was the lowest. Reducing the application of nitrogen fertilizer can significantly reduce the nitrate nitrogen content of the soil. P and K deficiency had little impact on nitrate nitrogen. The soil ammonium nitrogen in NE treatment was the highest in 0–40 cm, and most of it was adsorbed and fixed in soil particles after entering the soil. Only when the amount of nitrogen fertilizer was high, the soil adsorption of ammonium nitrogen reached saturation.

5 Conclusion

The application of fertilizer has a significant impact on the growth and yield formation of maize, especially the application of N fertilizer. A significant correlation between maize yield and AN, organic matter. The optimized fertilization treatment had a good effect on the growth indicators and yield components of maize, which could significantly reduce the bald tip length, increase the ear length of maize, and increase the yield. N deficiency could affect maize yield and its indicators seriously while P and K deficiency was not significant. Based on reducing the application of chemical fertilizer, the optimized fertilization could also improve the utilization rate of N, P, and K, and the yield increase effect was obvious. In summary, it was a reasonable practice to reduce fertilizer by combining local climate conditions and production conditions with fertilizer application technology, then and improving fertilizer utilization.

Acknowledgments. This work was supported by the National Key Research and Development projects (2021YFD1500202), Agricultural Science and Technology Innovation Leap project of the Heilongjiang Academy of Agricultural (CX23GG08, HNK2019CX13), Key Scientific and Technological Project of Heilleading project of the Chinese Academy of Sciences (XDA28100400), Project of soybean industry system (CARS-04-PS17).

References

1. Sun, W.T., Wang, R., An, J.W., Xing, Y.H., Wang, X.Z.: Study on the effect of balanced fertilization technology on the yield of corn. J. Maize Sci. **16**(3), 109 (2008)
2. Wang, Y.N., Mi, G.H.: Fertilizer application in maize production in northern China: current status and fertilization optimal potential. J. Maize Sci. **29**(3), 151 (2021)
3. Wang, L.L., Gu, H.J., Shi, Y.L., Wang, T.T.: Soil nitrogen transformation and corn yield as affected by a combination of urea and fertilizer additive NAM in San-Jiang Plain of China Soil and fertilizer sciences in China. **2**, 34 (2012)
4. Wang, Y.J., Lv, Y.J., Liu, H.T., Bian, S.F., Wang, L.C.: Integrated management of high-yielding and high nutrient efficient spring maize in northeast China. Scientia Agricultura Sinica **52**(20), 3533 (2019)
5. Cai, H.G., Mi, G.H., Zhang, X.Z., Ren, J., Feng, G.Z., Gao, Q.: Effect of different fertilizing methods on nitrogen balance in the black soil for continuous maize production in Northeast China. Plant Nutr. Fertilizer Sci. **18**(1), 89 (2012)
6. Wu, Q.H., Liu, X.B., Zhang, S.X., Yin, C.X., Li, G.H., Xie, J.G.: Application of 80% of routine phosphorus rate to keep high yield and P efficiency of maize and P balance in soil. Plant Nutr. Fertilizer Sci. **22**(6), 1468 (2016)
7. Fu, W., et al.: Spring maize yield and soil potassium balance under replacement of potassium with straw in karst peak cluster depression. Chin. J. Eco-Agric. **25**(12), 1823 (2017)
8. Gu, Y., Hu, W.H., Xu, B.J., Wang, S.Y., Wu, C.S.: Effects of nitrogen on photosynthetic characteristics and enzyme activity of nitrogen metabolism in maize under-mulch-drip irrigation. Acta Ecol. Sin. **33**(23), 7399 (2013)
9. Ji, J.H., Li, Y.Y., Liu, S.Q., Tong, Y.X., Liu, Y., Zhang, M.Y.: Optimal fertilization of maize production in black soil region of Heilongjiang Province-a case study in Shuangcheng City. Soil Crop **4**(2), 64 (2015)
10. Yang, H.M., et al.: Effects of the optimized fertilization on yield, nutrient balance, and eco-environmental benefits of wheat-maize rotation system. Chin. J. Eco-Agric. **31**(5), 699 (2023)
11. Cao, H.B., et al.: Optimization of nitrogen fertilizer recommendation technology based on soil test for winter wheat on Weibei Dryland. Scientia Agricultura Sinica **47**(19), 3826 (2014)
12. Berkhout, E.D., Malan, M., Kram, T., Islam, R.: Better soils for healthier lives an econometric assessment of the link between soil nutrients and malnutrition in Sub-Saharan Africa. PLoS One **14**(1), e0210642 (2019)
13. Wu, Y.L., et al.: Effects of nitrogen fertilizer on leaf chlorophyll content and enzyme activity at late growth stages in maize cultivars with contrasting tolerance to low nitrogen. Acta Pratacul. Sin. **26**(10), 188 (2017)
14. Ju, X.T., Gu, B.J.: Status-quo, problem and trend of nitrogen fertilization in China. J. Plant Nutr. Fertilizer **20**(4), 783 (2014)
15. Ji, J.H., et al.: Changes in yield and fertilizer use efficiency of spring maize in Heilongjiang over twenty years. J. Agric. Resour. Environ. **39**(06), 1099 (2022)
16. Ju, X.T., Zhang, C.: The principles and indicators of rational N fertilization. Acta Pedol. Sin. **58**(1), 1 (2021)

17. Wang, L.C., Zhao, L.P., Zhu, P., Gao, H.J., Peng, C.: Effects of different fertilizer application regimes on NO3–N and NH4+-N in black soil during spring maize growing season. J. Northeast Forestry University **37**(12), 85 (2009)
18. Zou, X.J., Zhang, X., An, J.W.: Effect of reducing and postponing of N application on yield, plant N uptake, utilization and N balance in maize. Soil Fertilizer Sci. China. (6) 25 (2011)
19. Zheng, L.F., Wu, S.D., Dang, T.H.: Effect of different fertilization modes on spring maize yield, water use efficiency and nitrate nitrogen residue. J. Soil Water Conserv. **33**(4), 221 (2019)

Research on the Clamping Force Performance of Cable Clamps on Long-Span Suspension Bridges Under Solar Radiation

Yue Cai[1]($\boxtimes$), Zhongchu Tian[1,2], Guibo Wang[3], Hao Zeng[4], and Cheng Liu[5]

[1] School of Civil Engineering, Changsha University of Science and Technology,
Changsha 410114, Hunan, China
caiyue_stu@163.com

[2] School of Civil Engineering, Fujian University of Science and Technology, Fuzhou 350118,
Fujian, China

[3] China Railway Construction Bridge Engineering Bureau Group First Engineering Co., Ltd.,
Dalian 116000, Liaoning, China

[4] China Railway Construction Port and Navigation Group Co., Ltd., Zhuhai 519070,
Guangdong, China

[5] Guizhou Provincial Transportation Construction Project Quality Supervision and Law
Enforcement Detachment, Guiyang 55002, Guizhou, China

Abstract. The anti-slip performance of cable clamps is one of the important structures to ensure the safety of suspension bridges, and there is currently limited research on the influence of temperature. by using validated meteorological data, an ABAQUS thermal coupling model is established to obtain the steady-state temperature field for a representative period and apply it as a mapped field to analyze the stress results of the cable clamp-main cable structure. The analysis concludes that the derived analytical solution in this study is close to the numerical solution of the finite element method, with a maximum error of 4.2%, meeting the requirements of practical engineering. The increase in temperature of the screw itself reduces the pre-tensioning force, while the increase in structural temperature, when the screw does not experience temperature variation, can actually increase the friction force of the cable clamp to some extent, which is beneficial for improving the friction resistance of the cable clamp. During the installation of cable clamps, it is recommended to perform it at low temperatures, and the surface of the screw can be thermally insulated to maximize the increase in friction force caused by structural temperature rise.

Keywords: Cable Clamp · Solar Radiation · Temperature Field · Thermal Coupling · Friction Resistance

1 Introduction

As a crucial connecting component between the hanger cable and the main cable, the cable clamp mainly improves its anti-slip ability by increasing the friction force with the main cable through bolt fastening [1, 2]. Once the cable clamp slips, it will cause

G. Feng (Ed.): ICCE 2023, LNCE 526, pp. 250–260, 2024.
https://doi.org/10.1007/978-981-97-4355-1_24

cracking and damage to the coating near the clamp and straight seams, redistribute the internal force of the main cable, change the line shape of the main beam, affect the safety of the bridge structure, and the process of change has a continuous divergent effect, which is irreversible [3].

In order to investigate the influencing factors of cable clamp slippage, Miao et al. [4, 5] proposed a new criterion for cable clamp sliding, which corrected the traditional Coulomb friction formula and analyzed the anti-slippage performance of cable clamps under different bolt tightening schemes through finite element modeling. Liu [6] and Ruan [7] conducted small-scale model tests on cable clamp anti-slippage, collected a large amount of test data, and obtained the influence laws of cable clamp tightening force, friction coefficient, and creep under different factors. Tang [8] and Zhang [9] mainly studied the reasons for the decay of cable clamp pre-tightening force. Shen [10, 11] studied the nonlinear relationship between the steel wires of the main cable at the cable clamp site and researched the anti-slippage friction resistance of the cable clamp based on a multi-scale model. Zhou [12], Zhao [13], and Zhao [14] based their research on actual engineering projects and produced large-scale models of main cable-cable clamp tests. The test data showed that the measured anti-slippage friction coefficient was greater than the specification requirement of 0.15, and the contact surface of the cable clamp was subject to nonlinear changes in force and had a complex stress composition. In addition to using traditional materials for the main cable in the above studies, Li [15], Zhuge [16], and Hou [17] also conducted related research on the sliding relationship between CFRP materials for the main cable and cable clamps, making it possible to further apply new material structures to suspension bridges.

The above research findings indicate that current studies have mostly focused on the influence of factors such as bolt pre-tensioning force, cable tension, and main cable creep. Suspension bridges are typical cable-supported structures, with the main load-bearing structure being made of steel. They are sensitive to temperature changes, and this effect is more pronounced in regions with large temperature differences between day and night [18]. Therefore, further research is needed to investigate the anti-slippage performance of cable clamps on suspension bridges under the influence of temperature.

2 Calculation of Temperature Effects on Cable Clamp Friction Resistance

For the main cable, temperature changes can cause variations in the cross-sectional area of the cable. This leads to the concept of the coefficient of thermal expansion for the main cable, which is expressed as follows:

$$\alpha_s = \frac{\pi (r_T^2 - r^2)}{\pi r^2 \Delta t} \tag{1}$$

In the equation, α_s represents the coefficient of expansion of the main cable surface; r_T represents the radius after temperature change; r represents the initial radius; Δt represents the temperature difference.

Then the change amount of the main cable radius before and after the temperature change is:

$$\Delta r = r(\sqrt{\alpha_s \cdot \Delta t + 1} - 1) \tag{2}$$

According to the coordination relationship between the main cable, cable clamp and screw deformation, it can be known:

$$\varepsilon_L = 2\Delta r / L \tag{3}$$

In the equation, ε_L represents the helical strain caused by the deformation of the main cable section; L represents the initial length of the helix.

Therefore, the influence relationship of the cable clamp tightening force under the temperature change of the main cable can be expressed by the following formula:

$$P_{mc} = \frac{2 \cdot r(\sqrt{\alpha_s \cdot \Delta t + 1} - 1)}{L} \cdot E_L A_L \tag{4}$$

In the formula, E_L, A_L are the elastic modulus and cross-sectional area of the screw respectively.

According to the principles of material mechanics, the tightening force changes under the temperature change of the screw itself are as follows:

$$P_{sc} = \alpha_L \Delta t \cdot E_L A_L \tag{5}$$

α_L is the screw linear expansion coefficient.

Therefore, the total influence expression of screw tightening force under the influence of temperature is:

$$P = \left[2r(\sqrt{\alpha_s \cdot \Delta t + 1} - 1)/L + \alpha_L \cdot \Delta t \right] \cdot E_L A_L \tag{6}$$

According to the classic Coulomb friction law, the ultimate anti-slip friction resistance between the cable clamp and the main cable can be expressed as F_{fr}.

$$F_{fr} = 2N_r = \frac{4P}{\mu_\theta}(1 - e^{-\mu_\theta \pi / 2}) \tag{7}$$

Then the friction formula after the influence of temperature is:

$$F_{fr} = 4 \frac{\left[2r(\sqrt{\alpha_s \cdot \Delta t + 1} - 1)/L + \alpha_L \cdot \Delta t \right] \cdot E_L A_L}{\mu_\theta}(1 - e^{-\mu_\theta \pi / 2}) \tag{8}$$

3 Solar Radiation Data

The Xi He Big Data platform [19] introduces various meteorological data sources, and based on artificial intelligence and machine learning algorithms, it performs downscaling calculations on existing meteorological elements, and optimizes fusion and calibration of meteorological data based on its own data grid. In order to verify the reliability of the data results, the platform's internal meteorological data and on-site radiation measurement data were compared for inspection.

The data collected by a solar radiation sensor installed at the construction site of the Qingshui River Bridge project is used as a comparison source for verification. The time span selected is August, when the sunlight is relatively strong, and the data collection time is from 8:00 to 17:00 (August), with a time interval of 1 h. Rainy days in each month are excluded. The data is shown below in Fig. 1.

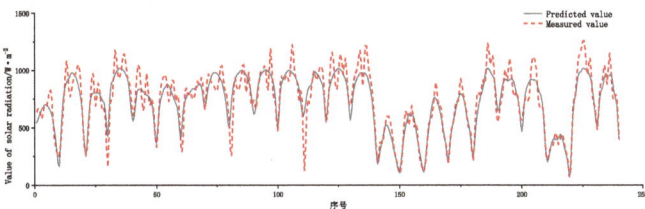

Fig. 1. Comparison between radiation prediction and actual measurement in August

Based on the data curve in the graph, it can be observed that the measured values and predicted values are well aligned. There is a relatively large deviation in the high radiation value area, with a maximum deviation of 15.2% in July and 17.6% in August. Overall, the measured values tend to be slightly higher. According to the calculation formula of the MAPE function, the M value for July is 9.95%, and for August it is 11.08%. These values meet the requirements for accurate estimation and can be used in practical engineering projects.

4 Project Overview and Model Construction

4.1 Project Overview

The Changshou Economic Development Zone Bridge is a single-span simply supported steel box girder suspension bridge with a main span of 739 m and a vertical span ratio of 1:9.11. The longitudinal spacing of the suspenders is 12 m, and the suspenders near the tower are 15.5 m away from the tower centerline. The bridge cable clamps mainly consist of upper and lower half clamps and M42 high-strength bolts. The design clearance between the two half clamps is set at 4 cm, and the effective length of the bolts is greater than 0.7 times the diameter of the main cable inside the clamps in Fig. 2.

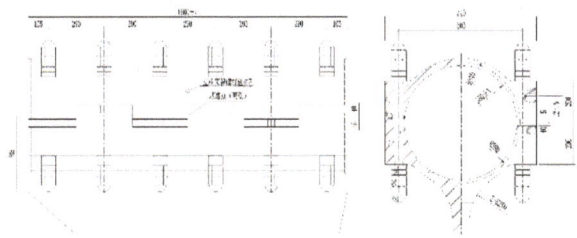

Fig. 2. Cable clamp design dimensions. (unit: mm)

4.2 Model Parameter Selection

Main Cable Properties. In reference [10], a large amount of actual field data on the tension forces of the cable clamps and the dimensions of the main cable were analyzed. The three-dimensional (axial, tangential, and radial) anisotropic equivalent material property relationship of the main cable was proposed, and its material adaptability was confirmed through experimental data. The fitting curve and relationship equation are shown in Table 1.

Table 1. Equivalent material parameter table

Elastic modulus (MPa)			Poisson's ratio		
Axial	Radial	Tangential	Axial	Radial	Tangential
195000	36000	22000	0.26	0.02	0.26

The author compared the main cable dimensions, construction methods, and design aspects of the experimental bridge in reference [10] with those of the Changshou Economic Development Zone Bridge, and found that except for slight differences in the quantity of galvanized steel wires used for the main cable, all other indicators are completely consistent. Therefore, disregarding factors such as construction errors, it can be considered that the three-dimensional anisotropic material for the main cable in reference [10] is equally applicable to the main cable model in this paper.

Thermophysical Parameters. Zhang [20] summarized the research results on the thermal properties of the main cable in the past. By fitting the corresponding thermal property parameter model of the main cable through laboratory and on-site main cable model test data, the correctness of the parameter model has been verified in the paper, and can be used for temperature field calculation and analysis of the main cable of suspension bridges. As shown in Table 2.

Table 2. Thermophysical parameters.

	Thermal Conductivity $(W{\cdot}(m{\cdot}°C)^{-1})$			Convective heat transfer coefficient $(W{\cdot}(m^2{\cdot}°C)^{-1})$	Thermal expansion coefficient	radiation heat transfer rate	Contact heat transfer
main cable	Axial	Radial	Tangential	3.14	1.13e-5	0.8	complete heat transfer
	40	1.2	25				
rope clip	38.53			46.52	1.14e-5	0.8	

4.3   Finite Element Model Establishment

A model was created using the ABAQUS software, a large-scale general-purpose finite element software. The main cable and cable clamps were simulated using C3D8T thermal-coupled hexahedral elements, while the high-strength bolts were simulated using spatial beam elements. The contact surfaces between the ends and the nuts on the cable clamps were coupled to simulate the transmission of bolt forces, with pre-tension applied through bolt loading. The axial friction coefficient was set to 0.15, the circumferential friction coefficient was set to 0.2, and the friction coefficient at the cable clamp stopper was set to 0.15. The ends of the main cable were fixed constraints, and a vertical displacement was applied at one end of the cable clamp. According to the principle of action and reaction, the frictional resistance of the cable clamps could be obtained by extracting the boundary reaction forces of one end of the main cable. The computational model is shown in Fig. 3.

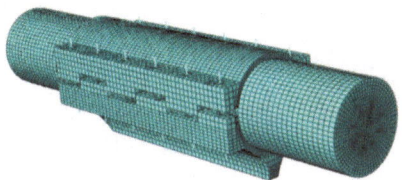

Fig. 3. Finite element model diagram

5   Result Analysis

5.1   Analysis of the Influence of Screw Temperature

Based on the equation for calculating the frictional resistance of the cable clamps and the finite element model, different values of bolt temperature were determined to analyze the differences between the finite element solution and the analytical solution. The results are shown in the table below. For different bolt temperatures, the maximum difference ratio between the finite element solution and the analytical solution is -4.2%, and the minimum is -0.3%. The comparison of the calculation results in these two cases shows small differences, indicating that the finite element calculation results are close to the analytical solution, further confirming the accuracy of the established finite element model (Table 3).

5.2   Force Analysis of Cable Clamp Under Steady Temperature Field

During daytime solar radiation, factors such as radiation intensity, radiation angle, and wind speed at different times can all affect the temperature field results. Therefore, to simplify the calculation, we take the example of the day with the highest radiation at the bridge site in mid-July 2023, with time set at 7 AM and 12 PM. The solar radiation

Table 3. Comparison of friction resistance at different screw temperatures[a]

Temperature change value (°C)	Finite element solution (N)	Analytical solution (N)	Difference percentage (%)
10	2139705	2162424	1.1
20	2043835	2037244	-0.3
30	1928565	1912064	-0.9
40	1823755	1786884	-2.0
50	1719035	1661704	-3.3
60	1604560	1536524	-4.2

[a]Note: The 500kN preload force model is used as the temperature effect calculation model/.

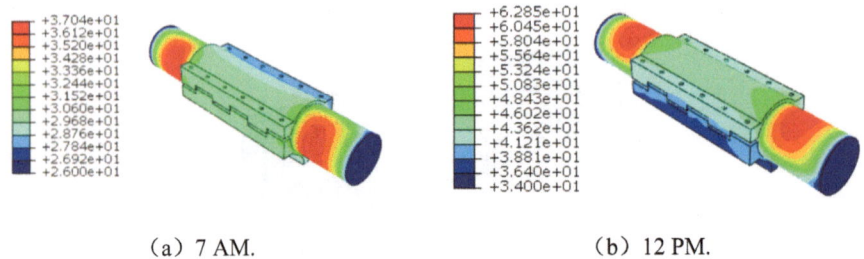

(a) 7 AM.   (b) 12 PM.

Fig. 4. Temperature field results. (unit:°C)

is assumed to be horizontal at 7 AM and vertically from top to bottom at 12 PM. The temperature field results are shown in Fig. 4.

The temperature results mentioned above were loaded into the model in the form of a mapping field, with the ambient temperature based on the data provided by the platform, and the bolt pre-tightening force set at 500 kN without applying temperature load to the bolt.

The tangential friction force at the contact point of the cable clamp was extracted, as shown in Fig. 5. According to the results in the figure, compared with the original model, at 7 AM, the numerical range of the inner friction force distribution of the upper cable clamp is basically consistent, and the distribution area of higher friction force is expanding. Overall, the frictional resistance of the cable clamp is increasing at this time. However, at 12 PM, although the maximum frictional force has increased, it is mainly concentrated in the corner area of the cable clamp and local positions near the screw hole, and the frictional force in a larger area at the top is decreasing. Combined with the temperature field results, it can be inferred that during the transition from low temperature to high temperature, the frictional resistance of the cable clamp will increase to a certain extent. However, after exceeding a certain temperature, the deformation of the cable clamp relative to the radial deformation of the main cable increases, causing some deformation inconsistency and a decrease in contact density, resulting in a decreasing trend in frictional resistance. Comparing with the results of cable clamp frictional force, it is basically consistent with the description above. The cable clamp frictional force at the side is greatly increased under high temperature, while most areas show a decreasing trend.

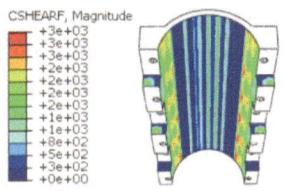

（a）Original model

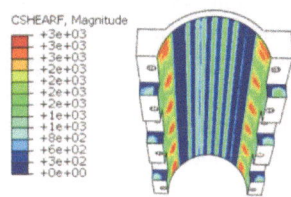

（b）Friction force of upper rope clamp at 7AM

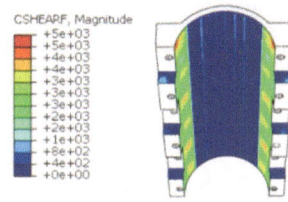

（c）Friction force of upper rope clamp at 12 pm

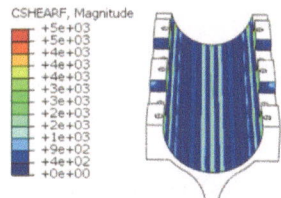

（d）Original model

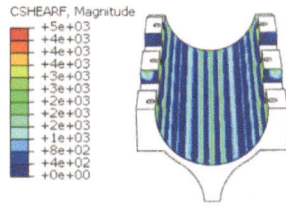

（e）Friction force of lower rope clamp at 7am

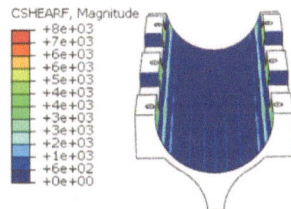

（f）Friction force of upper rope clamp at 12 pm

Fig. 5. Cable clamp friction force results. (unit: N)

5.3 Screw Force Analysis Under Steady Temperature Field

Observing the bolt axial stress results shown in Fig. 6, it can be found that the bolt axial stress increases correspondingly with the increase of temperature. At 7 AM, the increase in bolt axial stress on the left and right sides is not consistent. Considering the direction of radiation, the temperature on one side is higher than the other, and the axial stress on the side with higher temperature will be higher than that on the other side with lower temperature. In addition, the increase in bolt axial force caused by the uneven temperature difference on both sides of the cable clamp is greater than the effect of uniform temperature difference. Therefore, in actual bolt axial force testing, if the site cannot reach a constant temperature state, testing can also be performed under uniform temperature difference.

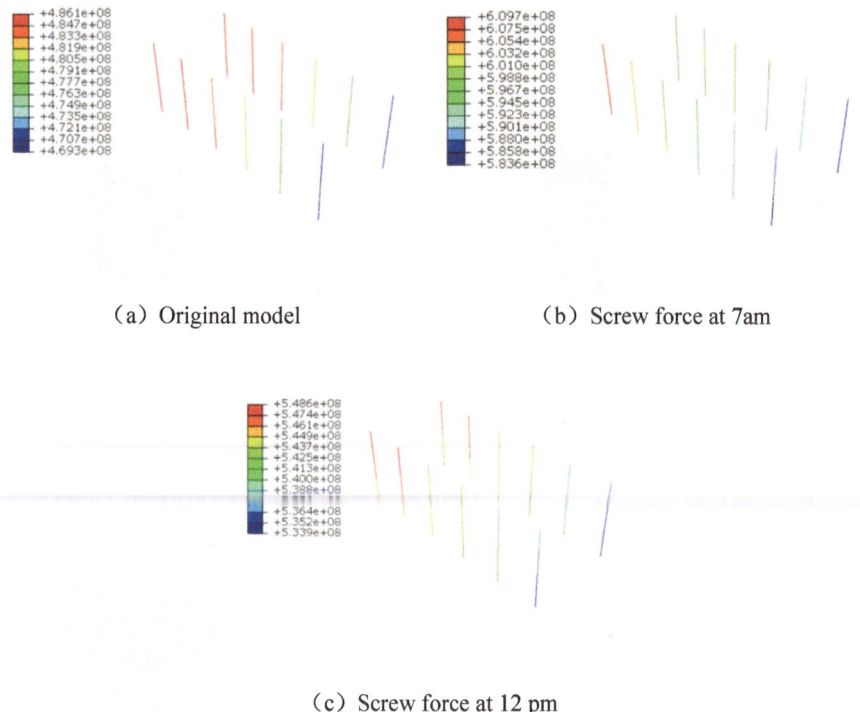

（a）Original model   （b）Screw force at 7am

（c）Screw force at 12 pm

Fig. 6. Screw shaft stress results. (unit: Pa)

6 Conclusion

Based on the validated platform radiation data, this article establishes a thermal-mechanical coupling model of the cable clamp-main cable using ABAQUS software, and compares the finite element solution with the analytical solution to obtain the effect of bolt temperature. In addition, the corresponding stress situation under steady-state temperature field is obtained for the frictional force of the cable clamp and the axial stress of the bolt. The following conclusions are drawn:

With the increase of temperature, the bolt pre-tightening force decreases, and the analytical solution is close to the finite element solution, with a maximum error of 4.2%, which can be used for actual on-site engineering calculations.

The increase of temperature causes structural deformation of the main cable and cable clamp under thermal variation, which increases the frictional force of the cable clamp to a certain extent, which is beneficial to reducing the probability of slippage during use.

Based on the above two analysis results, in actual projects, the cable clamp pre-tightening should be performed at low temperature to avoid pre-tightening force loss caused by cooling after high-temperature pre-tightening. In addition, thermal insulation treatment can be applied to the surface of the bolt to maximize the increase in frictional force caused by structural warming and improve the anti-slippage safety factor.

Acknowledgments. This research was funded by the National Nature Science Foundation of China, grant number (52078058); the Postgraduate Research and Innovation Project of Hunan, grant number (CX20200834); Research on Key Technology of Long Span Reinforced Concrete Suspension Arch Construction, grant number (DQJ-2021-B02).

References

1. Zhang, W.: Theory and Construction Control Technology of Suspension Bridge Design. Science Press, Beijing (2021)
2. Wu, S., et al.: Runyang Yangtze River Highway Bridge Construction - Volume III Suspension Bridge. People's Communications Press, Beijing (2006)
3. Ma, W., Liu, S., Wang, C., et al.: Study on main cable gap ratio and anti-sliding test of cable clamp for Liujiaxia Bridge. World Bridges 42(05), 59–62 (2014)
4. Miao, R., Shen, R., Wang, L., et al.: Theoretical and numerical studies of the slip resistance of main cable clamp composed of an upper and a lower part. Adv. Struct. Eng. 24(4), 691–705 (2021)
5. Miao, R., Shen, R., Tang, F., et al.: Nonlinear interaction effect on main cable clamp bolts tightening in suspension bridge. J. Constr. Steel Res. 182, 106663 (2021)
6. Liu, H., Wang, G., Huang, J., et al.: Experimental and numerical study on the anti-sliding performance of full-lock coil cable clamps. J. Constr. Steel Res. 187, 106957 (2021)
7. Ruan, Y., Luo, B., Ding, M., et al.: Theoretical and experimental study on the antisliding performance of casting steel cable clamps. Adv. Civ. Eng. 2019, 1–18 (2019)
8. Tang, M., Che, T., Song, X., et al.: Research on pre-tightening force of self-anchored suspension bridge cable clamp based on creep theory. J. South China Univ. Technol. (Nat. Sci. Ed.) 50(01), 59–68 (2022)
9. Zhang, P.: Analysis of the causes of decrease in pre-tightening force of cable clamps in suspension bridges and preventive maintenance measures. Highway 64(02), 101–105 (2019)
10. Ruili, S., Kai, H., Zhen, H.: Analysis model of nonlinear relationship between main cable wires at cable clamp position of suspension bridge. J. Archit. Civ. Eng. 35(01), 111–118 (2018)
11. Shen, R., He, K., Miao, R.: Analysis of ultimate anti-sliding frictional resistance of bolted cable clamps based on multiscale model. Bridge Constr. 48(05), 16–20 (2018)

12. Zhou, Z., Yuan, Q., Zhou, C., et al.: Test research on anti-sliding performance of zinc-aluminum alloy coating steel wire cable clamps in suspension bridges. World Bridges **43**(05), 40–43 (2015)
13. Zhao, C., Yan, M., Zhang, A., et al.: Test research on anti-sliding performance of hongdu bridge cable clamps. World Bridges **41**(01), 64–68 (2013)
14. Zhao, D.: Research on anti-sliding and stress testing of self-anchored suspension bridge cable clamps. Railw. Constr. Technol. **10**, 18–19 (2009)
15. Yang, L.: Calculation method for cable clamp design of CFRP cable suspension bridge. China J. Highw. Transp. **28**(10), 67–75 (2015)
16. Zhuge, P., Zhang, Z., Wang, S., et al.: Interfacial anti-sliding performance between CFRP main cable and cable clamp in long-span suspension bridge. J. Southwest Jiaotong Univ. **49**(02), 208–212 (2014)
17. Hou, S., Qiang, S., Liu, M., et al.: Experimental study on frictional properties between CFRP main cable and cable clamp. J. Shenzhen Univ. (Sci. Eng. Ed.) **29**(03), 16–21 (2012)
18. Wang, D., Zhang, W.: Calculation model construction and analysis of temperature field of suspension bridge main cable. Highw. Traffic Sci. Technol. **32**(08), 66–71 (2015)
19. Hersbach H., et al.: ERA5 hourly data on single levels from 1959 to present. Copernicus Climate Change Service (C3S) Climate Data Store (CDS) (2018)
20. Zhang, W., Zhang, L., Zhong, N.: Temperature field research based on experimental thermal properties parameters of suspension bridge main cable. J. Chongqing Jiaotong Univ. (Nat. Sci) **35**(01), 1–4 (2016)

Study on Selection of Excavation Methods and Spacing Distance for Ultra-shallow-Buried and River-Crossing Tunnel with Small Clearance

Zhiqun Gong[1], Hongyu Guo[2], Helin Fu[3(⊠)], Yibo Zhao[3], and Zetong Peng[3]

[1] China Construction Infrastructure Co., Ltd., Beijing 100044, China
[2] CCFEB Civil Engineering Co., Ltd., Changsha 410000, Hunan, China
[3] Department of Civil Engineering, Central South University, Changsha 410075, Hunan, China
fu.h.l@csu.edu.cn

Abstract. For the mined land section of the river-crossing tunnel with small clearance, the study should be carried out for excavation method and safe spacing before the tunnel construction, for engineering properties of surrounding rock is poor and the distance between two tunnel tubes is small. In this paper, in combination with the mined land section works of the river-crossing tunnel on Rongjiang Fourth Road, the finite difference numerical simulation software was adopted to calculate mechanical behaviors of three construction methods: three-bench method, CD method and CRD method under different excavation spacing (10 m, 20 m, 30 m and 40 m) on the basis of grouting and pre-reinforcement of surrounding rock in the early stage. The calculation results indicate that vault displacement and ground settlement can notmeet the specification requirements when the three-bench method is adopted for construction, even if the excavation spacing between the former and latter tunnels is 40m; When the CD method is adopted for construction, vault displacement and ground settlement meet the specification requirements when the excavation spacing between the former and latter tunnels is 40m; When the CRD method is adopted for construction, vault displacement and ground settlement meet the specification requirements when the excavation spacing between the former and latter tunnels is 30m; The control effect of initial support deformation and surrounding rock displacement is in the following sequence: CRD method > CD method > three-bench method. Considering various factors, it is found that the CD method is more reasonable and efficient. Therefore, the CD method is recommended for tunnel excavation and the tunnelling spacing should be controlled to 40m.

Keywords: River-Crossing Tunnel · Small Clearance · Numerical Simulation · Excavation Method · Safe Spacing

1 Introduction

Due to the small spacing between the two tunnels of the small clear distance, the construction of the rear tunnel have adverse effects on the existing lining and the stress-strain state of the surrounding rock of the first tunnel, seriously endangering the construction

© The Author(s) 2024
G. Feng (Ed.): ICCE 2023, LNCE 526, pp. 261–272, 2024.
https://doi.org/10.1007/978-981-97-4355-1_25

safety of the first tunnel. For tunnels excavated by the bench cut method, the invert closure distance has a great impact on the deformation and pressure of the surrounding rock.

Jin Xiaoguang et al. [1] analyzed the construction mechanical behavior of side wall heading and two-bench method, double wall heading method, and double hole and two-bench method by numerical simulation. Jiang Kun et al. [2] relied on the Kuiqi large-section and small clear distance tunnel project to reveal the influence of different construction methods on the settlement of the tunnel arch through numerical simulation. Yan Qixiang et al. [3, 4] used elastic-plastic numerical calculation methods to analyze the stress and deformation characteristics of the middle rock column in small clear distance tunnels. Li Yunpeng et al. [5] calculated the reasonable clear distance by numerical simulation under different levels of surrounding rock. Sun Zhigang [6] conducted a study on the excavation steps of the front and rear tunnels of small clear distance overlapping tunnels through three-dimensional numerical simulation. Wang Kang [7] relied on a large -section and small clear distance tunnel to reveal the influence of excavation sequence on the stability of surrounding rock through numerical simulation. Zhang Dingli et al. [8] established an analytical formula for the stress state of the intermediate rock wall in small clear distance tunnels using the bipolar coordinate method. Li et al. [9] analyzed the viscoelastic plastic deformation during the excavation process of the small spacing tunnel by benches method through numerical simulation. Chen et al. [10] revealed the distribution of seepage field and its impact on structural mechanical properties of small clean tunnels in water-rich areas through seepage model experiments. Li et al. [11] revealed the mechanical response laws of the construction process of large-span and small clear distance tunnels through model experiments and numerical simulations. Song et al. [12] compared the effects of different excavation methods and clearance on the stress characteristics and displacement of tunnel support through numerical simulation. Yao et al. [13] analyzed the deformation patterns and stress characteristics of small spacing tunnels under different levels of surrounding rock through numerical simulation. Tang et al. [14] established numerical simulation models of two-bench method, core soil method, and side wall heading method for a small clear distance highway tunnel. Su et al. [15] analyzed the plastic zone and internal forces of shallow buried small clear distance loess tunnels through numerical simulation.

Many scholars analyzed the stress and deformation effects of different excavation methods on small spacing tunnels through numerical simulation, model experiments, and other methods. However, few scholars studied the excavation methods and safe construction spacing of large cross-section and small clear distance underwater tunnels suitable for weak surrounding rock conditions. Therefore, this study combines the Rongjiang Tunnel Project with finite difference numerical simulation software to calculate the stress and displacement under different excavation intervals using three construction methods, namely the three step method, CD method, and CRD method, and reveal their evolution laws. It is of great significance for guiding the safe construction of large cross-section and small clear distance underwater tunnels in weak surrounding rocks.

2 General Situation of the Project

Rongjiang Tunnel's length of the main line is 2.412 km. The main works include the cross-river tunnel and the entrance/exit ram. The minimum distance from the tunnel cross-section is only 8.0m. The surrounding rock in this tunnel section has poor self-stabilizing ability, with an overlying layer of weathered rock less than 1 times the tunnel diameter. The landward segment of the tunnel is located near Chaoyang Road, requiring stringent control of settlement caused by construction activities. Excavating a small clearance tunnel in weak surrounding rock can have a significant impact on the surrounding environment. Implementing appropriate reinforcement measures and determining a reasonable excavation method and construction spacing are crucial for controlling settlement during the construction process.

3 Overview of Numerical Computational Models

In this paper, the finite difference software FLAC3D is used for three-dimensional numerical simulation calculations.

3.1 Assumptions for Numerical Simulations

Due to the complexity of engineering practices and the non-uniform nature of geological strata, it is impossible to fully encompass the actual conditions of the project in the model during model setup and numerical calculations. The following assumptions are made under the premise of meeting the needs of the project:

The soil and rock layers are assumed to have a homogeneous layered distribution. The surrounding rock is considered to be isotropic and a continuous elastic-plastic material. The weakening effects of groundwater on the surrounding rock and support structures are not taken into account. When simulating soil and rock material, only self-gravitational stresses are taken into account and tectonic stresses are ignored.

3.2 Model Dimensions

According to Saint Venant's principle, in order to minimize the influence of the model boundaries on the model calculation error, the length on both sides of the model is taken as 7 times the diameter of the tunnel, and the overburden layer was taken to be the most unfavorable conditions, and The thickness of the soil above the tunnel is 9.0m, which takes into account the layering effect of the soil. The numerical model established in this study has the following dimensions as shown in Fig. 1.

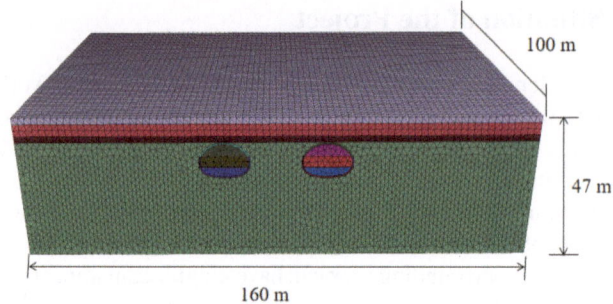

100 m

47 m

160 m

Fig. 1. Three-dimensional numerical model

4 The Simulation of Construction Methods and Safe Spacing for the Rongjiang Tunnel

Based on the results of on-site investigation, the surrounding rock of the tunnel in the land section is class V. This tunnel is a super small clearance and large cross-section tunnel. The commonly used construction methods for excavation of tunnels with super small clearance and large cross-section in soft rock include the three-bench method, CD method, and CRD method. This paper analyzes the minimum spacing between tunnels under different excavation methods using these three methods. The excavation step distance and tunnel step distance for the three excavation methods are shown in Fig. 2.

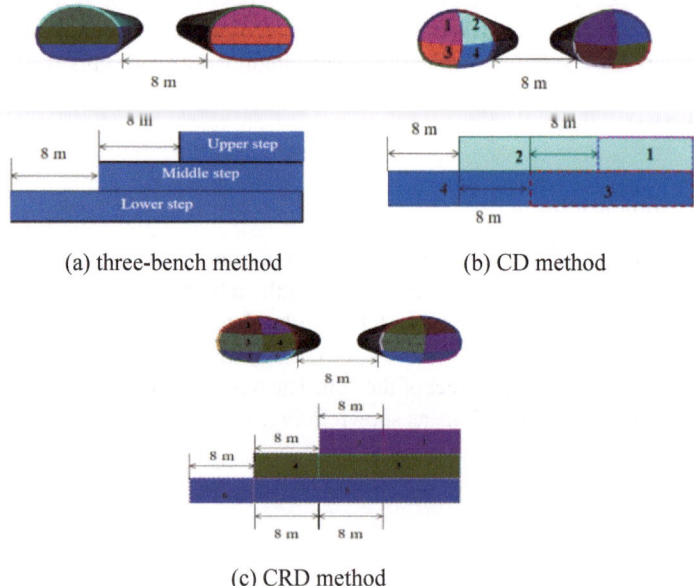

(a) three-bench method (b) CD method

(c) CRD method

Fig. 2. Excavation steps for different excavation methods

4.1   Material Properties and Boundary Conditions

The soil layer above the tunnel is mainly composed of soft soil layers such as sand layers, and is generally described using the Mohr-Coulomb constitutive model. While the tunnel is located in the hard weathered rock, which is generally described by D-P constitutive model (Table 1).

Table 1. Physical and mechanical parameters of rock and soil strata

Name	Density /(kg/m³)	Elastic modulus /MPa	Poisson's ratio	Cohesive force /kPa	Internal friction angle /(°)
Miscellaneous Fill	1760	30.9	0.32	32.5	10.5
Fine sand	1820	45.8	0.28	12.1	10.1
Round gravel soil	1860	32.5	0.3	19.1	14.1
Moderately weathered argillaceous sandstone	1920	1100	0.31	1040	43.3

For the advanced small conduit grouting support of the tunnel, a single layer of $\varphi42 \times$ 4 mm advance grouting pipes with a length of 4.5 m is used. The circumferential spacing is 0.4 m, the outer insertion angle ranges from 5° to 15°, and the longitudinal spacing is 4 m. A total of 46 pipes are used in a single ring. To simulate the advanced support, a numerical model is constructed using beam elements to represent the corresponding pipes, as shown in Fig. 3. The reinforcement area of the pipes is determined by calculating the pipe's range, and the strength of the soil within the reinforcement area is increased to simulate the implementation of advance support measures.

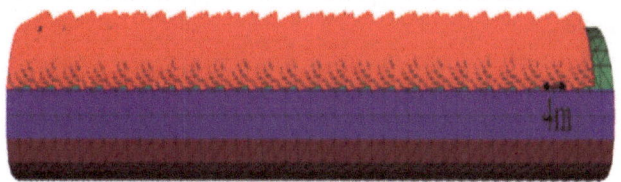

Fig. 3. Advance small conduit support

For the initial support, anchor rods and shotcrete support are used. The shotcrete thickness is 30 cm, and 6.0 m long hollow anchor rods are arranged within a 120° range of the arch crown, with a longitudinal spacing of 1 m and a circumferential spacing of 0.6 m.There are a total of 26 anchor rods in one ring. Cable elements are used in the modeling to represent the corresponding anchor bolts, as shown in Fig. 4.

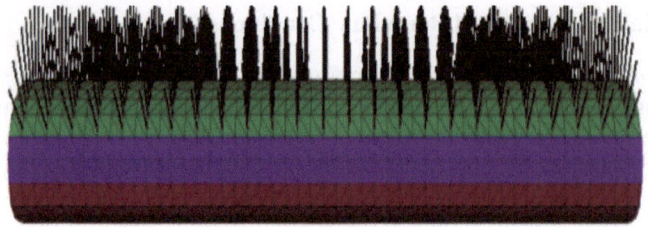

Fig. 4. Bolting support

The soil strength within the reinforcement area is increased to simulate the reinforcement measures. Existing research suggests that the soil strength in the advance reinforcement area is generally increased by a factor of 1.2, while the soil strength in the grouting area can be increased by a factor of 1.5. Rongjiang Tunnel's pre-reinforcement project in the mined land section of the tunnel adopts a high-pressure jet grouting method. This means grouting and pre-reinforcing from the tunnel roof to the ground surface. In the model calculations, it is assumed that the reinforcement area is located above the tunnel roof. When simulating the temporary invert arch support in the CD method and CRD method, Shell elements included in FLAC3D are used to represent the temporary support, as shown in Fig. 5.

Fig. 5. Temporary support

The physical and mechanical properties of the support measures used in the model are shown in Table 2.

Table 2. Support parameters

Name	Elastic modulus/MPa	Poisson's ratio
Advance small catheter	1.8×10^5	0.22
Anchor stock	2.2×10^5	0.22
Advance reinforcement area	1.32×10^3	0.28
Initial support	2.3×10^3	0.23
Shell	1.3×10^5	0.25
Grouting pre reinforcement area	1.98×103	0.26

4.2 Simulation

In the numerical simulation, the three-bench method of excavation was carried out according to the following steps:

Assign the properties of the advanced support material, excavating the upper bench of the first tunnel step by step, with a depth of 2 m each time, and assign the properties of the initial support material at the corresponding position, and assign the anchor strengths to the anchors, and then run the calculations;

After excavating the upper bench of the former tunnel for 8 m, the middle bench will be excavated at the same time, and after excavating the upper bench for 10 m, start excavating the later tunnel. At this time, the excavation distance between the two tunnels will be 10 m.

After 16 m of excavation for the upper bench of the former tunnel and 8 m for the middle bench, start excavating the lower bench. At this time, the excavation of the upper, middle and lower benches is the same, each time excavation is 2 m. After 18 m of excavation for the middle bench of the former tunnel, start excavating the middle bench of the later tunnel, and after 18 m of excavation for the lower bench of the former tunnel, start excavating the lower bench of the later tunnel. The upper and lower benches of the two tunnels keep the same speed, and the excavation work is completed after 5 steps of cyclic excavation, analyzing the surface settlement and the deformation of the initial support and other indicators, and changing the spacing between the former and later tunnels to 20 m, 30 m and 40 m, comparing the control indicators, and determining the reasonable spacing between the tunnels.

The excavation steps of CD and CRD methods are basically similar to those of the three-step method and will not be repeated here. The ultimate relative displacements of the initial support shall meet the provisions in Table 3. The maximum deformation of the initial support at the vault should not exceed 0.08%H ~ 0.16%H. The height of the crossing tunnel on Rongjiang Road 4 is H = 8.6 m, which means that the deformation of the initial support at the top of the arch must not exceed 14 mm.

Table 3. Limit relative displacement of initial support with width 7 m < B ≤ 12 m.

Surrounding rock level	Tunnel burial depth h/m		
	h ≤ 50	50 < h ≤ 300	300 < h ≤ 500
Relative horizontal displacement of arch foot/(%)			
IV	0.10~0.30	0.20~0.80	0.70~1.20
V	0.20~0.50	0.40~2.00	1.80~3.00
Relative settlement of the arch crown/(%)			
IV	0.06~0.10	0.08~0.40	0.30~0.80
V	0.08~0.16	0.14~1.10	0.80~1.40

The deformation of the structure was simulated using the three-bench method (10 m between the former and later tunnels) and is shown in Fig. 6. In Fig. 6, after the tunnel was excavated by the three-bench method, the deformation was that the top of the arch was sinking and the bottom of the arch was bulging to a certain extent, and as the excavation proceeded, the value of the top of the arch sinking and the value of the bottom of the arch bulging increased, and the size of the surface settlement and the range of the settlement increased as well.

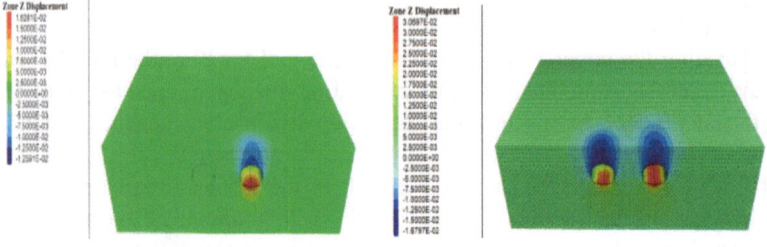

(a) excavation of upper bench for 10m   (b) excavation of upper bench for 30m

Fig. 6. Deformation by using three-bench method

Comparing the surface settlement data and vault settlement data under different excavation spacing. The monitoring statements of corresponding cross sections were compiled, and the changes of maximum surface settlement and vault settlement with the depth of excavation on the upper bench of the three-bench method under the excavation spacing of 10 m, 20 m, 30 m, and 40 m were plotted, as shown in Fig. 7.

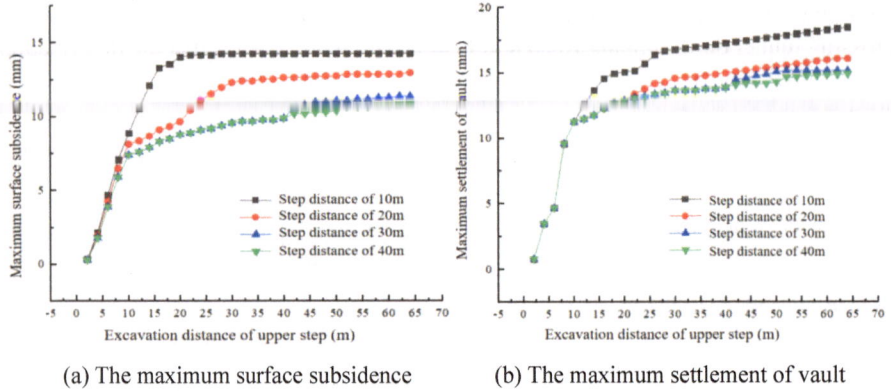

(a) The maximum surface subsidence   (b) The maximum settlement of vault

Fig. 7. Deformation of ground and vault during excavation by three-bench method

In Fig. 7, when the excavation distance between the first and the second tunnels are 10 m, 20 m, 30 m and 40 m, the maximum surface settlement are 14.3 mm, 13.1 mm, 12.2 mm and 11.7 mm, and the maximum vault settlement are 18.5 mm, 15.2 mm, 14.6 mm and 13.7 mm, respectively. As the excavation distance between the two tunnels increases, the maximum settlement of the vault and the maximum surface settlement decreases, and the excavation distance between the two tunnels is greater than 40 m, which still does not meet the requirements of the construction specification, and it is necessary to increase the excavation distance between the two tunnels.

The deformation of the structure was simulated using the CD method of excavation (10 m between the former and later tunnels) and the results are shown in Fig. 8.

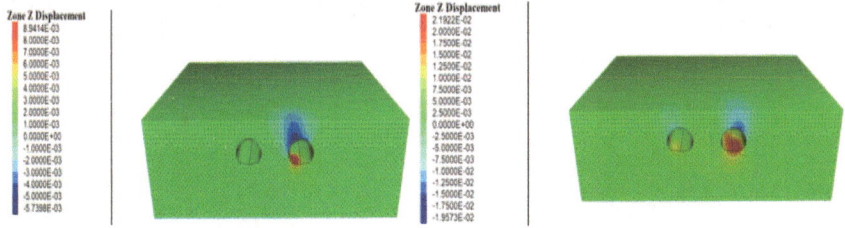

(a) Excavation of upper left section for 10m (b) Excavation of upper left section for 30m

Fig. 8. Deformation by using CD method

In order to compare the surface settlement and vault settlement data under different excavation spacing. The monitoring statements of corresponding sections were compiled, and the maximum surface settlement and vault settlement with the excavation depth of the upper-left part of the CD method were plotted under the excavation spacing of 10 m, 20 m, 30 m, and 40 m, as shown in Fig. 9.

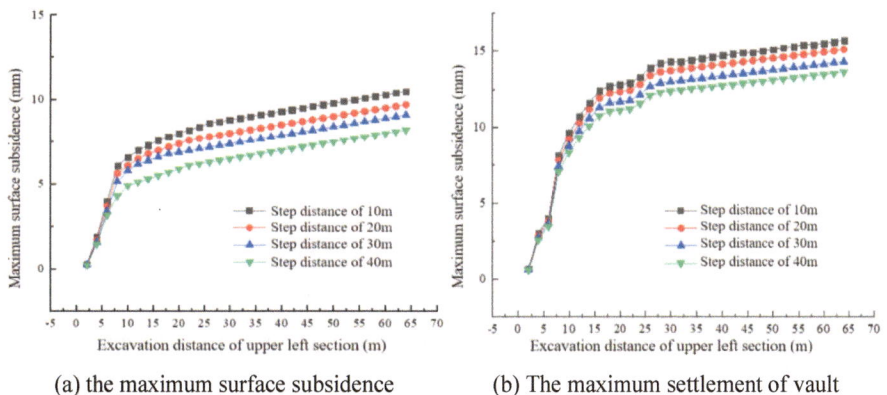

(a) the maximum surface subsidence   (b) The maximum settlement of vault

Fig. 9. Deformation of ground and vault during excavation by CD method

In Fig. 9, with the increase of excavation spacing between the former and later tunnels, the maximum arch settlement and surface settlement gradually decrease. When the excavation spacing between the two tunnels is 20 m, 30 m, and 40 m, the maximum arch settlement are 15.1 mm, 14.3 mm, and 13.6 mm, and the maximum surface settlement are 11.4 mm, 10.4 mm, and 9.7 mm, in that order. The settlement control value can be satisfied when the excavation distance between the former and the later tunnels is 40 m. The excavation spacing between the former and later tunnels must be at least three times the diameter of the tunnel in order to meet the construction specification. It was also demonstrated that the CD method was more effective in controlling the deformation of the tunnel and the ground settlement deformation than the three-bench excavation method.

The deformation of the structure was simulated using the CRD method of excavation, as shown in Fig. 10.

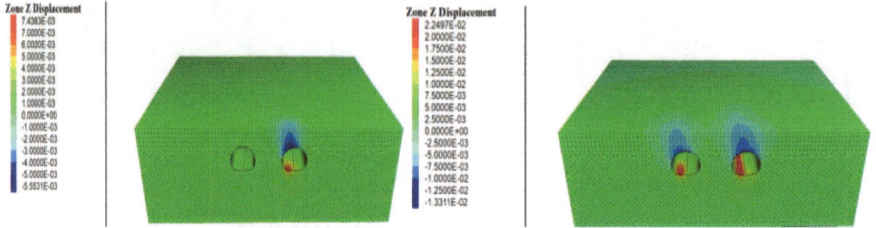

(a) Excavation of upper left section for 10m (b) Excavation of upper left section for 30m

Fig. 10. Deformation by using CRD method

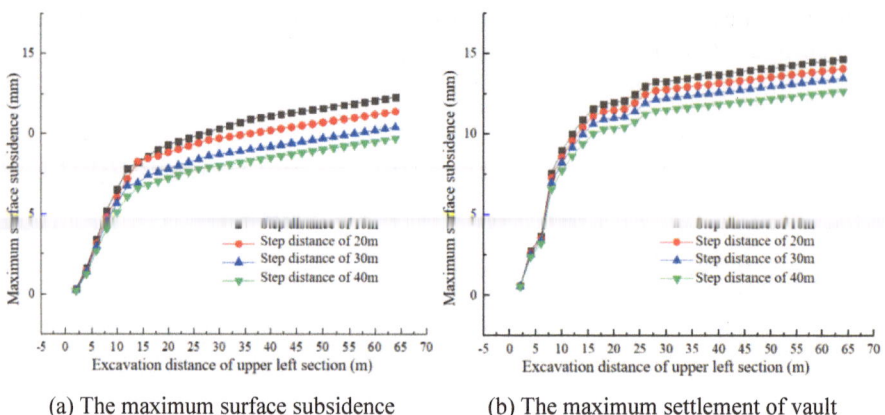

(a) The maximum surface subsidence   (b) The maximum settlement of vault

Fig. 11. Deformation of ground and vault during excavation by CRD method

Comparing with the surface settlement data and arch settlement data under different excavation spacing, the monitoring statements of the corresponding sections were compiled, and the changes of the maximum surface settlement and the maximum arch settlement with the depth of the upper left part of the CRD excavation under the excavation spacing·of 10 m, 20 m, 30 m, and 40 m were plotted, in Fig. 11.

In Fig. 11, with the increase of the excavation distance between the two tunnels, the maximum value of the arch settlement and the maximum surface settlement decrease gradually. When the excavation distance between the two tunnels are 20 m, 30 m and

40 m, the maximum arch settlement are 14.1 mm, 13.5 mm and 12.7 mm, and the maximum surface settlement are 9.7 mm, 9.1 mm and 8.2 mm in the order of the excavation distance between the two tunnels. When the excavation distance between the two tunnels is ≥30 m, the settlement control requirement can be met. If the CRD method is used on site, the excavation spacing between the former and later tunnels must be at least two times the diameter of the tunnels in order to meet the construction specification. In terms of the deformation of initial support and displacement control of surrounding rock, CRD method > CD method > three-bench method.

5 Conclusion

Through the study of the construction mechanical behavior of the mined land section of the Rongjiang Fourth Road river-crossing tunnel with different excavation spacing of the former and later tunnels using the three-bench method, CD method and CRD method, and the analysis of the surface settlement and vault displacement during the construction process with the aid of three-dimensional numerical simulation, the following conclusions were drawn:

(1) When the three-bench excavation method is used, the deformation of the tunnel and surface settlement cannot meet the construction requirements. When the excavation distance is 40 m, the maximum settlement value of the vault is 13.7 mm and the maximum settlement value of the surface is 11.7 mm, which cannot meet the requirements of the construction specification.
(2) Adopting the CD method of excavation can meet the construction specification when the excavation distance between two tunnels is greater than or equal to 40 m. Adopting the CRD method of excavation can meet the construction specification when the excavation distance between two tunnels is greater than or equal to 30 m.
(3) In terms of the control effect of initial support deformation and perimeter rock displacement, CRD method > CD method > three-bench method.
(4) Considering the factors of safety, economy and construction efficiency, it is recommended to use CD method of excavation for this project, and the spacing of tunnel excavation is controlled to be 40 m.

Acknowledgment. Funding: This study was financially supported by the Hunan Provincial Construction Science and Technology Plan Project (KY202109).

References

1. Jin, X., Liu, W., Qin, F., et al.: Excavation method analysis of little distance tunnel in freeway. J. Railw. Eng. (2), 3–68 (2004)
2. Jiang, K., Xia, C., Bian, Y.: Optimization analysis of construction schemes of small space tunnel with bidirectional eight traffic lanes in jointed rock mass. Rock Soil Mech. **33**(3), 841–847 (2012)
3. Yan, Q., He, C., Yao, Y.: Study on construction characteristic and dynamic behavior of soft rock tunnel. Chinese J. Rock Mech. Eng. **25**(3), 572–577 (2006)

4. Yan, Q., He, C., Yao, Y.: Study on mechanical effect of small-distance tunnels located at soft ground. Chinese J. Undergr. Space Eng. **1**(5), 693–697 (2005)
5. Li, Y., Wang, Z., Han, C., et al.: Simulating study on construction process of tunnels with small spacing for difference classes of surrounding rocks. Rock Soil Mech. **27**(1), 11–16+18 (2006)
6. Sun, Z.: Study on construction mechanical effect of overlapping metro tunnel with small interval. Chongqing University, Chongqing (2017)
7. Wang, K.: Spatial deformation and load release mechanism of surrounding rock in tunnel construction with super-large section and small spacing and its engineering application. Shandong University, Jinan (2017)
8. Zhang, D., Chen, L., Fang, Q., et al.: Research and application on central rock wall dike stability of small interval tunnel. J. Beijing Jiaotong University, **40**(1), 1–11 (2016)
9. Li, Y.P., Ai, C.Z., Wang, Z.Y., et al.: Research on the optimal interval of tunnel with small spacing considering visco-elastic plasticity. In: Pacific Rim Conference on Rheology. China University of Petroleum, Beijing, 102249, China (2005)
10. Chen, Z., Li, Z., He, C., et al.: Investigation on seepage field distribution and structural safety performance of small interval tunnel in water-rich region. Tunnelling and Underground Space Technology (2023)
11. Liping, L., Hongyun, F., Hongliang, L., et al.: Model test and numerical simulation research on the mechanical response law of lager span and small interval tunnels constructed by CD method. Tunnelling and Underground Space Technology incorporating Trenchless Technology Research, p. 132 (2023)
12. Song, Z., Tian, X., Liu, Q., et al.: Numerical analysis and application of the construction method for the small interval tunnel in the turn line of metro. Sci. Progr. **103**(3) (2020)
13. Yong, Y., Chuan, H.E., Zhuo-Xiong, X.: Study on mechanical behavior and reinforcing measures of middle rock wall of parallel tunnel with small interval. Rock Soil Mech. (2006)
14. Tang, M.M., Wang, Z.Y., Li, Y.P.: Study of construction methods for crossing bias small interval highway tunnel. Rock Soil Mech. **32**(4), 1163–1168 (2011)
15. Su, M.Z., Wang, F.X., Liu, R.: Type selection of arches in primary support of loess tunnel with small interval under shallow buried & unsymmetrical pressure. Appl. Mech. Mater. **1800**, 170–173 (2012)

Research on the Influence of Tunnel Invert Excavation on the Rheological Deformation of Different Levels of Surrounding Rock

Zhiqun Gong[1], Hongyu Guo[2], Helin Fu[3(✉)], Yibo Zhao[3], and Zetong Peng[3]

[1] China Construction Infrastructure Co., Ltd., Beijing 100044, China
[2] CCFEB Civil Engineering Co., Ltd., Changsha 410000, Hunan, China
[3] Department of Civil Engineering, Central South University, Changsha 410075, Hunan, China
fu.h.l@csu.edu.cn

Abstract. The surrounding rock in the closed section of the inverted arch creates a bearing ring that, when combined with the upper initial support, provides stable initial support. However, excavation of the inverted arch can disrupt the original balance and significantly impact the tunnel's stability. Using the classic Burgers creep constitutive model, we conducted numerical analyses of the construction process for tunnels with different closure and exposure distances of the inverted arch under varying levels of surrounding rock, using the three-step excavation method. We compared the maximum displacement of the arch crown to study the influence of the closure and exposure distances of the inverted arch on the stability of the initial support lining. Our results show that the displacement of the arch crown is primarily influenced by the strength of the surrounding rock; the lower the strength of the rock, the greater the displacement of the arch crown. Furthermore, the displacement of the arch crown increases as the closure distance of the inverted arch increases. Conversely, the exposure distance of the inverted arch has a minimal impact on arch displacement, and the longer the exposure distance, the greater the arch displacement. These findings can serve as a foundation for improving existing standards and adapting them to the spatial requirements of large-scale mechanized operations.

Keywords: Burgers Creep Constitutive Law · Numerical Calculation · Rheological Tunnel · Inverted Arch Excavation

1 Introduction

In the weak and broken surrounding rock, the inverted arch structure is often used to increase the stiffness of the tunnel support to improve the mechanical performance and control the overall deformation. It plays a very important role in the tunnel support structure. Wang et al. [1] effectively prevented the development of extreme deformation of loess tunnels by using temporary steel arch and the temporary inverted arch combination. Shreedharan et al. [2] found that the inverted arch tunnel may be more efficient in reducing roof sag and floor heave for the existing geo-mining conditions. Sung et al. [3]

© The Author(s) 2024
G. Feng (Ed.): ICCE 2023, LNCE 526, pp. 273–285, 2024.
https://doi.org/10.1007/978-981-97-4355-1_26

revealed the mechanical behavior characteristics of tunnel construction of weak rock by using finite difference software. Fang et al. [4] found that the integrity of inverted arch could effectively restrain the bottom deformation.

For tunnels excavated by the bench cut method, the invert closure distance has a great impact on the deformation and pressure of the surrounding rock. Sun et al. [5] discussed the characteristics of the influence of the closure time of the support on the stress state of the surrounding rock. Designing of the tunnel support in weak rock masses is the most time consuming task of the tunnel engineers [6]. The support design is done using the collected data. With realized optimum design, the loss of time and money is prevented [7]. C.O et al. [8] used numerical analysis to simulate the mechanical behavior of the tunnel rock mass more precisely. L. Cantieniet al. [9] evaluated the mechanical behavior of the face core through extrusion measurement. The excavation method and support system can be optimized by numerical simulation [10]. The combination of the two methods is widely used to solve the problem of extreme deformation of tunnel surrounding rock [11].

The study above put forward regular suggestions for the closure distance of the back arch, but did not propose specific values for the reasonable distance. Based on those analysis, Zhong [12] took a series of numerical analysis on the impact of disposable excavation length of loess tunnel on additional displacement around the tunnel, yielding the best disposable excavation length of loess arch. Meng [13] proposed the maximum back arch closed distance under large-section loess tunnel deformation and support force by field test measurement. Jin Baocheng [14] combined the numerical simulation of the loess tunnel to analyze the clearance convergence of the arch. However, under different surrounding rock conditions, the invert closure distance also shows greater differences in different construction methods. Therefore, the invert closure distance during tunnel excavation should be adapted to local conditions [15]. The surrounding rock has the characteristics of time-deformation. When the rheological properties are considered, the stability analysis of the tunnel will become particularly complicated, which puts forward new requirements for the distance between the excavation face and the invert closure.

There are few researches on the spatial influence of rheological tunnels on excavation step and invert closure distance. Based on this, relying on the Songshan Tunnel and Shimen Gang Tunnel projects of Jiangxi-Shenzhen High-speed Railway under construction, we study the impact of tunnel elevation arch excavation on tunnel deformation under different surrounding rock conditions based on numerical analysis software, aiming to further improve the relevant specifications and technical requirements and provide basic research on large-scale mechanized operation.

2   General Situation of the Project

GSSG-2 bid for Front Station Project of Ganzhou-Shenzhen Railway from Jiangsu-Guangdong Provincial Boundary to Tangxia Section, the length of the main line is 38.777 km. The Songgangshan Tunnel and Shimengang Tunnel are the controllable works along the whole line within the benchmark section and the most difficult ones. The total length of Songgangshan Tunnel is 9881 m, the first long tunnel along Gansu-Shenzhen Railway, and the total length of Shimengang Tunnel is 5759 m. In the area,

Class III surrounding rocks are constructed by two-step method and Class IV and V surrounding rocks are constructed by three-step method.

Both tunnels use large-scale mechanized operation, which makes it difficult to exert the mechanical effect during construction. Therefore, in this paper, a numerical analysis is conducted on the tunnel construction process with different closure distances and exposed distances of the inverted arch, considering the creep constitutive model of the surrounding rocks. Songgangshan Tunnel and Shimengang Tunnel Project is under construction. The results provide a technical basis for improving relevant specifications and large-scale mechanized operation.

3   Numerical Modeling Calculation

3.1   Numerical Model Establishment

Using FLAC3D6.0 software, we established a numerical calculation excavation model to investigate the influence of tunnel excavation on surrounding rocks, excavation mode, and support characteristics. The model followed the tunnel construction steps shown in Fig. 1, with dimensions of 80 m along the tunnel strike (longitudinal Y-axis) and 70 m in the horizontal direction orthogonal to the tunnel, a buried depth of 20 m, and a lower part depth of 30 m. The upper boundary of the calculation model adopted a free boundary condition, while the remaining sides and underside used normal restraint boundary conditions.

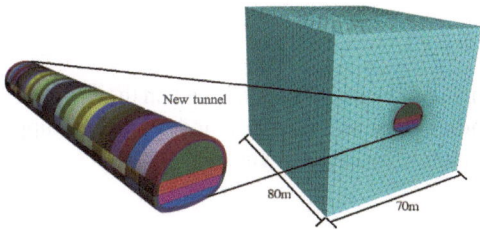

Fig. 1. Numerical analysis and calculation model

To simulate different conditions, we proposed various levels of surrounding rock conditions, using a two-step method in Class III surrounding rocks, with advance support and initial support in accordance with the construction conditions. To investigate the influence of different excavation steps and the length of the first excavation of the inverted arch on tunnel deformation, we varied the excavation steps at 30 m, 40 m, 45 m, 50 m, 55 m, and 60 m, and the length of the first excavation of the inverted arch at 3 m, 4 m, 5 m, and 6 m.

3.2   Numerical Simulation Steps

The three-dimensional simulation considers the three step excavation method under various levels of surrounding rock conditions, and the excavation diagram is shown in Fig. 2:

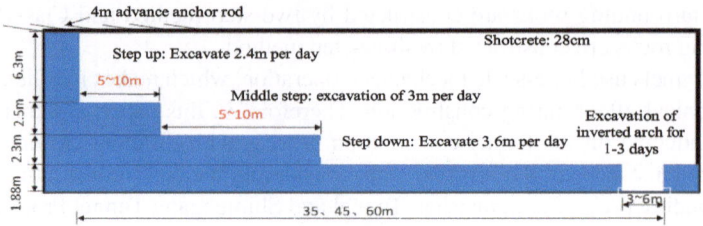

Fig. 2. Diagram of three-step excavation

To conduct the calculation, the model followed the following steps: (1) advanced support by increasing the mechanical strength parameters of rock and soil within the advanced support range; (2) excavation of upper steps; (3) first support of the upper steps; (4) excavation of 10 steps (24.0 m) on the upper bench followed by excavation of the middle bench; (5) primary support for the middle step; (6) construction of two excavation steps (7.2 m) for the middle step, followed by excavation of the lower step; (7) initial support provided for the lower steps; (8) excavation of the inverted arch after three excavation steps (10.8 m) of the lower step without initial support. The creep model calculation had an excavation cycle time of 36000 s (10 h) for each step. The excavation duration was 17 days for a 30 m closed distance, 22 days for a 40 m closed distance, 25 days for a 45 m closed distance, 29 days for a 50 m closed distance, 35 days for a 55 m closed distance, and 38 days for a 60 m closed distance.

3.3 Material Parameter Determination

In the modeling of tunnel excavation process, based on the experiment, the basic physical parameters of surrounding rock, initial support and secondary support materials, soil and support materials are mainly involved in Table 1.

Table 1. Basic parameters of simulated materials.

Material Science	Deformation modulus/GPa	Poisson's ratio	Cohesion/MPa	Internal friction angle/°	Density/kg/m³
V-grade surrounding rock	18	0.30	0.05	25	2000
Initial expenditure	1.6	0.4	/	/	2500
Secondary support	20	0.25	/	/	2500
IV-grade surrounding rock	30	0.25	0.35	30	2200
III-grade surrounding rock	40	0.22	0.50	35	2500

3.4 Creep Constitutive Law and Its Parameters

The classic Burgers creep model only considers rock viscoelasticity and cannot accurately depict the creep behavior of soft rocks, which typically exhibit instantaneous plasticity, elasticity, viscoplasticity, and viscoelasticity in Fig. 3.

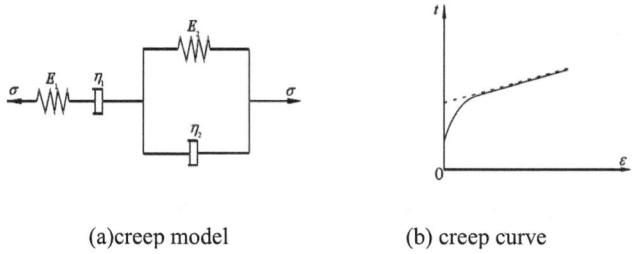

(a)creep model   (b) creep curve

Fig. 3. Typical burgers creep model and characteristic curve.

Under constant axial stress σ_0, the axial strain $\varepsilon(t)$ is:

$$\varepsilon(t) = \frac{\sigma_0}{9K} + \frac{\sigma_0}{3G_1} + \frac{\sigma_0}{3G_2} + \frac{\sigma_0}{3G_2}e^{-E_2/\eta_2 t} + \frac{\sigma_0}{3\eta_1}t \tag{1}$$

In Eq. (1), K represents the bulk modulus of the rock sample, G_1 represents the elastic shear modulus of the model, and G_2 represents the modulus of control delay elasticity, η_1 and η_2 determine the rate of viscous flow and delayed elasticity in the model, respectively.

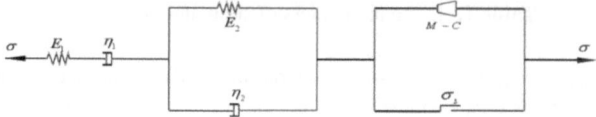

Fig. 4. Modified burgers creep model.

However, weak rocks often exhibit simultaneous properties of instantaneous plasticity, instantaneous elasticity, viscoplasticity, and viscoelasticity. To address this issue, using creep tests, we establish a new creep model that simulates the viscoelastic plastic properties of various rock samplesy. A new plastic element based on the Mohr Coulomb criterion, is added to the Burgers creep model as the basic model. Before the stress reaches the yield stress determined by the Mohr Coulomb criterion, the element's strain is 0. Once the stress is greater than or equal to the yield stress, it fully follows the Mohr Coulomb plastic flow law, in Fig. 4.

Table 2 presents the rheological parameters of the Kelvin and Maxwell bodies in the modified Burgers model for different levels of surrounding rock, which were determined based on on-site sampling and indoor experiments.

Table 2. Creep parameters of surrounding rocks of different grades.

Material Science	shear-maxwell	viscosity-maxwell	shear-kelvin	viscosity-kelvin
V-grade surrounding rock	3.333×10^9	7.2×10^{12}	3.8×10^9	1.8×10^{11}
IV-grade surrounding rock	3.67×10^9	7.2×10^{12}	4.5×10^9	1.8×10^{11}
III-grade surrounding rock	4.2×10^9	7.2×10^{12}	5.0×10^9	1.8×10^{11}

3.5 Numerical Simulation Based Analysis of Arch Settlement of V-Class Surrounding Rock Tunnel

Numerical simulations used the ideal elastic-plastic model based on the Mohr Coulomb criterion. The inverted arch had a closed distance of 30 m, and a one-time excavation of 6 m was conducted for the inverted arch. The three-dimensional simulation results can be observed in Fig. 5.

The maximum settlement curve of the arch crown is presented in Fig. 6, with a maximum settlement deformation of only 2.5 mm.

The modified Burgers creep model was employed to numerically calculate the excavation model of a V-level surrounding rock three-step tunnel, with a 30 m closed distance for the inverted arch and a one-time excavation of 6 m for the inverted arch. The simulation results in three dimensions are presented in Fig. 7.

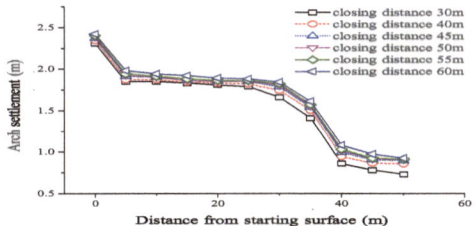

Fig. 5. Settlement curve with Mohr-Coulomb model

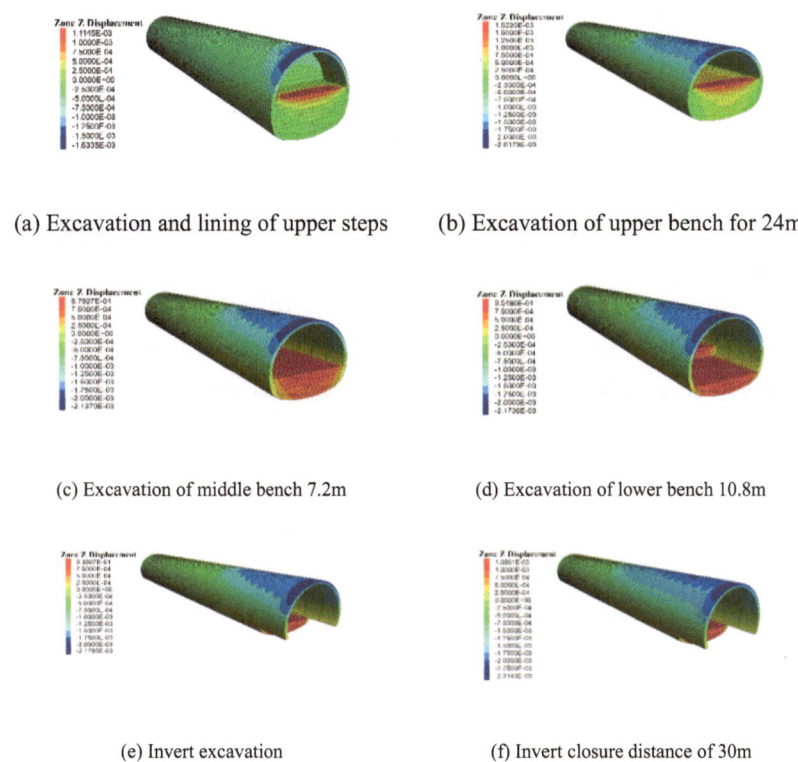

(a) Excavation and lining of upper steps   (b) Excavation of upper bench for 24m

(c) Excavation of middle bench 7.2m   (d) Excavation of lower bench 10.8m

(e) Invert excavation   (f) Invert closure distance of 30m

Fig. 6. Deformation nephogram of three-step excavation tunnel with Mohr-Coulomb criterion

Figure 8 presents the maximum settlement curve of the arch crown for various closure distances. The maximum settlement of the arch crown reaches 24 mm when the closure distance is 60 m. The settlement distribution is more realistic compared to the Mohr Coulomb criterion based three-step excavation model. The failure to consider the effect of stress release and timeliness on the mechanical and deformation properties of surrounding rocks and deformation of the arch crown during excavation. During excavation, the deformation at the arch waist of the unclosed section increases with excavation steps. However, the deformation at the arch waist of the unclosed section of the inverted arch remains stable, which differs significantly from the actual monitoring

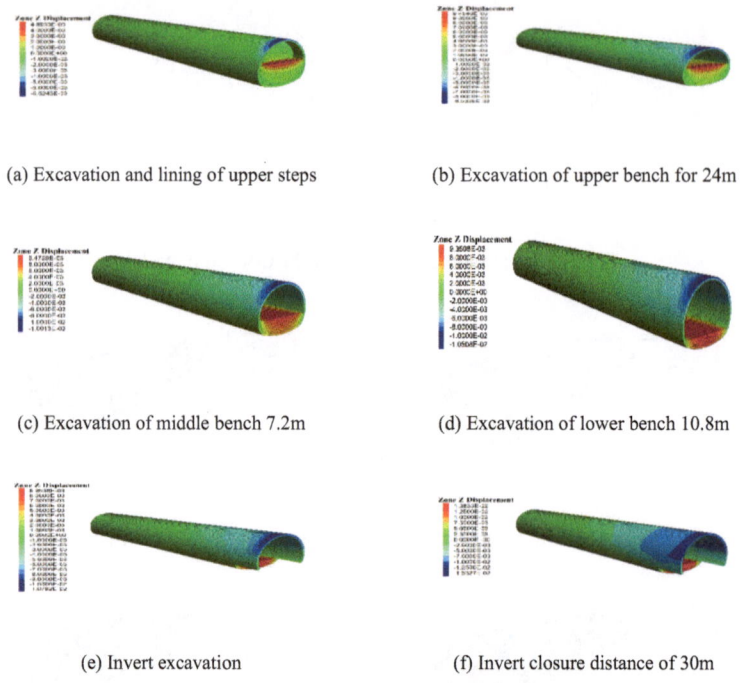

(a) Excavation and lining of upper steps   (b) Excavation of upper bench for 24m

(c) Excavation of middle bench 7.2m   (d) Excavation of lower bench 10.8m

(e) Invert excavation   (f) Invert closure distance of 30m

Fig. 7. Deformation nephogram of three-step excavation tunnel with modified Burgers model

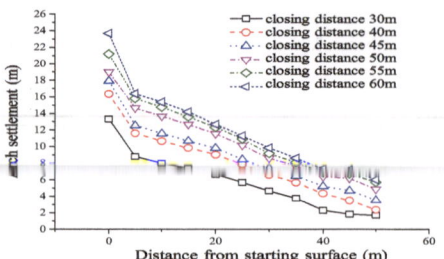

Fig. 8. Calculating the settlement curve of vault with modified Burgers creep model

situation. It is essential to consider the decrease in rock strength due to stress release after excavation and the time-dependent nature of excavation steps.

4   Analysis of the Influence of Tunnel Arch Subsidence

4.1   Impact of Invert Step Distance on Arch Crown Subsidence in V-Class Surrounding Rock Tunnel

Based on the above method, we calculated the settlement of the arch crown caused by different closure distances for V-grade surrounding rock in Table 3. As the closure distance increases from 30 m to 60 m, the maximum displacement of the arch crown increases from 13.33 mm to 25.1 mm.

Table 3. Maximum displacement of vault under different closure distances of inverted arch in V-grade surrounding rock

Closing distance (m)	Excavation time (day)	Maximum displacement of vault (mm)
30	17	13.33
40	22	16.38
45	25	17.93
50	29	21.71
55	35	23.9
60	38	25.1

4.2   Influence of Excavation Length of inverted arch of Class V Surrounding Rock Tunnel on Vault Settlement

The settlement of the arch crown caused by different lengths of the exposed inverted arch in V-grade surrounding rock was calculated.

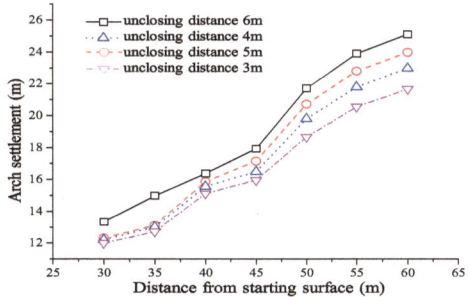

Fig. 9. Variation curve of vault settlement

The results are presented in Fig. 9 and Table 4. The calculation results show that as the exposed distance of the inverted arch decreases, the maximum settlement of the arch decreases, indicating that a smaller exposed distance of the inverted arch leads to stronger initial support stability and a more pronounced initial support force ring effect. However, the settlement curve shows that the maximum settlement of the arch is not sensitive to the exposed distance of the inverted arch, indicating that the effect of the exposed distance of the inverted arch on stability is not significant.

Table 4. Maximum displacement of inverted arch without closure of 3-6m in V grade rockmass

Closing distance (m)	Unenclosed inverted arch 6 m Maximum displacement of vault (mm)	Unenclosed inverted arch 5 m Maximum displacement of vault (mm)	Unenclosed inverted arch 4 m Maximum displacement of vault (mm)	Unenclosed inverted arch 3 m Maximum displacement of vault (mm)
30	13.33	12.3	12.2	11.98
35	14.98	13.1	13	12.7
40	16.38	15.85	15.55	15.1
45	17.93	17.13	16.47	15.94
50	21.71	20.7	19.78	18.64
55	23.9	22.78	21.77	20.54
60	25.1	23.96	22.95	21.64

4.3   Influence of Surrounding Rock Class on Arch Settlement

The stability of a tunnel is primarily influenced by the nature of the surrounding rocks. Therefore, investigating the relationship between the closure distance of the inverted arch and the maximum settlement of the initial support arch under different surrounding rock classes, the strength of the surrounding rocks was varied based on the aforementioned model.

The results obtained using the Burgers creep model are shown in Fig. 10 and Table 5. The maximum settlement value of a V-grade surrounding rock vault is larger than that of III and IV surrounding rocks, and the maximum settlement of a IV surrounding rock vault is slightly higher than that of a III surrounding rock vault at each closure distance. With an increase in the closure distance, the maximum settlement value of the vault also increases. An increase in the grade of surrounding rocks results in a significant reduction of the displacement of the tunnel vault. The stability of initial support is closely linked to the grade of surrounding rocks. When the surrounding rock conditions are good, the inverted arch closure distance can be increased, and the exposed distance can also be increased to facilitate the operation of large machinery. Conversely, when the surrounding rock conditions are poor, the inverted arch closure distance should be shortened, and the exposed distance of the inverted arch reduced, while initial support should be promptly applied to ensure the stability of the surrounding rock. Additionally, increasing the length of circular footage and reducing the time under good surrounding rock conditions can significantly enhance the stability of the tunnel.

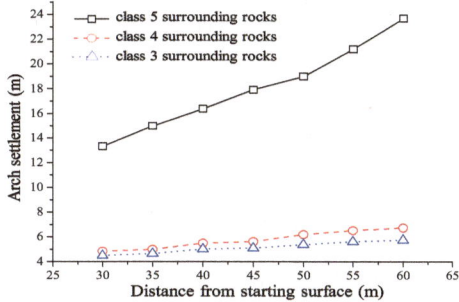

Fig. 10. Tunnel vault settlement curves caused by different levels of surrounding rock and different excavation methods

Table 5. Numerical results

Closing distance (m)	Maximum Settlement of class V Surrounding Rock Arch (mm)	Maximum Settlement of Class IV Surrounding Rock Arch (mm)	Maximum Settlement of Class III Surrounding Rock Arch (mm)
30	13.33	4.82	4.49
35	14.98	4.98	4.63
40	16.38	5.48	5.00
45	17.93	5.61	5.08
50	21.71	6.18	5.36
55	23.9	6.5	5.6
60	25.1	6.72	5.74

5   Conclusion

(1) A decrease in the closure distance of the inverted arch results in a reduction of the maximum settlement of the vault. The stability of the initial support is stronger with a smaller closure distance. The maximum settlement of the vault is not sensitive to the exposed distance of the inverted arch. So the influence of the distance of the inverted arch on the stability of the initial support is not significant. After mechanized operation, the sealing distance of V-grade surrounding rock inverted arch can be controlled within a range of 55 m, while that of IV-grade surrounding rock inverted arch can be controlled within a range of 70 m. For III-grade surrounding rock, the sealing distance can be controlled depending on the specific situation.

(2) The excavation length of V-grade surrounding rock inverted arch can be controlled within a range of 5 m at one time.

(3) The step method is suitable for excavation in Grade III, IV, and V surrounding rocks, and the length can be controlled based on the settlement of the vault caused by the excavation.

Acknowledgement. This study was financially supported by the Hunan Provincial Construction Science and Technology Plan Project (KY202109).

References

1. Ke, W., Sx, B., Yzb, E., et al.: Deformation failure characteristics of weathered sandstone strata tunnel: a case study. Eng. Failure Anal. (2021)
2. Shreedharan, S., Kulatilake, P.: Discontinuum–equivalent continuum analysis of the stability of tunnels in a deep coal mine using the distinct element method. Rock Mech. Rock Eng. **49**(5) (2016)
3. Choi, S.O., et al.: Stability analysis of a tunnel excavated in a weak rock mass and the optimal supporting system design. Int. J. Rock Mech. Min. Sci. (2004)
4. Fang, Y., Ding, K.: Mechanical effects analysis of inverted arch. In: IOP Conference Series: Materials Science and Engineering, vol. 490, no. 3, p. 032026 (6pp) (2019)
5. Sun, M., Zhu, Y., Li, X., et al.: Experimental study of mechanical characteristics of tunnel support system in hard cataclastic rock with high geostress. Shock Vibr. **2020**(5), 1–12 (2020)
6. Study on Mechanical Properties of Long Column for Steel Concrete under Axial Compression. Anhui Architecture (2012)
7. Aksoy, C.O., Onargan, T.: The role of umbrella arch and face bolt as deformation preventing support system in preventing building damages. Tunn. Undergr. Space Technol. **25**(5), 553–559 (2010)
8. Aksoy, C.O., Ogul, K., Topal, I., et al.: Reducing deformation effect of tunnel with non-deformable support system by jointed rock mass model. Tunnell. Underground Space Technol. Incorp. Trenchless Technol. Res. **40**(feb.), 218–227 (2014)
9. Cantieni, L., Anagnostou, G., Hug, R.: Interpretation of core extrusion measurements when tunnelling through squeezing ground. Rock Mech. Rock Eng. **44**(6), 641–670 (2011)
10. Jza, B., Zta, B., Xwa, B., et al.: Engineering characteristics of water-bearing weakly cemented sandstone and dewatering technology in tunnel excavation (2022)
11. Barla, G., Bonini, M., Semeraro, M.: Analysis of the behaviour of a yield-control support system in squeezing rock. Tunnell. Underground Space Technol. Incorp. Trenchless Technol. Res. **26**(1), 146–154 (2011)
12. Zhong, Z.L., Liu, X.R., Yuan, F., et al.: Effect of excavation length of inverted arch in one step on stability of multi-arch tunnels in loess. Chin. J. Geotech. Eng. (2008)
13. Meng, D., Tan, Z.: Deformation control technology and supporting structure stress of large section loess tunnel. China Civ. Eng. J. (2015)
14. Jin, B., Wang, X.: Research on reasonable distance of invert closure in loess tunnel by pre-cutting method. Railw. Eng. (2016)
15. Mingqing, D.U., Dong, F., Ao, L.I., et al.: Mechanism and failure mode of floor heave in tunnel invert of high speed railway under expansive surrounding rock. China Railw. Sci. (2019)

Study on Dynamic Response of Gas Insulated Line Pipe Gallery Under Vehicle Traveling Loads

Lin Li[1], Zhenzhong Wei[1], Yunan Jiang[1], Yong Liu[1], Junrong Gong[2], Kunjie Rong[2(✉)], and Li Tian[2]

[1] Shandong Electric Power Engineering Consulting Institute Co. Ltd., Jinan 250013, Shandong, China

[2] School of Civil Engineering, Shandong University, Jinan 250061, Shandong, China
kunjierong@sdu.edu.cn

Abstract. This study relies on the Huangbuling 500 kV transmission project to investigate the dynamic response law of gas insulated line (GIL) pipe gallery under vehicle traveling loads. A numerical model of GIL pipe gallery considering soil-structure interaction is developed using finite element software ABAQUS, and the effect of soil pressure on GIL is studied through static analysis. This study proposes a continuous step loading method for simulating vehicle traveling loads, and parametric analyses of different traveling directions and speeds are carried out to reveal the GIL's dynamic response law in depth. The results show that the maximum stresses in the concrete and steel reinforcement cage of the GIL under soil pressure are located at the root of the right-side wall (both are less than the strength design value), and the maximum value of displacement is 6.556 cm occurs in the middle of the top plate. The maximum vertical acceleration, velocity and displacement responses under vehicle load occur at the top midpoint of the pipe gallery, while the maximum stress response occurs at the lower left corner. When heavy vehicle passes over the pipe gallery at different speeds, the peak acceleration, velocity and stress of the pipe gallery tends to increase with the increase of vehicle speed, while the peak displacement does not change significantly.

Keywords: Vehicle Traveling Load · Gil Pipe Gallery · Soil Pressure Analysis · Dynamic Response · Parametric Analysis

1 Introduction

With the "Western Development" and "West-to-East Power Transmission" strategy, China has launched a large number of large-scale hydroelectric power stations, these large-scale projects are mostly arranged in the way of underground plant, and the construction of their supporting transmission projects is more difficult, and gas-insulated transmission lines (GIL) is one of the main approaches to solving this problem. The use of GIL pipe gallery in the Huangbuling 500 kV power transmission project alleviated the problem of coil out pipe gallery restriction, and improved the durability of engineering

© The Author(s) 2024
G. Feng (Ed.): ICCE 2023, LNCE 526, pp. 286–296, 2024.
https://doi.org/10.1007/978-981-97-4355-1_27

pipelines and reduced their maintenance costs. However, vehicle operation often causes undesirable vibrations in GIL, and there will be hidden dangers such as pipeline scattering and structural collapse, increasing the maintenance cost of the transmission project in later period.

Recently, scholars have conducted a large number of studies on the mechanical properties and dynamic response of integrated pipe galleries based on field tests and numerical simulations [1–6]. Yu [7] conducted numerical simulation analysis of a typical integrated pipe gallery system and nonlinear dynamic analysis of the integrated pipe gallery and its surrounding soil using finite element software.

When an integrated pipe gallery inevitably crosses traffic routes, the pipe gallery is subject to traffic loads generated by moving vehicles. In this regard, related scholars have developed a series of researches on the dynamic response of integrated pipe gallery under traffic load [8–10]. Jiang L et al. [11] discussed the effect of different speeds on the structure of the integrated pipe gallery under the same working condition, and analyzed the dynamic stress change rule of the integrated pipe gallery under the actual traffic load. The results show that the effect of vehicle speed on the left wall and axillary region of the longitudinal integrated pipe gallery is greater than that of the cross-section.

Moreover, relevant scholars have conducted a series of studies on GIL pipe gallery [12, 13]. Tang P et al. [14] analyzed in detail the seismic performance of the cross-river GIL integrated pipe gallery structure in terms of structural deformation, shield tunnel segment opening and structural seismic damage for the special characteristics of cross-river GIL integrated pipe gallery structure.

Based on the above, most of the current studies mainly focus on integrated pipe galleries, and some of them involve GIL pipe galleries, while there are even fewer studies on the dynamic response of GIL pipe galleries under vehicle loads.

The rest of this study is structured as follows: Sect. 2 develops a coupled numerical model of GIL considering soil-structure interaction. Section 3 carries out the analysis of the effect of soil pressure on the structural response of GIL. Section 4 proposes a simulation method for vehicle traveling loads and investigates the dynamic response law of GIL in depth by parametric analysis of traveling direction and speed. Finally, the systematic conclusions of this study are presented in Sect. 5.

2 Soil-GIL Coupling Numerical Model

2.1 Engineering Situations

The structure type of GIL pipe gallery of this project is reinforced concrete structure, the pipe gallery is made of C35 waterproof concrete, the cross-section size is 5.5 m × 5.2 m, and the top plate is 1.5 m from the ground surface. The cushion is made of 100-thick C20 concrete. Steel bars are of HPB300 and HRB400 grade. The GIL pipe gallery structure is shown in Fig. 1. The parameters of the steel bars used in the pipe gallery are shown in Table 1. According to the geotechnical engineering investigation report of this project, the soil layer around the pipe gallery mainly consists of silty clay, fully weathered volcanic breccia and strongly weathered volcanic breccia.

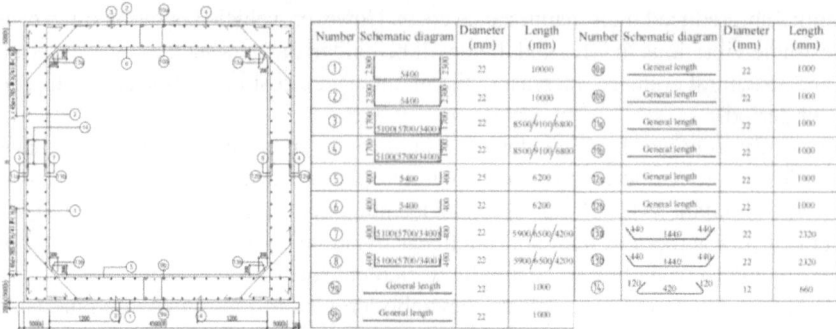

Fig. 1. Structure of GIL pipe gallery

2.2 Basic Assumptions

The following basic assumptions are considered in the numerical modeling process: 1) the underground pipe gallery structure can be simplified as an elastic structure for analysis. The soil body will be considered as an elastic material in the numerical simulation; 2) The soil is considered to be uniformly distributed in layers without considering the internal pore ratio of the soil and the effect of groundwater; 3) the materials are all isotropic and homogeneous.

2.3 Finite Element Modeling

The finite element analysis software ABAQUS is used to establish the numerical model of GIL pipe gallery considering soil-structure interaction, as shown in Fig. 2. The width of the road above the pipe gallery is 5.5 m, the direction of the length of the pipe gallery is perpendicular to the direction of the road, and the length of the pipe gallery is taken as 5.5 m. The model size in the vertical pipe gallery length direction is taken as 40 m. The final size of the finite element model is determined to be 40 m × 5.5 m × 20 m. In the finite element model, the soil body, GIL pipe gallery and concrete cushion are simulated using solid element (C3D8R); the reinforcement is simulated using wire element (B31). After the meshing is completed, the number of model nodes is 30096 and the number of cells is 33009.

The "Embedded" in "Constraints" is used between the concrete and the steel reinforcement cage inside the pipe gallery; "Surface-to-surface contact" is used between the pipe gallery, the concrete cushion and the silty clay, and the tangential behavior in the "Contact Properties" defines the friction coefficient $\mu = 0.194$, and the normal behavior is defined as "Hard Contact"; Only the normal behavior between the fully weathered volcanic breccia and the strongly weathered volcanic breccia is defined as "Hard Contact"; The pipe gallery and the concrete cushion are poured as a single entity, with "Tie" in "Constraints" between the two. The soil surrounding the pipe gallery has a large proportion of volcanic breccia, which are in rigid contact with each other, so the soil body boundary conditions can be defined as fixed constraints. The material parameters of the soil are shown in Table 1 and the material parameters of the GIL structure are detailed in Table 2.

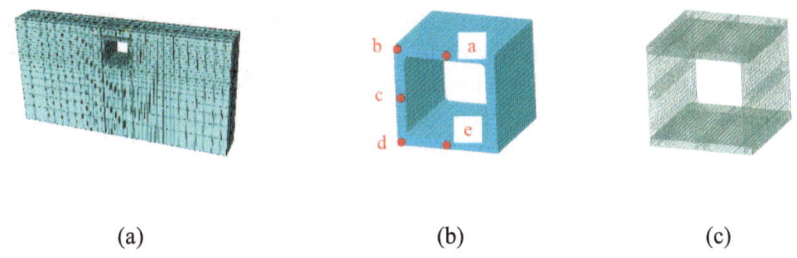

(a)   (b)   (c)

Fig. 2. GIL pipe gallery-soil body finite element model. (a) Overall finite element model; (b) GIL model; (c) Reinforcement cage model

Table 1. Soil material parameters.

No.	Soil thickness (m)	Density (kg/m³)	Friction coefficient	Cohesion (Pa)	Friction angle	Poisson's ratio	Elastic modulus
1	2.5	1930	0.194	49000	10.9°	0.25	31.25 MPa
2	0.95	2590	0	0	0°	0.2	52.65 MPa
3	16.55	2590	0	0	0°	0.2	135 MPa

Note: Soil layer numbers 1, 2 and 3 are silty clay, fully weathered volcanic breccia and strongly weathered volcanic breccia, respectively.

Table 2. GIL structure material parameters.

Material	Material strength	Elastic modulus (Pa)	Density (kg/m³)	Poisson's ratio	Yield stress (MPa)
Concrete	C40	3.15E10	2500	0.2	35
Steel bar	HRB400	2E11	7850	0.3	330
	HPB300	2.1E11	7850	0.3	270

3 Static Analysis of Soil Pressure in GIL Pipe Gallery

3.1 Soil Pressure Analysis

After geo-stress balance, a static general analysis step is created to analyze the effect of soil pressure on the pipe gallery. The stress and displacement clouds of the concrete and the reinforcement cage of the pipe gallery are shown in Fig. 3. It can be seen that overlying soil body gravity and pipe gallery's own gravity, region of high structural stress in the pipe gallery appears in the pipe gallery side wall root, the maximum value of stress is located in the pipe gallery right side wall root (1.814 MPa) under the effect of lateral soil pressure. The concrete strength used in the pipe gallery is C35 and the design value of strength is 16.7 MPa, which is much larger than 1.814 MPa, the pipe gallery concrete strength meets the requirements. Region of high stress in the reinforcement cage appears in the cage side root and the middle of the bottom surface, the maximum value of stress is located in the cage right side root (17.94 MPa). The longitudinal bars strength used in the cage are of HRB400 grade and the design value of tensile and compressive strength is 360 MPa, which is much larger than 17.94 MPa, the cage strength meets the requirements. From the displacement cloud map, it can be seen that overlying soil body gravity and pipe gallery's own gravity, the maximum displacement response of the concrete and the maximum displacement of the steel reinforcement cage appear in the middle of the top plate and are both 6.556 cm under the effect of lateral soil pressure.

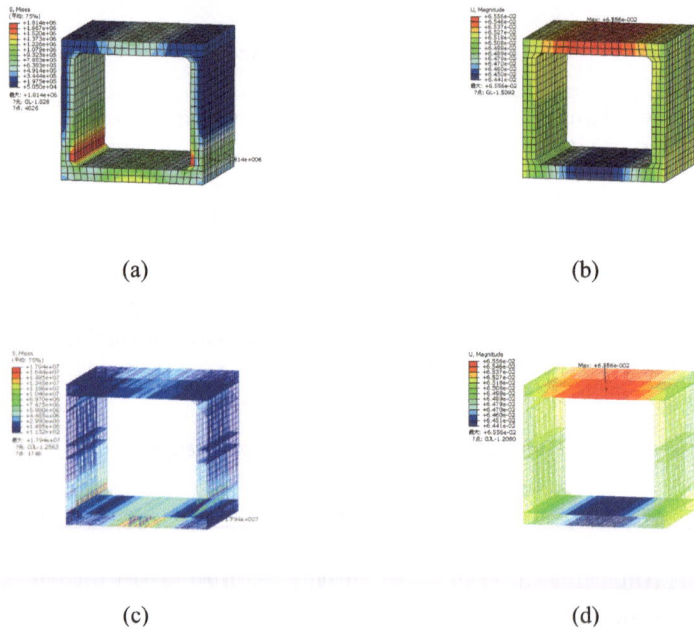

(a)   (b)

(c)   (d)

Fig. 3. The stress and displacement clouds of the concrete and the reinforcement cage of the pipe gallery. (a) Concrete stress cloud; (b) Concrete displacement cloud map; (c) Reinforcement cage stress cloud map; (d) Reinforcement cage displacement cloud map

4   Dynamic Response Analysis of GIL Under Vehicle Traveling Loads

4.1   Vehicle Traveling Load Simulation

In order to simulate the vehicle load in a more standardized way, the vehicle load passing through the road with buried GIL pipe gallery is now expressed according to the standard vehicle load on bridges in the Technical Specification of Urban Road Engineering (GB51286–2018), which is the Urban-Class B vehicle load, as shown in Table 3. Figure 4 shows a distribution diagram of the vehicle traveling load.

Table 3. Urban-Class B vehicle load

Axle number	Unit	1	2	3	4	5
Axle weight	KN	30	120	120	140	140
Wheel weight	KN	15	60	60	70	70
Longitudinal wheelbase	m		3.0	1.4	7.0	1.4
Transverse center distance per axle group	m	1.8	1.8	1.8	1.8	1.8
Wheel footprint area	m	0.3×0.2	0.6×0.2	0.6×0.2	0.6×0.2	0.6×0.2

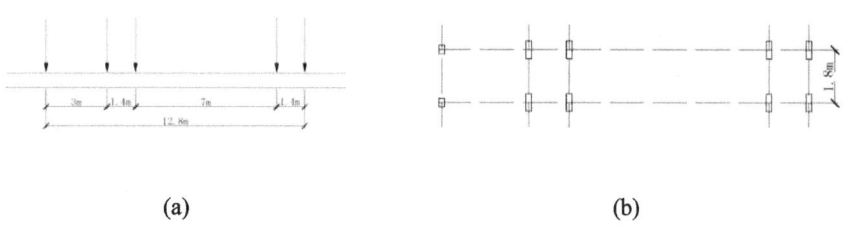

(a)   (b)

Fig. 4. Distribution diagram of the vehicle traveling load. (a) Elevation distribution diagram; (b) Plane distribution diagram.

In this study, the application of vehicle traveling loads is achieved by using the continuous step loading method, and the spacing between the vehicle loads applied by two adjacent steps is set to 3 m. In the model, the length of the soil body in the vertical pipe gallery length direction is 40 m, and the length of the selected standard heavy vehicle is 12.8 m, after calculation, a total of ten steps are required to move the vehicle from one side of the soil to the other in the direction of the vertical GIL to simulate the whole process of vehicle passing through the GIL.

4.2 Dynamic Response Analysis

Four speeds, 10 km/h, 20 km/h, 30 km/h and 40 km/h, are selected to analyze the dynamic response of the pipe gallery. The vehicle load conditions are shown in Table 4.

Table 4. Vehicle load conditions

Condition No.	Vehicle speed (km/h)	Individual step time (s)
1	10	1.08
2	20	0.54
3	30	0.36
4	40	0.27

In order to comprehensively analyze the effect of heavy vehicle passing through on various parts of the GIL pipe gallery, a total of five locations in the pipe gallery cross-section are selected for analysis: a) midpoint at the top of the pipe gallery, b) upper-left corner of the pipe gallery, c) midpoint of the pipe gallery side wall, d) lower-left corner of the pipe gallery, and e) midpoint at the bottom of the pipe gallery, as shown in Fig. 2(b). By extracting the dynamic response time travel curves of heavy vehicle traveling vertically in the direction of the pipe gallery in the above mentioned parts under different working conditions, the dynamic response of different locations of the pipe gallery to the heavy vehicle load transmitted from the soil body above is analyzed, and at the same time, the location of the pipe gallery structure most affected by the heavy vehicle load is summarized.

A comparison of the peak vertical and horizontal acceleration, velocity, displacement response and stress response of each part of the GIL pipe gallery when the vehicle speed is 10 km/h is shown in Fig. 5. It can be seen that the vertical acceleration, velocity and

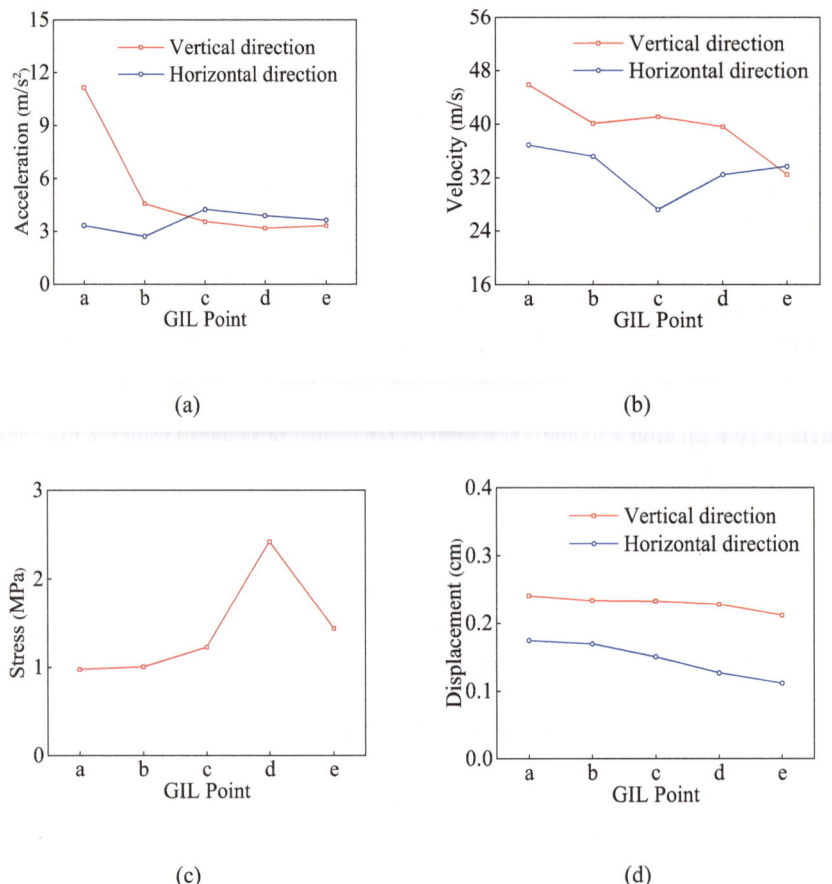

Fig. 5. Peak response of each measurement point in the GIL pipe gallery at a vehicle speed of 10 km/h. (a) Acceleration; (b) Velocity; (c) Displacement; (d) Stress

displacement responses of the midpoint at the top of the pipe gallery are the largest, and subsequently only the midpoint at the top of the pipe gallery is extracted as a characteristic point for vertical acceleration, velocity and displacement response analysis; the lower-left corner of the pipe gallery has the largest stress response, and subsequently only the lower-left corner of the pipe gallery is extracted as a characteristic point for stress response analysis.

Figure 6(a) shows the acceleration response of the GIL pipe gallery under each working condition, and it can be seen that when heavy vehicle passes over the pipe gallery at different speeds, the peak acceleration of the pipe gallery tends to increase with the increase of the vehicle speed, and the peak acceleration reaches a maximum of 18.99 m/s^2 when the vehicle speed reaches 40 km/h. The analysis results of velocity response are illustrated in Fig. 6(b), it can be seen that when heavy vehicle passes over the pipe gallery at different speeds, the peak velocity of the pipe gallery tends to increase with the increase of the vehicle speed, and the peak velocity reaches a maximum of 50 mm/s when the vehicle speed reaches 40 km/h. Figure 6(c) shows that when heavy vehicle passes over the pipe gallery at different speeds, the peak stress of the pipe gallery

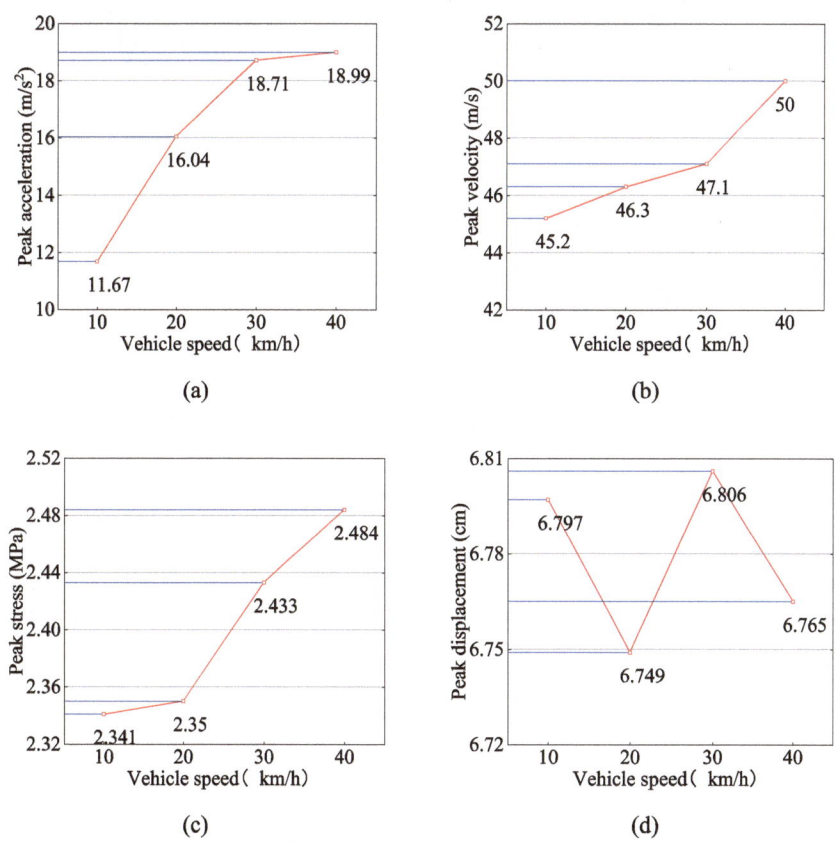

Fig. 6. Peak dynamic response of GIL in vertical direction under each working condition. (a) Acceleration; (b) Velocity; (c) Displacement; (d) Stress

tends to increase with the increase of the vehicle speed, and the peak stress reaches a maximum of 2.484 MPa when the vehicle speed reaches 40 km/h. For Fig. 6(d), when heavy vehicle passes over the pipe gallery at different speeds, the peak displacement of the pipe gallery does not change significantly and fluctuates within a small range as the speed increases, and the peak displacement reaches a maximum of 6.806 cm when the vehicle speed reaches 30 km/h.

Figure 7 illustrates the dynamic response cloud map of the GIL in vertical direction for the condition where the peak response is the largest, the maximum acceleration and velocity values occur at the GIL's top midpoint, the maximum stress value occurs at the GIL's lower-left corner, and the maximum displacement value occurs at the GIL's right top, and the dynamic response of the other parts is small.

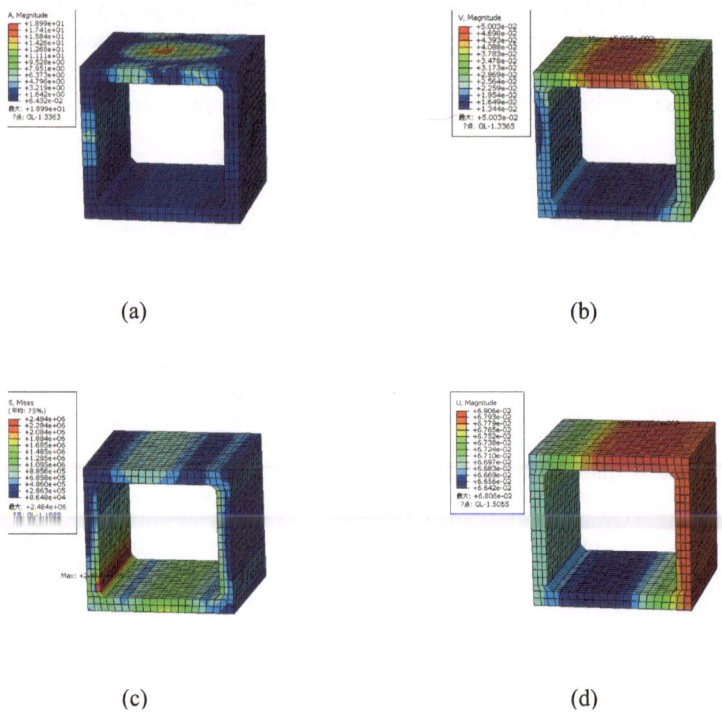

(a)   (b)

(c)   (d)

Fig. 7. Dynamic response cloud map of GIL in vertical direction for working conditions with maximum peak response. (a) Acceleration (Condition 4); (b) Velocity (Condition 4); (c) Displacement (Condition 4); (d) Stress (Condition 3)

5   Conclusions

This study relies on the Huangbuling 500 kV transmission project to investigate the dynamic response law of GIL pipe gallery under vehicle traveling loads. A numerical model of GIL pipe gallery considering soil-structure interaction is developed using finite element software ABAQUS, and the effect of soil pressure on GIL is studied through static analysis. A continuous step loading method for simulating vehicle traveling loads is proposed, and parametric analyses of different traveling directions and speeds are carried out to reveal the GIL's dynamic response law in depth. The following conclusions are mainly obtained:

(1) Based on soil pressure analysis, the Max. Stress value of pipe gallery is located in the root of the right-side wall (1.814 MPa), which is much smaller than the concrete strength design value. The Max. Stress value of the reinforcement cage is located in the root of the right-side face (17.94 MPa), much smaller than the tensile and compressive strength design value. The Max. Value of displacement occurs in the middle position of the top plate (6.556 cm).
(2) The vertical acceleration, velocity and displacement responses are largest at the midpoint of the top of the pipe gallery and the stress response is largest at the lower-left corner under vehicle load. When heavy vehicle passes over the pipe gallery at different speeds, the peak acceleration, velocity and stress of the pipe gallery tends to increase with the increase of vehicle speed, while the peak displacement does not change significantly.

References

1. Xie, J., Huang, N., Feng, J., et al.: Study on deformation of shallow buried underground pipe gallery. In: IOP Conference Series: Materials Science and Engineering, vol. 741, no. 1, p. 012058. IOP Publishing (2020)
2. Yu, J., Sang, L.: Seismic finite element analysis of underground integrated pipe corridor based on generalized reaction displacement method. Appl. Math. Nonlinear Sci. (2023)
3. Bo-Tuan, D., Pan, L.I., Xin, L.I., et al.: Mechanical behavior of underground pipe gallery structure considering ground fissure. J. Mt. Sci. **19**(2), 16 (2022)
4. Chen, J., Shi, X., Li, J.: Shaking Table test of utility tunnel under non-uniform earthquake wave excitation. Soil Dyn. Earthq. Eng. **30**(11), 1400–1416 (2010)
5. Qian, H., Zong, Z., Wu, C., et al.: Numerical study on the behavior of utility tunnel subjected to ground surface explosion. Thin-Walled Struct. **161**, 107422 (2021)
6. Wang, Y., Jin, X., Yang, Q.: Stress analysis of surrounding rock of rectangular pipe gallery in Mountain City. In: IOP Conference Series Materials Science and Engineering, vol. 739, p. 012001 (2020)
7. Yu, J.: Research on nonlinear dynamic analysis of seismic performance of comprehensive pipe gallery. In: 2nd International Conference on Applied Mathematics, Modelling, and Intelligent Computing (CAMMIC 2022), vol. 12259, pp. 1261–1267. SPIE (2022)
8. Yuan, Z., Cao, Z., Cai, Y., et al.: An analytical solution to investigate the dynamic impact of a moving surface load on a shallowly-buried tunnel. Soil Dyn. Earthq. Eng. **126**, 105816 (2019)
9. Yi, H., Qi, T., Qian, W., et al.: Influence of long-term dynamic load induced by high-speed trains on the accumulative deformation of shallow buried tunnel linings. Tunn. Undergr. Space Technol. **84**, 166–176 (2019)
10. Tang, L., Quan, Y., Zhu, Y., et al.: Application of improved calculation method considering the vehicle loads in branch utility tunnel. Geotech. Geol. Eng. **37**, 251–266 (2019)
11. Jiang, L., Xie, Z., Huang, Y.: Prototype test study on dynamic stress of utility tunnel under traffic load. J. Vibroeng. **22**(1), 170–183 (2020)

12. Huang, T., Mao, X., Yu, Y., et al.: Structural health monitoring of UHV GIL shield cross-river electric power pipe gallery. In: 2021 International Conference on Power System Technology, pp. 2357–2363. IEEE (2021)

13. Wang, S., He, N., Wang, H., et al.: Study on the characteristics and prediction of tunneling-induced ground movements in sutong GIL utility tunnel. Electr. Power Surv. Des. **05**, 39–45 (2022)

14. Tang, P., Yang, M., Zhuang, H., et al.: Lateral seismic performance of the utility tunnel crossing the Yangzi river. J. Disast. Prevent. Mitig. Eng. **42**(03), 507–515 (2022)

Mechanical Property Analysis of a Deployable Tape-Spring Boom Using for Aerospace Structures

Sicong Wang, Haizhen Sun$^{(\boxtimes)}$, and Lining Sun

School of Mechanical and Electrical Engineering, Soochow University, No. 8 Jixue Road, Suzhou 215137, Jiangsu, China
hzsun@suda.edu.cn

Abstract. Deployable coilable tape-spring booms have many advantages especially for use in space, such as light-weight, high folded-ratio and small storage volume. The boom's folding and deploying process is accompanied with the large-scaled deformation of thin-walled materials, which may damage the boom on stress concentration points. For the sake of avoiding failures during coiling and deployment process, this paper aimed at acquiring the critical points on a boom which were vulnerable to be destroyed. Since the interactions of the boom's infinitesimals were complicated, a numerical model was considered to be introduced. Meanwhile, the mechanical properties of the boom's deployed state were also analyzed for a better design, and both the deploying and deployed behaviors were further analyzed through parametric study. The research of this paper will give more guidance on the design of tape-spring booms and the selection of the key parameters.

Keywords: Deployable · Tape-spring · Mechanical Property · Aerospace

1 Introduction

Extendable thin-walled tape-spring boom is a new kind of deployable mechanism which has been widely used for many applications especially for aerospace technologies. This should thank to its advantages such as light-weight, high folded-ratio and self deployable property. The boom is usually used for combining with flexible membranes forming large-scale boom-membrane structures after wholly deployed, such as membrane antennas, solar arrays, solar sails, space-telescope star-shaders, etc.

A diagram is presented in Fig. 1 to illustrate the deployment process of the boom. The boom looks like a carpenter's tape, which is like a slit boom when fully deployed. Before the deployment, the boom is coiled and locked onto a cylindrical hub, and thus some strain-energy is restored in the boom material in the process. During the deployment, either the boom tip or the hub is selected to be fixed (depending on working conditions), and the other side would be released and deployed under the drive of the restored strain-energy. In some conditions the boom is needed to be placed into a restraint mechanism

G. Feng (Ed.): ICCE 2023, LNCE 526, pp. 297–308, 2024.
https://doi.org/10.1007/978-981-97-4355-1_28

for a better control (usually with a motor), while sometimes it can be self-deployed [1, 2].

The boom's first famous practical use in space was on Hubble Telescope as the supporting structures for the solar arrays. In this mission, the booms deployed and worked on orbit successfully at the beginning and were unfortunately out of work before long because of the bending and buckling failures [3]. Apart from the semi-circular style boom as shown in Fig. 1 (which was commonly called Storable Tubular Extendable Member, STEM), the boom with a lenticular cross-section was also used for engineering on a solar sail by DLR, Germany [4]. The lenticular boom (commonly called Collapsible Tubular Mast, CTM) was able to obtain relatively higher torsional stiffness because of its closed cross-section configuration. Besides, a boom with a triangular cross-section (commonly called Triangular Rollable And Collapsible, TRAC) was also invented and launched in NanoSail-D mission by NASA. The triangular boom could acquire higher bending stiffness and buckling load as its cross-section geometry was more dispersed, however, the torsional stiffness was relatively low just because of this [5, 10, 12].

The existing research concerning tape-spring booms was mainly concentrated on analyzing the boom's moving behaviors during deployment process, and upgrading or optimizing the boom's cross-section configurations for acquiring better mechanical properties. However, since the deployment process was accompanied by large-scale and complex deformations on the boom materials, the booms were easily to be damaged by extremely high stress caused by stress concentration [6]. Therefore, finding the stress concentration point which was the most vulnerable to be damaged during the whole deployment process was necessary to be investigated. As the interaction of the boom infinitesimals was too complicated to be analyzed, establishing a numerical model was considered for use. Section 2 built a finite element model of a tape-spring boom for analyzing the stress concentration during the deployment. For giving some verification and providing some support on the corresponding parametric study, the analytical method for driving force calculation was also carried out. Sec. 3 focused on investigating the mechanical properties of the boom's deployed state based on both numerical and theoretical methods. According to the analysis in Sects. 2 and 3, Sect. 4 studied the parametric impact on the boom's deploying and deployed states, and Sect. 5 concluded the whole paper. The research in this paper would provide more guidance on the design of extendable tape-spring booms and the parametric selections.

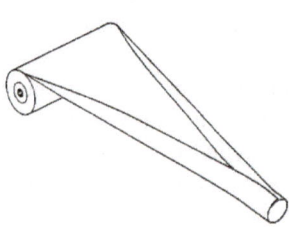

Fig. 1. Tape-spring boom diagram (STEM)

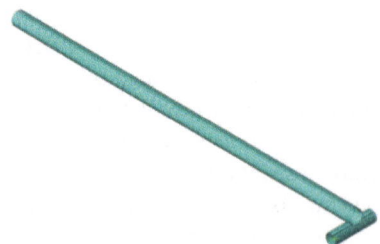

Fig. 2. Numerical model of a tip-spring boom

2   Deployment Process Analysis

Deployment process was one of the most important modes of a tape-spring boom, which directly determined the working success of a boom and even the corresponding membrane structures. From the experimental study previously, tearing-failure caused by stress concentration was easily to be found on a boom material because of the large deformations happening in this process [7, 8]. Therefore, finding the concentration points and acquiring the highest stress a boom needed to sustain were necessary for boom designs and material selections.

Since a boom's infinitesimal deformations were too complicated to study through analytical methods, a numerical model was considered to be used for the analysis. Meanwhile, the deployment process could be regarded as the inverse process of coiling at low moving speed, which was easier to be analyzed. From above, a numerical model of a tape-spring boom with a central hub was established in Abaqus which was presented in Fig. 2. The parameters used in the model were selected to mimic those used in InflateSail CubeSat mission which was launched in 2015 where the definitions and values of the parameters were listed in Table 1 [9], and the corresponding geometric diagram was shown in Fig. 3 (in which O_1 presented the boom's circular center, and I_{x2} marked the neutral surface of the cross-section). Based on the simulation experience, the S4R shell elements were selected and the rectangle meshes were used [9]. According to the element refinement, twenty meshes along the boom's cross-section were proved to be appropriate through considering calculation amount and simulation precision comprehensively, as the model with forty meshes output very similar results. For the sake of universality of the analysis, the isotropic material was introduced into the model. Furthermore, the material property of the central hub was set as rigid body since the hub deformation was able to be ignored during the deployment, while the mesh size was selected to be appropriate for the boom meshes. For obtaining the stress distribution caused by boom deformation accurately, the quasi-static analysis was used for all the steps in the simulation. The simulation steps and analysis details were shown as follows:

Table 1. Boom parameters

Name	Symbol	Units	Value
Opening-angle	θ	deg	10
Boom external radius	r	mm	180
Hub external diameter	R	mm	150
Boom length	l	m	20
Wall wall-thickness	t	mm	0.8
Modulus of elasticity	E	GPa	70

Step 1 introduced equal edge loads on the boom's both edges for making the boom flat; **Step 2** added a surface pressure acting on the boom root making the root attached and fixed on the central hub; **Step 3** released the edge loads and the root pressure, and

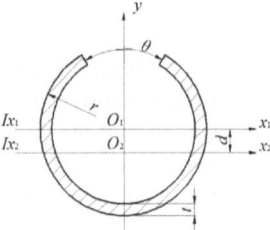

Fig. 3. Parameter diagram of the boom cross-section

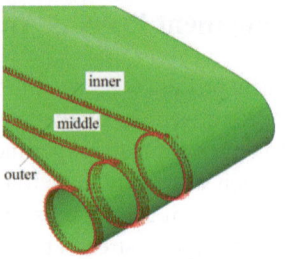

Fig. 4. Definition for outer, middle and inner longitudinal cross-sections of a coiled boom (1 turn)

spun the hub for coiling the boom on the hub (during this step, a slight concentrate force was introduced on the boom tip pointing at the opposite side of the hub for preventing the boom's uncoiled section rotating around the hub, and the force was removed at the end of the step). The stress distribution of the boom during the coiling process (also regarded as the deployment process) was obtained and recorded in Step 3.

2.1 Stress Distribution During Deployment

Based on the numerical model established in Abaqus, the stress distribution on the boom during coiling (deploying) could be acquired. Since the boom's configuration was symmetric during the whole process, the results from one side of the symmetry were recorded from the analysis. The forth strength theory was selected for analysis and Von Mises stress was acquired from the model, because plastic deformation of the materials was strictly prohibited during the whole process. Figure 4 showed the definition of the outer, middle and inner (longitudinal) cross-sections, while Fig. 5 presented Von Mises stress along the cross-sections respectively. The results listed in Fig. 5 were the distribution after the boom was coiled for one turn as a representative. In this figure, x-axis showed the actual path-length along the longitudinal cross-section, and 1 m boom-length from the root was present since this section with sharp stress change was easier to be failed.

According to the plots shown in Fig. 5, it could be found that the sharp stress change happened at the boom's root region and the connecting region between the coiled section and the transitional section, while the plots away from the two regions changed smoothly. Meanwhile, the highest stress of the inner cross-section was higher than that of the other two cross-sections. Moreover, the highest stress occurred at the inner cross-section near the boom root. From above, the highest stress point would appear on the middle of the boom root when coiling.

Furthermore, for checking the stress variation trend during coiling, the stress of the boom's transversal cross-section on the boom root was acquired from the model. Figure 6 showed the stress on the root when the boom was coiled for one turn to three turns as a representative for finding the changing regularity, while the simulation contour was also presented in Fig. 7 for providing more clarification. The definition of x-axis in Fig. 6

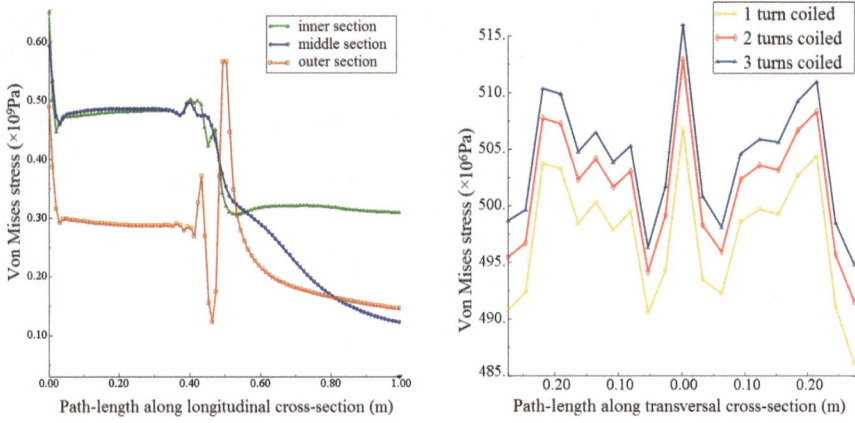

Fig. 5. Stress distribution along the boom's longitudinal cross-sections

Fig. 6. Stress distribution on boom root when coiling (1–3 turns)

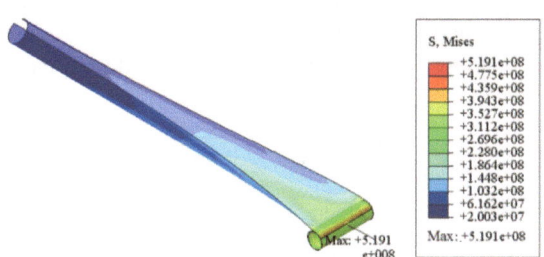

Fig. 7. Simulation contour

was the path-length along the boom's transversal cross-section, and the stress presented in Fig. 6 and Fig. 7 were both Von Mises stress.

From Fig. 6 and Fig. 7, it could be observed that the stress along the transversal cross-section rose with the number of the coiled turns, while the plots from different turns looked fairly similar. Also, all the plots were generally symmetric around $x = 0$ because of the boom's configuration. The stress plots in Fig. 6 ought to be continuously along the cross-section, while the broken lines listed was because of the mesh limit. Nevertheless, the plots in this figure was precise enough to acquire the variation regularity of the stress.

To sum up, the boom's most vulnerable point during deployment was always staying at the middle of the transversal cross-section on the boom root, and the value of this point rose gradually following the growing of the number of the coiled turns.

2.2 Driving Torque Analysis

Based on the numerical model established, the boom's deployment torque was also able to be obtained through adding another analysis step after Step 3 in the simulation as follows.

Step 4 released the restraints acted on the central hub for making the boom deploy automatically, and the driving torque on the hub was recorded during the deployment process (3 turns for deployment).

In order to avoid calculation abortion, explicit analysis was used for this step, and the corresponding numerical results were been given in Fig. 8. According to the simulation plots in this figure, the torque overall increased with the deployment progress. This was because the bending curvatures of the inner coil layers was smaller than those on the outer coil layers, which gave the boom higher spring-back ability. Meanwhile, the fluctuation of the plots was mainly caused by the boom's untight coiling resulted from mesh limit of the finite element models (see Fig. 9 for clarification), and the troughs appearing at the starting point of each turn were caused by the hump on the connecting part between the boom and the central hub.

For the sake of giving some verification for the numerical results and providing some support on the parametric study in the following sections, an analytical model for driving torque analysis was also established.

When coiled on the central hub, the boom's longitudinal cross-section could be approximately regarded as an Archimedes spiral (see Fig. 10 where a_1 presents the boom's coiling angle around the hub)[10–13]. According to the geometric analysis from Fig. 10, the boom's bending/coiling curvature when coiled at a_1 can be expressed as

$$\kappa_{\alpha_1} = \frac{1}{R + \frac{t}{2} + \frac{t}{2\pi}\alpha_1} \tag{1}$$

while, the boom's coiled length is given as:

$$l_1 = \int_0^{\alpha_1} \left[\left(R + \frac{t}{2} \right) + \frac{t}{2\pi}\alpha \right] d\alpha \tag{2}$$

Based on the elastic mechanical theory, the boom's driving torque when deploying at the point with the angle a_1 is shown as $M = \kappa_{\alpha_1} E I_z$ where

$$I_z = \frac{\left(r - \frac{t}{2} \right) \cdot \left(\pi - \frac{\theta}{2} \right) \cdot t^3}{6} \tag{3}$$

which is given through material mechanical method.

By introducing the results calculated from the theoretical analysis into Fig. 8, it could be found that the numerical results were generally scattering around the corresponding analytical results. Therefore, the numerical model was available for analyzing the mechanical properties of the boom during coiling/deploying process.

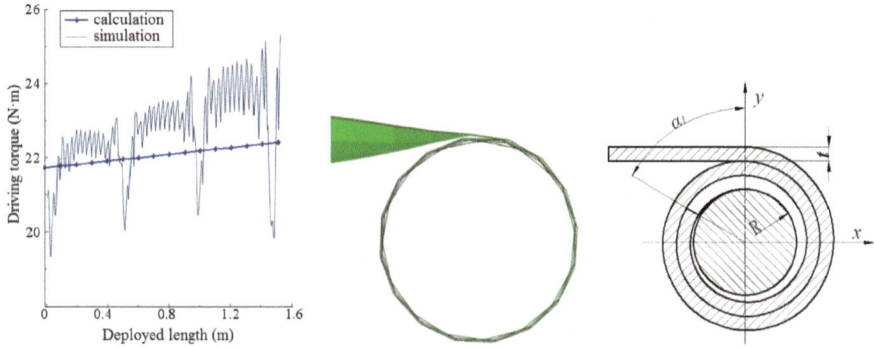

Fig. 8. Driving torque during deployment

Fig. 9. Untight coiling caused by mesh limit

Fig. 10. Diagram of longitudinal cross-section

3 Deployed State Analysis

Bending stiffness and buckling load are a boom's critical properties when at its deployed state. The numerical model built in Sect. 2 is able to be applied for deployed state analysis as well, while an analytical model will also be established for providing some verification and for the further parametric study.

3.1 Bending Stiffness Analysis

Based on the numerical model established in Sect. 2, the bending stiffness around x-axis and y-axis (axis definitions shown in Fig. 3.) could be acquired respectively. The numerical results were listed in Table 2.

For giving more verification and providing support on parametric study, the theoretical analysis for bending stiffness was presented as follows.

According to the material mechanical methods [11], the moment of inertia around x-axis I_x can be given as:

$$I_x = \int_A y^2 dA = 2 \int_{r-t}^{r} \gamma d\gamma \int_{-\frac{\pi}{2}}^{\frac{\pi}{2}-\frac{\theta}{2}} r^2 \sin^2 \vartheta \, d\vartheta \tag{4}$$

where A is the area of the boom's transversal cross-section which can be expressed as:

$$A = \pi \left[r^2 - (r-t)^2 \right] \frac{2\pi - \theta}{2\pi} \tag{5}$$

Based on Eq. (5) and geometric analysis in Fig. 3, the boom's inertia moment (deployed state) around x_1-axis can be shown as:

$$I_{x_1} = r^3 t \left[1 - \frac{3}{2}\left(\frac{t}{r}\right) + \left(\frac{t}{r}\right)^2 - \frac{1}{4}\left(\frac{t}{r}\right)^3 \right] \left(\pi - \frac{\theta}{2} - \frac{\sin\theta}{2} \right) \tag{6}$$

Table 2. Numerical and analytical results comparison

Bending stiffness	Numerical result	Analytical result	Relative error
Around x-axis ($\times 10^5$ N·m^2)	1.1551	1.2094	4.50%
Around y-axis ($\times 10^5$ N·m^2)	1.6677	1.7765	6.12%

Because $t < < r$, i.e. $t/r \to 0$, Eq. (6) can be simplified written as:

$$I_{x_1} = r^3 t \left(\pi - \frac{\theta}{2} - \frac{\sin\theta}{2} \right) \tag{7}$$

Meanwhile, the distance between x_1-axis (circular center) and x_2-axis (neutral surface) d is able to be found through:

$$d = \bar{y} = \frac{\iint_A y\, d\Lambda}{A} = \frac{2 \int_{r-t}^{r} y\, dy \int_{-\frac{\pi}{2}}^{\frac{\pi-\theta}{2}} r\sin\vartheta\, d\vartheta}{A} \tag{8}$$

According to the parallel-axis theorem, the inertia moment around x_2-axis is presented as:

$$I_{x_2} = I_{x_1} - d^2 A = I_{x_1} - \bar{y}^2 A \tag{9}$$

Therefore, the boom's (deployed state) bending stiffness around x-axis is acquired through $U_x = EI_{x_2}$.

In the similar way, the bending stiffness around y-axis can be shown by analogizing with Eqs. (6) and (9) as:

$$U_y = E \cdot I_y = E r^3 t \left(\pi - \frac{\theta}{2} + \frac{\sin\theta}{2} \right) \tag{10}$$

where y-axis is the boom's symmetric surface and neutral surface simultaneously in this case.

According to Eqs. (9) and (10), the boom's (deployed) bending stiffness around x-axis and y-axis are also listed in Table 2. By comparing the results from numerical model and theoretical analysis, it can be observed that the errors are relatively low (around 5%), and the results generally match with each other well. Besides, the analytical result is slightly higher that that from the numerical model, and this is because the boom's cross-sections are totally non-deformable in the theoretical analysis.

3.2 Buckling Load Analysis

Buckling is also a critical mechanical property for a boom's deployed state, which is researched by numerical analysis and analytical method in this subsection. For numerical analysis, a reference point was introduced at the circular center of the boom tip, which was rigidly connected with the tip edge for mimicking a closure head commonly used on the tip (especially for boom-membrane structures). The load was acted on the reference point aiming at the boom root which had been fully fixed before the analysis. By using linear perturbation step in Abaqus, the boom's first buckling load P_{cr} was acquired which was listed in Table 3. Similarly with the bending stiffness analysis, analytical method by Euler formula as Eq. (13) was also used to provide some analytical verification, whose result was presented in Table 3.

Table 3. Buckling load results comparison

	Numerical result	Analytical result	Relative error
Buckling load P_{cr} (N)	533.16	746.02	28.53%

According to the comparison in Table 3, the error between the numerical and analytical results was relatively high (up to 30%) and the analytical value was higher the numerical one. This was because the boom was not only bent but also twisted under the the pressure of the concentration force in the simulation, while the Euler formula was only used for calculating the cases under ideal conditions (flexural buckling). Therefore, the theoretical method was only able to be used for sketchy calculating the boom's buckling load in this case, and the numerical method was more precise for predicting.

4 Parametric Study

For the sake of providing some guidance for a boom design, based on the analytical methods in Sects. 2 and 3, a parametric study was carried out in this section.

4.1 Driving Torque Study

Based on the analytical method in Sec. 2, the impact of the boom radius (r), hub radius (R), boom wall-thickness (t) and half opening-angle ($\theta/2$) on the boom's driving torque was investigated (shown in Fig. 11). The figure was plotted when the boom was coiled for 2 turns as a representative. On account of the plots in Fig. 11, the driving torque rose with the increasing of boom radius and decreasing of hub radius. This was because higher boom radius made the boom have longer cross-section path-length, and the lower hub radius led to higher coiling curvature which provided higher strain-energy for drive during the deployment. Moreover, the driving torque also went up when the boom wall-thickness was rising or the opening angle was reducing with the similar reasons. Meanwhile, the plots changed non-linearly in the figures, and increasing the values of the boom wall-thickness and the hub radius was more efficient for improving the boom's driving torque.

4.2 Bending Stiffness Study

Furthermore, the parametric study on the boom's bending stiffness was also investigated based on the theoretical analysis in Sect. 3. The influences of boom wall-thickness (t), hub radius (R), boom wall-thickness (t) and half opening-angle ($\theta/2$) on the bending stiffness around x-axis (easier to be failed according to the study in Subsect. 3.1) was presented in Fig. 12.

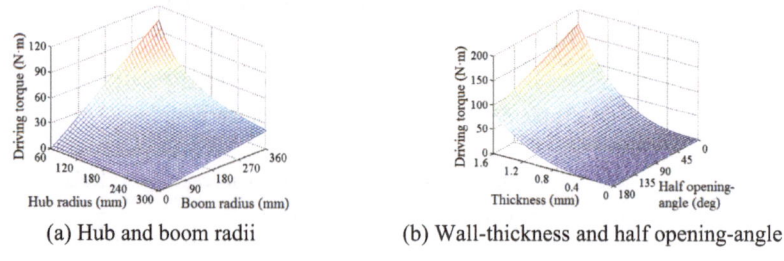

(a) Hub and boom radii (b) Wall-thickness and half opening-angle

Fig. 11. Driving torque parametric study

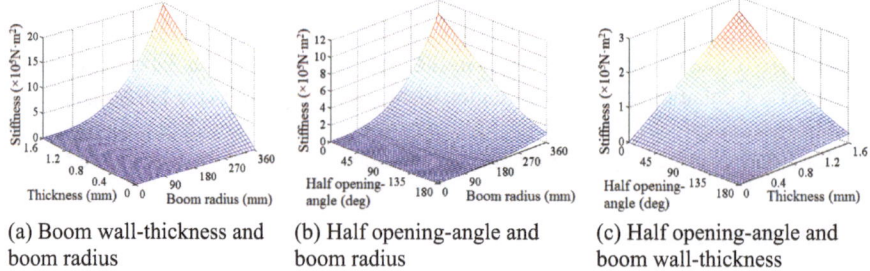

(a) Boom wall-thickness and boom radius (b) Half opening-angle and boom radius (c) Half opening-angle and boom wall-thickness

Fig. 12. Parametric study of boom bending stiffness around x-axis

According to the plots shown in Fig. 12, the bending stiffness rose linearly with the increasing of the boom wall-thickness, and the stiffness went up non-linearly with the deceasing of the half opening-angle and the increasing of the boom radius. Further, from the figure, changing the boom radius was the most efficient way to improve the boom's bending stiffness around x-axis.

5 Conclusion

Extendable tape-spring boom is a new kind of deployable mechanism which is widely using for many applications especially for aerospace technologies in recent years. As the boom is a large-deformation flexible mechanism during folding or deployment process, failure caused by stress concentration is one of the most common ways for boom damage. Therefore, acquiring the point with the highest stress on the boom during the whole process is necessary to be carried out. Since the interactions of the infinitesimals were fairly complicated during the deformation process, the numerical method was considered to be introduced into the investigation.

This paper established a numerical model of an isotropic tape-spring boom. Through analyzing the stress distributions along the boom's longitudinal and transversal cross-sections, the point with the highest stress during the deployment process was found.

Meanwhile, the boom's deployment driving force was also obtained from the simulation. For the sake of verifying the results from numerical analysis and providing support for further parametric study, the theoretical model was also built for acquiring the boom's driving force during deployment. From the comparison, the results from the two methods generally matched well. Furthermore, the bending stiffness and buckling load at the boom's deployed state were analyzed through both numerical model and theoretical analysis as well. Afterwards, a parametric study was carried out to find the most efficient method for improving the boom's driving torque and bending stiffness. From the investigation, increasing the boom wall-thickness was the most efficient way to improve the driving force, and increasing the boom radius was the best mode for promoting the bending stiffness. The research from this paper will provide more guidance on tape-spring boom designs and the corresponding parameter selections.

The authors would like to acknowledge that the research presented in this paper was carried out with the aid of the National Natural Science Foundation of China (Grant No. 52205027) and the Science and Technology Project of Jiangsu Province (Grant No. BK20220496).

References

1. Wang, S., Xu, S., Lu, L., Sun, L.: Roll-out deployment process analysis of a fiber reinforced polymer (FRP) composite tape-spring boom. Polymers **15**, 2455 (2023)
2. Wang, S., Schenk, M., Jiang, S., Viquerat, A.: Blossoming analysis of composite deployable booms. Thin-walled Struct. **193–194**, 141–151 (2020)
3. Lake, M.S., Francis, W.H.: Robust. Highly scalable solar array system. In: 3rd AIAA Spacecraft Structures Conference, San Diego, California (2016)
4. Leipold, M., Widani, D., Groepper, P.: The European solar sail deployment demonstration mission. In: 57th International Astronautical Congress, Valencia, Spain (2006)
5. Jeon, S.K., Footdale, J.N.: Scaling and optimization of a modular origami solar array. In: 2018 AIAA Spacecraft Structures Conference, Florida (2018)
6. Strauble, M., Block, J..: Deployable composite booms for various gossamer space structures. In: 52nd AIAA/ASME/ASCE/AHS/ASC Structures, Structural Dynamics and Materials Conference, Denver, Colorado (2011)
7. Cui, Y., Santer, M.: Characterisation of tessellated bi-stable composite laminates. Compos. Struct. **137**, 93–104 (2016)
8. Mallol, P., Tibert, G.: Deployment modeling and experimental testing of a bi-stable composite boom for small satellites. In: 54th AIAA/ASME/ASCE/AHS/ASC Structures, Structural Dynamics, and Materials Conference, MA (2013)
9. Mallol, P., Mao, H.N., Tibert, G.: Experiments and simulations of the deployment of a bi-stable composite boom. In: 54th AIAA/ASME/ASCE/AHS/ASC Structures, Structural Dynamics, and Materials Conference, MA (2013)
10. Wang, S., Schenk, M., Guo, H., Viquerat, A.: Tip force and pressure distribution analysis of a deployable boom during blossoming. Int. J. Solids Struct. **193**, 141–151 (2020)
11. Guest, S.D., Pellegrino, S.: Analytical models for bistable cylindrical shells. Proc. Math. Phys. Eng. Sci. **462**, 839–854 (2006)
12. Daton-Lovett, A.: Extendible Member. US Patent US6217975B1 (2001)
13. Timoshenko, S.P., Woinowsky-Krieger, S.: Theory of plates and shells (1959)

Reliability Analysis of Surrounding Rock Stability in Hydraulic Tunnels Based on Fuzzy Random Theory

Jiangyu Liu[1], Xiao Sun[1,2(✉)], Yan Liu[1], and He Wang[1]

[1] Nanchang Institute of Technology, Nanchang 330099, Jiangxi, China
512930251@qq.com
[2] Jiangxi Engineering Research Center of Water Engineering Safety and Resources Efficient Utilization, Nanchang 330099, Jiangxi, China

Abstract. The study of the stability of rock masses in hydraulic tunnels is a highly complex problem with significant uncertainty, involving a large amount of randomness and fuzziness. Traditional deterministic analysis methods are insufficient in capturing the aforementioned characteristics. Therefore, based on fuzzy random theory, this paper establishes a fuzzy random reliability calculation model for the stability of rock masses in hydraulic tunnels. The model constructs a functional relationship that represents the safety margin of the stability, taking into account the fuzziness and randomness of the failure events in hydraulic tunnels. The proposed model is applied to an engineering case study, and the computed results align well with the actual conditions. This model provides a more accurate reflection of the real state of the rock masses in hydraulic tunnels and offers a scientifically rational alternative for calculating the stability of rock masses in hydraulic tunnels.

Keywords: Hydraulic Tunnels · Rock Masses · Fuzzy Random Theory · Stability

1 Introduction

Hydraulic tunnels are crucial components in water infrastructure projects, serving important functions such as diversion and water conveyance. However, the stability of the rock masses directly affects the safety of tunnel structures. Due to the complex geological nature of rock masses, which consist of various types of rocks and structural planes, their stress states are also influenced by groundwater and in-situ stress conditions. Therefore, the stability analysis of rock masses in hydraulic tunnels is an uncertainty problem that involves fuzziness and randomness [1].Currently, in the analysis of rock mass stability in hydraulic tunnels, methods such as the safety factor method and finite element method are commonly used. These methods share the common characteristic of representing many design parameters with single values, often failing to reflect the actual stress conditions of the structures.

© The Author(s) 2024
G. Feng (Ed.): ICCE 2023, LNCE 526, pp. 309–315, 2024.
https://doi.org/10.1007/978-981-97-4355-1_29

Fuzzy random reliability theory considers the randomness of rock mechanics parameters, dimensions of structures, loads, and lining material properties. Engineering structural reliability regards these engineering variables as random variables [2, 3]. Zou Shanshan proposed a probability-based Miner's rule using fuzzy theory to predict the fatigue life of critical components under random load stress and obtained a fatigue life prediction model [4].

Currently, in water infrastructure projects, there are still many factors that impact hydraulic structures, and the design parameters vary greatly, while the number of statistically available parameters is limited. Therefore, reliability theory is less commonly used in hydraulic structures [5, 6]. This paper primarily focuses on the fuzzy random reliability theory and establishes a fuzzy random reliability model for the stability of rock masses in hydraulic tunnels. The calculation method of fuzzy random reliability for rock masses is considered when support structures are present.

2   Establishment of the Fuzzy Random Reliability Calculation Model for the Stability of Rock Masses in Hydraulic Tunnels

2.1   The Construction of Function Function

Establishment of safety reserve function:

In order to facilitate analysis, computation, and practical application, a functional relationship is constructed to represent the safety margin of rock masses in hydraulic tunnels. The safety margin is defined as:

$$Z = R - S \tag{1}$$

S—Load Effects on Rock Masses; R—Structural Resistance.

2.2   Basic Assumptions and Fundamental Mechanical Models.

Basic Assumptions and Fundamental Mechanical Models. In order to avoid excessive complexity in the established model and solution process, assumptions are adopted for the analytical process inclueded: the rock mass is considered as an isotropic continuum;the initial stress field in the rock mass is due to self-weight and is in a state of hydrostatic pressure ($\lambda = 1$);the tunnel is relatively long, and the analysis is conducted based on the assumption of plane strain, within the small deformation range;the tunnel cross-section is assumed to be circular; after yielding, the rock mass still satisfies the **Mohr-Coulomb** criterion; only the overall deformation and failure of the rock mass are considered.

Determination of Equivalent Strain Values around the Tunnel Periphery. For axisymmetric problems, when body forces are neglected, the equilibrium differential equation is given by:

$$\frac{\partial \sigma_r}{\partial r} + \frac{\sigma_r - \sigma_\theta}{r} = 0 \tag{2}$$

The stress within the plastic zone satisfies the plasticity criterion:

$$\sigma_{\theta p} = \frac{1 + \sin \varphi}{1 - \sin \varphi} \sigma_{rp} + \frac{2c \times \cos \varphi}{1 - \sin \varphi} \tag{3}$$

When the tunnel is freshly excavated without lining, the radial load at the inner edge of the plastic zone, which is the tunnel periphery, is zero. This serves as the boundary condition at this stage. By using Eqs. (2) and (3) along with the boundary condition, the stress field and displacement field of the surrounding rock can be determined:

$$\varepsilon_{\theta p} = -u \Big|_{Rp} \frac{Rp}{r^2} = \frac{1 + \mu}{E} \cdot (p \cdot \sin \varphi + c \cdot \cos \varphi) \frac{R_p^2}{r^2}; \bar{\varepsilon} = \sqrt{\frac{2}{3} \cdot e_{ij} \cdot e_{ij}}; \varepsilon_{1p} = \varepsilon_{rp}, \varepsilon_{2p} = 0, \varepsilon_{3p} = \varepsilon_{\theta p} \tag{4}$$

In the equation, R_p represents the radius of the plastic zone, c is the cohesive strength, φ is the internal friction angle, r is the radius of the tunnel.

Thus, the strain field of the surrounding rock in the plastic zone is obtained as:

$$\varepsilon_{rp} = u \Big|_{Rp} \frac{Rp}{r^2} = \frac{1 + \mu}{E} \cdot (p \cdot \sin \varphi + c \cdot \cos \varphi) \frac{R_p^2}{r^2}; \varepsilon_{\theta p} = -u \Big|_{Rp} \frac{Rp}{r^2} = \frac{1 + \mu}{E} \cdot (p \cdot \sin \varphi + c \cdot \cos \varphi) \frac{R_p^2}{r^2} \tag{5}$$

According to the plasticity theory, the equivalent shear strain is obtained as:

$$\bar{\varepsilon} = \sqrt{\frac{2}{3} \cdot e_{ij} \cdot e_{ij}} \tag{6}$$

As the tunnel is a plane strain problem, $\varepsilon_{1p} = \varepsilon_{rp}, \varepsilon_{2p} = 0, \varepsilon_{3p} = \varepsilon_{\theta p}$.
So the equivalent variation field is obtained as:

$$\bar{\varepsilon} = \frac{2\sqrt{3}}{3} \cdot \frac{1 + \mu}{E} \cdot \frac{R_0^2}{r^2} \cdot \left[\frac{(p + c \cdot ctg\varphi)(1 - \sin \varphi)}{d \cdot ctg\varphi} \right]^{\frac{1 - \sin \varphi}{\sin \varphi}} \cdot (p \cdot \sin \varphi + c \cdot \cos \varphi) \tag{7}$$

Since $(\bar{\varepsilon})_{p_i} = 0$ is maximum at the perimeter of the tunnel, decreasing with increasing radius r, In Eq. (7) let $r = R_0$ then we have:

$$\bar{\varepsilon} = \frac{2\sqrt{3}}{3} \cdot \frac{1 + \mu}{E} \cdot \frac{R_0^2}{r^2} \cdot \left[\frac{(p + c \cdot ctg\varphi)(1 - \sin \varphi)}{d \cdot ctg\varphi} \right]^{\frac{1 - \sin \varphi}{\sin \varphi}} \cdot (p \cdot \sin \varphi + c \cdot \cos \varphi) \tag{8}$$

The Fuzzy Random Reliability Calculation Process of Surrounding Rock Stability. Using the limit strain criterion, the limit state equation of surrounding rock stability can be obtained:

$$Z = \varepsilon_0 - \bar{\varepsilon} = \frac{2\sqrt{3}}{3} \cdot \frac{1 + \mu}{E} \cdot \frac{R_0^2}{r^2} \cdot \left[\frac{(p + c \cdot ctg\varphi)(1 - \sin \varphi)}{d \cdot ctg\varphi} \right]^{\frac{1 - \sin \varphi}{\sin \varphi}} \cdot (p \cdot \sin \varphi + c \cdot \cos \varphi) \tag{9}$$

The formula contains fuzzy random variables $E, c, \varphi, \varepsilon_0, \mu, p$, Considering the actual situation of the surrounding rock's Poisson ratio μ, the variation of bulk density γ is generally one order of magnitude smaller than c, φ, so only $E, c, \varphi, \varepsilon_0$ are taken as basic fuzzy random variables in the above formula. In geotechnical engineering, most of the variables related to geotechnical parameters are normally distributed, so $E, c, \varphi, \varepsilon_0, \bar{\varepsilon}$ are fuzzy random variables with normal distribution, The fuzzy random reliability is calculated by using the composite function derivation rule and Harlin method.

Then the limit state equation is:

$$Z(\bar{x}) = x_1 - \left(\frac{x_2}{x_3}\right)^{x_4} \cdot x_5 \tag{10}$$

The limit state equation contains five substitution variables $x_i (i = 1, 2, \cdots 5)$, and each substitution variable is a function of four variables $E, c, \varphi, \varepsilon_0$.

Let $E, c, \varphi, \varepsilon_0$ correspond to variables $y_i = (i = 1, 2, 3, 4)$ respectively, namely: $E = y_1, c = y_2, \varphi = y_3, \varepsilon_0 = y_4$.

then get:

$$\frac{\partial Z}{\partial y_i}\Big|_{y^*} = \sum_{j=1}^{5} \frac{\partial Z}{\partial x_j} \cdot \frac{\partial x_j}{\partial y_i}\Big|_{y^*} \tag{11}$$

As long as the reliability index β is obtained, the fuzzy random reliability can be obtained. Harlin method believes that the geometric meaning of the reliability index is the shortest distance from the origin to the limit state equation in the standard normal coordinate system,

If the basic fuzzy random variables are mutually independent, they all follow a normal distribution, and have mean values u_i and mean square deviations σ_i respectively, and the functional function is a linear combination of basic fuzzy random variables, namely:

$$Z(\bar{x}) = u_0 + \sum_{i=1}^{n} u_i, \quad \beta = \frac{a_0 + \sum_{i}^{n} a_i \cdot u_i}{\left(\sum a_i^2 \cdot \sigma_i^2\right)^{\frac{1}{2}}} \tag{12}$$

If the functional function is a nonlinear combination of basic variables, then the functional function can be expanded into a Taylor series at the verification point p^* as follows:

$$Z(\bar{x}) = Z\left(x_1^*, x_2^*, \cdots x_n^*\right) + \sum_{i=1}^{n} \left(\frac{\partial Z}{\partial x_i}\right)_{p^*} \cdot \left(x_i - x_i^*\right) \tag{13}$$

Thus, it can be obtained:

$$\beta = \frac{\sum_{i=1}^{n} \left(\frac{\partial Z}{\partial x_i}\right)_{p^*} \cdot \left(u_i - x_i^*\right)}{\left(\sum_{i=1}^{n} \left(\frac{\partial Z}{\partial x_i}\right)_{p^*}^2 \cdot \sigma_i^2\right)^{\frac{1}{2}}}; x_i^* = u_i + \sigma_i \lambda_i \beta; \tag{14}$$

$$\lambda_i = \frac{-\left(\frac{\partial Z}{\partial x_i}\right)_{p*} \cdot \sigma_i}{\left(\sum_{i=1}^{n} \left(\frac{\partial Z}{\partial x_i}\right)_{p*}^2 \cdot \sigma_i^2\right)^{\frac{1}{2}}}; \; Z\left(x_1^*, x_2^* \cdots, x_n^*\right) = 0; \qquad (15)$$

Since the verification point is unknown in advance, when expanding into a Taylor series, a point must be assumed in advance, such as the mean value point of each basic variable. In the calculation process, use the iteration method to gradually approach the real verification point and correct the obtained β value until the result converges satisfactorily. Assume the initial value of x_i^*, generally take the average value u_i.

Hydraulic tunnels are more complex than above-ground structures, making the working state of the surrounding rock very fuzzy. Therefore, using the exact judgment criterion of "either this or that" sometimes does not match the actual situation. If considering the fuzziness of the judgment criterion, it is necessary to fuzzify the limit strain judgment criterion and make it a fuzzy subset Ω, whose membership function is shown in the following formula:

$$u_\Omega(\varepsilon_0, \bar{\varepsilon}) = \begin{cases} 0 & \varepsilon_0 - \bar{\varepsilon} \prec -L \\ \frac{\varepsilon_0 - \bar{\varepsilon} + L}{2L} & -L \leq \varepsilon_0 - \bar{\varepsilon} \leq L \\ 1 & \varepsilon_0 - \bar{\varepsilon} \succ L \end{cases} \qquad (16)$$

The value of L in the formula can be taken according to the actual situation, and here L can be set to 0.05%.

3 Example

A circular hydraulic tunnel with a diameter of 6 m, a burial depth of 300 m, located in limestone strata, sprayed C30 concrete immediately after tunnel excavation, with a concrete thickness of 8 cm. Through the indoor test of rock samples, the mechanical parameters of surrounding rock samples and the mechanical parameters of concrete used in working state, after fuzzy random processing, the statistical characteristic values of normal distribution are shown in Table 1. Since the variation of bulk density γ, Poisson's ratio μ of rock and elastic modulus E_1 and Poisson's ratio μ_1 of concrete are much smaller than that of other parameters, they are considered as constants.

Table 1. Fuzzy random statistical values of mechanical parameters

Fuzzy random variable	Surrounding rock					concrete		
	$c(MPa)$	$\varphi(\circ)$	μ	$E(MPa)$	λ	$E_1(MPa)$	μ_1	R_C
average	1	40	0.34	20000	2.2	30000	0.17	20
Variance	0.11	4.2		4000				4.0

The vertical component of the initial ground stress is calculated from the above data as follows:

$$P_v = \gamma \cdot H = 2.2 \times 300 = 6.6 \text{ MPa} \quad \lambda = \frac{\mu}{1 - \mu} = \frac{0.34}{1 - 0.34} = 0.52 \quad (17)$$

The average ground stress is: $P = \frac{1+0.52}{2} \times 6.6 = 5.0 \text{ MPa}$.

According to Eq. (5), we get $D = 53.323$, $F = 3.602$, and rearrange to get.

$$1194.05 \left(\frac{R_p}{R_0} \right)^{5.602} + 6345.43 \left(\frac{R_p}{R_0} \right)^{3.602} - 11783.48 = 0 \quad (18)$$

Using Newton's iteration method, we get after 3 iterations: $\frac{R_p}{R_0} = 1.12$

Therefore, the plastic radius is: $R_p = 3.36 \text{ m}$, then: $P_r = \frac{706 \times (10 - c \cdot ctg\varphi)}{E + 2123}$.

According to Eq. (4), the tangential stress on the inner side of the lining is:

$$\sigma_\theta = \frac{26828 \times (10 - c \cdot ctg\varphi)}{E + 2123} \quad (19)$$

According to the limit strength criterion, the functional function is:

$$Z = R_c - \sigma_\theta = R_c - \frac{26828 \times (10 - c \cdot ctg\varphi)}{E + 2123} \quad (20)$$

This functional function contains four basic variables, namely R_c, E, c, φ. And this function is a nonlinear function of these four variables. Therefore, this time we use the Halin method to solve the reliability index β.

After iteration, the reliability index of the lining is finally determined to be $\beta = 3.949$, and the support structure is stable and reliable.

4 Conclusion

This paper mainly expounds the fuzzy stochastic reliability analysis method of surrounding rock stability of hydraulic tunnel. Based on the fuzzy stochastic reliability theory, a fuzzy stochastic reliability calculation model for surrounding rock stability of hydraulic tunnel is established, and the accuracy of the reliability calculation model is verified by practical engineering cases:

(1) On the basis of considering the fuzziness and randomness of surrounding rock parameters, a fuzzy stochastic model for surrounding rock reliability analysis of hydraulic tunnel is proposed.
(2) On the basis of the established fuzzy stochastic reliability model for surrounding rock stability, the influence of support structure on surrounding rock stability is considered, and the fuzzy stochastic reliability of support structure is calculated.
(3) The fuzzy stochastic reliability model of surrounding rock stability of hydraulic tunnel is applied to a circular hydraulic tunnel, and the surrounding rock reliability considering the support structure is calculated, and its reliability index $\beta = 3.949$ is obtained. The support structure is reliable and consistent with the actual situation. This model considers both the fuzziness and randomness of surrounding rock, which is more scientific and reasonable than the traditional reliability method.

Acknowledgment. This research is financially supported by the Graduate Innovation Special Fund Project of Jiangxi Province (Project No. YC2023-S998).

References

1. Liu, T.Y., Ma, N.J., Gao, Q.: Reliability Analysis of Underground Tunnel Engineering. China University of Mining and Technology Press, Xuzhou (1998)
2. He, S.Q., Wang, S.: Structural Reliability Analysis and Design. National Defense Industry Press, Beijing (1993)
3. Zhao, G.F., Jin, W.L., Gong, J.X.: Structural Reliability Theory. China Architecture and Building Press, Beijing (2000)
4. Zou, S.S., Jin, J.Q., Wang, X.R., Qi, J.Y.: Key components of the mechanism under random loads fatigue life prediction and analysis. Mach. Des. Manuf. (09), 9–13+7 (2022)
5. Zhao, H.X., Zhang, G.J., Zhang, S.J.: Reliability analysis of surrounding rock stability of unpressurized tunnel. J. Hefei Univ. Technol. **21**(01), 65–70 (1998)
6. Su, Y.H., He, M.C., Cao, W.G.: Convex set model analysis method for non-probabilistic reliability of surrounding rock stability of rock mass underground structure. Chin. J. Rock Mech. Eng. **24**(03), 373–383 (2005)

Research on Seismic Vibration Table Simulation of Large LNG Storage Tanks Based on Numerical Simulation

Juan Su, Guanghui Zhu$^{(\boxtimes)}$, Wenjie Li, Jinchi Zhang, and Jingwei Su

Offshore Oil Engineering Co., Ltd., Tianjin NJ 300461, China
zhugh10@cnooc.com.cn

Abstract. In order to study the dynamic characteristics and seismic performance of a large LNG storage tank, a finite element analysis model of the shaking table test model of a large LNG storage tank was established by using finite element calculation software with the background of 160,000 m^3 LNG storage tank. The seismic response of the model structure was studied from the three directions of acceleration amplification factor, relative displacement and strain, and the seismic performance of the storage tank model under different seismic waves was obtained. The research results are of great significance for the study of dynamic characteristics and overall seismic performance of large LNG storage tanks.

Keywords: Lng Storage Tank · Finite Element Analysis · Acceleration Amplification Factor · Relative Displacement · Strain · Seismic Wave

1   Introduction

As an important force to achieve China's "double carbon" goal, the consumption of natural gas is increasing year by year, and the construction of LNG receiving stations and storage tanks is developing rapidly [1–3]. At present, China has several LNG receiving stations and storage tanks under construction or have been put into operation.

The loss of property and life caused by earthquake [4, 5] disasters cannot be measured, especially when the gas storage tank and liquid storage tank in the energy industry are damaged and leaked in the earthquake disaster, which causes huge damage. Therefore, the study of earthquake response has important engineering significance for the safety design of storage tank. In 1969, Edwards used finite element method for the first time to conduct numerical simulation research on the coupled seismic response of tank and liquid [6]. W.a.ash and Haroun et al., from the United States, have conducted various researches on cylindrical liquid storage tanks by applying the finite element theory [7–10]. They apply the theory of potential flow to discrete the fluid into ring elements with rectangular sections without the free sloshing of the fluid surface, and transform the circumferential displacement into a one-dimensional problem assumed to be linear elastic deformation. The Sanders shell theory is used as a thin elastic shell on the wall of the tank and is discrete into ring elements. Finally, the coupling problem of tank and liquid

© The Author(s) 2024
G. Feng (Ed.): ICCE 2023, LNCE 526, pp. 316–329, 2024.
https://doi.org/10.1007/978-981-97-4355-1_30

is simplified into an empty tank vibration problem with mass. The dynamic response of empty tank and liquid storage depth, the influence of soil contact and geometrical initial defects were studied. It is reasonable in theory to use finite element theory to calculate the natural characteristics and dynamic response, and it can be used to simulate the vibration situation of storage tank. However, the finite element method is more suitable for solving the natural characteristics of storage tank because of its large calculation amount and the difficulty in compiling the calculation program. This simulation study is of great significance to the research of dynamic characteristics and overall seismic performance of the storage tank.

2 Storage Tank Geometry

Taking 160,000 m³ LNG storage tank as an example, the tank Liquefied Natural Gas is 51.00 m tall, and the largest diameter is 87.10 m. As shown in Fig. 1, it is a sealed double-wall structure with an inner tank made of 9% nickel steel and supported by anchoring bands, and an outer tank consisting of prestressed reinforced concrete walls and a combined reinforced concrete and steel plate dome structure.

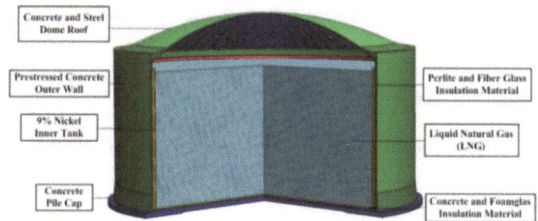

Fig. 1. Prototype structure of full containment LNG storage tank

2.1 Structural Unit Analysis Model

The basic components of the overall structure of LNG storage tanks can be divided into two categories: beam and column components and wall panel components. In this simulation, fiber beam element and layered shell element are used to simulate the composite beam-column component and wall panel component respectively. Figure 2 shows the fiber beam element model.

Figure 3 shows the model of the layered shell element. For reinforced concrete wall panels, the in-plane deformation and out-of-plane bending are described by plane stress element and layered shell element respectively. For the plane stress element (sometimes called the membrane element in structural engineering), the element model commonly used in the traditional finite element method can be used. The strain in the plane of the steel bar and the concrete is consistent, and the internal force of the plane element is the sum of the internal forces of the steel bar and the concrete.

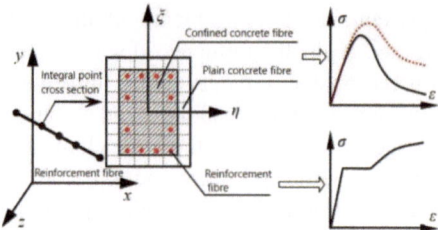

Fig. 2. The fiber beam element model.

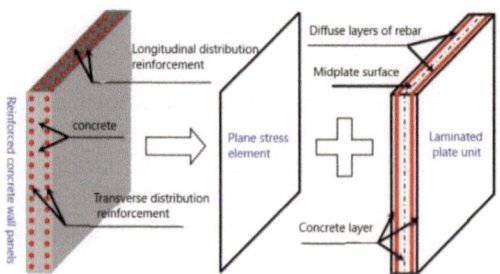

Fig. 3. Layered shell element model

2.2 Structural Analysis Model

Based on the finite element method, the spatial discretization of the continuous structure can form the following structural dynamic equation:

$$M\ddot{u} + C\dot{u} + Ku = f^{ext} \tag{1}$$

where M, C and K are the mass matrix, damping matrix and stiffness matrix respectively. Is the external force vector; u is the displacement vector. The solution of the above dynamic equation is generally based on the Newmark method. Here, the three-stage Newmark method expression is adopted.

2.3 Additive Mass Model

The additional mass method [11] is an approximate method to calculate the liquid-solid coupling problem. The basic idea is to equate the impact dynamic pressure of the liquid on a certain point of the tank wall with the inertia force of the additional mass moving together at the point, and simulate the impact effect of the liquid by applying the additional mass to the inner tank wall. This calculation model makes the calculation decouple, thus reducing the calculation amount.

2.4 Constitutive Relation of Concrete Damage

In order to simulate the whole process of cracking, crushing, damage and destruction of concrete after being stressed, the concrete damage model [12–14] is mainly adopted.

According to the Code for Design of Concrete Structures (GB50010-2015), the concrete damage model is introduced as follows:

By introducing the tensile damage variable, the uniaxial tensile stress-strain curve of concrete is expressed as:

$$\sigma = (1 - d_t)E_c\varepsilon$$

$$d_t = \begin{cases} 1 - \rho_t\left[1.2 - 0.5x^5\right]x \le 1 \\ 1 - \dfrac{\rho_t}{\alpha_t(x-1)^{1.7}+x}x > 1 \end{cases}$$

$$x = \frac{\varepsilon}{\varepsilon_{t,r}}$$

$$\rho_t = \frac{f_{t,r}}{E_c\varepsilon_{t,r}}$$

(2)

where: α_t is the uniaxial tension pressure-strain curve.

2.5 Main Material Parameters

The main mechanical parameters of concrete, steel and thermal insulation materials involved in the scale model of LNG storage tank are listed in Table 1 and Table 2.

Table 1. Concrete material parameters.

Concrete	Compressive strength/f_{cu} (MPa)	Average tensile strength/f_{ctm} (MPa)	Modulus of elasticity/E_c (GPa)
C50	32.4	2.64	34.5

Table 2. Other material parameters

Name	Elastic modulus(MPa)	Poisson's ratio	Density(kg/m^3)
130 thick foam brick	965	0.24	120
100 thick foam brick	900	0.24	200
Expanded perlite	12	0.2	66
Glass Mat	800	0.12	16

3 LNG Tank Structure ABAQUS Finite Element Model

The finite element software ABAQUS is used to analyze the seismic response of the LNG storage tank model. The basic idea of numerical simulation is as follows: firstly, the whole model is established, including prestressed tendons, ordinary rebar, concrete cap, concrete outer tank, steel dome, steel inner tank and thermal insulation layer. According to Housner theory [15, 16], the equivalent additional mass of hydraulic pressure is obtained, and the additional mass is distributed on the wall of the tank. The corresponding structural analysis of the seismic response results of the storage tank is carried out to evaluate its seismic performance.

3.1 ABAQUS Numerical Model for LNG Storage Tank

Outer and Inner Tank Models. The ABAQUS finite element analysis model of the LNG storage tank model structure was established using the aforementioned unit model. The geometric model and mesh division of the inner and outer tanks are shown in Fig. 4 respectively.

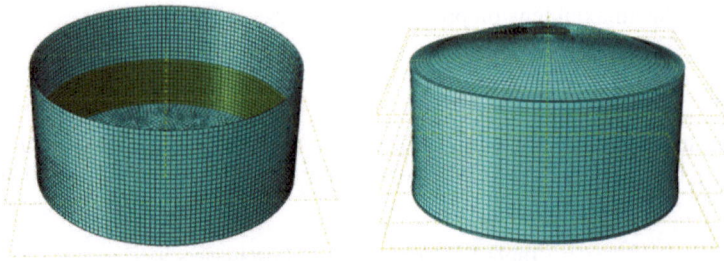

(a)ABAQUS model for steel inner tanks   (b)ABAQUS model for reinforced concrete outer tanks

Fig. 4. Geometric model of inner and outer tanks

Reinforcement and Prestressing. T3D2Truss unit is used for both common reinforcement bars and prestressed tendons of the storage tank, and the geometric model is shown in Fig. 5. In ABAQUS, a cooling method is used to apply prestressing force to the steel bars.

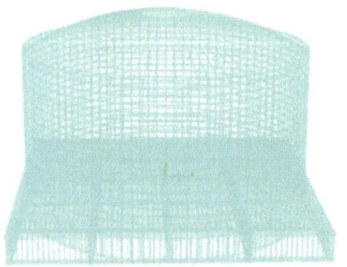

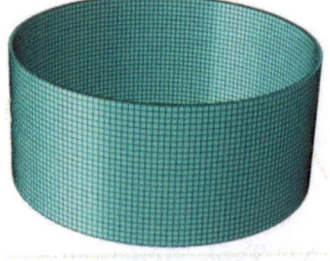

Fig. 5. ABAQUS Model of Steel Fabric   **Fig. 6.** ABAQUS model of insulation layer

Geometric Model of Insulating Layer. The insulating layer on the side wall of the storage tank is 46mm thick expanded perlite and 24mm thick glass fiber felt. ghABAQUS uses C3D8 solid unit to build the insulating layer, as shown in Fig. 6. The insulating layer is processed in different layers, and each layer is assigned with different material properties. The bottom insulating layer is modeled in the same way as the side wall insulating layer.

Additional Mass Distribution. As shown in Fig. 7 and Fig. 8, ABAQUS adopts the additional mass method to carry out dynamic response numerical simulation analysis of the inner tank. A layer of user-defined units is laid at the interface between the storage tank and the liquid and attached to the wall of the storage tank. One is tank wall element mesh, and the two types of finite element mesh share nodes.

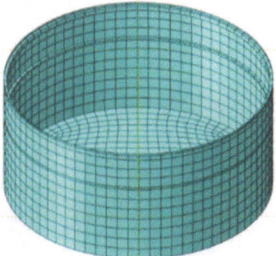 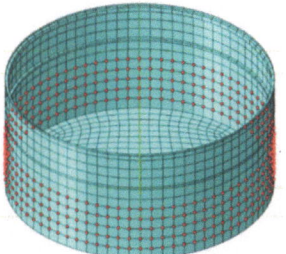

Fig. 7. Finite Element Model of Storage Tank   **Fig. 8.** Additional Mass Model

4   ANSYS Builds the Finite Element Model

The seismic response analysis of LNG storage tank model is carried out with ANSYS finite element software. The basic idea of numerical simulation is as follows: firstly, the whole model is established, including prestressed tendons, concrete cap, concrete outer tank, steel dome, steel inner tank and thermal insulation layer. Then, according to Housner theory, the equivalent additional mass of hydraulic pressure is obtained, and the additional mass is distributed on the wall of the tank. ANSYS-APDL is used to analyze the seismic response results of the tank and evaluate its seismic performance [17–20].

4.1   Establishment of LNG Storage Tank Model

Outer Tank Model. The external tank structure includes the external wall and the roof of the tank, as shown in Fig. 9. The external wall and roof structure adopts solid SOLID65 unit, which is a three-dimensional solid unit specially designed in ANSYS software to face concrete materials and can simulate the unique mechanical phenomena of concrete materials.

Steel Bar Model. As shown in Fig. 10, the prestressed tendon of the storage tank adopts Link180 unit. In the shaking table test, the prestressed steel bar imposes prestressed constraint on the external wall of concrete through the anchor hole installed on the external wall. For the convenience of modeling, the prestressed steel bundle is directly attached to the surface of the external wall, and shared with the external wall concrete.

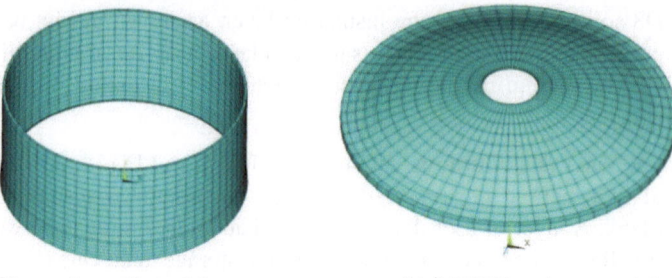

(a) ANSYS exterior wall model   (b) ANSYS tank top model

Fig. 9. Finite Element Model of Outer Tank

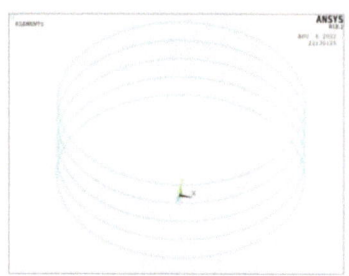   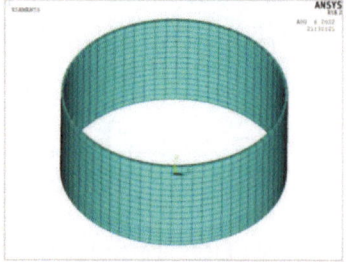

(a)ANSYS pre-stressed reinforcement model   (b)ANSYS insulation layer model

Fig. 10. Model of prestressed reinforcement and insulation layer for storage tanks

According to Housner's theory, when a storage tank is subjected to a horizontal acceleration $a_1(t)$ from the bottom of the tank, the liquid impact pressure acting on any point on the tank wall (θ, y) is shown in formula (3).

$$P_R = a_1 \rho h \left(y/h - \frac{1}{2}(y/h)^2 \right) \sqrt{3} \tan h \left(\sqrt{3}\frac{R}{h} \cos \sigma \right) \tag{3}$$

In the formula:
$a_1(t)$—horizontal acceleration (m/s^2).
ρ—Liquid density in the storage tank (kg/m^3).
r—radius of the storage tank (m).
h—Liquid level height (m).
σ—Azimuth angle of any point along the circumference (rad).
y—Height from the point to the bottom plate (m).
In ANSYS, a cooling method is used to apply prestress to steel bars. The temperature difference T between the initial and the temperature field that reaches the effective stress value is calculated by formula (4).

$$\sigma_{pe} = \lambda \cdot \Delta T \cdot E \tag{4}$$

In the formula:
σ_{pe}—effective tensile stress;

λ—Linear expansion coefficient of steel bars, taken as 1.2×10^{-5};

E—Elastic modulus of steel bars, taken as 2.3269×10^{11} N/m²;

When the initial temperature is 0 °C, the effective stress value can be reached by applying − 55.64 °C to the prestressed steel bar.

5   Comparative Analysis of Two Finite Element Simulation Results

5.1   Seismic Wave Input

In order to study the seismic performance of LNG storage tank under earthquakes with different spectrum and amplitude acceleration, two artificial seismic waves were synthesized by El Centro wave (El), and the site-specific seismic response spectrum. Artificial waves are denoting RG-1 respectively. The acceleration time history are shown in Fig. 11.

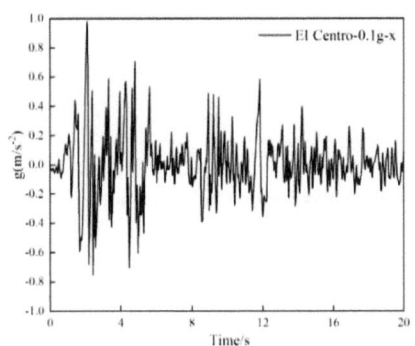 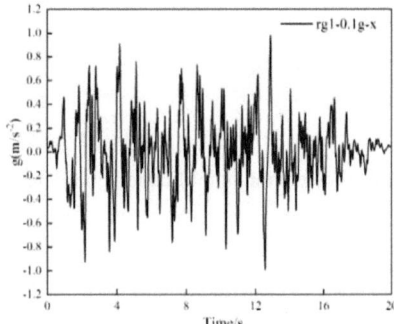

(a)El Centro wave acceleration history   (b)Artificial wave 1 acceleration time histor

Fig. 11. Acceleration time history of two seismic waves

5.2   Analysis of Test Results

As shown in Fig. 12 and Fig. 13, for seismic wave input, the amplitude of the acceleration dynamic amplification coefficient for the outer tank of the empty tank model is between 0.85 and 1.50, and the acceleration dynamic amplification coefficient for the inner tank is between 1.00 and 2.00; Under the same PGA and seismic wave conditions, the dynamic amplification coefficient amplitude of the half tank water outer tank model is slightly lower than that of the empty tank outer tank model, ranging from 0.90 to 1.60, while the dynamic amplification coefficient amplitude of the half tank water inner tank model is larger than that of the empty tank inner tank model, ranging from 1.30 to 2.10.

For different seismic wave inputs, the distribution of the acceleration dynamic amplification coefficient of the model along the height has a similar variation pattern, with the following characteristics:

(1) There is a significant turning point in the outer tank, and for different operating conditions and whether the interior is filled with water, there is a significant turning point that occurs at the transition between the outer tank sidewall and the dome.
(2) The acceleration dynamic amplification coefficient at the end of the model increases more significantly from above the outer tank ring beam to the observation opening of the dome compared to other tank side parts of the outer tank.
(3) For different working conditions and whether the interior is filled with water, the acceleration amplification coefficient of the inner tank increases with height.

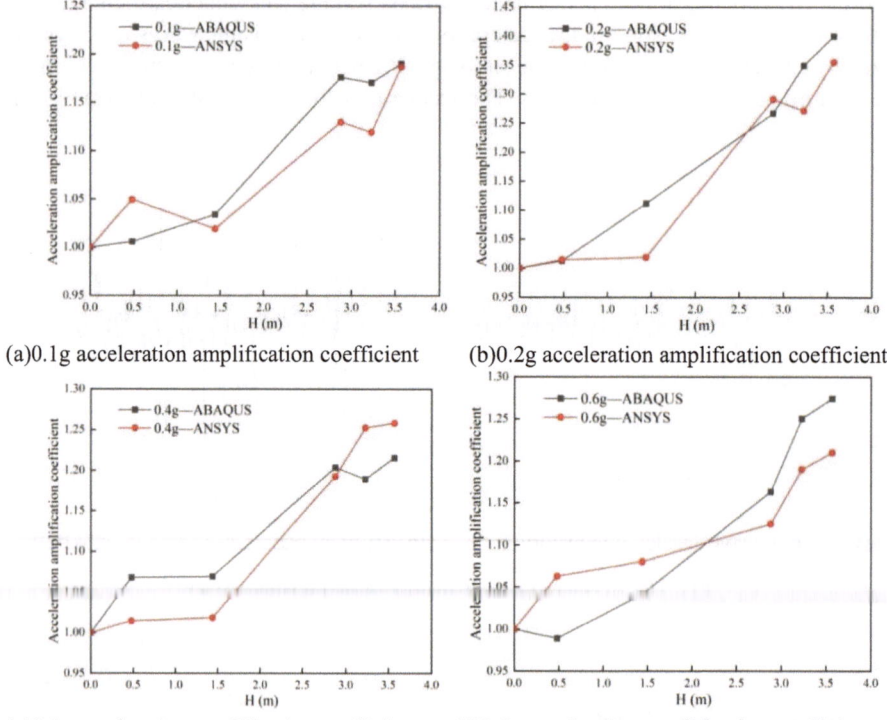

(a)0.1g acceleration amplification coefficient   (b)0.2g acceleration amplification coefficient

(c)0.4g acceleration amplification coefficient   (d)0.6g acceleration amplification coefficient

Fig. 12. Comparison of Acceleration Amplification Coefficients of Outer Tank under El Centro Wave Seismic Action

As shown in Fig. 14 and Fig. 15, in the maximum relative displacement of the outer tank, for the horizontal relative displacement within the height range, it can be observed that the values of the outer wall and dome increase as the height increases.

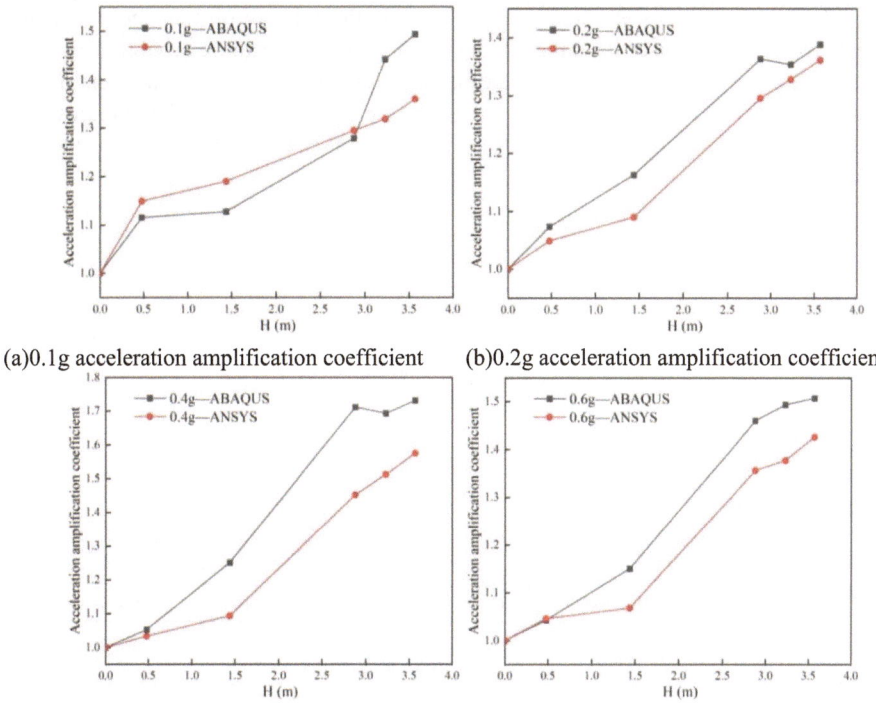

(a)0.1g acceleration amplification coefficient (b)0.2g acceleration amplification coefficient

(c)0.4g acceleration amplification coefficient (d)0.6g acceleration amplification coefficient

Fig. 13. Comparison of Acceleration Amplification Coefficients of Outer Tank under RG-1 Wave Seismic Action

For the acceleration amplification coefficient in the height range, when the design seismic wave is small, the lower part of the exterior wall has a good barrier effect on the seismic wave, resulting in a smaller acceleration value. However, when the design seismic wave is small, the acceleration amplification coefficient is larger, while when the design seismic wave value is large, the acceleration amplification coefficient is smaller. Within the height range, as the height increases, the overall amplification coefficient shows an increasing trend.

The relative displacement at the variable cross-section of the outer tank is relatively large, and the maximum interlayer displacement angle is significantly greater than other positions. Due to the large bending moment and shear force at the bottom of the structural model under earthquake action, it often has large relative displacement and interlayer displacement angle.

Below the middle of the outer tank, the maximum relative displacement increases with height, and decreases from the middle to below the ring beam. For vibrations dominated by the first mode of vibration, if the deformation mechanism is mainly bending, the relative displacement will increase with the increase of height, while for structures dominated by shear deformation, the relative displacement will decrease with the increase of height.

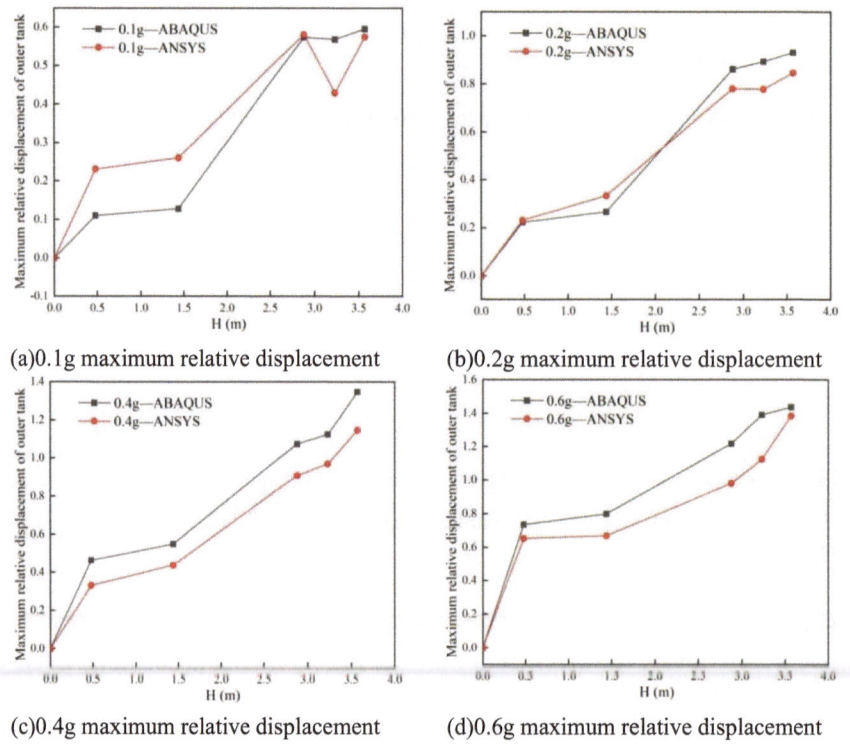

(a)0.1g maximum relative displacement   (b)0.2g maximum relative displacement

(c)0.4g maximum relative displacement   (d)0.6g maximum relative displacement

Fig.14. Maximum relative displacement of outer tank along height under El Centro wave seismic action

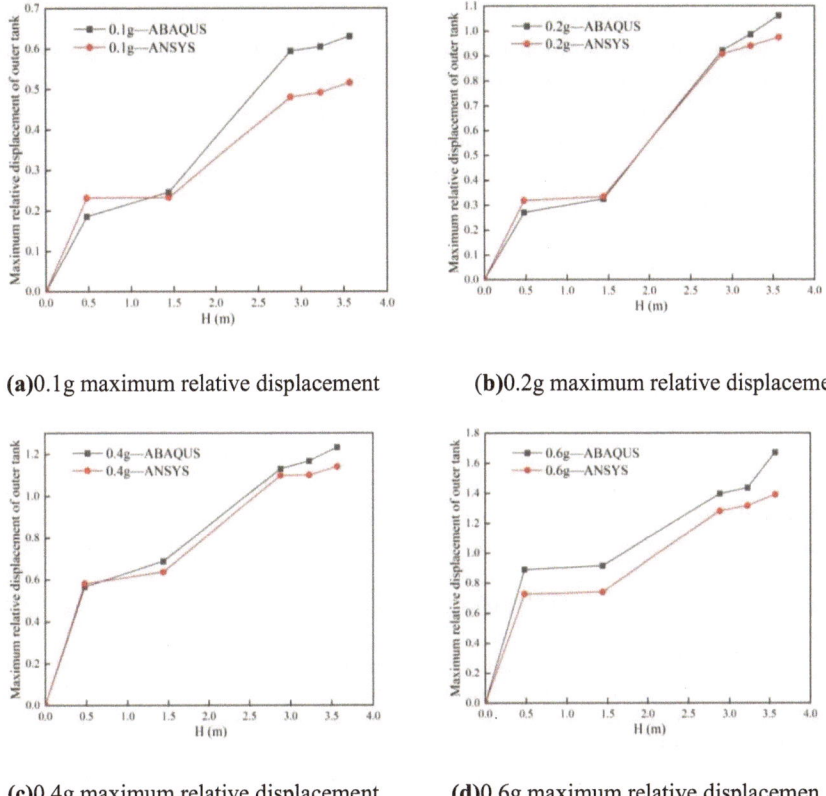

(a)0.1g maximum relative displacement   **(b)**0.2g maximum relative displacement

(c)0.4g maximum relative displacement   **(d)**0.6g maximum relative displacemen

Fig. 15. Maximum relative displacement of outer tank along height under RG-1 wave seismic action

6 Conclusion

This article establishes a finite element analysis model for the vibration table test model of large LNG storage tanks using ABABQUS and ANSYS finite element calculation software. The dynamic time history response analysis of the structure with empty and half tank water was conducted through a finite element model, and the results of dynamic time history analysis such as acceleration, displacement, and stress were obtained. The main conclusions are as follows:

(1) Under the same peak ground motion (PGA) condition, the acceleration amplification coefficient of the model along the height under the action of artificial waves is generally larger than that under the action of natural waves. Overall, there is no trend of decreasing the dynamic acceleration amplification coefficient with the increase of earthquake amplitude under the same seismic wave action.

(2) Under the action of seismic waves with the same waveform, as the peak ground motion increases, the relative displacement response of the model increases. At the same time, the sudden change in size at the variable cross-section of the outer tank

leads to stress concentration, making the stiffness at the variable cross-section less than other positions, resulting in the maximum displacement angle between the variable cross-section position of the outer tank and the root of the outer tank being greater than other positions; Due to the shaking of water during vibration, the relative displacement below the middle of the outer tank increases. As the distance from the water surface increases, it can be roughly assumed that the impact of water vibration on relative displacement is decreasing.

References

1. Luo, D.Y., Sun, J.G., Liu, C.G., Cui, L.F., Wang, Z.: Effect analysis of near-fault ground motion on large LNG storage tank. J. Nat. Disast. **29**(04), 102–110
2. Roetzer, J., Douglas, H., Maurer, H.: Hazard and safety investigations for LNG-Tanks. LNG J. **15**(6), 72–89 (2005)
3. Zhang, Y.F., Li, T.F., Teng, Z.C.: Research status and prospect of LNG storage tank. Contemp. Chem. Ind. **50**(07), 1662–1666 (2021). (in Chinese)
4. Tang, G.G.: Research on some problems of super large LNG storage tank. Harbin Institute of Technology (2017)
5. Chen, H.Y.: Seismic response and temperature field analysis of super large LNG storage tank. Guangzhou University (2020)
6. Edwards, N.W.: A procedure for dynamic analysis of thin walled liquid storage tanks subjected to lateral ground motions. Ph.D. dissertation, University of Michigan (1969)
7. Nash, W.A., et al.: Response of liquid storage tanks to seismic motion. Bull. Seismol. Soc. Am. **73**(2), 151–159 (1985)
8. Haroun, M.A.: Dynamic analysis of liquid storage tankd. In: EERL 80-04. California Institute of Technology, Pasadena (1980)
9. Haroun, M.A.: Assessment of seismic harzards to liquid storage tanks at port facilities. In: Proceedings of the 1990 Pola Seismic Workshop, pp. 100–105 (1991)
10. Haroun, M.A.: Wajdi Abou-izzeddine. parametricstudy of seismic soil-tank interaction. I: horizontal excitation. J. Struct. Eng. **118**(3) (1992)
11. Jadhav, M.B., Jangid, R.S.: Response of base-isolated liquid storage tanks. Shock Vibr. **11**(1), 33–45 (2004)
12. Zhuang, Z., Liu, H.Z., Li, Q.: Developing an additional mass model to analyze the dynamic response of storage tanks. Mod. Manuf. Eng. (01), 69–73 (2006)
13. Du, X.H.: Numerical simulation of fluid structure coupling for prestressed LNG storage tanks under earthquake action. Chinese society of mechanics mechanics and engineering applications. In: Chinese Society of Mechanics: Henan Provincial Society of Mechanics, vol. 5 (2012)
14. Han, F., Gong, X.C., Chen, G.J.: Numerical analysis of seismic resistance of vertical liquid storage tanks based on fluid structure coupling theory. J. Wuhan Univ. Sci. Technol. **34**(05), 359–363 (2011)
15. Housner, G.W.: Dynamic pressure on accelerated fluid container. Bull. Seism. Soe. Am **47**(1), 15–35 (1957)
16. Rostasy, F.S., Sprenger, K.H.: Strength and deformation of steel fibre reinforced condrete at very low temperature. Int. J. Cem. Compos. Lightweight Concr. **2**, 47–51 (1981)
17. Meinen, E.: LNG Storage enclosed in prestressed concrete safety walls. Oil Gas J. **5**, 117–120 (1979)

18. Law, B.: LNG storge tanks: concrete in an ultra-cold environment. Concr. Constr. **28**(6), 465–466 (1983)
19. Lebris, P.: Three safety factors included in LNG storage tank system. Pipe Line Ind. **49**(3), 45–46, 48, 50 (1978)
20. Shibata, H., Okamura, H., Takagi, T., Ukaji, K., Ito, H., Ouchi, H.: Ductility of prestressed concrete at extremely low temperature for PC-LNG-aboveground storage tank. Publ. Atomic Energy Soc. Jpn. **91** (1993)

Seismic Fragility Analysis of Space Special-Shaped Steel Box Arch Bridge

Donglian Tan and Yue Zhao[✉]

School of Railway Transportation, Shanghai Institute of Technology, 100 Haiquan Road,
Fengxian District, Shanghai, China
854886369@qq.com

Abstract. Seismic fragility analysis is of paramount significance in studying the seismic characteristics of bridges, especially for special structures like special-shaped bridges. Special-shaped bridges exhibit more complex geometric and mechanical properties. Therefore, it is even more essential to conduct seismic fragility analysis for special-shaped bridges. Analyzing the response of special structures to lateral seismic loads benefits the evaluation of their seismic performance and facilitates enhancements in seismic design. A finite element model was established using Midas/Civil to investigate the seismic fragility of a space special-shaped steel box arch bridge under lateral seismic loads. To assess the seismic fragility of space special-shaped steel box arch bridges, displacement data from bridge abutments, arch support platforms, and bearings were selected as damage indicators. The seismic fragility curves for bridge components were constructed using the incremental dynamic analysis (IDA) method. In the end, the weighting method was adopted to calculate the seismic fragility curve for the entire bridge system. The bearings and bridge abutments are more susceptible to seismic damage under lateral seismic loads, making them the vulnerable components of the bridge. Among the four bearings, the ones located near the arch ribs have poorer seismic performance and constitute vulnerable areas in the bearing layout. Through the case study of the seismic fragility analysis of space special-shaped steel box arch bridge, common issues in the fragility analysis of special bridges were explored. This further provides rational recommendations for enhancing and improving the seismic design of special-shaped bridge structures.

Keywords: Space Special-Shaped Steel Box · Seismic Performance · Bridge Structural System Fragility · Incremental Dynamic Analysis Method

1 Introduction

With the continuous development of technology, various aesthetically pleasing new types of bridges have emerged. Among these, spatial special-shaped arch bridges have garnered significant attention in existing literature, primarily focusing on their design and construction [1, 2]. However, there has been relatively limited research on the seismic fragility of spatial special-shaped steel box arch bridges. Analyzing a bridge's response to lateral seismic forces allows us to better assess and improve its seismic performance,

© The Author(s) 2024
G. Feng (Ed.): ICCE 2023, LNCE 526, pp. 330–337, 2024.
https://doi.org/10.1007/978-981-97-4355-1_31

ensuring that the bridge remains safe and stable during earthquakes. This is of critical importance for enhancing a bridge's earthquake resistance, as it helps reduce potential losses and risks associated with earthquakes.

Kim et al. [3] utilized the Monte Carlo simulation method and response spectrum analysis to establish fragility curves for bridges. The findings revealed that bridges considering multiple excitation factors are more susceptible to seismic damage compared to those considering uniform excitation. The fragility curves were represented using different seismic indicators such as PGA, PGV, and SA. It was concluded that PGA and SA are more suitable and efficient seismic indicators. Choi et al. [4] simultaneously considered the correlation between the failure modes of bearings and pier columns under seismic actions. They introduced the first-order boundary method to establish the upper and lower bounds of systemic fragility curves for bridges. Li et al. [5] considered the correlation between the failure modes of pier, bearing, and abutment components and established systemic fragility curves for medium-span reinforced concrete continuous beam bridges using a comprehensive second-order boundary method. Pan [6] conducted a seismic fragility analysis on a steel bridge in New York State and found that nonlinear regression outperforms traditional linear regression in fitting the bridge's response data.

Currently, most research on seismic fragility focuses on conventional bridges, with limited studies on the seismic fragility of irregular arch bridges. This paper takes a specific spatial irregular steel box arch bridge as an example and establishes a finite element model. By applying fragility theory and conducting incremental dynamic analysis (IDA), the fragility curves of bridge components are developed to evaluate their vulnerability. Additionally, the paper explores the systemic fragility curve of such bridge structures using the weight method. The findings of this study can serve as a reference for similar vulnerability research on bridges.

2 Bridge Analytical Model

To ensure the accuracy and reliability of the finite element numerical model, this paper used the finite element analysis software Midas/Civil to establish a complete spatial finite element model of the bridge. In this model, the main beam, main arch, pile foundation, and abutment were simulated using beam elements, while the cables were simulated using truss elements. The model consisted of 1,209 beam elements and 12 truss elements. The main beam is elastically connected to the abutments, whereas the main arch is rigidly connected to the foundation of the arch seat. The bearings are simulated using a combination of rigid and elastic connections. The connections between the main beam pile foundations and the abutments, as well as between the arch seat pile foundations and the foundation of the arch seat, are treated with master-slave constraints. The finite element model of the bridge is shown in Fig. 1.

There are No. 1 to No. 4 bearings on the main beam, and the specific arrangement is shown in Fig. 2. Among them, bearing 1 is a fixed bearing, bearing 2 is a longitudinal moving bearing, bearing 3 is a transverse moving bearing, and bearing 4 is a bidirectional moving bearing. The maximum allowable displacement of the bearing is 200 mm.

Fig. 1. Bridge Finite Element Model

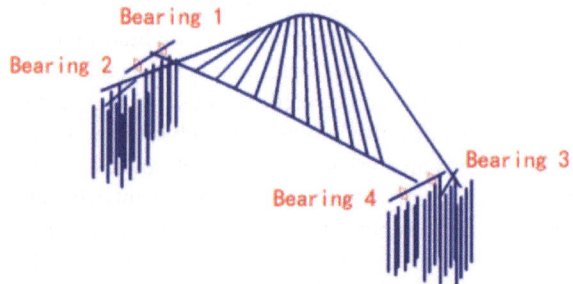

Fig. 2. Bearings Arrangement

3 Seismic Fragility Analysis of Bridge

3.1 Selection and Division of Damage Index

The damage indicators for the bearing components are calculated based on the maximum allowable displacement of the bridge's movable bearings and coefficients corresponding to different damage states. The damage indicators for the bridge abutments and arch bearings are determined based on the damage states proposed by the scholar Zheng Kaifeng [7]. The damage state and damage index of the bearing, abutment, and arch support platform are shown in Table 1.

3.2 Seismic Fragility Analysis of Components

The 'frequency statistics method' is used for the statistics of the failure state of the component under different PGA. When the ' frequency statistics method ' is used to draw the fragility curve, because the probability points corresponding to the failure state of each level under each level of PGA are directly connected, the obtained will be a set of broken lines. The current fitting methods for probabilistic points assume that the fragility curves of components conform to a lognormal distribution. These methods employ the

Table.1 Component failure index

Damage State	Displacement of Bearing	Displacement of Abutment	Displacement of Arch Support Platform
No Failure	$\mu \leq 60$	$\mu \leq 25$	$\mu \leq 25$
Slight Failure	$60 < \mu \leq 100$	$25 < \mu \leq 50$	$25 < \mu \leq 50$
Moderate Failure	$100 < \mu \leq 160$	$50 < \mu \leq 100$	$50 < \mu \leq 100$
Extensive Failure	$160 < \mu \leq 200$	$100 < \mu \leq 150$	$100 < \mu \leq 150$
Complete Failure	$\mu > 200$	$\mu > 150$	$\mu > 150$

probability density function of the lognormal distribution to perform regression analysis on the probabilistic points. This analysis enables the determination of the mean and standard deviation of the failure probability density function for each component at various levels. Subsequently, these computed mean and standard deviation values are inserted into the cumulative distribution function of the standard normal distribution. By doing so, the fragility curves for the different damage states of each component can be derive [8]. This article utilizes the lognormal distribution function and employs maximum likelihood estimation in the Matlab software to fit the lognormal distribution for structural fragility.

Assuming that the seismic fragility curves conform to a two-parameter lognormal distribution:

$$F_j(IM) = \Phi\left[\frac{ln(IM/c_j)}{\zeta_j}\right] \tag{1}$$

where, $F_j(\cdot)$ represents the fragility function of the damage state, whereas c_j and ζ_j represent the median and logarithmic standard deviation, respectively, corresponding to that damage state. The fragility curve of components is shown in Fig. 3.

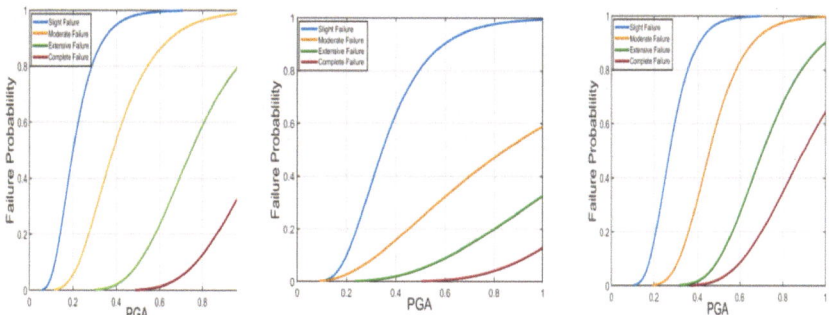

Fig. 3. Bridge Abutment and Bridge Arch Support Platform and Bridge Bearing Fragility Curve

The probability of failure states for the four bearings of the irregular arch bridge under different PGA values was analyzed statistically, as shown in Figs. 4 and 5.

Significant differences exist in the probability of complete failure among the four bearings, At a PGA of 1.0g, the probabilities of complete failure for bearings 2 and 3 are 70% and 90%, respectively. Bearing 4 exhibits the lowest probability of complete failure among the four bearings at a PGA of 1.0g, with a probability of 35%. This demonstrates its strong seismic resistance and higher safety performance.

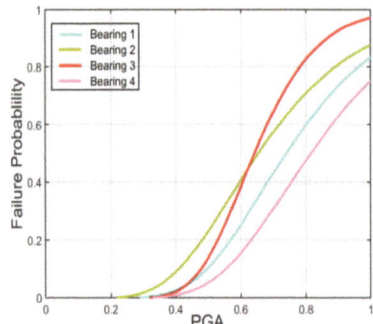

Fig. 4. Extensive Damage Fragility Curve of Bearing

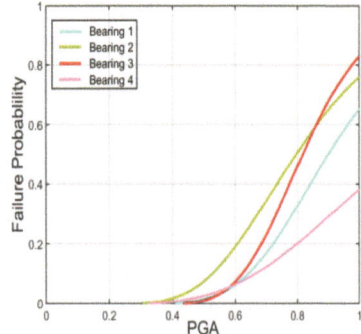

Fig. 5. Complete Damage Fragility Curve of Bearing

3.3 Seismic Fragility Analysis of System

The bridge structure is a complex system composed of a large number of different components. If only the fragility analysis of a certain component is carried out, the seismic performance of the whole bridge system cannot be fully reflected. The calculation results show that the damage level and failure probability of the components are quite different, so it is necessary to carry out the fragility analysis of the bridge system.

The weighting method takes into account the importance of each component in the bridge system. It categorizes the major components, which are the ones that, if damaged, would lead to the failure of the entire bridge system, as one group, and the remaining components as another group, considering their respective weights. The weighting method has been proven to be suitable for the analysis of bridge system fragility [9].

In this paper, the bridge abutment and arch support platform are divided into main components, and the bearing is divided into other components [10]. The weight of each component is calculated as shown in Table 2. The corresponding calculation weights are computed based on the provided weights of the components, as shown in Eq. (2):

$$\omega_\eta = \frac{\omega_q}{\omega_{mq}} \tag{2}$$

where, ω_η represents the calculated weight of the component, ω_q represents the weight of the component, and ω_{mq} represents the weight of the main component.

The weight method [9] is used to calculate the failure probability of the bridge system. The calculation formula of the weight method is shown in Eq. (3).

$$P_{sys} = \max_{i=1}^{m} P_{fi} + \eta \sum_{j=1}^{n} P_{fj} \le 1 \tag{3}$$

Table 2. Member weight calculation

Member	Weight	Calculation weight
Abutment	23	1
Arch Support Platform	23	1
Bearing	3	0.13

where, P_{sys} is the failure probability of the bridge system. P_{fi} is the probability of failure of the i th main member.m is the number of main components. P_{fj} is the probability of failure of the j th other component.n is the number of other components. η is the corresponding weight.

It is evident from Fig. 6 that the maximum bandwidths for the system and its main components, ranging from slight failure to complete failure, are 0.111, 0.121, 0.113, and 0.085, respectively. This indicates that the probability of failure at each level on the system's fragility curve is higher than that of the main components, aligning with the actual earthquake-induced damage observed in the project.

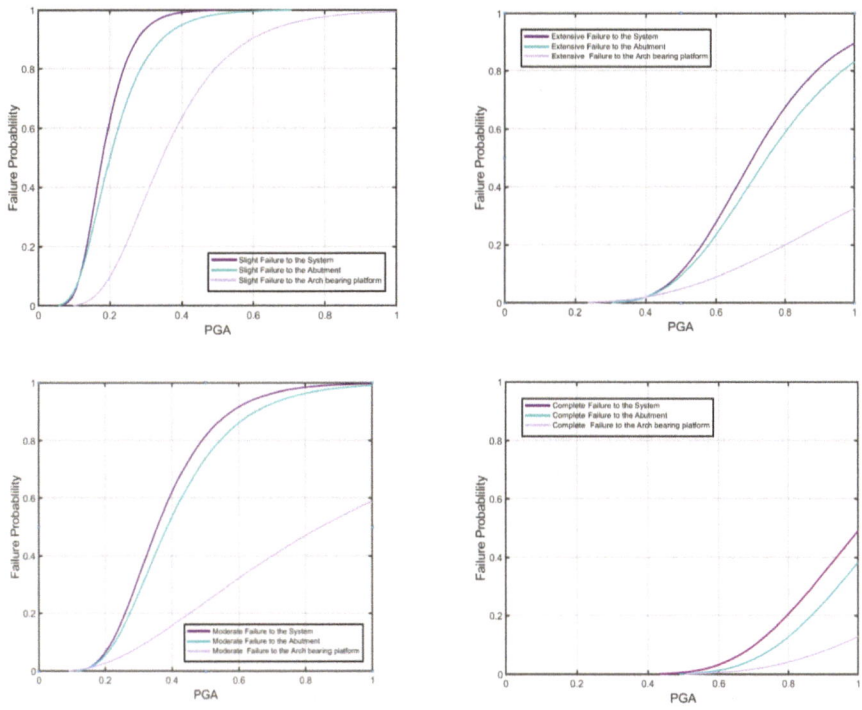

Fig. 6. Fragility Curve of System and Main Components in Failure State

4 Conclusions

The conclusions derived from this study are as follows:

A comparative fragility analysis of multiple components reveals that the maximum probability of complete failure for bearings is 70%, for the bridge abutment is 40%, and for the arch support platform is 10%. This indicates that the bridge abutment and bearings are susceptible to damage under seismic actions, while the arch support platform exhibits better seismic performance. It suggests that the bridge abutment and bearings are the vulnerable components of the bridge, and their seismic protection should be given priority.

References

1. Zhang, Y.W.: Study on construction optimization and construction control technology of CFST shaped tied arch bridge. Lanzhou Jiaotong University, China. https://doi.org/10.27205/d.cnki.gltec.2019.000363
2. Zhang, Z.J.: Construction monitoring technology of a through reinforced concrete special-shaped bowstring arch bridge. China Acad. J. Electron. Publ. House. **229**(05), 193–195+222+21 (2018). https://doi.org/10.16799/j.cnki.csdqyfh.2018.05.053
3. Kim, S.H., Shinozuka, M.: Fragility curves for concrete bridges retrofitted by column jacketing and restrainers. In: American Society of Civil Engineers Sixth U.S. Conference and Workshop on Lifeline Earthquake Engineering (TCLEE), 10–13 August 2003, Long Beach, California, United States, pp. 906–915 (2003). https://doi.org/10.1061/40687(2003)92
4. Choi, E., Des Roches, R., Nielson, B.: Seismic fragility of typical bridges in moderate seismic zones. Eng. Struct. **26**(2), 187–199 (2004). https://doi.org/10.1016/j.engstruct.2003.09.006
5. Li, L., Wu, W., Huang, J., et al.: Vulnerability study of medium-span RC continuous girder bridge system under seismic action. J. Civ. Eng. **45**(10), 152–160 (2004). https://doi.org/10.15951/j.tmgcxb.2012.10.005
6. Pan, Y., Agrawal, A.K., Ghosn, M.: Seismic fragility of continuous steel highway bridges in New York State. J. Bridg. Eng. 12(6), 689–699 (2007). https://doi.org/10.1061/(asce)1084-0702(2007)12:6(689)
7. Zheng, K., Chen, L., Zhuang, W., et al.: Bridge fragility analysis based on probabilistic seismic demand model. Eng. Mech. **30**(05), 165–171+187 (2013)
8. Gardoni, P.: Probabilistic models and fragility estimates for bridge components and systems. Ph.D. thesis, University of California Berkeley, CA, USA (2002)
9. Huang, F.H., He, P.X., Wu, T.F.: Seismic vulnerability analysis of through concrete-filled steel tube arch bridge. J. HeFei Univ. Technol. (Nat. Sci.) **45**(06), 801–807 (2022). https://doi.org/10.3969/j.1003-5060.2022.06014
10. JTG H11-2004. Specification for maintenance of highway bridges and culverts, People's Communications Publishing, Beijing, China

Prediction of Construction Water Inflow of Karst Water-Rich Tunnel Without Drainage Tunnel

Wei Li[1], Shiquan Hao[1], Ju Wang[2], and Junpeng Zou[2(✉)]

[1] China Railway Third Bureau Group Fifth Engineering Co., Ltd., Jinzhong 030600, Shanxi, China
[2] China University of Geosciences (Wuhan), Wuhan 430000, Hubei, China
zoujunpeng@cug.edu.cn

Abstract. The construction of infrastructure in China faces substantial challenges due to the complex geological conditions and groundwater in karst areas, particularly in tunnel construction. The construction of the Maoping Port railway tunnel in the Three Gorges Hub area has been accompanied by the emergence of caves and faults. The excavation process for the tunnel has been found to expose caves and cavities, leading to sudden surges of water and mud, which can pose a risk. Hence, there is a pressing need to precisely forecast the water inflow in the karst region of Yangmuling Tunnel, in order to furnish guidance for the tunnel's drainage and to avert surging water disasters in the absence of a relief hole. This study focuses on the water-rich karst issue in Yangmuling Tunnel. The research employed on-site hydraulic pressure monitoring and finite element numerical simulation to investigate the seepage and water influx properties of the tunnel without water relief hole. The outcomes of the study reveal that a low water pressure zone forms near the tunnel after excavation, and water pressure gradually increases outward. When the tunnel reaches the karst pipe, there is a sharp increase in water influx. Simultaneously, the low water pressure zone in the karst area sharply expands, resulting in an irregular shape. Excavation of the tunnel exposed the karst pipe, causing a rise in flow velocity in the karst zone towards the front of the tunnel face, but it subsequently declined rapidly. The alteration in flow velocity within the rock section surrounding the tunnel face was comparatively uniform.

Keywords: Karst Tunnel · Water Pressure Monitoring · Water Inflow Prediction · Numerical Simulation

1 Introduction

The development of China's infrastructure construction within the karst landscape of southwest China poses significant challenges. The complex geological conditions and higher instances of groundwater in karst areas result in immeasurable negative impacts on tunnel construction. These challenges seriously threaten construction safety and impede the progress of project1. Numerous tunnels fail to recognize the karst geological conditions and groundwater distribution characteristics, resulting in severe water and mud

© The Author(s) 2024
G. Feng (Ed.): ICCE 2023, LNCE 526, pp. 338–349, 2024.
https://doi.org/10.1007/978-981-97-4355-1_32

surges. Frequent disasters in water-rich karst tunnels require extensive and thorough research into key scientific issues, namely predicting tunnel water ingress and preventing and managing major water and mud surges. Such research can efficiently inform engineering practice and guarantee the safety of construction projects.

Although predicting water surges during tunnel construction is challenging, scholars at home and abroad have recently made significant progress in forecasting such phenomena in water-rich karst tunnels. Wang et al. devised an equation for computing the water surge at the tunnel face when exposed to fault fragmentation using the potential function's superposition method [2]. Zhang Qingsong used a model test system to investigate the changes in physical quantities such as seepage pressure, stress-strain displacement and surge material in the tunnel perimeter rock after fault exposure in the fault fragmentation zone [3, 4]. Fu Hailin et al. have established a simplified model of tunnel seepage near faults by equating faults with a certain dipping angle in the tunnel cross-section to a vertical dipping angle. They have deduced a formula for calculating the water influx in the tunnel [5]. However, traditional research oversimplifies the geological model, often neglecting the complex geological conditions and hydraulic properties of karst aquifers. In areas of strong karst development, the water-bearing medium exhibits highly non-homogeneous characteristics, resulting in extremely uneven spatial distribution of karst groundwater. The hydrogeological conceptual models and associated mathematical models can often provide an insufficient representation of the characteristics of karst groundwater movement [6]. As a result, current predictions of water inflow in karst tunnels continue to deviate significantly from observations. Therefore, it is crucial to identify water surge conditions systematically in karst tunnels to predict water surge effectively. To achieve this, a more efficient method of water surge prediction in the tunnels needs to be found, and a karst groundwater model should be established within the study area of the tunnels. This will enable a precise description of their structural features and water flow behavior [7, 8].

Currently, numerical simulation has become a widely used tool for predicting tunnel water infiltration in various complex geological environments. Chiu and Chia [9] employed a modular three-dimensional finite difference groundwater flow model (MODFLOW) in order to reproduce the groundwater seepage field whilst simultaneously utilising a drainage package with the aim of predicting the activity of the tunnel. Using the numerical simulation method, Fang Yong [10] conducted a simulation of surge water behaviour in tunnels located in complex tectonic areas.

This study focuses on the issue of water-rich karst present in the Yangmuling tunnel at Maoping Harbor Relief Railway in the Three Gorges Hub. Firstly, the monitoring results of the pore water pressure are analysed, and subsequently, a numerical model is created using the finite element method. The study examines the issue of sudden water influx in the Yangmuling tunnel due to karst formation in the absence of relief holes. The research incorporates numerical calculations to determine the extent of water influx in the karst area of the tunnel without relief holes, providing a robust foundation for simulating the surge water in the karst region. The study conducts a numerical calculation of water ingress in Yangmuling Tunnel in the absence of a water discharge hole. The findings offer valuable reference and guidance for estimating water inflow in karst tunnels and preventing water surge disasters.

2 Project Overview

The Yangmuling Tunnel spans 3648.3 m and is situated in Yiling District, Yichang City, Hubei Province, between Mouyang Village and Taojiaxi. Figure 1 shows the regional geographic location of the tunnel site area. The tunnel inlet's designed shoulder elevation is at 305.026 m, while the tunnel outlet's is at 292.743 m. The maximum depth of the tunnel reaches 428.934 m. The strata in the tunnel location are mainly composed of Shipai Formation ($\in_1 t$), Shuijingtuo Formation ($\in s$), and Tianzhushan Formation ($\in sh$) of the Lower Cambrian alongside Dolomite from the Upper Aurignacian Lampshade Formation ($Z_2 dn$). Dolomitic greywacke from the Upper Aurignacian Steep Mountain Tuo Formation ($Z_2 d$), conglomerate from the Lower Aurignacian Nantuo Formation ($Z_1 n$), and sandstone from the Lower Aurignacian Liantuo Formation ($Z_1 l$) can be found in the tunnel area.

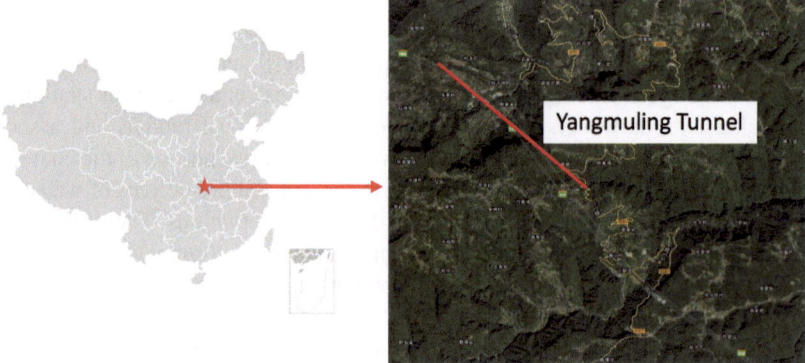

Fig. 1. Geographical location of the tunnel site area

Based on the excavation situation on site, the surrounding rock in the tunnel construction section predominantly comprises medium to weakly weathered sandstone. The rock body also features soft and weak interlayers, leading to poor rock stability. Additionally, the rock body exhibits a fragmented structure, with joints and fissures more developed, resulting in poor rock stability and high water content. Due to the non-soluble rock stratum in this section, the surrounding rock does not contribute to tunnel water surge. However, the rock body contains ample bedrock fissure water, resulting in water surges in the form of rain or strand surge, with a minor amount of water surge. Given that the tunnel excavation is about to penetrate the soluble rock layer, the rock mass displays a high degree of fragmentation and there is a plentiful groundwater system. Water disasters in this area often exhibit a high incidence and suddenness, and feature abundant water and sediment. Hence, it is imperative to anticipate the volume of water inflow in the tunnel of the karst water-rich area and present appropriate measures for prevention and management to ensure tunnel construction safety. Figure 2 illustrates the water leakage at the site of the Yangmuling Tunnel.

(a) water seepage and gushing from the tunnel face (b) Water surges in the sidewalls of the tunnel

Fig. 2. Water leakage in Yangmuling Tunnel site

3 Pore Water Pressure Monitoring in Tunnels

3.1 Instrument Introduction and Installation

For this water pressure test on the Yangmuling Tunnel, the JYKYJ-370 vibrating string-type permeameter is utilized to measure fluid pressure. Figure 3 shows a photograph of the vibrating string seepage manometer and a schematic diagram of the installation of the seepage manometer in the tunnel wall.

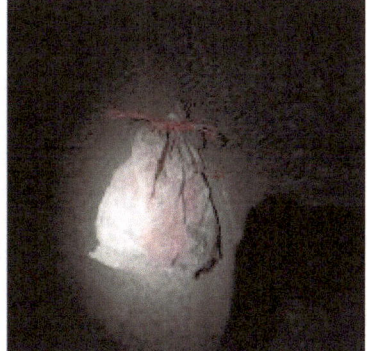

(a) Sinusoidal piezometers

(b) nstallation of seepage manometers on tunnel walls

Fig. 3. Photo of sinusoidal osmometer and schematic diagram of installation osmometer

3.2 Osmotic Pressure Analysis

The water pressure on the surrounding rock of the tunnel section is measured by the seep-age manometer, and the monitoring results of the Yangmuling Tunnel exit DK22+354 point 1 and DK22+370 point 2 are shown in Fig. 4.

At measurement point 1 of the Yangmuling Tunnel (Fig. 4a), data shows that as of July 31st, there is a maximum water pressure of 6.84 kPa on the tunnel's surface at the Yangmuling Tunnel point. This pressure corresponds to a groundwater depth of 0.68 m, and the pore water pressure remains stable at 6. Technical term abbreviations are explained when first used. At pressures ranging from 0 to 6.84 kPa, the pore pressure on the tunnel surface is low, and the water level only reaches 0.6 to 0.68 m. Although there is a consistent water influx in the section, the overall amount is not significant. The main water source comes from the bedrock fissures, which has a relatively minor impact on the tunnel's construction. The construction of the tunnel is minimally affected. Data from Point 2 of the Yangmuling Tunnel (Fig. 4b), as of 31st July, indicates that the maximum water pressure at the tunnel surface is 2.55 kPa, corresponding to a groundwater depth of 0.26 m. The pore water pressure is stable at 2.0–2.55 kPa, and the tunnel surface pore pressure is relatively low, with a head of only 0.2–0.26 m.

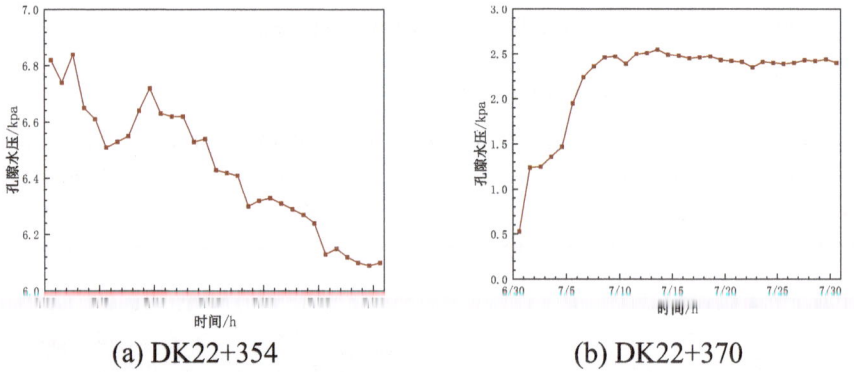

(a) DK22+354   (b) DK22+370

Fig. 4. Pore water pressure variation curve of surrounding rock at measuring points 1 and 2 of Yangmuling Tunnel

4 Numerical Calculation Model for Tunnel Seepage and Surge

4.1 Modelling

To investigate sudden water flow in a tunnel that has a karst pipe in front of the tunnel face, we utilized the Abaqus finite element software to develop a 150 m × 70 m × 70 m three-dimensional geological model based on the actual geological conditions of the Yangmuling Tunnel. Refer to Fig. 5 for the model's visual representation. The tunnel has a depth of 428.9 m and features a bottom section designed as a side wall, while the top section is shaped like a horseshoe. The tunnel is 6.5 m wide and 8.6 m high, with the side wall measuring 3.25 m in height. In this model, a karst pipeline with a diameter of 6 m is present in front of the tunnel face. The rock layer contains groundwater, with the water level 40 m above the tunnel base. The aim is to simulate the pore pressure distribution of the tunnel, and the seepage and surge of the tunnel water, over a 100-h period after excavation. The peripheral rock and tunnel surfaces serve as free drainage boundaries after tunnel excavation. The tunnel's left, right, front, and rear boundaries are set to hydrostatic pressure according to the actual groundwater level. The bottom boundary is a no-flow boundary.

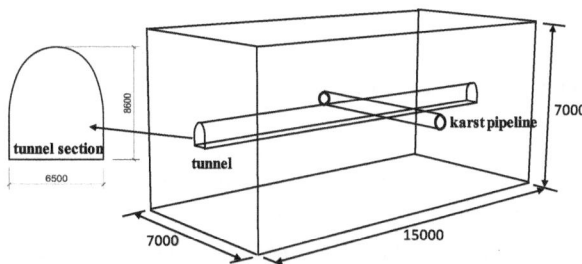

Fig. 5. 3D geological model and tunnel profile

4.2   Governing Equation

Effective Stress Principle. The overall stress upon any plane located inside a saturated geotechnical body can be split into two distinct entities: the effective stress and the pore water pressure [11]. Pore water pressure is defined as such.

$$\overline{\sigma} = \sigma + (\chi uw + (1 - \chi)ua)I \tag{1}$$

where σ is the total stress, $\overline{\sigma}$ is the effective stress, χ is related to the surface tension between the saturated geotechnical body and the liquid-gas, $\chi = 1.0$ for fully saturated geotechnical body and $\chi = 0.0$ for dry geotechnical body.

Seepage Equation. The permeability of a geotechnical body ought to be determined within a coupled fluid permeability/stress analysis. The permeability law is Forchheimer's permeability law [12], which defines the permeability coefficient $\overline{k}$ as:

$$\overline{k} = \frac{ks}{(1 + \beta\sqrt{vwvw})}k \tag{2}$$

where: k is the permeability coefficient of saturated geotechnical soil; β is a coefficient reflecting the effect of velocity on the permeability coefficient. When $\beta = 0.0$, Forchheimer's law of osmosis simplifies to Darcy's law. v_w is the velocity of fluid. k_s is the coefficient related to the degree of saturation, and $ks = 1.0$ when the degree of saturation $Sr = 1.0$.

The flow rate of pore fluid is related to pore pressure [13], i.e.:

$$vn = ks(uw - u_w^\infty) \tag{3}$$

where: v_n is the flow velocity in the direction normal to the boundary, k_s is the seepage coefficient, u_w is the pore water pressure at the boundary, and u_w^∞ is the reference pore pressure.

Calculation Parameter Selection Determination. According to the geological investigation report of Yangmuling Tunnel, the rock mechanical parameters of the peripheral rock, and the permeability parameters, are as selected and shown in Table 1 below. Due to the development of peripheral rock fissures in the specified section of the model and the high degree of rock fragmentation, the material parameters of the peripheral rock area of class V are chosen for calculation in this study.

Table 1. Rock mechanics parameters of tunnel surrounding rock

Perimeter rock grade	density (kg/m^3)	elastic modulus (GPa)	poisson's ratio	internal friction angle (°)	cohesion (MPa)	Porosity	permeability coefficient (m/s)
III	27	15	0.27	45	0.42	0.25	2×10^{-9}
IV	20	5	0.32	30	0.18	0.43	5×10^{-9}
V	18.5	0.02	0.3	23	0.03	0.54	5×10^{-7}

5 Results

5.1 Overall Analysis of the Seepage Field

Overall Analysis of the Water Pressure Field. When the tunnel is excavated into the karst region, Fig. 6 and Fig. 7 show water pressure cloud maps of various profiles ($XY_Z = 0$, $YZ_X = 0$).

The water pressure map of the tunnel profile after excavation is shown in Fig. 6. Figure 6(a–d) shows the water pressure map under different distance conditions from the tunnel face to the karst pipe. Before excavation, the maximum pore water pressure is 312.45 kPa, and after excavation, a low water pressure zone is formed in the vicinity of the tunnel, and the outward water pressure gradually increases, and the water pressure distribution is symmetrically distributed. With the discharge of pore water, the pore water pressure of the rock around the hole will eventually be reduced to 0 kPa, which corresponds to the monitoring results of the pore water pressure in the tunnel. At this time, the zone of low water pressure in the karst area expands sharply and shows an irregular shape, the shape of which is related to the distribution of karst caves and karst pipes.

Figure 7 illustrates the water pressure map of the $YZ_X = 0$ profile following the tunnel excavation. If the tunnel is not excavated to the karst pipe, a low water pressure zone forms near the cave following excavation, where the water pressure distribution is symmetrically distributed along the Z-axis. When excavating the tunnel to expose

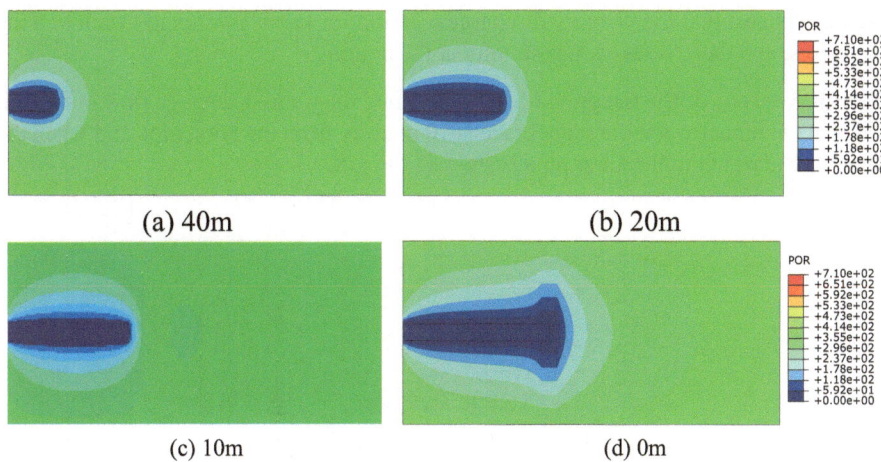

Fig. 6. Water pressure cloud map of section XYZ = 0 when the tunnel was excavated to the karst pipeline

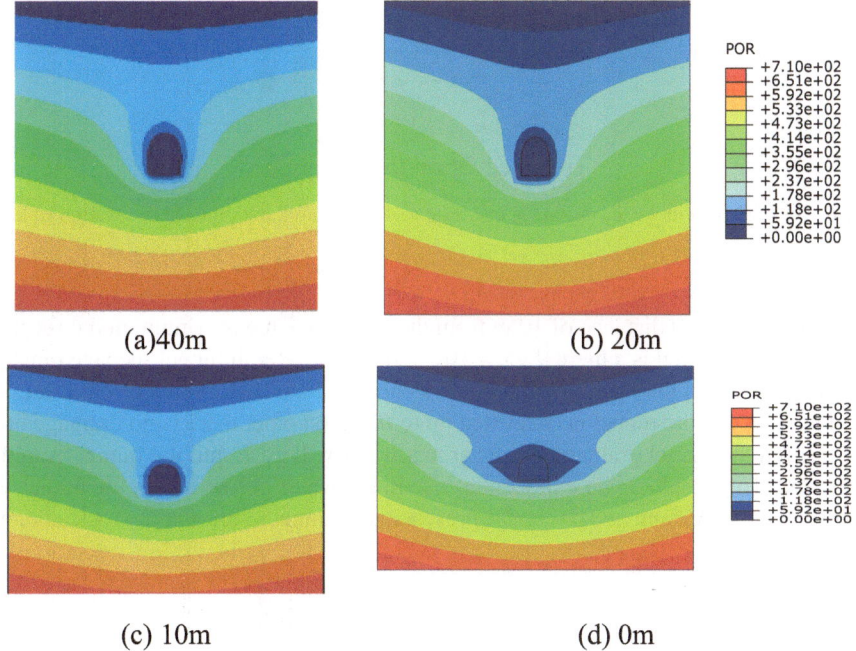

Fig. 7. Water pressure cloud image of YZX = 0 section when tunnel excavation to karst pipeline.

the karst pipe, the low water pressure area surrounding both the tunnel and the vault expands. This creates an inverted triangle water pressure distribution on either side of the tunnel, and there is no alteration in the water pressure at the base of the tunnel. These findings suggest that the primary discharge location for karst water is the vault and the two gangs, and there is less water surging at the bottom.

Overall Analysis of the Seepage Field. The cloud diagram of seepage flow in the unit nodes in the tunnel when the tunnel face is 10 m away from the karst pipe and when the tunnel is excavated to the karst pipe is shown in Fig. 8.

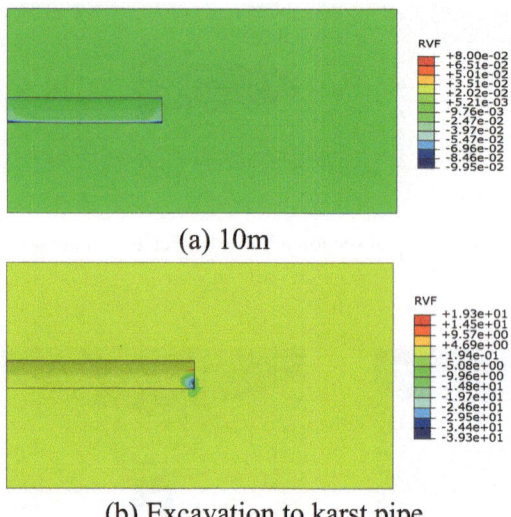

(a) 10m

(b) Excavation to karst pipe

Fig. 8. RVF after tunnel excavation (a: excavate to 10 m from karst pipe; b: excavation to karst pipe)

From Fig. 8, at a distance of 10 m from the karst pipe, the maximum nodal seepage recorded in the tunnel is a mere 9.95×10^{-2} m³/h. The overall tunnel seepage remains relatively low, with the maximum seepage occurring at the bottom foot of both sides of the tunnel. When the tunnel is excavated up to the karst pipe, there is no drainage hole to constrain the tunnel drainage. Thus, groundwater will gush into the tunnel from the karst pipe. This results in an exponential rise in the water inflow within the tunnel. At this point, the maximum nodal seepage present in the tunnel's surrounding rock amounts to 39.3 m³/h.

5.2 Water Pressure and Fluid Velocity Analysis

In order to visually represent the changes of the water pressure and velocity fields of the surrounding rock, four probe lines were selected in the range of 50 m in front of the tunnel face to be analysed and studied. The number and location of the selected probe lines are shown in Table 2.

For the four probe lines mentioned above, their corresponding profiles of infiltration water pressure and fluid velocity are displayed in Fig. 9.

Table 2. Select the number and position of detection lines

Serial number	Detection line range	Selection of detection points
1	X = 0–50 m, Y = 0 m (bottom of the tunnel)	22
2	X = 0–50 m, Y = 3.0 m (median line of a cave)	22
3	X = 0–50 m, Y = 6.0 m (top of the cave)	22
4	X = 0–50 m, Y = 8.6 m (top of arch)	22

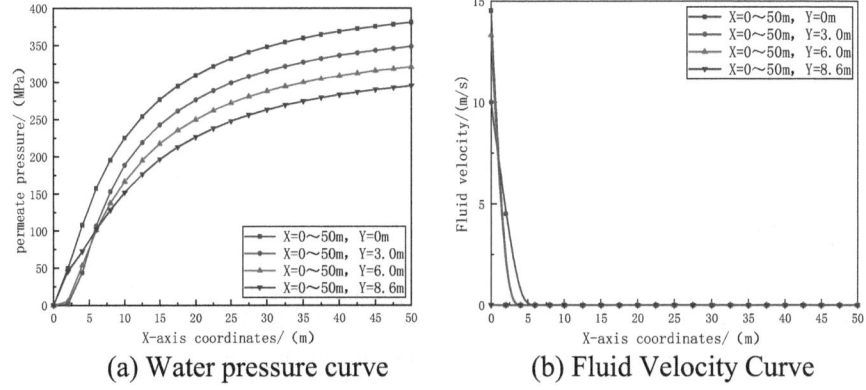

(a) Water pressure curve   (b) Fluid Velocity Curve

Fig. 9. Pore water pressure and velocity curves 50 m in front of the tunnel face

Based on the data presented in Fig. 9(a), it can be inferred that the water pressure is zero at the tunnel face for both the middle line and top line of the cavern. Additionally, the water pressure fluctuates slightly within a span of 3 m in front of the tunnel face before increasing abruptly. The growth rate then decreases at a distance of 20 m, ultimately stabilising at 348.54 kPa and 321.16 kPa. The water pressure at the tunnel face is 0 kPa for both the bottom line of the tunnel and the top line of the arch. Subsequently, the water pressure gradually rises, slowly increasing until it reaches the maximum values of 381.05 kPa and 295.37 kPa. The water pressure starts at 0 kPa, gradually increases, then slowly rises again, finally tending towards the aforementioned maximum values. Once passing through the karst pipe, water pressure rises sharply and then gradually continues to increase. The water pressure in the tunnel section close to the tunnel face is evidently lower compared to the exterior, and it increases as it moves farther away from the tunnel face.

From Fig. 9(b), it can be inferred that the excavation of the tunnel uncovered the karst pipe, resulting in the maximum fluid velocity of 14.5 m/s and 13.3 m/s at the bottom line and top line of the cavern, respectively, at the tunnel face. Subsequently, the fluid velocity experiences a rapid decrease within four meters in front of the tunnel face, eventually stabilising at 1.7×10^{-4} m/s. The maximum fluid velocity at the midline of the cavern opening is 10 m/s at the tunnel face. The velocity rapidly decreases within 5 m in front of the opening and then changes gradually. The karst pipe does not extend to the top of the tunnel, resulting in minimal variation of fluid velocity at the top. The maximum fluid velocity at the tunnel face is 6.7×10^{-3} m/s. Beyond a range of 5 m in front of the tunnel face, the flow velocity remains consistent among the lines. The highest fluid velocity is observed near the tunnel face according to the tunnel excavation which uncovered the karst pipe, causing a surge in velocity in the karst region immediately ahead of the tunnel face. Then, the velocity quickly drops off and eventually levels out in the surrounding rock segment.

5.3 Surge Analysis

Based on the model calculation, the sum of seepage volume can be acquired from the nodes of the tunnel face and the 2 m perimeter of the hole behind the tunnel face. If the tunnel traverses through the crushed rock body, the estimated tunnel water influx is 35.12 m^3/h. On the other hand, if the tunnel face passes through the karst pipeline, the highest influx of water in the tunnel is at 638.5 m^3/h.

6 Conclusion

Taking the Yangmuling water-rich karst tunnel boring project as a case study, we constructed a numerical model via the finite element method, which was then combined with the monitoring data on pore water pressure within the tunnel. The findings demonstrate that:

(1) The maximum water pressure on the surface of Yangmuling Tunnel is 6.84 kPa, which corresponds to a groundwater depth of 0.68 m. Additionally, the pore water pressure remains stable between 6.0 and 6.84 kPa, and the pore pressure on the surface of the tunnel is minimal.
(2) Prior to tunnel excavation, the highest recorded pore water pressure surrounding the tunnel was 312.45 kPa. Post-excavation, a zone of low water pressure develops around the tunnel, and water pressure gradually rises. As pore water is discharged, the rock surrounding the tunnel experiences a reduction in pore water pressure, eventually decreasing to 0 kPa, in accordance with the monitored pore water pressure results within the tunnel. When the excavation of the tunnel reveals the karst pipe, the outpouring of karst groundwater leads to a dramatic expansion in the low water pressure zone within the karst region.
(3) As the tunnel is excavated to the karst pipe, the tunnel water inflow increases rapidly, resulting in a maximum nodal seepage of 39.3 m^3/h and a maximum water inflow of 638.5 m^3/h.
(4) Once it passes through the karst pipe, the water pressure in front of the face increases rapidly, followed by a slow increase. The peak velocity of the fluid appears near the tunnel face. As a result of the tunnel excavation, the karst pipe that causes a high flow velocity in the karst area ahead of the tunnel face is revealed. Nonetheless, it then swiftly decreases towards the front of the tunnel face, and then the velocity of the flow starts to change gradually in the surrounding rock section. Afterward, the flow rate in the encircling rock section tends to stabilise.

References

1. Li, L.C., Tang, C.A., Liang, Z.Z., Ma, T.H., Zhang, Y.B.: Numerical simulation on water inrush process due to activation of collapse columns in coal seam floor. J. Min. Saf. Eng. **26**(2), 158–162 (2009)
2. Hwang, J.H., Lu, C.C.: A semi-analytical method for analyzing the tunnel water inflow. Tunn. Undergr. Space Technol. **22**(1), 39–46 (2007)
3. Zhang, Q.S., Wang, D.M., Li, S.C.: Development and application of water and mud burst model test system for tunnel in fault fracture zone. Chin. J. Geotech. Eng. **39**(03), 417–426 (2017)
4. Wang, D.M., Zhang, Q.S., Zhang, X.: Fault fracture zone tunnel water inrush disaster mud evolution model test research. Rock Soil Mech. **5**(10), 2851–2860 (2016). https://doi.org/10.16285/smj.r.2016.10.016
5. Fu, H.L., Li, J., Cheng, G.W.: Based on the conformal mapping of faults zone tunnel water inflow prediction. J. Huazhong Univ. Sci. Technol. (Nat. Sci. Ed.) **49**(01), 86–92 (2021). https://doi.org/10.13245/j.hust.210115
6. Li, F.H.: Groundwater system identification and inflow prediction of Gaojiaping tunnel. Chin. J. Underground Space Eng. **14**(01), 250–259 (2018)
7. Li, S.C., et al.: Gaussian process model of water inflow prediction in tunnel construction and its engineering applications. Tunn. Undergr. Space Technol. **69**, 155–161 (2017)
8. Zhang, P., Huang, Z., Liu, S., Xu, T.: Study on the control of underground rivers by reverse faults in tunnel site and selection of tunnel elevation. Water **11**(5), 889 (2019). https://doi.org/10.3390/w11050889
9. Chiu, Y.-C., Chia, Y.: The impact of groundwater discharge to the Hsueh-Shan tunnel on the water resources in northern Taiwan. Hydrogeol. J. **20**(8), 1599–1611 (2012). https://doi.org/10.1007/s10040-012-0895-6
10. Fang, Y., Wang, H.W., Zhou, C.Y.: Constraints tunnel pressure cavity in front of a small impact on the surrounding rock stability analysis. J. Hunan Univ. (Nat. Sci. Ed.) **44**(9), 137–145 (2017)
11. Lin, C.N., Li, L.P., Han, X.R.: Research on prediction method of tunnel water inrush in complex karst area. Chin. J. Rock Mech. Eng. (07), 1469–1476 (2008)
12. Huang, Y.Z., Wang, E.Z.: Experimental study on effective stress sensitivity coefficient of low permeability rock permeability. Chin. J. Rock Mech. Eng. (02), 410–414 (2007)
13. Fei, K., Zhang, J.W.: Application of ABAQUS in Geotechnical Engineering. China Water Resources and Hydropower Press, Beijing (2010)

The Effect of Modified Styrene Butadiene Rubber Latex on the Properties of Emulsified Asphalt and Mixture

Qiwei Zhou[1,2(✉)], Zhixin Shi[2], Shilun Zheng[3], Mengzhen Zhao[1], Mingyuan Yuan[1], Chunzhuang Xing[4], and Fu Xu[5]

[1] China Merchants Chongqing Communications Technology Research and Design Institute Co., Ltd., Chongqing, China
zhouqiwei@cqjtu.edu.cn
[2] Chongqing Jiaotong University, Chongqing, China
[3] Zunyi Highway Administration Bureau, Zunyi, Guizhou, China
[4] Guangxi Cenxing Expressway Development Co., Ltd., Guangxi, China
[5] Chongqing Zhixiang Paving Technology Engineering Co., Ltd., Chongqing, China

Abstract. In order to study the effect of styrene butadiene rubber (SBR) latex at various dosages on the properties of emulsified asphalt and its mixtures as well as to reveal SBR's modification mechanism and action, the conventional test, dynamic shear rheology test, fluorescence microscopy test, infrared spectroscopy test, contact angle test, wet wheel abrasion test and rutting deformation test were carried out to investigate the conventional properties, rheological properties, microphase structure, adhesive properties and the abrasion resistance of its mixtures, the resistance to water damage and rutting resistance of the modified SBR emulsified asphalt. The results show that SBR can make a significant improvement in ductility and softening point of emulsified asphalt with reduction in emulsified asphalt's penetration. SBR emulsified asphalt complex shear modulus and rutting factor increases and the phase angle decreases with the dosage of SBR The spatial structure of SBR can be stabilized in the emulsified asphalt, the linear correlation coefficients between the SBR doping and its area share were found to be high by binarized fluorescence microscopy image analysis. The SBR can make improvement in the contact angle for emulsified asphalt and in free energy on emulsified asphalt's surface. The physical modification between SBR and emulsified asphalt. And SBR features excellent road properties, such as abrasion resistance, water damage resistance and rutting resistance.

Keywords: Road Engineering · Emulsified Asphalt · Fluorescence Microscopy · Digital Image Processing · Styrene Butadiene Rubber Latex · Adhesion · Micro-Surfacing

G. Feng (Ed.): ICCE 2023, LNCE 526, pp. 350–364, 2024.
https://doi.org/10.1007/978-981-97-4355-1_33

1 Introduction

As of 2022, China's has realized over 170,000 km of highway mileage with 5 million km of road mileage, ranking first in the world. There has been a highway maintenance mileage of 5,350,300 km, accounting for 99.9% of highway mileage [1], China's highway maintenance industry is undergoing a period of rapid development. Emulsified asphalt features such advantages as excellent fluidity at room temperature, no heating during construction, green and low-carbon [2, 3]. It is widely used in highway maintenance projects, and it has just become one of the core materials for highway maintenance [4–6]. As improvement has been made in latex preparation process in recent years, polymer-modified emulsified asphalt has become a research hotspot in the industry. Such modified latex as styrene butadiene styrene [7, 8], SBR rubber [9, 10] and waterborne epoxy resin [11–13] have been widely used in emulsified asphalt. Of them, SBR latex has significantly improved the low-temperature resistance to cracking and adhesion of emulsified asphalt while improving the high temperature performance of emulsified asphalt, having become the most widely used modifier. Domestic and foreign counterparts have carried out relevant researches such as Meng [14], Wang [15] and Che [16] found that SBR latex can improve the storage stability and high temperature performance of emulsified asphalt, and further enhance the rutting and cracking resistance of the mixture. SBR latex blending amount is recommended to be 3%–5%. Yang GM [17], Yang TW [18], Gong R [19] and Hu FG [20] studied the conventional properties of SBR emulsified asphalt, rheological properties, fluorescence microscopy and road performance of SBR asphalt mixtures, and found that the high temperature resistance to deformation, low-temperature resistance to cracking and relaxation characteristics of the modified asphalt as well as the shear capacity of the mixture have been improved with the increase of SBR doping, and the SBR doping should not be more than 4%. However, there are few reports both at home and abroad which describe the microscopic phase structure and dichotomization analysis, surface energy and adhesion concerning emulsified asphalt modified with SBR latex.

This paper starts with the effect of SBR latex dosing on the properties of emulsified asphalt, and further reveals the role of SBR latex on the performance enhancement of emulsified asphalt and its mixtures through conventional test, DSR, fluorescence microscopy, contact angle and infrared spectroscopy so as to further reveal the role of SBR latex on the performance enhancement of emulsified asphalt and its mixtures, with a purpose to provide a theoretical basis at certain level and a basis of data for the application of real engineering.

2 Materials and Testing

2.1 Raw Materials

(1) Matrix asphalt

70 # matrix asphalt is employed. Asphalt specific conventional performance indicators are shown in Table 1.

(2) Modified SBR latex
   The modified SBR latex is a high solid content latex provided by a company in the United States with a solid content of 63%, of which the proportion of styrene is 24% and the proportion of butadiene is 76%.
(3) Aggregate
   The aggregate consists of 0–5 mm limestone and 5–10 mm basalt, and all its properties are in compliance with the existing specifications.

Table 1. The basic Properties of Matrix Asphalt.

Pilot project	Unit	Test results	Specification	Test Method
Penetration (25 °C, 100 g, 5 s)	0.1 mm	70.4	60–80	T0604-2011
Softening point	°C	50.2	≥45	T0606-2011
Power Viscosity (60 °C)	Pa · s	185	≥160	T0620-2011
Ductility (15 °C, 5 cm/min)	cm	>150	≥40	T0605-2011
Wax content (distillation method)	%	1.6	≤2.2	T0615-2011
Flash point	°C	280	≥260	T0611-2011
Density (15 °C)	g/cm³	1.030	Actual measurement	T0603-2011

2.2 Preparation of Modified Emulsified Asphalt

In this paper, the method where emulsification is followed by modification is employed. The emulsifier is of a slow-cracking and fast-setting cationic emulsifier with a dosage of 2%. SBR emulsified asphalt is prepared as follows: 0%, 1%, 2%, 3%, 4%, 5% SBR (and emulsified asphalt mass ratio) were added to the matrix emulsified asphalt respectively, and stirred well for later use; the mixing duration is 3 min with a mixing rate of 500 rpm. Then SBR modified emulsified asphalt was formed. The preparation process is shown in Fig. 1.

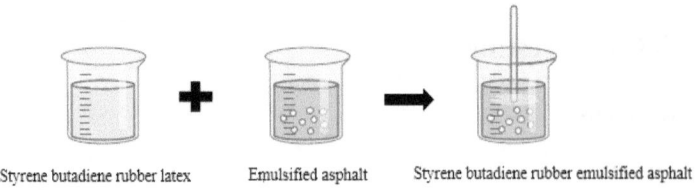

Styrene butadiene rubber latex   Emulsified asphalt   Styrene butadiene rubber emulsified asphalt

Fig. 1. SBR Modified Emulsified Asphalt Preparation Process.

2.3 Micro-surfacing Grading Design

The micro-surfacing grade composition is shown in Table 2.

Table 2. Micro-surfacing Grading.

Grading	Sieve size/mm								
	13.2	9.5	4.75	2.36	1.18	0.6	0.3	0.15	0.075
Synthetic grades	100.0	100	75.9	61.8	44.4	25.8	15.3	11.7	10.0
Lower limit of design gradation range	100.0	100	70.0	45.0	28.0	19.0	12.0	7.0	5.0
Higher limit of design grading range	100.0	100	90.0	70.0	50.0	34.0	25.0	18.0	15.0
Median value	100.0	100	80.0	57.5	39.0	26.5	18.5	12.5	10.0

The composition of the gradation at the micro-surfacing was finalized by the tests and the amount of water added was determined to be 5%, the amount of cement was 1% and the optimum oil/gravel ratio was 6.3%.

2.4 Test Methods

Conventional Performance Test. According to JTG E20-2011 "*Highway Asphalt and Mixture Test Procedure*", the evaporated residue of emulsified asphalt modified with SBR at varied dosages was tested for penetration at 25 °C, softening point and ductility at 5 °C in order to characterize the effect of SBR on the conventional properties of emulsified asphalt.

DSR Test (Temperature Scanning Test). The temperature scanning test was carried out on the emulsified asphalt modified with SBR using an AR 1500ex dynamic shear rheometer. The test conditions were described as follows: the angular velocity was 10 rad/s, the control strain was 12%, the scanning temperature was 30–66 °C, and the temperature interval was 6 °C.

Microscopic Phase Structure. The microstructure of SBR emulsified asphalt was observed by LW300LFT-LED type fluorescence microscope with a magnification of 400 times. And its fluorescence microscope pictures were subjected to digital image processing to convert the effective information in the figure into digital information. The area percentage of SBR was employed thus to further analyze the effect of SBR on the phase structure of emulsified asphalt.

A Therom Scientific Nicolet iS5 Fourier infrared spectrometer was used to test the changes in functional groups and chemical composition of SBR emulsified asphalt. The tested wave numbers range from 400 cm^{-1} to 4000 cm^{-1}, and the number of scans was 32 with a resolution of 4 cm^{-1}.

Adhesion Test. HARKE-CA contact angle tester was used to measure the contact angle of SBR emulsified asphalt by lay-drop method and calculate free energy on its surface, and thus to study the interfacial adhesion strength of SBR emulsified asphalt from the point of view of work and energy, and to gain the effect of SBR blending on the adhesion performance of emulsified asphalt.

Mixture Test. Mixable tests, wet wheel abrasion tests and rutting tests were carried out on the mixtures to analyze the effect of SBR on the abrasion resistance, water damage resistance and rutting resistance of modified emulsified asphalt micro-surfacing mixtures.

3   Results

3.1   Conventional Performance

The general performance test results of SBR emulsified asphalt are shown in Fig. 2. When the ductility is greater than 100 cm, it is recorded 100 cm. It can be seen from the figure that: With the increase of SBR mixing, the softening point and ductility of asphalt increase, and the degree of penetration decreases. The softening point of the original emulsified asphalt is low, the degree of penetration is high, and the asphalt is brittle in the 5 °C ductility experiment. When the dosage of SBR goes up to 3%, the softening point of the modified emulsified asphalt has increased by 11.4% compared to the original sample of emulsified asphalt. Ductility at 25 °C went up increased to 100 cm, the penetration decreased to 57.6 (0.1 mm). This indicates that the low-temperature cracking resistance of SBR-modified emulsified asphalt was enhanced and the high-temperature performance was improved. This is because the addition of SBR absorbs the light oil from asphalt and produces a swelling reaction, with the asphalt changing from the solgel type to the sol-gel type. A more stable mesh structure takes shape after the molecules inside the asphalt are restrained.

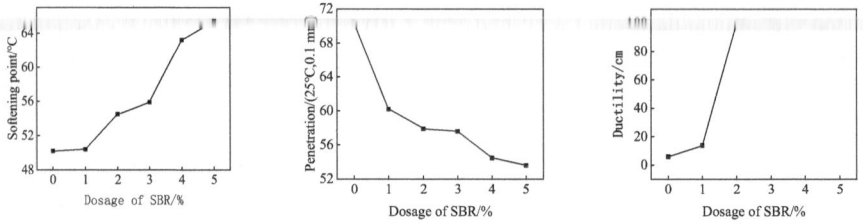

Fig. 2. Conventional performance of Emulsified Asphalt with Different SBR.

3.2   Rheological Performance

Temperature scanning tests were carried out on emulsified asphalt with varied SBR dosages. The results are shown in Fig. 3. When the temperature is determined, the addition of SBR induces reduction the phase angle (δ) of emulsified asphalt along with increment in both complex shear modulus (G^*) and rutting factor ($G^*/\sin \delta$). It shows that the addition of emulsified asphalt leads to a more stable spatial structure of SBR, and the reduction in light oil content lowers the temperature sensitivity of asphalt and improves the high-temperature deformation resistance and rutting resistance of asphalt as well. When SBR doping remains unchanged, the phase angle tends to increase and the complex shear modulus and rutting factor decrease with the increment in temperature. At this moment, the internal elastic component of asphalt is transformed into a viscous along with weakened deformation resistance of asphalt.

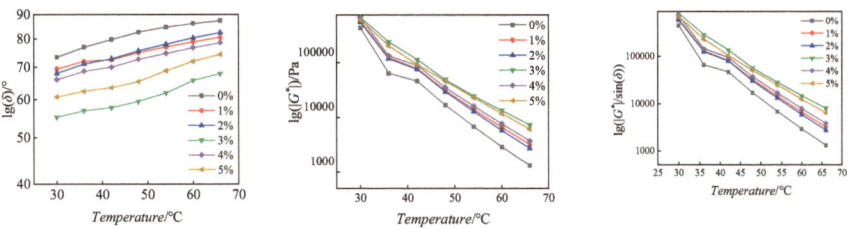

Fig. 3. Rheological Properties of Emulsified Asphalt with Varied SBR.

3.3   Microscopic Phase Structure

Fluorescence Microscope.   Fluorescence microscope with 400 times magnification was used to observe the distribution of SBR and asphalt. As shown in Fig. 4, fluorescence microscope image on the left hand side is darker than on the right hand side. But there is a similarity in proportion for fluorescent area which shows noticeable 3D characteristics. The emulsified asphalt image shows indistinct fluorescence emitted from the emulsifier, and the rest of the images show noticeable fluorescent material emitted from the SBR. When the dosage of SBR stands at 1%, SBR exists in the continuous phase of emulsified asphalt in the form of small particles that are well dispersed. When SBR is 3%, spatial network structure takes shape in modified emulsified asphalt along with high structural strength, improved high temperature performance and enhanced softening point as opposed to that for the conventional performance. When the dosage of SBR continues to increase to 4%, the latex begins to aggregate with a poor degree of dispersion. It shows from an overall point of view that it is best for SBR's content to stand 3%.

Digital Image Processing of fluorescence microscope images of emulsified asphalt modified with SBR was performed using MATLAB. First of all, the fluorescence microscope images of asphalt with SBR of varied dosage were converted to grayscale image. Due to noticeable difference in gray levels between the SBR region and the background region in the image, unequal threshold segmentation of asphalt images with SBR of varied dosages was carried out to accurately extract the pixels in the target region when the original image was converted to grayscale image. The ratio of the SBR region and the background region is finally obtained, and the data obtained for each sub-block region is averaged to obtain the data results of the global image. The segmentation is shown in Fig. 5, where the black part is asphalt and the white part is SBR latex.

The binarized image of SBR emulsified asphalt was analyzed at each dosage to gain the ratio of the number of SBR pixel points to the number of pixel points of the modified

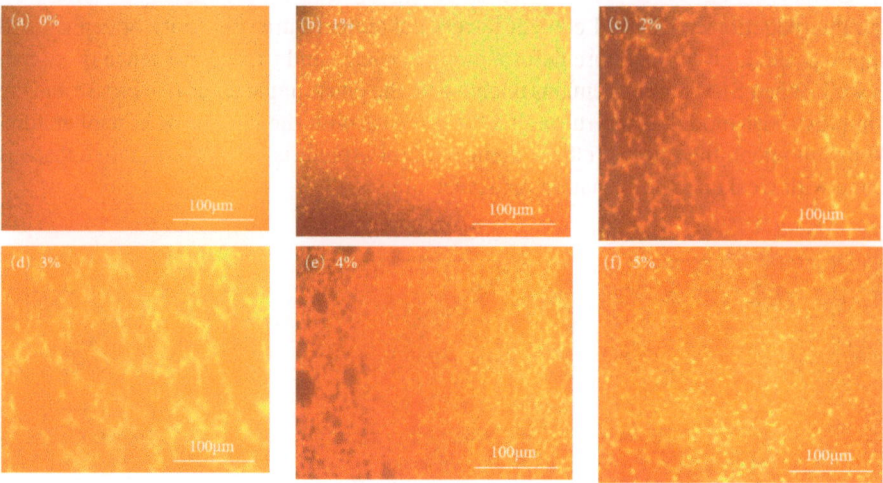

Fig. 4. Fluorescence Microscope Images of Emulsified Asphalt with SBR of varied dosages.

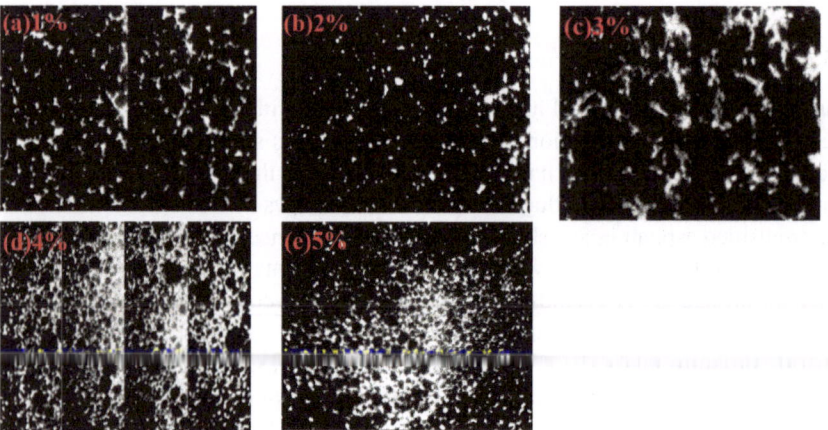

Fig. 5. Grayscale Plot and Binarized Image of Emulsified Asphalt with SBR of Varied Dosages.

emulsified asphalt in this image, namely the area percentage of SBR at each dosage. The results are shown in Fig. 6, and the formula is shown in Eq. (1) (2).

$$R = \frac{\sum\limits_{i=1}^{n} i}{\sum\limits_{j=1}^{m} j} \times 100\% \tag{1}$$

$$R^- = \frac{\sum\limits_{R=1}^{k} R}{K} \tag{2}$$

where i—pixel points of SBR in the sub-block;

 j—all pixel points in the sub-block;

 k—the number of sub-blocks;

 R—the area percentage of the sub-block;

 R^-—the area percentage of the total block, namely the average of the sub-block summation.

It is indicated in Fig. 6 that the area percentage of SBR is linearly distributed with the increase of butadiene doping. The area percentage is only 5.52% at 1% of SBR doping. As the dosage of SBR increases, the area percentage also increases linearly. The percentage of area reached 12.37% at 3% of SBR. The linear fitting of the data shows that the linear pattern of the dosage of SBR and its area percentage is good, and the correlation coefficient can go up to 98.9%, which indicates that the SBR and emulsified asphalt are highly compatible.

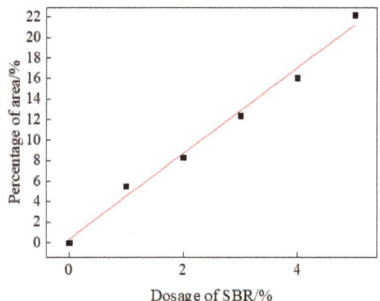

Fig. 6. The Area Percentage of SBR.

Infrared Spectrum. Infrared spectroscopic tests were performed on the evaporated residues of as-received emulsified asphalt and 3% SBR emulsified asphalt. These results are shown in Fig. 7. The same absorption peaks appeared in the two infrared spectral curves, including 2919 cm^{-1} and 2849 cm^{-1} for the absorption peaks of the stretching vibration of -CH$_2$ alkanes, and 1456 cm^{-1} and 1376 cm^{-1} for the absorption peaks of the bending vibration of -CH$_3$ alkanes. In addition to the above-mentioned absorption peaks, absorption peaks of butadiene and styrene components of the SBR composition appeared in the 3% SBR emulsified asphalt. These include 966 cm^{-1} for trans-butadiene, which is an out-of-plane bending vibration of trans-C-H olefin, and 699 cm^{-1} for styrene, which is an out-of-plane bending vibration of monosubstituted benzene ring C-H olefin. As a result, it can be seen that there is no new functional group in the emulsified asphalt of 3% SBR. Therefore, no change was found in the chemical composition of SBR and emulsified asphalt which was just subject to a physical reaction in Fig. 8.

3.4 Adhesion Performance

According to the first law of thermodynamics, the work consumed by the molecules inside a substance to migrate to the surface is the potential energy of the molecules on

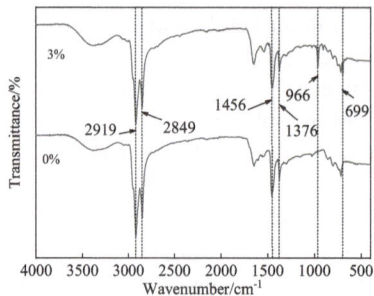

Fig. 7. Infrared Spectrums of SBR Emulsified Asphalt.

Fig. 8. Contact Angle Pictures.

the surface, called the surface free energy (γ). When asphalt adheres to the aggregate, the aggregate will adsorb the asphalt to reduce its own surface free energy under the influence of the force field. As a result, the surface energy is an intrinsic factor that affects the adhesion performance of asphalt. Surface energy by the polar component (also known as Lewis acid-base component, γ^P) and dispersion component (also known as van der Waals component, γ^d). The polar component in turn consists of Lewis acid (γ^+) and Lewis base (γ^-). The expression for the surface energy expression is show as follows:

$$\gamma = \gamma^d + \gamma^P = 2\sqrt{\gamma^+\gamma^-} \tag{3}$$

The relation between the surface energy parameters for the liquid and bitumen are expressed as follows:

$$\gamma_L(1 + \cos\theta) = 2\sqrt{\gamma_a^d \gamma_L^d} + 2\sqrt{\gamma_a^+ \gamma_L^-} + 2\sqrt{\gamma_a^- \gamma_L^+} \tag{4}$$

where γ_L—surface energy of the liquid;
   γ_a^d, γ_L^d—Dispersive component of bitumen and liquid;
   γ_a^+, γ_L^+—Lewis acid fraction of the bitumen and liquid;
   γ_a^-, γ_L^-—Lewis base fraction of the bitumen and liquid.
   The modified emulsified asphalt with SBR of varied dosages was subjected to contact angle tests using three test liquids: water, glycerol and formamide. The surface free energy parameters of each liquid at 25 °C are shown in Table 3 below:
   As shown in Fig. 9, the contact angle results of emulsified asphalt for each SBR dosage are slightly different. The contact angle results obtained for all three liquid

Table 3. Surface Free Energy of Liquids.

Liquid type	γ_L	γ_L^d	γ_L^P	γ_L^+	γ_L^-
Water	72.8	21.8	51.0	25.5	25.5
Glycerol	64.0	34.0	30.0	3.92	26.08
Formamide	58.0	38.0	19.0	2.28	16.72

reagents showed an increasing trend with the increase of SBR dosage, in which the contact angle of emulsified asphalt with water was the greatest, and the contact angle with formamide was the smallest. The coefficients of variation of the contact angles measured for the three liquids were around 2% with good reproducibility. The results of the linear analysis of the surface energy of the three liquids tested and its product with the cosine value of the contact angle are shown in Fig. 9. The figure shows that the correlation coefficients of the linear fits of the six SBR doped emulsified asphalt are all greater than 90%, indicating a good linear relationship. It indicates that the selected fluids are applicable to the surface energy testing of SBR modified emulsified asphalt.

The results of the contact angle test were used in Eqs. (3) (4), and the results are shown in Fig. 10: With the addition of SBR, the surface energy of the modified emulsified asphalt increases, and the surface free energy is more prone to decreasing when combined with the aggregate, and the asphalt-aggregate adhesion performance is better. This is due to the formation of a stable 3D network structure between the SBR and the asphalt, which enhances the adhesion of the asphalt.

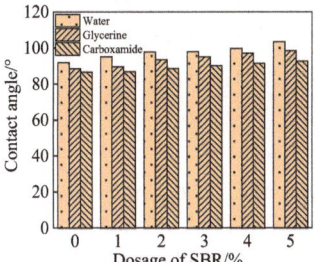 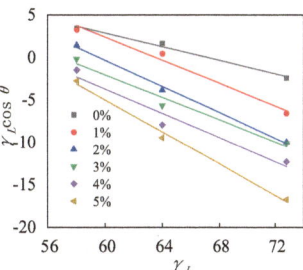

Fig. 9. Contact Angle Results of SBR Modified Emulsified Asphalt.

3.5 Mixture Test

Mixable Test. The performance of asphalt is subject to mixable duration when asphalt is used during a project. Required mixable duration is necessary when modified emulsified asphalt is applied to micro-surfacing. In this paper, emulsified asphalt dosage is 10%, water dosage is 5% and cement dosage is 1% in the mixable test. The above-mentioned materials were poured into 100 g of the graded aggregate and mixed to observe the mixing duration. The temperature during testing was 23 °C and the measured solid

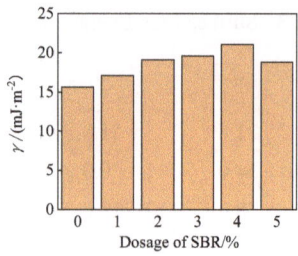

Fig. 10. Surface Free Energy of SBR Modified Emulsified Asphalt.

content of the modified emulsified asphalt was 63%. The mixing duration of the asphalt mixtures are shown in Table 4. With increasing amount of SBR, the mixing duration becomes longer along with increasing fluidity. This may be the result of formation of a more stable spatial network structure of the modified emulsified asphalt, which becomes stronger when combined with the mixture.

Table 4. Mixable duration of SBR emulsified asphalt Mixture.

Dosage of SBR (%)	Mixable duration (s)	Slurry state
0	122	Dry and harder to mix
1	150	Sizing general
2	175	Sizing general
3	>180	Good pulping
4	>180	Good pulping
5	>180	Good pulping

Abrasion Resistance and Resistance to Water Damage. The 1 h and 6 d wet wheel abrasion tests were conducted on modified emulsified asphalt mixtures with SBR of varied dosages using the above-mentioned mixing test formulations. The wet wheel abrasion values are shown in Fig. 11. From the 1 h and 6 d wet wheel abrasion results, it can be concluded that the wet wheel abrasion values of the mixes are higher when no SBR is added. As the dosage of SBR increases, the adhesion and consolidation of asphalt and aggregate become more noticeable, and the abrasion resistance and water damage resistance of the mixture increase. As shown in Fig. 12, the cohesion between aggregates is weak when SBR is not added, and the falling of stone in shape of large particles can be noticeably found. And when the SBR dosing is 3%, the emulsified asphalt and aggregate have strong adhesion ability and the mixture loss is less. This may be the result of the stable structure formed by SBR and asphalt, which inhibits the movement of molecules within the asphalt, thus reducing the effect of water immersion on the asphalt mixture. However, when the dosage of SBR was increased to 4%, the abrasion resistance and water damage resistance of the mixture decreased instead of increasing. Asphalt mixture's resistance to water damage was the highest at 3% SBR dosage.

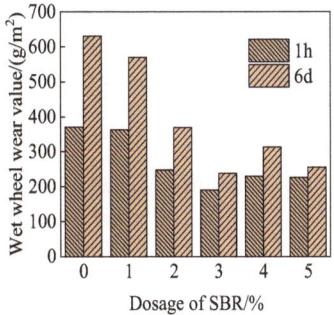

Fig. 11. Wet Wheel Wear Values of SBR Modified Emulsified Asphalt Mixture.

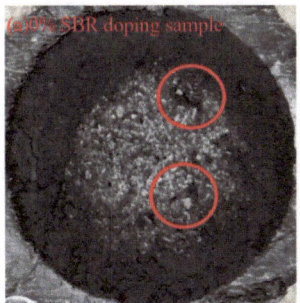 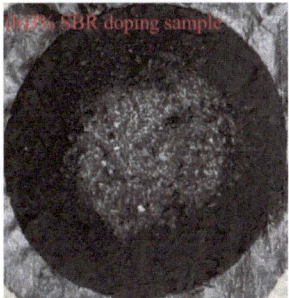

Fig. 12. Wet Wheel Abrasion Specimens of SBR Modified Emulsified Asphalt Mixture.3.5.3. Rutting resistance.

Rutting Deformation Test. The load wheel rolling test was conducted on the SBR emulsified asphalt mixture, and the rutting deformation test results are shown in Fig. 13. The width deformation rate and rutting depth rate of SBR modified emulsified asphalt mixture both become higher first and lower later with the increment SBR doping, which shows that the rutting resistance of modified emulsified asphalt mixture is enhanced first and weakened later. The improvement in the performance of the mixture is the result of the swelling reaction between SBR and asphalt, which restricts the movement of molecules within the asphalt, and the asphalt becomes harder, which enhances the rutting resistance after combining with the mixture. However, when the dosage of SBR is overly high, they will be aggregated. Consequently, the structure of modified emulsified asphalt becomes weaker and the performance of SBR-emulsified asphalt mixtures will deteriorate.

Rutting Test. A modified asphalt mixture rutting test was employed as a rutting test method. The 5 cm-thick rutting plate specimen spares set aside a top at a height of

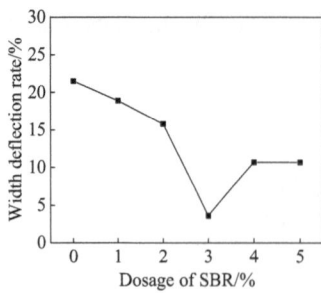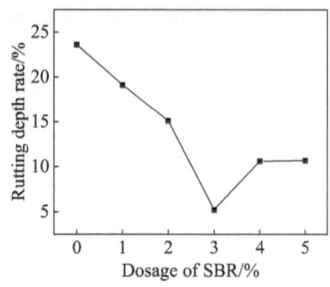

Fig. 13. Rutting Deformation Rate of Emulsified Asphalt of SBR Modified Emulsified Asphalt Mixture.

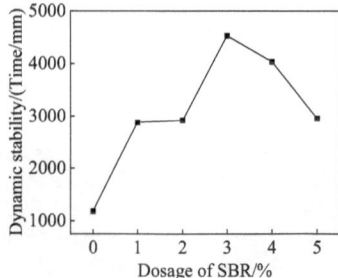

Fig. 14. Dynamic Stability of SBR Modified Emulsified Asphalt Mixture.

1 cm, and the SBR emulsified asphalt mixture of each dosage was put into the surface of other graded asphalt mixtures, simulating the asphalt structure at the time of micro-surfacing. The results of the dynamic stability of the rutting test at 60 °C are shown in Fig. 14. Figure 14 shows that the dynamic stability of the mixture increases as SBR dosage is increased, and the rutting resistance at the micro-surfacing is improved, and the dynamic stability of the mixture is higher and the rutting resistance is stronger when SBR dosage is increased to 3%. This is because the emulsified asphalt and aggregate with 3% of SBR has better combining ability, high structural strength and strong resistance to deformation. Similarly, the rutting resistance of SBR emulsified asphalt mixtures decreases when SBR is doped at 4% and higher.

4   Conclusion

In summary, the results of conventional test, rheological test, microscopic phase structure, adhesion test and Fourier infrared spectroscopy test performed on the SBR emulsified asphalt showed that the SBR can effectively improve the high and low temperature performance and adhesion properties of emulsified asphalt; the type of modification of SBR and emulsified asphalt is of physical modification; a stable spatial network structure can be formed via 3% SBR and emulsified asphalt, when the doping is overly high, SBR

Aggregation occurs. Thus, it is not recommended to overly dope SBR; after the fluorescence microscope images of SBR emulsified asphalt were digitized, it is concluded that the compatibility between SBR and emulsified asphalt is good.

Tests on SBR emulsified asphalt mixtures showed that the water damage resistance, abrasion resistance and rutting resistance of SBR emulsified asphalt are improved by increment in SBR. The mix performance of emulsified asphalt reached its optimum at 3% of SBR dosing. In an overall consideration of the results of test on SBR emulsified asphalt and its mixture, dosage of SBR is recommended to be at 3%.

Acknowledgments. This paper is one of the phase results of the Science and Technology Project of Chongqing Science and Technology Bureau (CSTB2023TIAD-GPX0003), the Science and Technology Project of Guizhou Provincial Department of Transportation (2022-122-002) and the Science and Technology Project of Jiangxi Provincial Department of Transportation (2021H0020).

References

1. Statistical bulletin on transportation industry development. China Water Transp. **07**, 29–33 (2023)
2. Meng, Y.J., Zhao, Q.X., Lu, Z.B., Liu, Z.R., Xu, R.G., Lei, Y.L.: Preparation and properties of waterborne epoxy resin/butadiene rubber composite modified emulsified asphalt Science. Technol. Eng. **21**(13), 5524–5531 (2021)
3. Yan, H.B., Chen, X.: Preparation and performance of nanometer TiO_2/SBR composite modified emulsified asphalt. Enterp. Technol. Dev. **12**, 61–63 (2022)
4. Kong, L., Li, J., Luo, Q.X., Shen, Z.Z., He, Z.Y.: Preparation and performance of waterborne epoxy resin emulsified asphalt Applied. Chem. Eng. **50**(08), 2076–2081 (2021)
5. Zhang, S., Zhang, S., Zhang, H.L., Wu, C.F., Wan, N.: Aging properties of waterborne epoxy emulsified asphalt residues Highway and Transportation. Sci. Technol. **40**(03), 1–7 (2023)
6. Lin, X., Li, Q., Lu, Y.: Highway maintenance: building a new generation of highway maintenance technology system. China Highway **05**, 22–25 (2022)
7. Li, Q., Chu, Z.S., Luo, X., Huang, S.L., Huang, X.L.: Preparation of high dosage SBS modified emulsified asphalt. J. Hubei Univ. (Nat. Sci. Ed.) **42**(02), 217–221 (2020)
8. Cong, Y.F., Xie, J.X., Liu, H., Wei, H.: Preparation and performance analysis of linear SBS latex modified emulsified asphalt. J. Liaoning Univ. Petrochemical Technol.s **39**(03), 17–21 (2019)
9. Zheng, J.P.: SBR modified emulsified asphalt and its micro-surface treatment road performance research. Shandong Transp. Sci. Technol. (03), 71–73 (2023)
10. Jia, Q.R., Liu, Q., Xu, Q.Y., Ma, J.Y.: Preparation and basic performance evaluation of emulsified asphalt modified with SBR styrene-butadiene latex. Enterp. Technol. Dev. **10**, 63–66 (2022)
11. Liu, H.H., Li, X.J.: High temperature rheological properties of waterborne epoxy emulsified asphalt. J. Chongqing Jiaotong Univ. (Nat. Sci. Ed.) **39**(10), 67–73 (2020)
12. Kezhen, Y., Junyi, S., Kaixin, S., Min, W., Goukai, L., Zhe, H.: Effects of the chemical structure of curing agents on rheological properties and microstructure of WER emulsified asphalt. Constr. Build. Mater. **347**, 128531 (2022)
13. Yin, Y.Y., Han, S., Kong, H.Y., Han, X., Guo, H.: Optimization and performance evaluation of waterborne epoxy resin modified emulsified asphalt micro-surfacing based on tunnel driving environment. Constr. Build. Mater. **315**, 125604 (2022)

14. Meng, Y., Chen, J., Kong, W., Hu, Y.: Review of emulsified asphalt modification mechanisms and performance influencing factors. J. Road Eng. **3**(2), 141–155 (2023). https://doi.org/10.1016/j.jreng.2023.01.006

15. Wang, D.C., et al.: Assessment of testing methods for higher temperature performance of emulsified asphalt. J. Clean. Prod. **375**, 134101 (2022)

16. Che, T.K., Pan, B.F., Li, Y.D., Ge, D.D., Jin, D.Z., You, Z.P.: The effect of styrene-butadiene rubber modification on the properties of asphalt binders: aging and restoring. Constr. Build. Mater. **316**, 126034 (2022)

17. Yang, G.M., Li, H., Chen, S., Pei, X.G., Wu, Y.: Study on the effect of styrene-butadiene rubber content on the high and low temperature performance of modified asphalt. New Mater. Chem. Ind. **47**(06), 248–251 (2019)

18. Yang, T.W.: Study on conventional and rheological properties of emulsified asphalt modified with styrene-butadiene latex. Adhesion **49**(12), 31–35 (2022)

19. Gong, R., Xu, P., Guo, Y.Q.: The effect of styrene-butadiene latex on the performance of emulsified asphalt modified with adhesive layer. Chin. Foreign Highw. **34**(06), 223–225 (2014)

20. Hu, F.G., Tian, X.G., Hu, H.L., Li, G.Y., Guo, C.H.: Effect of SBR latex dosage on the performance of modified emulsified asphalt. J. Constr. Mater. **24**(04), 895–900 (2021)

Optimization of Impact Resistant Structures with Negative Poisson's Ratio Based on Response Surface Methodology

Qiang Zeng[1], Shenqiang Feng[2], and Zhengkai Zhang[3](✉)

[1] Chengdu Xingchen Investment Group Co., Ltd., Chengdu, People's Republic of China
[2] Chengdu Construction Engineering Group Co., Ltd., Chengdu, People's Republic of China
[3] School of Mechanical and Electrical Engineering, Xi'an University of Architecture and Technology, Xi'an, Shaanxi, People's Republic of China
woodncy@163.com

Abstract. In order to enhance the blast resistance capabilities of critical structural elements in buildings, a blast-resistant honeycomb sandwich protective structure based on a negative Poisson's ratio structure was proposed and designed. The failure mechanisms of the protective structure under distributed impulse loads were investigated. The structural parameters of the protective structure were optimized using the response surface methodology, and simulation analysis was conducted using the finite element method. The results demonstrate that the optimized negative Poisson's ratio honeycomb sandwich protective structure exhibits excellent blast resistance performance.

Keywords: Negative Poisson Ratio · Auxetic Honeycomb Structure · Response Surface Methodology · Latin Hypercube Experimental Design Method

1 Introduction

Many modern architectures, such as factories, airports, train stations, theaters, etc., are large-span shell structures lacking internal supports. In the event of an accidental explosion, the impact of the explosion shock wave can easily lead to building collapse. Therefore, implementing effective blast engineering measures to reduce the explosive dynamic response of shelter structures and enhance the survivability of buildings is of paramount importance. Currently, buildings often employ methods such as reinforced concrete structures, steel structures, and the construction of blast walls as blast protection measures. However, these protection measures are time-consuming to deploy, construction-intensive, and limited in terms of size. There is a need for a structurally simple, easily deployable, cost-effective, and highly blast-resistant architectural structural protection scheme.

One of the protective measures is to limit the deformation of the building structure during impact to prevent further accidents. Therefore, a protective structure capable of absorbing impact energy has been developed to reduce the risk of accidents. The

© The Author(s) 2024
G. Feng (Ed.): ICCE 2023, LNCE 526, pp. 365–372, 2024.
https://doi.org/10.1007/978-981-97-4355-1_34

protective structure must be able to effectively absorb dynamic impact energy and should also be lightweight to reduce the load on the building and effectively lower costs, as shown in Fig. 1.

The Poisson's ratio of a material refers to the ratio of transverse strain to axial strain. Structures or materials with a negative Poisson's ratio are called auxetic structures or materials. Negative Poisson's ratio means that when an axial force is applied to the material, it tends to increase in the direction perpendicular to the applied force [1].

Under the impact load, materials with a positive Poisson's ratio will move outward from the impacted area in the direction perpendicular to the impact load, while materials with a negative Poisson's ratio will contract towards the impacted area. At this time, the local density of the material with a negative Poisson's ratio increases and the modulus increases, thus exhibiting better impact resistance and blast resistance compared to other honeycomb structures as shown in Fig. 2.

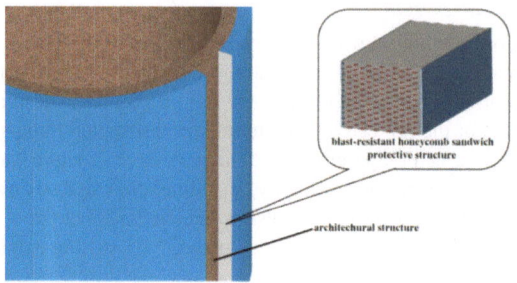

Fig. 1. Diagram of blast-resistant honeycomb sandwich protective structure.

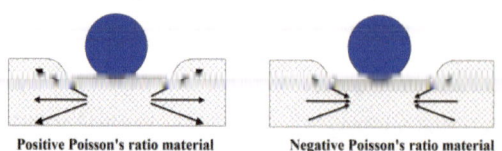

Fig. 2. The deformation of the material under impact compression.

Due to the unique properties of materials with a negative Poisson's ratio, which originate from their internal structure, there is a growing interest in optimizing methods to further enhance their performance [2–4]. Researchers are actively exploring various approaches to improve the properties of materials with a negative Poisson's ratio, such as optimizing the internal structure of the material to enhance its auxetic properties. This can involve designing specific patterns or arrangements of the structure's constituents to achieve desired mechanical behaviour.

By focusing on these research areas and utilizing optimization methods, it is possible to further enhance the performance of materials with a negative Poisson's ratio and unlock their potential for various applications, including in protective materials [5–7].

In order to enhance the blast-resistance capabilities of buildings, this paper proposes a honeycomb sandwich protective structure based on a negative Poisson's ratio structure.

The stress-strain relationship of the structure under impact loads was analysed using the finite element method. Furthermore, a combination of response surface methodology and genetic algorithm was employed to optimize the structural parameters of the protective structure [8].

2   Numerical Modelling and Validation of Auxetic Structure

The structural parameters of the auxetic structure depicted in Fig. 3 is presented. Preliminary modelling and simulation need to be conducted to establish precise parameters and validate the structure.

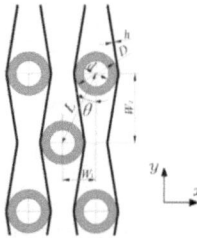

Fig. 3. Diagram of blast-resistant honeycomb sandwich protective structure.

The initial model chosen for this purpose is based on the work by references [9] and [10]. The geometric dimensions of the model can be found in Table 1.

Table 1. The geometric dimensions.

	Value	Description
d	8 mm	internal diameter of the tubes
D	10 mm	external diameter of the tubes
W_1	8 mm	centre distance of two adjacent tubes in the x-direction
W_2	20 mm	centre distance of two adjacent tubes in the y-direction
h	1 mm	the thickness of the corrugated sheets
l	–	length of the straight ligaments
θ	–	angle of the perpendicular line of the straight ligament to x-axis, which corresponds to the half wrapping angle formed between the corrugated sheet and tube

In this paper, a finite element simulation was conducted to investigate the deformation process of the proposed impact-resistant protective shell, as shown in Fig. 4. The numerical analysis of the in-plane impact characteristics of honeycomb materials was conducted using the ANSYS/Explicit Dynamic Finite Element software. In the calculations, the matrix material utilized was aluminium alloy, assumed to exhibit ideal elastoplastic behaviour. The material parameters for the aluminium alloy were set as follows: shear modulus Es = 27.6 GPa, yield stress σy = 680 MPa, density ρ = 2.7 × 103 kg/m³, and Poisson's ratio ν = 0.3.

Fig. 4. Finite element mode of the blast-resistant honeycomb structure.

The cell walls were modelled using the SHELL163 shell element. To ensure convergence, five integration points were defined along the thickness direction. Additionally, a single-sided automatic contact algorithm was applied in the calculations. The surfaces of the rigid plate and the external surface of the honeycomb specimen were considered to be smooth, with no friction between them.

The proposed impact-resistant protective shell subjected to an initial velocity of 50 m/s and a duration of 3 ms on its upper surface (Fig. 5).

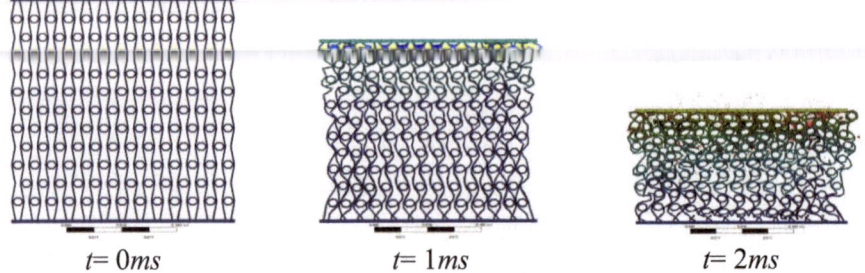

| $t = 0ms$ | $t = 1ms$ | $t = 2ms$ |

Fig. 5. Finite element mode of the blast-resistant honeycomb structure.

The process of deformation occurring in auxetic structure under low-speed impact can be divided into four stages (Fig. 6): elastic zone, plateau zone, enhancement zone, and densified zone. The graph below illustrates this, with the x-axis representing strain and the y-axis representing stress.

The force-deformation behaviour of an auxetic structure is initially unstable when subjected to impact. The elastic zone is also very brief. In the plateau zone, the stress of the structure fluctuates within a small range around a constant value. However, as the

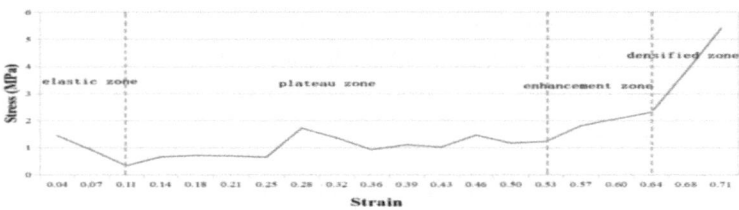

Fig. 6. Stress strain curve of auxetic structure.

strain increases, the stress no longer fluctuates around a constant value but gradually increases. The structure enters the plateau stress-enhanced zone, and the strain at this point is known as the plateau-enhanced strain. As the strain of the auxetic structure further increases, the honeycomb walls start to compress, and the slope of the stress-strain curve rapidly increases and approaches a constant value. The structure enters the densified zone, and the strain at this point is known as the densified strain. Densification occurs because as the deformation increases, the voids between the honeycomb cells gradually fill up. Ideally, densified strain should be reached when the voids are filled. However, in practical situations, the densification strain is generally smaller than the ideal case. This is because in real-world scenarios, some of the rods in the auxetic structure experience compression and friction between each other, leading to densified strain even when some voids still exist.

Due to the instability and transient nature of the elastic region of an auxetic structure under impact loads, the energy absorbed during the entire deformation process is relatively small. Therefore, when studying the theoretical model of how much energy a negative Poisson's ratio structure can absorb under impact loads, the focus is primarily on two stages: the platform region and the platform stress-enhanced region. The theoretical results are used to represent the energy absorbed by the negative Poisson's ratio structure from the onset of the impact load until densification. From the onset of the impact load until densification, the absorbed energy E can be expressed as the sum of the energy absorbed in the platform region E1 and the energy absorbed in the platform stress-enhanced region E2, i.e., $E = E1 + E2$. This can be represented on the equivalent stress-strain curve as the area enclosed by the equivalent stress-strain curve and the horizontal axis.

Clearly, in the case of constant corrugated plate thickness and tube wall thickness, the microstructure of the negative Poisson's ratio influences the shape of the equivalent stress-strain curve, thereby affecting the energy absorption capacity of the structure.

3   Response Surface Methodology and Genetic Algorithm Optimization

Based on this research foundation, it is necessary to further optimize the structure with the energy absorption-to-mass ratio as the optimization goal. In order to expedite the optimization process, this paper adopts an optimization method based on response surface models. The Latin hypercube experimental design method was used to randomly select 15 sample points, and a polynomial response surface model was used to fit the simulation results of these 15 sample points, in order to obtain the model parameters that result in the maximum energy absorption-to-mass ratio.

3.1 Style and Spacing

The response surface methodology involves fitting a polynomial regression equation to model the complex nonlinear relationship between the optimization objective and the design variables [8]. The regression equation for a multivariate quadratic response surface approximation model is given by:

$$y(x) = \beta_0 + \sum_{i=1}^{n} \beta_i x_i + \sum_{i=1}^{n} \beta_{ii} x_i^2 + \sum_{i<j}^{n} \beta_{ij} x_i x_j + \varepsilon \tag{1}$$

where y(x) represents the predicted response, x1, x2, ..., xn are the design variables, β_0, β_1, β_2, ..., β_n are the regression coefficients, β_{ii} are the quadratic coefficients, β_{ij} are the cross-product coefficients, and ε is the error term.

The negative Poisson's ratio structure introduced in this paper is determined by three independent variables: the lateral spacing W1, the longitudinal spacing W2, and the pipe diameter D. Any change in one of these variables will have a certain impact on the structural strength. However, in order to expedite the search for optimal results, this paper selects two of these variables, lateral spacing W1 and longitudinal spacing W2, as the influencing factors, while keeping the pipe diameter D, which has a relatively small impact on the structure, constant.

This paper uses the Latin hypercube sampling method for experimental design. The range of values for the lateral spacing W1 is set to [6.5, 8], and the range of values for the longitudinal spacing W2 is set to [16, 20]. A total of 15 random samples were taken.

Based on the previous analysis, the energy absorbed by the impact-resistant structure from the onset of impact to densification, denoted as E, can be calculated using the finite element method. After matrix calculation, the energy-mass ratio response surface model is obtained as follows:

$$y = -5422.13 + 920.08x_1 + 246.85x_2 - 32.03x_1^2 - 1.72x_2^2 - 25.41x_1x_2 \tag{2}$$

The response surface fitting model is shown in the Fig. 7, from which it can be seen that the optimization problem has a global optimal advantage. The two-dimensional variable optimization of the response surface model is carried out, and the optimal advantage is found to be x = 6.821 mm, y = 18.65 mm, and the maximum energy-mass ratio is 158.3J/kg. The optimized impact energy mass ratio is greater than the original structure's impact energy mass ratio of 137.2 J/kg.

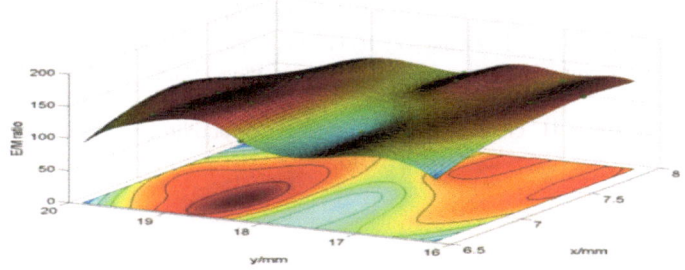

Fig. 7. Response surface fitting model.

4 Discussion and Conclusions

In this paper, theoretical analysis and optimization simulation of the energy absorption process of the new negative Poisson's ratio structural model under impact load are carried out, and the following conclusions are obtained:

(1) The impact calculation model of the new negative Poisson ratio structure is established, and the stress-strain curve of the impact-resistant structure is obtained from the finite element calculation results, so as to obtain the energy-absorption energy-mass ratio.
(2) By using the Latin hypercube experimental design method, the variables related to the structure parameters of negative Poisson ratio were randomly sampled, and 15 sample points were obtained. Through the finite element calculation results of 15 sample points, the response surface model of the new negative Poisson ratio structure was obtained by response surface fitting technology.
(3) According to the response surface model, the optimal impact model parameters of the negative Poisson ratio impact resistant structure were found with the energy-absorption-mass ratio as the optimization objective, and the impact resistance finite element calculation of the optimal model was re-performed. The calculation results verified the effectiveness of the optimization results.

Acknowledgments. The research was supported by Xi'an Science and Technology Plan Project (No. 2021JH-QCY7-0024).

References

1. Chan, N., Evans, K.E.: J. Cell. Plast. **35**, 166–183 (1999)
2. Wang, W., Dai, S., Zhao, W., Wang, C., Ma, T.: Compos. Struct. **275**, 114458 (2021)
3. Biharta, M.A.S., Santosa, S.P., Widagdo, D., Gunawan, L.: World Electr. Veh. J. **13**, 118 (2022)
4. Gao, J., Xue, H.P., Gao, L., Luo, Z.: Comput. Methods Appl. Mech. Eng. **1**, 211–236 (2019)
5. Lisiecki, J., Błażejewicz, T., Kłysz, S., Gmurczyk, G., Reymer, P., Mikułowski, G.: Phys. Status Solidi B **250**, 1988–1995 (2013)
6. Lowe, A., Lakes, R.S.: Cell. Polym. **19**, 157–167 (2000)

7. Zhang, J., Lu, G., You, Z.: Compos. Part B Eng. **201**, 108340 (2020)
8. Sosina, S., Remillard, E.M., Zhang, Q.: Technometrics **61**, 50–65 (2019)
9. Zhengkai, Z., Hong, H., Shirui, L., Binggang, X.: Phys. Status Solidi B **250**, 1996–2001 (2013)
10. Zhengkai, Z., Hong, H., Ginggang, X.: Smart Mater. Struct.Struct. **22**, 1–6 (2013)

Study on the Influence of Secondary Grouting on Soil Settlement Above Shield Tunnel

H. L. Wang[1], J. J. Ge[2], M. Wang[3], Y. X. Jiang[4], S. Feng[5], Y. D. Li[1], L. Ran[6], and T. Jin[7(✉)]

[1] Chengdun Suian Underground Engineering Co., Ltd., Shanghai 201108, China
[2] Zhejiang Hupingyan Intercity Rail Company Limited, Jiaxing 314200, China
[3] Polytechnic Institute, Zhejiang University, Hangzhou 310058, China
[4] Hangzhou Metro Group, Hangzhou 310017, China
[5] Shanghai Tunnel Engineering Co., Ltd., Shanghai 200032, China
[6] Department of Civil Engineering, Zhejiang University of Science and Technology, Hangzhou 310012, China
[7] Department of Civil Engineering, Zhejiang University, Hangzhou 310058, China
cetaojin@zju.edu.cn

Abstract. Secondary grouting can effectively control the ground settlement caused by shield construction and is widely used in urban shield tunnel construction. Therefore, it is necessary to study the grouting material and grouting effect of secondary grouting. Based on the background of below-passed Metro Line, this paper determines the optimal ratio of grouting materials by studying the effects of different cement content and superplasticizer addition on the fluidity, consistency, initial setting time and compressive strength of cement inert slurry, this paper also studies the influence of grouting on the settlement of soil above the tunnel by monitoring the settlement value of soil above the tunnel during grouting. This study contributes to the understanding of grouting techniques in urban shield tunnel construction. The results show that Grouting has a significant effect on controlling the settlement of shield machine in the later stage of crossing, and the ground settlement can be basically stable within 4 days.

Keywords: Secondary Grouting · Soil Settlement · Shield Tunnel

1 Introduction

With the continuous development of China's transportation industry, the living population of many large cities is increasing, and the urban area is expanding. In order to facilitate citizens' travel, reduce citizens' travel commuting time, and alleviate urban traffic pressure, subway tunnels may be in the process of construction or normal service. For some reasons (such as adjacent construction, vehicle load, etc.) have a certain uneven settlement. Many cities in China (such as Beijing, Shanghai, Guangzhou, Hangzhou, etc.) are increasing subway mileage.

The shield tunnel can well control the construction disturbance and has strong adaptability to different strata. However, in the process of shield tunnel construction, it will

G. Feng (Ed.): ICCE 2023, LNCE 526, pp. 373–381, 2024.
https://doi.org/10.1007/978-981-97-4355-1_35

inevitably cause surface subsidence, which will have a serious impact on nearby buildings and structures [1]. Controlling the ground settlement caused by shield tunnel excavation is the key to the safe construction of the tunnel, especially the below-passed tunnel project. The grouting strength and grouting pressure of the shield tunnel gap have a significant impact on the ground settlement. Kasper [2] analysed the influence of parameters such as grouting pressure behind the wall and balance pressure on the surface settlement during shield tunnel construction, and concluded that the surface settlement decreased with the increase of grouting pressure and balance pressure on the surface. In the past, the research on ground surface settlement caused by shield tunnelling was usually carried out from two aspects: horizontal ground surface deformation caused by shield tunnelling and ground surface deformation along the direction of shield tunnelling. In recent years, researches on surface settlement caused by shield construction mainly include field measurement analysis [3, 4], theoretical analysis [5], numerical simulation [6] and machine learning [7]. The use of suitable grouting materials and grouting parameters will improve the effect of grouting and better control the settlement of soil above the tunnel. Among the existing grouting materials, cement-water glass slurry has good injectability and high strength after solidification. However, it has poor water resistance and is not suitable for high water-bearing strata. In this paper, the optimal ratio of secondary grouting materials is obtained through laboratory tests. The influence of secondary grouting on the soil above the tunnel is studied by monitoring the settlement of the soil above the tunnel before and after secondary grouting.

2   Engineering Background

The U-shaped groove between the railway station and the test section has a length of 3126.220 m on the left line and 3126.555 m on the right line. The left line of the section corresponds to ring numbers 1890 to 1945, while the right line corresponds to ring numbers 1892 to 1951. The test section (as shown in Fig. 1) is located before the subway line and passes through a silty sand layer with a full-section designation of J-4 (as shown in Fig. 2). The depth of burial for the test section is 25.36 m. The horizontal section of the design axis is a straight-line segment, and it has a vertical slope of 15‰.

Fig. 1. Position diagram of test section

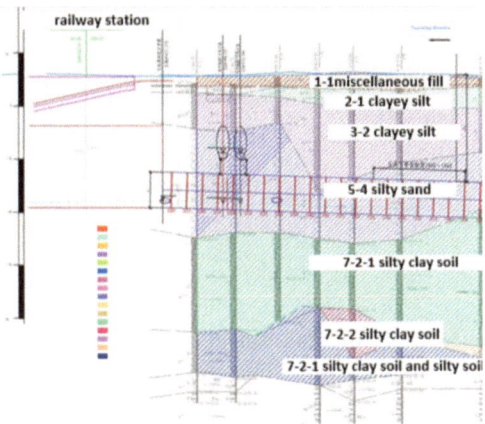

Fig. 2. Stratum distribution graph

The test section of the below-passed subway line is required to be located in an area with the same geological stratum and buried depth as the crossing section. Specifically, the left line test section is designated within the 1890 ring to 1949 ring. Both the test section and the crossing section are characterized by a full-Sect. 5–4 silty sand layer.

3   Indoor Test

The indoor test of secondary grouting was carried out for the silty sand stratum in the test section. According to the indoor test results, the relevant parameters and ratios of synchronous grouting in the test section were determined.

The test section is located in the silty sand layer. The silty sand layer is quite different from the clay layer in front of the excavation. The permeability coefficient of the silty sand layer is relatively large, while the hard slurry has the characteristics of good filling performance and good impermeability. Compared with the inert slurry, it is more suitable for the silty sand layer. However, the hard slurry is more prone to plugging in the actual grouting process, resulting in shutdown and delays in the construction period. Considering the good filling performance of the inert slurry and the difficulty of plugging, the project intends to add a small amount of cement to the inert slurry. In the case of ensuring fluidity, the formation of slurry strength is accelerated as much as possible, so that the slurry has early strength faster.

In order to improve the inert slurry obtained by optimization, different amounts of cement were added on the basis of the original site ratio to shorten the initial setting time and improve the compressive strength. Combined with the dosage of cement and water reducing agent of common hard slurry, the dosage of cement is 10kg/m3, and the dosage of 20kg/m3,30kg/m3,40kg/m3 is set respectively. The engineering performance of cement inert slurry with different cement content is studied. The research focuses on the fluidity, consistency, initial setting time and compressive strength of the slurry. The specific ratio is shown in Table 1.

Table 1. Proportioning Table.

test number	lime	coal fly ash	fine sand	bentonite	water	cement	additive
1	80	350	1050	80	420	0	2.5
2	80	350	1050	80	420	20	2.5
3	80	350	1050	80	420	30	2.5
4	80	350	1050	80	420	40	2.5

By studying the effects of different cement content and superplasticizer addition on the fluidity, consistency, initial setting time and compressive strength of cement inert slurry, the cement inert slurry with 20kg/m3 cement content was finally adopted. The fluidity is 28.3cm, the consistency of the slurry is 10.6cm, the initial setting time is 26 h, the compressive strength is 30–60 kPa, and the bleeding rate is reduced by 3%.

4 Monitoring System

For the test section of subway line, six layered settlement points of soil are buried in the test section, and the buried depth is 2 m above the tunnel vault to the ground. A monitoring section (ZD319-ZD329) is set up every 10 m within the test section, and an encryption axis point is added every five meters. The monitoring frequency is increased from the original 2 times/day to 4 times/day, and the monitoring points are arranged as shown in Fig. 3:

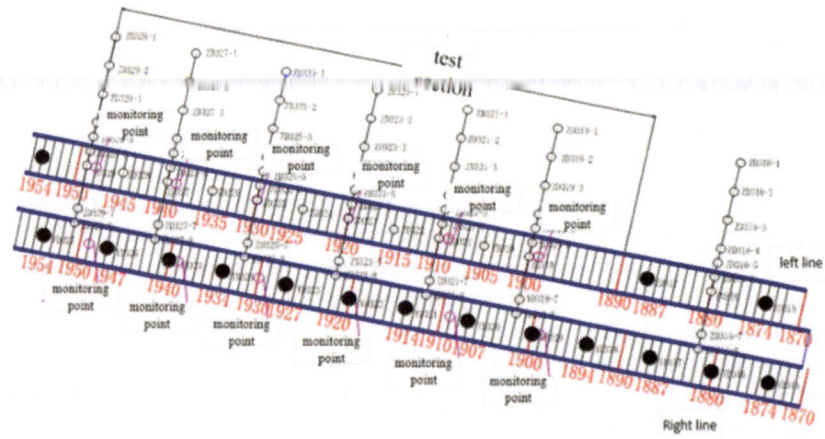

Fig. 3. Monitoring points graph

5 Data Analysis

5.1 Without Grouting

The 1920 ring was not grouted, and after the completion of the crossing, the single settlement value fluctuated significantly (as shown in Fig. 4). The maximum single settlement was −1.29 mm, and the maximum single uplift was 1.58 mm. The stable period of the ZD323 ground axis reached 6 days (as shown in Fig. 5).

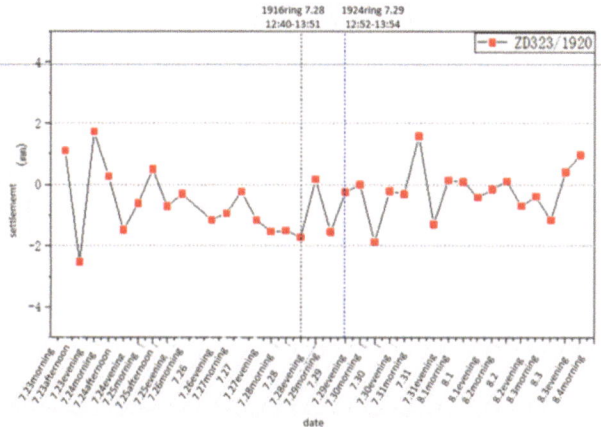

Fig. 4. Single settlement change at ZD323

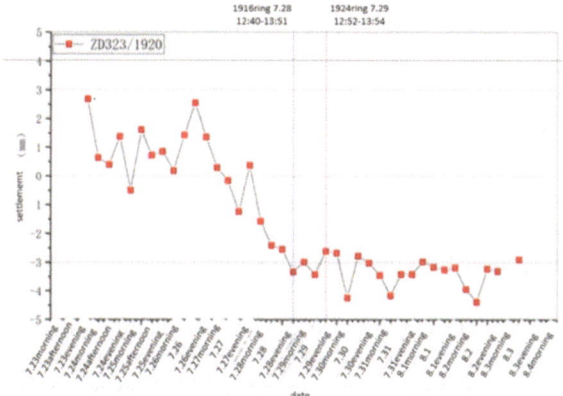

Fig. 5. Accumulated settlement change at ZD323

5.2 After Grouting

The 1900 ring grouting volume is 0.8 m³, the grouting pressure is 0.9 Mpa. The monitoring data is shown in the Fig. 6 and Fig. 7. After grouting, the data basically stabilized, and after 36 h, the monitoring data remained stable.

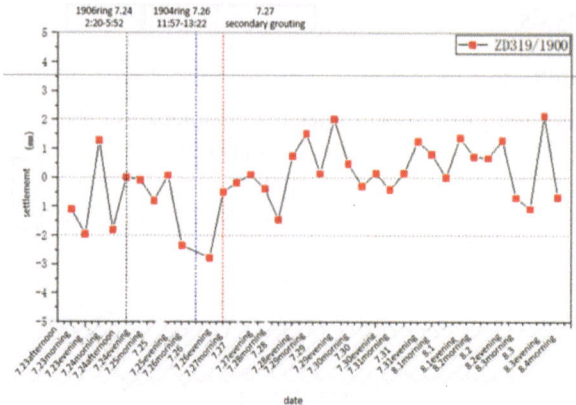

Fig. 6. Single settlement change at ZD319

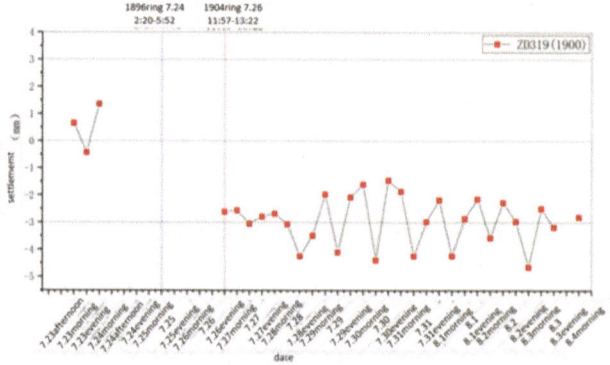

Fig. 7. Accumulated settlement change at ZD319

1915 ring grouting volume is 1 m³, the grouting pressure is 0.9 Mpa. After grouting is completed, the single variable is basically stable within 1 mm (as shown in Fig. 8), and the monitoring data can stabilize within 4 days after grouting is completed (as shown in Fig. 9).

The grouting volume for the 1930 ring is 1.5 m³, the grouting pressure is 0.9 Mpa. After grouting, the settlement value was small. The settlement can be stable within 2 mm (as shown in Fig. 10), and there was no obvious uplift. However, later monitoring data showed that there was still some settlement (as shown in Fig. 11).

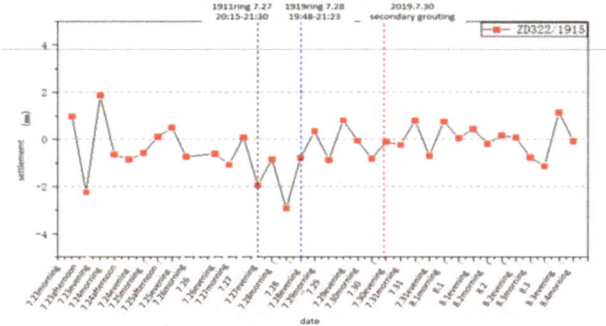

Fig. 8. Single settlement change at ZD322

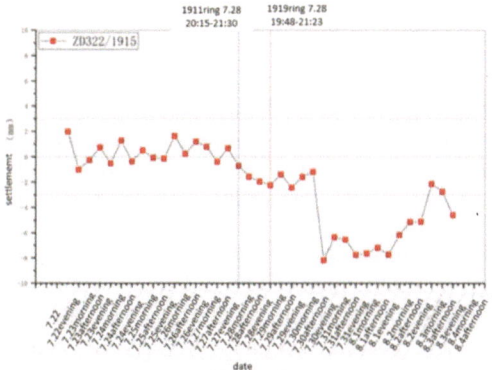

Fig. 9. Accumulated settlement change at ZD322

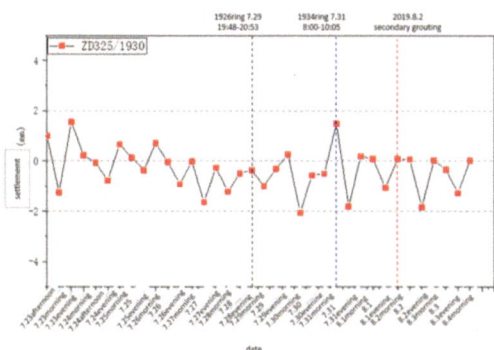

Fig. 10. Single settlement change at ZD325

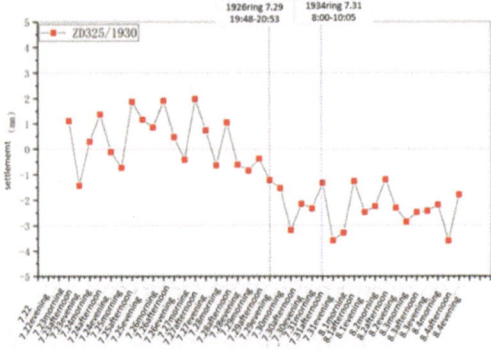

Fig. 11. Accumulated settlement change at ZD325

6 Conclusion

Based on the background of Metro Line, the optimum ratio of secondary grouting material is determined by experiment, and the influence of secondary grouting on the settlement of soil above the tunnel is studied by monitoring the settlement value of soil above the tunnel during secondary grouting. The following conclusions are obtained:

1. The secondary grouting makes the ground settlement significantly reduced after the crossing, and the ground settlement can be stabilized within four days.
2. After the secondary grouting is completed, there is no large variable in the monitoring data above the soil. The grouting volume is 0.8–1.5 m^3, and the grouting pressure is 0.9 Mpa and the ground stability period is different. When the grouting amount is 1 m^3 and the grouting pressure is 0.9 Mpa, the effect of controlling the late settlement of the surface is the most obvious, and the stability period is the shortest.

Acknowledgments. The work was supported by Shanghai Tunnel Engineering Co., Ltd. and the National Natural Science Foundation of China (No. 52308332).

References

1. Zhang, Z.G., Zhang, Z.X., Jiang, Y.J., Bai, Q.M., Zhao, Q.H.: Analytical prediction for ground movements and liner internal forces induced by shallow tunnels considering non-uniform convergence pattern and ground-liner interaction mechanism. J. Soils Found. **57**, 211–226 (2017)
2. Kasper, T., Meschke, G.: A numerical study of the effect of soil and grout material properties and cover depth in shield tunnelling. J. Comput. Geotech. **33**, 234–247 (2006)
3. Zhao, W., et al.: Analysis of the additional stress and ground settlement induced by the construction of double-O-tube shield tunnels in sandy soils. J. Appl. Sci. **9**, 1399 (2019)
4. Ding, Z., Wei, X.J., Wei, G.: Prediction methods on tunnel-excavation induced surface settlement around adjacent building. J. Geomech. Eng. **12**, 185 (2017)

5. Ding, Z., He, S.Y., Zhou, W.H., Xu, T., He, S.H., Zhao, X.: Analysis of ground deformation induced by shield tunneling considering the effects of muck discharge and grouting. J. Transp. Geotech. **30**, 1–10 (2021)
6. Yuan, L., Cui, Z.D., Tan, J.: Numerical simulation of longitudinal settlement of shield tunnel in the coastal city Shanghai. J. Marine Geotechnol. **35**, 365–370 (2017)
7. Ding, Z., Zhao, L.S., Zhou, W.H., Bezuijen, A.: Intelligent prediction of multi-factor-oriented ground settlement during TBM tunneling in soft soil. J. Front. Built Environ. **8**, 1–16 (2022)

Convergence Deformation of Existing Shield Tunnel Induced by Adjacent Shield Tunnelling Construction

Y. J. Song[1], C. R. Lu[1], X. J. Li[1], L. A. Zhao[1], X. W. Ye[2], and T. Jin[2(✉)]

[1] Shaoxing Rail Transit Group Co., Ltd., Shaoxing 312000, China
[2] Department of Civil Engineering, Zhejiang University, Hangzhou 310058, China
cetaojin@zju.edu.cn

Abstract. The subway system has benefited the transportation in a lot of modern cities with advantages such as convenience, rapidity, large volume, etc. However, with the construction of more and more subway systems, the underground space is increasingly crowded. According to engineering practice, existing tunnel is inevitably distorted by adjacent tunneling construction, reducing stability and safety, especially when the distance between two tunnels is small. To explore the convergence deformation of existing tunnels, field monitoring and analysis are conducted on the convergence deformation of the existing tunnel induced by the adjacent tunnel construction. This research finds that: (i) the convergence of each monitoring section gradually develops and stabilizes, and then the convergence of tunnel segments exhibits "horizontal duck egg" pattern and non-uniform convergence pattern; and (ii) after the tunnel shield machine passes, due to the lateral compression deformation of surrounding soil, horizontal convergence of the existing tunnel has an increase in horizontal convergence and a decrease in vertical convergence.

Keywords: Convergence Deformation · Shield Tunnel · Adjacent Shield Tunnelling

1 Introduction

The excavating of tunnel close to existing tunnel unavoidably impacts the nearby soil, causing the deformation and additional force of existing tunnel structures, which can pose hazards to the safety of the tunnel during normal operations. Common occurrences include uneven longitudinal settlement, structural cracks and damage, different degrees of opening in segment joints, as well as seepage and leakage [1–3]. Due to the complexity of the tunnel project, it is crucial to precisely ascertain the impact of shield tunnel construction on the existing tunnel structure, and appropriate and effective measures should be adopted to avoid structural damage.

Researchers have conducted studies on the influence of adjacent construction on existing tunnel structures using methods such as field monitoring, physical model testing, finite element method. For field monitoring, Shi et al. [4] based analysis on the

© The Author(s) 2024
G. Feng (Ed.): ICCE 2023, LNCE 526, pp. 382–389, 2024.
https://doi.org/10.1007/978-981-97-4355-1_36

monitoring data of the uplift displacement of the subway tunnel below a certain foundation pit excavation in Shanghai. The monitoring data showed that the existing tunnel in the pit bottom covering area showed a significant uplift, while the tunnel part not covered by the foundation pit settled; Liu et al. [5] conducted a comprehensive field monitoring of the deformation response of deep foundation pits and adjacent tunnels, including lateral displacement of underground continuous walls, tunnel crown settlement, elastic horizontal displacement, segment convergence, and opening width of segments. Based on the data, they analyzed the process of tunnel deformation; Li and Yuan [6] used a measurement system to implement displacement monitoring and analysis of the entire process of shield tunneling under existing tunnels; Zhang [7] analyzed the field monitoring data of Shanghai Metro Line 11, crossing Line 4 from above and below. By conducting the numerical simulation on the project, scientifically set shield parameters; Jin et al. [8] reported and analyzed the monitoring data of the construction of a shield tunnel under an existing tunnel in Shenzhen and compared the deformation of the existing tunnel and the ground caused by the construction of the shield tunnel underpass.

In terms of physical model testing, Huang et al. [9] used a centrifuge experiment to obtain the response the foundation pit project and proposed the concept of safe distance between the foundation pit project and the existing tunnel below; Kim et al. [10] conducted a scale model experiment to study the influence of tunnel gap, angle, and lining segment stiffness on the interaction during the process of shield tunneling under existing tunnels in clay; Ng et al. [11] combined the working conditions of shield tunneling under existing tunnels, used centrifuge experiments to analyze the changes in the impact of soil loss on existing tunnels in sandy soil layers.

About finite element method, Chakeri et al. [12] used a finite element software to simulate the construction process of the Torside Tunnel near Line 4 using the New Austrian Tunneling Method. The results showed that the settlement rate of Metro Line 4 was directly related to the tunnel spacing. At the same time, the simulation software calculated the stress and strain curves of Metro Line 4 during the underpass process; Lin et al. [13] took a double-line shield tunnel in Changsha obliquely crossing an existing tunnel as a support and analyzed the deformation response of the double-line tunnel obliquely crossing the existing tunnel through numerical simulation; Lai et al. [14] took a special case of a tunnel close to and approximately parallel to an existing tunnel as a research object and studied the deformation characteristics based on the tunnel based on monitoring data and FDM numerical simulation.

In this paper, field monitoring and analysis are conducted on the convergence deformation of the right tunnel induced by the excavation of the left tunnel, aiming to explore the patterns of influence on the deformation of the existing tunnel caused by the construction of the adjacent tunnel.

2 Engineering Background

2.1 Engineering Project

The newly excavated shield tunnel of metro line 2 in a specific city, was constructed using an earth pressure balance (EPB) shield machine. The tunnel was launched at Jinghu Station and completed at Houshu Road Station, traversing three passages (2 pedestrian lanes and 1 vehicular lane). Total length of tunnel is 1946 m. The MJS (Metro Jet System) method is employed to enhance the stability of the tunnel, reinforcing the shield tunnel structure both above and below the shield tunnel. The spacing distance between the axis of lines measures 14.2 m. The outer and inner diameter of shield is 6.7 m and 5.9 m. The segment has a thickness of 0.4 m, a ring width of 1.2 m, and is constructed using C50 concrete. As depicted in Fig. 1, the tunnel is primarily buried in silty clay from 20 m to 26.7 m.

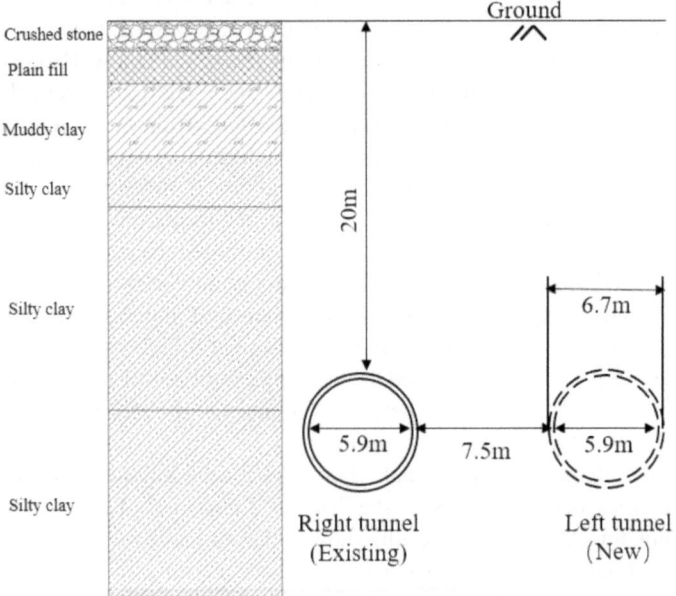

Fig. 1. Cross section of tunnels

The parameters of the soil are shown in Table 1.

Table 1. Soil parameters

Soil type	γ_{sat} (kN/m^3)	e	C (kPa)	ϕ (°)	E (MPa)	R$_{inter}$
Crushed stone	18.7	0.88	18	2	/	/
Plain fill	18.5	0.90	21.3	15.6	4.73	/
Muddy clay	16.9	1.37	13.5	9.7	2.86	4.34
Silty clay	19.0	0.83	13.5	32.8	4.80	3.50
Silty clay	18.5	0.90	28.1	18.6	5.49	3.04
Silty clay	18.2	1.04	21.0	13.0	3.95	3.75

2.2 Field Monitoring

To precisely investigate the impact of the left shield tunneling on the structure of the right tunnel, laser range finders are installed in right tunnel to monitor the convergence deformation of tunnel structures. During tunneling of left line, the convergence deformation of the right tunnel is monitored every ten minutes. The data acquisition and transmission system comprise a data acquisition instrument, an acquisition instrument and communication cables. The convergence monitoring layout are shown in Fig. 2.

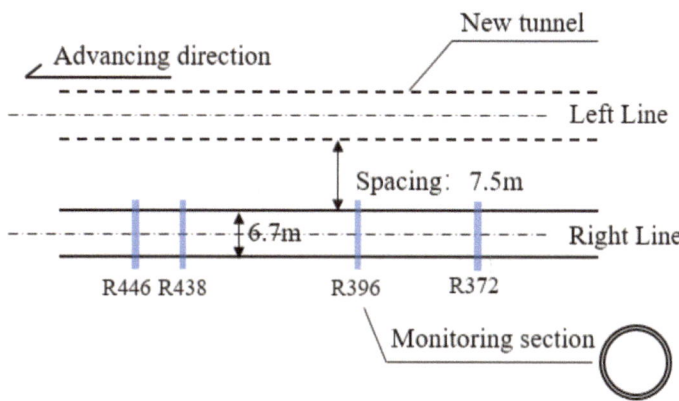

Fig. 2. Convergence monitoring section layout

Convergence monitoring sections are established at rings 372, 396, 438, and 446. Laser range finders are installed along the tunnel's waist at each monitoring section, dedicated to monitoring vertical and horizontal convergence. The installation position of monitoring instruments is shown in Fig. 3.

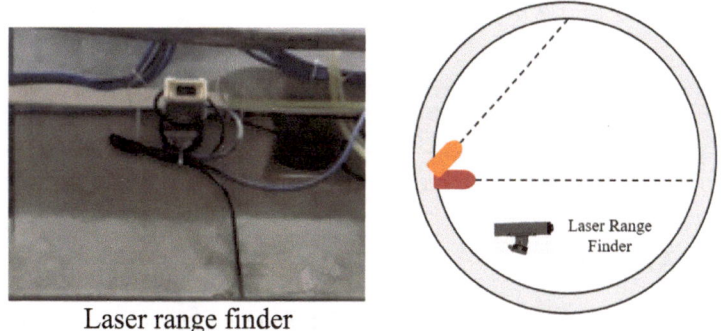

Fig. 3. Monitoring section layout design

3   Measurement Results

The monitoring sections for lateral convergence of tunnel segments during the right tunnel shield machine's passage through the area with pile foundation effects are rings 372, 396, 438, and 446. Convergence monitoring for ring 372 began on January 1, 2022, while monitoring for rings 396, 438 and 446 started on January 19, 2022.

   From Fig. 4, for ring 372 and ring 438, there is horizontal expansion and vertical convergence, representing a "horizontal duck egg" convergence deformation pattern. Both horizontal and vertical convergence occur in ring 396 and ring 446, indicating a non-uniform convergence deformation pattern. Horizontal and vertical convergence at monitoring sections are shown in Table 2.

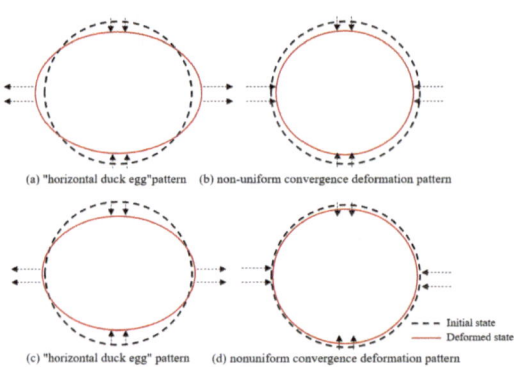

(a) "horizontal duck egg"pattern   (b) non-uniform convergence deformation pattern

(c) "horizontal duck egg" pattern   (d) nonuniform convergence deformation pattern

Fig. 4. Horizontal convergence patterns at monitoring sections: (a) R372, (b) R396, (c) R438, (d) R446

Table 2. Final convergence at monitoring sections

Monitoring section	Horizontal convergence (mm)	Vertical convergence (mm)
R372	6.6	−7.4
R396	−4.0	−6.7
R438	1.9	−9.8
R446	−1.5	−1.3

4   Result Analysis

In this project, the axial spacing between the left and right tunnel axes is 14.2 m, with the minimum distance between the two tunnels (distance from the boundary of the left tunnel to the boundary of the right tunnel) being 7.5 m. The right tunnel is excavated before the left tunnel. To investigate the impact of the new tunnel excavating on the existing tunnel structure, this section further analyzes the horizontal convergence when the left tunnel shield machine approaches the monitored tunnel segments.

The horizontal and vertical convergence for each monitoring section is shown in Fig. 5, where positive and negative values represent the expansion and convergence of segment, respectively. Figure 5 reveals that the convergence (expansion) of each monitoring section gradually develops and stabilizes. Before and after shield machine passing the position of monitoring section, due to the change of surrounding soil pressure, the deformation of tunnel gradually developed; when shield machine passing, the deformation suddenly changed.

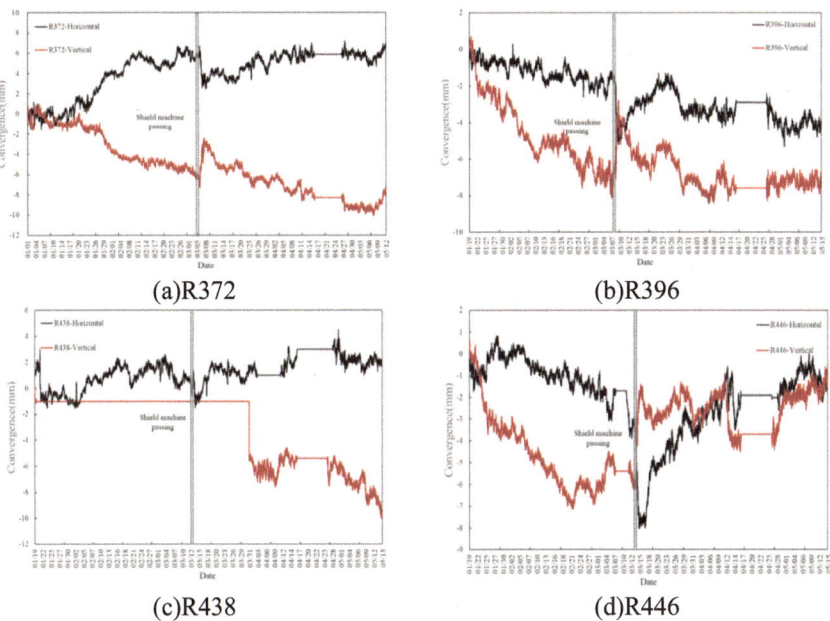

(a)R372   (b)R396

(c)R438   (d)R446

Fig. 5. Horizontal convergence at monitoring sections

During the excavation of the left tunnel while the right tunnel is already in place, it can be observed that as the left tunnel shield machine approaches the monitoring sections stage, there is minimal change in horizontal convergence of the right tunnel segments. However, after the tunnel shield machine passes through each monitoring section, the horizontal convergence of the right tunnel segments gradually increases, while the vertical convergence decreases. After the convergence changes stabilize, the vertical convergence for ring 372 is −3.8 mm, and the horizontal convergence is 3.4 mm. For ring 396, the vertical convergence is −5.3 mm, and the horizontal convergence is −4.1 mm. For ring 446, the vertical convergence is −2.1 mm, and the horizontal convergence is −7.6 mm.

Due to lateral compression, right tunnel segments experience horizontal convergence and vertical expansion. This is the result of the horizontal compression induced by the shield tail synchronous grouting pressure as it passes through each monitoring section of the left tunnel. This pressure causes the surrounding soil to expand outward, transmitting deformation to the right tunnel. Simultaneously, the lateral deformation of the tunnel segments in the left tunnel also applies a compressive force on the surrounding soil, finally causing the convergence of the right tunnel segments.

5 Conclusions

This paper analyses the convergence deformation of the right tunnel induced by the left shield tunneling. Based on the analysis of field monitoring result, the mainly conclusions drawn are as follows:

(1) The convergence (expansion) of each monitoring section gradually develops and stabilizes over time. For the upper part with the influence of pile foundations, the convergence of tunnel segments exhibits "horizontal duck egg" pattern and non-uniform convergence pattern. When reaching a stable state, the maximum horizontal convergences for rings 372, 396, 438, and 446 are 6.6 mm, −4.0 mm, 1.9 mm, and −1.5 mm; The maximum vertical convergences are −7.4 mm, −6.7 mm, −9.8 mm, and −1.3 mm, respectively.

(2) When the shield machine reaches and passes, there is no apparent change of horizontal convergence in the existing tunnel. However, after the tunnel shield machine passes, due to the horizontal compression caused by synchronous grouting pressure and lateral deformation of the left shield tunnel segment, surrounding soil experiences lateral compression deformation. This causes the horizontal convergence of the right tunnel to have an increase in horizontal convergence and a decrease in vertical convergence.

Acknowledgments. The work was supported by the National Natural Science Foundation of China (No. 52308332).

References

1. Sirivachiraporn, A., Phienwej, N.: Ground movements in EPB shield tunneling of Bangkok subway project and impacts on adjacent buildings. Tunn. Undergr. Space Technol. **30**, 10–24 (2012)
2. Gharehdash, S., Barzegar, M.: Numerical modeling of the dynamic behaviour of tunnel lining in shield tunneling. KSCE J. Civ. Eng. **19**, 1626–1636 (2015)
3. Yun, Y., Sun, Y., Lin, Z., Wang, J., Ye, Y.: Influence of lateral foundation pit excavation on forces and cracks of shield segment lining. J. Liaoning Tech. Univ. (Nat. Sci.) **41**, 337–344 (2022)
4. Shi, J., Liu, G., Huang, P., et al.: Interaction between a large-scale triangular excavation and adjacent structures in shanghai soft clay. Tunn. Undergr. Space Technol. **50**, 282–295 (2015)
5. Liu, B., Zhang, D., Yang, C., et al.: Long-term performance of metro tunnels induced by adjacent large deep excavation and protective measures in Nanjing silty clay. Tunn. Undergr. Space Technol. **95**, 103–117 (2020)
6. Li, X.G., Yuan, D.J.: Response of a double-decked metro tunnel to shield driving of twin closely under-crossing tunnels. Tunn. Undergr. Space Technol. **28**, 18–30 (2012)

7. Zhang, Z., Huang, M., Wang, W.: Evaluation of deformation response for adjacent tunnels due to soil unloading in excavation engineering. Tunn. Undergr. Space Technol. **38**, 244–253 (2013)
8. Jin, D., Yuan, D., Li, X., et al.: Analysis of the settlement of an existing tunnel induced by shield tunneling underneath. Tunn. Undergr. Space Technol. **81**, 209–220 (2018)
9. Huang, H., Huang, X., Zhang, D.: Centrifuge modelling of deep excavation over existing tunnels. Geotech. Eng. **167**, 3–18 (2015)
10. Kim, S.H., Burd, H.J., Milligan, G.W.E.: Model testing of closely spaced tunnels in clay. Géotechnique **48** (1998)
11. Ng, C.W.W., Lu, H., Peng, S.: Three-dimensional centrifuge modelling of the effects of twin tunnelling on an existing pile. Tunn. Undergr. Space Technol. **35**, 189–199 (2013)
12. Chakeri, H., Hasanpour, R., Hindistan, M.A., et al.: Analysis of interaction between tunnels in soft ground by 3D numerical modelling. Bull. Eng. Geol. Environ. **70**, 439–448 (2011)
13. Lin, X., Chen, R., Wu, H., et al.: Deformation behaviors of existing tunnels caused by shield tunneling undercrossing with oblique angle. Tunn. Undergr. Space Technol. **89**, 78–90 (2019)
14. Lai, H., Zheng, H., Chen, R., et al.: Settlement behaviours of existing tunnel caused by obliquely under-crossing shield tunneling in close proximity with small intersection angle. Tunn. Undergr. Space Technol. **97**, 103–118 (2020)

Numerical Simulation-Based Investigation of Fatigue Fracture Mechanisms in Hot In-Place Recycling of Asphalt Pavement

Kai Zhao[1,2], Yongli Zhao[1(✉)], and Yi Cao[1]

[1] School of Transportation, Southeast University, Nanjing 210096, Jiangsu, China
yonglizhao2016@126.com
[2] Shanxi Transportation Holdings Group Co., Ltd., Taiyuan 030032, Shanxi, China

Abstract. Hot in-place recycling (HIR) technology rejuvenates a certain depth range (typically not exceeding 6 cm) of old asphalt concrete pavement in a single step. Suitable for early-stage surface pavement distresses, it prevents them from progressing into deeper layers. Despite numerous experimental studies on the fatigue and fracture properties of hot recycled materials, limited research exists on numerical simulation of fatigue fracture in such materials. Numerical simulations, offering cost-effectiveness and overcoming size constraints, are the focus of this paper. The study uses a developed fatigue fracture co-simulation program to investigate fatigue fracture behavior in pavement structures under wheel load and temperature effects. It explores the influence of material properties, layer configurations, and initial crack positions of HIR materials on fatigue cracking occurrence. The research provides valuable insights into fatigue fracture patterns in HIR material pavement structures, offering a theoretical foundation for preventing fatigue fracture in such materials.

Keywords: Hot In-Place Recycling · Fatigue Fracture · Numerical Simulation

1 Introduction

In recent years, the recycling of asphalt pavement has gained considerable momentum due to its numerous advantages, including cost reduction during initial implementation, resource and waste area preservation, and positive environmental implications [1, 2]. The disposal of reclaimed asphalt pavement (RAP) not only causes air and water pollution but also results in substantial resource wastage [3, 4]. The production of new asphalt mixture involves extensive mining of sand and gravel, posing a threat to soil and vegetation preservation. To align with the country's environmental protection efforts and global sustainable development principles, the research on recycling old asphalt pavement has gained significant traction in China [5].

The regeneration process of asphalt pavement encompasses in-place recycling and central plant recycling, with in-place recycling further classified into cold in-place recycling (CIR) and hot in-place recycling (HIR) [6–8]. Similarly, central plant recycling

© The Author(s) 2024
G. Feng (Ed.): ICCE 2023, LNCE 526, pp. 390–407, 2024.
https://doi.org/10.1007/978-981-97-4355-1_37

is categorized into central plant cold recycling and central plant hot recycling [6, 9]. Currently, considerable investigation has been undertaken on the selection and action principles of regenerants, the aging mechanism of asphalt, performance characteristics of RAP, and the mixing ratio of new asphalt mixture in the HIR process of asphalt [10–12]. HIR technology is becoming increasingly prevalent in asphalt pavement maintenance engineering. The process involves three primary stages: pavement preheating, milling and loosening of the original surface, and paving and rolling. During the process, the original road surface is fully heated using a heating machine, milled and dispersed, and then mixed with recycled asphalt mixture in a continuous mixer [8, 13]. Appropriate amounts of regenerant, new asphalt, and new aggregate are added to achieve uniform mixing. The mixture is subsequently transported to the paver for paving and vibrating before being compacted using roller equipment. The geothermal regeneration technology offers six advantages: rapid construction with short cycle duration, minimal noise pollution, suitability for repairing pavement damage and cracks, the potential to improve aggregate gradation, reduce asphalt content, and enhance pavement performance [14].

HIR of asphalt pavement involves several key technologies that play crucial roles in the construction process [15]. Research in China has yielded promising results in this area. However, previous studies have primarily focused on the ratio and performance of regenerants and recycled asphalt mixtures, mainly at the basic theory and method research stage [10, 16, 17]. Few studies have explored the technology's construction applicability or its integration with field construction conditions. The influence of HIR on the propagation of reflection cracks has also received limited attention [18].

The pretreatment of the original pavement in the HIR process is of utmost importance [19]. Neglecting this stage can lead to the rapid occurrence of early diseases in the recycled pavement, leading to unsatisfactory results. In China, semi-rigid base asphalt pavement is a prevalent structural form, often accompanied by reflection cracks extending from the base layer to the asphalt surface layer [20–22]. Since HIR targets only a 4 cm surface layer, neglecting the pretreatment of reflective cracks can result in their penetration through the 4 cm HIR layer. Addressing reflection cracks before HIR can be labor-intensive and slow down construction progress.

Hence, it is imperative to investigate the propagation mechanism of reflection cracks in HIR of asphalt pavement [23, 24]. This analysis aims to determine the necessity of treating cracks in the original pavement during construction, considering economic and practical feasibility. The study assesses whether this treatment affects other pavement indicators. Employing a combination of physical engineering and theoretical analysis, the paper uses a fatigue fracture joint simulation program to numerically simulate the fatigue fracture process in pavement structures with HIR materials under wheel load and temperature conditions. Exploring the influence of material properties, layers, and initial crack positions of HIR materials on fatigue cracking, the paper seeks to provide insights into managing reflection cracks effectively and optimizing the performance of asphalt pavement structures.

2 Establishment of Numerical Model

2.1 Determination of Pavement Structure

Taking into consideration the common semi-rigid base asphalt pavement structure in China, the pavement structure combination comprises the following layers: a 4 cm hot in-place recycling AC-13 surface layer, a 6 cm AC-20 middle surface layer, an 8 cm AC-25 lower surface layer, a 30 cm cement stabilized macadam base, and a soil base (as illustrated in Fig. 1).

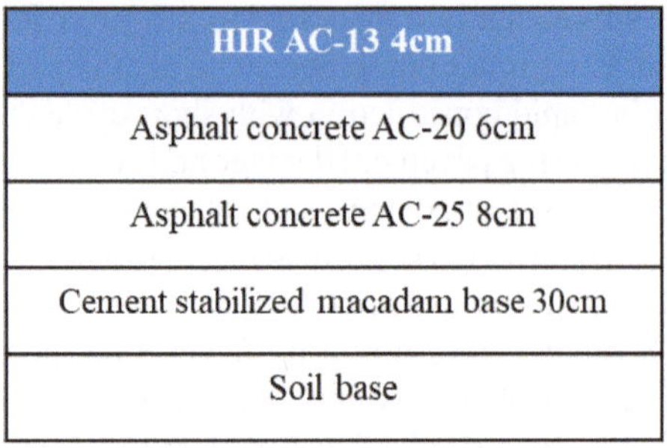

HIR AC-13 4cm
Asphalt concrete AC-20 6cm
Asphalt concrete AC-25 8cm
Cement stabilized macadam base 30cm
Soil base

Fig. 1. Pavement structure.

A 2D pavement structure model is established in finite element software with a width and depth of 4 m and 2 m, using linear plane strain elements. Applying a standard axle load of 100 kN, a pressure of 0.7 MPa, an equivalent circle diameter of 21.3 cm, and a center distance of 31.95 cm between wheels, a vertical pressure of 117371 Pa is exerted on the upper surface. This pressure is converted based on static equivalence principles, maintaining the same spacing and range of action as the standard load. Symmetrical boundary conditions are set on the left and right sides, while fixed boundary conditions are applied at the bottom. A meshing scheme (Fig. 2) enhances computational efficiency by increasing density from top to bottom and middle to sides, resulting in sparser meshes at the sides and lower part where deformation and damage typically occur within the specified wheel load range for each structural layer.

2.2 Calibration of Material Parameters

The mechanical response and fatigue accumulation of an asphalt pavement structure under load are significantly influenced by the material parameters of each structural layer. In this paper, particular attention is given to the impact of HIR asphalt mixture, which introduces notable changes to the material parameters of this layer. The detailed material parameters for each structural layer in the model are presented in Table 1. It is important to note that the analysis in this paper focuses solely on the influence of HIR asphalt mixture, and does not account for fatigue damage caused by the soil base.

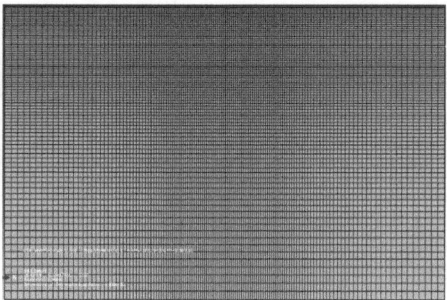

Fig. 2. Pavement structure mesh.

Table 1. Material parameters

Material Type	Modulus (MPa)	Poisson Ratio	a	p
HIR Asphalt Mixture	2000	0.3	0.1	4.5
AC-20	1500	0.3	0.06	4
AC-25	1500	0.3	0.06	4
Cement Stabilized Macadam	20000	0.25	6	9
Soil Base	60	0.4	—	—

3   Fatigue Analysis of Pavement Structure

3.1   Evolution Process of Pavement Structure Damage Field

If the fatigue damage of the pavement structure is disregarded, the initial application of an approximate load on the pavement generates the maximum tensile stress beneath the centerline of the two wheels at the bottom of the base, leading to damage initiation from this point. Figure 3 illustrates the damage distribution near this location for various loading cycles. The warmer colors in the figure represent greater damage. The initial damage predominantly occurs at the center of the two wheels on the bottom of the base, where the maximum tensile stress is concentrated. This damage gradually diminishes towards the surrounding area, with the surface structure experiencing primarily compressive stresses, resulting in lower damage.

As loading cycles accumulate, the fatigue damage field in the base layer extends laterally and upwards, exhibiting a horn-like distribution. The damage pattern at the bottom of the base layer becomes evident, with the highest damage concentration near the center of the two wheels and decreasing progressively towards the periphery. The damage distribution in the surface layer differs from that in the base layer. Notably, the prominent damage under the load is relatively small, and the maximum damage does not occur near the centerline of the two-wheel load. Instead, it is situated closer to the edge of the wheel load, displaying an approximate V-shaped distribution from top to bottom.

Figure 4 presents the damage distribution of the upper, middle, and lower layers after 3.25 million loading cycles, considering the pavement and load's symmetrical structures

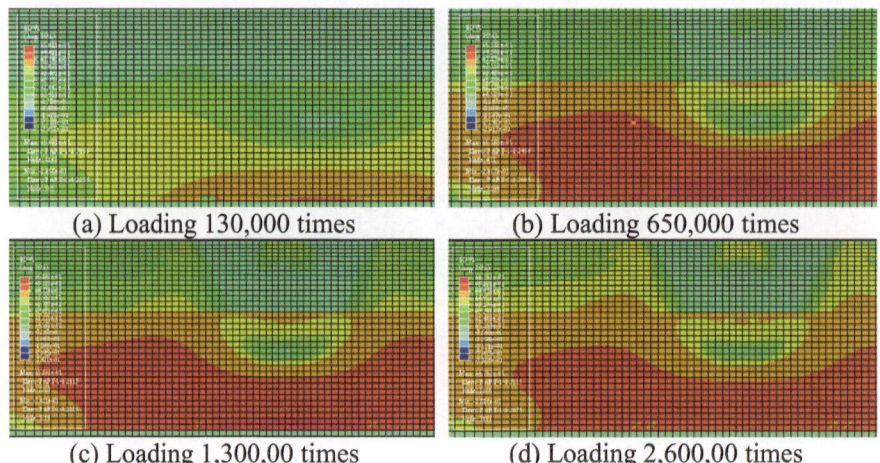

(a) Loading 130,000 times	(b) Loading 650,000 times
(c) Loading 1,300,00 times	(d) Loading 2,600,00 times

Fig. 3. Damage field evolution of pavement structure.

(only one side of the change is shown). For the bottom of the upper layer (Fig. 4a), some damage occurs near the center of the double wheel due to the proximity to the wheel load acting on the surface, aligning with slight uplift and cracking phenomena observed in the actual road surface's wheel track belt's center. However, as one moves further from the wheel load, the damage rapidly decreases to almost zero due to the pressure at the bottom of the layer directly below the wheel load, which is less prone to damage. Once the area directly beneath the wheel load is surpassed, the damage increases significantly, peaking at a distance of 1–3 cm from the wheel's edge and then declining. For the bottom of the middle layer (Fig. 4b), there is minimal damage near the center of the two wheels. As one approaches the edge of the wheel load, damage gradually initiates and grows rapidly, reaching a peak at approximately 1–3 cm outside the wheel's edge. The bottom layer (Fig. 4c) exhibits a similar pattern to the middle layer, with minimal damage near the center of the two wheels but starting at a slightly smaller range and showing some damage earlier. Beyond the wheel's edge, the range of higher damage is more extensive, and the reduction is less apparent after reaching the peak value at 1–3 cm from the outer edge of the wheel load.

The center line of the two wheels at the bottom of the base layer and the outer edge of the wheel load at the bottom of each layer are chosen, with approximately 1.5 cm outward from these points serving as the observation point. The resulting variation laws of the damage degree for each layer are presented in Fig. 5.

As the number of loading cycles increases, the maximum damage degree of each layer shows a rising trend. The early base layer and lower layer exhibit significantly greater damage compared to the middle and upper layers, attributed to force distribution within the pavement structure. In the later stage, damage in the middle and upper layers progresses at a faster rate. This can be partly attributed to the materials of the base and lower layers having accumulated considerable damage, making it difficult for them to sustain the same load-carrying capacity. Consequently, the main body of load-bearing shifts towards regions with less damage or unaffected material in the middle and upper

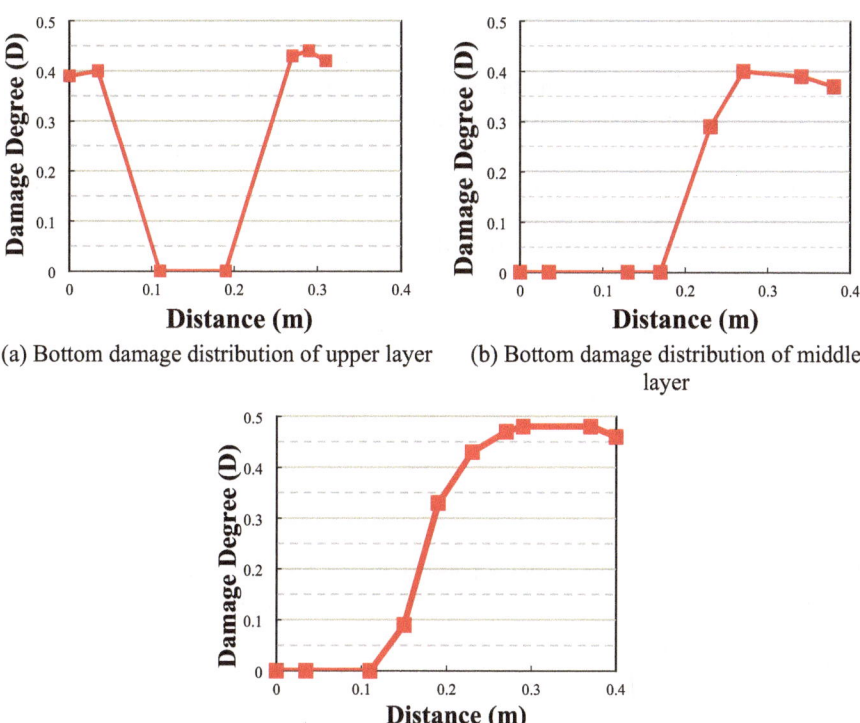

(a) Bottom damage distribution of upper layer

(b) Bottom damage distribution of middle layer

(c) Bottom damage distribution of bottom layer

Fig. 4. Horizontal distribution of surface layer bottom damage.

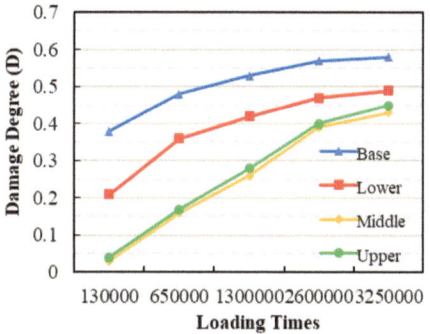

Fig. 5. The damage of each layer of pavement structure.

layers, allowing for good bearing capacity diffusion and transfer. Notably, the upper layer, consisting of hot in-place recycled material, exhibits faster damage accumulation compared to fresh asphalt concrete, as observed in related research. The later stage shows a slightly higher damage degree in the upper layer than in the middle layer.

3.2 Sensitivity Analysis

Material parameters of HIR asphalt mixture. The variation in RAP content and the use of different recycling agents in various HIR projects lead to differences in the nature and dosage of RAP. Consequently, the changes in pavement damage may also vary. This section takes into account the relationship between the material parameters obtained in the third chapter and the RAP content. Material parameters for HIR mixtures with two different RAP contents are set as presented in Table 2.

Table 2. Material parameters of HIR mixture with different RAP content

RAP Content	Modulus (MPa)	Poisson Ratio	a	p
Low	1800	0.3	0.06	4.5
High	2200	0.3	0.2	4.5

The variation laws of the damage degree for each layer are analyzed in both cases using the same method as in the previous section, as shown in Fig. 6.

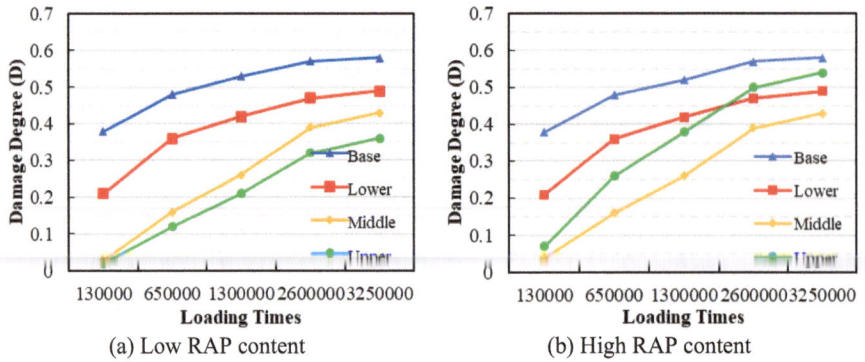

(a) Low RAP content   (b) High RAP content

Fig. 6. Pavement damage with different material parameters.

Comparing Fig. 6(a) and (b), it becomes evident that the material parameters of the HIR mixture significantly influence the damage in the upper layer, and the fatigue damage of the hot recycled mixture with higher RAP content accumulates faster. When the RAP content is low, the damage degree of the upper layer remains at a minimum level. However, with an increase in RAP content, the damage degree of the upper layer exceeds that of the middle and lower layers after 2.6 million loading cycles, becoming second only to the damage degree of the base layer.

HIR asphalt mixture layer. In practical HIR projects, a thin surface layer is sometimes applied over the thermal regeneration layer, which is positioned similarly to the middle surface layer. To investigate the damage changes in the pavement structure under this scenario, the aforementioned pavement structure is modified. Specifically, the upper layer is composed of SMA-13, the middle layer is the hot recycled mixture, and the remaining layers remain unchanged. The elastic modulus of SMA-13 is 1800 MPa, while other parameters are the same as those of AC-13. Figure 7 illustrates the pavement damage changes of the HIR mixture in different layers.

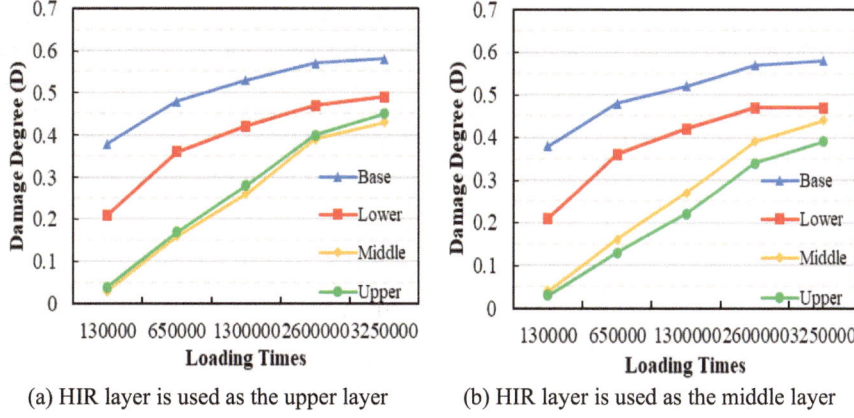

(a) HIR layer is used as the upper layer   (b) HIR layer is used as the middle layer

Fig. 7. Pavement damage of different layers of HIR mixture.

Observing Fig. 7, it becomes apparent that when the HIR layer is used as the middle layer and a mixture with better fatigue performance, such as SMA-13, is employed as the upper layer, the damage in the upper layer is reduced to a certain extent. However, the damage of the HIR structure layer itself does not show significant reductions.

Middle surface layer and base layer modulus. It is essential not only to use asphalt concrete with strong anti-rutting capability on the surface layer but also to enhance the strength of the middle layer. In this section, the modulus of the middle layer is increased from 1500 MPa to 1800 MPa, and the change in pavement structure damage is reanalyzed, as depicted in Fig. 8.

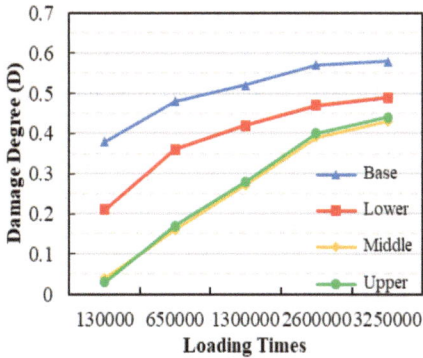

Fig. 8. Pavement damage of different layers of HIR mixture.

Upon comparing Fig. 8 with Fig. 5, it is evident that the increase in the middle layer's modulus results in the upper layer's damage degree becoming 0.44 after 3.25 million loading cycles, which is slightly lower than the value of 0.45 before the increase. However, the change is negligible. The enhancement of the middle layer's stiffness exhibits little effect on reducing damage in the upper layer and has minimal influence on the overall damage of the pavement structure.

The base layer constitutes the thickest part of the pavement structure and serves as the primary load-bearing component. Consequently, any changes in the material parameters of the base layer often lead to significant alterations in the overall pavement structure's damage. As accurate research results regarding the influencing factors and variation rules of base material fatigue parameters are lacking, this section focuses on changing the base modulus alone. The overall pavement structure's damage under different base modulus values is obtained and depicted in Fig. 9.

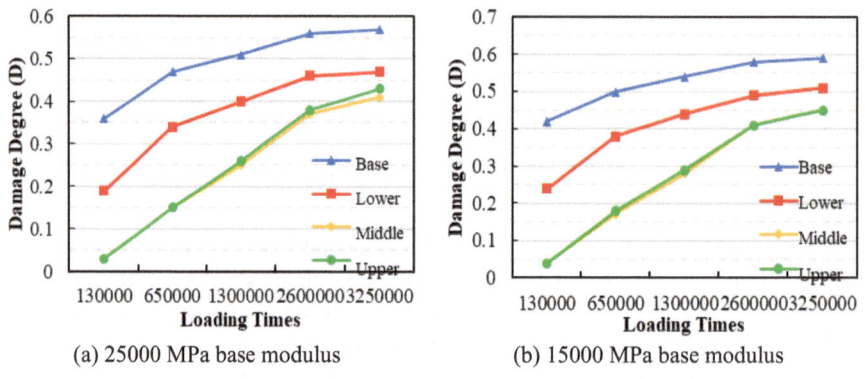

(a) 25000 MPa base modulus   (b) 15000 MPa base modulus

Fig. 9. The influence of different base modulus on pavement damage.

Observing Fig. 9, it becomes evident that changes in the base modulus do not substantially affect the overall damage development pattern. However, the damage degree in each layer does exhibit considerable variation. Table 3 presents the damage results of each layer after 3.25 million loading cycles.

Table 3. Different base modulus of each layer damage

Base Modulus (MPa)	Base Layer	Lower Surface Layer	Middle Surface Layer	Upper Surface Layer
15000	0.59	0.51	0.45	0.45
20000	0.58	0.49	0.43	0.45
25000	0.57	0.47	0.41	0.43

Remarkably, when solely increasing the modulus of the base layer, not only does the damage in the base layer decrease, but the damage in other layers also reduces. This phenomenon may be attributed to the fact that under actual wheel loads, the pavement experiences deformation. An increase in the base layer's modulus, owing to its larger thickness, significantly enhances the overall pavement structure's stiffness, thereby reducing the deformation generated under the same load. This mechanism is akin to a loading method that controls stress, leading to a reduction in the cumulative damage rate.

Overload. Currently, road transportation in China still faces a serious problem of overloading. The repetitive impact of overloading significantly degrades the pavement surface quality. This section primarily investigates whether overloading affects the fatigue behavior of the in-situ thermally regenerated pavement structure. The overload is set to 1.3 times the standard axle load, and the applied pressure in the model is changed from 117,371 Pa to 152,582 Pa. The resulting pavement structure damage development is presented in Fig. 10.

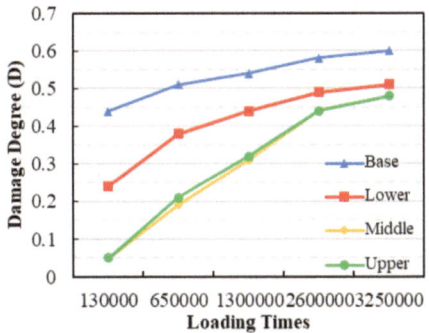

Fig. 10. Pavement structure damage under overload.

Comparing Fig. 10 with Fig. 5, noticeable changes are observed in the growth trend of the middle layer and the upper layer. The curves exhibit steeper inclines as a whole, and the gap between the upper and middle layers and the lower layer reduces during the later stage. Table 4 summarizes the damage degrees of each layer after the standard axle load and overload are respectively applied for 3.25 million loading cycles.

Table 4. Different base modulus of each layer damage

Damage Degree	Base Layer	Lower Surface Layer	Middle Surface Layer	Upper Surface Layer
Standard Axle Load	0.58	0.49	0.43	0.45
Overload	0.6	0.51	0.48	0.48
Growth Rate	3.45%	4.08%	11.63%	6.67%

The base layer shows the smallest growth rate in damage degree, while the middle layer experiences the largest growth rate, reaching 11.63%. The upper layer also exhibits a significant growth rate. Consequently, it is evident that increasing the axle load will invariably escalate the damage degree of each layer in the pavement under the same number of loading cycles, with a more pronounced effect in the middle and upper layers. Hence, the adverse impact of overloading on pavement structure damage should be carefully considered during actual usage.

4   Cracking Analysis of Pavement Structure

4.1   Cracking Analysis of Pavement Without Initial Cracks

Assuming the pavement surface has no initial crack, the distribution of the damage field in the pavement structure just before crack initiation (at 3.3 million loading cycles) is depicted in Fig. 11. The damage primarily occurs at the bottom of the semi-rigid layer and on both sides of the tire's grounding position. Concurrently, the main stress field of the pavement before crack initiation is presented in Fig. 12. The large modulus of the semi-rigid base results in the maximum principal stress occurring at the bottom of the semi-rigid layer. The main strain distribution on the road surface before crack initiation is illustrated in Fig. 13. The main strain concentrates primarily on both sides of the tire's grounding position and the semi-circular arc area beneath the tire.

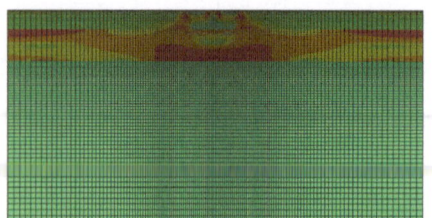
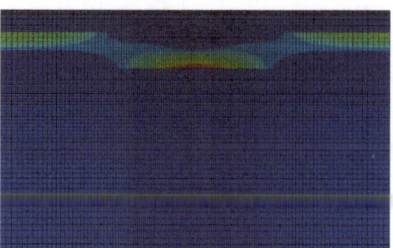

Fig. 11. Damage distribution.   **Fig. 12.** Master stress field distribution.

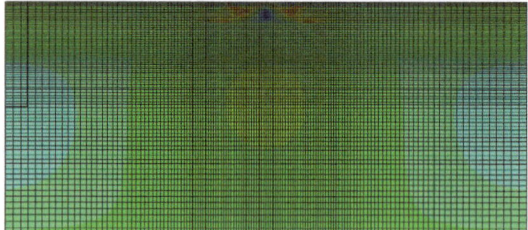

Fig. 13. Master strain distribution.

Considering damage, principal stress, and principal strain, there are two potential locations for crack propagation. From the perspective of damage and principal stress, the bottom of the semi-rigid layer is the primary candidate for crack propagation. However, regarding principal strain and damage, the two sides of the tire on the asphalt surface and the semi-circular arc area below the tire are also possible locations for crack expansion. The initiation point of crack expansion predominantly depends on the cracking criterion of the pavement material.

Based on the maximum principal stress criterion. If the maximum principal stress criterion of the semi-rigid base reaches the critical value, crack initiation occurs, and the crack starts propagating upward from the bottom of the semi-rigid base. Numerical simulation results reveal the initiation of two symmetrical cracks propagating from the location of maximum principal stress at the bottom of the semi-rigid layer. As the crack propagates, the principal stress concentration at the crack tip gradually diminishes. However, when the crack extends to approximately 70% of the thickness of the semi-rigid base (corresponding to 4.11 million loading cycles), the crack propagation comes to a halt, despite the increase in load cycles. Based on the horizontal stress distribution map in the final state of crack propagation, it is observed that when the crack propagates to about 70% of the thickness of the semi-rigid base, the crack tip nears the compression zone above the semi-rigid layer. As per the maximum principal stress criterion, crack propagation can only occur under tensile stress conditions. At this length of crack extension, the crack reaches the compressive zone, rendering further extension unfeasible in Fig. 14.

Based on the maximum principal strain criterion. If the crack resistance strength of the semi-rigid base is sufficiently high, it is possible for cracks to initiate due to the maximum principal strain criterion of the asphalt layer reaching the critical value. Figure 15(a) illustrates the distribution of the maximum principal strain in the asphalt layer. As evident from Fig. 15(b), when the number of loading cycles reaches 3.67 million, cracks form at the location of maximum principal strain concentration within the asphalt layer. With the progression of loading cycles, some cracks in the middle area of the tire extend to the surface of the asphalt layer, while cracks in other regions either remain unchanged or expand within a limited local area. Upon reaching 4.28 million loading cycles, the crack state stabilizes, and further increases in load cycles do not result in additional crack propagation.

In the aforementioned scenarios, the cracking of the semi-rigid base occurs due to the maximum principal stress criterion, while the cracking of the asphalt layer bottom is caused by the maximum principal strain criterion. Given the distinct material properties of the semi-rigid base and asphalt surface, their respective cracking criteria also differ. The semi-rigid layer, characterized by significant stiffness and brittleness, is more susceptible to the maximum principal stress fracture criterion. Conversely, the asphalt layer, featuring lower stiffness and greater ductility, is more inclined towards the fracture criterion controlled by the maximum principal strain. Considering the material differences and potential for distinct cracking criteria, it is important to acknowledge that both cases may occur simultaneously in real pavement structures, leading to the initiation of cracks in both the asphalt layer and the semi-rigid base layer. This understanding contributes to a comprehensive analysis of crack formation mechanisms and aids in the design and maintenance of asphalt pavement structures to enhance their longevity and performance.

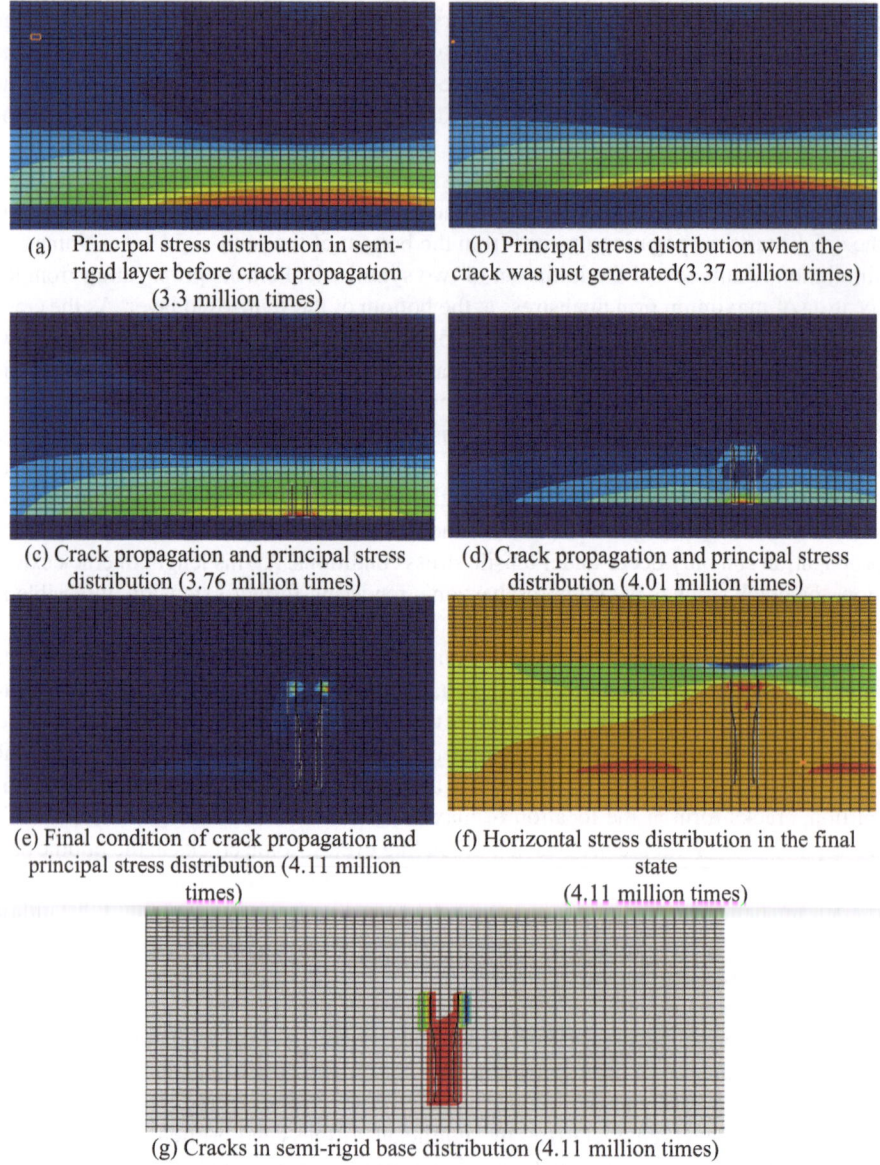

(a) Principal stress distribution in semi-rigid layer before crack propagation (3.3 million times)

(b) Principal stress distribution when the crack was just generated(3.37 million times)

(c) Crack propagation and principal stress distribution (3.76 million times)

(d) Crack propagation and principal stress distribution (4.01 million times)

(e) Final condition of crack propagation and principal stress distribution (4.11 million times)

(f) Horizontal stress distribution in the final state (4.11 million times)

(g) Cracks in semi-rigid base distribution (4.11 million times)

Fig. 14. Fracture analysis based on maximum principal stress criterion.

4.2 Cracking Analysis of Pavement Without Initial Cracks

When the initial crack is located at the bottom of the semi-rigid base, directly below the center of the two wheel loads, the crack propagates vertically upward along the load centerline. However, it ceases to expand when it approaches the compression zone. The crack propagation is thus confined within the semi-rigid layer, and after reaching a certain length (at 310,000 load cycles), further expansion is hindered due to the presence

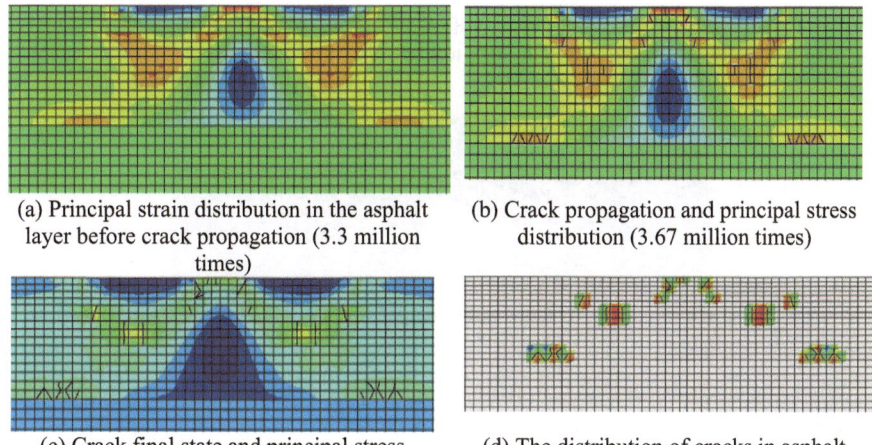

(a) Principal strain distribution in the asphalt layer before crack propagation (3.3 million times)

(b) Crack propagation and principal stress distribution (3.67 million times)

(c) Crack final state and principal stress distribution (4.28 million times)

(d) The distribution of cracks in asphalt surface layer (4.28 million times)

Fig. 15. Fracture analysis based on maximum principal strain criterion.

of compressive stress at the crack tip, in accordance with the maximum principal stress fracture criterion (Fig. 16).

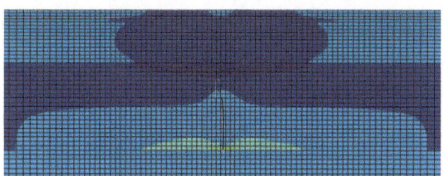

Fig. 16. Crack propagation status and principal stress distribution (310,000 times).

When the crack deviates 20 cm from the semi-rigid base below the wheel load center, similar to the previous case, the crack extends to a specific position and stops expanding. The crack's propagation path is upward and tilts toward the center of the wheel load (Fig. 17).

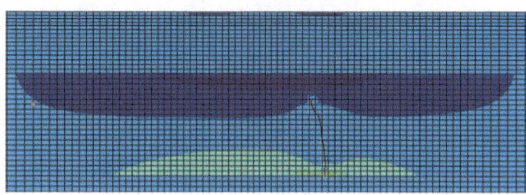

Fig. 17. Crack propagation status and principal stress distribution (430,000 times).

When the crack deviates 40 cm from the semi-rigid base below the wheel load center, the crack still extends to a certain position and halts expansion, following a path inclined towards the center of the wheel load. Compared with the 20 cm deviation case, the crack propagation path shows a greater inclination towards the wheel load center (Fig. 18).

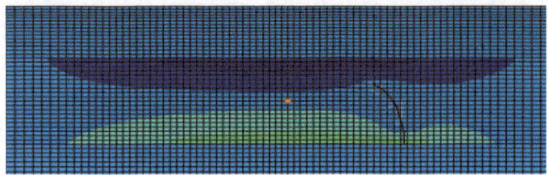

Fig. 18. Crack propagation status and principal stress distribution (450,000 times).

When the crack deviates 20 cm from the asphalt layer below the center of the wheel load, the crack expands upwards by two unit grid lengths but then stops as the load continues. Similar to the previous case, the crack cannot continue to expand when the crack tip approaches the compression zone (Fig. 19).

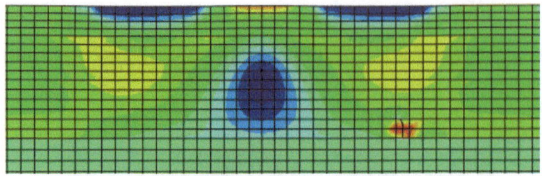

Fig. 19. Crack propagation status and principal stress distribution (220,000 times).

When the crack deviates 40 cm from the asphalt layer below the center of the wheel load, the crack propagates upwards, inclined towards the center of the wheel load. Similar to the 20 cm deviation case, the crack reaches a certain extent and ceases further expansion, but the propagation distance is longer (Fig. 20).

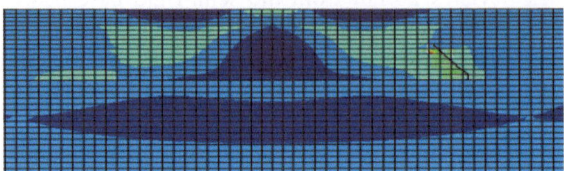

Fig. 20. Crack propagation status and principal stress distribution (270,000 times).

The following figure presents a statistical diagram of the initial crack position and the number of load cycles. It shows that pavement structures without initial cracks have a life 10–20 times longer than those with initial cracks. In the absence of initial cracks, the fatigue stage accounts for about 80% of the total life. With initial cracks, the farther the crack is from the centerline, the longer the fatigue life of the structure. The pavement structure's life is greater when the initial crack is located at the bottom of the semi-rigid layer compared to being at the bottom of the asphalt layer (Fig. 21).

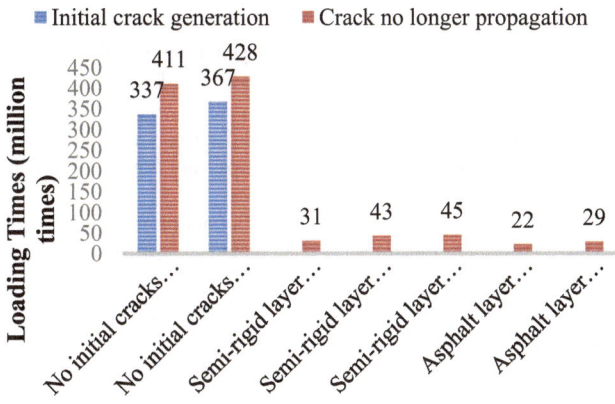

Fig. 21. Statistics of crack propagation.

5 Conclusion

This paper simulates the fatigue fracture process of a pavement structure with hot in-place recycling material under wheel load and temperature. Key findings include:

1. Initial damage occurs at the center of the two wheels at the base, extending as the number of load actions increases, with the surface layer showing a V-shaped distribution pattern.
2. Maximum damage generally increases from top to bottom, influenced by hot in-place recycling layer properties. Increased RAP content raises cumulative damage in later stages, with minimal impact on other layers.
3. Base modulus significantly affects overall damage; increasing it reduces damage in each layer. Lowering the modulus of the intermediate layer increases deformation and damage. Overloading exacerbates damage, particularly in upper and middle layers.
4. Cracking patterns vary based on the presence of initial cracks. Without initial cracks, pavement structures have a lifespan 10–20 times longer. Crack propagation direction is influenced by the initial crack's location and distance from the wheel load center.
5. Pavement structures without initial cracks have longer lifespans, with fatigue stages contributing around 80% of the total life. The location of the initial crack affects the structure's life.
6. Regardless of the presence of an initial crack, fatigue load under wheel load prevents crack extension to the pavement surface due to force distribution.

Acknowledgments. This work was supported by the National Natural Science Foundation of China [No. 52078132 and No. 52208448], and the Fundamental Research Funds for the Central Universities [No. 2242022k30058 and No. 300102212514]. The authors gratefully acknowledge their financial support.

References

1. Zaumanis, M., Mallick, R.B., Frank, R.: 100% recycled hot mix asphalt: a review and analysis. Resour. Conserv. Recycl. **92**, 230–245 (2014). https://doi.org/10.1016/j.resconrec.2014.07.007
2. Kalantar, Z.N., Karim, M.R., Mahrez, A.: A review of using waste and virgin polymer in pavement. Constr. Build. Mater. **33**, 55–62 (2012). https://doi.org/10.1016/j.conbuildmat.2012.01.009
3. Zaumanis, M., Mallick, R.B.: Review of very high-content reclaimed asphalt use in plant-produced pavements: state of the art. Int. J. Pavement Eng. **16**, 39–55 (2015). https://doi.org/10.1080/10298436.2014.893331
4. Huang, B.S., Li, G.Q., Vukosavjevic, D., Shu, X., Egan, B.K.: Laboratory investigation of mixing hot-mix asphalt with reclaimed asphalt pavement. In: Bitum. PAVING Mix. 2005, Transportation Research Board Natl Research Council, Washington, pp. 37–45 (2005). https://doi.org/10.3141/1929-05
5. Zheng, X., Easa, S.M., Yang, Z., Ji, T., Jiang, Z.: Life-cycle sustainability assessment of pavement maintenance alternatives: methodology and case study. J. Clean. Prod. **213**, 659–672 (2019). https://doi.org/10.1016/j.jclepro.2018.12.227
6. Xiao, F., Yao, S., Wang, J., Li, X., Amirkhanian, S.: A literature review on cold recycling technology of asphalt pavement. Constr. Build. Mater. **180**, 579–604 (2018). https://doi.org/10.1016/j.conbuildmat.2018.06.006
7. Giani, M.I., Dotelli, G., Brandini, N., Zampori, L.: Comparative life cycle assessment of asphalt pavements using reclaimed asphalt, warm mix technology and cold in-place recycling. Resour. Conserv. Recycl. **104**, 224–238 (2015). https://doi.org/10.1016/j.resconrec.2015.08.006
8. Ma, Y., Polaczyk, P., Park, H., Jiang, X., Hu, W., Huang, B.: Performance evaluation of temperature effect on hot in-place recycling asphalt mixtures. J. Clean. Prod. **277**, 124093 (2020). https://doi.org/10.1016/j.jclepro.2020.124093
9. Aravind, K., Das, A.: Pavement design with central plant hot-mix recycled asphalt mixes. Constr. Build. Mater. **21**, 928–936 (2007). https://doi.org/10.1016/j.conbuildmat.2006.05.004
10. Zhong, H., et al.: Investigating binder aging during hot in-place recycling (HIR) of asphalt pavement. Constr. Build. Mater. **276**, 122188 (2021). https://doi.org/10.1016/j.conbuildmat.2020.122188
11. Li, J., Ni, F., Jin, J., Zhou, Z.: A comparison of rejuvenator and sryrene–butadiene rubber latex used in hot in-place recycling. Road Mater. Pavement Des. **18**, 101–115 (2017). https://doi.org/10.1080/14680629.2016.1142465
12. Bouraima, M.B., Zhang, X.H., Rahman, A., Qiu, Y.: A comparative study on asphalt binder and mixture performance of two traffic lanes during hot in-place recycling (HIR) procedure. Constr. Build. Mater. **223**, 33–43 (2019). https://doi.org/10.1016/j.conbuildmat.2019.06.201
13. Liu, Y., Wang, H., Tighe, S.L., Zhao, G., You, Z.: Effects of preheating conditions on performance and workability of hot in-place recycled asphalt mixtures. Constr. Build. Mater. **226**, 288–298 (2019). https://doi.org/10.1016/j.conbuildmat.2019.07.277
14. Cao, R., Leng, Z., Hsu, S.-C.: Comparative eco-efficiency analysis on asphalt pavement rehabilitation alternatives: hot in-place recycling and milling-and-filling. J. Clean. Prod. **210**, 1385–1395 (2019). https://doi.org/10.1016/j.jclepro.2018.11.122
15. Li, X., et al.: Homogeneity evaluation of hot in-place recycling asphalt mixture using digital image processing technique. J. Clean. Prod. **258**, 120524 (2020). https://doi.org/10.1016/j.jclepro.2020.120524
16. Ma, Y., Polaczyk, P., Hu, W., Zhang, M., Huang, B.: Quantifying the effective mobilized RAP content during hot in-place recycling techniques. J. Clean. Prod. **314**, 127953 (2021). https://doi.org/10.1016/j.jclepro.2021.127953

17. Yao, Y., et al.: Strategy for improving the effect of hot in-place recycling of asphalt pavement. Constr. Build. Mater. **366**, 130054 (2023). https://doi.org/10.1016/j.conbuildmat.2022.130054

18. Xia, X., Han, D., Zhao, Y., Xie, Y., Zhou, Z., Wang, J.: Investigation of asphalt pavement crack propagation based on micromechanical finite element: a case study. Case Stud. Constr. Mater. **19**, e02247 (2023). https://doi.org/10.1016/j.cscm.2023.e02247

19. Pan, Y., et al.: Field observations and laboratory evaluations of asphalt pavement maintenance using hot in-place recycling. Constr. Build. Mater. **271**, 121864 (2021). https://doi.org/10.1016/j.conbuildmat.2020.121864

20. Gao, Y.: Theoretical analysis of reflective cracking in asphalt pavement with semi-rigid base. Iran. J. Sci. Technol.-Trans. Civ. Eng. **43**, 149–157 (2019). https://doi.org/10.1007/s40996-018-0154-8

21. Xia, X., Han, D., Ma, Y., Zhao, Y., Tang, D., Chen, Y.: Experiment investigation on mix proportion optimization design of anti-cracking stone filled with cement stabilized macadam. Constr. Build. Mater. **393**, 132136 (2023). https://doi.org/10.1016/j.conbuildmat.2023.132136

22. Idris, I.I., Sadek, H., Hassan, M.: State-of-the-art review of the evaluation of asphalt mixtures' resistance to reflective cracking in laboratory. J. Mater. Civ. Eng. **32**, 03120004 (2020). https://doi.org/10.1061/(ASCE)MT.1943-5533.0003254

23. Dave, E.V., Buttlar, W.G.: Thermal reflective cracking of asphalt concrete overlays. Int. J. Pavement Eng. **11**, 477–488 (2010). https://doi.org/10.1080/10298430903578911

24. Laboratory and Field Investigation of Reflective Crack Mitigation in Layered Asphalt Concrete Pavements – ProQuest (n.d.). https://www.proquest.com/openview/923d4391b0a90bbc2daa6c42df1f30fc/1?pq-origsite=gscholar&cbl=18750. Accessed 3 Mar 2023

Hybrid Noise Eliminating Algorithm for Radar Target Images Based on the Time-Frequency Domain

Xiaoyu Ma[✉], Kunmei Li, Zhiwei Wang, Wei Xu, Zheng Bian, Zicheng Du, and Longbo Deng

Xi'an Electronic Engineering Research Institute, Weiqu Fengqi East Street, Chang'an District, Xi'an 710100, Shaanxi Province, China
maxiaoyu0824@gmail.com

Abstract. The radar target imaging effect directly affects the resolution of the radar target, which affects the commander's decision. However, the hybrid noise composed of speckle and Gaussian noise is one of the main affecting factors. The existing methods for image denoising are hard to eliminate the hybrid noise in radar images. Hence, this paper proposes a new hybrid noise elimination algorithm for the radar target image. Based on the strong correlation between wavelet coefficients, this algorithm first uses the wavelet coefficient correlation denoising algorithm (WCCDA) to filter the high-frequency information and high-frequency part of low-frequency information for different directions of the three channels of the image. Then, an improved adaptive median filtering algorithm (IAMF) is proposed to perform fine-grained filtering on each re-constructed channel. Finally, the radar target image is reconstructed. The results show that the proposed algorithm outperforms the comparison approaches in the peak signal-to-noise ratio (PSNR) and mean-square error (MSE) indexes with better denoising effects.

Keywords: Radar Target Images · Hybrid Noise · Adaptive Median Filter · Wavelet Coefficient Correlation

1 Introduction

With the rapid development of radar technology, the application fields of radar are becoming more wide, especially in the military. Hence, the requirement of the radar performance is gradually increased, and the effect of noise elimination for radar target images is one of the crucial indicators for testing the radar performance. Radar target images are mainly affected by speckle noise caused by the coherence of scattering phenomena and Gaussian noise caused by complex environments. The radar imaging effect may directly affect the resolution of radar target recognition, which will delay the time for commanders to make effective strategies, Hence, an effective method for radar target image processing is necessary.

The difficulty in image denoising is that noise, edges, and textures all belong to high-frequency components, and it is difficult to distinguish them during the denoising process.

© The Author(s) 2024
G. Feng (Ed.): ICCE 2023, LNCE 526, pp. 408–420, 2024.
https://doi.org/10.1007/978-981-97-4355-1_38

The denoised image inevitably loses some edge and detailed information. Hence, this is also a vital issue in the image processing field today. The image denoising technology is mainly divided into denoising methods based on spatial domain and denoising algorithms based on frequency domain [1].

Spatial filters primarily declare noise occupying the spectrum higher regions by low-pass filtering groups of pixels. Hence, many methods filter out noise information by processing images with high frequency in the spatial domain. However, these methods can eliminate noise to a certain extent but may cause blur, loss of edge, and detailed information about the image. The frequency domain methods convert the image into a specific transformation domain and decompose the signal. Then, the suitable filter is designed for noise elimination based on the different statistical characteristics of the natural features of the image [2]. The frequency domain methods mainly include the method based on Fast Fourier Transformation (FFT) [3], Discrete Cosine Transform (DCT) [4], and Wavelet Transform (WT) [5]. Since WT can perform time-frequency analysis simultaneously and quickly discover the features of data mutation points, the algorithms based on WT are also one of the vital research contents in the image and signal processing field.

However, the WT only has information in the horizontal, vertical, and diagonal directions because it lacks directionality, so it cannot optimally represent two-dimensional images containing singular lines or surfaces. As a result, WT produces distortion and cannot reconstruct the original image well. In order to overcome these drawbacks, this paper proposes a hybrid noise elimination algorithm based on improved adaptive median filtering and wavelet coefficient correlation algorithm.

The proposed algorithm first uses the improved Wavelet Coefficient Correlation Denoising Algorithm (WCCDA) to perform wavelet transformation on the three channels R, G, and B of the image. Then, the high-frequency information and the high-frequency part of the low-frequency information are extracted in the horizontal, vertical and diagonal directions for each channel. After that, coarse-grained filtering on all high-frequency information is performed. To solve the color and line/surface distortion problems caused by image reconstruction, an improved adaptive median filtering algorithm (IAMF) is proposed to perform fine-grained filtering on each reconstructed channel and calibrate the reconstructed image. Finally, the original radar target image is reconstructed.

2 Wavelet Coefficient Correlation Denoising Algorithm

2.1 Wavelet Decomposition and Reconstruction of Radar Target Images

The two-dimensional discrete wavelet decomposition of digital images $f(x, y)$ can be expressed as the following formulas (1)~(5) [6]:

$$cA_0 = f(m, n) \tag{1}$$

$$cA_{j+1}(m, n) = \sum_k \sum_l h_{k-2m} \times h_{l-2n} \times cA_j(k, l) \tag{2}$$

$$cH_{j+1}(m, n) = \sum_k \sum_l h_{k-2m} \times g_{l-2n} \times cA_j(k, l) \tag{3}$$

$$cV_{j+1}(m, n) = \sum_k \sum_l g_{k-2m} \times h_{l-2n} \times cA_j(k, l) \tag{4}$$

$$cD_{j+1}(m, n) = \sum_k \sum_l g_{k-2m} \times g_{l-2n} \times cA_j(k, l) \tag{5}$$

where j is the wavelet decomposition scale, $\{h_k\}$ and $\{g_k\}$ represent low-pass and high-pass filters, the j-scale layer image cA_j is decomposed by one layer of wavelets to: low frequency coefficient cA_{j+1}, horizontal high frequency coefficient cH_{j+1}, vertical high frequency coefficients cV_{j+1} and diagonal high frequency coefficients cD_{j+1}.

After that, the system can process different information according to actual requirements. Finally, the image is reconstructed based on the processed high-frequency information in each direction and low-frequency information to reconstruct the original image. The reconstruction process is as follows:

$$cA_j(m, n){=}A_{j+1} + H_{j+1} + V_{j+1} + D_{j+1} \tag{6}$$

where A_{j+1}, H_{j+1}, V_{j+1} and D_{j+1} are sub-images of j scale images reconstructed by the low frequency coefficients and high frequency coefficients in three directions.

2.2 Radar Target Image Denoising Based on Wavelet Coefficient Correlation

There is a strong correlation between the upper and lower layers of the wavelet coefficients but the noise has no such correlation. The WCCDA is to compare the normalized correlation coefficients at each location for each layer and determine whether each data is a pixel or a noise point [7, 8]. The specific process can be expressed through Eqs. (7) and (8):

$$CW_{j,k} = W_{j,k} W_{j+1,k} \tag{7}$$

$$\widetilde{W_{j,k}} = CW_{j,k} \sqrt{\frac{PW_j}{PCW_j}} \tag{8}$$

where $W_{j,k}$ is the wavelet coefficient of the high-frequency, $CW_{j,k}$ is the correlation coefficient at point k of scale j, $\widetilde{W_{j,k}}$ is the normalized wavelet correlation coefficient, $PW_j = \sum_k W_{j,k}^2$ represents the energy of wavelet coefficients of scale j, $PCW_j = \sum_k CW_{j,k}^2$ represents the correlation coefficients energy of scale j.

Then, calculate the wavelet coefficient of the high-frequency $W_{j,k}$ for each layer first and compare $W_{j,k}$ and $\widetilde{W_{j,k}}$. If $\widetilde{W_{j,k}} \geq W_{j,k}$, the system considers that the point is the real pixel, takes $\widetilde{W_{j,k}} = W_{j,k}$, and sets $W_{j,k} = 0$. If $\widetilde{W_{j,k}} < W_{j,k}$, the system thinks that the pixel is controlled by noise, then leaving $W_{j,k}$, set $\widetilde{W_{j,k}} = 0$. Then, the system repeats the above process and recalculates $\widetilde{W_{j,k}}$ on each scale. Finally, the real pixel points are kept in $\widetilde{W_{j,k}}$, and the noise points are kept in $W_{j,k}$.

Normally, for single noise, most of the noise information is concentrated in the high-frequency part of the signal. However, when there has hybrid noises in the image, the low-frequency information is likely to contain residual large-amplitude noise. On the other hand, low-frequency and high-frequency information have different features. If they are processed together, their respective features are destroyed, thereby reducing the image denoising efficiency, increasing errors. Hence, the denoising process for low-frequency information is necessary. The proposed algorithm improves the traditional WCCDA, not only denoising the high-frequency information but also denoising the high-frequency part of the low-frequency information to achieve a more comprehensive noise elimination.

3 Improved Adaptive Median Filtering Algorithm

The median filter algorithm is widely used in the image processing fields. It sorts the pixels in the neighborhood by grayscale and selects the middle value of the group as the output pixel value to remove noise. The two-dimensional median filtering algorithm [9] is mainly used, as shown in Eq. (9):

$$P(i, j) = \text{median}(s(m)) \tag{9}$$

where m is the number of pixels, s is a gray sequence which is already sorted, i and j are the horizontal and vertical coordinates of the pixels respectively. The output pixels are determined by the image median, which makes the median filter less sensitive to the limit pixel value than the mean. Hence, it not only eliminates the isolated noise points but also gets better image clarity.

However, the window type and size directly affect the algorithm's performance. In theory, the larger the window size, the image processing effect is better, but if the window size is unrestricted to increase, it may damage the details of the image. Hence, one of the main problems is the selection of the window. In addition, the hybrid noise is with a large amplitude, and the image edge burr and distortion may occur. It is because the median point may take the noise point as the real pixel when the noise amplitude is large, which leads to the algorithm being invalid.

Many researchers have proposed the improved schemes. For instance, González et al. [10] propose a dynamic weighted adaptive median filtering algorithm for impulse noise, which can adaptively adjust the window size according to the noise amplitude. However these methods cannot restore the color and edge information well. Hence, this paper proposes an improved adaptive median filtering algorithm (IAMF). By selecting an appropriate filter window for each pixel point, the system can accurately determine whether the point is a noise point. The specific steps are as follows.

3.1 Expand the Pixel Matrix of the Original Image Pixel Matrix

To ensure the data in the filter window is not empty and valid, the original image pixel matrix needed to be expanded. The system sets a matrix expansion radius N for the algorithm. The overall performance of the system can be optimal when the matrix expansion radius N does not exceed 10 after experimental verification. The system assumes the original image matrix is $X = m \times n$. Then, the original image pixel matrix is

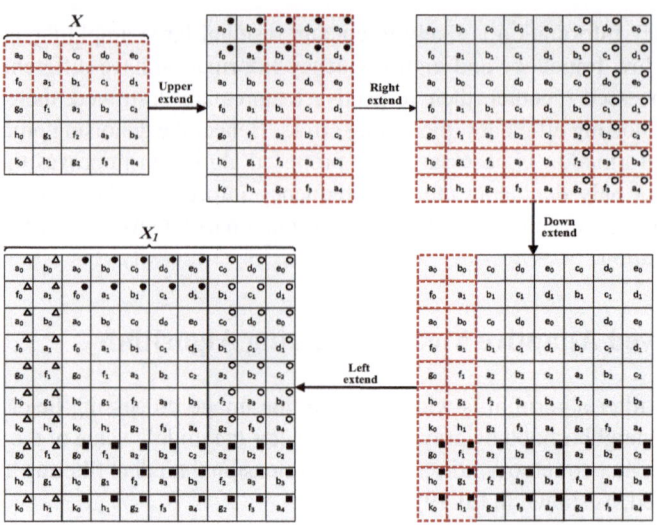

Fig. 1. Pixel matrix expansion process of the image

expanded from the upper, right, lower, and left directions with the expansion ampli-tudes of N, $N + 1$, $N + 1$ and N. The expanded image matrix dimension reaches $X_1 = (m + 2 * N + 1, \ n + 2*N + 1)$, where m and n represent the number of rows and columns of the original image pixel matrix.

The specific process is shown as follows: Assumes $m = n = 5, N = 2$, the expanded pixel matrix is $X_1 = 10 \times 10$ as shown in Fig. 1. The data enclosed by dashed circles rep-resents the expanded part. The expanded pixel data are represented by different markers. After 4 expansions, a new image pixel data matrix is obtained.

3.2 Adaptive Filter Window Size Computation

After the pixel matrix is expanded, the system selects an appropriate filter window size for each point in the original pixel matrix. For instance, the process of the filter window size for point $(7, 7)$ is shown in Fig. 2.

The system first sets filter radius $r = 1$ and calculates the neighborhood I of each pixel in the original image along the row direction of the matrix, as shown in Eq. (10).

$$I = X_1(i - r : i + r, j - r : j + r) \tag{10}$$

where i and j represent the coordinates of the pixel point. Then, the pixels in the neigh-borhood I are re-sorted to abtain $I\prime$, makes the pixels in each column are arranged in ascending order. Finally, the maximum $I_{max}\prime$ and minimum $I_{min}\prime$ are found, and the cen-tral pixel is used as the median point $I_{med}\prime$. If the center pixel value is 0, set $I_{med}\prime = 1$, and then determine whether $I_{med}\prime$ belongs to the range of $(I_{min}\prime, I_{max}\prime)$. If the above conditions are satisfied, the neighborhood size is considered to meet the actual filtering requirements. Otherwise, the system continues to expand the filter radius, let $r = r + 1$ $(r \leq N + 1)$, and repeats the above calculation process until a suitable filter window

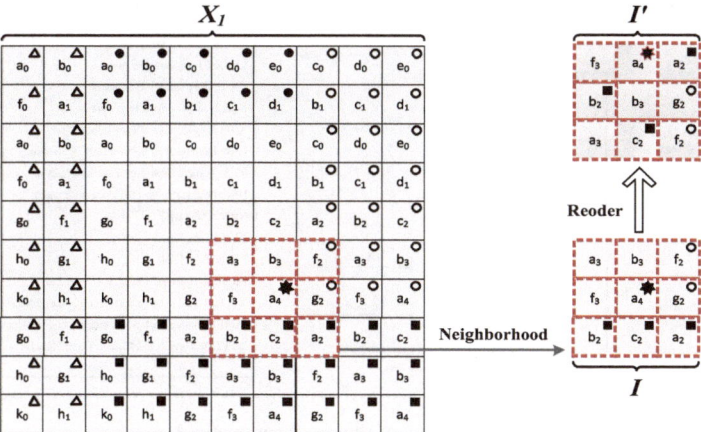

Fig. 2. Size calculation of filtering window

size is found. In addition, the system can adjust the size of the matrix expansion radius N according to the features of the noise.

3.3 Noise Point Judgement

First, determine whether the pixel in the original image is between the maximum and minimum value points within the filter window range. If this condition is met, the point is considered to be a real pixel and is retained. On the contrary, it means that the pixel is controlled by noise and needs to be replaced by the median point $I_{med}/$ within the filter window. Finally, the system repeats the above process to complete the image filtering.

4   Radar Target Image Denoising Algorithm Based on WCCDA and IAMF Algorithm

For the hybrid noise, the denoising methods for the single noise is not ideal and may easily cause image edges missing and color distortion. Hence, this paper combines the WCCDA and the IAMF algorithm to denoise the image. This algorithm can eliminate the noise while retaining the edges, details, and color information and can effectively prevent image distortion. The specific process is shown as Fig. 3.

The proposed algorithm is mainly divided into the following 5 steps:

1) Decompose the radar target image into R, G, and B channels, then extract the high-frequency information of each channel. In this step, the system extracts the high-frequency information in the horizontal, vertical, and diagonal directions and the high-frequency part of the low-frequency information.
2) Perform WCCDA on the high-frequency part of the low-frequency information of each channel, as well as the high-frequency information of each channel in three directions. This way can better retain the data features of different parts [11].

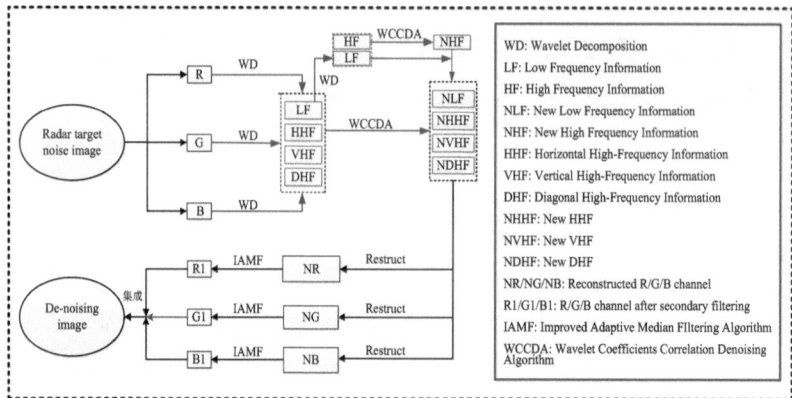

Fig. 3. Flow of the proposed method

3) Use the new low-frequency and high-frequency information to reconstruct the three channels of the image to complete the first coarse-grained filtering.
4) The IAMF algorithm is used to perform secondary fine-grained filtering on the three channels of the reconstructed image.
5) Integrate the denoised three channels to reconstruct the radar target image.

5 Experiments and Evaluation

5.1 Experimental Parameter Settings

To verify the effectiveness of the proposed algorithm, this paper adds the hybrid noise composed of Gaussian white noise with variances $\sigma_1 = 0.01$, $\sigma_2 = 0.02$, and $\sigma_3 = 0.03$ and multiplicative speckle noise with noise densities $d_1 = 10\%$, $d_2 = 20\%$, and $d_3 = 30\%$ to the radar target image, and conducts multiple experiments.

5.2 Wavelet Function Selection

There is another parameter that directly affects the system performance, which is the choice of wavelet function. Since the WCCDA algorithm mainly uses the correlation between wavelet coefficients of image data to eliminate noise, a suitable wavelet function can optimize the algorithm's performance. This experiment uses different wavelet functions in the WCCDA algorithm to process images, then calculates the correlation coefficients between the high-frequency wavelet coefficients before and after denoising. The larger the correlation coefficient value, the denoising effect is better [12]. The calculation process is shown in Eq. (11):

$$CW_{j,k\prime} = W_{j,k} \cdot W_{j,k\prime} \tag{11}$$

where $CW_{j,k\prime}$ is the correlation coefficient at point k of scale j, $W_{j,k}$ and $W_{j,k\prime}$ are the high-frequency wavelet coefficient before and after denoising, respectively.

Specifically, a hybrid noise composed by Gaussian white noise with variance $\sigma_2 = 0.02$ and multiplicative speckle noise with density $d_2 = 20\%$ is added to the image. Then, the three wavelet functions 'haar', 'symlet', and 'dbn' are selected for testing. After experiments verification, it is found that the three wavelet functions can obtain the maximum correlation coefficient when the wavelet scale is 8. Hence, the scale of the wavelet function in this paper is set to 8. The experiment results are shown in Table 1.

Table 1. Comparison of correlation coefficients under different wavelet functions

parameter	Wavelet function		
	haar	dbn	sym
scale	8	8	8
decomposition layer	3	3	3
wavelet correlation coefficient	0.9321	0.9215	0.9039

As Table 1 shown, the correlation coefficient value reaches the highest when using the 'haar' wavelet. It also proves the algorithm gets the best denoising effect when using the 'haar' wavelet function. Hence, the system uses the 'haar' wavelet function in the following experiments.

5.3 Hybrid Noise Eliminating

To verify the denoising effectiveness of the proposed algorithm (M4), the denoising effect for hybrid noise should be evaluated. This experiment adds Gaussian white noise and multiplicative speckle noise of different intensities to the radar target image and compares the denoising performance with different methods such as M1 [5], M2 [13], and M3 [14] in related fields.

Performance of the Algorithm When the Intensity of Gaussian Noise and Speckle Noise are $\sigma_1 = 0.01$ **and** $d_1 = 10\%$. After using different algorithms to denoise radar target images, the details, edges, and color recovery and the overall reconstruction effect are demonstrated, as shown in Fig. 4.

It can be seen that under the low-intensity condition of hybrid noise, all methods can effectively eliminate noise. M1 and M2 can eliminate noise and better retain the edge information. However, the image color has a distortion problem, and detailed information is seriously missing. In addition, the target orientation information and part of the image marks have been blurred. Though M3 is better than M1 and M2 in retaining image edges and detail information, the image color is completely distorted after denoising. The proposed method can better retain the image edges, details, and color information while eliminating noise.

Performance of the Algorithm When the Intensity of Gaussian Noise and Speckle Noise are $\sigma_2 = 0.02$ **and** $d_2 = 20\%$. To better verify the above analysis and the proposed algorithm, this experiment increased the noise intensity to $\sigma_2 = 0.02$ and $d_2 = 20\%$, and the results are shown in Fig. 5.

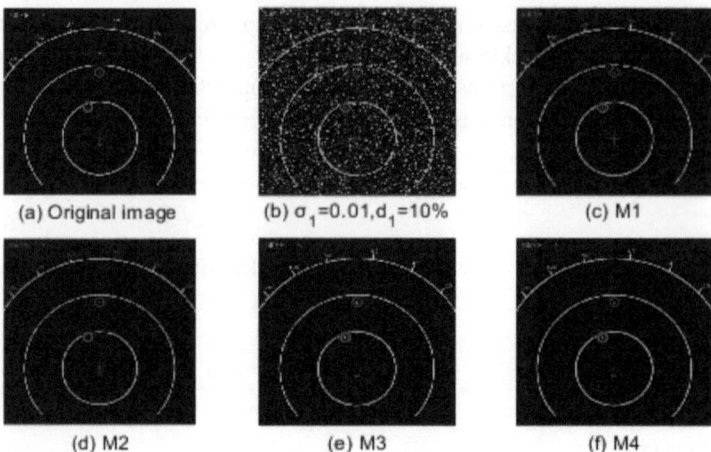

Fig. 4. Results of each method when the hybrid noise intensity is $\sigma_1 = 0.01$ and $d_1 = 10\%$

It can be seen that as the noise increases, M1 and M2 algorithms can still better eliminate noise and retain the edge information. However, the image blur is intensified, and the detailed information is seriously lost. Especially for M1, the resolution are seriously reduced, and the target with weak signal strength already cannot be identified from the image. Though the M3 algorithm retains the details and edge information well, it cannot eliminate noise absolutely, and the color distortion is serious. The proposed algorithm can retain the image edges, details, and color information better, which restores all the features of the original image with high clarity. In summary, compared with the other methods, the performance of the proposed method is better than other methods.

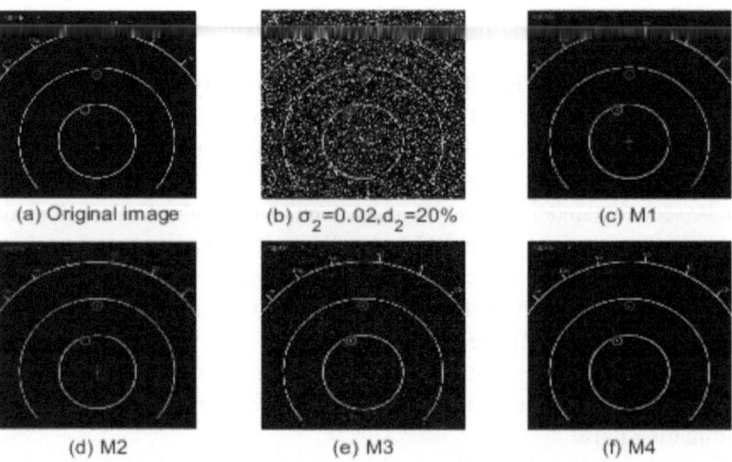

Fig. 5. Results of each method when the hybrid noise intensity is $\sigma_2 = 0.02$ and $d_2 = 20\%$

Performance of the Algorithm When the Intensity of Gaussian Noise and Speckle Noise are $\sigma_3 = 0.03$ **and** $d_3 = 30\%$. In order to test the maximum denoising performance of the proposed algorithm, this experiment continues to increase the intensity of the hybrid noise to $\sigma_3 = 0.03$ and $d_3 = 30\%$. The experiment results are shown in Fig. 6.

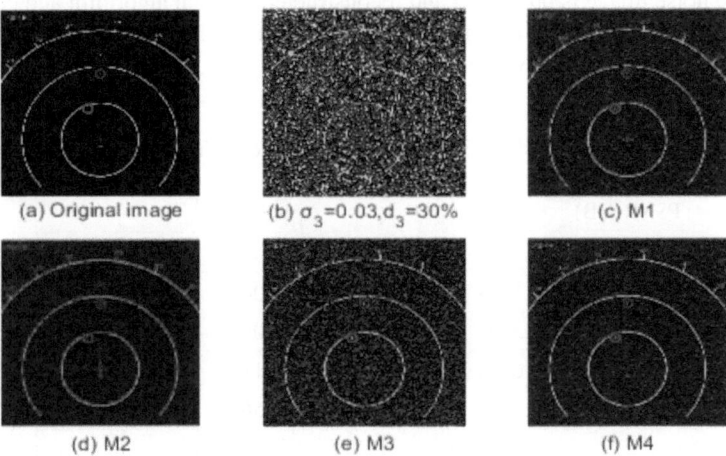

(a) Original image   (b) $\sigma_3 = 0.03, d_3 = 30\%$ (c) M1

(d) M2   (e) M3   (f) M4

Fig. 6. Results of each method when the hybrid noise intensity is $\sigma_3 = 0.03$ and $d_3 = 30\%$

As the noise further increases, the performance of each algorithm declines. The image processed by the M1 and M2 algorithms has a high degree of blur, while the detailed and edge information in the image cannot be distinguished. In addition to the above issues, the image processed by M3 has more serious color distortion problems, and the algorithm has failed. Though the denoising performance of the proposed method begins to decline, compared with the other three algorithms, the denoising effect is the best, in which the edge and detailed information of the image are well preserved and the clarity is high.

Evaluation of Denoising Performance for Each Algorithm. The above experiments show the results and analysis of each algorithm's processing of hybrid noise images with different strengths from an intuitive visual perspective. To more accurately display the denoising effect of each algorithm and verify the analysis results of the above experiments. This paper uses peak signal-to-noise ratio (PSNR) and mean square error (MSE) [15] as the measurement criteria for objective evaluation. The equations for MSE and PSNR are as follows:

$$MSE = \frac{1}{M \times N} \sum_{i=1}^{M} \sum_{j=1}^{N} [f(i,j) - g(i,j)]^2 \qquad (12)$$

$$PSNR = 10 \log_{10}(\frac{255^2}{MSE}) \qquad (13)$$

where, $f(i,j)$ is the reconstructed image after denoising, $g(i,j)$ is the original image without noise, M and N are the number of rows and columns of the image matrix. The larger the PSNR, the smaller the MSE, indicating that the image after denoising is closer to the original image.

Based on the above three experiments, the intensity of Gaussian noise is varied between $\sigma = 0.01 \sim 0.03$, and the speckle noise is kept at $d_2 = 20\%$ in this experiment. Then, the noise image is processed and reconstructed by each algorithm and compared with the original image to obtain the relevant PSNR and MSE values. The specific experimental results are shown in Table 2.

Table 2. The PSNR and MSE values after image processing under different noise

Method	PSNR (dB)			MSE		
	σ_1, d_2	σ_2, d_2	σ_3, d_2	σ_1, d_2	σ_2, d_2	σ_3, d_2
1	75.8813	73.1388	71.0691	33.0087	43.5519	47.6124
2	75.0383	74.5201	72.3904	36.0066	38.1646	39.0929
3	69.6129	67.6534	66.8925	64.1149	77.0323	82.6417
4	77.0052	76.3922	76.0383	29.5672	31.6469	32.9926

It can be seen from Table 2 that the proposed method in this paper gets the largest PSNR and the smallest MSE value compared to the other three algorithms, which also proves the consistency with the results and analysis of the above experiments. To more intuitively analyze the varies of PSNR and MSE with the intensity of hybrid noise, this paper gives the curves of PSNR and MSE values changing with noise intensity. As the intensity of Gaussian noise changes, while keeping the intensity of speckle noise stayed, the changes in PSNR and MSE can be more clearly observed. The results are shown in Figs. 7a and 7b.

Figure 7 can more intuitively show the varying of PSNR and MSE values for the four methods after processing noisy images. Even when the noise intensity reaches a, the method proposed in this article still maintains the maximum PSNR and minimum MSE value. It proves that the noise image processed by the proposed algorithm is closest to the original image and has high reliability. In addition, the performance of each algorithm decreases as the noise intensity increases, which is consistent with the above experimental results and analysis.

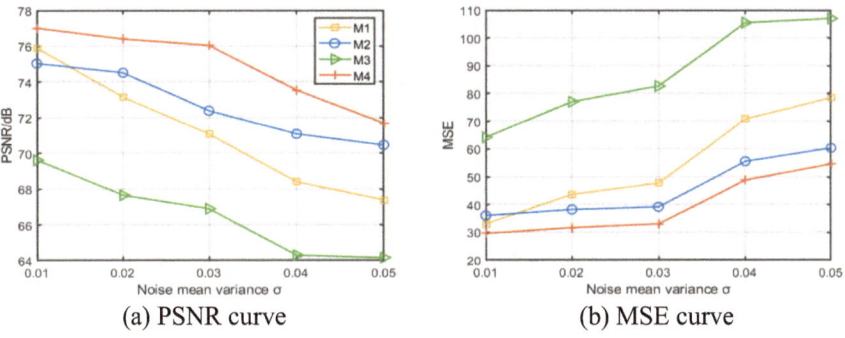

(a) PSNR curve   (b) MSE curve

Fig. 7. Curves of PSNR and MSE values with the various noise intensity

6 Conclusion

This paper proposes a new denoising algorithm for the radar target image based on wavelet transform to solve the problems of image blur, loss of edge and detail information, and image color distortion in the hybrid noise processing of radar target images. The proposed algorithm first uses the WCCDA algorithm to perform coarse filtering on the noisy image in different directions of R, G, and B channels to eliminate most of the noise in the image. Then, the IAMF algorithm is proposed to perform secondary fine-grained filtering on each channel after denoising. The IAMF algorithm expands the original target image matrix and adaptively selects an appropriate filter window size for each pixel in the image. It can quickly and accurately identify whether the pixel is a real pixel point and eliminate the remaining noise in the image to the maximum extent, which can better retain the image edges, details, and color information. Many simulation experiments prove that the proposed method in this paper has better denoising effect, higher image information recovery, and stronger stability than other related algorithms, which is suitable for radar target images denoising.

References

1. Fan, L., Zhang, F., Fan, H., et al.: Brief review of image denoising techniques. J. Vis. Comput. Ind. Biomed. Art **2**, 1–12 (2019)
2. Liu, X., Pedersen, M., Wang, R.: Survey of natural image enhancement techniques: classification, evaluation, challenges, and perspectives. J. Digit. Signal Process. **127**, 103547 (2022)
3. Kataoka, S., Yasuda, M.: Bayesian image denoising with multiple noisy images. Rev. Socionetwork Strateg. **13**, 267–280 (2019)
4. Padmapriya, R., Jeyasekar, A.: Blind image quality assessment with image denoising: a survey. J. Pharm. Negat. Results 386–392 (2022)
5. You, N., Han, L., Zhu, D., et al.: Research on image denoising in edge detection based on wavelet transform. Appl. Sci. **13**(3), 1837 (2023)
6. Song, Q., Ma, L., Cao, J.K., et al.: Image denoising based on mean filter and wavelet transform. In: 2015 4th International Conference on Advanced Information Technology and Sensor Application (AITS), pp. 39–42. IEEE (2015)

7. He, Z.: Wavelet Analysis and Transient Signal Processing Applications for Power Systems. John Wiley & Sons, Hoboken (2016)
8. Guo, T., Zhang, T., Lim, E., et al.: A review of wavelet analysis and its applications: challenges and opportunities. IEEE Access **10**, 58869–58903 (2022)
9. Jebur, R.S.: Image denoising using mean filter. Al-Salam J. Eng. Technol. **2**(2), 173–181 (2023)
10. González-Hidalgo, M., Massanet, S., Mir, A., et al.: Impulsive noise removal with an adaptive weighted arithmetic mean operator for any noise density. J. Appl. Sci. **11**(2), 560 (2021)
11. Deng, G., Liu, Z.: A wavelet image denoising based on the new threshold function. In: 2015 11th International Conference on Computational Intelligence and Security (CIS), pp. 158–161. IEEE (2015)
12. Li, W., Auger, F., Zhang, Z., et al.: Newton time-extracting wavelet transform: an effective tool for characterizing frequency-varying signals with weakly-separated components and theoretical analysis. J. Signal Process. **209**, 109017 (2023)
13. Deng, G., Liu, Z.: A wavelet image denoising based on the new threshold function. In: 2015 International Conference on Computational Intelligence and Security (CIS), pp. 158–161. IEEE (2015)
14. Faragallah, O.S., Ibrahem, H.M.: Adaptive switching weighted median filter framework for suppressing salt-and-pepper noise. AEU-Int. J. Electron. Commun. **70**(8), 1034–1040 (2016)
15. Tian, P., Li, Q., Ma, P., et al.: A new method based on wavelet transform for image denoising. J. Image Graph. **13**(3), 395–399 (2008)

Study on Strain Characteristics of Long Longitudinal Slope Asphalt Pavement Surface

Xu Li[1]([✉]), Weiqin Liu[1], Jinxi Qin[1], Xiuxing Zhao[1], and Jie Chen[2,3]

[1] Guangxi Xinfazhan Communications Group Co., Ltd., Guangxi, Nanning 530029, China
610635936@qq.com
[2] Guangxi Key Lab of Road Structure and Materials, Guangxi, Nanning 530007, China
[3] Guangxi Transportation Science and Technology Group Co., Ltd., Guangxi, Nanning 530007, China

Abstract. To study the changes in shear and tensile strains of asphalt pavement under vehicle moving loads on long longitudinal slopes, a structural model of asphalt pavement was established using Abaqus finite element calculation software. A single factor analysis was conducted on different slopes, driving speeds, temperatures, and braking coefficients. The calculation results show that the maximum shear strain increases with the increase of road slope, temperature, and braking coefficient, but decreases with the increase of driving speed; The maximum tensile strain increases with the increase of road slope, driving speed, and braking coefficient, but decreases with the increase of temperature. When the vehicle is driving smoothly, the maximum shear strain occurs at a distance of about 5cm from the road surface, and the maximum tensile strain occurs at a distance of 6cm from the road surface, both of which occur in the middle layer. In the design phase, targeted improvements can be made to the shear and tensile properties of the asphalt surface layer in the middle layer, in order to enhance the road performance of the asphalt surface layer in long and long longitudinal slopes. When the vehicle is braking, when the braking coefficient is high, the road surface will generate significant shear strain. In the design stage, it is necessary to improve the shear resistance of the upper layer of asphalt concrete in a targeted manner.

Keywords: Road Engineering · Long Longitudinal Slope · Asphalt Pavement · Finite Element Analysis · Dynamic Response

1 Introduction

With the accelerating process of China's transportation construction, the choice of highway continues to extend to the plateau and high altitude areas. Due to the limitation of terrain conditions, it is inevitable to have long longitudinal slope section, and even the slope length and slope exceed the limit specified in the existing specifications [1]. Compared with the small ramp pavement, the diseases such as rutting and displacement in the long longitudinal slope section are very serious, this is mainly because when the vehicle is driving on the long longitudinal slope section, the vehicle speed is low and

© The Author(s) 2024
G. Feng (Ed.): ICCE 2023, LNCE 526, pp. 421–430, 2024.
https://doi.org/10.1007/978-981-97-4355-1_39

there will be frequent start-up and braking acceleration. The asphalt concrete in the long longitudinal slope section will not only produce serious rut, but also produce shear failure under the action of driving, this is because the slow start of the uphill section and the deceleration braking of the downhill section produce a larger horizontal shear stress on the top and inside of the asphalt pavement surface than on the flat section. The reasons for this destruction are manifold. It is not only related to traffic volume and vehicle overload, but also related to technical conditions such as route line type, pavement structure type and construction quality.

In recent years, many scholars have used the finite element method to study the mechanical properties of asphalt pavement with long longitudinal slope.

Shi Tingwei et al. [2] established a three dimensional finite layer analysis model of asphalt pavement with long longitudinal slope by using the finite layer software 3DMove Analysis. It is found that the acceleration or braking of the vehicle will lead to a large increase in the maximum shear stress peak in the asphalt pavement. The maximum shear stress peaks appear at 0–4 cm below the road surface, so it is particularly important to improve the anti-rutting performance of the upper part of the long longitudinal slope asphalt surface. Zhou Taohong et al. [3] used Aansys software to model typical structures based on the definition of heavy load conditions and the analysis of common diseases, calculated and analyzed the surface deflection, tensile stress, compressive stress and shear stress of the typical structure under different axle loads, and analyzed the damage of the pavement structure. Yang Zhenzi et al. [4] used ANSYS software to quantitatively analyze the influence of high temperature and heavy load traffic on surface deflection and structural stress of asphalt pavement. Li Yanchun et al. [5] established a three dimensional finite element model using Ansys finite element software. By applying pulse load, the strain variation rule of large longitudinal slope of asphalt pavement under different conditions is obtained. Jun Fu et al. [6] discussed the relationship between shear stress and slope angle, load, pavement depth, interlayer contact condition and modulus by three dimensional finite element model. Zhou Yaxin et al. [7] simplified the load distribution model by calculating the equilibrium speed of heavy duty vehicles in the long longitudinal slope section under the equilibrium state, according to the commonly used pavement structure in china, a three dimensional finite element model of asphalt pavement was established to compare and analyze the mechanical response of asphalt pavement under different slope, temperature and asphalt layer thickness. Ruan Luming et al. [8], firstly, the traffic conditions in Chongqing are analyzed, and the typical heavy-duty vehicles and their climbing speed characteristics are obtained. The contact characteristics between the tire and the road surface of the heavy-duty vehicle are studied in depth, and a simplified model of tire grounding of heavy-duty vehicles is given. Then, the factors affecting the response indexes of asphalt pavement structure in long longitudinal slope section are studied. Finally, the fatigue damage variation rule of asphalt layer in high temperature month is analyzed based on Miner fatigue rule.

The research of the above scholars shows that the finite element method can be used to study the mechanical response of asphalt pavement with long longitudinal slope and its influencing factors. Therefore, this paper will use Abaqus calculation software to establish a finite element model of long longitudinal slope asphalt pavement, and discuss the influence of longitudinal slope, driving speed, temperature and braking coefficient on shear strain and tensile strain of asphalt pavement.

2 Analysis of Influencing Factors of Mechanical Model of Long Longitudinal Slope Section

The vehicle is subjected to various resistances when driving on the asphalt pavement. These resistances include rolling resistance F_1, slope resistance F_2, air resistance F_3 and acceleration or deceleration resistance F_4. In order to achieve a stable operating state, the traction of the car must be equal to the sum of the resistance encountered by the car during driving, that is:

$$F = F_1 + F_2 + F_3 + F_4 \tag{1}$$

The driving state of the vehicle on the longitudinal slope can be divided into two types: gradually decelerating to a stable speed and maintaining a uniform speed after entering the slope; the vehicle gradually accelerates to a stable speed and maintains a uniform speed after entering the slope.

3 Mechanics Model and Calculation Parameters of Asphalt Pavement in Long Longitudinal Slope Section

3.1 Finite Element Model Establishment

Considering the computational efficiency and accuracy, this paper uses a three dimensional model for mechanical response analysis. The model size is 8-node hexahedral element. The boundary conditions are assumed as follows: the bottom surface of the model is completely constrained, there is no lateral displacement on the left and right sides, there is no longitudinal displacement on the front and rear sides, and the contact state between the layers is completely continuous. The calculated load is a double circular load: the standard load is the tire ground pressure of 0.7MPa, the load circle radius is 106.5 mm, and the center distance of the two wheels is 319.5 mm. Among them, X is the lateral direction of the road, Y is the driving direction, and Z is the vertical direction. The size of the model is 5 m × 10 m × 5 m, and its structure is shown in Fig. 1. Vertical moving load and horizontal moving load are applied by ABAQUS's own subroutines DLOAD and UTRACLOAD.

3.2 Determination of Asphalt Pavement Surface Material Parameters

The parameters of asphalt mixture under different temperature conditions are shown in Table 1 (Table 2).

3.3 Pavement Computational Structural

The material and thickness of the pavement structure layer from top to bottom are shown in Table 3.

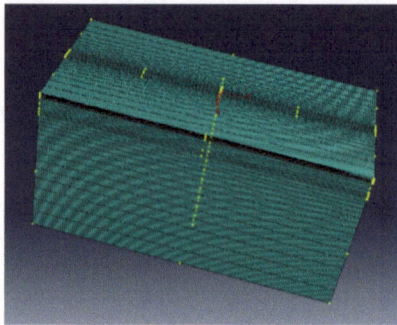

Fig. 1. Finite element model of pavement structure.

Table 1. Asphalt mixture parameters under different temperature conditions.

Mixture type	Temperature/°C	Elastic parameter		Density/(kg/m³)
		Modulus of resilience E/MPa	Poisson ratio μ	
Fine-grained bituminous concrete	20	870	0.25	2430
	30	620	0.30	
	40	554	0.35	
	50	530	0.40	
	60	526	0.45	
Medium grain bituminous concrete	20	910	0.25	2440
	30	752	0.30	
	40	600	0.35	
	50	440	0.40	
	60	380	0.45	
Coarse graded bituminous concrete	20	1031	0.25	2450
	30	900	0.30	
	40	710	0.35	
	50	500	0.40	
	60	390	0.45	

Table 2. Elastic parameters of base and soil materials.

Material	Compressive modulus of resilience E/MPa	Poisson ratio μ	Density/(kg/m^3)
Cement stabilized macadam CTB	15000	0.225	2700
Graded crushed stone GAB	400	0.35	2500
Soil SG	80	0.40	2000

Table 3. Asphalt pavement structure.

Layer of asphalt pavement	Thickness/mm
Fine-grained bituminous concrete	40
Medium grain bituminous concrete	60
Coarse graded bituminous concrete	80
Cement stabilized macadam	200
Graded crushed stone	200
Soil	-

4 Analysis of Pavement Mechanical Response of Long Longitudinal Slope Section

In the mechanical calculation and analysis, the shear stress on the driving direction of the vertical road table position of the wheel load center is analyzed and calculated. In the analysis and calculation of uniform speed, only 0.7MPa vertical stress of road surface is considered, and the influence of friction force in parallel direction is ignored.

4.1 Analysis of Mechanical Response Changes with Depth

Under the conditions of running speed of 60 km/h and temperature of 60 °C, the variation of shear stress of pavement structure with depth is calculated, and the results are shown in Fig. 2 and Fig. 3.

It can be seen from Fig. 2 and Fig. 3 that the shear strain and tensile strain of asphalt surface first increase and then decrease with the increase of depth, both the maximum tensile strain and the maximum shear strain are located in the middle surface layer, and the depth of the maximum tensile strain is greater than that of the maximum shear strain.

4.2 Influence of Longitudinal Slope Degree on Mechanical Response

When the driving speed is 60 km/h, the temperature is 60 °C, and the slope is 0%, 2%, 4%, 6% and 8% respectively, the maximum mechanical response and the change law of the position under different slope conditions are analyzed.

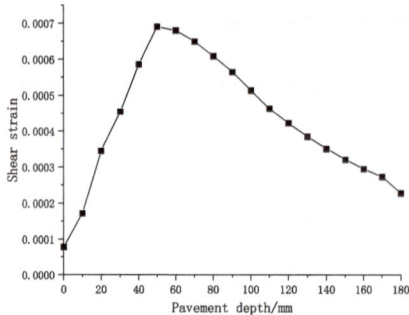

Fig. 2. Variation of shear strain with pavement depth.

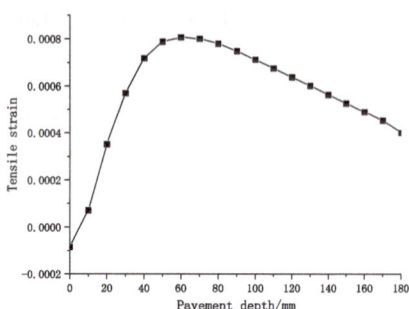

Fig. 3. Variation of tensile strain with pavement depth.

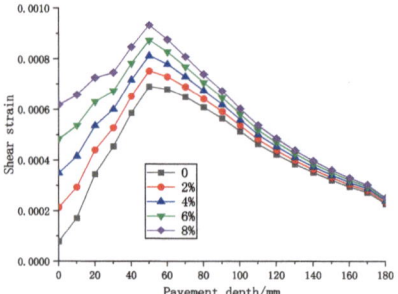

Fig. 4. Variation of shear strain under different slope conditions.

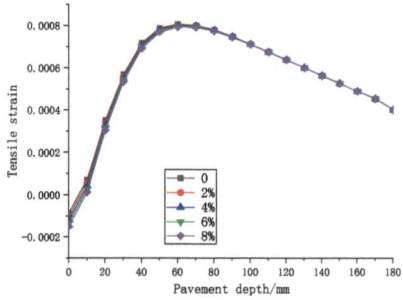

Fig. 5. Variation of tension strain under different slope conditions.

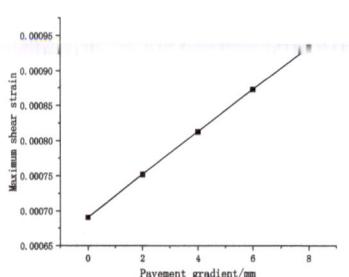

Fig. 6. Variation of maximum shear strain under different slope conditions.

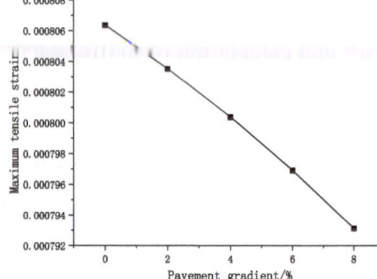

Fig. 7. Variation of maximum tensile strain under different slope conditions.

It can be seen from Fig. 4 and Fig. 5 that under different slope conditions, the shear strain and tensile strain of asphalt surface layer increase first and then decrease with the increase of pavement depth. It can be seen from Fig. 6 and Fig. 7 that the maximum shear strain increases and the maximum tensile strain decreases with the increase of road slope.

4.3 Influence of Running Speed on Shear Stress

Under the conditions of temperature 60 °C and longitudinal slope 4%, the maximum mechanical response and the change law of position were analyzed under different driving speeds of 40 km/h, 60 km/h, 80 km/h, 100 km/h and 120 km/h.

It can be seen from Fig. 8 and Fig. 9 that under different driving speeds, the shear strain and tensile strain of asphalt surface both increase first and then decrease with the increase of pavement depth. It can be seen from Fig. 10 and Fig. 11 that the maximum shear strain and maximum tensile strain both decrease with the increase of driving speed.

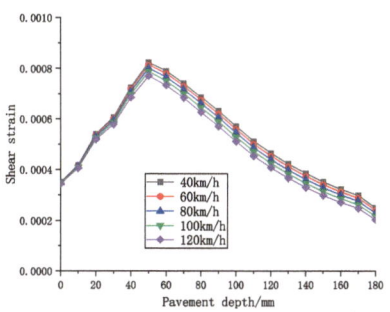

Fig. 8. Shear strain variation diagram at different running speeds.

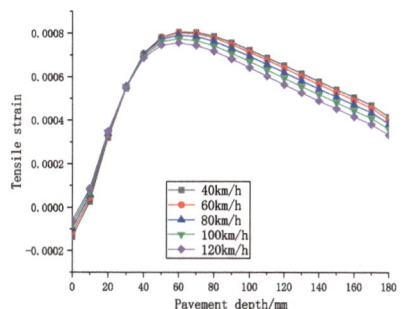

Fig. 9. Variation diagram of tension strain at different running speeds.

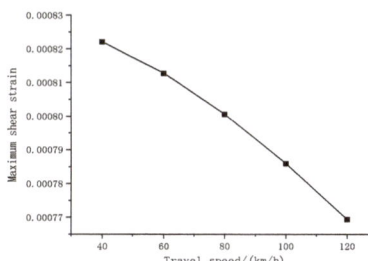

Fig. 10. Maximum shear strain variation at different running speeds.

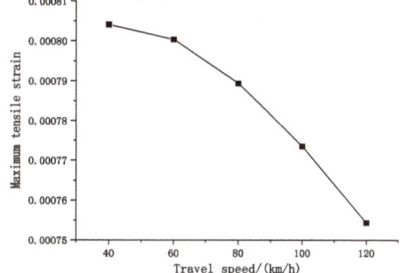

Fig. 11. Maximum tensile strain variation diagram at different running speeds.

4.4 Influence of Temperature on Shear Stress

Under the conditions of driving speed of 80 km/h and longitudinal slope of 4%, the maximum mechanical response and the change law of position at different temperatures of 20 °C, 30 °C, 40 °C, 50 °C and 60 °C were analyzed.

It can be seen from Fig. 12 and Fig. 13 that under different temperature conditions, the shear strain of asphalt surface increases first and then decreases with the increase of pavement depth. When the temperature is less than or equal to 40 °C, the tensile strain of asphalt surface increases first and then decreases with the increase of pavement depth. When the temperature is greater than 40 °C, the tensile strain of asphalt surface increases with the increase of pavement depth. It can be seen from Fig. 14 and Fig. 15 that the maximum shear strain and maximum tensile strain both increase with the increase of temperature.

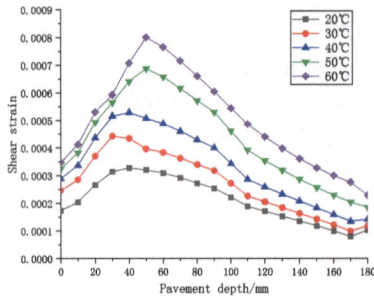

Fig. 12. Variation of shear strain at different temperatures.

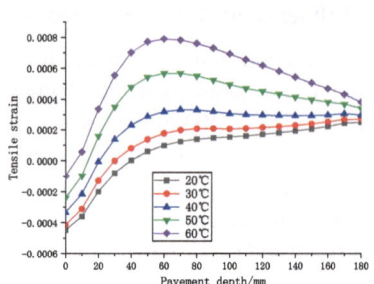

Fig. 13. Variation of tensile strain at different temperatures.

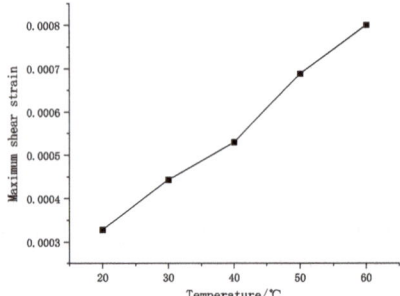

Fig. 14. Maximum shear strain variation at different temperatures.

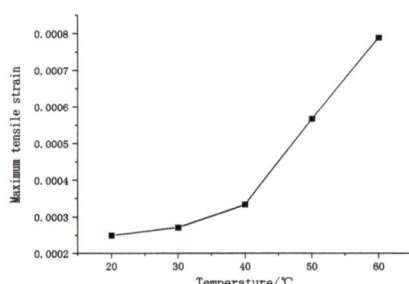

Fig. 15. Maximum tensile strain variation at different temperatures.

4.5 Influence of Braking Coefficient on Shear Stress

Under the conditions of running speed of 80 km/h, temperature of 60 °C and longitudinal slope of 4%, the maximum mechanical response and the change law of position under different braking coefficients f of 0.1, 0.3, 0.5, 0.7 and 0.9 were analyzed (Fig. 19).

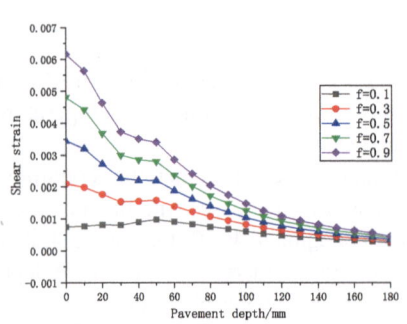

Fig. 16. Variation of shear strain under different braking coefficients.

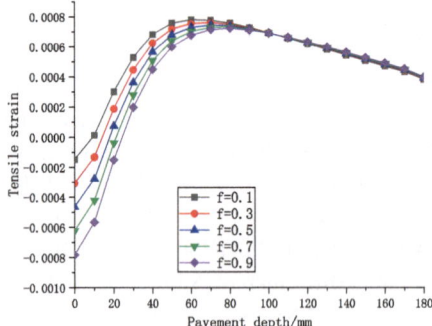

Fig. 17. Variation of tensile strain under different braking coefficients.

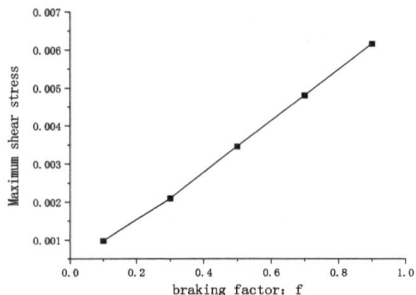
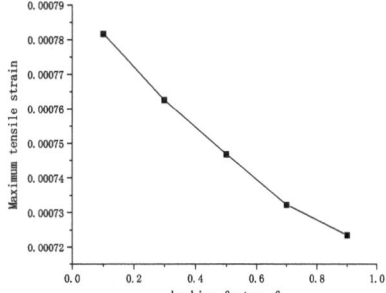

Fig. 18. Maximum shear strain variation under different braking coefficients.

Fig. 19. Maximum tensile strain variation under different braking coefficients.

It can be seen from Fig. 16 and Fig. 17 that under the conditions of different braking coefficients, when the braking coefficient is less than 0.1, the shear strain of asphalt surface increases first and then decreases with the increase of pavement depth. When the braking coefficient is 0.3, the shear strain of asphalt surface decreases first, then increases and then decreases with the increase of pavement depth. When the braking coefficient is greater than 0.3, the shear strain of asphalt surface decreases with the increase of pavement depth. The tensile strain of asphalt surface increases first and then decreases with the increase of pavement depth. It can be seen from Fig. 17 and Fig. 18 that the maximum shear strain increases and the maximum tensile strain decreases with the increase of the braking coefficient.

5   Conclusions

The asphalt pavement calculation model was established by finite element calculation software Abaqus, and the shear strain and tensile strain of the asphalt layer of the asphalt pavement with long longitudinal slopes were analyzed under the conditions of different slopes, travel speeds, temperatures and braking coefficients. The above analysis leads to the following conclusions:

(1) Maximum shear strain increases with increasing pavement gradient, temperature and braking factor and decreases with increasing travel speed. The maximum tensile strain increases with increasing pavement gradient, travel speed and braking factor and decreases with increasing temperature.

(2) Vehicles in the smooth running, the maximum shear strain and the maximum tensile strain appear in the middle surface layer. In the design stage, need to target to improve the middle surface layer of asphalt concrete shear resistance and tensile properties, in order to enhance the long longitudinal slope section of the asphalt surface layer of road performance.

(3) Vehicle braking coefficients have a small effect on tensile strains and a large effect on shear strains. When the braking coefficient is large, the road surface will produce a large shear strain. In the design phase, the top layer of asphalt concrete needs to be targeted to improve the shear resistance of asphalt concrete.

Acknowledgments. This research was funded by the Guangxi Key R&D Plan (NO. AB22080034), Guangxi Science and Technology Project (NO. ZY21195043).

References

1. Yang, D.Y.: Research on construction technology of highway asphalt pavement in long and large longitudinal slope section. Eng. Tech. Stud. **8**(9), 53–55 (2023)
2. Shi, T.W., Yan, K.Z., Zhao, X.W.: Analysis on shear stress in asphalt pavement with moving load of long and steep upgrade section. J. XiangTan Univ. (Nat. Sci. Edit.) **38**(4), 26–33 (2016)
3. Zhou, T.H.: Structural mechanics response and damage analysis of asphalt pavement under heavy load. Constr. Des. Eng. **3**, 87–90 (2019)
4. Yang, Z.Z.: Mechanical response analysis of asphalt pavement structure under the condition of high temperature and heavy-loaded. Technol. Highw. Transp. **36**(2), 20–26 (2020)
5. Li, Y.C., Xin, Y.Y.: The dynamic response of large longitudinal slope of asphalt road pavement under heavy impulsive load. Adv. Mater. Res. **1065–1069**, 806–813 (2014)
6. Fu, J., Qin, Y., Ding, Q.J.: Research on influence of longitudinal gradient to rigid base composite road of cross river (Sea) tunnel under large longitudinal slope. Adv. Mater. Res. **189–193**, 1621–1624 (2011)
7. Zhou, Y.X., Wu, C.: Dynamic response analysis of asphalt pavements with long longitudinal slopes. TranspoWorld **16**, 61–63 (2023)
8. Ruan, L.M.: Research on Fatigue Damage of AsphaltPavement Structure in Long and Steep Slope of Mountain Highway. Chongqing Jiaotong University (2018)
9. Guo, C.C., Ding, T.T., Lv, X., et al.: Study on rutting development and axle load conversion correction factor of long longitudinal slope asphalt pavement. J. China Foreign Highw. 1–15

Study on Insulation and Anti-frost Heave Effect of Polystyrene Board Under the Condition of Mold Bag Concrete

Fuqiang Guo[1](✉), Gangtie Li[1], Shichao Chen[1], Xia Yang[1], Yizhen Huo[2], and Jinxiang Bai[3]

[1] Inner Mongolia Agricultural University, Hohhot 010018, China
guofu101@163.com
[2] Hetao College, Bayannur 015000, China
[3] Ulanhot Water Conservancy Development Center, Ulanhot 137499, China

Abstract. As a new lining technology, mould bag lining technology has been widely used in recent years. In this paper, the channel frost heave in Hetao irrigation area of Inner Mongolia was studied and a field in-situ test platform was established. The experimental study on the insulation and anti-freeze of polystyrene board under the condition of concrete bag with different thickness was carried out by setting the concrete bag with different thickness and the test block of polystyrene board with different thickness. The research shows that the total accumulated temperature increases by 3.93%~9.22% and the frost heave rate decreases by 18.28%~55.44% by adding 2–5 cm mold bag concrete on the basis of 10 cm mold bag concrete. Laying 4~8 cm polystyrene board, the total accumulated temperature increased by 207.63%~272.25%, and the frost heave rate decreased by 71.43%~96.6%. The absolute slope of the curve fitting the frost heave rate and soil temperature decreased by 44.6%~58.7%. Laying polystyrene plate can inhibit the migration of soil water during freezing-thawing period, reduce the formation of frozen front ice intercalation, and thus reduce the damage of channel frost heave.

Keywords: Mold Bag Concrete · Polystyrene Board · Heat Preservation And Anti-Freezing Technology · Rate Of Frost Heave

1 Introduction

Bag concrete is an integral structure formed by pouring fluidity concrete into the bag with high pressure pump and solidifying after the excess water oozes from the fabric gap. The bag is made of geosynthetics woven into two layers with a certain thickness. This technology has the characteristics of fast construction speed, save labor and time, easy operation and strong terrain adaptability. It can be widely used in road, reservoir, river channel, seawall and other projects. In the aspect of lining technology of concrete channel for mold bag at home and abroad, researches have been carried out mainly on the concrete structure, material and application effect of mold bag. Li yatong [1] studied the bag concrete channel in hetao irrigation area of Inner Mongolia from the

© The Author(s) 2024
G. Feng (Ed.): ICCE 2023, LNCE 526, pp. 431–440, 2024.
https://doi.org/10.1007/978-981-97-4355-1_40

perspective of mechanical properties. Wang jun, huo yizhen et al. [2] proposed the optimized mix ratio of mold bag concrete. Zhang hailing, huo yizhen et al. [3] discussed the concrete roughness of mold bag and its influencing factors. Gong jibin [4] introduced the selection of mold bag in the concrete construction of mold bag, the technical index of concrete, the construction method of pouring concrete and the matters needing attention, based on the concrete laying and protecting project of the seven-star canal mingsha section. Wang jun, guo yanfen et al. [5] analyzed and studied the freeze resistance and macroscopic mechanical properties of mold bag concrete by measuring the mass loss rate and relative dynamic elastic modulus of test blocks with different mixing ratios. At present, the research on thermal insulation and antifreeze technology of polystyrene board used in precast concrete and cast-in-place concrete channel is relatively common [6–15], while the thermal insulation and antifreeze technology of polystyrene board laid in concrete channel of mold bag has not been studied yet. Therefore, combined with the characteristics of soil and climate in Hetao irrigation area of Inner Mongolia, this paper carried out an experimental study on the laying of polystyrene board with different thickness under the conditions of mold bag concrete.

2 Materials and Methods

In hetao irrigation area, Inner Mongolia, a platform for concrete treatment of different mold bags was set up. Each test platform has an area of 4 m × 4 m, and a total of 6 die bag concrete treatment platforms are set. Treatment 1~3 laid 10 cm, 12 cm and 15 cm mold bag concrete without heat preservation. Polystyrene insulation boards with thickness of 4 cm, 6 cm and 8 cm were laid under the concrete of 10 cm mold bags from 4 to 6. The 6 cm thick polystyrene board was laid vertically around each treatment block at a depth of 1 m to prevent interference from horizontal frost heaving.

The schematic diagram of the frost heave test platform is shown in Fig. 1, and the test processing is shown in Table 1. The test was conducted from November 2015 to April 2018, with a total of three freezing cycles.

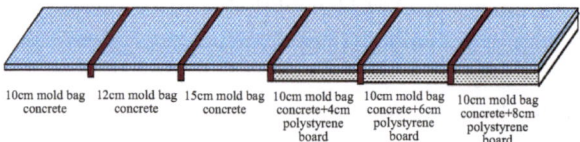

Fig. 1. Sketch map of frost heave test platform of different molded bag concrete treatment

In hetao area, the concrete construction thickness of the main channel mold bag is generally 10~15 cm, so considering the economic efficiency, the insulation board is laid under the 10 cm mold bag concrete in the test design, and the effect of insulation and anti-frost heave of the lining structure of mold bag concrete + insulation board is discussed.

Table 1. Design test of different molded bag concrete treatment

Experimental treatment	Thickness of mould bag concrete/cm	Heat preservation treatment
Treatment 1	10	No insulation
Treatment 2	12	No insulation
Treatment 3	15	No insulation
Treatment 4	10	4 cm polystyrene board
Treatment 5	10	6 cm polystyrene board
Treatment 6	10	8 cm polystyrene board

3   Results and Analysis

3.1   Different Bag Concrete Treatment Insulation Effect

The variation law of daily mean temperature of concrete treated with different mold bags.Statistical analysis was conducted on the daily mean temperature of concrete treatment with different mold bags (buried 30 cm deep) in the test site from 2015 to 2016, and the daily mean temperature variation process of concrete treatment with six different mold bags was drawn, as shown in Fig. 2.

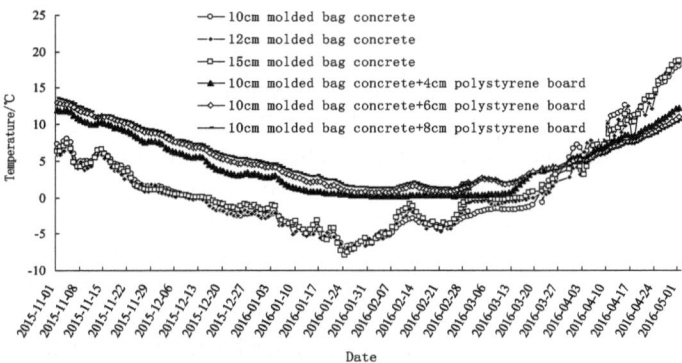

Fig. 2. Curve of daily average temperature variation of concrete treated with different mould bags in 2015~2016

As can be seen from the Fig.above, the daily mean temperature of the six treatments decreased first and then increased with time. The changes of the three treatments of 10 cm, 12 cm and 15 cm mold bag concrete are basically the same, indicating that the increase of the thickness of mold bag concrete does not significantly increase the insulation effect. Under the concrete condition of 10 cm mold bag, the changes of the three treatments of laying 4 cm, 6 cm and 8 cm polystyrene boards are basically the same, indicating that the difference of daily average temperature between laying 4 cm, 6 cm and 8 cm polystyrene boards is also small.

The total accumulated temperature and the effect of increasing the temperature of the concrete with different thickness.The daily average temperature of concrete treatment of 6 different mold bags in 3 freezing-thawing periods in the test site was accumulated to obtain the total accumulated temperature value of each treatment, and the average total accumulated temperature and average warming effect value of each treatment were calculated, and the average total accumulated temperature and average warming effect of concrete treatment of different mold bags were drawn, as shown in Fig. 3.

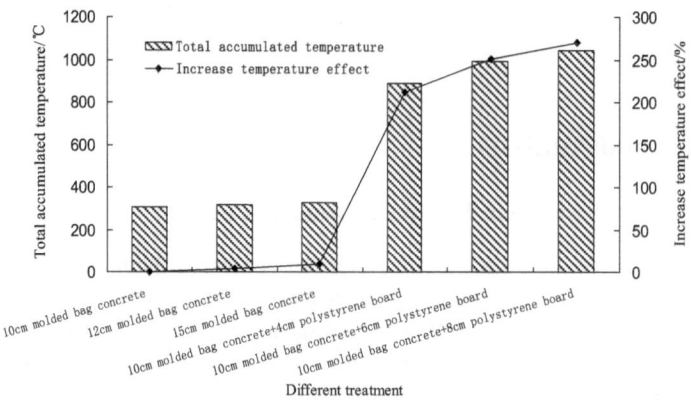

Fig. 3. Curve of Total accumulated temperature and increasing temperature effect of different Mould bags

As can be seen from Fig. 3, the average total accumulated temperature of concrete in mold bags with no insulation treatment in 10 cm, 12 cm and 15 cm is 278.35 °C, 289.30 °C and 304.01 °C, and there is no significant difference between them. The average total accumulated temperature of concrete in the three types of treatment with insulation board was 856.29 °C, 955.10 °C and 1036.17 °C, respectively, which were significantly higher than those without insulation. On the basis of 10 cm mold bag concrete, add 2~5 cm mold bag concrete, the heating effect is 3.93%~9.22%, and laid 4~8 cm polystyrene board, the heating effect is 207.63%~272.25%, indicating that mold bag concrete does not have the heat preservation effect, and the heat preservation board has a good insulation effect.

Effect of thermal insulation plate on soil temperature at different depths. The average ground temperature distribution diagram of 0~50 cm soil layer without heat preservation (comparison section) is shown in Fig. 4.

According to the analysis of the ground temperature at different depths, with the increase of the thickness of the insulation board, the ground temperature of the 0–10 cm soil layer on the surface increased less, indicating that the laying of the insulation board had less influence on the ground temperature of the 0–10 cm soil layer on the surface. With the increase of the thickness of the insulation board, the ground temperature of the 10–40 cm soil layer changed greatly, especially without the insulation treatment and the laying of 4 cm polystyrene board, the ground temperature of the 10–40 cm soil layer changed most significantly. The results showed that the insulation board laid on the

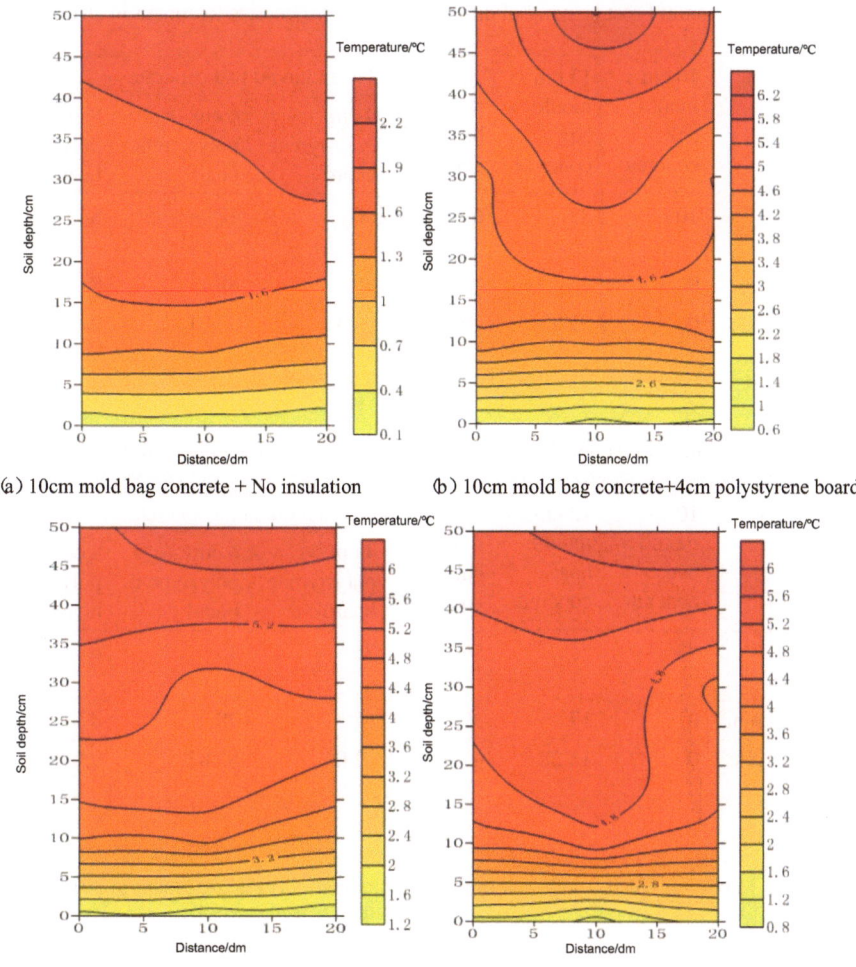

(a) 10cm mold bag concrete + No insulation   (b) 10cm mold bag concrete+4cm polystyrene board

(c)10cm mold bag concrete+6cm polystyrene board   (d)10cm mold bag concrete+8cm polystyrene board

Fig. 4. Soil temperature distribution map of different treatments

concrete foundation of mold bag had the greatest influence on the ground temperature of 10–40 cm soil layer.

3.2 Concrete Treatment of Different Mold Bag Anti - frost Heaving Effect

The maximum frost heaving capacity and the reduction of frost heaving capacity are handled by different mold bags of concrete. Table 2, 3 and 4 shows the maximum frost heave and reduction rate of concrete treated with different mold bags during 3 complete freezing-thawing periods from 2015 to 2018.

As can be seen from the above Table, the maximum frost heave of the three treatments of mold bag concrete + insulation board are all below 3 cm, while the frost heave of the

Table 2. Maximum Frost heaving of different mold bags concrete treatment in 2015~2016

Treatment of mould bag concrete	10 cm mould bag concrete	12 cm mould bag concrete	15 cm mould bag concrete	10 cm mould bag concrete		
				4 cm polystyrene board	6 cm polystyrene board	8 cm polystyrene board
Maximum value/mm	105	82	63	30	17	8
Reduction/mm	0	23	42	75	88	97
Reduction rate (%)	0	21.90	40.00	71.43	83.81	92.38

Table 3. Maximum Frost heaving of different mold bags concrete treatment in 2016~2017

Treatment of mould bag concrete	10 cm mould bag concrete	12 cm mould bag concrete	15 cm mould bag concrete	10 cm mould bag concrete		
				4 cm polystyrene board	6 cm polystyrene board	8 cm polystyrene board
Maximum value/mm	93	76	54	26	15	6
Reduction/mm	0	17	39	67	78	87
Reduction rate (%)	0	18.28	41.94	72.04	83.87	93.55

Table 4. Maximum Frost heaving of different mold bags concrete treatment in 2017~2018

Treatment of mould bag concrete	10 cm mould bag concrete	12 cm mould bag concrete	15 cm mould bag concrete	10 cm mould bag concrete		
				4 cm polystyrene board	6 cm polystyrene board	8 cm polystyrene board
Maximum value/mm	103	69	45.9	11	6	3.5
Reduction/mm	0	34	57.1	92	97	99.5
Reduction rate (%)	0	33.01	55.44	89.32	94.17	96.60

three treatments of mold bag concrete without insulation board is 5.4 cm~10.3 cm. On the basis of 10 cm mold bag concrete, add 2~5 cm mold bag concrete, reduce the frost heave rate of 18.28%~55.44%; And laying 4~8 cm polystyrene board, the frost heave reduction reached 71.43%~96.6%. It can be seen that the laying of insulation board under the concrete mold bag can significantly reduce the frost heave of the foundation soil, and the thicker the laid insulation board is, the more frost heave will be reduced, and the frost heave effect of polystyrene board is significantly greater than that of concrete mold bag.

3.3   Variation Law of Freezing Depth of Concrete Treatment with Different Mold Bags

Variation law of freezing depth in concrete treatment of different mold bags. Figure 5 shows the freezing depth variation process lines of the subsoil treated with 10 cm mold bag concrete, 12 cm mold bag concrete, 15 cm mold bag concrete and 10 cm mold bag concrete +4 cm polystyrene board, 10 cm mold bag concrete +4 cm polystyrene board, 10 cm mold bag concrete +4 cm polystyrene board in the south of the test site during a complete freezing-thawing period from 2015 to 2016.

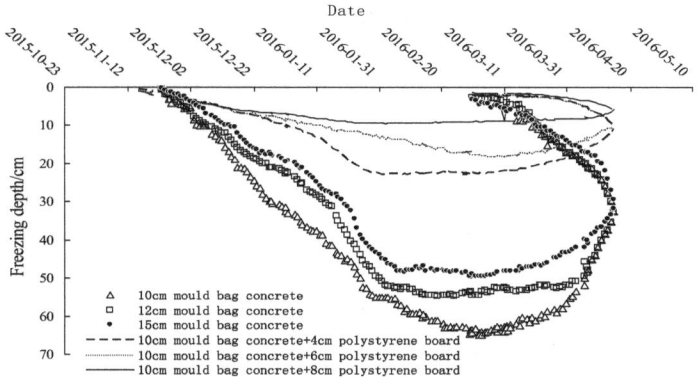

Fig. 5.   Freeze depth variation of different mold bag concrete treatment in 2015~2016

From the above analysis, it can be seen that during the complete freezing-thawing period from 2015 to 2016, the freezing depth of concrete treatment for each mold bag has the same change rule. All treated subsoil began to freeze in mid-November, reached its maximum depth in mid-late February, began to melt in early march, and melted in mid-April. The freezing depth of polystyrene board was significantly lower than that of concrete bag without heat preservation, and the freezing depth of concrete bag with 8 cm polystyrene board had the most gentle change, indicating that the freezing depth had the least change.If the concrete in mold bags of 10 cm is increased by 2~5 cm, the reduction rate of freezing depth is 17.74%~29.15%. Adding 4~8 cm polystyrene plate, the reduction rate of freezing depth reached 71.24%~85.08%. The results show that increasing the thickness of concrete in mold bag does not significantly reduce the freezing depth, while laying polystyrene board under the concrete in mold bag can significantly reduce the freezing depth of foundation soil.

Change law of frost heaving rate in different mold bag concrete treatment. Based on the analysis of the freezing depth and ground temperature data of different bag concrete

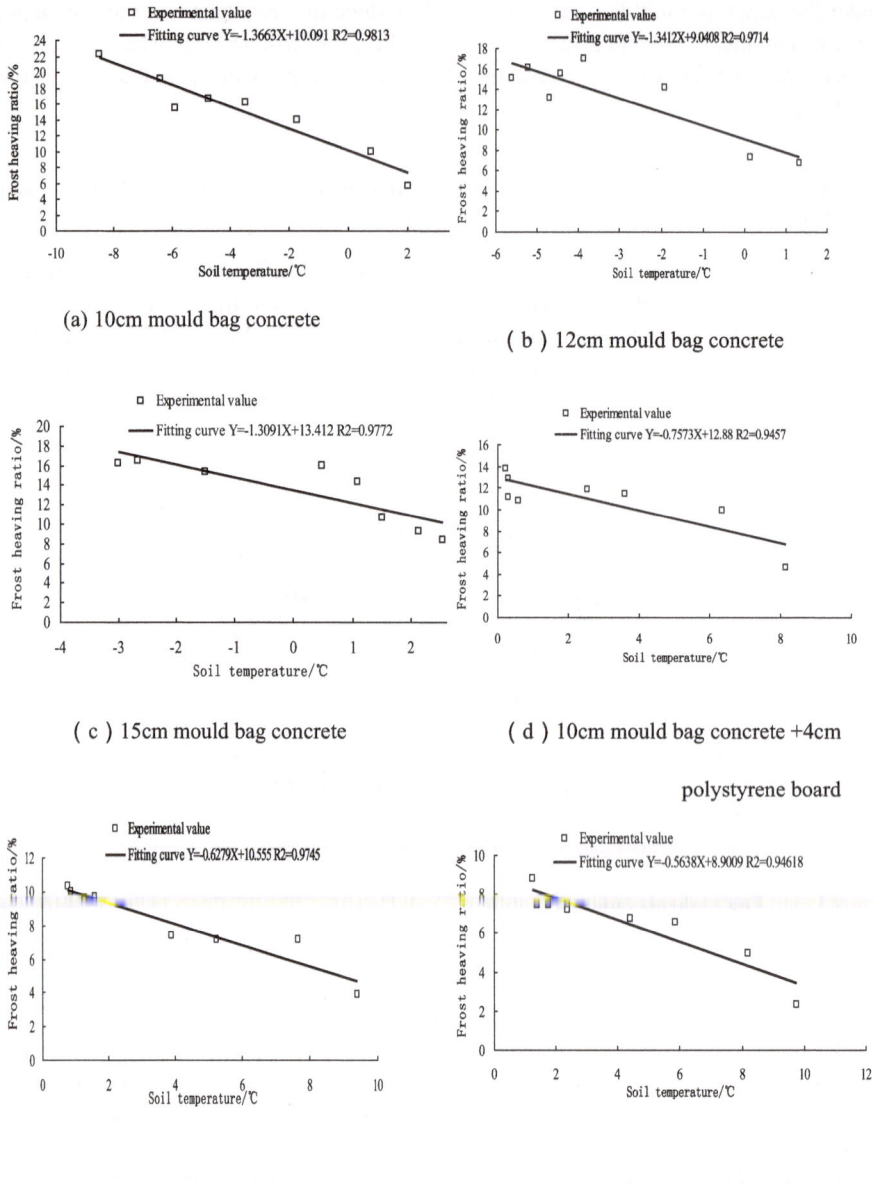

(a) 10cm mould bag concrete

(b) 12cm mould bag concrete

(c) 15cm mould bag concrete

(d) 10cm mould bag concrete +4cm

polystyrene board

(e)10cm mould bag concrete +6cm polystyrene board   (f)10cm mould bag concrete +8cm

polystyrene board

Fig. 6. Relationship between frost heave ratio and soil temperature in different mold bag concrete treatment

treatment from 2015 to 2016, the frost heave rate of each treatment in the process of frost heave and melting was calculated. The fitting curve of frost heave rate changing with temperature was fitted, as shown in Fig. 6.

As can be seen from Fig. 6, there is a linear relationship between frost heave rate of foundation soil and soil temperature under different mold bag concrete treatment, and the frost heave rate decreases with the increase of soil temperature. Under the condition of no insulation, the maximum frost heave rate of the 10 cm mold bag concrete is 22.2%. While the increase of 2–5 cm mold bag concrete, the maximum frost heave rate is 16% above, which decreases by 28%. The maximum frost heave rate of laying 4~8 cm polystyrene board treated was 14%~9%, which decreased by 38%~60%. It can be seen that the polystyrene board can significantly reduce the frost heaving rate of subsoil.

The slope of the curve fitting frost heave rate and soil temperature reflects the slowness of soil freezing process. The higher the absolute slope, the faster the freezing rate. When increasing 2–5 cm mold bags concrete on the foundation of 10 cm mold bags concrete, the absolute slope of the curve fitting the frost heave rate and soil temperature does not change. The results show that increasing the thickness of concrete does not delay the freezing process of foundation soil. When 4~8 cm polystyrene board was laid on the foundation of 10 cm mold bag concrete, the absolute slope of the curve fitting the frost heave rate and the soil temperature decreased by 44.6%~58.7%. The results showed that the change range of frost heave rate was small and the freezing duration was long after laying insulation board. That is, the laying of polystyrene board significantly slows down the freezing speed of the subsoil and prolongs the freezing duration, thus playing a thermal insulation effect.

4   Conclusion and Discussion

Laid bag concrete almost does not have the thermal insulation effect, the thermal insulation effect is far lower than the polystyrene board thermal insulation effect. Polystyrene insulation board is laid on the mold bag concrete, which has the greatest influence on the ground temperature in 10–40 cm soil layer.

The frost heaving reduction rate is only 18.28%~55.44% when 2~5 cm mould bag concrete is laid on the basis of 10 cm mould bag concrete, while the frost heaving reduction rate is 71.43%~96.6% when 4~8 cm polystyrene board is laid. The frost heaving reduction effect of polystyrene board is significantly greater than that of mould bag concrete. Increasing the thickness of the mold bag concrete does not significantly reduce the freezing depth, while laying polystyrene board under the mold bag concrete can significantly reduce the freezing depth of the foundation soil. Laying polystyrene board can significantly delay the freezing speed of foundation soil and prolong the freezing time, thus achieving the effect of heat preservation.

Acknowledgments. This study was funded by the Project of : Inner Mongolia Agricultural University Young Teachers Research Ability Enhancement Project (BR230136); Inner Mongolia Autonomous Region Science and Technology Plan Project (2022YFHH0088); Inner Mongolia Autonomous Region Natural Science Foundation Project (2023LHMS03056); Inner Mongolia Agricultural University High-level Talent Introduction Scientific Research Start-up Project (NDYB2020-19); Ordos City Science and Technology Major Project (2022EEDSKJZDZX012-3).

References

1. Li, Y.T.: Testing of channel mechanical properties of active duty lining mold-bag-concrete in large irrigation areas. China Rural Water Conserv. Hydropower **1**, 105–108 (2016)
2. Wang, J., Huo, Y.Z.: Study on mix proportion optimization and performance of formwork bag concrete. Sichuan Build. Mater. **45**, 7–9 (2019)
3. Zhang, H.L., Huo, Y.Z., Guo, Y.F.: Numerical simulation study on influencing factors of roughness of formwork bag concrete channel. Yellow River **41**, 157–160 (2019)
4. Gong, J.B.: Application of formwork bag concrete in channel construction of Qixing canal in Songsha section. Hous. Real Estate **31**, 134 (2018)
5. Wang, J., Guo, Y.F.: Analysis of macroscopic mechanical properties of formwork bag concrete under freeze-thaw cycle test. Value Eng. **37**, 116–117 (2018)
6. Guo, J., Lou, Z.K.: Application and numerical simulation of polystyrene insulation board in concrete lining canal. Yangtze River **44**, 57–60 (2013)
7. Li, S.N.: Experiments and application of EPS planks to freeze proof and leak proof irrigation ditch. J. Tarim Univ. **17**, 53–55 (2005)
8. Cheng, M.J., Shen, L.G., Bu, F.H.: Research on application of polystyrene insulation board in anti-freezing expansion of lining channel. J. Irrig. Drain. **30**, 22–27 (2011)
9. Zhang, W.Z.: Application of polystyrene foam plastic board in channel anti-freezing expansion. J. Water Conserv. Constr. Eng. **1**, 56–58 (2003)
10. Zhao, B., Li, J.W., Meng, C.: Experimental study on evolution law of thermal insulation performance of channel insulation materials. J. China Acad. Water Resour. Hydropower Sci. **5**, 28–33 (2017)
11. Wang, X.P., Cheng, Q.M., Zhang, Y.X.: Testing and analysis of thermal insulation performance of polystyrene foam for building. Plast. Technol.. Technol. **37**, 65–67 (2009)
12. Song, L.O.Y.F., Yu, S.C.: Checking calculation of frost heave and frost heave resistance of concrete anti-seepage channel in winter water delivery operation. Trans. Chin. Soc. Agric. Eng. **31**, 114–120 (2015)
13. Yin, Y.Z., Jia, Q., Zhao, Q.: Influence of water content on insulation durability of channel insulation board and thickness design of insulation board. Water Resour. Hydropower Eng. **2**, 170–174 (2017)
14. Zhao, B., Li, J.W., Men, C.: Experimental study on evolution law of thermal insulation performance of channel insulation materials. J. China Acad. Water Resour. Hydropower Sci. **5**, 28–33 (2017)
15. Guo, Z.H., Zhang, X.Q., Zhang, D.C.: J. Experimental study on full stress-strain curve of concrete. J. Archit. Eng. **3**, 1–12 (1982)

Numerical Analysis of High-Performance Reinforced Concrete Columns Under Eccentric Loading

Yuxuan Peng, Dezhang Sun$^{(\boxtimes)}$, and Junwu Dai

Institute of Engineering Mechanics, China Earthquake Administration, Key Laboratory of Earthquake Engineering and Engineering Vibration of China Earthquake Administration, Harbin 150080, China
ppp112517219@163.com

Abstract. The article is based on the eccentric compression test of HRB650E reinforced high-strength concrete columns. ABAQUS finite element software is used to conduct finite element analysis on HRB650E reinforced high-strength concrete column specimens. Comparing the simulation results with the experimental results, it can be found that the two fit well. By observing the patterns of experimental and simulation results, it can be concluded that the bearing capacity of HRB650E grade reinforced high-strength concrete columns is good, and improving the concrete strength can enhance the bearing capacity of the specimens.

Keywords: Eccentric Compression Test · HRB650E · Finite Element Analysis · ABAQUS

1 Introduction

In recent years, with the development of the economy, a large amount of infrastructure construction has consumed a huge amount of steel. The rapid development of steel smelting technology has laid the foundation for the research and promotion of high-strength steel bars. High strength steel bars can effectively reduce the use of steel, reduce costs, save resources, and maintain stable mechanical properties. Developed countries such as Europe and America have used over 95% of 400-600MPa steel bars, and countries such as Germany, Japan, and Australia have also achieved an adoption rate of 80–90% for steel bars above 400MPa [1, 2].

In 1996, Lloyd and other scholars conducted a study on the eccentric compression performance of 400MPa reinforced concrete specimens, focusing on the influence of factors such as longitudinal strength and stirrup strength on the stress performance of specimens. Based on the experimental results, a constitutive model of high-strength concrete and a calculation theory of high-strength steel reinforced concrete eccentric compression columns were established. The results indicate that the calculation theory is reasonable and the calculation results are relatively accurate [3].

In 2005, Tan and other scholars conducted axial and eccentric compression tests on 30 high-strength concrete column specimens, focusing on the influence of concrete

© The Author(s) 2024
G. Feng (Ed.): ICCE 2023, LNCE 526, pp. 441–452, 2024.
https://doi.org/10.1007/978-981-97-4355-1_41

strength and volume stirrup ratio on the mechanical performance of the specimens. The stirrup strength is divided into two types: 455MPa and 636MPa, and longitudinal bar strength is 595MPa. Based on the experimental results, researchers have proposed a reliable, simple, and fast calculation method for obtaining concrete stress [4].

From 2004 to 2009, Liu lixin and other scholars carried out a series of studies on HRB500 reinforced concrete members. The results show that the stress characteristics of HRB500 reinforced concrete members are similar to those of ordinary reinforced concrete members, and can improve the performance in terms of deflection and bearing capacity compared with ordinary reinforced concrete members [5–9]. Li hongyan carried out eccentric compression test on 500MPa high-strength reinforced concrete members to verify whether the domestic concrete code at that time was applicable to the calculation of concrete members in the test. The results show that the failure characteristics of 500MPa reinforced concrete members under eccentric compression are basically similar to those of ordinary concrete members, and the bearing capacity of the specimens calculated in accordance with design code for concrete structure (GB50010-2002) [10] is in good agreement with the measured bearing capacity, It is proved that the calculation method of ultimate bearing capacity of eccentrically loaded members specified in design code for concrete structure (GB50010-2002) [10] is suitable for 500MPa reinforced concrete members [11].

At present, the reinforcement above HRB600 level has not been included in design code for concrete structure (GB50010-2010) [12] in China, and the research on the specimens of high-strength reinforcement combined with high-strength concrete is relatively small. For the reinforcement with strength of 600MPa and above, Luo Shaohua conducted eccentric compression test research on 11 600MPa reinforced concrete columns, and proposed the design value of 600MPa reinforcement [13]; Zhang Jianwei and other scholars have studied the bending, axial compression and eccentric compression properties of HRB600 reinforced high-strength concrete specimens [14–16]; Rong Xian and other scholars carried out a series of tests on the axial compression performance, eccentric compression performance, bond anchorage, etc. of HRB600 and HRB600E reinforced high-strength concrete specimens [17–20]; Liu Chengtao studied the axial compression and eccentric compression performance of HRB600 reinforced concrete columns, focused on the applicability of crack calculation formulas in different national codes, and combined with finite element analysis, studied the influence of Longitudinal bar strength, reinforcement ratio and other factors. The results showed that the specimens with 600MPa high-strength reinforced concrete had better performance in terms of ultimate bearing capacity, deflection and so on, The failure characteristics are similar to those of ordinary reinforced concrete specimens [21]. Li Zhipeng studied the axial compression and eccentric compression properties of HRB635 reinforced high-strength concrete specimens [22]; Lin Wei and other scholars have studied the eccentric compression performance and bearing capacity of HRB635 grade hot-rolled ribbed high-strength reinforced concrete short columns, and combined with finite element analysis, put forward a simplified calculation formula for evaluating the eccentric compression ultimate bearing capacity of HRB635 grade hot-rolled ribbed high-strength reinforced concrete short columns [23].

At present, there is no research on the mechanical properties of 650MPa steel bars in China. Based on the eccentric compression performance test of HRB650E reinforced high-strength concrete columns, this paper uses ABAQUS to carry out finite element analysis, and compares the test results with the experimental results to study the variation of eccentric compression performance of specimens with the parameters such as steel bar strength, concrete strength, stirrup ratio and eccentricity.

2   Experimental Overview

2.1   Experimental Design

Five HRB650E reinforced high-strength concrete columns and two HRB400E reinforced high-strength concrete columns were designed, including two small eccentric compression specimens and five large eccentric compression specimens. The number and main parameters of the test piece is presented in Table 1, and the size and reinforcement arrangement of the specimens is presented in Fig. 1. The size of each specimen is the same, and the concrete grade, eccentricity, longitudinal bar strength, reinforcement ratio and eccentricity are the main variation parameters of the specimen.

Table 1. The main design parameters of eccentric compression specimens.

Specimen number	Designed concrete grade	Eccentricity e/mm	Longitudinal bar	Distributed steel[a]	Stirrup	Stirrup ratio
XP1	C55	100	HRB 400E	12D22	8@100	0.912%
DP2	C50	240	HRB 650E	12D22	8@100	0.912%
DP3	C55	240	HRB 650E	12D22	8@100	0.912%
DP4	C60	240	HRB 650E	12D22	8@100	0.912%
DP5	C55	240	HRB 650E	12D22	8@150	0.608%
DP6	C55	240	HRB 400E	12D22	8@100	0.912%
XP7	C55	100	HRB 650E	12D22	8@100	0.912%

[a] 12D22 refers to 12 longitudinal bars with a diameter of 22 mm.

2.2   Test Setting and Loading Scheme

Figure 2 shows the layout rules for strain measuring points. The middle section of the test piece is segmented into three layers, denoted as A, B, and C, with a 250 mm spacing between them. For the longitudinal bar, four strain measuring points are positioned in a clockwise manner at the corners of each layer. Additionally, four strain measuring points are situated at the four sides of the outer stirrup, as well as at the midpoint of the two long sides of the inner stirrup, resulting in a total of eight strain measuring points for

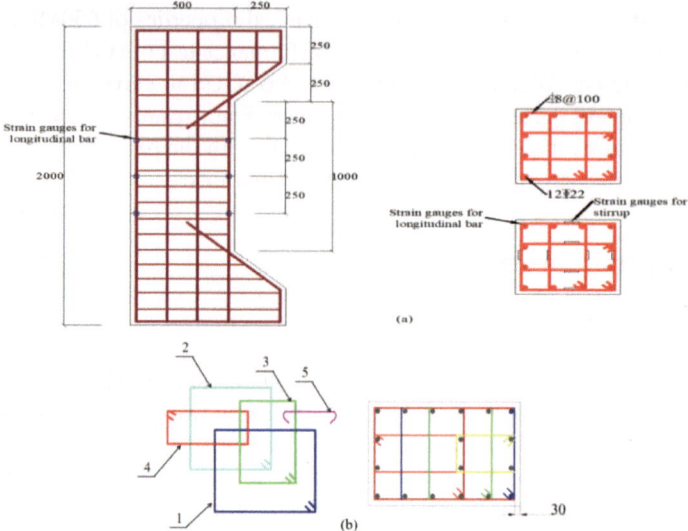

Fig. 1. Specimen size and reinforcement layout.

the stirrup. Regarding the longitudinal bar, Z-1 and Z-4 are situated on the compression side, while Z-2 and Z-3 are located on the tension side.

Symmetrically arranged at the middle height of the compression side and two sides of the concrete outer surface of the specimen, with a 125 mm spacing, are three strain measuring points. Furthermore, two strain measuring points are symmetrically arranged on the tension surface, with a 250 mm spacing. In total, there are 11 concrete strain measuring points.

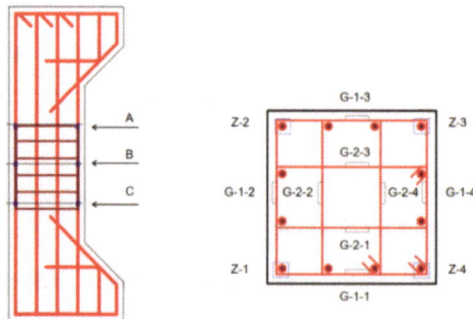

Fig. 2. Number of strain gauges for longitudinal bars and stirrups of the specimens.

For displacement measurement, five displacement meters are symmetrically arranged along the height direction at the tensile side of the central test piece, with a 200 mm spacing between them to measure the deflection of the test piece.

The test employs a 40000 kN multifunctional loading system for loading, with the vertical load value being directly read by the loading system. The loading scheme is formulated in accordance with the standard for test methods of concrete structures (GB50152-2012) [24]. Prior to the test, a steel base plate is placed at the lower support, and the test piece is lifted using a crane. After lifting, a steel base plate with the same size is placed on the upper part of the test piece. During the lifting and placing of the steel base plate, the eccentric position of the test piece and the center of the steel base plate are positioned on the connecting line between the center of the upper and lower supports. The laser broom is used to ensure that the connecting line is vertical.

Before the formal start of the test, preloading is carried out to clear the gap between the test piece and the loading part. After the formal loading commences, the large bias test piece is loaded to 500 kN, while the small bias test piece is loaded to 1000 kN. Subsequently, the load of each level is increased by 500 kN, with a load holding time of 60 s. After loading to approximately 50% of the ultimate bearing capacity, the load of each level of the large bias test piece is reduced to 300 kN, while the load of each level of the small bias test piece is reduced to 500 kN, and the load holding time is increased to 90 s. Upon reaching the peak value, the test switches to displacement control. The experiment is terminated when the bearing capacity drops to 85% of the peak value or when the component is severely damaged.

2.3   Mechanical Properties of Steel Bars and Concrete

The mechanical properties of reinforcement measured by the tensile test is presented in Table 2.Three grades of concrete (C50, C55, and C60) were used in the test. Simultaneously with making the test pieces, three samples of each grade were cast with the same concrete, resulting in a total of nine standard cube concrete test pieces, consistent with the curing conditions of the test pieces. The test is conducted in accordance with the standard for test methods for mechanical properties of ordinary concrete (GB50081-2002) [25], and the average compressive strength f_{cu}^0 of concrete is measured. Based on the formula (1) in Sect. 4.1 of the article description of design code for concrete structure (GB50010-2010) [12], the standard value of compressive strength $f_{cu,k}$, the standard value of axial compressive strength f_{ck}, and the standard value of axial compressive strength f_c of cubic concrete are calculated. The mechanical properties of concrete materials is presented in Table 3. The actual measured concrete strength in the test is different from the expected, and the measured compressive strength of C55 and C60 is relatively small.

$$\begin{cases} f_{cu,k} = f_{cu}^0 - 1.645\sigma \\ f_{ck} = 0.88\alpha_{c1}\alpha_{c2}f_{cu,k} \\ f_c = f_{ck}/1.4 \end{cases} \tag{1}$$

Table 2. Mechanical properties of reinforced bars.

Reinforcement grade	d/mm	F_y/MPa	F_u/MPa
HRB400E	22	445	625
HRB650E	22	675	878
HRB400	8	440	650

Table 3. Mechanical properties of concrete.

Design concrete strength grade	f_{cu}^0/MPa	$f_{cu,k}$/MPa	f_{ck}/MPa	f_c/MPa
C50	57.15	52.1	33.73	23.26
C55	55.83	50.8	32.74	22.58
C60	58.60	53.6	34.40	23.72

3   Establishment of Finite Element Model

3.1   Constitutive Relation

Constitutive Relation of Reinforcement. The constitutive relation of reinforcement adopts the bilinear model. The two straight lines are used to represent the elastic section and yield strengthening section of the reinforcement.

Constitutive Relation of Concrete. The CDP model is used in this simulation test, and the parameters in the CDP model are determined by using the concrete constitutive relationship provided in code for design of concrete structures (GB50010-2010) [12]. The stress strain curve of the model can be determined according to the following formulaı

$$\sigma = (1 - d_t)E_c\varepsilon$$
$$d_t = \begin{cases} 1 - \rho_t[1.2 - 0.2x^5] & x \le 1 \\ 1 - \frac{\rho_t}{\alpha_t(x-1)^{1.7}+x} & x > 1 \end{cases}$$
$$x = \frac{\varepsilon}{\varepsilon_{t,r}}$$
$$\rho_t = \frac{f_{t,r}}{E_c\varepsilon_{t,r}}$$

(2)

d_t is the uniaxial damage evolution parameter of concrete, α_t is the parameter value of the descending section of the concrete uniaxial tensile stress-strain curve, $f_{t,r}$ is the representative value of uniaxial tensile strength of concrete, $\varepsilon_{t,r}$ is the peak tensile strain of concrete corresponding to the representative value $f_{t,r}$ of uniaxial tensile strength.

$$\sigma = (1 - d_c)E_c\varepsilon$$

$$d_t = \begin{cases} 1 - \frac{\rho_c n}{n-1+x^n} & x \leq 1 \\ 1 - \frac{\rho_c}{\alpha_c(x-1)^2+x} & x > 1 \end{cases}$$

$$\rho_c = \frac{f_{c,r}}{E_c\varepsilon_{c,r}}$$

$$n = \frac{E_c\varepsilon_{c,r}}{E_c\varepsilon_{c,r}-f_{c,r}}$$

$$x = \frac{\varepsilon}{\varepsilon_{c,r}}$$

(3)

d_c is the damage evolution parameter of concrete under uniaxial compression, α_c is the parameter value of the descending section of the stress-strain curve of concrete under uniaxial compression, $f_{c,r}$ is the representative value of uniaxial compressive strength of concrete, $\varepsilon_{c,r}$ is the peak compression strain of concrete corresponding to the representative value $f_{c,r}$ of uniaxial compression strength.

The symbols and parameter values provided in the specifications align with their intended meaning. The measured values are substituted into the formula to determine the constitutive relationship, which is subsequently input into ABAQUS for analysis.

3.2 Modeling

The numerical simulation of the quasi-static test was conducted using ABAQUS with a finite element analysis model, and the results were compared with the experimental test results. The finite element model of the reinforced high-strength concrete column was established, as depicted in Fig. 3. The concrete column was represented using solid elements, while the reinforcement cage was modeled using truss elements. Additionally, a steel pad was included at the top area of the concrete column to match the actual test setup. The boundary condition at the bottom surface of the column constrained the displacement in XYZ directions. Vertical displacement was applied at a specific location on the top surface to simulate the loading conditions.

Fig. 3. The model of concrete columns and steel reinforcement cage.

4 Numeric Simulation Results and Analysis

4.1 Comparison Between Simulation Results and Test Results

The load-displacement curve obtained from ABAQUS numerical simulation is illustrated in Fig. 4. Using the test piece DP3 as a case study, a comparison between the finite element simulation results and the observed failure behavior is depicted in Fig. 5. It can be observed that the Mises stress in the upper section of the mid-span is notably high. This high Mises stress is near the compressive strength of the concrete when it is crushed, aligning with the observed behavior in the actual tests. Upon reaching the peak

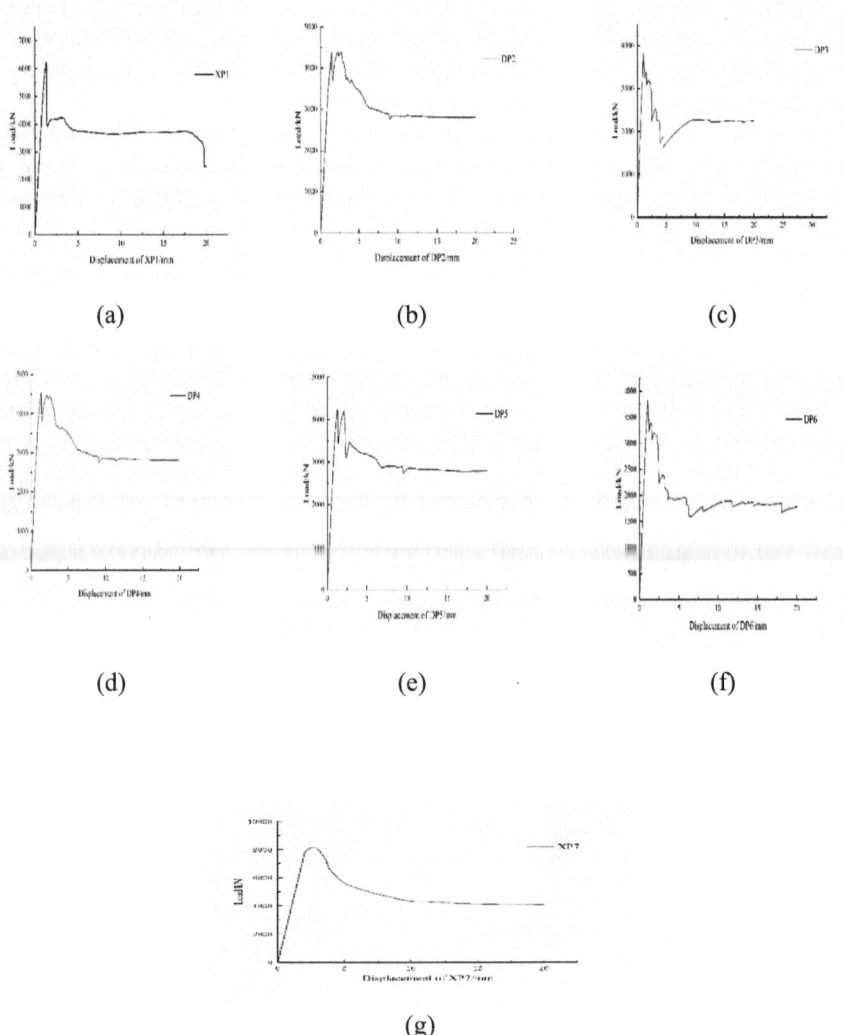

Fig. 4. Load-displacement curve.

load, a significant concrete collapse initially occurs in the upper section of the mid-span, subsequently spreading to both sides of the lower section. A comparison between the peak load simulation result, N_m, and the actual peak load result, N_u, is illustrated in Table 4. N_m is calculated using the measured data. The average ratio of N_u/N_m is 1.001, with a variance of 0.085. This suggests a strong agreement between the simulation and the actual results.

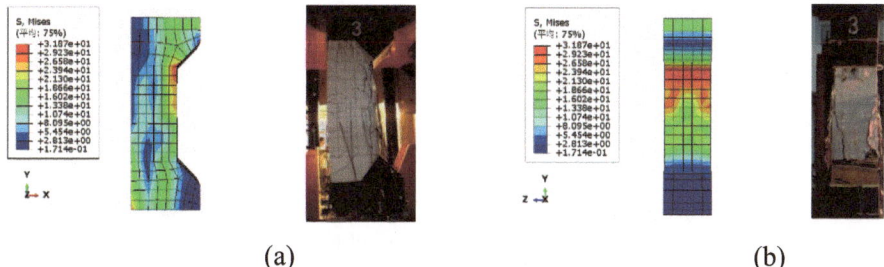

(a) (b)

Fig. 5. Comparison between simulation results and experimental results of specimen failure phenomenon using DP3 as an example.

Table 4. Comparison between the actual value of peak load N_u and the simulated value of peak load N_m

Specimen number	N_u/kN	N_m/kN	N_u/N_m
XP1	6697	6227	1.075
DP2	4167	4382	0.9509
DP3	4135	3833	1.0788
DP4	4085	4555	0.8968
DP5	3868	4237	0.9129
DP6	3807	3832	0.9934
XP7	9005	8170	1.1022

4.2 The Influence of Concrete Strength

In addition, ABAQUS was utilized to simulate the conditions outlined in Table 5, which involved concrete grades C70 and C80, represented as test pieces KZ1 and KZ2. The concrete strength, derived from the standard value of concrete compressive strength specified in the Code for Design of Concrete Structures (GB50010-2010) [12], was integrated into the CDP model for calculation purposes, resulting in the determination of the concrete constitutive relationship. The load-displacement curve obtained from ABAQUS is illustrated in Fig. 6. A comparison between the actual and simulated peak load values for test pieces KZ1, KZ2, and DP2 to DP4 is provided in Table 5. It is important to note that the differences in concrete strength among DP2, DP3, and DP4 are not significant. However, upon the inclusion of KZ1 and KZ2, the data presented in Table 6 unequivocally demonstrates that increasing the concrete strength enhances the load-bearing capacity of the concrete columns.

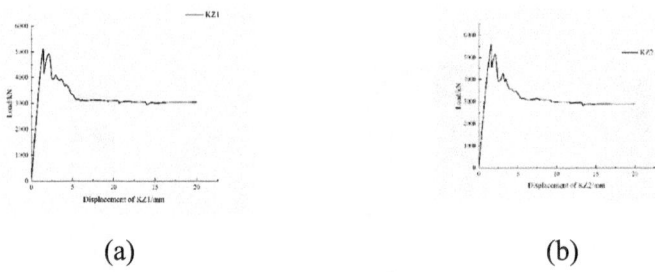

(a)　　　　　　　　　　　　　　　　　　　(b)

Fig. 6. Load-displacement curve of KZ1 and KZ2.

Table 5. Design parameters of specimens with concrete of C70 and C80.

Specimen number	Designed concrete grade	Eccentricity e/mm	Longitudinal reinforcement	Reinforcement mode	Stirrup reinforcement mode
KZ1	C70	240	HRB 650E	12D22	8@100
KZ2	C80	240	HRB 650E	12D22	8@100

Table 6. Comparison of peak loads of KZ1 and KZ2 with DP2-4

Specimen number	f_{ck}/MPa	N_u/kN	N_m/kN
DP2	33.73	4167	4382
DP3	32.74	4135	3833
DP4	34.40	4085	4555
KZ1	44.5	/	5111
KZ2	50.2	/	5589

5　Conclusion

The simulation results for the bearing capacity of HRB650E reinforced high-strength concrete columns closely align with experimental findings, demonstrating a strong fitting effect.

The finite element analysis model of HRB650E reinforced high strength concrete column is established by using ABAQUS. Numerical simulation analysis was conducted using test data, and the results were compared with experimental findings. The analysis revealed that increasing concrete strength extends the working condition and significantly enhances the bearing capacity of HRB650E reinforced high-strength concrete columns, as evidenced by the comparison of experimental and simulated data.

References

1. Zhai, Z.Q.: J. Shanxi Archit. **39**(25), 102–3 (2013)
2. Fan, Z., Xu, L., Feng, Y.: J. Struct. Eng. **29**(06), 169–76 (2013)
3. Loyd, L., Anne, N.: J. ACI Struct. J. **93**(6), 631–8 (1996)
4. Tan, T.H., Nguyen, N.B.: J. ACI Struct. J. **102**(2), 198–205 (2005)
5. Liu, L.X., Zhang, Y.L., Li, Q.: J. Zhengzhou Univ. (Eng. Edit.) **27**(4), 1–5 (2006)
6. Mao, D.L., Liu, L.X., Fan, L.: J. Ind. Build. **12**(90), 67–9 (2004)
7. Liu, L.X., Li, H.Y., Zhang, Y.L.: J. Zhengzhou Univ. (Engineering Edition) 2007 02, 30–4 (2007)
8. Liu, L.X., Yu, Q.B., Wang, X.L.: J. Zhengzhou Univ. (Eng. Edit.) **29**(S1), 161–6 (2008)
9. Yu, Q.B., Liu, L.X., Hu, D.D.: J. Build. Struct. **39**(S1), 527–30 (2009)
10. GB 50010-2002 Specifications for Design of Concrete Structures S
11. Li, Y.H.: Experimental Study on Mechanical Properties of 500MPa Grade Reinforced Concrete Eccentric Compression Members 2007. Zhengzhou University
12. GB 50010-2010 S. Specifications for Design of Concrete Structures
13. Luo, S.H.: Experimental Study on Mechanical Properties of 600MPa Grade Reinforced Concrete Eccentric Compression Members 2013. Southeast University
14. Zhang, J.W., Xia, D.R., Qiao, Q.Y.: J. Build. Struct. **40**(04), 74–80 (2019)
15. Zhang, J.W., Xia, D.R., Qiao, Q.Y.: J. Ind. Build. **47**(11), 77–83 (2017)
16. Zhang, J.W., Xia, D.R., Qiao, Q.Y.: J. Ind. Build. **47**(06), 6–12 (2017)
17. Rong, X., Du, H.X., Zhang, J.X.: J. Silicate Bull. **38**(01), 60–4 (2019)
18. Rong, X., Gao, Z.Y., Liu, C.T.: J. Build. Sci. **39**(07), 101–9 (2023)
19. Li, Q., Rong, X., Li, Y.Y.: J. Build. Struct. **46**(02), 8–11 29 (2016)
20. Zhang, J.X., Rong, X., Liu, P.: J. Build. Struct. **24**(82), 89–92 (2017)
21. Liu, C.T.: Experimental Research and Finite Element Analysis on Mechanical Performance of HRB600 Reinforced Concrete Compression Columns 2022. Hebei University of Technology
22. Li, Z.P.: Research On Axial Compression and Eccentric Compression Performance of Short Concrete Columns Confined with HRB635 Grade High-strength Steel Bars 2021. Hefei University of Technology
23. Lin, W., Shen, Q.H., Wang, J.F.: J. Hefei Univ. Technol. (Nat. Sci. Edit.) **46**(04), 500–11 (2023)
24. GB/T 50152-2012 Standard for Testing Methods of Concrete Structures S
25. GB/T 50081-2002 Standard for Standard for Test Methods for Mechanical Properties of Ordinary Concrete S

Hydraulic Characteristics of Undular Hydraulic Jumps Over Different Bed Roughness

Xinyu Lan, Jingmei Zhang, and Hang Wang[✉]

State Key Laboratory of Hydraulics and Mountain River Engineering, Chengdu, Sichuan, China
wanghang39@scu.edu.cn

Abstract. Occurrence of undular hydraulic jumps has impact on sediment transport and bank erosion in estuarine areas. The hydraulic properties vary when it forms on a rough bed, leading to modification of turbulent mixing processes. This paper reports on a study of undular hydraulic jumps generated with different types of bed roughness including smooth, rough rubber-matted and grated beds. The experiment was conducted with a flow rate of 241 m^3/h using acoustic displacement meters (ADMs) and an acoustic Doppler velocimeter (ADV). The obtained results reveal a discernible descending order in the undulations of the first three waves: smooth, grated, and rough rubber mat. In instances where roughness deviates to either extreme—being excessively small or large—the undulation of the first three waves intensifies. Through both free surface wave and turbulent flow field evolution experiments, it was determined that the roughness of the bed primarily influences the first wave, exerting the most pronounced impact on the crest of the initial wave. As the wave progresses, this influence gradually diminishes. Therefore, this article further posits that heightened bed roughness corresponds to an increased fluctuation in velocity at the bottom of the first peak, resulting in a weakened impact of toe jump oscillation on the flow field. Consequently, this diminishes the likelihood of negative velocity occurrences.

Keywords: Undular Hydraulic Jump · Free-Surface Undulation · Bed Roughness · Shallow Flow

1 Introduction

Unlike the classical hydraulic jump accompanied by free surface breaking and rolling, the undular hydraulic jump has a coherent and smooth free surface with subtle fluctuations, and this fluctuation will persist downstream over an extended distance. Undular hydraulic jumps generally occur when Froude numbers approach 1, which is a transitional state from supercritical flow to subcritical flow. Therefore, this special phenomenon can often be observed on coasts and estuaries.

Presently, the predominant focus in undular hydraulic jump research revolves around the examination of factors such as Froude numbers, aspect ratios, Reynolds numbers, and their collective impact on the hydraulic jump characteristics. In a study by Chanson and Montes (1995) [1], experiments were conducted within smooth rectangular channels,

© The Author(s) 2024
G. Feng (Ed.): ICCE 2023, LNCE 526, pp. 453–462, 2024.
https://doi.org/10.1007/978-981-97-4355-1_42

varying Froude numbers from 1.05 to 3.0. The investigation delved into the velocity, pressure, and energy distribution along the centerline longitudinal profile and specific positions in the transverse profile. Notably, within the Froude number range of 1.5 to 2.9, the wavelength and amplitude of free surface waves were identified as functions of both the Froude number and the aspect ratio (yc/W). Subsequently, Dunbabin (1996) [2] extended the exploration by investigating the velocity and pressure distribution upstream of the first crest under three distinct conditions: Fr = 1.41, 1.52, and 1.63.

Ohtsu et al. (2003) [3] divided undular hydraulic jumps into two types, I and II, based on the ratio H1/B, representing the upstream depth to river width. The primary distinction lies in whether the shock wave on the sidewall crosses at the first crest. At the same time, the variation of flow conditions with Reynolds number was delineated, and an analysis was conducted on the influence of Reynolds number on undular hydraulic jumps. In a study by Montes and Chanson (1998) [4], pitot tubes were employed for velocity and pressure measurements in the first three crests and two troughs, elucidating certain three-dimensional flow characteristics resulting from the development of lateral shock waves. Ohtsu et al. (1996) [5] found that shock waves were always observed before the first crest when Fr ≥ 1.2. When the intersection point of the shock wave is upstream of the first crest, the impact of the shock wave on the undular hydraulic jump is significant. If the intersection point is downstream, the impact of the shock wave can be ignored.

Chanson (1995a) [6] conducted experiments maintaining consistent upstream inflow conditions but varying sidewall roughness compared to Chanson and Montes (1995) [6]. The findings revealed distinguishable flow patterns from those established by the latter on smooth sidewalls. Pasha and Tanaka (2017) [7] mentioned that when vegetation is inserted into a rectangular flume, there will be a proportional decrease in speed and energy with higher vegetation density.

This article conducted two series of experiments on undular hydraulic jumps generated by different types of bed roughness (including smooth, rough rubber-matted, and grated beds) in a 0.3m wide channel, such as Fig. 1. The investigations were dedicated to studying the influence of bed roughness on both the free surface characteristics and the evolution of the turbulent flow field associated with undular hydraulic jumps. Non-intrusive measurements of the free surface were carried out using an acoustic displacement meter (ADM) along the centerline of the sink, while invasive velocity measurements were conducted using an acoustic Doppler velocimeter (ADV). This study demonstrates the main locations and causes of different turbulence characteristics induced by bed roughness, providing a theoretical basis for solving sediment deposition issue [8].

Fig. 1. Experimental phenomena of rough rubber-matted bed

2   Experimental Setup and Instrumentation

2.1   Facility and Flow Conditions

Experiments were conducted in a rectangular channel equipped with a circulating water supply system. As shown in the Fig. 2(a), the entire system comprises a high-level tank, a horizontal channel, and a circulation pool. Water is pumped into a high-level tank with dimensions of $2 \times 2 \times 2$ m^3 from the circulating pool. The flow is controlled by Siemens Micromaster 430 frequency converter and measured by LDG-200 electromagnetic flowmeter. At the same time, the flow is verified by a adjustable triangular overflow weir located at the tail of the flume, with a measurement accuracy of 2%. A movable baffle is positioned at the high-level tank's outlet to control the gate opening (inflow depth), while a triangular weir at the flume's tail regulates the toe position and tailwater depth. Rigorous checks are conducted on the tank's water level and gate opening to ensure uniform water flow into a horizontal channel measuring 12.5 m in length, 0.5 m in width, and 0.7 m in depth. Ultimately, the tail water is discharged into the circulating pool.

The experiment was conducted at $Q = 241$ m^3/h, Fr $= 1.75$, and Re $= 1.34 \times 105$, three sets of experiments were conducted on bed with different roughness. Nonintrusive measurements were performed using acoustic displacement meters on the free surface contour from the toe of the hydraulic jump to the end of the fourth wave, and invasive measurements were made using acoustic Doppler velocimeters on the velocity at the characteristic positions of the first two waves. The measurement cross-section selected for the experiment is the longitudinal section of the centerline of the flume. For recording convenience, a fixed point preceding the jump is designated as the starting point of the x-axis, with the bed denoted as $y = 0$. Figure 2(b) shows the geometric parameters measured in the experiments.

(a)

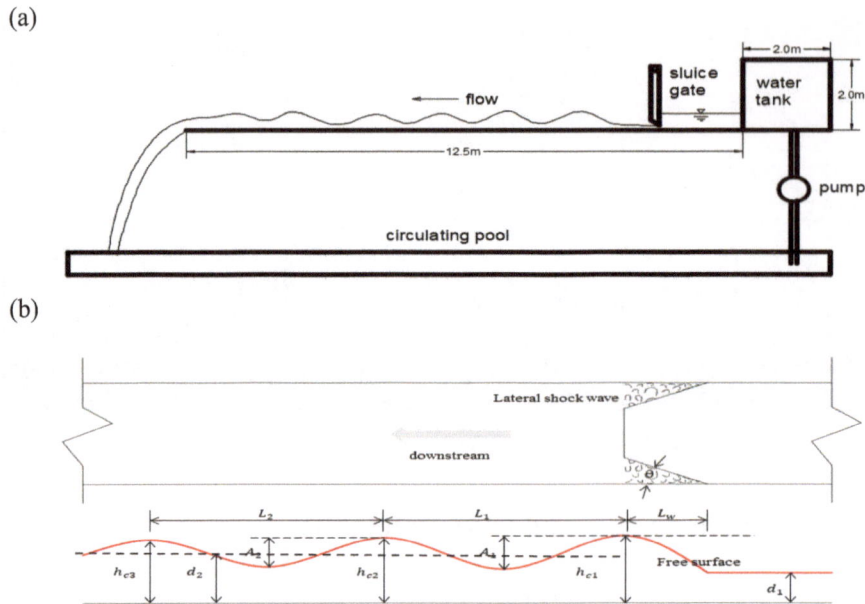

(b)

Fig. 2. Experimental facility and undular hydraulic jump flow: (a) experimental facility, (b) top view photo of flume and definition sketch of geometric parameters

2.2 Equivalent Roughness Height

The velocity profile shape in the boundary layer above the rough bed in the measured subcritical clear-water flow conditions was compared with the Law of the Wall in the log region, and the equivalent roughness heights corresponding to the rubber mat surface and grille surface were calibrated. Choose the roughness that best fits the theoretical curve. Compare the measured centerline free surface profile with the theoretical backwater calculation to verify the accuracy of roughness calculation. It is ultimately believed that the results of using the log-law estimation to characterize the two types of roughness are reliable. The results show that the the roughness of smooth bed is ks = 0.5 mm, the rubber bed is ks = 11.2 mm, and the grated-bed is ks = 18.4 mm.

3　Results: Free-Surface Characteristics

Our observation of general flow patterns shows in Fig. 1 and Fig. 3 indicates that under smooth bed conditions, transverse shock waves intersect at the first crest and do not intersect when the bed is rough.

The waveform in Fig. 4 shows the average free surface water depth from the toe of the hydraulic jump to the end of the fourth wave within 1 min. The horizontal axis represents the distance from the starting point of the hydraulic jump, and the vertical axis represents the water depth, both of which have been dimensionless. It can be seen that the waveform characteristics of the first two waves are the most obvious, and the fluctuation of the subsequent waves gradually weakens.

(a)   (b)

Fig. 3. Experiments under different bed roughness: (a) smooth bed (b) grated bed

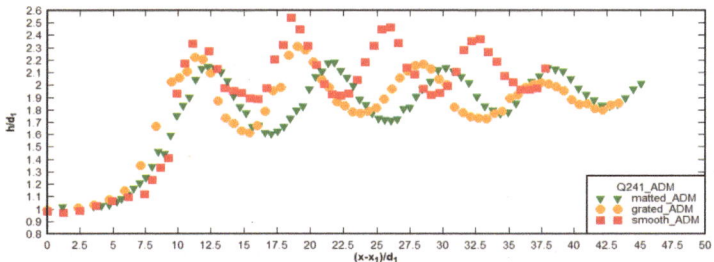

Fig. 4. Time-averaged free-surface profiles of undular hydraulic jump at centerline.

Figure 5 illustrates the water depths of the crests and troughs for the first four waves. Based on the waveform analysis, it is evident that the smooth bed exhibits the highest water depth at both the crest and trough of the wave. The grille and rough rubber mat show similar water depths at the trough, with slight variations at the crest. Specifically, at the first and second crests, the water depth of the grille is greater than that of the rough rubber mat. However, from the third crest onward, the water depth of the grille decreases sharply, only slightly surpassing that of the rough rubber mat, and eventually falling below the rough rubber mat at the fourth crest. This observation indicates that the waveform development of the grille is the most unstable, while the waveform of the rough rubber mat is the most stable. According to Fig. 5(a), the water depth of the second wave crest is the highest under all three operating conditions, and gradually decreases thereafter. In Fig. 5(b), the depth of the trough under smooth conditions is the most stable, while the grille is the least stable, showing a slow increasing trend overall. Across the first four waves, increased roughness corresponds to greater variation in trough depth as the waves develop. Simultaneously, as the depth of the wave crest gradually decreases, the trough depth gradually increases, and the undulating characteristics of the undular hydraulic jump gradually weaken. Comparing the Fig. 5(a) and(b), due to the difference in roughness, the difference in the dependent variable of the former is more significant, and the influence of roughness on the crest water depth is greater than that on the trough water depth.

(a)

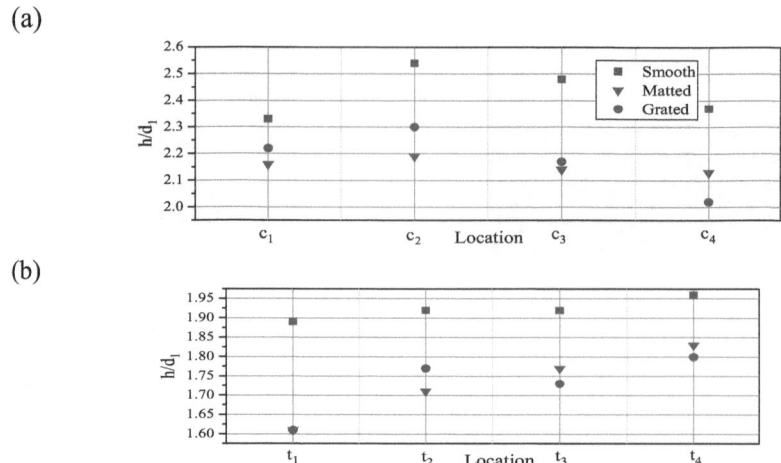

(b)

Fig. 5. The water depth at the first four wave characteristic positions: (a) the water depth at crest, (b) the water depth at trough

In Fig. 6, "L" represents the distance between two adjacent crests, for example, at first wave it represents the distance between the first and second crests, and the vertical axis represents the dimensionless crest distance value. In contrast to the pattern observed in crest and trough water depth, the distance between crests on a smooth bed is the smallest. Specifically, at the first crest distance, the rough rubber mat exceeds the grille, and in subsequent second and third crest distances, the grille exceeds the rubber mat. This pattern reaffirms that under grille conditions, the waveform characteristics of the first two waves are more pronounced. From the third wave onward, there is a significant increase in the distance between crests, a substantial decrease in crest depth, and a rapid weakening of waveform characteristics. This phenomenon may be attributed to the dissipation of most energy in the first two waves under grille conditions, leading to a notable attenuation of waveform characteristics at the onset of the third wave. Therefore, Fig. 6 also demonstrates that as the wave progresses downstream, the crest distance of the grille exhibits the most fluctuation. Additionally, the average value of crest distance is arranged in ascending order as follows: smooth, rough rubber mat, and grille.

Figure 7 illustrates the undulation of the first three waves, due to the significant difference in water depth between the upstream and downstream of the first wave crest, the calculation of the relief of the first wave crest has been improved. Across different roughness levels, the undulations of the first three waves follow a descending order: smooth, grille, and rough rubber mat. Notably, excessively small or large roughness intensifies the undulation of the first three waves, with the rough rubber mat exhibiting undulation between the other two and having the weakest undulation. The most significant difference in undulation occurs in the first wave, followed by a reduction in the disparity under various working conditions. The undulation of the rough rubber mat and grille becomes very close in the second and third waves.

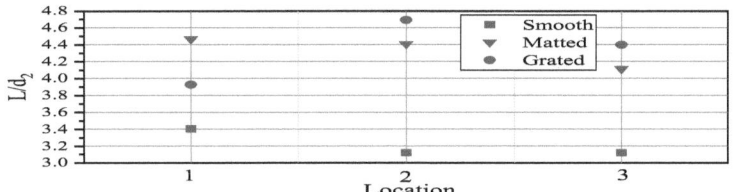

Fig. 6. Dimensionless crest distance of the first three waves: The abscissa represents the number of waves. L/d2 means the distance between the crests.

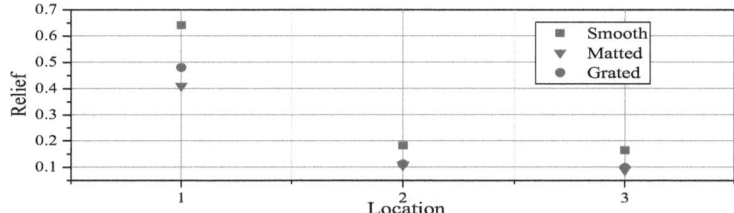

Fig. 7. Undulation of the first three waves. The horizontal axis represents the first, second, and third waves, and the vertical axis relief is defined as: 2Ai/Li. Furthermore, the undulation of the first crest is (A1 + A0)/2Lw

4   Results: Turbulent Velocity Characteristics

In the study of free surface wave characteristics, it was found that the waveform characteristics of the first two waves were the most prominent, and the turbulent flow field evolved more violently. After the second wave, the waves gradually weakened. In the examination of turbulent flow field evolution characteristics, the turbulent flow field information of the first two waves was explored, encompassing velocity, turbulence intensity, and turbulent kinetic energy at the crest and trough of the waves. Figure 8 depicts the distribution of velocity along the water depth at the crests and troughs of the first two waves. The smooth bed exhibits the smallest flow velocity and negative flow velocity appears near the bottom of the first wave crest. In contrast, the other two rough beds do not manifest negative flow velocity. This phenomenon is attributed to the minimal shear force and turbulence of the water flow at the bottom of the smooth condition, resulting in the emergence of a 'static water zone' at the bottom. With the oscillation of the jump toe, negative velocity appears at the bottom of the first wave crest. As the bottom roughness increases, the shear force of the bottom water flow intensifies, and turbulent kinetic energy rises (as depicted in Fig. 9), leading to the formation of numerous vortices in the static water zone. The streamline in the static water zone is no longer parallel to the free surface, and a small amount of mainstream may traverse the static water zone, disrupting the stability of the first crest bottom water flow. Consequently, the static water zone disappears, preventing the occurrence of negative velocity.

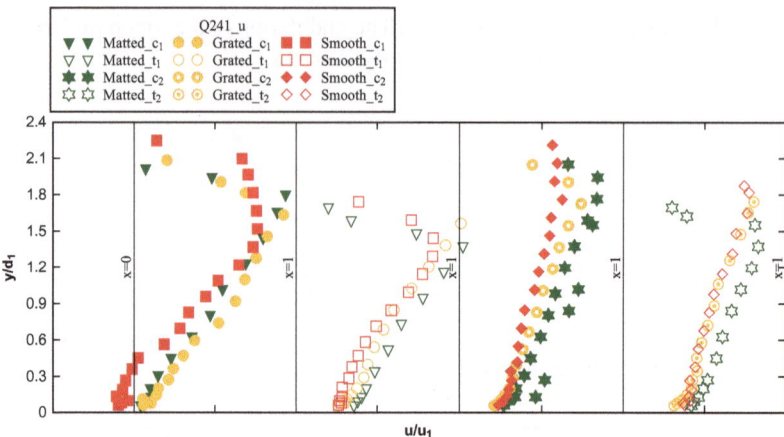

Fig. 8. Dimensionless longitudinal velocity distributions at different characteristic positions: c1, the first crest. t1, the first trough. c2, the second crest. t2, the second trough.

At the first crest, the velocity gradient near the surface of the rough bed is greater than that of the smooth bed. This observation may be attributed to the fact that under rough conditions, the shock wave intersects at the first crest, resulting in significant turbulence in the water flow near the free surface, as depicted in Fig. 9, with pronounced shear effects. Conversely, under smooth conditions, the shock wave does not intersect at the first crest. The flow velocities at the first and second crests exhibit a trend of increasing and then decreasing, and as the hydraulic jump develops, the gradient of flow velocity changes becomes smaller. This pattern aligns with observations made by Montes and Chanson (1998) in the velocity profile measured at the first crest of the undular hydraulic jump centerline. The decrease in speed near the water surface is attributed to surface rolling and breaking.

Figure 9 shows the distribution of turbulent intensity along the water depth. As the height increases, the turbulence intensity first increases and then decreases. Furthermore, as the hydraulic jump develops, this change diminishes gradually until it eventually disappears.

Roughness exerts a substantial impact on the first wave, with the most pronounced influence observed at the first crest. However, as it progresses to the second crest, the influence of roughness on turbulence intensity diminishes significantly, and the three curves closely approach overlapping. Across different roughness levels, the maximum turbulence intensity of the first two waves is situated in the lower half of the first wave crest. Specifically, at the bottom of the first crest ($y/d1 < 0.5$), the smooth bed exhibits the lowest turbulence intensity, while the grille displays the highest, indicating that greater bed roughness corresponds to increased velocity fluctuation at the bottom of the first crest.

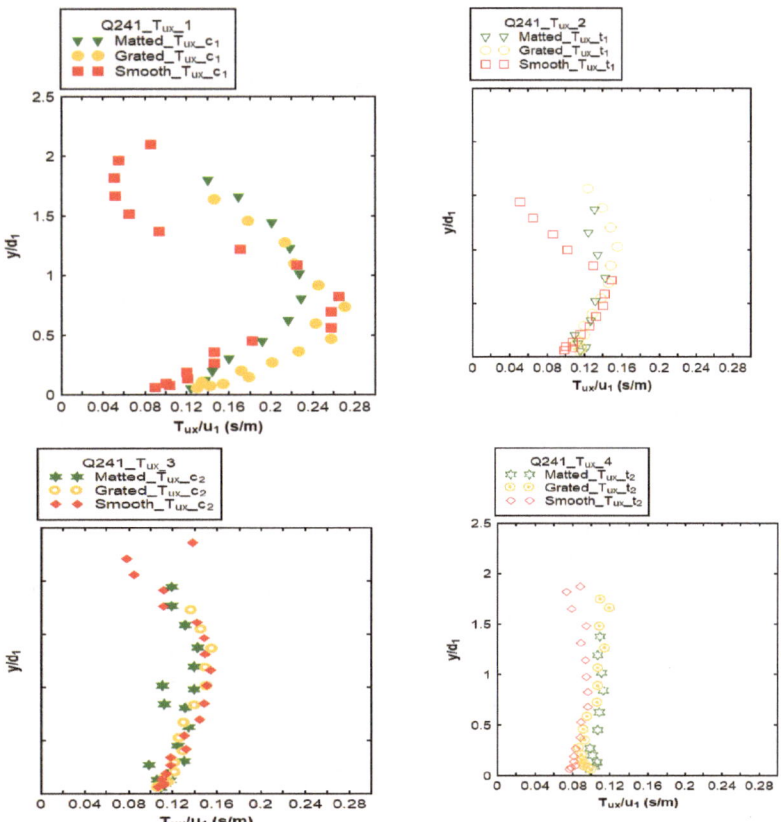

Fig. 9. Distribution of turbulent intensity in the mainstream direction along the water depth at the characteristic positions of the first two waves

5   Conclusion

An experimental study was conducted on a undular hydraulic jump in a 0.5m wide flume. The surface undulations of four standing waves were recorded. Under three different roughness conditions of smooth bed, rough rubber mat, and grille, different wave characteristics are displayed.

In the study of turbulent flow characteristics, the influence of roughness is mainly on the first wave, with the greatest impact on the first crest. The greater the roughness of the bed, the greater the fluctuation of the velocity at the bottom of the first crest.

At the first crest of the smooth bed, negative velocity was observed near the bottom, while the other two rough beds did not exhibit negative velocity. This phenomenon is attributed to the small shear force and turbulence of the water flow at the bottom of the smooth bed, resulting in the formation of a "static water zone" at the bottom. Simultaneously, with the oscillation of the toe, negative velocity emerges at the bottom of the first wave crest.

Under various roughness levels, the undulations of the first three waves follow a descending order: smooth, grille, and rough rubber mat. Excessive small or large roughness intensifies the undulation of the first three waves. The most substantial difference in undulation occurs in the first wave.

References

1. Chanson, H., Montes, J.S.: Characteristics of undular hydraulic jumps: experimental apparatus and flow patterns. J. Hydraul. Eng. **121**(2), 129–144 (1995)
2. Dunbabin, R.: Velocity Distributions within an Undular Hydraulic jump. University of Tasmania (1996)
3. Ohtsu, I., Yasuda, Y., Gotoh, H.: Flow conditions of undular hydraulic jumps in horizontal rectangular channels. J. Hydraul. Eng. **129**(12), 948–955 (2003)
4. Reinauer, R., Hager, W.H.: Non-breaking undular hydraulic jump. J. Hydraul. Res. **33**(5), 683–698 (1995)
5. Chanson, H.: Flow Characteristics of Undular Hydraulic Jumps. Comparison with Near-Critical Flows. Report CH45/95, Dept. of Civil Engineering, University of Queensland, Australia, June, p. 202 (1995a). (ISBN 0 86776 612 3)
6. Ohtsu, I., Yasuda, Y., Gotoh, H.: Discussion of Non-breaking undular hydraulic jumps, by R. Reinauer and W. H. Hager. J. Hydraul. Res. **34**(4), 567–572 (1996)
7. Pasha, G.A., Tanaka, N.: Undular hydraulic jump formation and energy loss in a flow through emergent vegetation of varying thickness and density. Ocean Eng. **141**, 308–325 (2017)
8. Montes, J.S., Chanson, H.: Characteristics of undular hydraulic jumps: experiments and analysis. J. Hydraul. Eng. **124**(2), 192–205 (1998)
9. Wang, H., Chanson, H.: Experimental study of turbulent fluctuations in hydraulic jumps. J. Hydraul. Eng. **141**(7), 04015010 (2015)
10. Hu, H., et al.: Free-surface undulation and velocity turbulence in shallow undular hydraulic jumps. Ocean Eng. **269**, 113566 (2023)
11. Liu, M., Rajaratnam, N., Zhu, D.Z.: Turbulence structure of hydraulic jumps of low Froude numbers. J. Hydraul. Eng. Hydraul. Eng. **130**(6), 511–520 (2004)

Comparative Analysis of Vibration Characteristics Testing for Different Pumped Storage Power Station Plants

Shengjun Wang[1], Lianghua Xu[2(✉)], Wei Xiao[3], and Guoqing Liu[2]

[1] State Grid Xinyuan Group Co., Ltd., Beijing 100032, China
[2] China Institute of Water Resources and Hydropower Research, Beijing 100048, China
`shepherd2008@126.com`
[3] Pumped Storage Technology and Economy Research Institute of State Grid Xinyuan Company Ltd., Beijing 100052, China

Abstract. The excitation force of the plant structural vibration of pumped storage power station mainly comes from hydraulic excitation, which is obviously related to the cascade combination formed by the number of runner blades and the number of movable guide vanes of the unit. The main frequency of the excitation force is closely related to the minimum pitch diameter number of the unit. The analysis of the vibration test data of the typical pumped storage power station plants showed that the structural vibration spectrum caused by the cascade combination unit with the minimum pitch diameter number equaled 1 or 2, generally manifests as a single main frequency and a large vibration amplitude of the plant. While, the structural vibration spectrum caused by the cascade combination unit with the minimum pitch diameter number equaled 4 is relatively dispersed, and the vibration amplitude of the plant is relatively low. The above research indicates that the vibration amplitude of the plant has a significant decrease by using the cascade combination unit with the minimum pitch diameter number equaled 4, which is very beneficial for the vibration reduction of the plant.

1 Introduction

Pumped storage is a green, low-carbon, clean and flexible regulating power supply with the most mature technology, the best economy and the most large-scale development conditions [1]. It has a good coordination effect with wind power, solar power, nuclear power and thermal power. Accelerating the development of pumped storage is an urgent requirement for building a new type of power system with new energy as the main body, is an important support for ensuring the safe and stable operation of the power system, and is an important guarantee for the large-scale development of renewable energy. Due to the long length of the diversion channel, the high pressure on the pressure side, and the design of the unit biased towards the runoff type, the vaneless area space between the runner inlet and the active guide vane is greatly limited. The vaneless area of the pump turbine forms a high-frequency and strong-amplitude dynamic and static interference pressure pulsation. The propagation, interference and superposition of the first and higher harmonics of the

© The Author(s) 2024
G. Feng (Ed.): ICCE 2023, LNCE 526, pp. 463–472, 2024.
https://doi.org/10.1007/978-981-97-4355-1_43

pulsation in the flow channel of the water guide mechanism, the volute flow channel and the pressure steel pipe may cause vibration, noise and various instability problems in various parts. For the plant vibration of hydropower station, there are three kinds of vibration sources, namely hydraulic vibration source, mechanical vibration source and electromagnetic vibration source [2]. Related studies show that the main vibration source of pumped storage power station unit and plant is hydraulic vibration source [3, 4], which is pressure pulsation in the flow channel of the unit. The common hydraulic vibration sources include hydraulic excitation and resonance caused by the dynamic and static interference in the vaneless area of the pump turbine, as well as pressure pulsation in the draft tube. Some of the pumped storage power stations that have been put into operation in China have relatively strong plant vibration. In the later period, the plant vibration is reduced to a certain extent by taking some measures such as replacing the runner [5–8]. The measured statistics show that the main vibration frequency of the pumped storage power station with strong plant vibration is basically the dynamic and static interference frequency of the pump turbine and its multiple frequency.

In this paper, the vibration characteristics of the measured plant structure are calculated and counted, and the specific influence of the pressure pulsation generated by the operation of typical different cascade combination units on the vibration characteristics of the plant structure is analyzed.

2 Investigation on Vibration Characteristics of Unit and Plant of Pumped Storage Power Station

When the unit operates under different loads, the main frequency of plant vibration response is multiple frequency of the overcurrent frequency of the unit. The main excitation source of the plant vibration is pulsating pressure in the flow channel of the unit.

The matching combination of the number of runner blades and the number of active guide vanes of the unit is called cascade combination. The cascade combination of early pumped storage power station units is mainly 9 + 20 or 7 + 20. With the improvement of research and development capacity of domestic units, the types of pump turbines with new cascade combinations continue to emerge, such as (5 + 5)/16, 9/22, 13/22, 11/20 and so on, among which (5 + 5)/16, 9/22 type have been the main type of medium and high head section. The statistical analysis of the plant vibration data obtained from the test of the operating power station shows that the main frequency of the plant vibration is mainly 2 or 3 times the overcurrent frequency, which is closely related to the cascade combination type of the unit.

The on-site vibration test analysis results of multiple research institutions are summarized, and the main frequency characteristics of the plant vibration of multiple pumped storage power stations are obtained, as shown in Table 1.

Table 1. Statistics on vibration frequency of some pumped storage power plants in China

No.	Power station	Single machine capacity (MW)	Speed (r·min^{-1})	Number of runner blades	Number of active guide vanes	Blade over current frequency (Hz)	Main frequency of plant vibration (Hz)
1	Shisanling	200	500	7	16	58.33	116.7
2	Baishan	150	200	7	20	23.33	70
3	Xilongchi	306	500	7	20	58.33	175
4	Tongbai	306	300	7	20	35	105
5	Liyang	250	300	7	20	35	105
6	Zhanghewan	250	333.3	9	20	50	100
7	Xianju	375	375	9	20	56.25	112.5
8	Dunhua	350	500	9	20	75	150
9	Fengning(I)	306	428.6	9	20	64.29	128.6
10	Shenzhen	306	428.6	9	20	64.29	128.6
11	Hongping	306	500	9	20	75	150
12	Xianyou	306	428.6	9	20	64.29	128
13	Huhehaote	306	500	9	20	75	150
14	Heimifeng	306	300	9	20	45	90
15	Baoquan	306	500	9	20	75	150
16	Huzhou	306	500	9	20	75	150
17	Guangzhou(I)	306	500	9	20	75	150
18	Xiangshuijian	250	250	9	20	37.5	75
19	Bailianhe	306	250	9	20	37.5	75
20	Pushihe	306	333.3	9	20	50	100
21	Qiongzhong	204	375	9	20	56.25	112.5
22	Taian	250	300	9	22	45	90/180
23	Tianhuangping	306	500	9	26	75	225
24	Yixing	250	375	9	26	56.25	168.75
25	Heimifeng(4#)	306	300	6 + 6	20	60	120
26	Jinzhai	300	333.3	13	22	72.2	144.4

The power stations with cascade combination of 9 runner blades and 20 guide vanes of the unit include Pushihe pumped storage power station, Heimifeng pumped storage power station (Unit 1#~3#), and Xianyou pumped storage power station. At present, there are many units of this type of cascade combination. When the unit operates at full load, the main frequency of the plant vibration is mainly reflected as 2 times the overcurrent frequency of the runner blade.

The power stations with cascade combination of 7 runner blades and 16 guide blades of the unit mainly include Shisanling pumped storage power station. When the unit of this type of cascade combination operates at full load, the main frequency of the plant vibration is mainly reflected as 2 times the overcurrent frequency of the runner blade.

The power stations with cascade combination of 7 runner blades and 20 guide blades of the unit mainly include Baishan pumped storage power station. When the unit of this type of cascade combination operates at full load, the main frequency of the plant vibration is mainly reflected as 3 times the overcurrent frequency of the runner blade.

The power stations with cascade combination of 9 runner blades and 26 guide blades of the unit include Tianhuangping pumped storage power station and Yixing pumped storage power station. When the unit of this type of cascade combination operates at full load, the main frequency of the plant vibration is mainly reflected as 3 times the overcurrent frequency of the runner blade.

3   Hydraulic Excitation Law of Different Cascade Combination Units

The vibration source of pumped storage power station units and plants is mainly hydraulic excitation. The hydraulic excitation characteristics of units with different cascade combinations are closely related to the number of guide vanes and the number of runner blades [9].

The main reason for the hydraulic excitation produced by the unit operation is that the outlet edge of the movable guide vane of the pump turbine is relatively thick. When the unit is running, the water flow passes through the movable guide vane, and the wake effect leads to a uneven flow field at the outlet of the guide vane. Under the action of water pressure, the water flow entering the rotating runner also produces regular potential flow disturbance at the inlet of the runner. The vaneless area between the runner and the movable guide vane will produce dynamic and static interference, and the pressure pulsation in the volute will show a periodic law as a whole [10]. The pressure pulsation in the volute causes vibration and noise of the plant, and the plant vibration and structural noise also show a similar periodic law.

The pressure field formed by the interaction between the runner blade and the guide vane can be described by the following equation [11]:

$$f(x) = \frac{B}{2}\cos[mZ_r\omega_n t - (mZ_r - nZ_g)\theta_s + \varphi_n - \varphi_m] + \frac{B}{2}\cos[mZ_r\omega_n t - (mZ_r + nZ_g)\theta_s - \varphi_n - \varphi_m] \quad (1)$$

where, B is the average amplitude of pressure pulsation. m and n are integers. θ_s is an angular coordinate, which is closely related to the angular coordinates of the static system (volute, guide vane and top cover) and the rotating system (runner). Z_g is the number of active guide vanes. Z_r is the number of runner blades. ω_n is the rotating speed of the unit.

Equation (1) shows that pressure pulsation is a function of time t and spatial angle θ_s. It has different low-order and high-order hydraulic excitation modes, which is determined by the cascade combination composed of the number of runner blades and the number of guide vanes.

The pitch diameter number k is described as

$$k = mZ_r \pm nZ_g \tag{2}$$

The minimum pitch diameter number k_1 and the maximum pitch diameter number k_2 are described as

$$k_1 = mZ_r - nZ_g \tag{3}$$

$$k_2 = mZ_r + nZ_g \tag{4}$$

$$k_{min} = Min(|k|) = Min(|k_1|) \tag{5}$$

It can be seen From Eq. (1) that there are many hydraulic excitation modes in the flow channel of the unit. The higher the harmonic order of the hydraulic excitation mode is, the larger the corresponding pitch diameter number $|k|$ is, and the smaller the vibration amplitude of the hydraulic excitation force is. On the contrary, the lower the harmonic order of the hydraulic excitation mode is, the smaller the corresponding pitch diameter number $|k|$ is [12], and the larger the vibration amplitude of the hydraulic excitation force is. Therefore, the harmonic corresponding to the minimum pitch diameter number k_1 is the main vibration excitation force of pressure pulsation, and the hydraulic excitation force corresponding to k_{min} is the max predominant vibration excitation force of pressure pulsation.

During the operation of the pumped storage unit, the flow components such as volute and guide vane are forced to vibrate in the hydraulic excitation mode generated by the dynamic and static interference. The frequency and vibration amplitude of the hydraulic excitation force are determined by the cascade combination mode and the unit speed.

The frequency of the hydraulic excitation force on the runner is the overcurrent frequency of the active guide vane f_r and its multiple frequency:

$$f_r = nZ_g f_n \tag{6}$$

The frequency of the hydraulic excitation force on the guide vane, head cover and volute is the overcurrent frequency of the runner blade f_s and its multiple frequency:

$$f_s = mZ_r f_n \tag{7}$$

where, f_n is the unit rotation frequency.

For the pumped storage power station that has been put into operation, when the unit speed and cascade combination are known, the amplitude and frequency characteristics of the hydraulic excitation force can be predicted and evaluated according to Eqs. (1)~(7).

The main excitation force of the plant vibration of pumped storage power station is hydraulic excitation. The pressure pulsation is transmitted to the peripheral concrete support through the head cover and volute, and then is transmitted to the plant structure, causing plant structure vibration and structural noise. Equations (1)~(7) can reflect the plant vibration characteristics of pumped storage power station to a certain extent.

4 Vibration Characteristics of Measured Pumped Storage Power Plant

The authors have obtained a lot of vibration data of the plant structure of pumped storage power station with various cascade combinations through measured tests. The typical hydropower stations tested include Baishan hydropower station, Heimifeng hydropower station, Shisanling hydropower station, Pushihe hydropower station and Jinzhai hydropower station. The minimum pitch diameter numbers of the unit in the above typical pumped storage power stations are also counted, as shown in Table 2. At the same time, the vibration data of these plant structures are analyzed and compared. It should be noted that, due to the obvious vibration of 4# unit in Heimifeng pumped storage power station, the runner has been replaced. The cascade combination of the replaced 4# unit has been transformed from the original 9/20 to the current $(6 + 6)/20$[8].

Table 2. Statistics for minimum pitch diameter of typical pumped storage power plant units

Power station & cascade combination	f_n (Hz)	n	m	k_1	f_s (Hz)	f_s/f_n
Pushihe	5.555	1	2	−2	100.0	18
9/20		2	4	−4	200.0	36
Shisanling	8.333	1	2	−2	116.7	14
7/16		2	5	3	291.7	35
Baishan	3.333	1	3	1	70.0	21
7/20		2	6	2	140.0	42
Heimifeng	5.000	1	2	4	120.0	24
(6 + 6)/20		2	3	−4	180.0	36
Jinzhai	5.555	1	2	4	144.4	26
13/22		2	3	−5	216.6	39

4.1 Analysis of Vibration Spectrum Characteristics of Measured Pumped Storage Power Station Plant

The frequency spectrum analysis of the measured data of the plant vibration of typical pumped storage power stations was carried out. From the calculated frequencies of the hydraulic excitation mode corresponding to the minimum pitch diameter number (see Table 2), it can be seen that the main frequencies of the actual plant vibration are very consistent with the frequencies of hydraulic excitation mode corresponding to k_1, which shows that the hydraulic excitation force of different cascade combination units is mainly the hydraulic excitation mode with $n = 1$ or $n = 2$ and $k = k_{min}$.

For the power stations with the minimum pitch diameter $k_{min} = 1$ of the unit, such as Baishan pumped storage power station, and the power stations with the minimum pitch diameter $k_{min} = 2$, such as Pushihe pumped storage power station and Shisanling pumped storage power station. The vibration spectrum of the plant structure of the three pumped storage power stations shows obvious dominant frequency, and the dominant frequency is m times the overcurrent frequency of the runner blade. Vibration fourier spectrograms of Baishan power station are shown in Fig. 1.

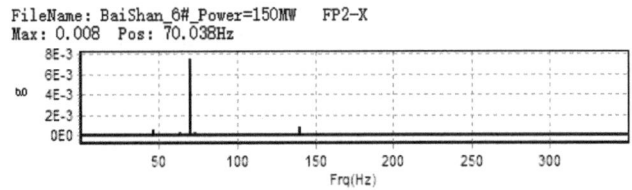

(a) X-direction vibration fourier spectrogram

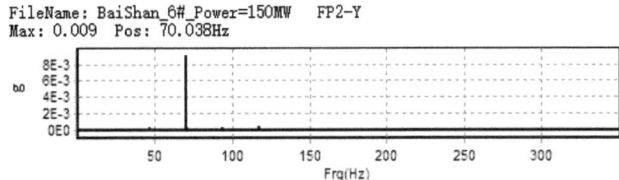

(b) Y-direction vibration fourier spectrogram

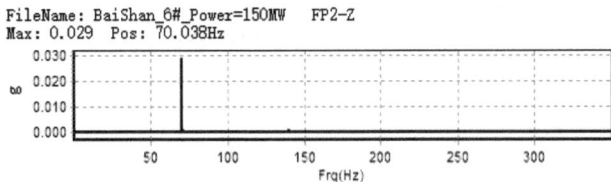

(c) Z-direction vibration fourier spectrogram

Fig. 1. Three orthogonal direction's vibration frequency spectrograms of a measuring point in Baishan power station

For the power stations with the minimum pitch diameter $k_{min} = 4$ of the unit, such as Jinzhai pumped storage power station and Heimifeng pumped storage power station (4# unit), although the main vibration frequency of the plant structure includes m times frequency of the overcurrent frequency of the runner blade, the vibration amplitude corresponding to the frequency is not significant. Vibration fourier spectrograms of Jinzhai power station are shown in Fig. 2. The plant vibration energy is distributed in a wide frequency band. The vibration amplitude of the plant structure of the two pumped storage power stations are significantly smaller than that of other power stations.

4# unit of Heimifeng power station adopts the cascade combination type of (6 + 6)/20. 6 + 6 is a runner type with six long blades and six short blades. Its hydraulic excitation mode is more complex and diverse than that presented by a single length runner blade. The main frequency characteristics of the hydraulic excitation force contain the

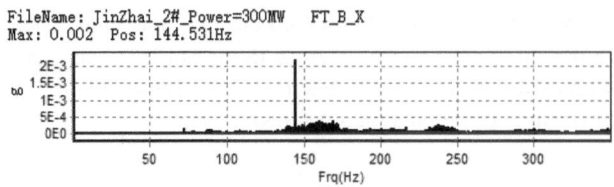

(a) X-direction vibration fourier spectrogram

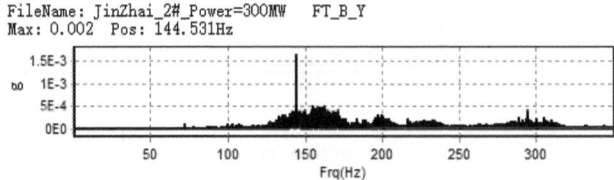

(b) Y-direction vibration fourier spectrogram

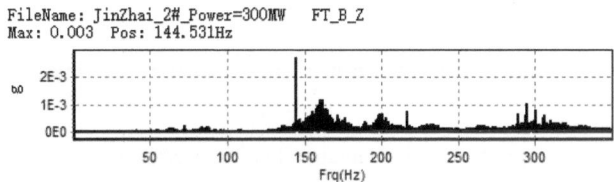

(c) Z-direction vibration fourier spectrogram

Fig. 2. Three orthogonal direction's vibration frequency spectrograms of a mseasuring point in Jinzhai power station

main frequency characteristics of the hydraulic excitation mode of the cascade combination of 12/20 and 6/20. There are many characteristic frequencies in the spectrum, but the vibration amplitude is small.

4.2 Statistical Comparison of Measured Vibration Response of Pumped Storage Power Station Plant

There are obvious differences in the vibration amplitude of the plant structure in several power stations with different cascade combinations that are investigated. It can be seen from the measured statistical results (Table 3) that the plant structural vibration of Heimifeng pumped storage power station (the cascade combination is $(6 + 6)/20$) and Jinzhai pumped storage power station (the cascade combination is 13/22) is significantly small. The common feature of these two combinations is $k_{min} = 4$.

From the perspective of spectrum characteristics, the plant vibration energy distribution characteristics of Heimifeng pumped storage power station and Jinzhai pumped storage power station are that the dominant frequency is not obvious, the vibration energy is distributed in a wide frequency band, and the vibration amplitude of the plant structure is significantly reduced. It shows that the plant vibration amplitude is obviously reduced

Table 3. Statisticals for plant vibration of different pumped storage power stations during full load generating operation of units

Location	Maximum vibration acceleration root mean square value (g)				
	Pushihe	Shisanling	Baishan	Heimifeng	Jinzhai
Floor slab of generator layer	0.133	0.339	0.061	0.019	0.015
	Z-direction	Z-direction	Z-direction	Z-direction	Z-direction
Floor slab of busbar layer	0.259	0.092	0.016	0.027	0.017
	Z-direction	Z-direction	X-direction	Z-direction	Z-direction
Floor slab of water turbine	0.046	0.292	0.016	0.062	0.021
	X-direction	Z-direction	Z-direction	Z-direction	Z-direction
Column of busbar layer	0.246	0.242	0.045	0.024	0.021
	Y-direction	X-direction	X-direction	Y-direction	Y-direction
Column of water turbine layer	0.114	0.135	0.038	0.034	0.024
	Y-direction	X-direction	Y-direction	Y-direction	Y-direction

by using the cascade combination mode with the number of pitch diameter $k_{min} = 4$, which is beneficial for the vibration reduction of the plant.

5 Conclusions

The excitation force amplitudes of pressure pulsation of the dynamic and static interference of the units with different cascade combinations are significantly different. The frequency spectrum analysis results of the measured vibration data in the typical pumped storage power plant structures also reflect the frequency characteristics of the dynamic and static interference, indicating that the main vibration source of pumped storage power plant vibration is hydraulic excitation force.

(1) When the minimum pitch diameter number of the unit is 1 or 2, the vibration spectrum of the plant structure shows a significant single dominant frequency, which is m times the overcurrent frequency of the runner blade.
(2) When the minimum pitch diameter number of the unit is 4, although the main vibration frequency of the plant structure contains m times frequency of the runner blade, the amplitude of the main vibration frequency is not large. The plant vibration energy is distributed in a wide frequency band, and the vibration amplitude of the plant structure has a significant decrease.

Acknowledgments. This study was supported by the State Grid Corporation of China Headquarters Management Technology Project (Grant no. 5419-202243054A-1-1-ZN).

References

1. Prospective Industry Research Institute. Panoramic map of China's pumped storage industry in 2022. Electr. Ind. 2021(12), 64–68
2. Ma, Z.Y., Dong, Y.X.: Vibration and corrective actions of water turbine generator set and powerhouse. China Water & Power Press, Beijing (2004)
3. Ouyang, J.H., Geng, J., Xu, L.H., et al.: Analysis on strong vibration cause of the powerhouse of a large-scale pumped-storage power station in China and study on its vibration reduction measure. J. Hydraul. Eng.Hydraul. Eng. **50**(8), 1029–1037 (2019)
4. Shang, Y.L., Li, D.Y., Ouyang, J.H.: Review on powerhouse self-oscillation characteristics of a large-scale power station. J. China Inst. Water Resour. Hydropower Res. **14**(1), 78–81 (2016)
5. Fan, Y.M., He, Q.Y., Wang, M.K., et al.: Application and research on vibration improvement of long and short blade runners. Water Conserv. Hydropower Technol. **53**(7), 58–68 (2022)
6. Guo, P., Liu, D.H., Li, Y.L., et al.: Vibration analysis and experience of pumped storage power plant. Hydropower Pumped Storage **8**(6), 99–104, 109 (2022)
7. Xie, W.X., Zhang, T., Qin, D.Q., et al.: Research and practice of hydraulic stability reconstruction of No. 1 unit runner of Pushihe pumped-storage power station. Large Electr. Mach. Hydraul. Turbine **2022**(3), 69–76 (2022)
8. Tang, Y.J., Fan, Y.L.: Analysis and treatment of violent vibration of powerhouse for Zhanghewan pumped storage power station. Water Resour. Power **37**(5), 149–151, 158 (2019)
9. Jia, W., Liu, J.S., Pang, L.J., et al.: Analysis on rotor-stator interaction and vibration of pump turbinein pumped storage power station. J. Vibr. Eng. **27**(4), 565–571 (2012)
10. Wuibaut, G., Bois, G., Caignaert, G., et al.: Experimental analysis of interactions between the impeller and the vaned diffuser of a radial flow pump. In: International Association on Hydraulic Research 2002 Symposium, Switzerland, pp. 1–11 (2002)
11. Nicolet, C., Ruehonnet, N., Avellan, F.: One-dimensional modeling of rotor stator interaction in Francis pump-turbine. In: 23th IAHR Symposium, pp. 1–15 (2006)
12. Pang, L.J., Lü, G.P., Zhong, S., et al.: Vortex shedding simulation and vibration analysis of stay vanes of hydraulic Turbin. J. Mech. Eng. **47**(22), 159–166 (2011)

Stability Analysis of Tailrace Outlet Slope at Right Bank of Kala Hydropower Station

Bing Pan[1,3](✉), Shuhong Hu[2], Weiwei Wu[1], Yongjin Cheng[3], and Zhen Jiang[3]

[1] PowerChina Huadong Engineering Corporation Limited, Hangzhou, Zhejiang, China
pan_b@hdec.com
[2] YALONG River Hydropower Development Company, Ltd., Chengdu, Sichuan, China
[3] HydroChina ITASCA Research and Development Center, Hangzhou, Zhejiang, China

Abstract. The tailrace outlet slope located on the right bank of the Kala hydropower station represents a typical bedding slope characterized by a moderate to large dip angle. The excavation at the toe of the slope create favourable conditions for rock mass deformation and failure. In order to evaluate the slope's stability, a thorough investigation and analysis of the engineering geological conditions are conducted. Additionally, the stress-strain characteristics resulting from the slope excavation are simulated utilizing the discrete element software 3DEC. Finally, the strength reduction method is used to evaluate the safety factor of the slope. The research results show that the stability of the slope is notably influenced by the presence of medium and large dip structural planes, the cutting excavation at slope toe can lead to deformation and localized block instability. The implementation of the system pre-stressed cables effectively increases the safety factor of the slope and ensures the stability of the tailrace outlet slope. These findings will provide valuable references for the design of similar rock slopes.

Keyword: Kala Hydropower Station · Stability Analysis · Strength Reduction Method · Bedding Slope

1 Introduction

The construction of large-scale hydropower projects in southwestern China has triggered considerable concerns regarding the stability of high and steep rock slopes. In response to this issue [1], Zheng Yingren, Song Shengwu, and Huang Runqiu have conducted a comprehensive study on high rock slope stability and have identified several common failure mechanisms associated with these slopes [2–4]. However, it is worth noting that rock slopes exhibit diverse characteristics based on their specific geological environments. In mountainous regions, rock slope deformation and failure caused by excavation is a significant geological challenges. Amongst these challenges, layered rock slopes are highly susceptible to collapse and failure as a result of excavation under the influence of weak structural planes such as joints, bedding planes and faults [5]. Therefore, the stability evaluation of the layered rock slope is particularly important.

The stability analysis of slopes typically utilizes two main methods: the strength reduction method and the limit equilibrium method (LEM) [6]. The LEM has been

© The Author(s) 2024
G. Feng (Ed.): ICCE 2023, LNCE 526, pp. 473–482, 2024.
https://doi.org/10.1007/978-981-97-4355-1_44

widely employed in civil projects due to its simplicity and effectiveness in assessing slope stability. However, this method still has certain limitations. For example, it does not take into account the deformation and stress characteristics of rock masses within slopes. To address these limitations, the strength reduction method based on numerical simulation can be employed, which provides a more comprehensive assessment of slope stability [7].

The Kala Hydropower Station, situated in the middle reach of the Yalong River, is the seventh cascaded development station. It serves primarily for power generation, with an installed capacity of 1020 MW. The reservoir maintains a normal water level of 1,987 m and has a storage capacity of $2.4 \times 10^9 m^3$. The right bank is a prototypical bedding slope with a dip angle 55~65°. Due to intricate geological and tectonic processes, various unfavorable geological features are present, including faults (f_{124}, f_{198}, f_{199}, f_{201} et al.), carbonaceous slate interlayer, and other weak geological structural planes within the tailrace outlet slope. Such unfavorable features contribute to the poor stability conditions of the slope, and excavation activities in the lower section of the slope further intensify the likelihood of rock mass deformation and failure [8]. Therefore, it becomes essential to evaluate the stability of the tailrace outlet slope. This study utilizes the strength reduction method to analyze the stability of the tailrace outlet slope at the Kala Hydropower Station. The findings of this analysis can provide valuable insights for the design and construction of the slope.

2　Geological Background

2.1　Engineering Geology

The outlet of the tailrace tunnels is characterized by a natural slope that exhibits significant height and steepness, as well as a complex geological composition. The bedrock in the study area is composed of metamorphic rock belonging to the Zagunao group of Upper Triassic(T3Z2). The exposed rock layers consist mainly of sandy slate, metamorphosed sandstone, marble, and carbonaceous slate structures with interbedded characteristics. The rock mass within the slope generally exhibits weak weathering, and the strongly unloaded rock mass extends horizontally to a depth of approximately 10~20 m, as shown in Fig. 1.

The rock formation within the slope exhibits a dip direction ranging from N55°E to N65°E and a dip angle of 55~65°, while the slope on the right bank follows a dip direction of N50°E to N60°E with a slope angle ranging from 30° to 45°. Within the slope, numerous faults have developed, with those striking NW to NWW being the most prominent and extensive, including f_{124}, f_{198}, f_{199}, f_{201} and others. Additionally, two primary sets of joints can be observed: the first set features orientations ranging from N65°E to N75°E with a southeast inclination angle of 85° to 90°, while the second set exhibits orientations from N30°W to N35°W with a southwest inclination angle of 25° to 30°.

2.2　Design of the Tailrace Outlet Slope

The tailrace outlet slope consists of two main parts: the cut slope and the natural slope. The cut slope has a maximum height of 112.5 m, extending from an elevation of 2015

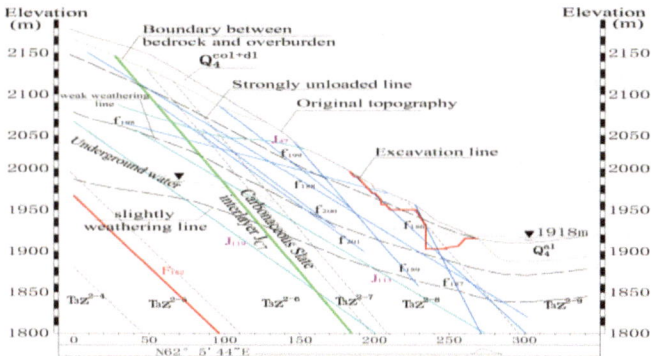

Fig. 1. The geological profile of the tailrace outlet slope.

m to 1902.5 m, as shown in Fig. 2(a). To ensure stability, different slope ratios are adopted. For elevations above 1950 m, a slope ratio of 1:0.5 is used. From 1950 m to 1930 m, a slope ratio of 1:0.3 is applied. Below 1930 m, the slope ratio becomes vertical. Additionally, benches, each 3 m wide, are established every 20 m in the cut slope. The natural slope above the cut slope has a height exceeding 500 m.

To safeguard and reinforce the slope, a designed protection and reinforcement scheme comprising three key elements is implemented, as illustrated in Fig. 2(b). Firstly, a layer of 15 cm of shotcrete is uniformly applied to all newly excavated surfaces. Secondly, rock bolts are placed at 2-m intervals to provide shallow support, with the length of the bolts ranging from 6 to 9 m. And thirdly, pre-stressed cables are employed to control slope deformation and ensure overall stability. These cables are designed with a capacity of 2000 kN, with lengths ranging from 40 to 50 m, and spacing set at 4 m × 4 m.

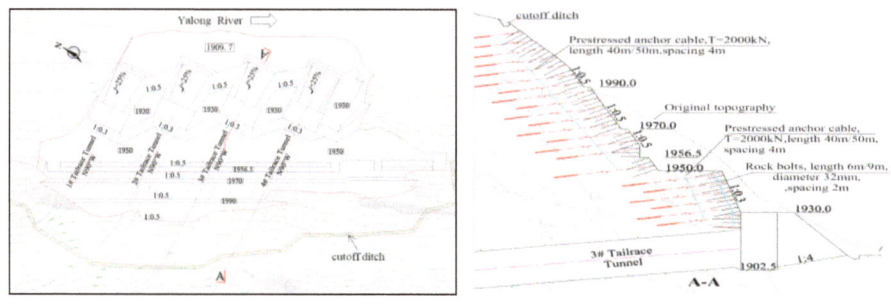

(a) The layout of excavation   (b) The support scheme

Fig. 2. The excavation plane view and support scheme of the tailrace outlet slope.

3   Numerical Models and Boundary Conditions

3.1   The Discrete Element Model

Based on the actual construction sequence, a simulation of the construction process is carried out using the 3DEC software, as shown in Fig. 3. The model is designed with dimensions of 660 m in the X direction, perpendicular to the slope strike, 300 m in the Y direction, and a vertical direction height of 650 m. Normal displacement constraints are applied on boundaries of left, right, frontal, and back, the bottom boundary is limited by a fixed constraint, and the top boundary is free.

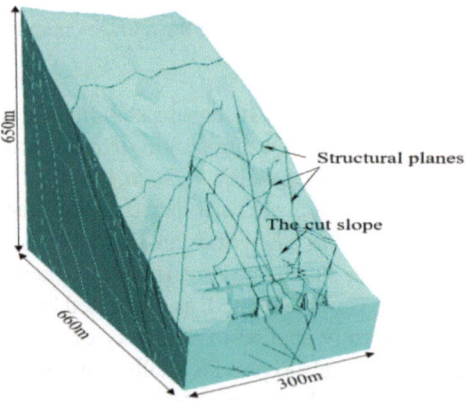

Fig. 3.   Numerical calculation model.

3.2   Calculation Parameters

Joint elements are utilized to simulate the geological structural planes, such as the carbonaceous slate interlayer J_{C7}, joints J_{47}, J_{48}, J_{119}, and faults f_{189}, f_{198}, f_{199} et al. The constitutive model employed for these structural planes follows a linear Coulomb shear-strength criterion, which sets limits on the shear force acting at each interface node. The rock mass is considered as an elastic-plastic material, employing a Mohr-Coulomb model.

Accurate estimation of the mechanical parameters of both the rock mass and structural features is very important for the stability analysis of the slope. In this study, the rock mass of the tailrace outlet slope is classified into four grades based on weathering characteristics, as depicted in Fig. 4. The strength parameters of the rock mass and faults are determined through in-situ testing and similar construction experience, which are listed in Table 1 and Table 2 [9].

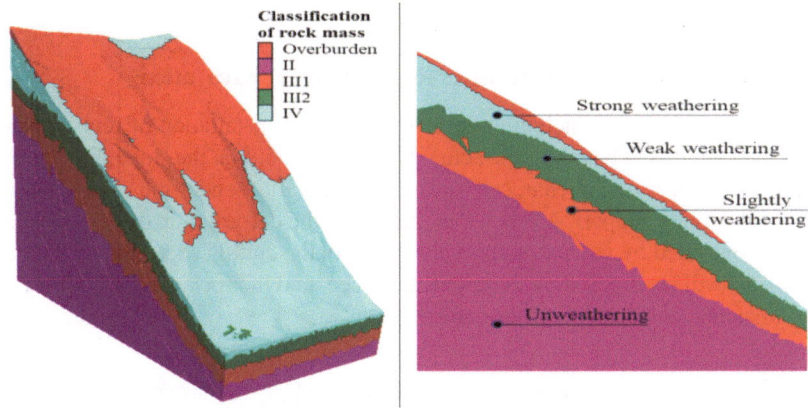

Fig. 4. Classification of rock mass.

Table 1. Mechanical parameters for the rock mass.

Classification of rock mass	Density $\rho(\text{kg/m}^3)$	Elastic modulus E (GPa)	Poisson's ratio	Internal friction angle $\varphi(°)$	Cohesion c (MPa)
II	2780	21.5	0.19	49.0	1.35
III$_1$	2770	13	0.23	43.5	1.15
III$_2$	2760	5	0.25	40.4	0.95
IV	2670	2.25	0.30	33.0	0.43
Overburden	2200	0.1	0.45	31.0	0.075

Table 2. Mechanical parameters for structural planes.

Structural feature		Friction angle $\varphi_s(°)$	Cohesion c_s(MPa)	Tensile strength T_s(MPa)
Joint, Bedding plane		26.5~31.0	0.1~0.15	0
Faults	filled with detrital material	24.2~28.8	0.1~0.15	0
	filled with detrital material and gouge	21.8~24.2	0.08~0.10	0
	filled with gouge	14.0~19.3	0.03~0.05	0
Carbonaceous slate interlayer		26.5~31.0	0.15~0.20	0

4 Analysis Results

4.1 The Deformation Characteristics Resulting from Excavation

The construction process, including excavation and the installation of reinforcements, is accurately simulated based on the actual work carried out on the slope. Fig shows the comparison of deformation characteristics after excavation under the conditions of no support and applied support. As shown in the Fig. 5, the deformation of the rock mass is primarily controlled by weak structural planes. With the cutting excavation at slope toe, the fault f_{198} will be exposed on the excavation face, and the rock mass on both sides of the fault has obvious discontinuous deformation.

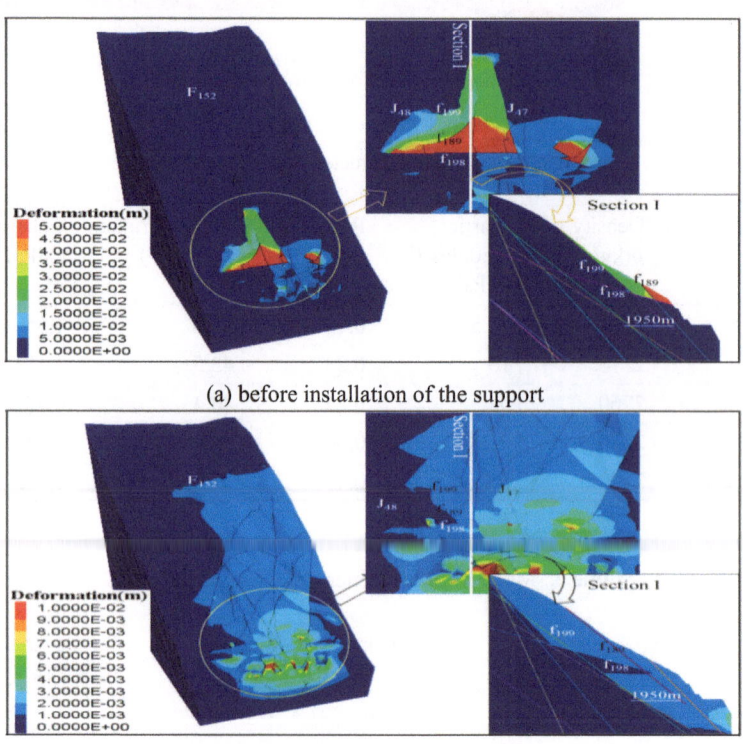

(a) before installation of the support

(b) after installation of the support

Fig. 5. The deformation Characteristics after excavation.

From Fig. 5(a), there is an unstable block, the bottom slip plane is fault f_{198}, the rear edge is fault f_{199}, and the upstream and downstream side boundaries are joint J_{47} and J_{48} respectively. Without considering any support measures, this block presents significant risks of instability and failure, thereby compromising construction safety. However, by implementing the reinforcement measures depicted in Fig. 2(b), the shear deformation along the fault f_{198} is effectively controlled, and the maximum deformation induced by excavation is generally within 10 mm.

4.2 Slope Stability Analysis

The strength reduction method [10] is used to analyze the stability of the slope after reinforcement with pre-stressed cables:

$$\left. \begin{array}{l} c_t = c/F_t \\ \varphi_t = \tan^{-1}(\tan\varphi/F_t) \end{array} \right\} \tag{1}$$

where Ft is strength reduction factor.

The strength reduction method employed in this study involves adjusting the shear strength of the structural plane by modifying the reduction factor Ft. Subsequently, a stability analysis of the slope is conducted using the Discrete Element Method. The reduction factor serves as the safety factor when the slope reaches a critical failure state. In this paper, the critical failure state is determined based on the plastic yielding zone connection criterion and the displacement mutation criterion [11].

Figure 6 shows the deformation velocity of the slope under critical instability, its show that the deformation velocity of the huge potential block composed of structural planes f_{198}, J_{47} and J_{48} is significantly higher than that of other parts. To analyze the stability of this block, three observation points (P1, P2, and P3) are set along its boundaries. Figure 7 presents the relationship curves between the displacements of the observation points and the strength reduction factor. Notably, the displacement of P2 experiences a sharp increase when the strength reduction factor exceeds 1.25. Therefore, according to the displacement mutation criterion, the safety factor of the block composed of structural planes f_{198}, J_{47} and J_{48} can be defined as 1.25.

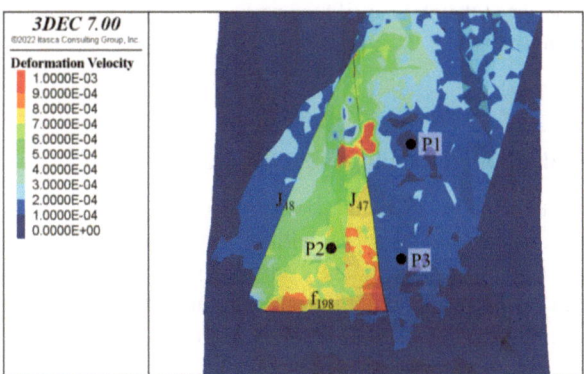

Fig. 6. The deformation velocity of the slope under critical instability.

Figure 8 shows the development of the plastic yielding zone on the structural planes throughout the process of strength reduction. Examination of the Fig. 8 reveals that plastic yielding failure initially occurs on the structural planes near the rear edge of the block. As the reduction factor increases, the plastic yielding zone gradually extends towards the front edge of the block. At a strength reduction factor of 1.30, the plastic

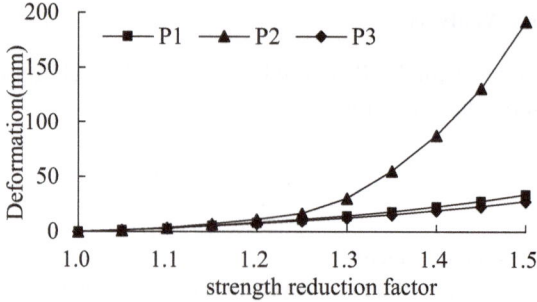

Fig. 7. Relation curve of displacements of the observation points and strength reduction factors.

zone on the bottom slip plane f_{198} becomes fully connected. Therefore, based on the criterion of plastic yielding zone connection, the reduction factor of 1.25 from the previous calculation step is defined as the safety factor for the block.

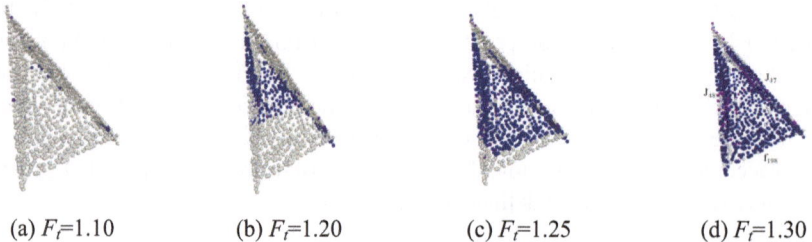

(a) F_r=1.10   (b) F_r=1.20   (c) F_r=1.25   (d) F_r=1.30

Fig. 8. The development of the plastic yield zone (shown in blue) on the structural planes with the strength reduction factors.

4.3 Stability Evaluation and Guiding to Construction

The stress-strain characteristics of slope excavation indicate that cutting the slope toe will cause sliding failure along fault f_{198} without any support. However, implementing pre-stressed cable support can significantly enhance the stability of the slope and effectively restrict shear displacement within weak structural planes. With the installation of this support system, the safety factor of the tailrace outlet slope reaches 1.25, thereby satisfying the minimum safety factor requirement specified in the code [12]. Therefore, the current reinforcement scheme can be considered rational and dependable. Nevertheless, it is crucial to maintain a vigilant monitoring of geological conditions throughout the slope excavation process and carefully assess their potential impacts on slope stability.

5 Conclusions

This study aims to investigate and analyze the deformation characteristics and stability of a complex bedding slope located at the outlet of the tailrace tunnels in the Kala Hydropower Station. The following conclusions can be drawn:

(1) The overall stability of the tailrace outlet slope mainly depends on the huge potential block formed by structural planes f_{198}, J_{47} and J_{48}. The excavation of the slope toe has resulted in the removal of the rocks that previously acted as a stabilizing factor, leading to a significant deterioration in slope stability.

(2) On the basis of ensuring the safety of the supporting structure itself, the pre-stressed anchor cables can effectively restrict the shear deformation of the block along the fault f198, thereby improving the overall stability of the slope.

(3) The stability of the slope is assessed using the strength reduction method, with a comparison made between the calculation results obtained from two failure criteria: the plastic yielding zone connection criterion and the displacement mutation criterion. The results indicate that the safety factor of the huge block reinforcement by pre-stressed anchor cables is 1.25, and the stability of the block meets the requirements of the code, which can ensure the safety of the slope.

References

1. Xu, N.W., Wu, J.Y., Dai, F., et al.: Comprehensive evaluation of the stability of the left-bank slope at the Baihetan hydropower station in Southwest China. Bull. Eng. Geol. Env.Env. **77**, 1567–1588 (2018)
2. Zheng, Y.R., Zhao, S.Y., Deng, W.D.: Numerical simulation on failure mechanism of rock slope by strength reduction fem. Chin. J. Rock Mech. Eng. **22**, 1943–1952 (2003)
3. Song, S.W., Feng, X.M., Xiang, B.Y., et al.: Research on key technologies for high and steep rock slopes of hydropower engineering in southwest china. Chin. J. Rock Mech. Eng. **30**, 1–22 (2011)
4. Huang, R.Q.: Geodynamical process and stability control of high rock slope development. Chin. J. Rock Mech. Eng. **27**, 1525–1544 (2008)
5. Wang, J.L., Yang, J., Chen, Y.H., et al.: Study on deformation and stability of complex layered high and steep rock slope. J. Hydraul. Eng.Hydraul. Eng. **46**, 1414–1422 (2015)
6. Li, N., Guo, S.F., Yao, X.C.: Further study of stability analysis methods of high rock slopes. Rock Soil Mech. **39**, 397–406 (2018)
7. Jiang, Q.H., Qi, Z.F., Wei, W., Zhou, C.B.: Stability assessment of a high rock slope by strength reduction finite element method. Bull. Eng. Geol. Env.Env. **74**, 1153–1162 (2015)
8. Wei, Y.F., Nei, D.X.: Analysis on progressive slipping-shear failure mode of bedding slope of hard rock with medium or large dip angle. Appl. Mech. Mater. **438**, 1232–1237 (2013)
9. Zhang, C.S., Zhi-gang, S.A.N., Shi, B., et al.: Feasibility study report of kala hydropower project located on yalongjiang river stage (Hangzhou: Power China Huadong Engineering Corporation Limited) (2015)
10. Dawson, E.M., Roth, W.H., Drescher, A.: Slope stability analysis by strength reduction. Geotechnique **49**, 835–840 (1999)
11. Tu, Y., Liu, X., Zhong, Z., Li, Y.: New criteria for defining slope failure using the strength reduction method. Eng. Geol. **212**, 63–71 (2016)
12. Yao, S.X., Zhou, H., Cai, Y.P., et al.: Code for slope design of hydropower projects NB/T 10512-2021 (Beijing: China Water Power Press) p5 (2021)

Temperature Control Simulation of Rock Wall Crane Beam Based on Cooling Water Pipe

Chuncheng Ma[1], Haoming Zhang[2]([⊠]), Shuangquan Xu[1], and Xiji Li[1]

[1] Grid Xinyuan Shandong Weifang Pumped Stroage Co., Ltd., Nanjing, China
[2] College of Water Conservancy and Hydropower Engineering, Hohai University, Nanjing, China

844691059@qq.com

Abstract. Currently, the problem of concrete cracks is one of the common quality problems that is difficult to solve in the field of water conservancy and hydropower engineering. In order to study the distribution law of concrete temperature field during the construction period of underground cavern rock wall crane beams, different cooling water pipe arrangements were adopted to simulate and analyze the internal temperature and stress of concrete. The results show that the three-layer parallel cooling water pipe arrangement has the best cooling effect and can effectively control concrete cracking.

Keywords: Crane Beam · Cooling Water Pipe · Water Pipe Layout · Temperature Field

1 Introduction

In the process of concrete pouring and hardening, the hydration reaction of cement emits large heat. This chemical reaction leads to the change of the concentration of each component (such as the hardening of cement gel) and causes the structure to heat up. Conversely, an increase in temperature will exacerbate the rate of hydration reaction. The hydration reaction will rapidly increase the temperature of concrete to 70 °C. Subsequently, as the hydration rate slows down and the temperature decreases, thermal shrinkage occurs, leading to the generation of thermal stress. Under the dual driving forces of chemistry and thermodynamics, the physical characteristics of temperature, strength, stiffness, creep, and shrinkage of newly poured concrete will undergo significant changes, which can easily lead to cracking [1]. A large number of engineering crack treatment and investigation results show that 80% to 90% of cracks in concrete structures are caused by the tensile stress generated during the cooling process of concrete construction exceeding the tensile strength of concrete.

The rock wall crane beam of the underground powerhouse of a hydropower station is a structure that uses grouting long anchor rods to anchor the reinforced concrete beam body on the rock wall. Its inclined surface is conducive to forming a wide upper and narrow lower factory structure, which can effectively narrow the span of the factory,

G. Feng (Ed.): ICCE 2023, LNCE 526, pp. 483–494, 2024.
https://doi.org/10.1007/978-981-97-4355-1_45

reduce the excavation amount of the factory, reduce the engineering cost, and maintain the stability of the surrounding rock. During the construction period, temperature stress is prone to cracking. Currently, China mainly focuses on three aspects of crack prevention and resistance of concrete: ① improving the crack resistance performance of concrete from the material aspect, increasing its tensile strength and ultimate tensile value, reducing its hydration heat and shrinkage value; ② Propose reasonable structural types and dimensions from the perspective of structural design; ③ Take corresponding temperature control measures during construction to reduce adverse factors that cause concrete cracking [2].

The thermal conductivity of concrete is poor, making it difficult to dissipate in a short period of time and form a nonlinear temperature field inside. In the initial temperature rise stage, it is in a plastic flow state, and the uneven temperature distribution will not generate significant tensile stress in the concrete; However, during the process of temperature decrease, concrete has already reached a certain strength. The changes in temperature field and other constraints limit the free shrinkage of concrete, resulting in a certain amount of temperature stress. As the surface temperature is often lower than its internal temperature, significant tensile stress will be generated on the surface of concrete. If it exceeds the ultimate tensile strength of concrete, surface cracks will occur, which will further develop into penetrating cracks [3, 4]. Years of research have shown that the main cause of such cracks is the combined effect of concrete shrinkage (self generated volume strain) and environmental temperature changes and constraints. Therefore, the accuracy of temperature field simulation is crucial for the study of crack formation in concrete.

In recent years, research on concrete temperature and stress has been continuously deepening, and different scholars have made their research on this issue more specific and targeted. Tasri conducted research on thermal stress and temperature gradient caused by the space between the cooling pipes and cooling water temperature of post cooling components [5]. Myers and Charpin treat the heat transfer problem between cooling water pipes and concrete as plane strain and analyze the effects of cooling water flow rate and pipe material on concrete temperature [6]. Zeng has studied the effect of pouring speed on temperature stress [7]. Qiang has achieved certain results in the calculation of water pipe cooling algorithms [8].

Previous research on the effect of cooling water pipes on the temperature and stress fields of concrete has been relatively limited. This study mainly compares the layout of different cooling water pipes, Propose the best method to meet the requirements of temperature control and crack prevention.

2 Theory and Method

2.1 Theory of Temperature Field Calculation

Heat conduction is a specific heat transfer method, in which heat is transferred from the high-temperature area to the low-temperature area inside an object, satisfying the heat conduction equation:

$$\frac{\partial T}{\partial t} = \frac{\lambda}{\rho c}\left(\frac{\partial^2 T}{\partial x^2} + \frac{\partial^2 T}{\partial y^2} + \frac{\partial^2 T}{\partial z^2}\right) + \frac{\partial \theta}{\partial t} \tag{1}$$

In the formula: λ is the thermal conductivity of concrete, c is the heat capacity of concrete, ρ is the density of concrete, θ is the heat of concrete hydration, and T is the temperature of concrete.

In the heat conduction of solids, the heat flow rate (the amount of heat per unit area per unit time) is directly proportional to the temperature gradient, but the direction of the heat flow is opposite to the direction of the temperature gradient. The heat flow rate transmitted by concrete through the boundary of water pipes can be expressed as:

$$q = -\lambda \frac{\partial T}{\partial x} \tag{2}$$

Considering the variation of water temperature along the cooling water pipe, the heat change per unit length of cooling water per unit time can be expressed as:

$$T_{out} = T_{in} - \frac{\lambda}{q_w \rho_w c_w} \iint\limits_{s} \frac{\partial T}{\partial x} ds \tag{3}$$

The heat transferred from the pipe wall to the water flow per unit time can be expressed as: due to hydration, under adiabatic conditions, the temperature rise rate of concrete is:

$$\frac{\partial \theta}{\partial \tau} = \frac{Q}{c\rho} = \frac{Wq}{c\rho} \tag{4}$$

In the formula: is the adiabatic temperature rise of concrete, W is the amount of cement per unit of concrete, q is the hydration heat released per unit weight of cement per unit time.

2.2 Calculation Theory and Solution Method for Concrete Creep Stress

The theory of elastic creep stress can effectively describe the constitutive relationship of concrete. The strain increment of concrete during the time period Δt is:

$$\{\Delta \varepsilon_n\} = \left\{\Delta \varepsilon_n^e\right\} + \left\{\Delta \varepsilon_n^c\right\} + \left\{\Delta \varepsilon_n^T\right\} \tag{5}$$

Among them: $\left\{\Delta \varepsilon_n^e\right\}$ represents the increment of elastic strain; $\left\{\Delta \varepsilon_n^c\right\}$ is incremental creep strain; $\left\{\Delta \varepsilon_n^T\right\}$ is increment of temperature strain.

Zhu Bofang pointed out that the sum of elastic strain increment and creep strain increment can be written as:[9]

$$\left\{\Delta \varepsilon_n^e\right\} + \left\{\Delta \varepsilon_n^c\right\} = \{\eta_n\} + \frac{1 + C(t_n, \overline{\tau}_n)}{E(\overline{\tau}_n)}[Q]\{\Delta \sigma_n\} \tag{6}$$

Among them: $\{\eta_n\} = \sum_{s=1}^{m}\left[1 - \exp(-r_s \Delta \tau_n)\right]\{\omega_{sn}\}$, $\{\omega_{sn}\}$ is the state variable that changes over time.

In summary, the stress-strain relationship of concrete can be obtained as follows:

$$\{\Delta \sigma_n\} = \frac{E(\overline{\tau}_n)}{1 + E(\overline{\tau}_n)C(t_n, \overline{\tau}_n)}[Q]^{-1}(\{\Delta \varepsilon_n\} - \{\eta_n\} - \left\{\Delta \varepsilon_n^e\right\} + \left\{\Delta \varepsilon_n^c\right\}) \tag{7}$$

The basic idea of using sequential coupling method to solve the stress field of concrete is to first use the HETVAL subroutine to solve the temperature field of the structure, and then import the temperature field to solve the stress field using the UMAT subroutine. Calculation model and calculation parameters.

2.3 Simulate Cooling Water Pipes

The diameter of the cooling water pipe is set to 28 mm, which is much smaller than the size of the concrete structure. The key to the accuracy of the calculation results lies in the precise establishment of a large volume concrete model containing the cooling water pipe. Therefore, the substructure method can effectively solve the complex structural problems mentioned above. The substructure method, also known as superelement technique, divides complex structures into several small modular substructures, each of which is connected by boundaries [10]. When solving, first fix the common boundary, calculate the stiffness matrix of each substructure relative to the common boundary, and then use the finite element balance equation to assemble the stiffness matrix of the substructure through the common boundary, forming the overall balance equation of the entire structure.

Finally, use the temperature of the common boundary node solved as the specified temperature to solve the internal temperature of the substructure. This study used an octagonal water pipe structure to simulate, and the grid diagram of the division is shown in Fig. 1. And take one layer of units as the cooling water pipe, as shown in Fig. 2.

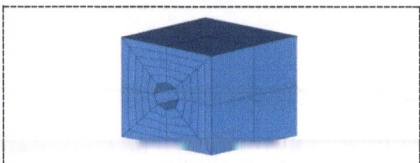

Fig. 1. Schematic diagram of the sub-structure

Fig. 2. Simulation diagram of cooling water pipe

3 Finite Element Simulation Analysis

3.1 Calculation Model

Due to the spatial structural characteristics of the rock wall crane beam, it is advisable to use the three-dimensional nonlinear finite element method for calculation. Meanwhile, the finite element method can effectively simulate the nonlinear temperature field formed during the pouring and hardening process of rock wall crane beams. Extend horizontally from the side wall to the interior of the surrounding rock to three times the anchoring depth of the system anchor rod, that is, take the thickness of the surrounding rock as 27.0 m; Vertically, the section height of the rock wall crane beam extends 4 times upwards and downwards from the top and bottom of the rock wall crane beam, that is, 9.0 m upwards and downwards respectively; The axial direction is taken as 1.5 times the width of a crane, which means the longitudinal length is taken as 11.0 m. In order to ensure high computational accuracy, the grid size of the rock wall crane beam and its surrounding area should not exceed 1/16 of the height of the rock wall crane beam, with a grid size of 0.19 m. Among them, the Mohr Coulomb yield criterion is used for the concrete of the rock wall crane beam and the nearby rock mass. A thin layer element with a thickness of 10cm is used to characterize the contact surface between the rock wall crane beam and the surrounding rock. The elastic modulus is taken as 1/10–1/2.5 of the surrounding rock and the Mohr Coulomb yield criterion is used. In three-dimensional finite element calculations, the rock mass and rock wall crane beam both use spatial eight node hexahedral elements. The overall coordinate origin of the model is selected in the middle of the rock wall crane beam, with the X-axis transverse to the river, the Y-axis longitudinal to the river, and the Z-axis vertical upwards. The specific overall structural model and the detailed three-dimensional finite element calculation mesh of the rock wall beam are shown in Figs. 3 and 4.

Fig. 3. Structural model   **Fig. 4.** Finite element mesh of rock wall beam

3.2 Boundary Conditions and Material Parameters

The thermal parameters of rock wall beam concrete include: concrete specific heat, thermal conductivity, surface heat dissipation coefficient, and concrete adiabatic temperature rise;

The adiabatic temperature rise model of concrete adopts an exponential function model, as shown in the following equation:

$$\theta = \theta_0 \left(1 - e^{-a\tau^b}\right) \tag{8}$$

Among them: θ_0 is final adiabatic temperature rise, τ is Time, a, b is time related parameters. The parameters of the concrete adiabatic temperature rise model are taken as: 46.0, 0.499, and 1.068.

Concrete surface heat dissipation coefficient: 960 kJ/(m2 · d · °C) for the horizontal plane, multiplied by 1.08 for the vertical plane. When there is a template on the concrete surface, take 1/3 of the value without the template. The surface heat dissipation coefficient of the surrounding rock is taken as 500 kJ/(m2 · d · °C).

The growth curve of concrete elastic modulus with age adopts a composite exponential model form, namely:

$$E(\tau) = E_0(1 - e^{-c\tau^d}) \tag{9}$$

In the formula: E_0 represents the final elastic modulus, it is recommended to take 36.00 MPa; c, d is two parameters related to the growth rate of the elastic modulus, with values of 0.28 and 0.52, respectively. Table 1 lists the calculation parameters of the material.

Table 1. Calculation Parameters

	Thickness (kg m^{-3})	specific heat kJ/(kg °C)	Poisson's ratio (GPa)	elastic modulus	thermal conductivity kJ/(m d °C)	expansion coefficient
concrete	2500	0.96	0.617	36	220	1.05×10^{-5}
rock	2600	0.91	0.23	13	86	9×10^{-6}

The following formula is used for the creep degree of concrete:

$$C(t, \tau) = \frac{0.23}{E_0}\left(1 + 9.2\tau^{-0.45}\right)\left[1 - e^{-0.3(t-\tau)}\right]$$
$$+ \frac{0.52}{E_0}\left(1 + 1.7\tau^{-0.45}\right)\left[1 - e^{-0.005(t-\tau)}\right] \tag{10}$$

The stress relaxation coefficient of concrete is expressed by the following equation:

$$K(t, \tau) = 1 - \left(0.4 - 0.6e^{-0.021\tau^{0.10}}\right)\left(1 - e^{-(0.2+0.2/\tau^{0.09})}(t - \tau)^{0.30}\right) \tag{11}$$

The underground powerhouse of the pumped storage power station is located in a deep area above 200 m underground. In the absence of a heat source, the temperature inside the cave should change according to a regular pattern around the average temperature throughout the year. The average temperature in Weifang is 12.8 °C. Considering that there is a significant difference between the temperature inside the cave and the temperature due to the influence of construction machinery. Based on past experience, it is assumed that the temperature inside the cave is approximately between 24 and 29 °C, and varies with the four seasons according to a regular pattern.

3.3 Calculation Results

Four types of cooling water pipe arrangements are shown in Fig. 5. Scheme 1: No cooling water pipe as control group. Scheme 2: Adopting a double-layer parallel cooling water pipe layout method. Scheme 3: Adopting a single-layer vertical cooling water pipe layout method. Scheme 4: Adopting a three-layer vertical cooling water pipe layout.

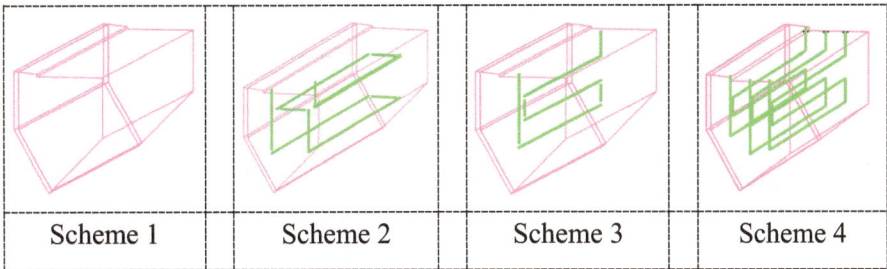

| Scheme 1 | Scheme 2 | Scheme 3 | Scheme 4 |

Fig. 5. Diagram of different cooling water pipe layout schemes

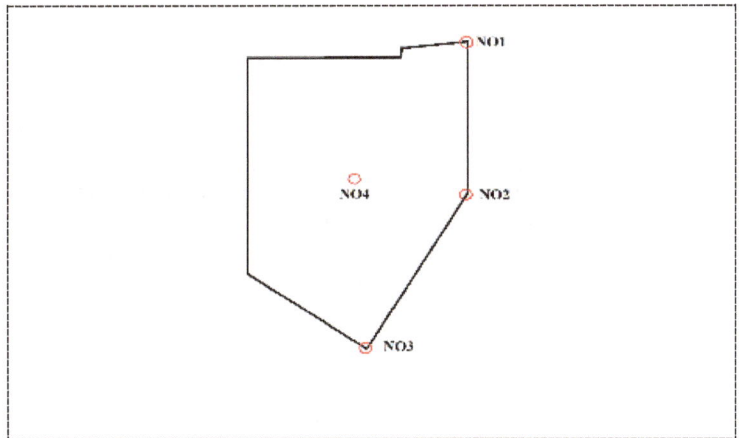

Fig. 6. Schematic diagram of key points on rock wall beams

Temperature Field Calculation. Select four key points of the rock wall beam for analysis, as shown in Fig. 6.

A three-dimensional calculation model of the rock wall beam of a pumped storage power station was constructed based on the actual design scheme, using thermal calculation parameters of concrete and foundation. At the same time, based on the actual construction situation and corresponding boundary conditions, a three-dimensional transient temperature field simulation calculation was carried out during the construction period of the rock wall beam. Specific analysis was conducted on the temperature of the cave, the temperature calculation results after pouring the rock wall beam, and the stress results at key points.

Figure 7 and 8 show the temperature variation patterns of the four characteristic points in Scheme 1 and Scheme 4 over time, respectively. Comparing the various schemes, it can be seen that the cooling effect of the three row vertical cooling water pipe arrangement in Scheme 4 is the best, so only the variation diagrams of the control group and Scheme 4 are shown. Throughout the entire calculation cycle, the temperature at the center position N04 without cooling water pipes was higher than the temperature at the other three key points. When the template was removed on the 7th day, there was a significant change

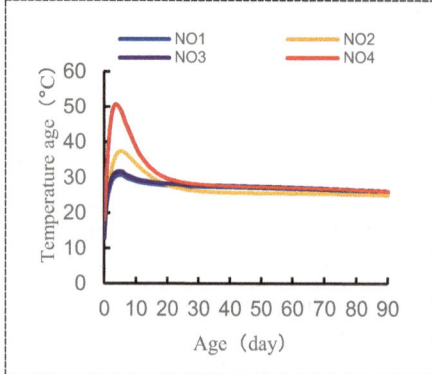
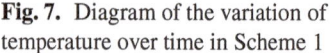
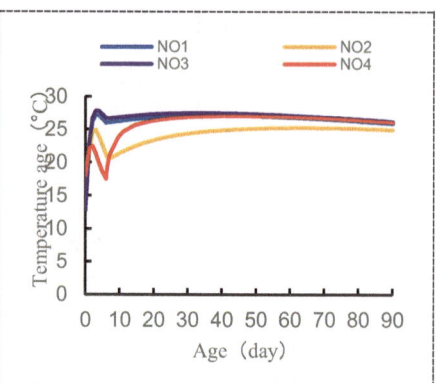

Fig. 7. Diagram of the variation of temperature over time in Scheme 1

Fig. 8. Diagram of the variation of temperature over time in Scheme 4

in temperature at each key point, and then the temperature gradually decreased. In the later stage of pouring, the temperature at each key point was significantly affected by the temperature and gradually tended to be consistent. And the center point NO4 in scheme 4 has the lowest temperature, and the other key points have not exceeded 30°C, indicating that the cooling effect is very good. After pouring, the temperature gradually decreases and the hole temperature gradually decreases.

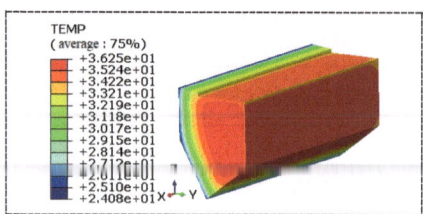

Fig. 9. Temperature distribution map of the middle section of the rock wall beam wall after one days of pouring

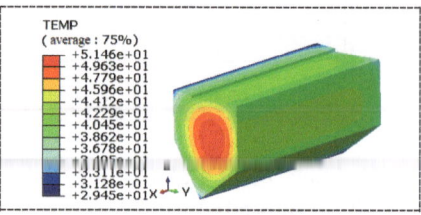

Fig. 10. Temperature distribution map of the middle section of the rock wall beam wall after 3.5 days of pouring

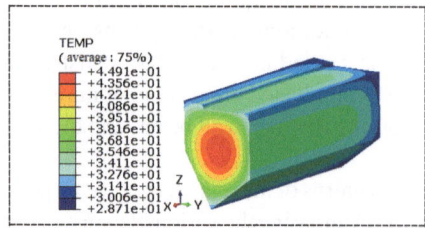

Fig. 11. Temperature distribution map of the middle section of the rock wall beam wall after 7 days of pouring

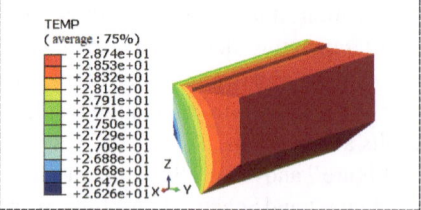

Fig. 12. Temperature distribution map of the middle section of the rock wall beam wall after 28 days of pouring

Figure 9, 10, 11 and 12 shows the temperature cloud maps of the middle section of the rock wall beam on the 1st, 3.5th, 7th, and 28th days of scheme 1.The internal temperature of the rock wall beam concrete rises rapidly three days before pouring. In the condition of no cooling water pipe, the highest temperature is about 36.25 °C on the first day, and reaches the highest value on the 3.5th day, about 51.46 °C. The highest temperature is located at the core of the rock wall beam concrete beam, and the concrete temperature gradually diffuses and decreases from the inside out; As time went on, the rate of temperature decrease at each location slowed down. After removing the template, the highest temperature reached 28.74 °C at 28 days.

In working condition two of the cooling water pipe, the temperature reached its highest value on the first day, about 33.88 °C. The high-temperature part is located at the corner of the rock wall beam surface where the water pipe is not buried and in contact with the air. On the third day, the temperature reached its highest value, about 41.50 °C. After removing the template, the temperature slowly decreased and tended towards the temperature inside the cave. At 28 days, the highest temperature reached 28.73 °C.

In working condition three of the cooling water pipe, the temperature reached its highest value on the first day, about 34.13 °C. The concrete temperature gradually diffused and increased from the inside out with the water pipe as the center. On the third day, the temperature reached its highest value, about 42.99 °C. After removing the template, the temperature slowly decreased and tended towards the temperature inside the cave. On the 28th day, the highest temperature reached 28.73 °C.

The cooling effect of the cooling water pipe is most obvious in the fourth working condition of the cooling water pipe. On the third day, the temperature no longer rises, and the peak appears one day earlier, with a temperature peak decrease of 15.66 °C. On the first day, the temperature reached its highest value, about 31.92 °C, and the concrete temperature gradually increased from the inside out with the water pipe as the center. On the second day, the temperature reached its highest value, about 35.80 °C, and the high-temperature part was only distributed on the surface of the rock wall beam and the front end in contact with the air. On the sixth day, the temperature inside the cave had dropped to nearly 28.80 °C. In the first 6 days, due to the formwork covering the rock wall beam, the heat dissipation of the concrete was slow. After 7 days of pouring and removing the formwork, the heat dissipation speed significantly accelerated. After 28 days of pouring, the temperature remained basically constant. Afterwards, the temperature decreased and was basically consistent with the temperature inside the cave. After 28 days, the highest temperature reached 28.73 °C. The final temperature stabilizes around 28 °C in the construction environment.

Figure 13 and 14 show the variation of the first principal stress over time for the four feature points in Scheme 1 and Scheme 4, respectively. The tensile stress of rock wall beam concrete is mainly caused by the self weight of the concrete and the temperature tensile stress. In the early stage of concrete pouring, the surface of the rock wall beam bears tensile stress and the interior bears compressive stress. And the surface tensile stress and internal compressive stress almost reach their maximum values when the concrete temperature reaches its peak, after three days of pouring. At this time, the maximum tensile stress on the outer surface is 1.38 MPa, and the maximum tensile stress is located at the contact position between the rock wall beam concrete and the crane beam rail. Due to the constraint of the lateral surrounding rock on the rock wall beam, the maximum compressive stress is located at the corner point of the transition zone between the rock

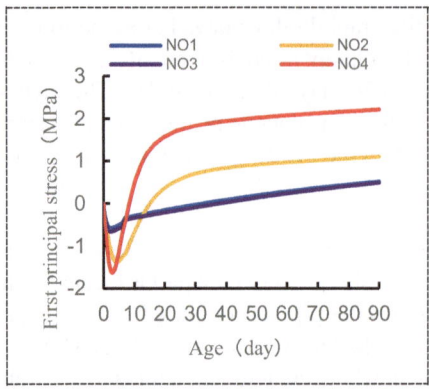

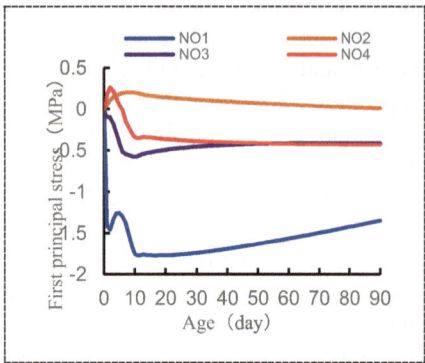

Fig. 13. Diagram of the variation of the first principal stress over time in Scheme 1

Fig. 14. Diagram of the variation of the first principal stress over time in Scheme 4

wall beam and the surrounding rock, which is 2.32 MPa. The maximum compressive stress inside the concrete of the rock wall beam is 1.63 MPa, located at the core of the concrete beam. As time increases, the tensile stress on the outer surface transforms into compressive stress. After the removal of the template on the 7th day, the internal compressive stress gradually decreases and transforms into tensile stress.

The compressive stress inside the concrete of the rock wall beam in working condition five of the cooling water pipe is mostly less than 0.90 MPa, and the maximum compressive stress is located at the corner point of the transition zone between the rock wall beam and the surrounding rock, which is 1.77 MPa. The central part of the beam is located near the cooling water pipe, with a maximum compressive stress of 0.33 MPa on the first day. During the cooling water period, the compressive stress gradually decreases in the first three days. After stopping the water supply and removing the mold, the internal temperature of the concrete rises, and the outer surface temperature is basically consistent with the hole temperature. The internal compressive stress gradually increases, reaching a maximum of 0.78 MPa on the tenth day after the mold is removed. Only at the contact between the cooling water pipe and the concrete is subjected to significant tensile stress, with a maximum tensile stress of 1.17 MPa. The maximum tensile stress on the outer surface reaches 0.62 MPa at the outlet of the cooling water pipe, while the maximum tensile stress in other areas is below 0.35 MPa. In the later stage of pouring, the temperature at each location is significantly affected by air temperature, with the internal temperature slightly lower than the external surface temperature, and the maximum tensile stress on the external surface reaching 1.35 MPa.

4   Conclusion

The maximum temperature of the unconnected cooling water pipe is 51.46 °C. The peak temperature in Scheme 2 decreases by 9.96 °C, the peak temperature in Scheme 3 decreases by 8.47 °C, and the peak temperature in Scheme 3 decreases by 15.66 °C. It can be seen that the arrangement of three rows of vertical cooling water pipes has the best cooling effect. The temperature of each key point in Plan 4 did not exceed 30 °C throughout the entire time cycle, and the temperature at the center point of the beam decreased most significantly.

After the removal of the non cooling water pipe mold, as the heat dissipation on the surface of the rock wall beam concrete increases, the hydration heat of the concrete continues to release heat, and the temperature difference between the inside and outside of the concrete increases, resulting in compressive stress on the surface of the concrete and tensile stress on the inside. The key point where the cooling water pipe is not connected in Plan 1 is subjected to compressive stress before demoulding and tensile stress after pouring. The key points of Plan 4 experience a decrease in the peak compressive stress before demolding, while NO1 experiences a smaller tensile stress. The tensile stress decreases in the later stage of pouring. The maximum tensile stress at the core of the concrete beam in the rock wall beam without cooling water pipes is 2.50 MPa. It can be seen that the tensile stress of scheme four also decreased the most significantly, with a peak tensile stress decrease of 0.21 MPa before demolding and a maximum tensile stress decrease of 1.15 MPa in the later stage of pouring.

References

1. Yu-Shuang, W., Xian-Bin, Y., Hao, Z.: An energy-based chemo-thermo-mechanical damage model for early-age concrete 2024 J. Eng. Fract. Mech. **295**, 109758 (2024)
2. Gu, G.l., Xu, W., Wu, Y.H.: Optimization of temperature control technology for cooling large volume concrete water pipes. J. Concr. **11**, 141–145 (2021)
3. Wang, Y., Xiong, G.: Experimental and simulation of internal temperature field in large volume concrete based on water pipe cooling. J. Civil Eng. Manage. **40**(03), 69–76 (2023)
4. Hua, C.R.: On the types and causes of early cracks in concrete. J China's new Technol. Prod. **15**, 158 (2011)
5. Tasri, A., Susilawati, A.: Effect of cooling water temperature and space between cooling pipes of post-cooling system on temperature and thermal stress in mass concrete. J. Build. Eng. **24**, 100731 (2019)
6. Myers, T., Fowkes, N., Ballim, Y.: Modeling the cooling of concrete by piped water. J. Eng. Mech. **135**(12), 1375 (2009)
7. Zeng, J.Q., Li, G.R., Chen, X.C.: Analysis of temperature creep stress in concrete foundation blocks using anisotropic thermal parameters of bedrock. J. Chengdu Univ. Sci. Technol. **05**, 1–6+71 (1994)
8. Qiang, S., Zhu, Y.M., Ding, B.Y.: Research on temperature control of gravity dams based on discrete algorithm of cooling water pipes. J. Hydroelectric Energy Sci. **05**, 93–95+155 (2008)
9. Zhu, B.F.: Equivalent heat conduction equation for concrete considering the cooling effect of water pipes. J. Hydraulic Eng. **3**, 28–34 (1991)
10. Liu, N., Liu, G.N.: Finite element substructure simulation technique for water pipe cooling effect. J. Hydraul. Eng. **12**, 45–50 (1997)

Conflict Between Fracture Toughness and Tensile Strength in Determining Fracture Strength of Rock Samples with a Circular Hole

Ali Lakirouhani[✉] and Somaie Jolfaei

Department of Civil Engineering, Faculty of Engineering, University of Zanjan, Zanjan, Iran
`rou001@znu.ac.ir`

Abstract. In this article, using a two-dimensional numerical model based on the finite element method, the fracture pressure of rock samples with a circular hole in the center is obtained. In this model, a mixed criterion is used to determine the fracture pressure, which is a combination of the tensile strength criterion and the fracture toughness criterion. The superiority of this model over the tensile strength criterion or the fracture toughness criterion is that it shows well the effect of the size of the hole in the middle of the sample on the fracture stress. The main and innovative finding of this article is that if the radius of the hole in the center of the sample is more than 70 mm, the fracture strength of the sample is dependent on the tensile strength of the rock material, and if the radius of the hole is less than 70 mm, the fracture strength is dependent on the fracture toughness. According to other results, the fracture strength decreases with the increase of the radius of the hole in the center of the sample, the decrease rate is high at first and then decreases with the increase of the radius of the hole.

Keywords: Stress Intensity Factor · Hole Radius · Size Effect · Brittle Rock · Fracture Toughness

1 Introduction

In rock mechanics laboratories, for various reasons, fracture resistance tests are performed on rock samples with a hole in the center. For example, in hydraulic fracturing tests, to determine the fracture pressure, rock samples with a circular cavity are subjected to triaxial compressive stresses. But it has been observed that the fracture pressure in rock samples, although it depends on the mechanical properties of the rock and the fluid injection rate, but it also depends on the hole radius. In hydraulic fracturing tests, the fracture pressure is also dependent on the length of the initial crack in the cavity wall. The dependence of the fracture pressure on the size of the borehole is also very important in the field, where the breakdown pressure of the hydraulic fracture is supposed to be used in the estimation of the in situ stresses.

In rock and rock masses, failure is mainly caused by shear and tensile stresses caused by compressive stresses [1–7], in other words, when the tensile stresses exceed the tensile

© The Author(s) 2024
G. Feng (Ed.): ICCE 2023, LNCE 526, pp. 495–503, 2024.
https://doi.org/10.1007/978-981-97-4355-1_46

strength of the rock, tensile failure occurs [8–13]. This is the traditional classic view of tensile failure, against which there is the view of energy instability. The theory of energy instability, which today has led to the linear elastic fracture mechanics (LEFM) method, states that when the stress intensity factor at the crack tip exceeds the fracture toughness, a crack starts to propagate [14, 15]. On the other hand Leguillon (2002) showed that both the energy criterion and the stress criterion are necessary conditions for failure, but neither of them alone is sufficient [15]. This is while the energy criterion provides the lower bound for crack length and the stress criterion leads to the upper bound for admissible crack length.

If there is a hole in the center of the sample, due to the concentration of tensile stress, tensile failure occurs in the cavity wall and along the maximum in situ stress. Although, due to the use of linear elastic theory, the hole size is not included in the classical criterion based on tensile strength, but the fracture mechanics criterion is able to consider the effect of the initial crack length on the fracture pressure.

Some laboratory studies showed that the fracture pressure is dependent on the tensile strength and the fracture toughness of the material [7, 15], in such conditions, the effect of the size of the hole in the center of the sample can be seen in the fracture pressure. Recently, the authors obtained the breakdown pressure of hydraulic fracture by combining the tensile strength criterion and the fracture mechanics criterion and investigated the effect of borehole radius and initial crack length on the breakdown pressure [16].

This study investigates fracture pressure of brittle rock samples with a circular cavity using a plane strain two-dimensional finite element model. The mixed failure criterion is based on a combination of tensile strength (stress criterion) and fracture toughness (energetic criterion) of the material. The superiority of this model over the previous models is that it examines both the tensile strength of the material and the fracture mechanics at the same time, and by using it, the effect of the radius of the hole in the center of the sample can be seen on the fracture resistance. Basically, the mixed criterion in this paper covers the gap between the tensile strength criterion and the toughness criterion. This model, while being simple, is very effective in determining the fracture stress of brittle materials. In the next section, the geometry of the problem and the solution method are explained, and in the Sect. 3, the results of the numerical analysis are given.

2   Finite Element Numerical Model, Mixed Criterion

As seen in Fig. 1, a sample with a circular cavity of radius $R_o = 5$ cm is considered. This sample is under compressive stress p_o. Here, the dimensions of the model are considered to be 2×2 m^2. The purpose of this article is to determine the pressure p_o so that tensile failure occurs in the wall of the cavity and along p_o. In this article, the mixed criterion is used to find the fracture pressure, which means that material tensile strength criterion and the fracture mechanics criterion must both be satisfied, i.e.:

$$\begin{cases} K_I(p_0, c) = K_{IC} \\ \sigma_\theta(p_0, c) = \sigma_t \end{cases} \rightarrow \begin{cases} p_f = p_0 \\ c_f = c \end{cases} \tag{1}$$

In this relationship, p_f is the fracture pressure and c_f is the fracture initiation length. σ_θ is the maximum tangential stress along the imaginary crack with length c. σ_t is the tensile strength of the rock material and K_{Ic} is the fracture toughness. K_I is the stress intensity factor at the crack tip with a length of c. To find these values, we draw the graphical form of the first and second criterion in one plot. The point of intersection of two criteria indicates the fracture pressure and the length of fracture initiation. Figure 2 shows an example of a mixed criterion.

After discretizing the geometry of the problem and applying the boundary conditions, the problem is solved by the finite element method and the fracture pressure is obtained. For this, a code is written in the MATLAB program.

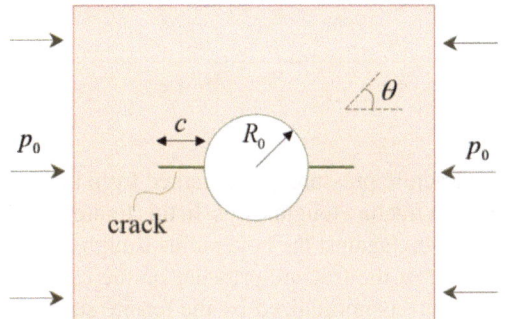

Fig. 1. The geometry of 2D numerical model

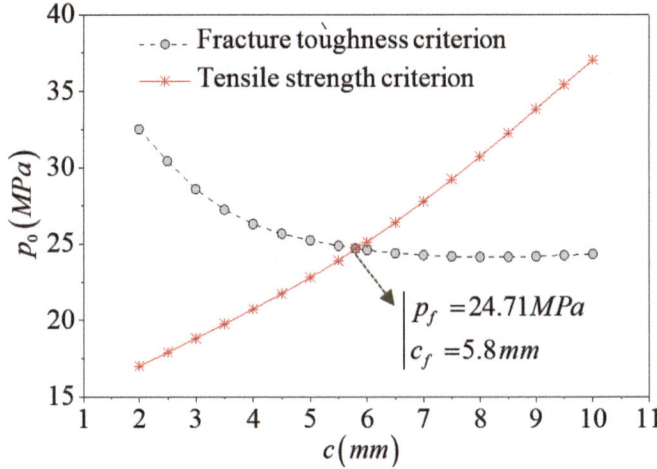

Fig. 2. The mixed criterion, $R_0 = 50\,mm$, $\sigma_t = 14\,MPa$, $K_{Ic} = 2.5\,MPa\,\sqrt{m}$

3   Numerical Analysis Results

For the characteristics of the materials presented in Table 1 and using the mixed criteria explained above, various numerical analysis are performed, the results of which are given below.

Table 1. The mechanical properties of model

Rock properties	Symbol	Unit	Value
Young's modulus	E	GPa	40
Poisson's ratio	v	---	0.22
Tensile strength	σ_t	MPa	8, 10, 12,14
Fracture toughness	K_{Ic}	$MPa\sqrt{m}$	1.5, 2, 2.5, 3

Figure 3 shows the fracture pressure (p_f) obtained from the mixed criterion against the hole radius for different fracture toughnesses. In this figure, it can be seen that the rate of change of fracture pressure against the hole radius is high at first and then decreases. Therefore, the dependence of the fracture pressure on the cavity radius can be clearly seen. If the fracture pressure is normalized by the tensile strength of the material and plotted against the radius of the hole, Fig. 4 is obtained. The important point that can be obtained from this Fig. 4 is that for $R_o > 70$ mm, $\frac{p_f}{\sigma_t} \simeq 1$, that is, for holes with a large radius, the tensile strength of the rock determines the fracture pressure, while for holes with a small radius, the fracture mechanics criterion plays an essential role in determining the fracture pressure of the samples. Thus, the mixed criterion captures the size effect well. Figures 3 and 4 show that with increasing toughness, the fracture pressure increases. But for $R_o > 70$ mm, the fracture pressure for different fracture toughnesses is almost the same, and therefore toughness has no effect on the fracture pressure of the rock samples, this is because for $R_o > 70$ mm as mentioned, it is the tensile strength of the rock that is effective in the fracture of the model (Fig. 3). As the radius of the hole decreases, the distance between the curves with different toughness increases, because for $R_o < 70$ mm, the fracture toughness is effective in the fracture of the model.

Figure 5 shows the fracture pressure against the hole radius for different values of tensile strength. In this figure, it can be seen that for $R_o < 70$ mm, the sensitivity of the model to the tensile strength of the rock is low, but for $R_o > 70$ mm, the fracture pressure increases slightly with the increase of tensile strength.

Figures 6 and 7 also show the fracture initiation length (c_f) versus the hole radius. It can be seen in Fig. 6 that for samples with the same hole radius, the length of fracture initiation increases with the increase of fracture toughness. But Fig. 7 illustrates that, for samples with the same hole radius, as the tensile strength of the sample increases, the fracture pressure decreases. Also, Figs. 6 and 7 illustrate that as the radius of the hole increases, the initiation fracture length increases, the rate of increase of the initiation length of the fracture is high at first, but then it decreases.

Figure 8 shows the comparison between the results of the numerical model presented in this paper and the experiments conducted by Carter et al. (1992) on sandstone [7]. This Fig. Shows that there is a good agreement between the two sets of results.

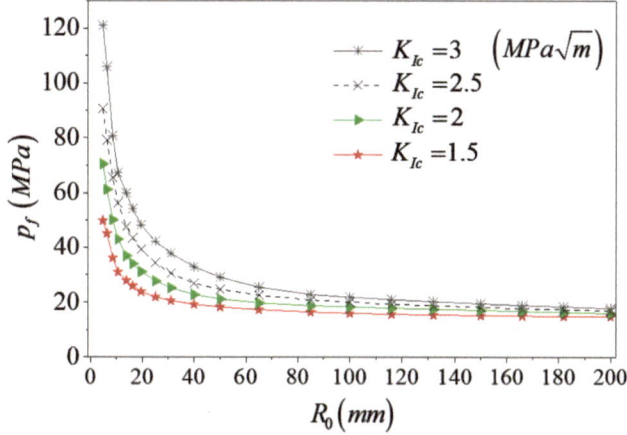

Fig. 3. Fracture pressure versus cavity radius ($\sigma_t = 14\ MPa$)

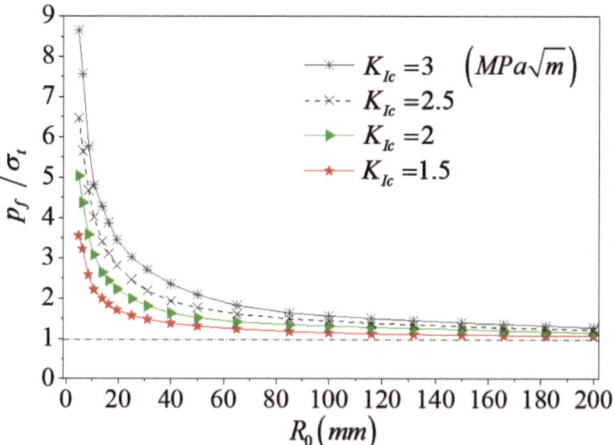

Fig. 4. Normalized fracture pressure obtained from the mixed criterion versus cavity radius

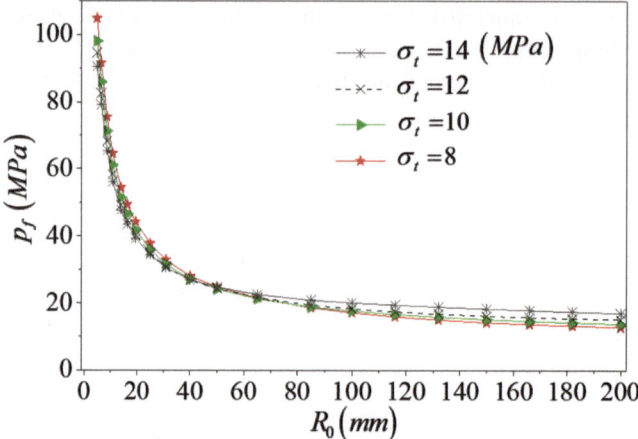

Fig. 5. Fracture pressure versus cavity radius $\left(K_{Ic} = 2.5\,MPa\sqrt{m}\right)$

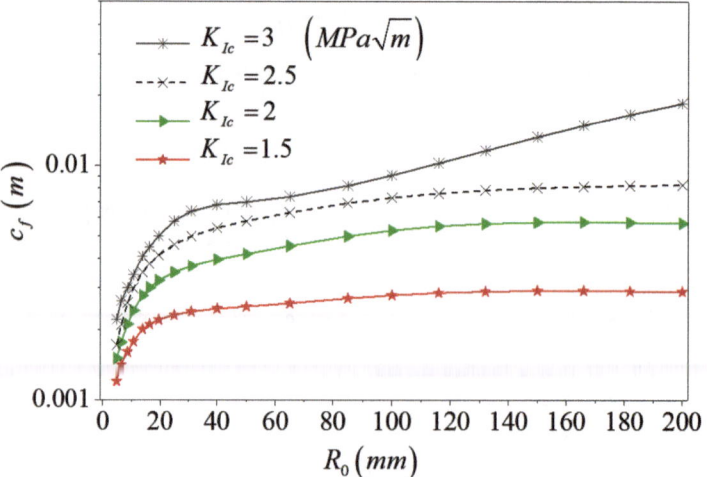

Fig. 6. Fracture length versus cavity radius $\left(\sigma_t = 14\,MPa\right)$

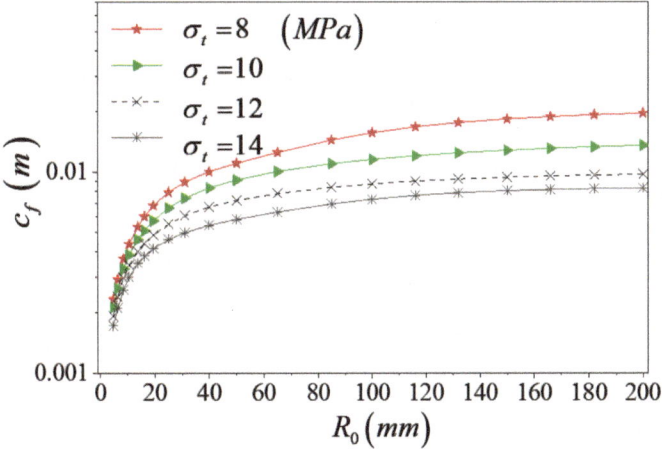

Fig. 7. Fracture length versus cavity radius $\left(K_{Ic} = 2.5\,MPa\sqrt{m}\right)$

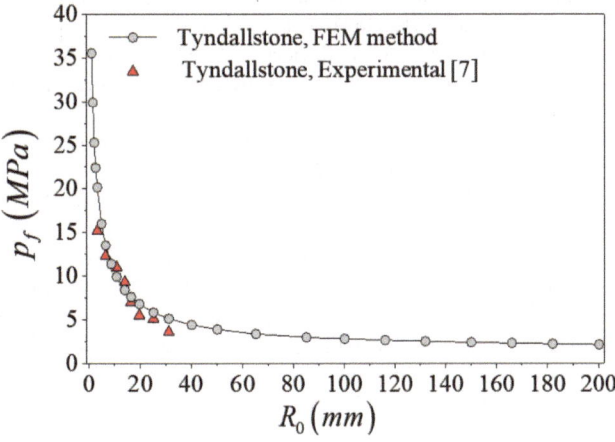

Fig. 8. Comparison between finite element numerical analysis and laboratory results

4 Conclusion

In this article, using a numerical model based on the finite element method and mixed tensile fracture criterion technique, the fracture pressure of rock samples with a hole in its centre was obtained. Using the technique of mixed fracture criterion, the effect of the size of the hole in the middle of the sample was observed on the fracture pressure. The obtained results show that when $R_o > 70$ mm, the tensile strength criterion plays a role in determining the fracture pressure of rock samples, while for $R_o < 70$ mm, the fracture mechanics criterion determines the fracture pressure of the model. Also, as the radius of the hole increases, the fracture pressure decreases, the fracture pressure reduction rate is high at first and then decreases, and the fracture pressure converges to the tensile strength of the rock.

References

1. Hoek, E.: A photoelastic technique for the determination of potential fracture zones in rock structures. In: ARMA US Rock Mechanics/Geomechanics Symposium, pp. ARMA-66. ARMA (1966)
2. Bahrehdar, M., Lakirouhani, A.: Evaluation of the depth and width of progressive failure of breakout based on different failure criteria, using a finite element numerical model. Arab. J. Sci. Eng. **47**(9), 11825–11839 (2022)
3. Bahrehdar, M., Lakirouhani, A.: Assessment of interplay of mud cake and failure criteria on the lower limit of safe borehole pressure. Indian Geotech. J. 1–13 (2024). https://doi.org/10.1007/s40098-023-00856-8
4. Lakirouhani, A., Hasanzadehshooiili, H.: Review of rock strength criteria. In: Proceedings of the 22nd World Mining Congress & Expo, pp. 473–482 (2011)
5. Lakirouhani, A., Jolfaei, S.: Hydraulic fracturing breakdown pressure and prediction of maximum horizontal in situ stress. Adv. Civil Eng. **2023**, 1–14 (2023)
6. Lakirouhani, A., Bahrehdar, M., Medzvieckas, J., Kliukas, R.: Comparison of predicted failure area around the boreholes in the strike-slip faulting stress regime with Hoek-brown and Fairhurst generalized criteria. J. Civ. Eng. Manag. **27**(5), 346–354 (2021)
7. Carter, B.J., Lajtai, E.Z., Yuan, Y.: Tensile fracture from circular cavities loaded in compression. Int. J. Fract. **57**, 221–236 (1992)
8. Lakirouhani, A., Ghorbannezhad, S., Medzvieckas, J., Kliukas, R.: Failure analysis around oriented boreholes using an analytical model in different faulting stress regimes. J. Civ. Eng. Manag. **29**(4), 360–371 (2023)
9. Jolfaei, S., Lakirouhani, A.: Sensitivity analysis of effective parameters in borehole failure, using neural network. Adv. Civil Eng. **2022**, 1–16 2022
10. Hubbert, M.K., Willis, D.G.: Mechanics of hydraulic fracturing. Trans. AIME **210**(01), 153–168 (1957)
11. Zhang, X., Lu, Y., Tang, J., Zhou, Z., Liao, Y.: Experimental study on fracture initiation and propagation in shale using supercritical carbon dioxide fracturing. Fuel **190**, 370–378 (2017)
12. Bahrehdar, M., Lakirouhani, A.: Effect of eccentricity on breakout propagation around noncircular boreholes. Advances in Civil Engineering. **2023**, 6962648 (2023)
13. Jolfaei, S., Lakirouhani, A.: Initiation pressure and location of fracture initiation in elliptical wellbores. Geotech. Geol. Eng. **41**, 1–20 (2023)
14. Atkinson, C., Thiercelin, M.: The interaction between the wellbore and pressure-induced fractures. Int. J. Fract. **59**, 23–40 (1993)
15. Leguillon, D.: Strength or toughness? A criterion for crack onset at a notch. Eur. J. Mech.-A/Solids **21**(1), 61–72 (2002)
16. Lakirouhani, A., Jolfaei, S.: Assessment of hydraulic fracture initiation pressure using fracture mechanics criterion and coupled criterion with emphasis on the size effect. Arab. J. Sci. Eng. **49**, 1–12 (2023)

Hydrogeology-Induced Retrogressive Slope Toe Failure Initiation: A Coupled Physical and Numerical Modeling Approach

Biruk Gissila Gidday[1](✉) and Bisrat Gissila Gidday[2]

[1] Civil Engineering Department, Addis Ababa Science and Technology
University, Addis Ababa, Ethiopia
birukgidday2000@gmail.com
[2] Civil Engineering Department, Arba Minch Institute of Technology, Arba Minch University,
Arba Minch, Ethiopia

Abstract. The conceptual premise of the current study is that the chronological effect of geologic parameters and hydrologic variation has been enhancing the toe slope since then, and hence the natural slope has remained stable. To capture the phenomenon, in addition to an intensive literature review and the subsequent findings and rationales, comprehensive laboratory testing, physical flume experiments, and numerical modeling were performed. The flume experiments were extensively instrumented (pore pressure transducers, suction sensors, strain transducers, and tracer chemicals) to observe the retrogressive failure mechanism. The results show that the abrupt change in slope toe stress caused by an increase in slope angle, rise in pore water pressure, and dissipation of soil suction were responsible for radical change in strain and sudden failure. Findings show that the slope toe section was the spot of failure initiation and was subjected to sudden failure associated with accelerated retrogressions. Physical flume assessment and numerical simulations both show that modifying the slope toe could increase overall slope stability. Hence, slope toe contributes far more to slope stability than previously thought. Even though the soil type range and slope angle at which the retrogressive failure occurs remain a source of considerable ambiguity, proper slope toe treatment can be regarded as a critical remedy measure.

Keywords: Hydrogeology · Retrogressive Failure · Slope Toe · Stress · Flume Experiment · Soil Suction

1 Introduction

Significant progress has been made in the study of retrogressive and progressive slope failure in both saturated and unsaturated soils over the last few decades. In general, progressive and retrogressive slides that grow in the same or opposite direction of motion are the two recognized types of successive slope failures. Due to the obvious removal of the nether section of ground support, retrogressive slides were thought to be a series of simple columnar or circular cylindrical slides [1, 2]. Several laboratory-based flume

© The Author(s) 2024
G. Feng (Ed.): ICCE 2023, LNCE 526, pp. 504–513, 2024.
https://doi.org/10.1007/978-981-97-4355-1_47

experiments coupled with numerical models have been conducted over the last two decades to capture retrogressive failure phenomena [3, 4]. A recent study reported that an increase in slope toe saturation on river sand and residual granite could have resulted in a dominant failure mode of shallow retrogressive sliding [4]. Several papers have reported limited comparative studies of physically measured and numerical model responses, implying that more work is needed to capture the mechanisms of failure in natural slopes [5, 6]. Recent studies have developed theoretical and empirical models for complex infiltration analysis solutions [7, 8]. The empirical model developed by [7], based on the concept of wetting front advancement, has been the most widely used. Infiltration, combined with the assumption of uniform porous media, plays a significant role in slope instability, which is less likely to be true for natural soils [9, 10]. It has often been assumed that failures in many natural slopes begin at the toe as a consequence of the fact that, toe stress concentration is likely to be higher [11]. Classical research findings revealed that failure began far from the toe; however, it is invalid to assert that failure must always begin at the upper parts of a natural slope [13].

2 Background and Rationale

Since hillslope stability analyses include so many space-time variables, it may be difficult to fully address the effects of seepage and infiltration in variable saturation conditions. However, the geologic time series influence of infiltration around the toe slope was investigated in the current work with some conceptualized premises as one of the major controlling variables. When a physical-based model is combined with numerical analysis, the premises may be able to aid in understanding the mechanism that triggers retrogressive failure in natural slopes. Hence, the current study focused primarily on the true cause of slope toe failure and subsequent retrogression, which is not adequately addressed in the existing literature and research findings. The author was intrigued by the previously mentioned frequent failure initiation of the toe slope and decided to look into it further.

The coarser material in the toe portion not only filters and dissipates seepage energy but also acts as back support, enhancing stability under higher toe shear stress. According to research, cutting or eroding the slope toe support from a long natural slope can result in slope failure According to a recent review of the literature on the subject, reducing river erosion can help to slow the progression of slope-toe retrogressive landslides. However, when excessive deformation is applied, retrogression accelerates and the soil mass detaches irreversibly. As a result, it was fair to conclude that a mode of slope failure at the toe has received little attention thus far. Because of the prolonged hydrogeologic environment, the material stiffness in the toe soil may be greater than in the upslope portion, and the failure mode may be elastic.

3 Material and Methods

3.1 Instrumentation

Virtual-hydromet digital soil suction recorder: The virtual-hydromet is a microcontroller digital soil moisture-temperature recorder that embodies the cutting-edge microcontroller instrumentation design to evaluate volumetric water content and the corresponding soil suction. The assessment of moisture variation due to rainfall application was carried out using the Virtualhydromet moisture sensor product.

Pore pressure transducers:Positive and negative pore pressure measurements using piezometers and miniature tensiometers have been used in both field and laboratory-scale experiments to assess slope stability. In the current study, strain gauge type pore pressure transducers with 350 Ω resistance having a measuring capacity of 200 kPa and accuracy of 1.6mV/V (3200×10^{-6} strain) were used. Matric suction dissipation and the subsequent increment of pore water pressure due to rainfall infiltration were measured using tensiometers and pore pressure transducers located at different points in the flume (Table 1). The number of available sensors and the depth of the flume dictated the general location of the embedded pore pressure and suction sensors (Table 1).

Table 1. Location matrix of embedded sensors

Suction Sensors (S)	S1	S2	S3	S4	S5	S6	S7
Points (X, Y, Z)	(71, 11,4)	(26, 22, 17)	(52, 21, 12)	(63, 22, 6)	(41, 21, 1)	(22, 31, 21)	(11, 21, 4)
Pore water transducers		PWT_C4	PWT_C2	PWT_C3	PWT_C1		
Points (X, Y, Z)	NA	(26, 22,17)	(52, 21,12)	(63, 22, 6)	(41, 21, 1)	NA	NA
Strain sensors (ST)	ST (1, 2)	ST (3, 4)	ST (5, 6)	ST (7, 8)	ST (9, 10)	ST (11 − 13)	ST (14 − 16)
Points (X, Y, Z)	(75, 10, 5); (75, 20, 7);	(25, 20, 15); (35, 25, 10);	(50, 20, 10) (70, 10, 20);	(65, 20, 5); (55, 30, 10)	(55, 5, 10); (50, 20, 15)	(20, 30, 20); (40, 30, 20); (45, 15, 30)	(35, 20, 20); (30, 20, 20); (25, 25, 25)

Strain transducers and cameras:In civil engineering, strain gauge technology has long been used to measure deformation and compare it to analytical models. In the current study, 120-Ω strain transducers were used to capture the deformation characteristics during an increase in slope angle and rainfall infiltration. Strain gauges were embedded in both vertical and horizontal configurations to capture the real-time deformation and subsequent stress accumulation. The transducers were attached to a 1 mm steel rod and vertically embedded in various locations in the soil during compaction. Since the current study was focused on toe retrogressive failure initiation, most of the strain transducers were embedded near the toe of the slope. Even though stress sensors were not employed, the deformation characteristics and the pore pressure transducer observation could be enabled to estimate the location of stress accumulation.

This inorganic compound is a purplish-black crystal that dissolves when it comes into contact with infiltrated rainwater, producing intensely pink to purple color intensity gradients. The initial moisture content dissolved some of the tracing crystals that had been left for 24 h for moisture equilibration (Fig. 1). To observe the boundary of the tracer, the photo image was processed using ImageJ software, and the gray image is shown in Fig. 2b. Thus, the purple-colored area coverage could serve as a reference point in the future. The initial linear schematics of the tracing color scheme, shown in Fig. 1a, were compared to the dilated pattern of the color during rainfall infiltration.

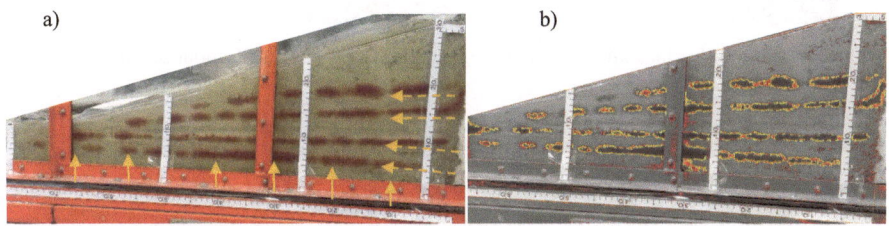

Fig. 1. (a) Linear orientation of trace element; (b) Boundaries of tracer (Gray image analysis)

4   Measurements and Observations

Rainfall characteristics: As can be seen from Table 1, the rainfall scheme began with a smooth transition from antecedent to main rainfall, with an intensity ranging from 4.4 mm/hr to 14.2 mm/hr applied on the slope of 25°. Figure 2a shows the impact of antecedent rainfall on the surface and Fig. 2b, c on side soils (tracer) during the first 48 h of periodic rainfall applications. To achieve a consistent distribution of rainfall, the spray nozzle locations (vertical and lateral distance) and subsequent raindrops on the soil surface were examined in a controlled environment. The initial tracer extent in Fig. 2was monitored after 24 h to account for moisture equilibration time; however, an additional 48 h were required to observe the effect of antecedent rainfall (Fig. 2b, c). When compared to the initial color extent depicted in Fig. 2, there was an increase in both pinkish area coverage and color signal strength at distances of 55 cm and 70 cm in the X-direction (Fig. 2b, c).

Slope failure: It was observed in both densities that the finer particles eroded from the uppermost part of the slope settled around the base-toe intersection and contributed to the stability of the slope temporarily. Given the significance of erosion, it is reasonable to conclude that the top surface fine particles were steadily eroded, resulting in the surface being fully degraded. The increase in slope angle caused a significant decrease in infiltration on the upper portion of the slope. During test label FMT4, the rainfall intensity increased to 43.3 mm/hr and the slope angle reached 40°. For in-situ testing conditions, the toe slope failure continued, accompanied by gradual retrogressive failure. The modified density testing, on the other hand, did not fail at label FMT4 but rather revealed a complete change in surface morphology.

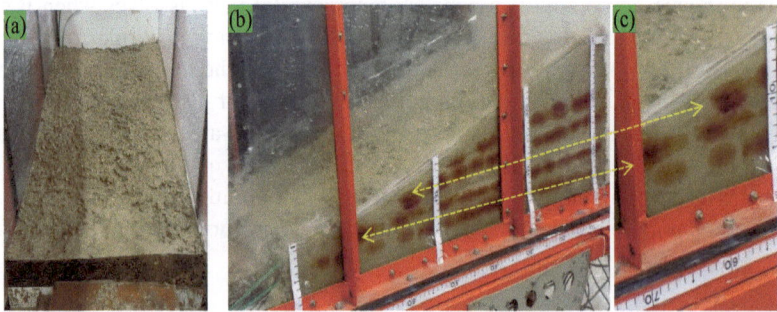

Fig. 2. (a and b) Surface and side view; (c) Enlarged side view at X about 55 cm – 70 cm

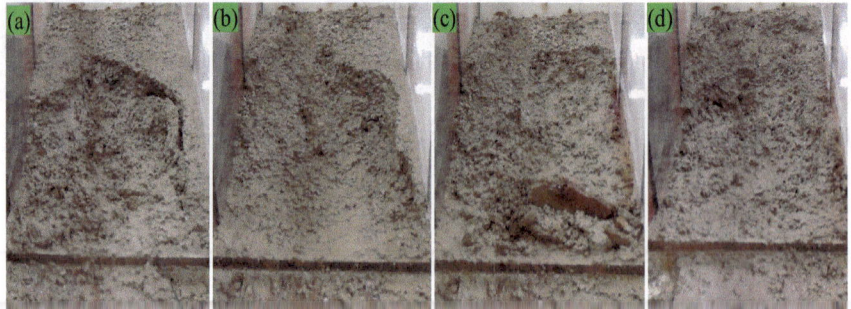

Fig. 3. Retrogressive failure for a modified density state

As shown in Fig. 3a, b, soil suction measurements were grouped based on their characteristics, namely early and late dissipation of soil suction. It was observed that the matric suction of the soil for suction sensors namely SS3, SS4, and SS6 became nil at about 45 h, 35 h, and 80 h, respectively (Fig. 3a). However, SS4 showed extended near-zero reading oscillation after 80 h with a maximum magnitude of 7 kPa. It was also discovered that sensors positioned at shallower depths, namely SS2 and SS6, recorded suction changes in response to infiltration (Fig. 3a, b). Damping of the suction shown in Fig. 3 was observed due to intermittent application of rainfall. The first two-day rainfall application, illustrated in Table 1, significantly reduced the soil suction as shown in Fig. 3.

Tracing chemical: This inorganic compound ($KMnO4$) had a purplish-black crystal when it was placed as a single-line orientation in the flume during soil compaction. It can be seen from Fig. 1a that the initial dissolved color was purple when in contact with field moisture content during compaction. A subsequent increment of the purple area coverage images during infiltration was taken by a continuous camera record and compared with the initial image (Reference area coverage). Finally, ImageJ software was used to process and analyze sequential side photo images to determine area coverage by the diluted tracer. Gray image analysis was used to calculate the area and perimeters of the individual tracer images, shown in yellow cross-dots and red boundaries, and these are used to calculate, the area perimeter ratio. The area perimeter ratio represents the merging of the neighborhood tracer area during further infiltration, resulting in a single larger tracer area. Based on the observations, the area perimeter ratio for the second day was greater than the first day, indicating that infiltration has increased at the lower 2/3 section (Fig. 4a, b). Some intermediate image analysis results were not shown as the first and last results captured the effect of infiltration.

5 Numerical Modeling

In the current study, slope stability analyses were performed under both in-situ and modified density states of saturated steady seepage conditions using GeoStudio 2021 V3 software's SEEP/W and SLOPE/W. The analyses also considered the groundwater table condition in a similar region to the soil sampling region. The slope geometry and rainfall characteristics shown in Fig. 4a were nearly identical to the laboratory physical flume experiments. The modified and enlarged portion of the slope toe has different observation sections, as shown in Fig. 4b. At the lower one-third slope portion, section profile j is parallel to the slope surface with elevation-distance coordinates of 4.56 to 2.61 and 8.57 to 11.78, respectively (Fig. 4b). Until it reaches the modified bottom section, the stress profile in Fig. 5a is nearly constant.

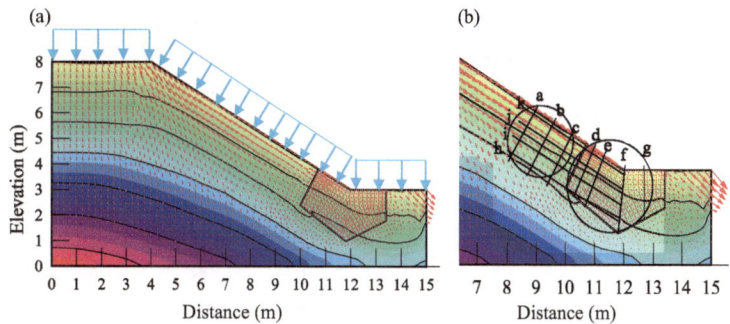

Fig.4. (a) Model geometry, rainfall, and flow gradients; (b) Modified slope toe section

Because the profile section was parallel to the slope face, the stress profile remained unchanged. However, an abrupt change in stress was observed beginning around a modified section and continuing to the end of the profiling line. This revealed that higher stresses were accumulated at the toe section, not only as a result of increased compactive effort but also as a result of slope angle and the resulting increase in driving gravity stress. Higher slope inclination may exacerbate this phenomenon in response to rainfall infiltration. Similarly, until they reached the modified section, the corresponding strains remained constant. Following that, the strain changes were increased and accumulated at a specific location and near zero strain value. Figure 4a depicts how the stress fields increase in depth. However, up to 1.2 m depth, constant strains were observed, followed by abrupt changes in strain increment at a distance of 11.9 m. (Fig. 4b). This layer above could be regarded as the zone of maximum shear mobilization.

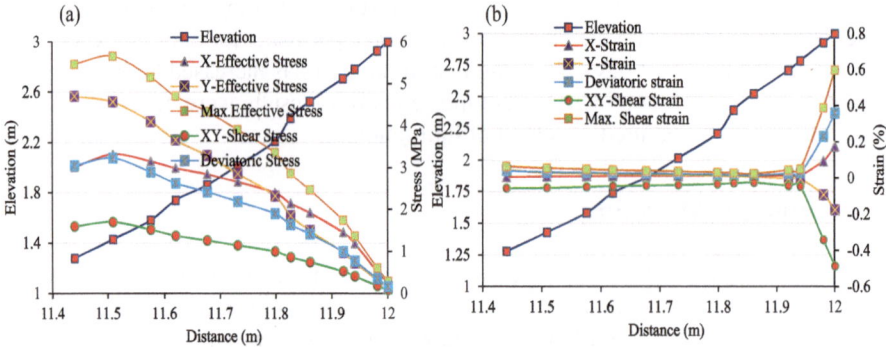

Fig. 5. Observation section f: (a) Stress profile; (b) Strain profile

In Fig. 5b, the in-situ/uniform compaction profile was compared to the modified section profile. Constant changes in strain were observed, as shown in Fig. 5b, in contrast to a modified section, which had an abrupt change in strain for an equal section length. Even though the soil was modeled with a 10% density increase, the unique property escalation of c' indicates that the region was more consolidated than the entire slope. Similarly, a sudden increase in shearing resistance was observed at approximately 10.5 m along the slip surface (Fig. 6). A 10% increase in density at the slope toe results in a 10.69 unit increase in shear resistance, according to Fig. 6b. This demonstrated that a minor change near the slope's toe could potentially improve slope stability and thus be used as a mitigation measure.

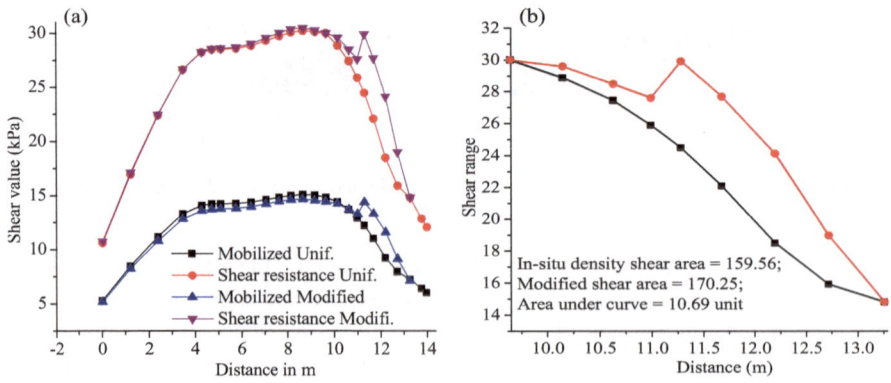

Fig. 6. (a) Variation in shear on in-situ and modified compaction; (b) Calculated shear area

6   Results and Discussions

As shown in Fig. 5a, the matric suction for SS3, which was placed in the lower third of the slope with PWT_C2, steadily dissipated suction and reached zero within 45 h. The corresponding pore pressure at PWT_C2 steadily increased due to superficial infiltration. Image processing and area-perimeter ratio analysis results verified that the infiltration and subsequent wetting front reached PWT_C2 (Table 2). Given this sensor (PWT_C3) was so close to the slope's toe, the suction should have dissipated as soon as feasible. In contrast, the pore pressure sensor PWT_C3, which was positioned alongside SS4, had continuously raised. In general, suction and pore pressure sensors have been expected to produce inverse results. However, a rise in PWT_C3 was found while soil suction climbed, which might be attributed to the combined influence of moist soil weight and pore air pressure. Despite this, image analysis revealed that the wetting front advancement had not reached PWT_C3 and had even shown modest infiltration during failure (Figs.4). As a result, the premise for the observation of increased pore pressure at PWT_C3 was accepted. Suction measurements for SS1, SS2, and SS7 were retained until the slope collapsed, and values for SS7 were retained even after the slope failed (Fig. 3b). Since the suction sensor SS1 was placed around the slope toe, the soil suction should have dissipated sooner, but it had a magnitude of around 15 kPa (Fig. 3b). The phenomenon reveals how a smaller proportion of finer soil can significantly impede infiltration. It can be deduced that slope toe modification using a finer proportion and proper drainage can improve soil suction and thus maintain slope stability. Otherwise, improper drain installation or trench excavation at the slope toe could remove the resisting soil mass and potentially trigger slope failure.

Numerical modeling (SEEP/W and SLOPE/W) sectional observation results revealed that stress fields crossing the modified slope toe are increasing and subjected to overstress (Fig. 6a). The subsequent changes in strain were increasing and also concentrating towards the assumed exit slip surface (Fig. 6a). When stresses exceed the maximum shear stresses, the under drained shear strength getting lower and hence progress of failure most likely initiated at the slope toe and ends following a potential slip surface. This revealed that modifying the slope toe in conjunction with appropriate drainage can improve slope stability and thus be used as a mitigation measure. Studies show that slope failures can be triggered by slope toe excavation for any use of engineering activity and are also believed as a hazardous procedure. The premises dictated in the introduction above, in natural slopes, the initial stress field at slope toe had been subjected to an additional change in stress due to geological slope formation. A subsequent time-induced hydrogeological and morphological phenomenon had also been superimposed on the original stresses. This significant phenomenon could be due to the fact that the overstressed slope toe had been already enhanced in stress-bearing through hydrogeologic changes.

7   Conclusions

The results of both physical experiments and numerical modeling show that the slope toe section contributes significantly to slope stability and can be more than theoretically confirmed. The abrupt change in slope toe stress due to an increase in slope angle and pore water pressure has resulted in a drastic change in strain which led to sudden failure. Since the stresses and the corresponding strains were accumulated at the slope toe, excessive deformation was the final stage of slope toe triggering. Thus, the retrogression accelerated along the slip surface, and the soil mass detached irrevocably and irreversibly. It may not be concluded that landslides often occur abruptly, however, natural slopes, particularly overstressed slope toes most often fail suddenly. Therefore, hillsides and toes associated with intensive public activities should be investigated in the manner that the slope toe material has a potential sudden risk or not. The premises and the subsequent results of this study applied not only to the slope toe but can also be extended upslope since the toe stress resistance capacity ascends. It can be concluded from a coupled physical and numerical modeling that a smaller enhancement of slope toe material, i.e., installation of the displacement pile, slope toe reinforcement, anchor, and jet grouting might be cost-effective and feasible. In addition, the slope toe modification using a finer proportion and proper drainage can improve soil suction and thus maintain slope stability.

References

1. Chandel, A., Singh, M., Thakur, V.: Retrogressive failure of reservoir rim sandy slopes induced by steady-state seepage condition. Indian Geotech J **51**, 705–718 (2021). https://doi.org/10.1007/s40098-021-00502-1
2. Wang, B., Vardon, P.J., Hicks, M.A.: Investigation of retrogressive and progressive slope failure mechanisms using the material point method. Comput. Geotech. **78**, 88–98 (2016). ISSN 0266–352X, https://doi.org/10.1016/j.compgeo.2016.04.016
3. Puzrin, A.M., Gray, T.E., Hill, A.J.: Retrogressive shear band propagation and spreading failure criteria for submarine landslides. Géotechnique **67**(2), 95–105 (2017)
4. Tohari, A., Nishigaki, M., Komatsu, M.: Laboratory rainfall-induced slope failure with moisture content measurement. J. Geotech. Geoenviron. Eng. **133**(5), 575–587 (2007). https://doi.org/10.1061/(ASCE)1090-0241133:5(575)
5. Sako, K., Kitamura, R., Fukagawa, R.: Study of slope failure due to rainfall: a comparison between experiment and simulation. In: Proceedings of the 4th International Conference on Unsaturated Soils, 2–6 April, Carefree, AZ, ASCE, pp. 2324–2335 (2006)
6. Hamdhan, I.N., Schweiger, H.F.: Finite element method–based analysis of an unsaturated soil slope subjected to rainfall infiltration. Int. J. Geomech. **13**(5), 653–658 (2013)
7. Green, W.H., Ampt, C.A.: Studies on soil physics: flow of air and water through soils. J. Agric. Sci. **4**, 1–24 (1911)
8. Beven, K., Horton, R.E.: Robert E. Horton's perceptual model of infiltration processes. Hydrol. Process. **18**(17), 3447–3460 (2004)
9. Li, J.H., Zhang, L.M.: Geometric parameters and REV of a crack network in soil. Comput. Geotech. **37**, 466–475 (2010)
10. Wang, G., Sassa, K.: Factors affecting rainfall-induced flowslides in laboratory flume tests. Géotechnique **51**(7), 587–599 (2001)
11. Chowdhury, R., Flentje, P., Bhattacharya, G.: Geotechnical slope analysis. In: Geotechnical Slope Analysis (2009)
12. Garg, S.K.: Irrigation Engineering And Hydraulic Structures Volume 2 of Water resources engineering; Ed. 23; Publisher, Khanna (2009). ISBN: 8174090479, 9788174090478
13. Blight, G.E., Brackley, I.J., van Heerden, A.: Landslides at amsterdamhoek and bethlehem – an examination of the mechanics of stiff fissured clays. Civ. Eng. South Africa, June, 129–140 (1970)

Evaluating the Performance of Reinforced Unpaved Roads Using Plate Load Tests

Bisrat Gissila Gidday[1]([✉]) and Biruk Gissila Gidday[2]

[1] Arba Minch Institute of Technology, Arba Minch University, Arba Minch, Ethiopia
bisratgissila@gmail.com
[2] Civil Engineering Department, Addis Ababa Science and Technology University, Addis Ababa, Ethiopia

Abstract. This study presents the strength of geosynthetic-reinforced unpaved road sections over soft subgrade utilizing plate load Test through a rectangular model box. Geotextile, geogrid, and geonet were used as a reinforcing component. The bearing capacity determined by the plate load test from the laboratory was compared with an analytical solution. The primary objective behind the reinforcement is to decrease the structural section without changing the traffic capacity and to increase the durability of the pavement. The test results showed the implications of a geosynthetic position on the strength of unpaved test sections, through a double reinforcement position always yielded the best improvement. It is seen from the results that there is a substantial increment in the Bearing Capacity of unpaved sections due to single reinforcements attaining a rate of about 34.60% with bi-axial geogrid, about 29.81% with geonet, and about 32.50% with nonwoven geotextile when reinforced within top one-third of subgrade layer.

Keywords: Bearing Capacity · Unpaved Roads · Geosynthetic Reinforcement · Soft Subgrade · Plate Load Test

1 Introduction

Weak subgrades are a widespread challenge in a temporary access road or a permanent road constructed over a weak subgrade road construction. The paved or unpaved surface can be deteriorated due to deformation of the subgrade. The benefit of geosynthetics in unpaved roads constructed over a weak subgrade is known to give a reinforcing benefit to the roadway sections. Geogrids, geonets, and geotextiles assist in sharing the loads more effectively and raise the efficient bearing capacity of the subgrade. Geosynthetic materials have been commonly utilized as reinforcement/ stabilization in structures through boundless materials, such as slopes, roads, embankments, and retaining walls. Geosynthetic stabilization and reinforcement is a mechanical process. Geosynthetics located either on top of the subgrade or within the subgrade/base course layer work with the soil and granular material to make a reinforced section through separation, confinement, and/or reinforcement functions.

© The Author(s) 2024
G. Feng (Ed.): ICCE 2023, LNCE 526, pp. 514–523, 2024.
https://doi.org/10.1007/978-981-97-4355-1_48

Geo-grids are widely accepted as reinforcement for enhancing engineering strength [1]. Reinforced soil technology is one of the mainly successful fields of civil engineering and has gained broad popularity because of its functional, constructional, and cost-effective benefits [2]. The geosynthetic reinforcement is mostly located between the sub-base and base layers at the interface between the subgrade and sub-base layers or within the base course layer of the flexible pavement [3, 4]. The placement location of reinforcement is the major factor influencing the bearing capacity of reinforced granular soil and maximum bearing capacity is examined when the depth of placement of reinforcement is lowered. Thus, geotextile assists in decreasing the vertical stress acting on the subgrade than in unreinforced pavements [5]. The use of geosynthetic reinforcement is effective in a weak subgrade with prominence for higher rut depths which can mobilize the 'tension membrane effect' of the geotextile [6]. The existence of the reinforcement layer raises lateral restraint or passive resistance of the fill material, raising the rigidity of the system and decreasing the vertical and lateral pavement deformation [7–9]. Reinforcement positioned high up in the granular layer hinders the lateral movement of the aggregate due to frictional interaction and interlocking between the fill material and the reinforcement which increases the apparent load-spreading capacity of the aggregate and decreases the required fill thickness [10, 11].

Geosynthetics have been in permanent use over the previous few decades as a reinforcing material in the cross-sections of pavements. Geogrids are mainly effective for reinforcement purposes and therefore, are mainly utilized in the design of road cross-sections. The primary objective behind the reinforcement is to decrease the structural section without changing the traffic capacity of the pavement or the durability of the pavement. An experimental test setup is needed to investigate the strength characteristics of both unreinforced and reinforced sections to investigate the behavior and to realize how and to what extent geosynthetics affects the engineering characteristics of subgrade soil. The function of geosynthetics in the design of a flexible pavement system has been calculated by performing laboratory tests on similar pavement sections, both in unreinforced and reinforced situations.

Many investigations in the field of geosynthetics application in pavement design have depicted that there is and decrease in base course thickness for specified structural capacity [12–14] and extension in the working life of pavement [15, 16, 17, 18] Performed tests on unpaved sections with varying base course thicknesses and depicted that a geotextiles-reinforced section with a 350 mm thick base layer performed the same as an unreinforced section with a 450 mm thick base layer. [19] Examined the construction of field-reinforced sections that contained a base course that was 50 mm thinner than that of unreinforced sections. Geogrid is extremely firm in tension, if compared to the subgrade material or aggregates, hence lateral stress is minimized in the reinforced base aggregate, resulting in less vertical deformation at the road surface. Due to shear interaction between the geogrid and material, the shear strength and thus the load distribution capacity of the utilized base course material is significantly improved [20].

This relation supports the decrease of reinforced aggregate layer thickness in comparison to the un-reinforced aggregate layers. There is a 40% reduction in the base course thickness after reinforcement in comparison to the unreinforced section for similar load-carrying capacity.

The main aim of this research study is to assess the performance of unpaved road sections reinforced with geosynthetics at different depths thereby also determining the optimal position of reinforcement. For this reason, extensive small-scale in-box static PLTs were conducted on several geosynthetic reinforced and unreinforced unpaved test sections. The improvement due to reinforcement has been assessed in terms of bearing capacity and base course reduction. The unpaved sections have been reinforced by applying one and a double layer of geosynthetics at different locations in the cross-section. Three types of geosynthetics (geogrids, geonets, and geotextiles) have been used in this study and their performance has been compared.

2   Material Used in the Study

2.1   Laboratory Model Tests

Figure 1 shows a typical cross-section of the testing set of a model device used in this study. The test box is rectangular-shaped, having inside dimensions of 1000 mm × 1000 mm and 800 mm in depth, and the walls of the tank have a thickness of 6 mm.

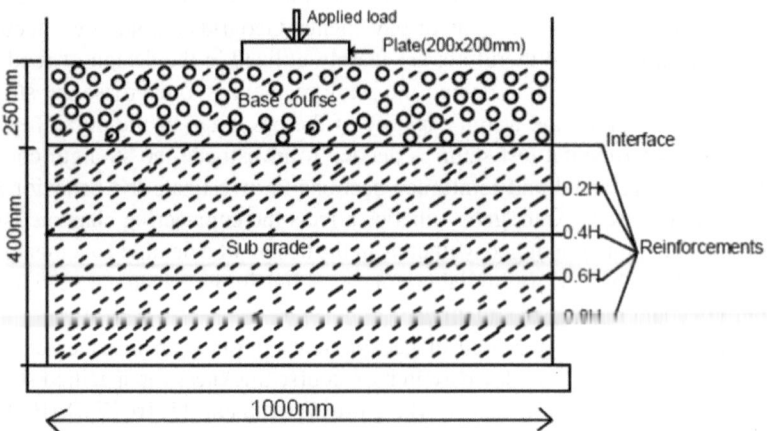

Fig. 1. Typical cross-section showing positions of reinforcements and loading configuration

2.2   Model Footing

The model footing was prepared from a steel square plate with a dimension of 200 mm × 200 mm, and 25 mm thickness. The load is transmitted to the footing via a ball bearing which is located between the footing and the proving ring. A proving ring of capacity 100 kN was applied.

2.3 Test Material

Important physical properties of the clay soil are shown in Table 1 and it was categorized as CH according to the Unified Soil Classification System (USCS), and A-7-5 according to the American Association of State Highway and Transportation\Officials (AASHTO) classification systems.

Table 1. Properties of clay soil

Parameter	Value
Specific gravity	2.75
Liquid limit (%)	67.88
Plastic limit (%)	34.69
Shrinkage limit	19.5
Plasticity index (%)	33.19
Optimum moisture content, (%)	26
Maximum dry density, (KN/m^3)	15.33

The size of aggregates ranges between 10–20 mm along with 10% stone dust from total weight were utilized to make the base course compacted to a unit weight of 20.61 kN/m3. This particular blend of materials in the base course has been chosen to determine the lowest voids and optimum compaction filled with stone dust, briefly stated by [21].

2.4 Reinforcement

Three types of geosynthetics (as shown in Fig. 2 during placement) with varying tensile strengths have been utilized in this study. The properties of these geosynthetics are given in Table 2.

Fig. 2. Geosynthetics creep testing machine, a) Nonwoven geotextile; b) Biaxial geogrid

The tensile strength of the geotextile and geogrid, obtained by the wide-width strip method [1] with a creep testing machine was illustrated in Fig. 2.

Table 2. Properties of Geosynthetics

Property	Biaxial geogrid	Non woven geotextile	Geonet
Material	Polypropylene	Polypropylene	High-density polyethylene
Mesh aperture size(mm)	34.6 × 34.6	0.1 × 0.1	7.6 × 7.6
Thickness(mm)	0.96	1.50	3.26
Mass per unit area(g/m^2)	200.00	334.00	250.00
Ultimate tensile strength (kN/m)	25.45	20.5	21.86

3   Model Constructions and Testing Procedure

The typical cross-section of the testing setup is illustrated in Fig. 1. The test setup includes a 250 mm aggregate base layer overlying subgrade clay of thickness 450 mm contained in a rectangular tank. The subgrade layer is set in three layers compacted to attain a unit weight of 13.80 kN/m^3 which is 90% of its maximum dry unit weight as determined by the Modified Proctor test. The gravel aggregates were located on top of the subgrade layer and blended with stone dust were located in 250 mm thickness compacted to attain a maximum dry unit weight. The load was used on the top manually by a mechanical jack in slight increases till getting a failure. The settlement of reinforced and unreinforced clay soil was measured utilizing two dial gauges located on the different sides of the footing. The small-scale in-box static plate load test during the placement of reinforcement in the laboratory is shown in Fig. 3.

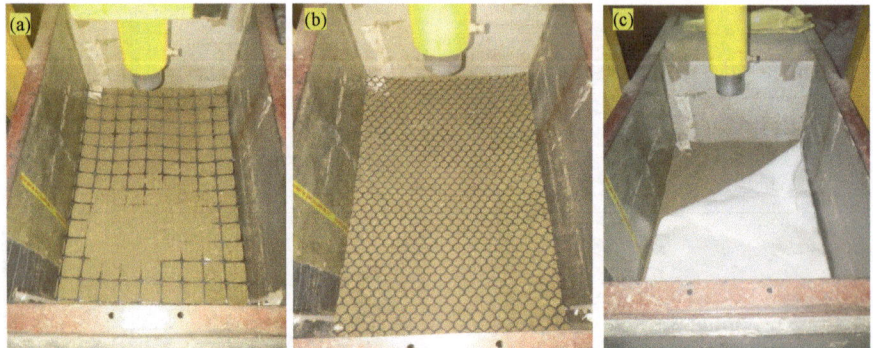

Fig. 3. The test set up under progress for plate load test a) Biaxial geogrid; b)Geonet; c) Geotextile

4 Results and Discussion

Results obtained from different laboratory tests by plate load to observe the load intensity versus settlement curves in both the unreinforced and reinforced conditions as illustrated in Figs. 4, 5, and 6 at different depths. The higher load intensities were recorded for a total settlement of 12.5 mm in all the test conditions. The position of geosynthetic reinforcement within the subgrade layer is one essential factor in the strength of unpaved sections. The following reinforcement configurations were selected to study this effect: placing geosynthetic at the base/subgrade interface, placing geosynthetic at 0.2 H, 0.4 H, 0.6 H, and 0.8 H of the subgrade layer, and placing double reinforcing layer at the upper (0.2 H and 0.4 H) and lower at (0.6 H and 0.8H) of the subgrade layer thickness. The effectiveness of reinforcement for the specified type of clay soil has been determined in the top one-third of the subgrade layer. [16] Examined that utilizing a geogrid at the top of the third layer in a soil sample by varying the plasticity index causes a significant rise in the CBR value compared with unreinforced soil in both soaked and unsoaked conditions. [22] Investigated the same trend while examining the result of geogrid reinforcement on the ultimate bearing capacity of sand.

It is examined from the results that there is a substantial increase in the load-carrying capacity of reinforced unpaved sections as compared to unreinforced conditions. The BCR value at the 12.5 mm settlement range for biaxial geogrid is 1.21–1.53, for geonet, it is 1.10–1.42 and from 1.05–1.51 for geotextile by moving the location of single layer reinforcement.

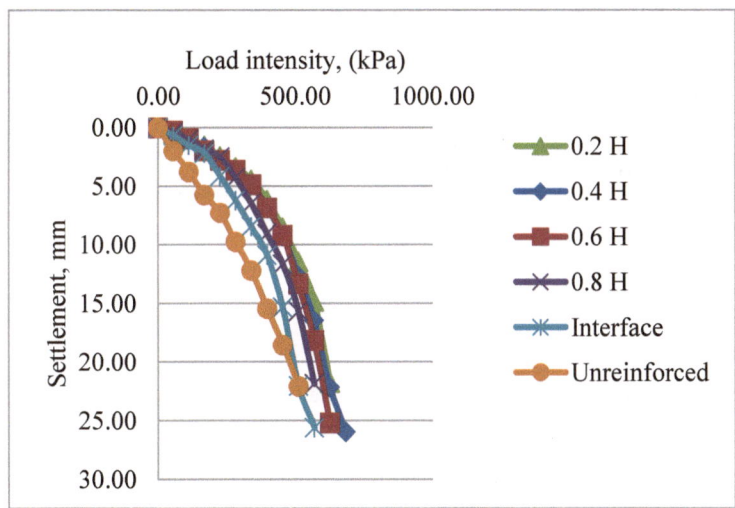

Fig. 4. Load intensity versus settlement curve for single biaxial geogrid reinforcement

Figure 7 depicts load intensity versus settlement curves for the tests performed in both the unreinforced and double-reinforced conditions at varying depths. As predicted, the BCR rises as the number of reinforcement layers increases. The BCR is defined as the ratio of the bearing capacity of reinforced unpaved sections to that of the unreinforced

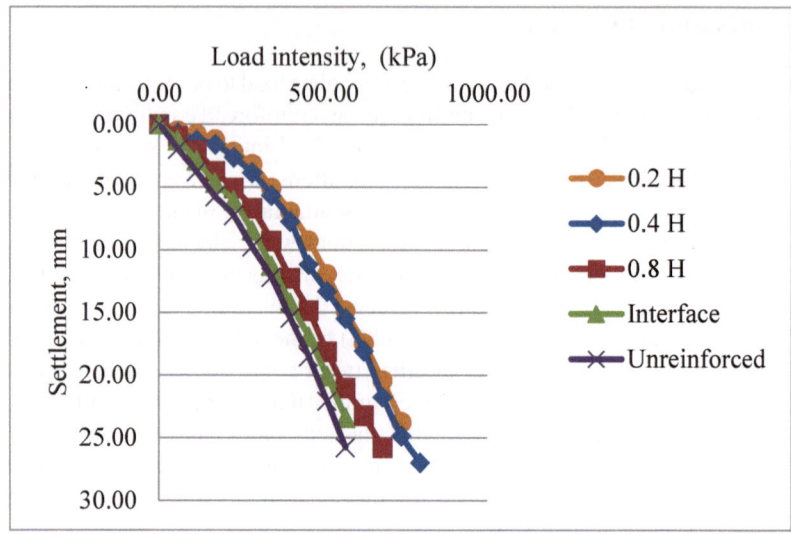

Fig. 5. Load intensity versus settlement curve for single nonwoven geotextile reinforcement

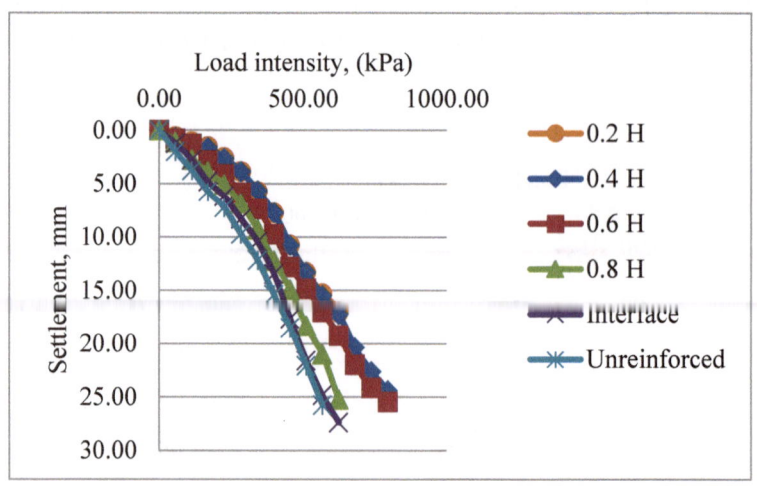

Fig. 6. Load intensity versus settlement curve for single geonet reinforcement

unpaved section. Among the two-layer geosynthetics, biaxial geogrid performed better than geonets and geotextiles. [11] Explained that bearing capacity rises with an increase in the number of geogrid layers from 33.33% with a single layer to 44.44% with a double layer, though double-layer reinforcement may become too expensive in road construction.

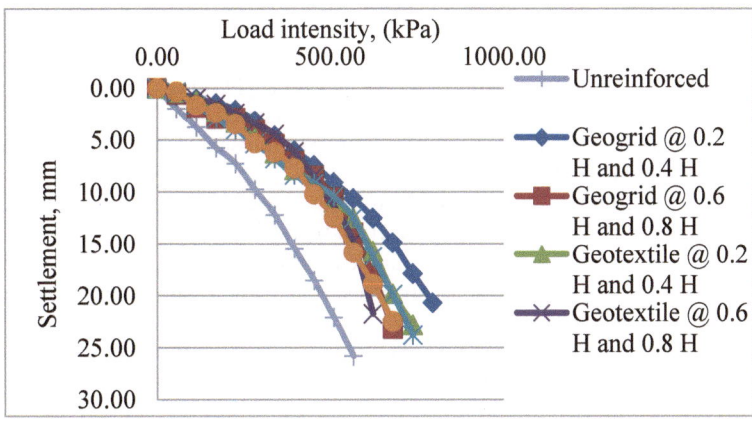

Fig. 7. Load intensity versus settlement curve for double-layer geosynthetic reinforcement

4.1 Ultimate Bearing Capacity of Reinforced Unpaved Sections

The experiment part of this study is considered to be a two-layer system. Since the performance of the granular aggregates base layer is much bigger than that of the underlying subgrade soil layer, a punching shear failure will exist in the granular aggregates layer followed by a general shear failure in the underlying soil layer. This kind of failure mode was primarily investigated by [13] for stronger soil overlying weaker soil. The cohesion and friction angle of the aggregate (base course) and the soft clay layer soil (subgrade) have been obtained from the experimental test as aggregate: $C_t = 8.82$ kPa (due to dust particles), $\Phi_t = 42°$, and subgrade: $C_b = 18$ kPa, $\Phi_b = 19°$. The adhesion, C_a, the punching shear coefficient, K_s, and the mobilized friction angle, δ, can be obtained by the graph given by Meyerhof and Hanna as 7.23 kPa, 5.2, and $30°$, respectively.

5 Conclusions

The following major conclusions are drawn from the results presented from the in-box static plate load experimental tests carried out on several geosynthetic reinforced and unreinforced unpaved test sections.

1. For a single layer of reinforcement, the upper one-third position gave the maximum enhancements under static loading conditions.
2. Laying the geosynthetic reinforcement in double locations yielded the largest enhancement. However, double-layer reinforcement may become uneconomical in road construction.
3. The highest BCR value of about 1.53, with bi-axial geogrid, about 1.51 with geotextile, and about 1.42 with geonet were investigated when reinforced within the top one-third of the subgrade layer.
4. From the test result, the bearing Capacity of unpaved sections due to single reinforcements attained a rate of about 34.60% with bi-axial geogrid, about 29.81% with geonet, and about 32.50% with nonwoven geotextile.

References

1. Rajesh, U., Sajja, S., Chakravarthi, V.K.: Studies on engineering performance of geogrid reinforced soft subgrade. In: 11th Transportation Planning and Implementation (2014)
2. Giroud, J.P., Noiray, L.: Geotextile-reinforced unpaved road design. J. Geotech. Eng. Div. **107**(9), 1233–1254 (1981)
3. Benjamin, C., Bueno, B., Zornberg, J.: Field monitoring evaluation of geotextile-reinforced soil-retaining walls. Geosynthetics Int. **14**(2), 100–118 (2007)
4. Bueno, B.S., Benjamin, C., Zornberg, J.G.: Field performance of a full-scale retaining wall reinforced with non-woven geotextiles slopes and retaining structures under seismic and static conditions. In: ASCE, pp 1–9 (2005)
5. Cancelli, A., Montanelli, F.: In-ground test for geosynthetic reinforced flexible paved roads. In: Proceedings of Twelfth European Conference on Soil Mechanics and Geotechnical Engineering (1999)
6. Carroll, R.G., Walls, J.C., Haas, R.: Granular base reinforcement of flexible pavements using geogrids, In Proceeding of Geosynthetics' 87, pp. 46–57. IFAI, New Orleans (1987)
7. ASTM D4595 Standard Test Method for Tensile Properties of Geotextiles by the Wide-Width Strip Method, ASTM International, West Conshohocken, PA, USA (2023)
8. Perkins, S.: Constitutive modeling of geosynthetics. Geotext. Geomembr. **18**(5), 273–292 (2000)
9. Som, N., Sahu, R.: Bearing capacity of a geotextile-reinforced unpaved road as a function of deformation: a model study. Geosynthetics Int. **6**(1), 1–17 (1999)
10. Omar, M.T., Das, B.M., Puri, V.K., Yen, S.C.: Ultimate bearing capacity of shallow foundations on the sand with geogrid reinforcement. Can. Geotech. J. **30**(3), 545–549 (1993)
11. Singh, A.K., Mittal, S.: Analysis of reinforced unpaved roads by modified structural number method. Int. J. Geosynthetics Ground Eng. **4**(1), 1–8 (2018). https://doi.org/10.1007/s40891-017-0115-5
12. Anderson, P., Killeavy, M.: Geotextiles, and geogrids: cost-effective alternate materials for pavement design and construction, Proceeding of Geosynthetics'89, IFAI, vol. 2, pp. 353–360. Sand Diego, USA (1989)
13. Chen, Q.: An experimental study on characteristics and behavior of reinforced soil foundation (Ph.D. dissertation), Louisiana State University, Baton Rouge, USA (2007)
14. Dey, A., Meena, S.: Geosynthetic reinforced unpaved road resting on c-ϕ subgrade. In: Indian Geotechnical Conference, Roorkee (2013)
15. Koerner, R., Soong, T.Y., Koerner, R.M.: Earth retaining wall costs in the USA, Geosynthetic Research Institute Report, Folsom, Pennsylvania, USA (1998)
16. Kolay, P.K., Kumar, S., Tiwari, D.: Improvement of bearing capacity of shallow foundation on geogrid reinforced silty clay and sand, Journal of the Transportation Research Board No. 2462, Transportation Research Board of the National Academies, Washington, pp. 98–108 (2014)
17. Leng, J.: Characteristics and behavior of geogrid reinforced aggregate under cyclic load, PhD thesis, North Carolina State University, Raleigh, USA (2002)
18. Meyerhof, G.G., Hanna, A.M.: Ultimate bearing capacity of foundations on layered soils under inclined load. Can. Geotech. J. **15**(4), 565–572. https://doi.org/10.1139/t78-060 (1978)
19. Miura, N., Sakai, A., Taesiri, Y.: Polymer grid reinforced pavement on soft clay grounds. Geotext. Geomembr. **9**, 99–123 (1990)
20. Montanelli, F., Zhao, A., Rimoldi, P.: Geosynthetics reinforced pavement system: testing & design. In: JN Paulson (ed) Geosynthetics'97 Proceedings, Long Beach, CA. vol. 2. Industrial Fabrics Association International, Roseville, MN, pp. 619–632 (1997)

21. Naeini, S.A., Moayed, R.: Effect of plasticity index and reinforcement on CBR value of soft clay. Int. J. Civil. Eng. **7**(2), 124–130 (2009)
22. Perkins, S.W., Edens, M.Q.: A design model for geosynthetics reinforced pavements. Int. J. Pavement Eng. **4**(1), 37–5024 (2003)

The Effect of Glass Fiber on the Notched Izod Impact Strength of Polybutylene Terephthalate/Glass Fiber Blends'

Luong Quoc Khanh[1], Tran Hoang Phuc[1], Nguyen Dinh Quang[1],
Pham Thi Hong Nga[1(✉)], Pham Quan Anh[1], Nguyen Thanh Tan[1],
and Ho Thi My Nu[2]

[1] Ho Chi Minh City University of Technology and Education, No. 1 Vo Van Ngan Street.,
Thu Duc, Ho Chi Minh City, Vietnam
hongnga@hcmute.edu.vn
[2] Ho Chi Minh City University of Industry and Trade, 140 Le Trong
Tan Street, Tay Thanh, Tan Phu, Ho Chi Minh City, Vietnam

Abstract. Improving the fire resistance of polymer and construction materials
is an urgent requirement today. This article aims to study the incorporation of
Polybutylene terephthalate (PBT), a versatile polymer used in most glass fiber
(GF) combination fields, to study their fire resistance and impact resistance for
manufacturing battery housings. We conducted injection molding and scanning
electrode microscopy (SEM) and had the expected results. The impact strength of
PBT/GF samples measured according to ASTM D256 is 4.69, 4.43, 3.71, 4.67,
and 6.28 kJ/m^2. The PBT/25% GF sample achieved the highest average impact
range of 6.28 kJ/m^2. The suitable GF content in the mixture is from 20–30%.
Combined with SEM microstructure, it was found that the content and density of
GF distribution have a positive effect on the mixture, making the bond between
them remarkably tight and robust. Thanks to the property of GF, which is non-
flammable, it makes the material more effective fire resistant. This research is the
basis for scientific developments that help solve the problem of impact resistance
of this new plastic in production and is a premise for other research purposes in
the future.

Keywords: Polybutylene Terephthalate · Glass Fiber · PBT/GF Blend · Impact
Strength · Polymers

1 Introduction

The optimization of battery safety, efficiency, and shatter resistance was the inspiration
for starting this research. Polybutylene terephthalate (PBT) is a good candidate for the
purposes mentioned. PBT is a thermoplastic engineering polymer with high strength and
impact strength, low moisture absorption, water and chemical resistance, high thermal
stability, hardness, abrasion, short molding time, and good surface [1].

© The Author(s) 2024
G. Feng (Ed.): ICCE 2023, LNCE 526, pp. 524–530, 2024.
https://doi.org/10.1007/978-981-97-4355-1_49

PBT is widely used in electrical systems, including lamp holders, switches, circuit breakers, and motor housings. It has high heat resistance, mechanical strength and water resistance, and excellent insulating properties [2–4]. However, pure PBT is flammable. Glass fiber (GF) is added to PBT to prevent dripping and help maintain the integrity of combustion. Phosphorus-based compounds are added to PBT to improve fire resistance [5]. Melamine derivatives have been shown to synergize with phosphorus flame retardants to enhance the flame retardants of mixtures further [6, 7].

PBT is a suitable material for making battery cases. However, in this area, even if PBT has high rigidity to a certain extent, this property can be improved, and one way to do this is to add GF to reinforce PBT [8]. GF-modified PBT can manufacture electronic components that require operation under prolonged high-temperature conditions with high dimensional stability [9–11] and materials with high fire resistance [12, 13]. The effect of GF length and content has been reported in some polymers, such as polypropylene [14–17]. However, no publication has yet mentioned research on the effects of GF on its stiffness resistance to PBT.

The advantage of PBT is its resistance to deformation over time with a stable temperature. The hardness and bearing capacity are good, but poor fire resistance still has has disadvantages, so it cannot be widely used in some fields. GF is a suitable choice to improve the fire resistance of the material. If PBT is reinforced with GF, the impact toughness increases, contributing to the optimal properties.

2 Materials and Methods

Table 1. The composition of the sample (wt.%).

Sample	PBT (wt.%)	GF (wt.%)
GF5	95	5
GF10	90	10
GF15	85	15
GF20	80	20
GF25	75	25

This study used plastic materials PBT-30GF and PBT. PBT-30GF is produced by TA COMA CO., LTD, and PBT resin is supplied from Lanxess (Germany). PBT and PBT-30GF were mixed in the proportions in Table 1 and dried at 110 °C for about 4 h, reaching a humidity of less than 0.03%, using Toshiba's 100-ton injection molding method.

After injection molding, samples obtained at each ratio were used for impact strength measurement on the AG-X plus material tester, according to ASTM D256–10. Before proceeding with sizing, ensure the measuring medium is at a temperature of 23 ± 2 °C and a relative humidity of $50 \pm 5\%$ in the air. In addition, laboratory samples need to be stored for at least 40 h. Measurement: First, enter the measurement parameters into the machine at a speed of 5 mm/min. Next, fasten the V-notched sample to the device. After that, start measuring the impact length. Allowing the device to apply force to the upper part of the sample will cause the sample to break later. Finally, record the test data of the sample and remove the piece. Proceed to repeat the sequence of steps for the following models.

AG-X Plus is a PBT/GF blend impact strength tester. This device uses a method of converting energy from potential energy to kinetic energy to determine the impact resistance of a sample. The mass, length, and speed of movement of the striking hammer will affect the accuracy of the results when measuring. The AG-X Plus can compensate for errors caused by air friction resistance and hammer drop angle with the electronic control system. The ASTM D256-10 measurement method was used in this experiment's instructions and explanations in Fig. 1.

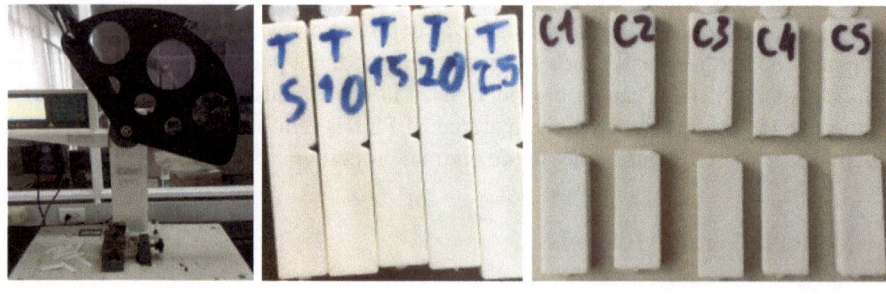

a) AG-X plus machine   b) Sample before testing   c) Sample after testing

Fig. 1. V-notch impact strength testing

3   Results and Discussion

Table 2. Average impact strength parameter of the test specimen

Sample	Impact strength of samples (kJ/m²)				
	GF5	GF10	GF15	GF20	GF25
1	4.89	4.23	3.85	4.56	5.92
2	4.58	3.86	4.25	4.97	5.18
3	4.71	5.07	3.28	5.14	5.56
4	4.66	5.31	3.46	4.61	7.57
5	4.63	3.68	3.73	4.08	7.18
Average impact strength (kJ/m²)	4.69	4.43	3.71	4.67	6.28
Standard deviation	0.11	0.73	0.37	0.41	1.04

The data in Table 2 show that the impact strength for each measurement is irregular. The GF5 prototype will gradually decrease from 4.69 kJ/m^2 to GF15 of 3.73 kJ/m^2 and progressively increase to 6.28 kJ/m^2. This result may be due to the effect of the distribution density of GF on the test specimen. The standard deviation is in the range of 0.11-0.73, which is relatively small.

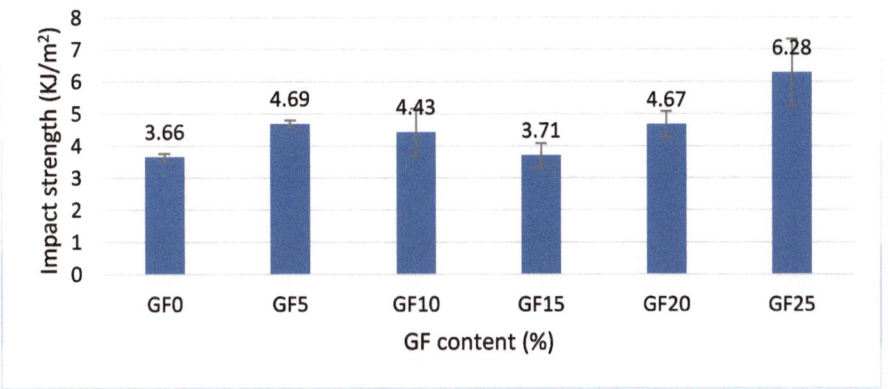

Fig. 2. Impact strength of PBT/GF blend

Figure 2 describes the impact strength of the PBT/GF blend. Looking at the column chart, the impact strength of the sample containing 25% GF has a higher average impact strength than other samples. The sample containing 15% GF is the lowest. This result is due to the GF content present in this mixture. It is more flexible in GF5 and GF10 models with low GF content. The GF content is higher from GF20 and GF25 models and above, so it will be more rigid and less flexible. The GF15 model is shallow because the impact strength and flexibility are low, so breaking with a manageable force is easy. The impact strength will likely increase if the amount of GF increases. However, the material is very brittle when GF reaches its maximum. Our team will adjust below 30% GF in the mixture to limit defects and improve impact strength at the appropriate level.

Impact strength is influenced by GF content, which has been studied for the mechanical properties of PBT. The dispersion density of GF helps the mixture reduce plastic shrinkage and retain the shape of the sample. The impact strength of the mix has increased significantly with the addition of GF [15]. Although the impact strength is uneven, it is still 100% higher than PBT. The impact strength of PBT of 3.66 kJ/m^2 is the lowest, lower than PBT/15GF of 3.71 kJ/m^2. The reason for the increase in impact toughness is the presence of GF. The GF content further enhances the impact intensity. After measuring the impact strength, we will conduct a microscopic scan to observe compatibility between GF and PBT.

Figure 3 shows the SEM microstructure of the fracture surface of the PBT/GF mixture samples corresponding to the percentage of GF present in the PBT mixture. The increased density of GF on the PBT matrix shows the compatibility between PBT and GF in the mix. PBT stands out on GF surfaces as fault lines, meaning GF is highly compatible with PBT. Figure 3 shows the effect of GF content on the impact strength of PBT and

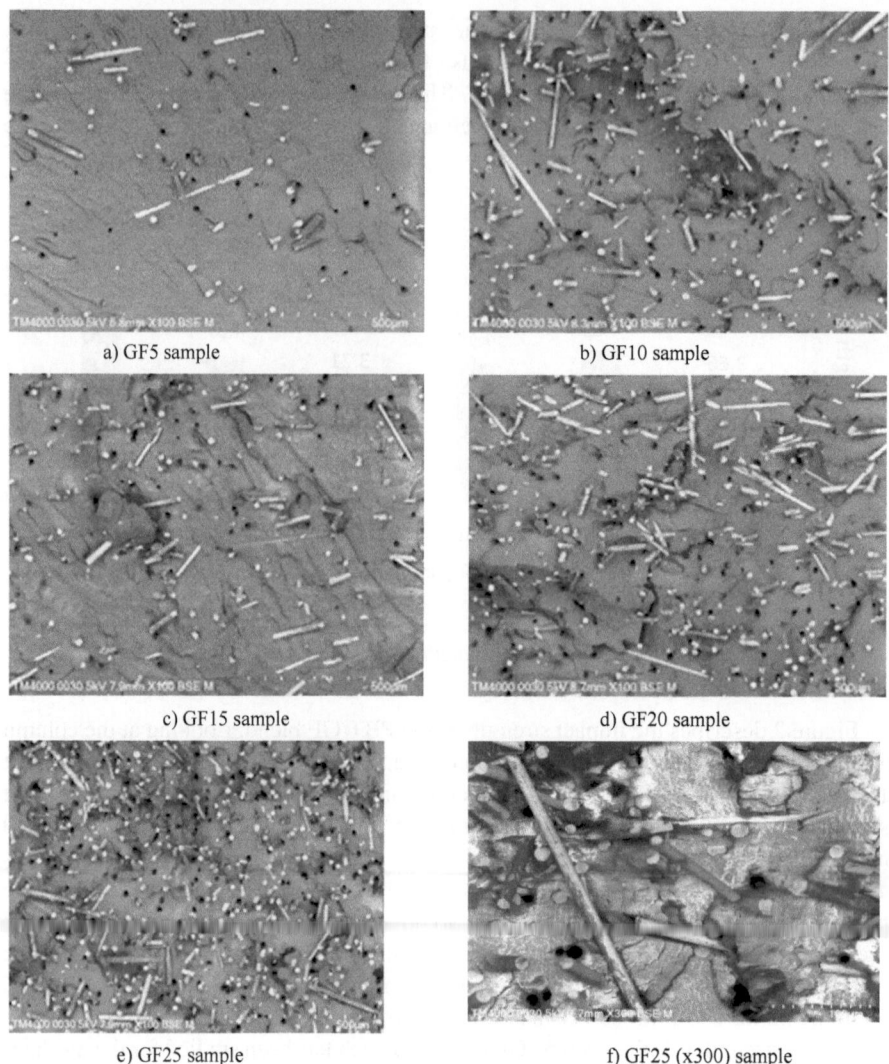

a) GF5 sample

b) GF10 sample

c) GF15 sample

d) GF20 sample

e) GF25 sample

f) GF25 (x300) sample

Fig. 3. SEM microstructure of the PBT/GF blend

PBT/GF mixtures. Compared with the neat PBT, the impact strength of the PBT/GF mixture moderately rises as the GF percentage increases. The nature of the hard GF relative to the matrix dominates the increase in impact strength. They can be caused by high dispersion and adhesion between GF and PBT matrices. Therefore, it can be seen that adding GF to PBT is an effective means of enhancing the properties of PBT.

At fiber content greater than 5 wt.%, the fiber length of the mixture is an essential factor for the mechanical strength of PBT/GF mixtures. From the results of Fig. 3, we have seen that the GF length of mixtures made with lower molecular weight PBT [16] is longer than that of mixtures produced with higher molecular weight PBT [17]. This behavior is probably due to the more severe decomposition of GF in the dense matrix than during mixing PBT and GF in plastic injection machines.

4 Conclusions

The mechanical properties are improved compared to 100% PBT sample. The slightest impact strength is 3.71 kJ/m^2, and the highest is 6.28 kJ/m^2. The impact strength decreases when GF reaches 15% but is still higher than 100% PBT. The additional filling increases friction between the elements of GF and uniform distribution throughout the mix, which helps to create strong bonds that increase the rigidity of the material; plus, the uniform distribution of GF throughout the mixture through SEM images has known the impact of GF. Flame retardant properties are also further improved due to the addition of GF scattered in the mix. PBT resin is additionally reinforced with GF, which can prevent the shrinkage of the mixture and create uniformity, which is a factor that clarifies the effect of the toughness of the improved material. This result makes the material less susceptible to environmental influences. This result is ideally suited for manufacturing battery cases without using expensive plastics while still having good properties and durability that save on fees. It even opens up new applications in sectors that require strength and resilience, such as the automotive or electronics industries. Overall, our research expands our understanding of how GF affects the mechanical properties of PBT and shows new prospects for developing materials with similar mechanical properties, which can be advanced development, as a premise for promoting rich applications in the manufacturing industry.

Acknowledgments. This work belongs to the project grant No: SV2024-231 funded by Ho Chi Minh City University of Technology and Education, Vietnam. We acknowledge Ho Chi Minh City University of Technology and Education, and Material Testing Laboratory. We also acknowledge the assistance of the ICCE2023Committee in sharing our study. With their appreciated support, it is possible to conduct this research.

References

1. Xie, Z., Wu, X., Jeffrey Giacomin, A., Zhao, G., Wang, W.: Suppressing Shrinkage/Warpage of PBT Injection Molded Parts with Fillers, Polymals. Compos, pp. 2–4 (2016)
2. Gao, F., General, L., Fang, Z.: Effect of a novel phosphorous-nitrogen containing intumescent flame retardant on the fire retardancy and the thermal behaviour of poly(butylene terephthalate). Polymers. Degrad. Stab. **91**(6), 1295–1299 (2006)
3. Braun, U., Bahr, H., Sturm, H., Schartel, B.: Flame retardancy mechanisms of metal phosphinates and metal phosphinates in combination with melamine cyanurate in glass-fiber reinforced poly (1,4-butylene terephthalate): the influence of metal cation. Polymers. Adv. Technol. **19**(6), 680–692 (2008)
4. Camino, G., Costa, L., di Cortemiglia, M.P.L.: Overview of fire retardant mechanisms. Polymers. Degrad. Stab. **33**, 131–154 (1991)
5. Xiao, J., et al.: Fire retardant synergism between melamine and triphenyl phosphate in poly (butylene terephthalate). Polymers. Degrad. Stab. **91**, 2093–2100 (2006)

6. Gallo E., Braun U., Schartel B., Russo P., Acierno D.: Polymdegrad and stabil **94**, 1245–1253 (2009)
7. Mohd Ishak Z.A., Ariffin, A., Senawi, R.: Eur Polymers. J. **37**, 1635–1647 (2001)
8. Threepopnatkul, P., Sukprasul, N., Santiuiparat, P., Prom-Oh, W., Kulsetthanchalee, C.: **27**(1), 31–34 (2017)
9. Zhang, J.: Study of poly (trimethylene terephthalate) as an engineering thermoplastics material. J. Appl. Polymers. Sci. **91**, 1657–1666 (2004)
10. Bergeret, A., Ferry, L., Ienny, P.: Influence of the fibre/matrix interface on ageing mechanisms of glass fibre reinforced thermoplastic composites (PA-6,6, PET, PBT) in a hygrothermal environment. Polymers. Degrad. Stab. **94**, 1315–1324 (2009)
11. Brehme, S., Köppl, T., Schartel, B., Altstädt, V.: Competition in aluminium phosphinate-based halogen-free flame retardancy of poly (butylene terephthalate) and its glass-fibre composites. e-Polymers **14**, 193–208 (2014)
12. Yang, W., Hu, Y., Tai, Q., Lu, H., Song, L., Yuen, R.K.: Fire and mechanical performance of nanoclay reinforced glass-fiber/PBT composites containing aluminum hypophosphite particles. Compos. Part A Appl. Sci. Manuf. **42**, 794–800 (2011)
13. Hajiraissi, R., Parvinzadeh, M.: Preparation of polybutylene terephthalate/silica nanocomposites by melt compounding: evaluation of surface properties, August 2011, **257**(2), 8443–8450 (2001)
14. Hashem, S.: Polymer i. TEST **30**, 801–810 (2011)
15. Mohd Ishak, Z.A.: Short glass fibre reinforcedpoly (butylene terephthalate) part 2 –effect of Hygrothermal aging Onmechanical properties **67**, 271–277 (2013)
16. Nise, M., Langer, B., Schumacher, S., Grellmann, W.J.: Appl. Polym. Sci. **111**, 2245 (2009)
17. Kim, Y.H., Kim, D.H., Kim, J.M., Kim, S.H., Kim, W.N., Lee, H.S.: Macromol Res. **17**, 110 (2009)

Numerical Investigation of the Aerodynamic and Thermal Behavior of a Flow Around the Blades of an Axial Gas Turbine

S. Haddout[1]($\boxtimes$) and Jian Zhang[2]

[1] Faculty of Science, Ibn Tofail University, Kenitra, Morocco
soufian.haddout@gmail.com
[2] Jiangsu University of Science and Technology, Zhenjiang, China

Abstract. Gas turbines have emerged as integral components in various industrial applications, particularly in the generation of electrical power. The widespread adoption of gas turbines has sparked increased interest among researchers and designers to explore and enhance various aspects of this machinery. This study is dedicated to simulating the flow of compressible transonic fluid through a configuration of eight blades, akin to those found in gas turbines. The primary objective is to analyze and determine the pressure and temperature distribution surrounding each blade. The configuration under investigation aligns with the one previously studied by T. Arts through experimental means. The numerical results generated by the Fluent code will be examined and discussed, shedding light on the intricacies of fluid dynamics within this specific turbine blade arrangement. This research aims to contribute valuable insights for further refining the performance and efficiency of gas turbines in practical applications.

Keywords: Fluent Code · Gas Turbine · Blades · Performance

1 Introduction

In the contemporary landscape of power generation, gas turbines have evolved into indispensable components, playing a pivotal role in diverse industrial sectors, especially in the production of electrical energy [1–3]. The increasing prevalence of gas turbines has spurred a heightened interest among researchers and designers, propelling them to delve deeper into every aspect associated with this machinery, all with the overarching goal of effecting necessary improvements [4, 5]. This comprehensive study is dedicated to the intricate simulation of the flow of compressible transonic fluid through a configuration consisting of eight blades, mirroring the design commonly found in gas turbines. The focal point of our investigation revolves around the meticulous analysis and determination of the pressure and temperature distribution near each turbine blade. The chosen configuration for examination aligns with the experimental study conducted by T. Arts [6], providing a foundation for our numerical simulations. By leveraging the Fluent code, we aim to generate accurate and insightful numerical results. These results will then be subjected to thorough discussion, offering a nuanced understanding of the fluid

G. Feng (Ed.): ICCE 2023, LNCE 526, pp. 531–537, 2024.
https://doi.org/10.1007/978-981-97-4355-1_50

dynamics at play within this specific turbine blade arrangement. This research aspires to contribute not only to the theoretical understanding of gas turbine performance but also to provide practical insights that can inform the ongoing efforts to enhance the efficiency and effectiveness of gas turbines in real-world applications. Through this exploration, we anticipate uncovering valuable knowledge that will contribute to the continuous advancement of gas turbine technology.

2 Problem Statement and Equations

This work focuses on the numerical modeling of transonic flow around the VKI-CT2 8-profiles located in the stator of an axial turbine. The numerical simulation of the flow is conducted using Fluent software based on the Navier-Stokes equations. Turbulence is accounted for using first-order and second-order turbulence models. The discretization method employed is the finite volume method, and an unstructured tetrahedral mesh is adopted for a generalized Cartesian coordinate system.

The system of averaged Navier-Stokes equations can be expressed in the following conservative form [Refer to [6]]:

$$\frac{\partial w}{\partial t} + div(Fc - Fd(w, wx, wy, wz)) = S(w) \tag{1}$$

3 Meshing and Boundary Conditions

In the simulation process, meshing and defining appropriate boundary conditions are crucial aspects. The mesh serves as a discretized representation of the computational domain, while boundary conditions prescribe the behavior of the flow at the domain boundaries. For this study, an unstructured tetrahedral mesh of the computational domain is utilized. The choice of mesh type and refinement is particularly important near the walls (Fig. 1), and it is influenced by the Reynolds number and turbulence model applied. In cases where viscous effects are significant, such as at a Reynolds number of approximately $Re = 105$, a refined mesh is employed to capture the details of the flow near the surfaces accurately.

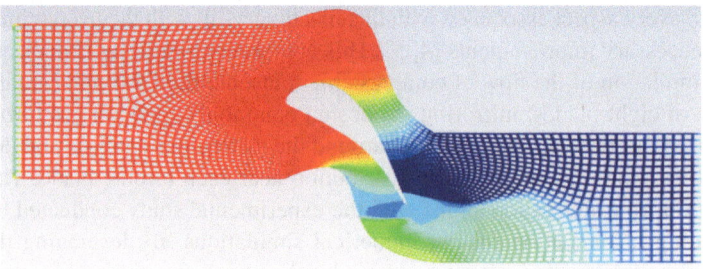

Fig. 1. Mesh of the domain studied

Additionally, specifying proper boundary conditions is essential to mimic real-world scenarios. These conditions might include the inflow and outflow conditions, as well as the treatment of solid surfaces like the turbine blades. The accuracy and reliability of the simulation results depend on the careful consideration and application of these meshing and boundary condition parameters.

The flow is assumed to be steady, compressible, and viscous. At the inlet of the domain, both the velocity and static temperature of the fluid are considered as specified boundary conditions. At the outlet, the static pressure of the fluid is captured. On the walls, it is assumed that the temperature is known. The obtained results are validated against the wall 251 (test) [1], providing a confirmation of the simulation outcomes. This validation against a known case helps ensure the accuracy and reliability of the numerical model in replicating real-world fluid dynamics, especially under the given assumptions and boundary conditions.

4 Results and Discussions

4.1 Mach Number

The Mach number is a dimensionless quantity that represents the ratio of the speed of an object, in this case, the fluid flow, to the speed of sound in that fluid. It is a crucial parameter in aerodynamics and fluid dynamics, providing insight into the compressibility effects of the fluid. In the context of the results obtained using the Shear-Stress Transport (SST) model, the Mach number gives us valuable information about the speed of the fluid at different locations within the computational domain (Fig. 2).

Inlet of the Domain: The fluid enters the domain at a low velocity. This initial condition is crucial as it sets the baseline for the subsequent flow behavior.

Leading Edge of the 8-Blade Configuration: As the fluid encounters the leading edge of the 8-blade configuration, it undergoes deceleration. This deceleration is likely influenced by the presence of the blades, causing changes in the flow pattern and velocity distribution.

Inter-Blade Space (especially on the Upper Surface): The fluid then begins to accelerate in the inter-blade space, with a particular emphasis on the upper surface or extrados. This acceleration could be attributed to the aerodynamic shape of the blades and the consequent pressure differences on the upper surface, leading to increased flow velocity.

Trailing Edge: At the trailing edge of the blades, there is a sudden decrease in velocity. This reduction in speed might be a result of the interaction between the fluid from the extrados and intrados, causing a difference in momentum (QM) between the two regions.

Swirling Zone Formation: The sudden decrease in velocity at the trailing edge is accompanied by the formation of a swirling zone. This swirling motion is likely induced by the difference in momentum between the fluid from the upper surface (extrados) and that from the lower surface (intrados).

Shockwave Near the Trailing Edge: Additionally, a shockwave is noted near the trailing edge. Shockwaves often occur when there is a sudden change in flow conditions, such as a rapid decrease in velocity. The presence of a shockwave indicates a significant change in the flow dynamics at this location.

Understanding the Mach number distribution and its variations at different points along the flow path provides valuable insights into the aerodynamic characteristics of the 8-blade configuration. It helps in optimizing the design and predicting the performance of the system, considering compressibility effects and shockwave formations.

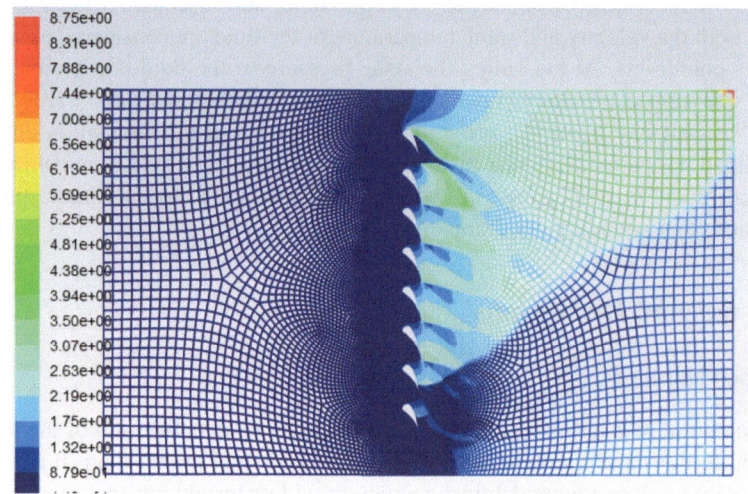

Fig. 2. Variation of the Mach number obtained by the SST-model (case MUR251)

Pressure/Temperature. The interplay between pressure and temperature is crucial for understanding the behavior of a system, such as the flow around turbine blades (Fig. 3 and 4).

High Inlet Pressure: The excessively high pressure at the inlet of the domain could be attributed to various factors, including upstream conditions, geometry of the inlet, or operational parameters. Understanding the source of this high pressure is essential for optimizing the system's performance and preventing potential damage.

Pressure Degradation in Inter-blade Region: The decrease in pressure within the inter-blade region suggests that there might be interactions between the fluid and the blades leading to energy losses. This phenomenon could be caused by the complex aerodynamic interactions occurring between adjacent blades or the presence of turbulent flows. Investigating and mitigating these pressure losses are vital for improving overall efficiency.

Different Pressure Changes on Extrados and Intrados: The discrepancy in pressure changes between the extrados and intrados indicates an asymmetry in the flow field around the blades. This could be a result of uneven blade loading, non-uniformities in the incoming flow, or blade design considerations. Identifying the cause of this asymmetry is essential for achieving a more uniform distribution of pressure and optimizing the aerodynamic performance of each blade.

Temperature Increase at Shockwave Location: The sudden increase in temperature at the shockwave location near the trailing edge is indicative of compression heating. As the flow accelerates and encounters a shockwave, there is a conversion of kinetic energy to thermal energy, resulting in a temperature rise. This temperature spike should be carefully monitored, as excessive heating can lead to material degradation and affect the overall durability of the blades.

Inconsistent Effect Among Blades: The variability in the impact of the shockwave among the eight blades suggests that there might be differences in their individual aerodynamic loading or geometrical characteristics. Investigating these variations and their influence on the pressure and temperature distribution will help in optimizing the design and ensuring uniform performance across all blades.

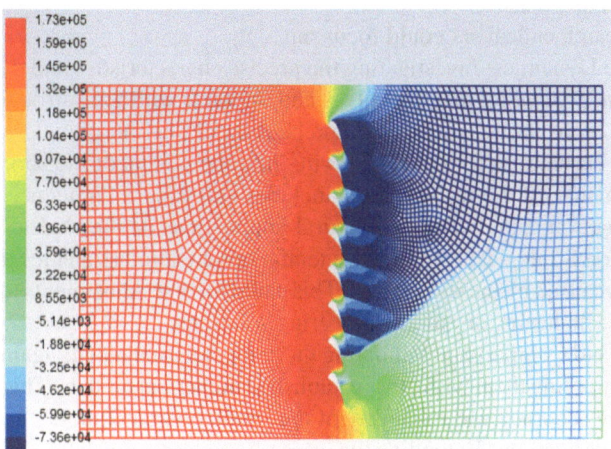

Fig. 3. Pressure variation (P) obtained by the SST-model

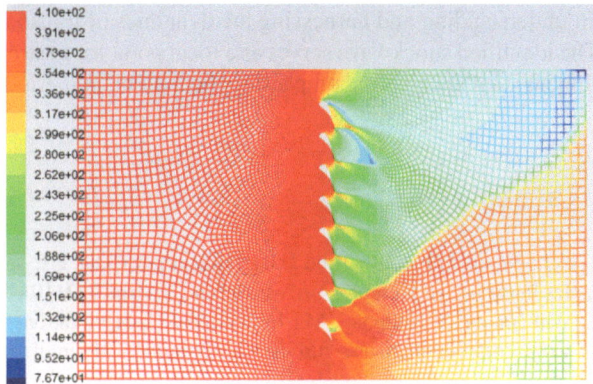

Fig. 4. Temperature variation (T) obtained by the SST-model

5 Conclusions

The preliminary results obtained in the course of this research work are promising. The viscous calculation of transonic flow around the 8 blades has revealed the presence of a shockwave and its varying effects. This observation prompts the need for further exploration, both in terms of theoretical understanding and practical implications, in future research. These early findings offer a glimpse into the complex behavior of transonic flow around the turbine blades. The identification of a shockwave introduces a layer of intricacy to the study, necessitating a deeper dive into the physical and practical aspects associated with its presence.

Future research endeavors could focus on:

Shockwave Dynamics: Investigating the precise characteristics and dynamics of the shockwave. Understanding how it evolves and interacts with the surrounding flow will contribute to a more comprehensive comprehension of the transonic flow phenomena.

Impact on Turbine Performance: Delving into the effects of the shockwave on the overall performance of the turbine. This includes assessing its influence on efficiency, pressure distribution, and potential structural implications on the blades.

Optimization Strategies: Exploring potential optimization strategies to mitigate any adverse effects caused by the shockwave. This could involve adjustments to blade design, flow control mechanisms, or other engineering solutions.

Validation and Comparison: Validating the computational results against experimental data and comparing them with other established numerical models. This step is crucial for ensuring the accuracy and reliability of the simulation outcomes.

Practical Applications: Extending the research to address practical applications, such as turbine design improvements or the development of guidelines for handling transonic flows in similar contexts.

In summary, while the current results provide a foundation, there is a rich landscape for exploration in understanding and harnessing the dynamics of transonic flow around turbine blades. The identified shockwave serves as a focal point for future investigations, inviting a multidimensional analysis encompassing theoretical, numerical, and practical considerations.

References

1. Choi, M.G., Ryu, J.: Numerical study of the axial gap and hot streak effects on thermal and flow characteristics in two-stage high pressure gas turbine. Energies **11**(10), 2654 (2018)
2. Nakhchi, M.E., Win Naung, S., Rahmati, M.: Influence of blade vibrations on aerodynamic performance of axial compressor in gas turbine: direct numerical simulation. Energy **242**, 122988 (2022)
3. Burberi, E., Massini, D., Cocchi, L., Mazzei, L., Andreini, A., Facchini, B.: Effect of rotation on a gas turbine blade internal cooling system: numerical investigation. J. Turbomach. **139**(3), 031005 (2017)
4. Schobeiri, M.T., Nikparto, A.: A comparative numerical study of aerodynamics and heat transfer on transitional flow around a highly loaded turbine blade with flow separation using RANS, URANS and LES. In: Turbo Expo: Power for Land, Sea, and Air, vol. 45738, p. V05CT17A001. American Society of Mechanical Engineers, June 2014

5. Fan, X., Du, C., Li, L., Li, S.: Numerical simulation on effects of film hole geometry and mass flow on vortex cooling behavior for gas turbine blade leading edge. Appl. Therm. Eng. **112**, 472–483 (2017)
6. Arts, T., de Rouvroit, L., Rutherford, A.W.: Aero thermal investigation of a highly loaded transonic turbine guide vane cascade. Von Kerman Institute–Belgium, September 1990

Leveraging Variational Autoencoder for Improved Construction Progress Prediction Performance

Fatemeh Mostofi[1(✉)], Onur Behzat Tokdemir[2], and Vedat Toğan[1]

[1] Department of Civil Engineering, Faculty of Engineering, Karadeniz Technical University, 61080 Trabzon, Türkiye
`fatemee.mostofi@gmail.com`
[2] Department of Civil Engineering, Faculty of Civil Engineering, Istanbul Technical University, 34469 Istanbul, Türkiye

Abstract. The imbalanced construction dataset reduces the accuracy of the machine learning model. This issue that addressed by recent construction management research through different sampling approaches. Despite their advantages, the utilized sampling approaches are reducing the reliability of the prediction model, while posing the risk of artificial bias. The objective of this study is to address the challenge of imbalanced datasets in construction progress prediction models using a novel variational autoencoder (VAE) that generates synthetic data for underrepresented classes. The VAE's encoder-decoder architecture, along with its latent space components, is optimized for this task. A comparative analysis using decision tree-based ML models, including grid search optimization, substantiated the effectiveness of the VAE approach. The results indicate that the hybrid dataset benefited the ML models from the addition of the synthesized dataset, showing 2% improvements in performance metrics across most models. The synthetic data generated by VAEs contributes to the construction of more balanced datasets, which, in turn, can lead to more reliable and accurate predictive models. The enhanced accuracy of the VAE-ML model addresses the class imbalance problem and improves the reliability of construction productivity predictions and related resource allocation plans.

Keywords: Generative Model · Variational Autoencoder (VAE) · Imbalanced Construction Dataset · Machine Learning (ML)

1 Introduction

The construction industry often grapples with the complexity of data, particularly when it comes to monitoring and predicting project progress. Placing effective project control is essential for the success of construction projects (Ezzeddine et al. 2022), reducing the rate of construction budget and schedule failures. However, the increasing complexity of projects coupled with inefficiencies in project control systems accelerates the rate of budget and schedule failures (Ezzeddine et al. 2022). Earned value analysis (EVA),

G. Feng (Ed.): ICCE 2023, LNCE 526, pp. 538–545, 2024.
https://doi.org/10.1007/978-981-97-4355-1_51

employing indicators such as cost performance index (CPI) and schedule performance index (SPI), is a widely accepted tool for comprehensive construction performance analysis (Kim and Kim 2014). SPI along with CPI provides crucial insights into the health and progress of a project, being the bird's-eye view of the performance triangle of a project (Kim and Kim 2014). The effectiveness of EVM is often limited by the quality of input data, particularly from traditional forecasting methods like S-curves. Recognizing the limitations of traditional methods, the construction sector has increasingly adopted ML solutions (Candaş and Tokdemir 2022; Kazar et al. 2022; Koc 2023; Mammadov et al. 2023; Mostofi et al. 2022; Mostofi and Toğan 2023; Toğan et al. 2022). However, the efficacy of these models is contingent on the quality and size of the underlying datasets, which remains a challenge in the construction industry (Althnian et al. 2021; Li et al. 2017; Sordo and Zeng 2005).

While ML models have shown potential in forecasting construction productivity, their performance heavily relies on the quality and size of the underlying datasets (Althnian et al. 2021; Aroyo et al. 2021; Barbierato et al. 2022). A critical issue faced in the construction sector is the collection of ample, relevant, and reliable data, a challenge that is aggravated by the dynamic and heterogeneous nature of construction projects. Data augmentation has been recognized as a critical strategy to enhance the performance of ML models (Barbierato et al. 2022; Mostofi et al. 2023). Variational autoencoders (VAEs) stand out as a powerful tool in this context. VAEs are generative models capable of learning complex data distributions and generating new data samples. Research demonstrated the potential of VAE in improving the accuracy of ML by 8% (Islam et al. 2021). (Mostofi et al. 2023) explored the use of VAEs in construction management to improve the prediction accuracy of graph attention networks for CPI productivity prediction, achieving considerable improvement in prediction accuracy. However, this study did not evaluate the VAE on other ML models, particularly on SPI prediction. As a result, the present research delves into the application of VAEs on a comprehensive construction progress dataset, with the aim of generating synthetic data that addresses the imbalance issue. In addition, our study details the components of the VAE, including its encoder-decoder architecture, latent space, and hyperparameters, which are crucial for its effective functioning. The objective is to leverage the generative capabilities of VAEs to address the challenges of underperforming ML solutions due to training on imbalanced datasets, thereby aiding in more accurate and reliable construction project forecasting and control. A comparative analysis using decision tree-based models, such as random forest, AdaBoost, gradient boosting, LightGBM, XGBoost, extra trees classifier, and bagging models was performed.

2 Methodology

The methodology of using VAE for data generation and then aggregating it with a collected dataset involves several key steps. Figure 1 displays the research flow of this study.

The original dataset comprises 1,342 progress records collected from a construction project site, detailing the resources used for the execution of different construction activities, whereby their performance was reported by CPI and SPI. Followingly, the data

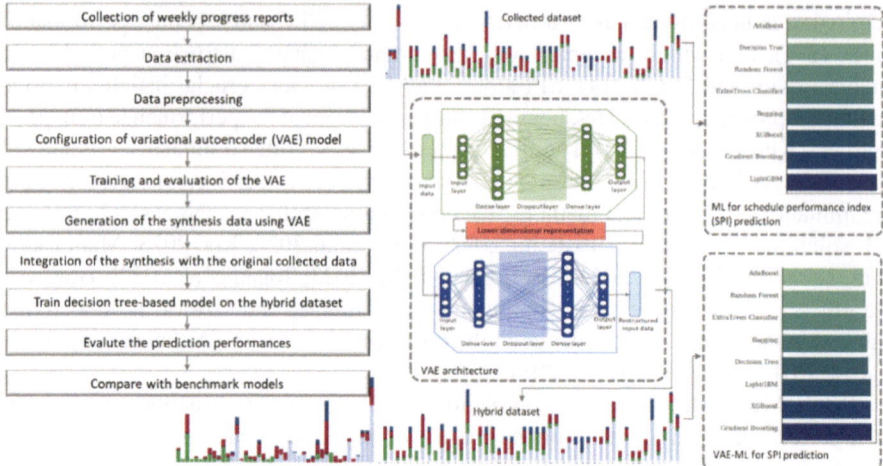

Fig. 1. Research flow.

preparation included handling records with inconsistencies and missing values, adjusting the format, and selecting the features for SPI prediction. Next the median instances per class were determined for the identification of underrepresented classes within the dataset. Subsequently, the underrepresented portion of data was used for targeted data generation and addressing the issue of class imbalance. Here a VAE model was configured with an encoder and decoder network. The encoder maps input data to a latent space representation, while the decoder reconstructs the data from the latent space. At this stage, a sampling function was implemented within the VAE architecture to facilitate the generation of new data points from the learned distribution. VAE was trained on the 80% of data related to underrepresented class data to learn the distributions specific to these classes, while the rest 20% was used for model validation. The training of VAE was guided through a reconstruction loss and the Kullback–Leibler divergence for the latent loss while being optimized based on the mean squared error. Figure 2 displays the performance of the utilized VAE.

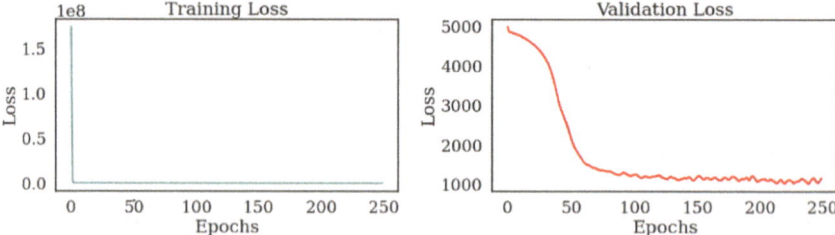

Fig. 2. Performance of the VAE proposed.

Generated data encompasses 315 new records created by the VAE, which aimed to address the imbalance in the original dataset. The generated synthesized data was then used to create a balanced dataset by combining it with the original dataset. The aggregated dataset, combining original and generated data, totals 1,496 observations. The combined dataset provides a more substantial and diverse training base, which could potentially lead to more robust models that generalize better to unseen data. Next, a variety of tree-based models were considered, namely decision trees, random forest, AdaBoost, gradient boosting, LightGBM, XGBoost, extra trees classifier, and bagging, ranging from simple decision rules to complex ensemble methods that aggregate multiple weak learners into a stronger predictive model.

3   Results

This study evaluated the proposed VAE-ML on various ML models while comparing their performance when applied to a dataset comprising synthesized and original data and the original collected data. For each model type, we conducted hyperparameter optimization using grid search, systematically working models through multiple combinations of parameter values and cross-validating them. This allows the determination of the tunes that give the best performance according to a specified metric. The models configured with their best hyperparameters were further evaluated based on accuracy, F1 score, precision, recall, and area under the curve (AUC) metrics. Figure 3 displays the prediction performance of the ML models trained over the original and hybrid datasets based on accuracy and F1 score metrics.

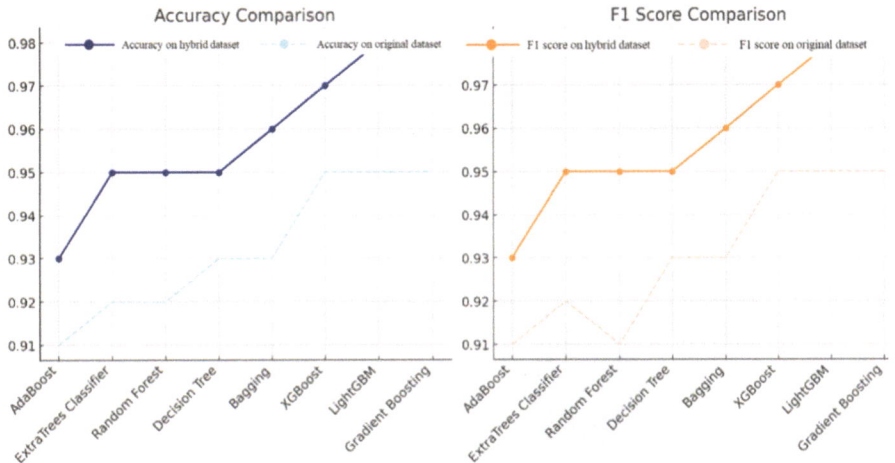

Fig. 3. Comparison of accuracy and F1 score performance of prediction models.

The synthesized dataset consistently exhibited superior performance compared to the original dataset, notably, models like LightGBM and gradient boosting achieved an accuracy of 0.98. The integration of VAE in ML models improved their prediction accuracies by about 2%. Figure 4 compares the prediction performance of these models.

The VAE-ML proposed in this research demonstrated a robust predictive capability using post data augmentation. Previous studies employed undersampling, oversampling,

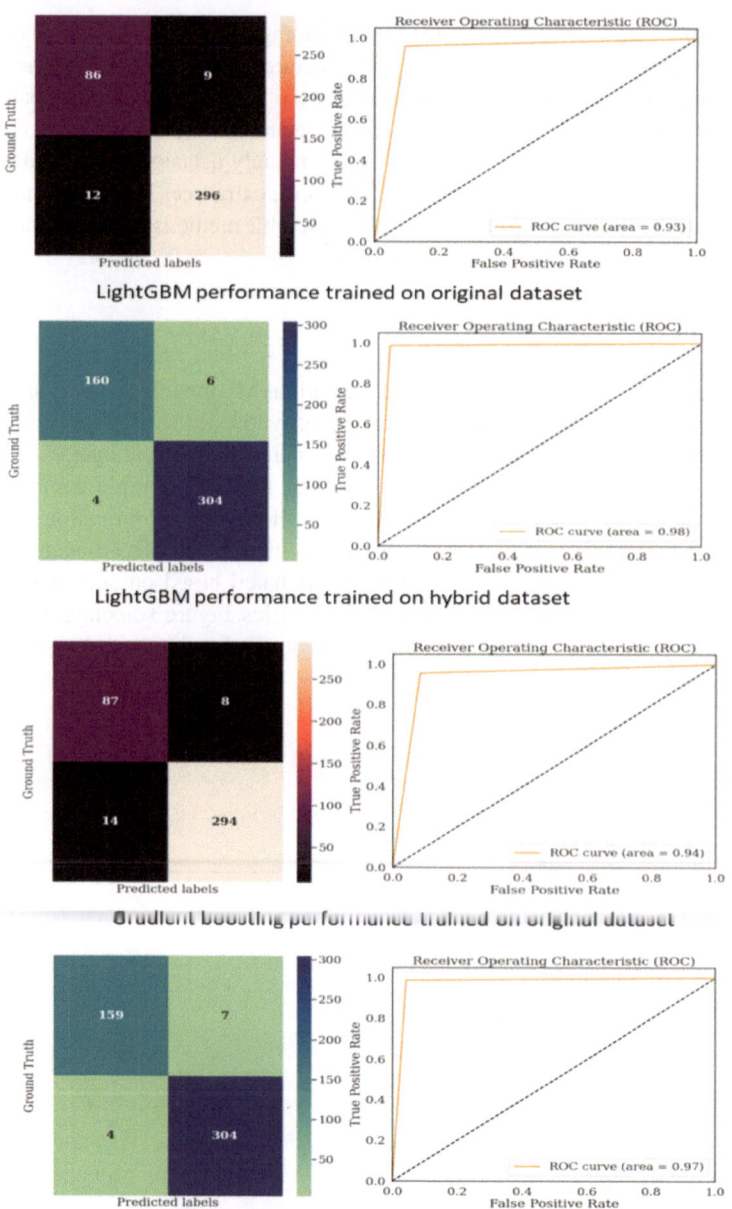

LightGBM performance trained on original dataset

LightGBM performance trained on hybrid dataset

Gradient boosting performance trained on original dataset

Gradient boosting performance trained on hybrid dataset

Fig. 4. Comparison of prediction performance in each SPI class using both datasets.

and synthetic minority oversampling technique (SMOTE) approaches. Undersampling removes the records related to the majority class (Guo et al. 2018; Mishra and Singh 2021) and thus loses important information from the utilized dataset (Taha et al. 2021). On the other hand, oversampling replicates the underrepresented class (Guo et al. 2018;

Mishra and Singh 2021), where the introduced repetition in the dataset poses the risk of overfitting (Taha et al. 2021).

SMOTE creates artificial samples using nearest-neighbor approaches over the minority class limits a synthetic sample from the minority class has been pivotal in enriching datasets, thereby fostering better classifier performance across various domains (Bogner et al. 2018; Chawla et al. 2002). However, the research raised a question about the reliability of the SMOTE as its over-generalization of the minority class can result in artificial biases (Bao and Yang 2023) and inclination of the prediction towards the minority class (Blagus and Lusa 2012, 2013).

Considering the drawbacks mentioned, there exists a gap in methodologies that not only address the class imbalance but also ensure the authenticity and reliability of prediction models is paramount.

The incorporation of VAE-generated synthetic data appears to be a potent strategy for enhancing ML model performance. This approach has demonstrated substantial benefits in accuracy, precision, recall, and AUC measures without incurring prohibitive computational costs. These findings suggest a promising direction for future research and applications. Future studies could focus on the application of VAEs across a wider range of predictive tasks and explore the impact of different types of generative models on the predictive accuracy of various machine learning algorithms.

While the results are promising, caution must be exercised in interpreting these findings. The synthetic data's performance boost must be validated against more extensive datasets from different construction projects to ensure the ability of VAE to improve the prediction performance of the ML model.

4 Conclusion

There exists a gap in construction management research for the methods that improve the prediction accuracy of the real-life construction dataset, with an imbalanced prediction class. The objective of this study was to address the challenges associated with data imbalance in construction management datasets and enhance the performance of machine learning models using data augmentation through a VAE model. The research entailed a comprehensive analysis of multiple decision tree-based ML models, evaluating their ability to predict construction project outcomes effectively, considering the SPI metric.

The obtained results highlighted several pivotal findings. Firstly, the incorporation of VAE-generated data led to an enhancement in the accuracy of SPI predictions, improving the prediction performance of the eight investigated ML approaches by up to 2%. The implementation of the VAE-ML model has improved the prediction accuracy of construction management datasets.

This suggests that addressing class imbalance through synthetic data generation can effectively improve model performance. Secondly, the results underscored the need for rigorous data pre-processing and augmentation to improve the robustness of ML models in construction management.

Overall, throughout the research, it was evident that data imbalance impacts the predictive capability of traditional models. The use of VAEs for data generation proved to be a promising approach to mitigate this issue. The generative capability of VAEs allowed for the creation of synthetic yet realistic samples, which balanced the dataset and provided a more uniform learning environment for the ML models.

However, the study also revealed gaps that warrant further investigation. The application of VAE was primarily focused on SPI prediction using different decision tree-based ML models, and its effectiveness on deep learning approaches remains unexplored, pointing to an area ripe for future research.

References

Althnian, A., et al.: Impact of dataset size on classification performance: an empirical evaluation in the medical domain. Appl. Sci. **11**(2), 796 (2021). https://doi.org/10.3390/app11020796

Aroyo, L., Lease, M., Paritosh, P., Schaekermann, M.: Data excellence for AI: why should you care (2021)

Bao, Y., Yang, S.: Two novel SMOTE methods for solving imbalanced classification problems. IEEE Access **11**, 5816–5823 (2023). https://doi.org/10.1109/ACCESS.2023.3236794

Barbierato, E., Della Vedova, M.L., Tessera, D., Toti, D., Vanoli, N.: A methodology for controlling bias and fairness in synthetic data generation. Appl. Sci. **12**(9), 4619 (2022). https://doi.org/10.3390/app12094619

Blagus, R., Lusa, L.: Evaluation of SMOTE for high-dimensional class-imbalanced microarray data. In: 2012 11th International Conference on Machine Learning and Applications, pp. 89–94. IEEE (2012)

Blagus, R., Lusa, L.: SMOTE for high-dimensional class-imbalanced data. BMC Bioinformatics **14**(1), 106 (2013). https://doi.org/10.1186/1471-2105-14-106

Bogner, C., Seo, B., Rohner, D., Reineking, B.: Classification of rare land cover types: Distinguishing annual and perennial crops in an agricultural catchment in South Korea. PLoS ONE **13**(1), e0190476 (2018). https://doi.org/10.1371/journal.pone.0190476

Candaş, A. B., Tokdemir, O.B.: Automated identification of vagueness in the FIDIC silver book conditions of contract. J. Constr. Eng. Manag. **148**(4) (2022). https://doi.org/10.1061/(ASCE)CO.1943-7862.0002254

Chawla, N.V., Bowyer, K.W., Hall, L.O., Kegelmeyer, W.P.: SMOTE: synthetic minority oversampling technique. J. Artif. Intell. Res.Artif. Intell. Res. **16**, 321–357 (2002). https://doi.org/10.1613/jair.953

Ezzeddine, A., Shehab, L., Lucko, G., Hamzeh, F.: Forecasting construction project performance with momentum using singularity functions in LPS. J. Constr. Eng. Manag. **148**(8) (2022). https://doi.org/10.1061/(ASCE)CO.1943-7862.0002320

Guo, H., Diao, X., Liu, H.: Embedding undersampling rotation forest for imbalanced problem. Comput. Intell. Neurosci.. Intell. Neurosci. **2018**, 1–15 (2018). https://doi.org/10.1155/2018/6798042

Islam, Z., Abdel-Aty, M., Cai, Q., Yuan, J.: Crash data augmentation using variational autoencoder. Accid Anal. Prev. **151**, 105950 (2021). https://doi.org/10.1016/J.AAP.2020.105950

Kazar, G., Doğan, N.B., Ayhan, B.U., Tokdemir, O.B.: Quality failures–based critical cost impact factors: logistic regression analysis. J. Constr. Eng. Manag. **148**(12), 04022138 (2022). https://doi.org/10.1061/(ASCE)CO.1943-7862.0002412

Kim, B.-C., Kim, H.-J.: Sensitivity of earned value schedule forecasting to s-curve patterns. J. Constr. Eng. Manag. **140**(7), 04014023 (2014). https://doi.org/10.1061/(ASCE)CO.1943-7862.0000856

Koc, K.: Role of national conditions in occupational fatal accidents in the construction industry using interpretable machine learning approach. J. Manag. Eng. **39**(6) (2023). https://doi.org/10.1061/JMENEA.MEENG-5516

Li, D.-C., Lin, W.-K., Lin, L.-S., Chen, C.-C., Huang, W.-T.: The attribute-trend-similarity method to improve learning performance for small datasets. Int. J. Prod. Res. **55**(7), 1898–1913 (2017). https://doi.org/10.1080/00207543.2016.1213447

Mammadov, A., Kazar, G., Koc, K., Tokdemir, O.B.: Predicting accident outcomes in cross-border pipeline construction projects using machine learning algorithms. Arab. J. Sci. Eng. 1–19 (2023). https://doi.org/10.1007/s13369-023-07964-w

Mishra, N.K., Singh, P.K.: Feature construction and smote-based imbalance handling for multi-label learning. Inf. Sci. (N Y) **563**, 342–357 (2021). https://doi.org/10.1016/j.ins.2021.03.001

Mostofi, F., Toğan, V.: Explainable safety risk management in construction with unsupervised learning, pp. 273–305 (2023)

Mostofi, F., Toğan, V., Ayözen, Y.E., Tokdemir, O.B.: Predicting the impact of construction rework cost using an ensemble classifier. Sustainability (Switzerland), **14**(22) (2022). https://doi.org/10.3390/su142214800

Mostofi, F., Toğan, V., Tokdemir, O.B.: Enhancing construction productivity prediction through variational autoencoders and graph attention network. In: Proceedings of 3rd International Civil Engineering and Architecture Congress (ICEARC 2023), pp. 120–128 (2023). Trabzon: Golden light Publishing

Sordo, M., Zeng, Q.: On sample size and classification accuracy: a performance comparison, pp. 193–201 (2005)

Taha, A.Y., Tiun, S., Abd Rahman, A.H., Sabah, A.: Multilabel over-sampling and under-sampling with class alignment for imbalanced multilabel text classification. J. Inf. Commun. Technol. **20** (2021). https://doi.org/10.32890/jict2021.20.3.6

Toğan, V., Mostofi, F., Ayözen, Y.E., Behzat Tokdemir, O.: Customized AutoML: an automated machine learning system for predicting severity of construction accidents. Buildings **12**(11) (2022). https://doi.org/10.3390/buildings12111933

Selection of Construction Materials for a Thermal Insulation Layer of a Road

A. F. Galkin[1]([✉]), N. A. Plotnikov[1], and V. Yu Pankov[2]

[1] Melnikov Permafrost Institute, 36 Merzlotnaya Street, 677010 Yakutsk, Russia
`afgalkin@mail.ru`
[2] North-Eastern Federal University, 58 Belinsky Street, 677027 Yakutsk, Russia

Abstract. Safety of the roads in the permafrost region is largely determined by their thermal regime. The aim of the present work was to discover a function to determine the thermal conductivity coefficient of materials used to construct a thermal insulation layer of a road to prevent foundation soils from thawing over a permitted thawing depth. Two cases were surveyed: when the natural temperature of the soil is equal to ice melting temperature and when it is not equal to ice melting temperature. Engineering formulas permitting to quickly select the required thermal resistance property of an insulation material based on the known Biot number were derived. An expedient regularity is observed: the thermal resistance of the thermal insulation layer is roughly proportional to the dimensionless thawing depth. Correspondingly, when selecting the construction materials for a thermal insulation layer it can be considered that the increase in permissible thawing depth increases proportionally to the increase of the thermal conductivity coefficient of the insulation material. Considering that the physical and mechanical properties of the soil are not constant along the road length, the thermal resistance of the thermal insulation layer should be determined for individual sections of the road rather than for the entire route. Accordingly, the construction materials can also vary depending on the selected solutions for the road construction.

Keywords: Permafrost · Roads · Thermal Insulation · Materials · Thermal Conductivity Coefficient · Permissible Thawing Depth

1 Introduction

Safety of the roads in the permafrost region is largely determined by their thermal regime, especially when the road is built on soil with high ice content [1–5]. As the soil thaws, its strength deteriorates and melting of the inner ice layer may cause the road to collapse [6–8] as the strength of the foundation soil depends on the phase state of the water in the pores [9–11]. A way to manage the thawing depth is to add an insulation layer. The insulation layer can be composed of a single material, such as polystyrene, or a combination of thermal accumulation and thermal insulation materials [12–14]. One notable insulation material is foam glass ballast [15]. The properties of a thermal insulation layer are usually selected to fulfill the main requirements of preserving the foundation soil in a frozen

G. Feng (Ed.): ICCE 2023, LNCE 526, pp. 546–554, 2024.
https://doi.org/10.1007/978-981-97-4355-1_52

state throughout the whole period of the road operation and allowing the soil to thaw up to a given depth while maintaining its load-bearing strength. The aim of this research is finding a function determine the thermal conductivity coefficient of insulation materials of the road that prevents the soil from thawing over the permissible depth.

2 Methodologies

An algorithm proposed in [16] is used to obtain the functions to determine the optimal properties of the insulation layer. The algorithm searches for a Biot number, yielded as a function of Fourier and Stefan numbers, guaranteeing that the thawing depth of the foundation soil will not exceed the depth allowed in the road design. The equations are presented in a dimensionless (criterion) form. The required thermal resistance of the insulation layer is determined using the obtained Biot number. The thawing depth with a thermal insulation layer applied can be determined using the formula [17, 18] resulting from solving a one-dimensional Stefan problem at boundary conditions of the third kind. The problem, in a dimensionless form, is:

$$h = \sqrt{2Fo/St + 1/Bi^2} - 1/Bi \tag{1}$$

where $Bi = h_0/(R \cdot \lambda_\Pi)$, $R = \delta_i/\lambda_i$, $Fo = a\tau/h_0^2$, $St = Lw/tC_p$, $h = H/h_0$

h is the thawing depth of the foundation soil, m. h_0 is the typical dimension (size), m. R is the thermal resistance, m^2K/W. λ_Π is the thermal conductivity coefficient of the thawed soil, W/mK. δ is the thickness of the insulation layer, m. λ_i is the thermal conductivity coefficient of the insulation layer, W/mK. L is the latent heat of ice thawing, J/kg. w is the ice content in the soil, unitless. C_p is the total heat capacity of the soil, J/kgK. a is the thermal diffusivity of the soil, m^2/s. t is the air temperature, °C. Bi is the Biot number. Fo is the Fourier number. St is the Stefan number.

Using the Eq. (1), a Biot number guaranteeing that the soil will not thaw beyond the permitted thawing depth over a given time period will be found:

$$Bi = (2h \cdot St)/\left(4Fo - h^2 St\right) \tag{2}$$

In the Eq. (1) it is adopted that the temperature of the active layer of soil is equal to ice thawing temperature. As demonstrated in [19], in most typical cases this assumption is expedient for engineering calculations. The Eq. (1) can be further specified by including the concept of effective heat capacity of the rocks [19]:

$$C = Lw + C_p|T_e| \tag{3}$$

where T_e is the temperature of the frozen active layer of the soil, °C. In this case, the Eq. (1) is transformed into the form:

$$h = \sqrt{2TFo/(St + 1) + 1/Bi^2} - 1/Bi \tag{4}$$

And the Biot number is found from the expression:

$$Bi = 2h \cdot (St + 1)/\left(4TFo - h^2(St + 1)\right) \tag{5}$$

The Stefan number notation changes. It will be equal to $St = Lw/C_p|T_e|$. The dimensionless temperature simplex T, included in the Eqs. (4) and (5) that describes the relationship between the air and soil temperatures, will be equal to $T = t/|T_e|$. The methods of determination of the Fourier and Stefan numbers when solving problems of heat exchange of atmospheric air with thawing or freezing rocks are considered in [20, 21].

Using the known Biot number, determined using the Eqs. (2) or (5), it is possible to quickly find the thermal resistance and choose a suitable material for the insulation layer of the road. The Biot number can be found using the equation:

$$Bi = 2h \cdot (St + 1)/\left(4TFo - h^2(St + 1)\right) \qquad (6)$$

where $R_\Pi = h_0/\lambda_\Pi$; $R_i = \delta/\lambda_i$; $Bi = \alpha R_\Pi$. In a physical sense, R_Π is a unit of thermal resistance of a thawed layer of the road foundation in $m^2 K/W$. The heat transfer coefficient for a flat surface a is determined by the formula [22]:

$$\alpha = 1/(R_\text{и} + 1/\alpha_0) \qquad (7)$$

where α_0 is a convective heat transfer coefficient that depends on the average air speed over a given time interval and can be determined using the Perlstein formula [23]. Should the thermal resistance of all layers of the road be considered, the thermal resistance of the insulation layer can be found from the expression:

$$R_i = \frac{R_i}{Bi} - \frac{1}{a_0} + \sum_{i=1}^{n} Ri \qquad (8)$$

Here R_i is the thermal resistance of an i-th individual layer of the road comprising n layers, excluding the thermal insulation layer, in $m2/W°C$.

The second member of the Eq. (8) is much smaller than the first one, Eq. (3) will be used for further analysis. This approach will provide some margin of safety as the excluded second member of the Eq. (8) decreases the thermal resistance of the insulation layer. Knowing the thermal resistance and thickness of the insulation layer, a material with the required thermal conductivity coefficient can be selected. A method to select an economically efficient material is proposed in [24]. The criterion of economic efficiency is equal to the product of the thermal conductivity coefficient of the material and the cost of one cubic meter of the material.

The physical and mechanical qualities of the foundation soil differ along the route and thus the thermal resistance of the insulation layer should be determined for individual sections of the road rather than for the entire road. Indicator m of the degree of change in the thermal resistance of the insulation layer m ($m = R_{i2}/R_{i1}$) and an indicator k ($k = h_2/h_1$) of the change in permissible thawing depth of the foundation soil is introduced. The indicators on the baseline section of the road (index 1) and specific section of the road (index 2) are determined using the formula (6). It includes a Biot number dependent on the corresponding parameter h. After several transformations, the relationship between parameters m and h is found:

$$m = k\left(\frac{2Fo}{St} - \frac{h_1^2}{2}\right)/\left(\frac{2Fo}{St} - \frac{k^2 h_1^2}{2}\right) \qquad (9)$$

Here h_1 is the permissible value of a dimensionless thawing depth of the foundation soil on the baseline section of the road.

3   Results and Discussion

An analysis of the equations shows that some of them reflect relationships relevant for practical applications. From the expression (6) a conclusion can be made that the thermal resistance of the insulation layer is directly proportional to the unit of thermal resistance of thawed foundation soil, soil and inversely proportional to the Biot number that restricts thawing to a maximum permitted depth. Calculations with varying data were done and their results are presented as charts in the Figs. 1, 2, 3 and 4. Figure 1 shows the dependence of the Biot number (Bi) on the dimensionless thawing depth of the foundation soil (h) at varying values of the complex 2Fo/St for A) small and B) large permissible dimensionless thawing depths. The shape of the charts shows that for small permissible depths (1-B), the relationship is close to linear. Independently of the value of the complex 2Fo/St, the degree of increase in the Biot number at change in the value of the Biot number is an almost constant value. For example, when the dimensionless thawing depth changes by 2.5x - from 0.2 to 0.5 - the degree of increase in Biot number for both a 2Fo/St complex equal to 4.0 and 2.5 remains a constant quantity equal to roughly 1.6. The shape of the curves describing the relationship of the Biot criterion on the dimensionless thawing depth for high thawing depth values (h > 0.5) is non-linear (Fig. 1A). For example, when the thawing depth increases twice, but within the interval 0.8 - 2.0, the degree of increase in the Biot number for the values of 2Fo/St complex of 4.0 is almost 2.0, and when the value of the complex is 2.5, the increase is 4.0. That is, almost twice as large, even though the degree of change in the dimensionless thawing depth remained the same and equal to 2.5, as in the first example.

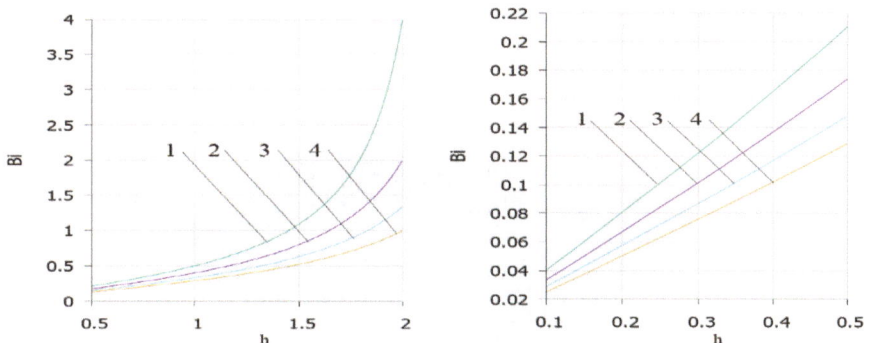

Fig. 1. Biot number depending on the dimensionless thawing depth of the foundation soil at various values of the complex 2Fo/St. 1 – 2.5, 2 – 3.0, 3 – 3.5, 4 – 4.0. A – for large permissible dimensionless thawing depths, B – for small permissible dimensionless thawing depths.

Figure 2 shows a 3D chart displaying the dependence of the Biot number on the Fourier and Stefan numbers at various values of the permissible thawing depths of the foundation soil.

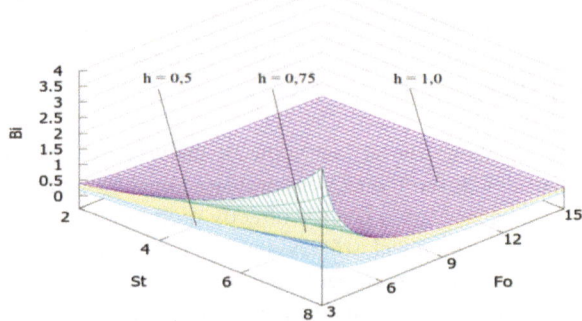

Fig. 2. Biot numbers depending on the values of the Fourier and Stefan numbers at different permissible thawing depths of the foundation soil (h): 1 – 0.5, 2 – 0.75, 3 – 1.0.

With the decrease in the dimensionless thawing depth, the dependence of the Biot number on the Fourier and Stefan criteria decreases. The greater the Stefan number and the smaller the Fourier number, the weaker the dependence. At small Stefan numbers and large Fourier numbers, the degree of change in the Biot number is insignificant, indicated by the planes merging in the figure. The greater the ice content in the foundation soil and the shorter the duration of the warm period during which the soil thaws, the smaller the thermal resistance of the insulation layer can be. When the layer is being designed, materials with higher thermal conductivity may be selected.

Figure 3 shows a dependence of the thermal resistance of the thermal insulation layer on the Biot number and the thermal resistance of thawed soil.

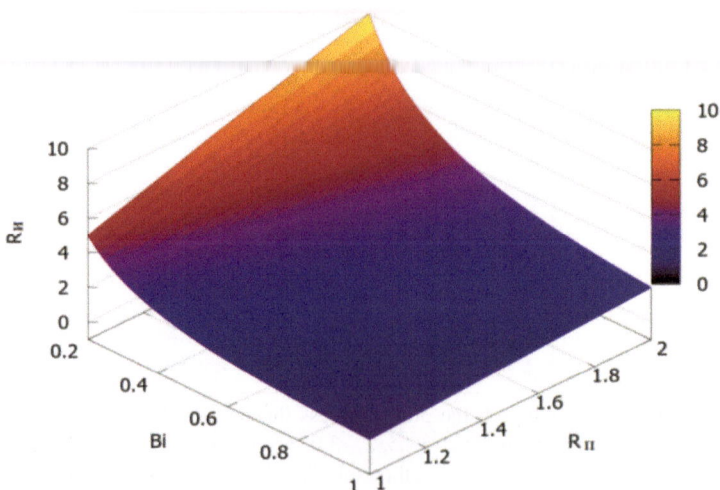

Fig. 3. The change in thermal resistance of the thermal insulation layer depending on the Biot number and the thermal resistance of a thawed soil layer.

The chart indicates that when the Biot number increases, the influence of the thermal resistance of a thawed soil layer on the thermal resistance of a thermal insulation layer decreases. Compare the edges of the chart at Biot numbers of 0.2 and 1.0. The edge that is closer in the picture (Bi = 1.0) is almost parallel with the R axis. At small Biot numbers the curve rises sharply. In the first case, the range of thermal resistance of the thermal insulation layer varies from 2 to 4 (blue), in the second case from 2 to 10 (yellow).

Figure 4 demonstrates the change in thermal conductivity coefficient depending on the Biot number set in the design and the thermal conductivity coefficient of thawed soil layer at varying thickness of the thermal insulation layer.

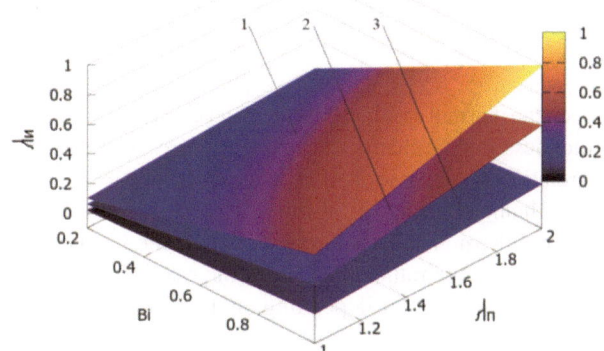

Fig. 4. Thermal conductivity coefficient of a thermal insulation layer material (λ_η, W/mK) depending on the Biot number set in the design and thermal conductivity coefficient of the thawed soil (λ_π, W/mK) at varying thickness of the thermal insulation layer. $1 - 0.5$ m, $2 - 0.2$ m, $3 - 0.1$ m.

The selection of thermal insulation material depends on the restrictions on total thickness of the road. The greater the permissible thickness, the more likely regular materials, such as burnt rocks or dry sand, whose thermal conductivity coefficients are in the range of $0.4 - 1.0$ W/mK could be sufficient as thermal insulation, depending on specific geocryological conditions. The Biot numbers must be in the ranges that do not permit soil thawing beyond the allowed depth. Figure 5 shows charts built on the basis of calculations done using the Eq. (9).

Figure 5 shows the dependence of the parameter m on k at various initial values of the base dimensionless quantity of the permissible thawing depth h_1 for two values of the 2Fo/St complex. The charts show that in the considered range of data, the relationship between m and k is almost linear. The gradient $\Delta m/\Delta k$ is close to one. This indicates that the thermal resistance of the insulation layer changes in a roughly equal proportion to the dimensionless thawing depth. If at some section of the road the permissible thawing depth can be increased twice, it means that the insulation layer can have a twice as large thermal conductivity coefficient as the material used for the rest of the road.

As the demonstrated by the chart, in the range of k (1, $0 \leq k \leq 2$, 0) the two planes almost merge, even though the 2Fo/St complex increases twofold, and the thawing depth changes by 2.5x. An analysis of the charts confirms the possibility of applying the

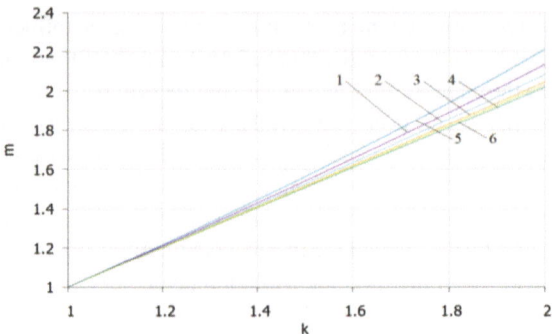

Fig. 5. The degree of change in thermal resistance m depending on the change in permissible thawing depth k at various values of the basic quantity h_1. $1 - 0.5, 2 - 0.4, 3 - 0.3, 4 - 0.2, 5 - 0.5, 6 - 0.2$ (1–4 when the 2Fo/St complex $= 6$, 5–6 when the 2Fo/St complex $= 4$).

relationship between the thermal conductivity of materials used for the thermal insulation layer of a given thickness and the change in permissible thawing depth at various sections of the road. This relationship is described by the function $\lambda_2 = k\lambda_1$. Considering that the physical and mechanical properties of the soil are not constant along the route, in the design the required thermal resistance coefficient of the insulation layer should be calculated for parts of the route rather than the whole road. Correspondingly, materials applied at different sections can vary depending on the construction solutions adopted.

4 Conclusion

Engineering equations allowing to assess the main heat exchange factors and thermal physical properties of the foundation soils on the choice of construction materials for the roads in the permafrost area were devised. In particular, it was demonstrated that the thermal resistance of the thermal insulation layer of the road is directly proportional to a unit of thermal resistance of the thawed layer of the foundation soil and is inversely proportional to the Biot number that restricts soil thawing to a required depth. A possibility of applying the relationship between the thermal conductivity coefficient of the construction materials of the thermal insulation layer of a given depth and a change in the permissible soil thawing depth at varying sections of the road. Further research should be focused on surveying the influence of modification of design parameters of the roads and the thermal physical properties of the foundation soil on the road safety and reliability.

References

1. Zheleznyak, M.N., Shesternev, D.M., Litovko, A.V.: Problems of stability of automobile roads in cryolithic zone. In: Materials of the XIV Russian Applied Science Conference and Exhibition "Prospects of Engineering Developments in the Russian Federation". Geomarket, Moscow, pp. 223–227 (2018)

2. Veli, Y.Y., Dokuchayeva, V.V., Fedorova, N.F.: Handook of Permafrost Soils Construction. Stroyizdat, Leningrad (1977)
3. Shatz, M.M.: The current state of the city infrastructure of Yakutsk and ways to increase its reliability. Georisk **2**, 40–46 (2011)
4. Kondratyev, V.G, Kondratyev, S.V.: How to protect the Federal highway "Amur" Chita – Khabarovsk from Geocryological Threats and Phenomena. Eng. Geol. 40–47 (2013)
5. Grechischev, S.E., Chistotnikov, L.V., Shur, Y.L.: Cryogenic Phyiscal Geological Processes and Their Forecasting. Nedra, Moscow (1984)
6. Pankov, V.Y.: The problem of mechanical loads on pavement of roads on cryolithic zone. E3S Web Conf. **363**, 01039 (2022)
7. Isakov, A., et al.: Modeling the operation of road pavement during the thawing of soil in the subgrade of higways. MATEC Web Conf. TransSiberia **239**, 05001 (2018)
8. Xu, G., Qi, J., Wu, W.: Temperature effect on the compressive strength of frozen soils: a review. In: Wu, W. (ed.) Recent Advances in Geotechnical Research. Springer Series in Geomechanics and Geoengineering. Springer, Cham (2019). https://doi.org/10.1007/978-3-319-89671-7_19
9. Crepeau, J., Siahpush, A.S.: Solid–liquid phase change driven by internal heat generation. C.R. Mec.Mec. **340**(7), 471–476 (2012)
10. Votyakov, I.N.: Physical Mechanical Properties of Frozen and Thawing Soils in Yakutia. Nauka, Novosibirsk (1975)
11. Zhang, X., Feng, S.G., Chen, P.C.: Thawing settlement risk of running pipeline in permafrost regions. Oil Gas Storage Transp. **6**, 365–369 (2013)
12. Bessonov, I.V., et al.: An analysis of construction solutions depending on the type of insulation materials in roads in the permafrost soils. Transp. Constr. 14–17 (2022)
13. Yartsev, V.P., Ivanov, D.V., Andrianov, K.A.: Forecasting the service time of extruded polystyrene on road constructions. Bull. Voronezh State Tech. Univ. Constr. Arch. 99–104 (2010)
14. Galkin, A.F., Pankov, VYu.: Thermal protection of roads in the permafrost zone. J. Appl. Eng. Sci. **20**(2), 395–399 (2022). https://doi.org/10.5937/jaes0-34379
15. Klochkov, Y.V., Nepomnyaschikh, E.V., Lineyvev, V.Y.: Application of foamglass to regulate thermal regime of the soil in complicated climatic conditions. Bull. Zabaykalsk State Univ. 9–15 (2015)
16. Galkin, A.F.: Efficiency evaluation of thermal insulation use in cryolithic zone mine openings. Metal. Mining Ind. **10**, 234–237 (2015)
17. Galkin, A.F.: Controlling the thermal regime of the road surface in the cryolithic zone. Transp. Res. Procedia **63**, 1224–1228 (2022)
18. Galkin, A.F., Pankov, Y.: Precision of determination of thawing depth of the frozen rocks. J. Phys. Conf. Ser. **2131**, 052079 (2021)
19. Galkin, A.F., Kurta, I.V.: Impact of temperature on the depth of thawing of frozen rocks. Mining Inform. Anal. Bull. (Sci. Tech. J.) **2**, 82–91 (2020)
20. Galkin, A.F.: Calculation of the fourier criterion when predicting the thermal regime of thawed and frozen dispersed rocks. Arctic Antarctic 1–10 (2022). https://doi.org/10.7256/2453-922. 2022.3.38555
21. Galkin, A.F., Yu Pankov, V.: Heat capacity of dispersed rocks. J. Phys. Conf. Ser. **2131**(5), 052076 (2021). https://doi.org/10.1088/1742-6596/2131/5/052076
22. Isachenko, V.P., Osipova V.A., Sukomel, A.S.: Heat Transfer. Energoizdat, Moscow (1981)
23. Perlstein, G.Z.: Water-thermal Reclamation of Frozen Rocks in the North-east of The USSR. Nauka, Novosibirsk (1979)
24. Galkin, A.F., Zheleznyak, M.N., Zhirkov, A.F.: Criterion of selection of building materials for thermal insulation layers of roads and road foundations. Achievements Mod. Nat. Sci. 108–113 (2022) https://doi.org/10.17513/use.37875

Thermal Performance Improvement of Hollow Fired Clay Bricks Embedding Phase Change Materials

Yassine Chihab[✉]

Department of Physics, Faculty of Sciences Dhar El Mahraz, Sidi Mohamed Ben Abdellah University, Fez, Morocco
yassine.chihab@usmba.ac.ma

Abstract. The construction industry in Morocco constitutes a substantial share, surpassing 33%, of the total energy consumption in the country, positioning it as one of the industries with the highest energy intensity. In spite of its considerable impact on energy consumption, the architectural design of buildings in Morocco frequently overlooks essential aspects such as the time lag and decrement factor linked to walls. This work aimed to examine the influence of incorporating phase change materials (PCMs) with three distinct melting temperatures on the dynamic thermal characteristics of bricks. The PCM is encapsulated in cylinders and integrated into the brick's solid matrix. The numerical simulations were conducted using COMSOL software, employing finite element and the apparent heat capacity methods. The results derived from the study indicate that integrating a 20% mass fraction of PCM with a melting point at 32 °C Results in a notable decrease of 34% in the heat flux swings. Additionally, there is a 2.5 h delay in the infiltration time of the external thermal temperature into the indoor space when comparing it to conventional bricks. Moreover, the application of PCM contributes to mitigating a significant portion of fluctuations in interior temperature. Consequently, incorporating PCMs into hollow clay bricks emerges as an effective approach for diminishing the energy demand associated with cooling buildings.

Keywords: Hollow Clay Bricks · Phase Change Material · Thermal Inertia · Building Energy Efficiency · Energy-Savings In Buildings

1 Introduction

Within Morocco, the construction industry comprises 33% of the overall final energy consumption and faces a considerable yearly increase in energy needs [1]. This upswing is predominantly ascribed to the considerable external thermal fluctuations experienced by building exteriors in warm climates, resulting in substantial energy consumption to maintain a satisfactory internal thermal comfort level. Despite the construction sector occupies a crucial role in the broader energy consumption scenario, it is also offers substantial opportunities to enhance thermal efficiency. This improvement can be realized by elevating the energy performance of construction materials within building envelopes.

© The Author(s) 2024
G. Feng (Ed.): ICCE 2023, LNCE 526, pp. 555–563, 2024.
https://doi.org/10.1007/978-981-97-4355-1_53

Furthermore, fired clay bricks continue to be the predominant construction elements in contemporary buildings [2]. This material offers numerous advantages, such as an improved thermal insulation properties attributed to the good insulation capacity of the air confined in the hole [3]. Nevertheless, its effectiveness is constrained in areas characterized by extreme climatic conditions [4]. To address this limitation, the incorporation of PCMs into bricks is suggested. To achieve this, the principal goal of the study is to evaluate the advantages of utilizing PCMs in brick walls. Exploring this integration is being studied as a strategy to decrease the energy needed for cooling buildings in the specific climate of Fez.

Several published papers [5–11] have highlighted the positive effects of incorporating PCMs into bricks, showcasing their ability to mitigate internal temperature fluctuations. Jia et al. [5] investigated the simultaneous incorporation of Insulation Material (IM) and PCM in bricks. Their results highlighted the distinct thermal adjustment mechanisms of IM and PCM, showcasing improvements in both steady and transient thermal performance, respectively. Furthermore, the outcomes outlined in [6] indicated that the optimal improvements in thermal inertia characteristics and thermal resistance of hollow clay walls were achieved by filling the central holes of hollow bricks with PCM32, coupled with the application of low emissivity paint or the introduction of IM into both internal and external holes. Rehman et al. [7] presented a dual-layer PCMs arrangement designed for walls in diverse climatic conditions, encompassing both hot and cold environments. Their study revealed that using double PCMs effectively upheld a stable inner temperature swings throughout both summer and winter, resulting in decreased energy needs in Islamabad.

In this paper, the main objective is to identify the appropriate type of PCM for integrating into bricks, with the aim of reducing overall cooling building loads in the climate of Fez.

2 Numerical Analysis of the Thermal Behavior of the PCM-Brick

2.1 Physical Model

The numerical study concentrated on evaluating the thermal characteristics of hollow bricks with 12 perforations incorporated with PCMs. The clay brick under examination is equipped with twelve air holes, each surrounded by cylindrical PCM capsules with a diameter of four millimeters and a mass fraction of 20% [8]. Figure 1 depicts the 2D configuration of the modeled brick, encompassing both PCM capsules and air holes. To fulfill the goal of this study, specific simplifications were incorporated.

- The examination of heat transfer is confined to a two-dimensional analysis.
- Convection is excluded from consideration within the PCM capsule.
- The Boussinesq formulation [12] is utilized to approximate the air density.
- The energy equation neglects the contribution of viscous heat dissipation.
- In this study, The thermal characteristics of the materials remain unchanged, except for air density (refer to Table 1).

The equations below delineate the mathematical model grounded on the previously mentioned assumptions:

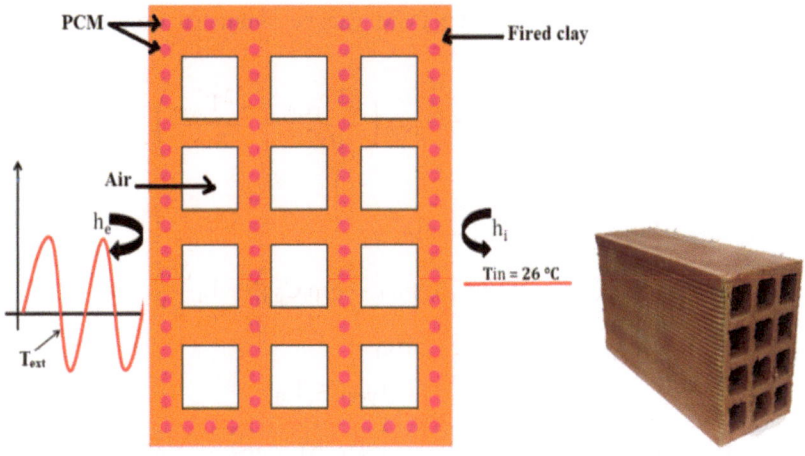

Fig. 1. Schematic illustration of the brick embedding PCMs

Table 1. The thermal characteristics of the utilized materials

Material	T_m(°C)	λ (W/(m.K))		C_p (J/(kg.K))		ρ (kg/m^3)		L_f (kJ/k)
		Solid	Liquid	Solid	Liquid	Solid	Liquid	
clay [2]	–	0.54	–	775	–	1810	–	–
Paraffin [13] (PCM24)	24	0.21	0.21	2100	2100	900	760	144
Capric acid [14] (PCM32)	32	0.149	0.149	2096	2088	1004	886	152
n-Eicosane [15] (PCM37)	37	0.15	0.15	2010	2040	778	856	241

Coupled heat transmission, involving conduction and convection, is described as follow:

$$\frac{\partial u}{\partial x} + \frac{\partial v}{\partial y} = 0 \tag{1}$$

$$\frac{\partial u}{\partial t} + u \cdot \frac{\partial u}{\partial x} + v\frac{\partial u}{\partial y} = -\frac{\partial P}{\partial x} + \mu_f\left(\frac{\partial^2 u}{\partial x^2} + \frac{\partial^2 u}{\partial y^2}\right) \tag{2}$$

$$\frac{\partial v}{\partial t} + u \cdot \frac{\partial v}{\partial x} + v\frac{\partial v}{\partial y} = -\frac{\partial P}{\partial y} + \mu_f\left(\frac{\partial^2 v}{\partial x^2} + \frac{\partial^2 v}{\partial y^2}\right) - \rho_f G\beta_T(T - T_0) \tag{3}$$

$$\frac{\partial T}{\partial t} + u \cdot \frac{\partial T}{\partial x} + v\frac{\partial T}{\partial y} = \frac{\lambda_f}{\rho_f c_f}\left(\frac{\partial^2 T}{\partial x^2} + \frac{\partial^2 T}{\partial y^2}\right) \tag{4}$$

The heat transmission through the PCM can be simulated using the following equations:

$$\rho_{pcm}C_{p,pcm}\frac{\partial T_{pcm}}{\partial t} = \lambda_{pcm}\left(\frac{\partial^2 T_{pcm}}{\partial x^2} + \frac{\partial^2 T_{pcm}}{\partial y^2}\right) \tag{5}$$

where:

$$\begin{cases} \rho_{pcm} = \theta\rho_s + (1-\theta)\rho_l \\ C_{p,pcm} = \frac{1}{\rho}\left(\theta\rho_s C_{p,s} + (1-\theta)\rho_l C_{p,l}\right) + L_f\frac{\partial\alpha_m}{\partial T} \\ \lambda_{p,pcm} = \theta\lambda_s + (1-\theta)\lambda_l \end{cases} \tag{6}$$

$$\begin{cases} \theta = 1; & \text{if } T_{pcm} \leq T_m - \frac{\Delta T}{2} \\ \theta = \frac{1}{\Delta T}\left(T_m + \frac{\Delta T}{2} - T_{pcm}\right); & \text{if } T_m - \frac{\Delta T}{2} \leq T_{pcm} \leq T_m + \frac{\Delta T}{2} \\ \theta = 0; & \text{if } T_m + \frac{\Delta T}{2} \geq T_{pcm} \end{cases} \tag{7}$$

$$\alpha_m = \frac{1}{2}\frac{(1-\theta)\rho_l - \theta\rho_s}{\theta\rho_s + (1-\theta)\rho_l} \tag{8}$$

Now, let's establish the initial and boundary conditions to solve the mathematical model:

$$\lambda_M\frac{\partial T(x,y,t)}{\partial x}|_{x=0} = h_1 \cdot \left[-T(x=0,y,t) + T_{sa}(t)\right] \quad \forall y \in [0,H] \tag{9}$$

$$\lambda_M\frac{\partial T(x,y,t)}{\partial x}|_{x=L} = h_2\left[T(x=L,y,t) - T_{in}\right] \quad \forall y \in [0,H] \tag{10}$$

With

$$T_{in} = 26\,°C \tag{11}$$

and

$$T_{sa}(t) = T_{amb}(t) + \frac{\alpha_s}{h_1}I(t) - \frac{\xi\Delta R}{h_1} \tag{12}$$

For vertical elements, $\frac{\xi\Delta R}{h_1} = 0$ [16].

In warmer day in the Fez region, the ambient temperature $T_{amb}(t)$ is written as follows:

$$T_{amb}(t) = 9.8 \times Sin\left(\frac{2\pi t}{86400} - \frac{\pi}{2}\right) + 35.5 \tag{13}$$

The solar irradiation I(t) is formulated as follows:

$$I(t) = 695.e^{-0.06\left(\frac{t}{86400} - 12\right)^2} \tag{14}$$

Ultimately, $T_{sa}(t)$ is given by the following equation:

$$T_{sa}(t) = 35.5 + 9.8 \times Sin\left(\frac{2\pi t}{86400} - \frac{\pi}{2}\right) + 28 \times \alpha_s \times e^{-0.06\left(\frac{t}{86400} - 12\right)^2} \tag{15}$$

The numerical solution of the mathematical model was conducted using the Galerkin finite element approach with the assistance of COMSOL software.

2.2 Numerical Validation

A comparative analysis is conducted, employing results extracted from the work [17], to assess the precision of the adopted solution procedure. Lachheb et al. [17] computed the temperature swings of the plaster-PCM composite using the finite volume method to solve the energy equation.

Table 2 offers a quantitative comparison of the peaks in the temperature swings of the plaster-PCM composite. The difference between the numerical data extracted from the literature [17] and the findings of this study is observed to be less than 1%.

Table 2. Transient temperature for different wallboard-PCM thickness

Thickness	2 cm		3 cm	
T (°C)	Tin, min	Tin, max	Tin, min	Tin, max
Present work	21.26	27.61	21.52	27.00
[17]	21.20	27.70	21.60	26.80
Deviation (%)	0.29	0.32	0.37	0.75

Furthermore, to affirm the precision of the numerical approach employed in this study for investigating conjugate heat transmission, an assessment was conducted a physical model encompassing natural convection flows in a square cavity [18].

Figure 2 illustrates the qualitative comparison of the dimensionless temperature distribution within the cavity. Notably, the data from the literature [18] and the results obtained through our simulations exhibit a remarkable similarity, as depicted in Fig. 2. Thus, the numerical technique utilized in this study demonstrates sufficient precision.

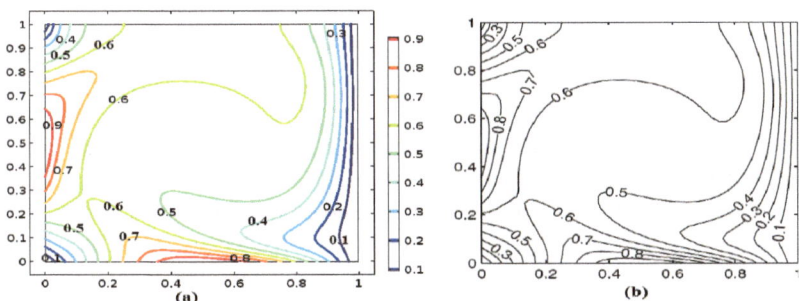

Fig. 2. The dimensionless temperature for Ra = 105 and Pr = 0.2: (a) current simulation. (b) the outcomes documented in [18].

3   Results and Discussion

To evaluate the impact of PCM type, numerical simulations were conducted on PCMs with three distinct melting temperatures (32 °C, 37 °C, and 42 °C). Figure 3 illustrates the transient internal temperature of the brick for various PCM type.

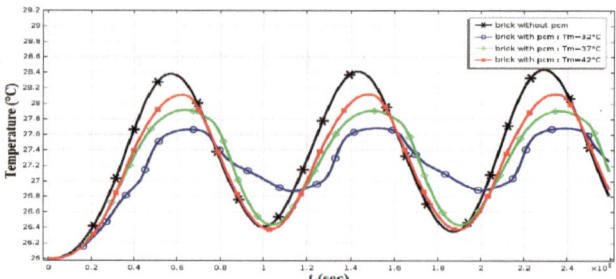

Fig. 3. Temperature swings for different PCM types

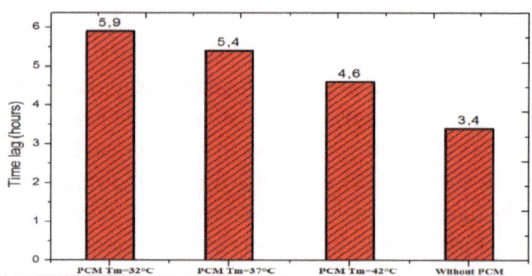

Fig. 4. The time lag for different PCM types

The findings reveal that the energy efficiency of the three PCM-based bricks surpasses that of the standard hollow brick without PCM. In comparison to the brick without PCM, bricks featuring a PCM mass fraction of 20% (PCM37 and PCM42) exhibit a minor decrease in temperature peak and a low phase shift. The PCM32-based brick shows a significant decrease in peak temperature and a substantial time lag (see Fig. 4) when compared to bricks without PCM. Considering indoor boundary conditions, the optimal Tm should closely align with the average outdoor thermal temperature. This alignment boost the benefit of a PCM's latent heat capacity, enhancing the potential to minimize total cooling loads by augmenting the wall's thermal inertia.

In conclusion, under the studied climatic conditions, PCM32 emerges as the most suitable choice when compared to the other two PCMs (PCM37 and PCM42). It effectively reduces the oscillations of the indoor thermal wave while maintaining a satisfactory level of comfort.

Figure 5 depicts the heat flux swings for the three types of PCM integrated into the bricks. It is evident that incorporating any PCM into brick permits attenuate the inner heat flux swings. Nevertheless, this decrease remains below 12% for the brick with PCM42 when compared to the brick without PCM. This outcome can be attributed to the low molten fraction of PCM42, as the external temperature consistently stays below its melting point for most of the time, rendering its latent heat capacity practically inactive.

On the other hand, it is evident that utilizing PCM32 permits attenuate the inner heat flux swings, amounting to approximately 34% compared to the brick without PCM.

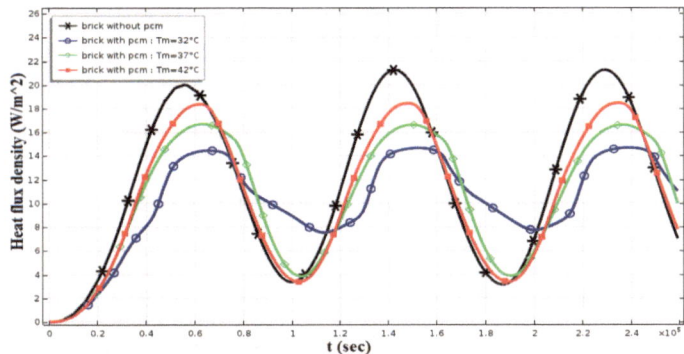

Fig. 5. Heat flux swings for different PCM types

Analyzing the variations in total liquid fraction enhances the comprehension of the results obtained in this section. Figure 6 illustrates that PCM42 has a low molten fraction (total liquid fraction less than 0.15), rendering its latent heat storage practically inactive. Consequently, incorporating PCM42 into bricks yields the lowest improvement compared to the other two PCMs. In contrast, PCM32 and PCM37 remain partially solid (partially liquid), activating their capacity for latent heat storage. Moreover, the molten percentage of PCM32 surpasses that of PCM37, indicating that the stored latent heat in the PCM32-based brick is higher.

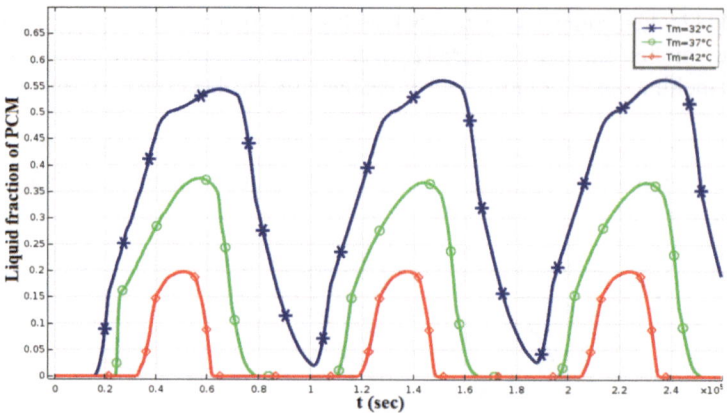

Fig. 6. Liquid fraction for different PCM types

4 Conclusion

This paper underscored the benefits of incorporating a PCM into bricks, with a specific PCM recommendation tailored to the Fez climate. The findings demonstrated that integrating a 20% mass fraction of PCM with a Tm of 32 °C resulted in a substantial 34% reduction in inner heat flux swings and a 2.5-h delay in the infiltration time of the external temperature swings into the indoor space compared to the brick without PCM. Furthermore, the outcomes suggested that the utilization of PCM significantly mitigates fluctuations in interior temperature. Consequently, the incorporation of PCMs in hollow bricks proves effective in maintaining satisfactory internal thermal comfort while concurrently reducing the energy required for cooling buildings.

References

1. Ministère de la Transition Énergétique et du Développement Durable. https://www.environne ment.gov.ma. Accessed 28 Sep 2021
2. Laaroussi, N., Lauriat, G., Raefat, S., Garoum, M., Ahachad, M.: An example of comparison between ISO Norm calculations and full CFD simulations of thermal performances of hollow bricks. J. Build. Eng. **11**, 69–81 (2017)
3. Aditya, L., et al.: A review on insulation materials for energy conservation in buildings. Renew. Sustain. Energy Rev. **73**, 1352–1365 (2017)
4. Leo Samuel, D.G., Dharmasastha, K., Shiva Nagendra, S.M., Prakash Maiya, M.: Thermal comfort in traditional buildings composed of local and modern construction materials. Int. J. Sustain. Built Environ. **6**(2), 463–475 (2017). https://doi.org/10.1016/j.ijsbe.2017.08.001
5. Jia, C., Geng, X.Y., Liu, F.D., Gao, Y.N.: Thermal behavior improvement of hollow sintered bricks integrated with both thermal insulation material (TIM) and phase-change material (PCM). Case Stud. Therm. Eng. **25**, 100938 (2021)
6. Chihab, Y., Bouferra, R., Garoum, M., Essaleh, M., Laaroussi, N.: Thermal inertia and energy efficiency enhancements of hollow clay bricks integrated with phase change materials. J. Build. Eng. **53**, 104569 (2022)

7. Rehman, A.U., Sheikh, S.R., Kausar, Z., McCormack, S.J.: Numerical simulation of a novel dual layered phase change material Brick Wall for human comfort in hot and cold climatic conditions. Energies **14**, 4032 (2021)
8. Hamdaoui, S., Bouchikhi, A., Azouggagh, M., Akour, M., Msaad, A.A., Mahdaoui, M.: Building hollow clay bricks embedding phase change material: thermal behavior analysis under hot climate. Sol. Energy **237**, 122–134 (2022)
9. Rai, A.C.: Energy performance of phase change materials integrated into brick masonry walls for cooling load management in residential buildings. Build. Environ. **199**, 107930 (2021). https://doi.org/10.1016/j.buildenv.2021.107930
10. Chihab, Y., Bouferra, R., Bouchehma, A.: Transient thermal behavior of clay walls integrated with phase change materials. J. Energy Storage **73**, 109246 (2023)
11. Bachir, A., Taieb, N.: Numerical analysis for energy performance optimization of hollow bricks for roofing. Case study: hot climate of Algeria. Constr. Build. Mater. **367**, 130336 (2023). https://doi.org/10.1016/j.conbuildmat.2023.130336
12. Théorie, B.: Analytique de la Chaleur. Gauthier Villars, Paris (1903)
13. Heat storage materials: Abhat a low temperature latent heat thermal energy storage. Sol. Energy **30**, 313–332 (1983)
14. Chuah, T.G., Rozanna, D., Salmiah, A., Thomas Choong, S.Y., Sa'ari, M.: Fatty acids used as phase change materials (PCMs) for thermal energy storage in building material applications. Jurutera **2006**(July), 8–15 (2006)
15. Alawadhi, E.M., Alqallaf, H.J.: Building roof with conical holes containing PCM to reduce the cooling load: numerical study. Energy Convers. Manage. **52**(8–9), 2958–2964 (2011)
16. Ulgen, K.: Experimental and theoretical investigation of effects of wall's thermophysical properties on time lag and decrement factor. Energy Build. **34**(3), 273–278 (2002). https://doi.org/10.1016/S0378-7788(01)00087-1
17. Lachheb, M., Younsi, Z., Naji, H., Karkri, M., Nasrallah, S.B.: Thermal behavior of a hybrid PCM/plaster: a numerical and experimental investigation. Appl. Therm. Eng. **111**, 49–59 (2017)
18. Roy, S., Basak, T.: Finite element analysis of natural convection flows in a square cavity with non-uniformly heated wall (s). Int. J. Eng. Sci. **43**(8–9), 668–680 (2005)

Analysis the Causes of the Slip Embankment Protecting the Slope of Lien Chieu Aviation Petroleum Depot

Hong Nam Nguyen[✉]

Division of Geotechnical Engineering, Faculty of Civil Engineering, Thuyloi University, 175
Tay Son Street, Dong Da District, Hanoi, Vietnam
hongnam@tlu.edu.vn

Abstract. The slope protection embankment of Lien Chieu aviation petroleum
depot was built in 2001. The embankment is responsible for ensuring safety for the
oil tank area and dispensing area, with the capacity of dispensing 13,800 m^3 of fuel
serving airports in the central region, from Hue to Binh Dinh provinces. In 2008,
due to the occurrence of floods, a landslide has destroyed the entire 59.7 m middle
segment of longitudinal embankment, leaving the foundation inert. The rest of the
embankment body in the North was cracked on the entire surface; the embankment
body fell inward; the weight relief plate was cut entirely from the embankment
body and collapsed downwards. The collapse of the slope embankment led to the
destruction of the oil tanks, causing the operation to stop. It was urgent at that
time to investigate the causes of the slope failure and suggest the safe remedial
design for the structure. The site investigation and numerical analyses show that
the reasons for the failure could be due to the insufficiency of drainage holes
to drain groundwater during the flood season characterized by heavy rain, which
made the lateral pressure acting on the wall increase significantly, possibly causing
the wall structure to be damaged. The proposed design which consists of a low
retaining wall and upper slope embankment, both reinforced by geotextiles, could
be stable according to numerical modelling with many loading cases. Attention
must be paid to the stability of the excavation slope during construction and the
difficulties of narrow space for the excavation.

Keywords: Embankment · Sliding · Stability · Rain Flood · Slope Protection

1 Introduction

The aviation fuel depot construction project in Lien Chieu, Da Nang, is located in Hoa
Hiep Bac ward, Lien Chieu district, Da Nang city (Fig. 1a). The construction area is
situated on the southern slope of Hai Van Pass down to the sea. To the west, it borders
National Road 1A, to the east is Kim Lien Bay in Da Nang Bay, located at $16^0 08' 00''$
north latitude and $108^0 09' 00''$ east longitude, in Hoa Hiep ward, Lien Chieu district, Da
Nang city. The project was built in 2001 with a capacity of 13,800 m^3 of various fuels to
serve airports in the Central region, from Hue to Binh Dinh. The slope protection project
has the task of ensuring the safety of the oil storage area and distribution area with a
high difference of 18 m.

© The Author(s) 2024
G. Feng (Ed.): ICCE 2023, LNCE 526, pp. 564–574, 2024.
https://doi.org/10.1007/978-981-97-4355-1_54

In 2008, during the period from October to November, due to the occurrence of floods landslides destroyed the entire 59.7-m-long central segment. The remaining embankment body in the south (about 4 m long) was still quite intact. The rest of the embankment body in the North (about 7 m) was cracked on the entire surface; embankment body fell inward; the load relief plate was severed from the embankment body and collapsed downwards. The collapse of the slope embankment led to the destruction of the oil tanks, causing the operation to stop (Fig. 1b).

To restore normal operations and long-term stability of the Lien Chieu oil storage tanks, it is necessary to rebuild the central segment of the slope embankment and to repair other damages to ensure the safety of the entire structure. Therefore, investment in the restoration of the tank foundation of the Lien Chieu aviation oil storage tank was very urgent and decisive in ensuring timely, effective and efficient supply of fuel to meet the increasing demand of the central region's airports.

The article analyzes the causes of the slope failure and suggests safety measures when reviewing some design proposals of relevant institutions as requested.

Fig. 1. The aviation fuel depot in Lien Chieu, Da Nang.

2   Method of Study

In this study, we conducted a survey at the site to investigate the failure. Our subsequent task involved analyzing and numerically assessing designs proposed by various relevant institutions, as well as recommending safety measures. For the numerical simulations, we utilized both the finite element method (FEM) and the limit equilibrium method (LEM). We employed Plaxis and GeoStudio software to analyze the stress-strain relationship and the factor of safety of the retaining walls, utilizing plane problem models.

3 Results and Discussion

3.1 Project Site Investigation

According to the topographic survey [4], the project location is on the southern mountain slope of Hai Van Pass with a slope elevation of +57 m (fuel dispensing station platform) to +39 m (oil tank farm). It faces the straight direction of National Highway 1A, with a steep slope and a slope height of 70 ° compared to the flat surface. Along National Highway 1A, the slope direction is North-South, which is also the direction of water flow from the mountain slope and outflows through the road trench. The water in the fuel dispensing station platform area flows towards the foot of the slope. The foot of the slope is an existing oil tank farm, in which the foot-to-tank 1 distance is about 1.5 m and the foot-to-tank 2 distance is about 2.5 m, which are very narrow and fixed distances.

According the original design [9], the length along the ridge is 101 m. The slope structure is made of reinforced concrete with a grade of 250, 13 m high. The thickness changes from 40 cm (at the foot of the roof) to 30 cm (at the top of the slope). The slope angle is 70 °. The foundation is 50 cm thick and 1.5 m wide, placed on natural ground. The top of the ridge is at elevation +48 m, while the foot of the ridge is at elevation +35 m. The load-bearing horizontal part is at elevation +44 m. The top of the ridge features a horizontal flange 1m wide and 25 cm thick for locking the roof. The inside of the ridge is filled with sand and gravel material with a compaction ratio (K value) of 0.9. From elevation +48 m to the distribution platform (elevation +53.6 m) is the soil slope, which is 70 °, reinforced by 11 layers of watered TT060, 3.5 m wide, with a spacing of 0.5 m.

Da Nang is situated in an area influenced by monsoon activity, thus falling within the monsoon climate zone. The Truong Son Mountain range to the west significantly influences the type and characteristics of the climate in the region. This area experiences a 2-season climate pattern: a dry season extending from January to September and a rainy season from September to December. The interplay of the monsoon regime and the Truong Son range generates distinct variations between the rainy and dry seasons across the project area.

The soil strata of the surveyed area (Fig. 2), from top to bottom, are as follows [10]:

Layer 1a: The yard consists of concrete, sand, crushed stone, and clay, with a poorly structured composition. This layer is distributed across the surface and encompasses most of the surveyed area. It has been identified in boreholes HK09, HK05, and HK03, with thicknesses ranging from 0.20 m (HK05) to 1.40 m (HK01). The average thickness of this layer is 0.67 m.

Layer 1b: consists of filling soil characterized by mixed clay with yellow-brown, yellow-gray, gray-brown, red-brown, mixed with grit, weathered rock in hard plastic state. This layer is distributed throughout the survey range, found in all boreholes (from HK01 to HK09), the thickness varies from 0.40 m (HK08) to 5.90 m (HK05).

Layer 2: comprises clay with yellow-brown, yellow-gray, and gray-brown hues, mixed with gravel and weathered rock, presenting a semi-hard to hard state. This layer is partially distributed across the survey area and has been encountered at boreholes HK02, HK04, and HK08. The thickness of this layer varies from 2.60 m (HK08) to 7.80 m (HK04).

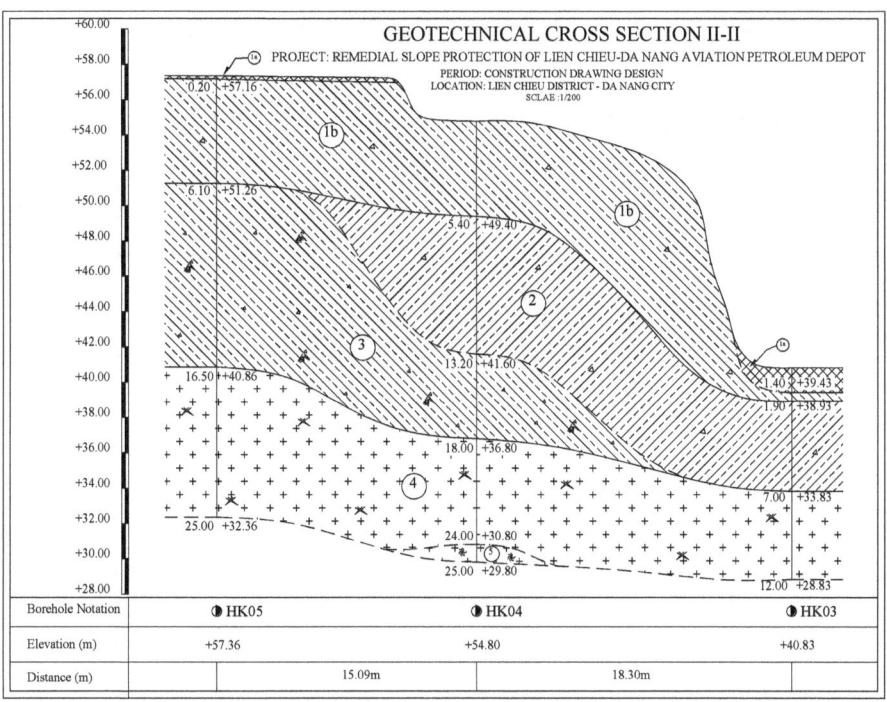

Fig. 2. Typical geotechnical cross section [10].

Layer 3: consists of intensely weathered granite, clay mixed with yellow-brown, gray-brown, and gray-white tones, combined with grit and soft weathered seams, presenting a semi-hard to hard state. This layer is identified in boreholes HK01, HK02, HK04, HK05, and HK06. The thickness of this layer varies from 0.50 m (HK06) to 13.80 m (HK02).

Layer 4: is characterized by strongly weathered granite, displaying yellow-grey and white-gray hues, with cracks and strong fragmentation, often exhibiting significant weathering. This layer is distributed and identified in boreholes HK02, HK03, HK04, HK05, HK07, and HK09. The thickness of this layer within the survey range is unknown, reaching depths of up to 25.00 m, as boreholes HK02 and HK05 terminate within this layer.

Layer 5 consists of yellow-grey, white-gray, and green-gray granite exhibiting moderate to light weathering, characterized by a block structure and cracks. This layer is partially distributed within the survey range and is only identified in boreholes HK01, HK04, HK06, HK07, HK08, and HK09. The thickness of this layer is unknown within the survey depth range, varying from 5.50 m (HK07) to 25.00 m (HK04), as all boreholes terminate within this layer.

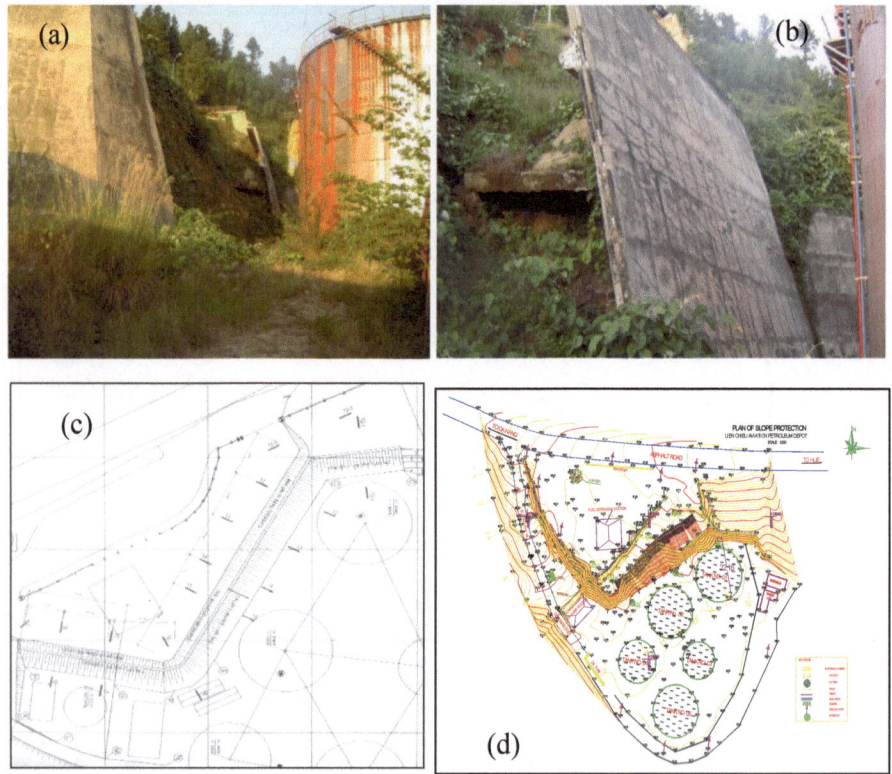

Fig. 3. Existing condition of the work at site: (a) General status of damaged slope embankment, (b) Cracked face plate and remaining broken offload plate (CD segment), (c) Slope protection embankment before 2008, (d) Layout of boreholes of this study.

In 2008, floods triggered landslides, resulting in the destruction of the entire section spanning 59.7 m along the embankment (from point D to point E, as shown in Figs. 3a, b, c). The remaining structure is now exposed, revealing a foundation that is 50 cm thick and 1.5 m wide. The southern section of the embankment (approximately 4 m long in the EF segment) remains largely intact. However, the northern portion of the embankment (around 7 m in the CD segment) is extensively cracked across its surface, causing the embankment body to collapse inward. The offload plate has been separated from the embankment body and has collapsed downwards. Consequently, the collapse of the slope embankment resulted in the destruction of the oil tanks, prompting the depot to cease all operations.

Based on the survey, the failure of the slope embankment wall could be attributed to the following reasons [3]:

1. Due to the absence of drainage holes on the slope embankment wall, heavy rainfall saturated the backfill soil, leading to increased lateral pressure on the wall. This heightened pressure likely contributed to structural damage within the wall. The failure primarily occurred at the base of the wall, where the highest shear stress was

experienced. Consequently, the wall underwent significant displacement, ultimately collapsing under its own weight and the pressure from the sliding soil. As a result, the embankment wall collapsed, and the embankment behind it slipped, exacerbating damage to concrete structures and adjacent components.

2. The placement of the relief plate on the embankment surface resulted in its transformation into a large-span cantilever (3 and 4 m) when the embankment subsided. The maximum shear stress occurred at the connection point between the relief plate and the wall panel. Due to the considerable weight of the relief plate and the soil above it, the plate was severed at the joint and collapsed. This phenomenon is evident in the northern section, specifically the beginning of the CD section, where the remaining part of the embankment wall became fractured (Fig. 3).

3. The destruction of the embankment wall originated in the middle near the beginning of the CD segment (Fig. 3), where the backfill was most extensive, resulting in the collapse of the entire segment. Subsequently, the backfill soil mass at the outset of the CD segment slid along the embankment, creating a significant gap between the excavated slope and the embankment wall surface. This displacement caused the embankment wall panel to move inward, in the opposite direction to its original position when the wall body was destroyed, resulting in surface-wide cracking. Consequently, the load relief plate was severed from the wall panel and collapsed. Although this segment of the embankment wall did not collapse entirely, its ability to support force was compromised, even though the reinforcements were still capable of maintaining its current state.

4. The damage to the embankment wall resulted in the landslide of the upper reinforced embankment from elevation +48 m to +54 m. The embankment block situated atop the embankment slipped, causing settlement and fracturing of the reinforced concrete layer of the fuel dispensing station platform, which subsequently collapsed.

5. In the original design [9], two types of embankment wall sections were utilized. Essentially, these two types are identical, differing only in the size of the load relief plate. However, observations from the actual site post-failure indicate that the efficacy, if any, of load reduction is minimal. For instance, at the termination of the longitudinal line (from position Sect. 15 to E; Fig. 3), the foundation consisted primarily of solid rock, visibly close to the embankment wall surface. Consequently, one end of the relief plate (3 m wide) likely rested on a stable stone foundation. However, during instances of increased lateral pressure on the wall due to soil saturation from groundwater, this load reduction plate did not perform adequately, potentially resulting in the undermining of the wall base.

6. The remnants of the bottom footing of the embankment wall were nearly intact, indicating that the embankment wall did not experience overall sliding, and the bottom foundation was situated on a solid rock foundation.

3.2 Numerical Simulations of the Structure at Site

To assess the impact of groundwater on the stability of the retaining wall numerically, we conducted simulations under various loading scenarios, including construction, basic, and special cases such as water drain pipe clogging (Fig. 4). The selected cross-section was at station 13 (KM: 0 + 94.37), corresponding to the geological cross-section passing through boreholes LK2 and LK5, associated with Sect. 2 of the embankment (Figs. 3c and 3d). At this location, the natural ground elevation is −1.95 m. The top elevation of the wall is +57.0 m, with a wall height of 18 m from the base. The ground soil comprises layers 1, 3, 4, 5A, and 5. For simplicity, the backfill and ground soil were simulated using the Mohr-Coulomb model [7], with parameter values as shown in Table 1.

Table 1. The Mohr-Coulomb material model parameters for ground and backfill soils.

ID	Soil layer	Type	γ_{unsat} [kN/m³]	γ_{sat} [kN/m³]	k_x [m/day]	k_y [m/day]	ν [-]	E_{ref} [kN/m²]	C_{ref} [kN/m²]	φ [°]	ψ [°]
1	Backfill	Drained	19.3	19.7	0.0432	0.0430	0.30	5000.0	17.0	19.0	0.0
2	3	Drained	19.5	20.2	0.0215	0.0215	0.30	7498.0	20.0	23.0	0.0
3	4	Drained	20.3	20.9	0.0215	0.0215	0.30	9000.0	22.0	26.0	0.0
4	5	Drained	21.0	21.0	8.64×10^{-4}	8.64×10^{-4}	0.20	1.0×10^5	100.0	35.0	0.0
5	5a	Drained	21.0	21.0	0.0215	0.0215	0.20	10000.0	20.0	30.0	0.0
6	Filter sand	Drained	17.0	19.0	1.0000	1.0000	0.25	10000.0	1.0	30.0	0.0

The construction process can be divided into three main stages: excavating the foundation pit and constructing the wall bottom slab; constructing the retaining wall; and backfilling soil behind the wall.

In the basic case, the groundwater level was assumed to be at the bottom of the foundation (elevation +37 m). The distributed load on the top surface was assumed to have values of q = 0 and 60 kPa. However, in the special case, we considered the scenario of a clogged water drain pipe, causing the groundwater level behind the wall to rise to elevation +45 m, while the water level in front of the wall was assumed to be at elevation +39.0 m. In this case, the distributed load on the top surface was assumed to be q = 55 kPa. It is important to note that the earthquake effect was not considered in these analyses.

The Plaxis software, version 8.6 [7], was utilized to analyze the stress-strain behavior using the Finite Element Method (FEM). The Factor of Safety (FOS) was calculated using the shear strength reduction technique [1]. Table 2 presents the results of deformation calculations for both the basic and special cases.

In the basic case, the wall demonstrated stability under two different surface load values: FOS = 1.42 (q = 0) and FOS = 1.24 (q = 60 kPa). However, in the special case of water drain pipe clogging, the rising groundwater level increased the horizontal pressure acting on the wall due to the water pressure. Consequently, the reduction of FOS to 1.11 (q = 55 kPa) did not meet the allowable FOS = 1.13 specified by TCXDVN 285–2002 [8]. Hence, there is a necessity to revise the remedial design plan to enhance the stability of the slope protection embankment.

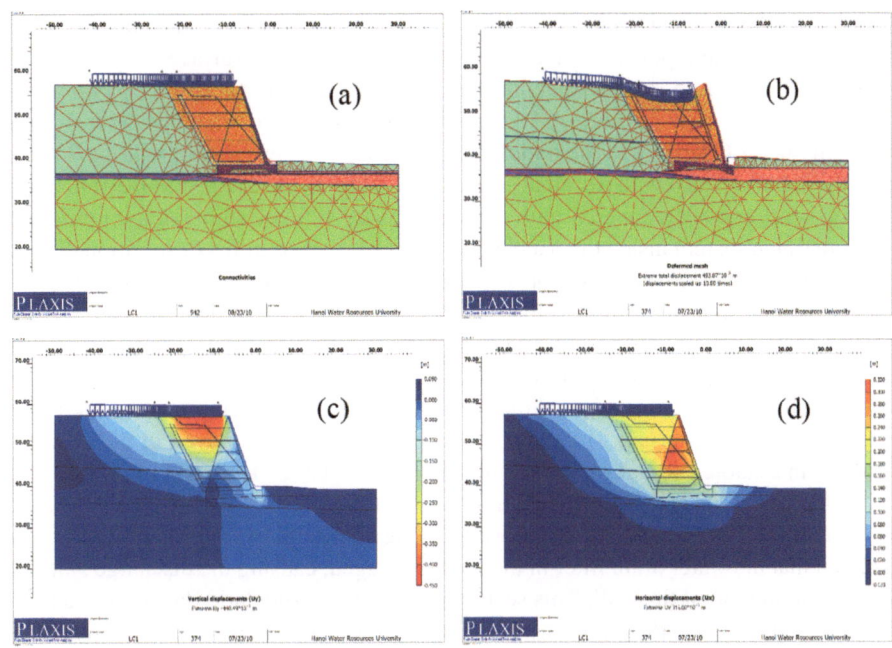

Fig. 4. Slope analysis of the retaining wall in the special case of clogged water drain pipes.

Table 2. Calculation results of deformation in the basic and special cases.

Case of calculation	Basic case		Special case
	q = 0	q = 60 kPa	q = 55 kPa
Maximum settlement when filling to the top (m)	0.24	0.47	0.44
Maximum horizontal displacement (m)	0.20	0.29	0.32

3.3 Checking Proposed Remedial Designs

In consideration of the actual topographical conditions, constructing a wall higher than 10 m would require extensive excavation, posing challenges due to limited terrain availability. Therefore, a design featuring a lower wall height of 6.0 m (from elevations + 39.80 m to +45.80 m) and an upper slope embankment (from elevations +45.80 m to +57.00 m) was proposed [2].

Given the steepness of the excavated slope (Fig. 5a), which fails to meet slope stability Factor of Safety (FOS) standards, an adjusted design [6] was proposed for the protection of the northern segment, as depicted in Fig. 5b. This structure comprises two main components:

Lower Part: A reinforced concrete (RC) wall, 6.5 m high, with a bottom plate 600 cm wide and 80 cm thick, featuring a vertical wall plate 80–50 cm wide. Triangular middle counterforts, 50 cm thick, and rectangular side counterforts, 50 cm thick, are included. Two rows of plastic pipes for drainage are arranged on the wall surface.

Upper Part: A 13.5 m high soil embankment with a slope factor of 1:1.5. Two berms are included: the slope leg berm, 250 cm wide with a drainage ditch (40-40 cm) made of 20 cm thick RC #200, and the middle berm placed at an elevation of +47 m, 700 cm from the leg berm, also with a drainage ditch (40-40 cm) made of RC #200, 20 cm thick. Geogrid reinforcement, covered with concrete mortar #200 or employing a reinforced concrete frame system inside the panel of protective stones, is utilized. Additionally, the embankment is reinforced with geotextile reinforcement, approximately 30 cm high.

The new design underwent rigorous slope stability analysis, considering various loading cases such as during construction, normal operation, and special cases. The GeoStudio software, version 6.02 [5], was utilized for analysis. A grid of centers of potential slip surface (13 × 13) and radii were implemented to determine the minimum FOS. The limit equilibrium Morgenstern-Price method was employed in the analysis, resulting in a stable excavation slope with an FOS of 1.19, as depicted in Fig. 6a.

In the case of the end of construction, it was assumed that the backfill soil is dry, and the groundwater level appeared at the wall footing elevation. In the special case, it was assumed that the water drain filter in the wall is clogged, causing the discharge water to overflow to the top of the wall. This scenario presents a particularly dangerous case.

All results satisfied the relevant standards when the wall was reinforced with geotextiles, with FOS values of 1.707, 1.858, and 1.591 for the construction, normal operation, and special cases, respectively (Fig. 6). It's noteworthy that for the design without slope reinforcement, the FOS value was 1.175, which did not meet the required standard in the case of construction. Clearly, the geotextile reinforcement proved effective in increasing the embankment stability.

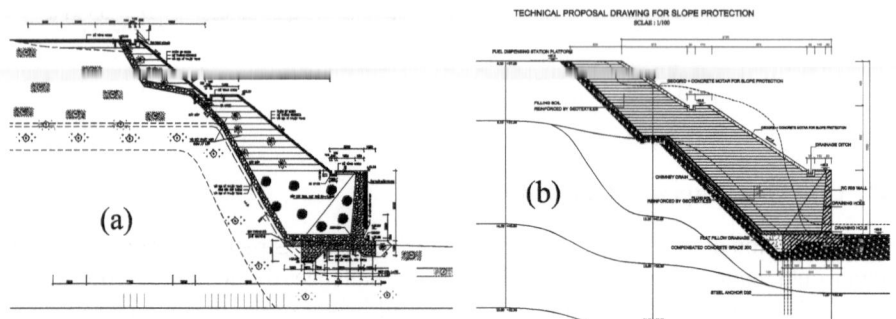

Fig. 5. Proposed remedial designs: (a) steep excavation, (b) gentle excavation.

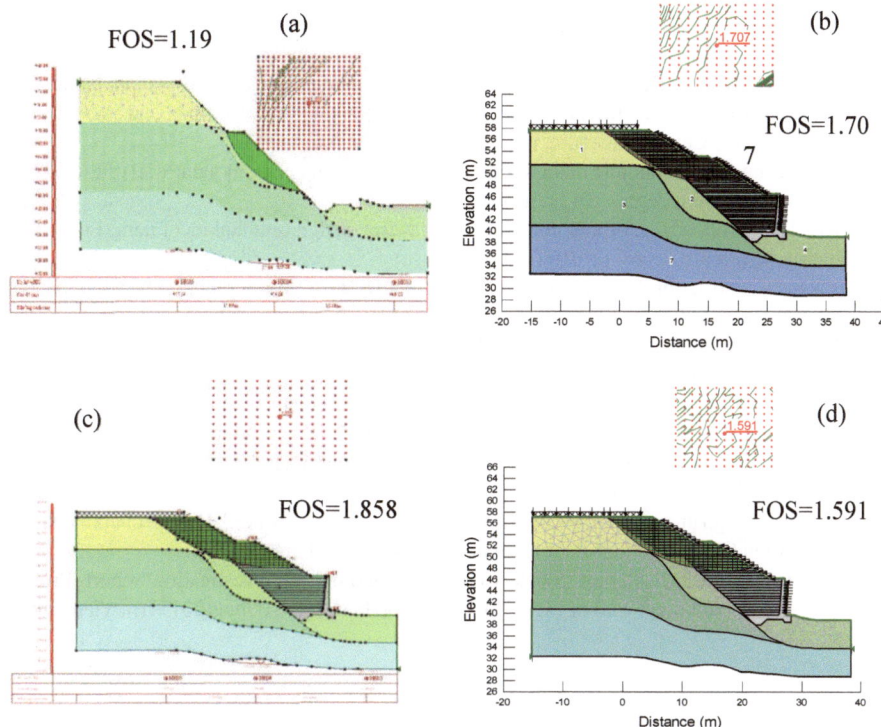

Fig. 6. Slope stability analyses of the new design.

4   Conclusion

The failure of the slope protection embankment at Lien Chieu aviation petroleum depot underwent thorough investigation. The reasons for the failure could be attributed to the insufficient drainage holes, which failed to effectively drain groundwater during the flood season with heavy rain. This resulted in a significant increase in lateral pressure acting on the wall, potentially causing structural damage.

To address these issues, a proposed design comprising a low retaining wall and upper slope embankment, both reinforced by geotextiles, was developed. Numerical modeling with various loading cases suggests that this design could ensure stability. However, particular attention must be given to the stability of the excavation slope during construction.

Considering the challenges posed by limited traffic space, a combination of soil nailing and geotextiles could prove effective for slope embankment protection in narrow excavation areas. This approach enhances stability while accommodating space constraints.

References

1. Brinkgreve, R.B.J., Bakker, H.L.: Non-linear finite element analysis of safety factors. In: Proceedings of the 7th International Conference on Computer Methods and Advances in Geomechanics, Cairns, Australia, pp. 1117–1122 (1991)
2. CRA: Technology economic main Report Project: Repairing embankment wall of Lien Chieu Aviation Petroleum depot (2011) (in Vietnamese)
3. CRA: Report on verification of construction drawing design (embankment item, Lien Chieu aviation petroleum depot) (2010) (in Vietnamese)
4. Da Nang Construction Technical Consulting Center DCTCC: Topographic survey report (Lien Chieu Aviation Petroleum depot) (2010) (in Vietnamese)
5. GEO-SLOPE: Stability Modeling with SLOPE/W 2007—An Eng. Method, 3rd edn. GEOSLOPE, Alberta (2008)
6. Nguyen Tien Dat: Remedial design of repairing embankment wall of Lien Chieu aviation petroleum depot. Bachelor thesis, Thuyloi University (2011) (in Vietnamese)
7. Plaxis BV: Plaxis 8 Professional version. The full manual. Delt, The Netherlands (2006)
8. TCVN 285:2002: Hydraulic works - The basic stipulation for design (2002) (in Vietnamese)
9. Transport design and technology transfer consulting center TDTTCC: Design of embankment wall of Lien Chieu Aviation Petroleum depot (2001) (in Vietnamese)
10. Viet Delta Consulting Joint Stock Company: Geological Investigation report Project: Repairing embankment wall of Lien Chieu Aviation Petroleum depot (2011) (in Vietnamese)

Dynamic Response and Reliability Assessment of Reinforced Concrete Members Under Vehicle and Train Collisions: Quantifying Impact Performance and Damage

Khalil AL-Bukhaiti[1], Yanhui Liu[1](✉), Shichun Zhao[1], and Daguang Han[2,3]

[1] School of Civil Engineering, Southwest Jiaotong University, Chengdu, Sichuan, China
eng.khalil670@hotmail.com, yhliu@swjtu.edu.cn
[2] Department of Civil Engineering and Energy Technology, Faculty of Technology, Art and Design, Oslo Metropolitan University, Pilestredet, Oslo, Norway
[3] Faculty of Civil Engineering, Southeast University, Nanjing, China

Abstract. This study analyzes the behavior of reinforced concrete (RC) members impacted by vehicles or trains by reviewing existing experimental and numerical work. RC structures are susceptible to such collisions, which can cause structural failure or collapse. The damage response depends on the energy transferred during impact, ranging from cosmetic effects to full structural failure. To properly characterize damage, indices are defined considering concrete and steel strain rates during and after plastic deformation. Concrete spalling is also discussed as an indicator of localized or global failure severity. Experimental results from other studies are presented and compared to theoretical models. The latest provisions for incorporating strain rate sensitivity into material models based on various standards are reviewed. This analysis aims to provide insights for better assessing the reliability of RC members after impacts. Understanding reliability is important for evaluating the safety of damaged structures and mitigating risks. The findings can help practitioners evaluate damaged structures and guide future design to avoid severe outcomes from vehicle and train collisions onto critical infrastructure.

Keywords: Unequal Impact · Shear Failure · Strain Rate · Crack Propagation · Numerical Analysis · Bending Failure

1 Introduction

Lateral impact loads such as earthquakes, explosions, vehicle collisions, train impacts, and ship impacts can cause significant damage to reinforced concrete (RC) structures if not properly addressed in the design [1, 2]. RC structures are composed of concrete and steel reinforcement and are designed to resist compressive and tensile forces. However, lateral loads induce complex stress states that can exceed the strength of materials [3, 4]. Understanding how RC structures behave under these dynamic loads is important for safety and performance. Several factors influence the response to lateral impacts. The strength of the concrete and steel reinforcement is critical, as higher

© The Author(s) 2024
G. Feng (Ed.): ICCE 2023, LNCE 526, pp. 575–584, 2024.
https://doi.org/10.1007/978-981-97-4355-1_55

material strengths improve impact resistance [5, 6]. The compressive strength of concrete and tensile strength of reinforcing steel allows the structure to withstand greater loads. Larger cross-sectional dimensions and more reinforcement provide better protection against impacts [7, 8]. The intensity and duration of lateral loads also matter [9, 10]. Short, high intensity impacts from explosions often cause the most damage through cracking and spalling of concrete cover [11, 12]. Proper design must consider load characteristics [13, 14]. In comparison, longer earthquake loads typically induce less severe cracking. Proper design must consider load characteristics. Research into RC structural behavior under lateral loads occurs through experimentation and numerical modeling. Full-scale and small-scale impact tests using methods like drop-weight impart dynamic loads [15, 16]. Observations from these experiments enhance understanding of RC response. Numerical simulations employ tools like finite element analysis to replicate complex loadings in a controlled virtual environment [17, 18]. Blast loading is an extensively researched lateral impact. Studies show explosives induce cracking and reinforcement failure. Parameters like explosive type/size/location and structure properties yield valuable data. High-strength concrete and reinforcement performed better. Train and ship collisions are also under examination. Varying cross-sectional geometries, vehicle velocities, material properties, and impact zones alter the structural response. Cracking and spalling patterns emerge depending on parameters. Advancing knowledge aims to refine modeling accuracy and develop improved designs. Fiber-reinforced polymers show promise as alternative reinforcement. Innovations target handling impacts without failure to safeguard infrastructure. As lateral impacts endanger the integrity, ongoing work strives to comprehend failure mechanisms better. The goal is to design RC members efficiently to withstand dynamic hazards through resistance strategies. Deeper insight supports assessing damaged structures and mitigating life safety risks from impact incidents. Continued research in this vital field fortifies resilient construction.

2   Collision Incidents

Vehicle-structure and train-structure collisions are unfortunate but relatively common incidents that can lead to loss of life, injury, and property damage if not properly addressed. As transportation and infrastructure development continues globally, protecting against these hazards grows more important. Collisions can occur in various scenarios. Due to driver error or vehicle malfunction, cars or trucks may crash into bridges, buildings, or guardrails. Derailed trains can impact a rail network's bridge piers, walls, or other fixed concrete assets. Ships navigating waterways also pose collision risks with bridges under some conditions. Several high-profile incidents demonstrate the destructive potential. In 2007, the cargo ship Cosco Busan struck a San Francisco Bay pier, resulting in an oil spill and over $70 million in cleanup costs. In China, fatal train wrecks between 2011–2018 involving reinforced concrete bridges caused hundreds of casualties [2]. Other cases internationally include trains hitting walls in Spain, the USA, and Turkey with loss of life. Factors contributing to accidents span both human factors and structural/design elements. Poor driving, train operations errors, mechanical failures, and unexpected natural events can all trigger impact scenarios. Infrastructure layout, bridge/wall placements, and wind/sea conditions in maritime settings also influence collision likelihood. Impacts create considerable damage depending on vehicle/vessel mass and velocity.

3 Methodology

The methodology used in a literature review is critical to its success and credibility. The methods and techniques used in a literature review determine the quality and relevance of the research sources reviewed and, thus, the validity and reliability of the findings. This paper describes the methods and techniques used in the literature review on RC structures subjected to lateral impact loads. The paper will include the databases, search terms, and inclusion and exclusion criteria. This paper used several databases, including the Web of Science, ScienceDirect, and Scopus. The databases were searched using the following terms: "reinforced concrete, lateral impact loads, blast loads, vehicle impact loads, train impact loads, ship impact loads, seismic loads, behavior, performance, damage." The article was limited to the last twenty years to ensure that the most recent research was included in the review. The paper included studies published in peer-reviewed journals, conference proceedings, and dissertations. The studies had to be in English and focused on the behavior of RC structures under lateral impact loads. The studies also had to include experimental or numerical analysis of the behavior of the structures. The studies included in this paper were analyzed systematically and comprehensively. The studies were first grouped into categories based on the type of lateral impact loads they addressed (blast, vehicle, train, ship, seismic). Then, the studies were reviewed, and their main findings were extracted and summarized. The studies were also compared to identify similarities and differences in their conclusions.

4 Investigating the Results of Impact Resistance on RC Members: A Study of Experimental, Numerical, and Analytical Methods

Reinforced concrete members are prone to severe damage from impact accidents, which can cause partial failure or collapse of the structure. Scholars have studied impact resistance through experiments, numerical analysis, and analytical methods.

4.1 Experimental Analysis

Research on the dynamic response of reinforced concrete members to impact loads has provided valuable insights but is limited by testing constraints. Most studies focus on mid-span impact, but parameters like boundary conditions, impactor shape, and impact position require further examination. Hughes and Speirs [3] conducted impact tests with varying support conditions and local stiffness. Members failed in bending concentrated at supports and mid-span. Stiffness had more influence on response than support. Demartino tested cantilevers and simply supported beams, finding boundary conditions and velocity greatly affected response and damage. Measured inertial forces indicated they resist over 2/3 of peak impact load. Pham and Hao [4] proposed models considering impact, support, and inertial forces. The failure occurred via bending and shear beyond concrete grades of 46MPa or collapse between 60-100MPa. Stress wave propagation generates local response, while global behavior is more important per another research. Impact body shape had little effect on overall resistance, according to some [5]. Axial load positively influenced minimum deformation but caused catastrophic failure at high velocity with low reinforcement. The shear failure occurred below static capacity. Varying impact velocity produced differing crack patterns - oblique cracks near impact (Type I) for high velocity, like static tests for inclined cracks (Type II). Type I was unique to impact. Reinforcement ratio and member grade influenced failure mode and response [6]. Further studies aim better to characterize dynamic reinforced concrete behavior under impact loads.

4.2 Numerical Analysis

Finite element software like ABAQUS, LS-DYNA, and ANSYS have become increasingly popular among scholars since the 1990s. Numerical simulation methods are widely used to simulate explosion and impact behavior in civil engineering [7–9] and to obtain essential conclusions about members' responses under impact load. The constitutive relationship under dynamic material load plays a vital role in numerical analysis. Constitutive relationship under a dynamic load of materials. Steel bars exhibit different mechanical properties under dynamic and static loading. The strain rate effect is significant under dynamic load and increases the yield strength more than the ultimate strength. Other models have been proposed to describe this effect, such as the Cowper-Symonds and Johnson-Cook models, which scholars widely recognize [10, 11].

$$\sigma = \left[1 + \left(\frac{\dot{\varepsilon}}{C} \right)^{\frac{1}{p}} \right] \sigma_0 \tag{1}$$

where σ is the stress of the steel under static loading, $\dot{\varepsilon}$ is the plastic strain rate, σ_0 is the stress of the steel under static loading, C and p are the parameters models related to the type of material. This model can estimate the dynamic yield strength and ultimate strength of steel at a given strain rate and is suitable for describing the strain rate effect of steel at lower strain rates. Johnson-Cook calculation formula of the model is as follows:

$$\sigma = \left(A + B\varepsilon^N \right) \left(1 + C \ln \dot{\varepsilon}^* \right) \left(1 - T^{*m} \right) \tag{2}$$

$\dot{\varepsilon}^* = \dot{\varepsilon}/\varepsilon_0$, $\varepsilon_0 = 1 S^{-1}$, T^* is a function of temperature, and A, B, C, n, and m are model parameters related to material properties. The first bracket in the formula indicates the stress-strain relationship at $\dot{\varepsilon}^* = 0$, $T^* = 0$, the second bracket indicates the strain rate effect, and the third bracket indicates the temperature effect. Early studies showed concrete strength increases with higher strain rates, though the degree depends on factors like static strength and rate magnitude. Abrams [12] found strength rose from 2×10^{-4}/s to 8×10^{-6}/s testing in 1917. Bischoff and Perry [13] analyzed data, finding strength grows more gradually below a critical rate but rapidly exceeds it. Atchley and Furr [14] found ultimate strength plateaued at high rates, possibly due to different strain measurement locations. Jia et al. [15] found low temperatures boosted strength with rising rates, s but high heat weakened it, as cracks formed beforehand, preventing rate effects. Cotsovos and Pavlovic [16] showed via modeling that local regions experience differing mechanical states in rapid tests, implying rate impact's structure rather than just material performance. Debates remain around whether strength boosts arise from direct rate impacts or confinement pressures in experiments. The most common design strength calculation comes from the CEB-FIP [17] Model Code fitted to test data. While mechanical properties under rapid loading are well-researched, controversies persist around strain rate effect mechanisms on concrete strength. Further studies could provide clarification.

$$\frac{\sigma_d}{\sigma_s} = \begin{cases} \left(\frac{\dot{\varepsilon}_d}{\dot{\varepsilon}_s}\right)^{1.026\alpha} & \dot{\varepsilon}_d \leq 30s^{-1} \\ \gamma\left(\frac{\dot{\varepsilon}_d}{\dot{\varepsilon}_s}\right)^{1/3} & \dot{\varepsilon}_d > 30s^{-1} \end{cases} \tag{3}$$

σ_d is the compressive strength of concrete under dynamic loading, σ_s is the compressive strength of concrete under static loading, $\dot{\varepsilon}_d$ is the strain rate value under static loading. The parameters α, γ values are specified in the specification. Research on concrete dynamic tensile properties also shows strength increases with higher strain rates. Hopkinson bar testing is commonly used. Malvar and Ross [18] found tensile strength grows more gradually below a critical rate before rapidly rising above it, similar to compression. Ross [19] determined strain rate had a greater impact on tensile versus compressive strength. Yan and Lin [20] studied how properties like strength, modulus, peak strain, and energy absorption changed with loading rate, temperature, and moisture content. Saturated concrete strength increased more substantially. Some debates exist around whether increased tensile strength stems from inertia effects or actual material behavior. Lu and Li [21] proved through modeling that increased strength was not due to higher simulated rates, indicating actual material behavior causes strengthening. The commonly used CEB-FIP [17] Model Code proposes a calculation formula to estimate dynamic tensile strength based on static properties and strain rate. Research continues to enhance understanding of concrete behavior under rapid tensile loading.

$$\frac{f_{ct,imp}}{f_{ctm}} = \begin{cases} \left(\frac{\dot{\varepsilon}_{ct}}{\dot{\varepsilon}_{ct0}}\right)^{1.026\alpha} & |\dot{\varepsilon}_{ct}| \leq 10s^{-1} \\ 0.0062\left(\frac{\dot{\varepsilon}_{ct}}{\dot{\varepsilon}_{ct0}}\right)^{1/3} & |\dot{\varepsilon}_{ct}| > 10s^{-1} \end{cases} \tag{4}$$

$f_{ct,imp}$ is the tensile strength of concrete under dynamic loading, f_{ctm} is the tensile strength of concrete under static loading, $\dot{\varepsilon}_{ct}$ is the material strain rate, and $\dot{\varepsilon}_{ct0}$ is the strain rate value under static loading.

4.3 Response of Members Under Impact Load

Cai et al. [22] used ABAQUS to simulate the dynamic response of 7 reinforced concrete members with a section size of 150 mm × 150 mm under low-speed horizontal impact loads. The authors studied the effect of impact mass and velocity on the failure mode of members. It is found that the inertia effect has a significant impact on the impact resistance of the member. By finite element analysis, Yu et al. [23] proposed simultaneously considering the effect of the impact body mass and velocity on the reinforced concrete member. Once the impact body mass and velocity are small, the maximum and residual deflection of the member is greater. EL-Tawil et al. [24] used the finite element simulation method to study the dynamic response of the bridge pier after being hit by a vehicle. The finite element method simulates the impact of heavy trucks' reinforced concrete members with square and circular cross-sections. The members' heights are 16.3 and 9.9 m, and the impact velocities are 1.35 m. The effects of different impact masses and velocities on reinforced concrete members' dynamic responses are analyzed. The results show that the equivalent static force calculation method proposed by AASHTO-LRFD [25] is not conservative in estimating the design force of an overweight truck hitting a bridge pier. To evaluate the vulnerability of reinforced concrete members to vehicle collisions. Thilakarathna et al. [26] established a numerical model of full-scale reinforced concrete members with circular cross-sections under an impact force. The mid-span deflection and bearing reaction force test data verify the model's accuracy. A method that can quantify the degree of damage to reinforced concrete members is proposed. Study the reinforced concrete members' dynamic response and failure mode under impact with and without CFRP layers. The experimental method and finite element analysis show that the CFRP reinforcement can change members' failure modes. The failure mode of reinforced concrete members that undergo shear failure under unequal span lateral impact load hitting by a derailed train is transformed into flexural failure after being wrapped by CFRP layers, as shown in Fig. 1 [27–35].

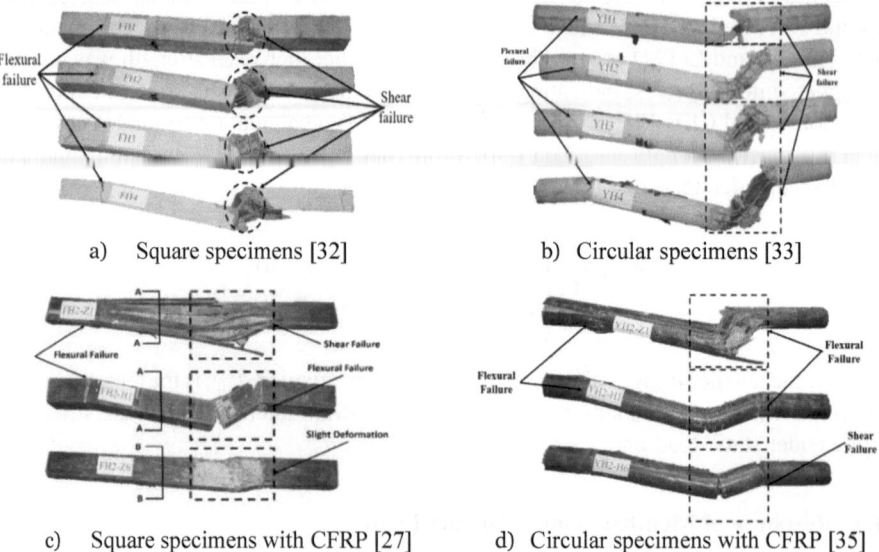

a)   Square specimens [32]   b)   Circular specimens [33]

c)   Square specimens with CFRP [27]   d)   Circular specimens with CFRP [35]

Fig. 1. Failure modes after the end of the impact scenario

5 Discussion

This section will discuss the key findings from the literature review in more detail and highlight important themes. One of the most prominent findings is that steel reinforcement and high-strength concrete are effective techniques for mitigating impact loads on RC structures. Several studies demonstrate their abilities to improve impact resistance by reducing cracking, spalling, and reinforcement failure [3, 4, 6]. However, the degree of benefit depends on factors like reinforcement ratio and impact parameters. Further research is still needed to optimize design guidelines around these techniques. Numerical simulation methods like finite element analysis were commonly used and provided valuable insights. However, the accuracy of such models relies on the constitutive relationships used to define dynamic material behavior under high strain rates. While extensive research has enhanced our understanding of rate effects, controversies remain around the mechanisms influencing concrete properties. More work is needed to clarify these issues and improve constitutive models. Regarding structure types, buildings, and bridges received the most attention due to their vulnerability. However, other structures like retaining walls are also at risk but lack comparable study. Combined loads and long-duration impacts were identified as prominent gaps. Considering multiple hazards simultaneously is important for robust design but challenging to reproduce experimentally. Classification systems for different failure patterns aim to aid damage evaluation but require further validation and standardization. Correlating measured response to specific performance limits would strengthen their practical application.

The discussion section of this paper presents the main findings, including the types of RC structures most studied and the most effective techniques for mitigating the effects of lateral impact loads. This section also highlights any major gaps in the literature and identifies areas for future research. In addition, this review aims to determine whether it has filled any knowledge gaps not addressed by previous studies. This paper aims to present the main findings of the literature review on RC structures subjected to lateral impact loads and discuss any major gaps in the literature. Buildings and bridges are the most studied RC structures concerning lateral impact loads. These structures are particularly susceptible to damage from blasts, vehicles, trains, ships, and seismic loads. Furthermore, it suggested that steel reinforcements and high-strength concrete effectively mitigate lateral impacts on RC structures. Several studies have found that using high-strength concrete and steel reinforcement improves the ability of the structures to withstand lateral impact loads, reducing cracking and spalling of the concrete and failure of the steel reinforcement. This review paper also revealed that numerical analysis, such as finite element analysis, is a commonly used technique for studying the behavior of RC structures under lateral impact loads. These techniques allow for the simulation of different loading scenarios, providing valuable insights into the behavior of the structures under different types of loads and conditions. The article revealed several gaps in the current research on RC structures subjected to lateral impact loads. One major gap is the lack of research on the behavior of RC structures under combined loads, such as simultaneous blast and seismic loads. In addition, there is a lack of research on the behavior of RC structures under long-duration loads, such as those caused by prolonged vehicle or ship impact. Another gap in the literature is the lack of research on the behavior of RC structures under different types of lateral impact loads, such as those caused by aircraft impact. Additionally, there is a lack of research on the behavior of RC structures caused by derailed train impact.

6 Conclusion

The conclusion of this paper is crucial in summarizing the main findings of the research and providing recommendations for future research on the topic. This paper summarized the main findings of the literature review on RC structures subjected to lateral impact loads and provided recommendations for future research. The article also emphasized the importance of continued research to improve the design and performance of RC structures subjected to lateral impact loads. The specific contributions outlined damage responses, failure patterns, real-world incidents, material behavior insights, and remaining research gaps in this field. Providing a comprehensive review of the damage responses of RC members to various lateral impact loads such as vehicle, train, and ship collisions based on existing experimental and numerical studies. Various performance-based studies were examined to find the best way to predict damage. Classifying different failure patterns of RC members under impact loads based on damage severity can help evaluate structural integrity after a collision event. Highlighting real-world collision incidents to demonstrate the destructive potential of such accidents and the need to improve impact resistance of infrastructure. Summarizing the latest understandings of material behavior under high strain rates from the literature to inform more accurate simulation of impact load scenarios. Identifying remaining gaps in research, such as the behavior of combined loads, to guide future work towards more robust assessment and design for impact resistance.

Two alternative configurations of RC columns were used in each case. According to the findings of the research topic, the impactor speed is related to the duration for which the peak impact force is generated. This article also includes experimental and numerical findings on the effects of drop hammer impact weight on columns and beams. Finally, when comparing dynamic and impulsive testing to quasi-static testing, it seems that concrete materials exhibit an apparent increase in strength. Much test data documented in the article regarding strain rate sensitivity has been discussed. Moreover, more modeling studies and field testing are needed to develop the dynamic response methodology that precisely evaluates the damage of RC members by impact force and improves decision-making on what must be done to protect the RC members from impact force damage.

References

1. Zhang, C., Gholipour, G., Mousavi, A.A.: Blast loads induced responses of RC structural members: State-of-the-art review. Compos. B Eng. **195**, 108066 (2020)
2. Zhang, C., Gholipour, G., Mousavi, A.A.: State-of-the-art review on responses of RC structures subjected to lateral impact loads. Arch. Comput. Methods Eng. **28**(4), 2477–2507 (2021)
3. Hughes, G., Speirs, D.M.: An Investigation of the Beam Impact Problem. Cement and Concrete Association, p. 117 (1982)
4. Pham, T.M., Hao, H.: Effect of the plastic hinge and boundary conditions on the impact behavior of reinforced concrete beams. Int. J. Impact EngEng **102**, 74–85 (2017)

5. Kishi, I.K., Ikeda, K., Mikami, H., Kanie, S.: Effects of nose shape of steel weight on impact behavior of RC beams. In: Proceeding of the Japan Concrete Institute (2000)
6. Yi, W., Zhao, D.: Simplified approach for assessing shear resistance of reinforced concrete beams under impact loads. ACI Struct. J. **113**(4), 747–756 (2016)
7. Priestley, M.J.N., Seible, F., Calvi, G.M.: Seismic Design and Retrofit of Bridges. Wiley (1996). https://doi.org/10.1002/9780470172858
8. Sharma, H., Gardoni, P.: Probabilistic demand model and performance-based fragility estimates for RC column subject to vehicle collision. Eng. Struct.Struct. **74**, 86–95 (2014)
9. Bertrand, D., Kassem, F., Delhomme, F., Limam, A.: Reliability analysis of an RC member impacted by a rockfall using a nonlinear SDOF model. Eng. Struct.Struct. **89**, 93–102 (2015)
10. Soroushian, P., Choi, K.: Steel mechanical properties at different strain rates. J. Struct. Eng.Struct. Eng. **113**(4), 663–672 (1987)
11. Malvar, L.J.: Review of static and dynamic properties of steel reinforcing bars. Mater. J. **95**(5), 609–616 (1998)
12. Abrams, D.A.: Effect of rate of application of load on the compressive strength of concrete. J ASTM Int. **17**, 364–377 (1917)
13. Bischoff, P.H., Perry, S.H.: Compressive behaviour of concrete at high strain rates. Mater. Struct.Struct. **24**(6), 425–450 (1991). https://doi.org/10.1007/BF02472016
14. Atchley, B.L., Furr, H.: Strength and energy absorption capabilities of plain concrete under dynamic and static loadings. undefined **64**(11), 745–756 (1967)
15. Jia, B., Tao, J., Li, Z., Wang, R., Zhang, Y.: Effects of temperature and strain rate on dynamic properties of concrete. Trans. Tianjin Univ. **14**(S1), 511–513 (2008). https://doi.org/10.1007/s12209-008-0087-6
16. Cotsovos, D.M., Pavlović, M.N.: Numerical investigation of concrete subjected to compressive impact loading. Part 1: a fundamental explanation for the apparent strength gain at high loading rates. Comput. Struct. **86**, 145 (2008)
17. Bulletin, Model Code 2010 - First complete draft, Volume 1, vol. 2. (2010)
18. Malvar, L.J., Ross, C.A.: Review of strain rate effects for concrete in tension. Mater. J. **95**(6), 735–739 (1998)
19. W, T.J., et al.: Effects of strain rate on concrete strength. ACI Mater. J. **92**(1), 37–47 (1995)
20. Yan, D., Lin, G.: Dynamic properties of concrete in direct tension. Cem. Conc. Res. **36**(7), 1371–1378 (2006)
21. Lu, Y.B., Li, Q.M.: About the dynamic uniaxial tensile strength of concrete-like materials. Int. J. Impact EngEng **38**(4), 171–180 (2011)
22. Cai, J., Ye, J., Wang, Y., Chen, Q.: Numerical study on dynamic response of reinforced concrete columns under low-speed horizontal impact loading. Procedia Eng. **210**, 334–340 (2017)
23. Yong Jae, Y., Kim, C.-H., Cho, J.-Y.: Investigation of behavior of RC beams subjected to impact loading considering combination of mass and impact velocity. Procedia Eng. **210**, 353–359 (2017). https://doi.org/10.1016/j.proeng.2017.11.088
24. El-Tawil, S., Severino, E., Fonseca, P.: Vehicle collision with bridge piers. J. Bridg. Eng.Bridg. Eng. **10**(3), 345–353 (2005)
25. American Association of State Highway and Transportation Officials, AASHTO LRFD Bridge design specifications. Washington, DC (2012)
26. Thilakarathna, H.M.I., Thambiratnam, D.P., Dhanasekar, M., Perera, N.: Numerical simulation of axially loaded concrete columns under transverse impact and vulnerability assessment. Int. J. Impact EngEng **37**(11), 1100–1112 (2010)
27. AL-Bukhaiti, K., et al.: Experimental study on existing RC circular members under unequal lateral impact train collision. Int. J. Concr. Struct. Mater. **16**(1), 1572–1596 (2022). https://doi.org/10.1186/s40069-022-00529-5

28. AL-Bukhaiti, K., Yanhui, L., Shichun, Z., Abas, H., Aoran, D.: Dynamic equilibrium of CFRP-RC square elements under unequal lateral impact. Materials **14**(13), 3591 (2021). https://doi.org/10.3390/ma14133591

29. Liu, Y., Al-Bukhaiti, K., Abas, H., Shichun, Z.: Effect of CFRP shear strengthening on the flexural performance of the RC specimen under unequal impact loading. Adv. Mater. Sci. Eng. **2020**, 1–18 (2020)

30. Liu, Y., Dong, A., Zhao, S., Zeng, Y., Wang, Z.: The effect of CFRP-shear strengthening on existing circular RC columns under impact loads. Constr. Build. Mater. **302**, 124185 (2021)

31. AL-Bukhaiti, K., Yanhui, L., Shichun, Z., Abas, H.: CFRP strengthened reinforce concrete square elements under unequal lateral impact load. In: Pellegrino, C., Faleschini, F., Zanini, M.A., Matos, J.C., Casas, J.R., Strauss, A. (eds.) EUROSTRUCT 2021. LNCE, vol. 200, pp. 1377–1387. Springer, Cham (2022). https://doi.org/10.1007/978-3-030-91877-4_157

32. Abas, H., Yanhui, L., Al-Bukhaiti, K., Shichun, Z., Aoran, D.: Experimental and numerical study of RC square members under unequal lateral impact load. Struct. Eng. Int.. Eng. Int. **33**(1), 147–164 (2021). https://doi.org/10.1080/10168664.2021.2004976

33. AL-Bukhaiti, K., et al.: Experimental study on existing RC circular members under unequal lateral impact train collision. IJCSM 2022 **16**(1), 1–21 (2022)

34. AL-Bukhaiti, K., et al.: Failure mechanism and static bearing capacity on circular RC members under asymmetrical lateral impact train collision. Structures **48**, 1817–1832 (2023)

35. Yanhui, L., et al.: Failure mechanism analysis of circular CFRP components under unequal impact load. Structurae **22**, 1668–1676 (2022)

Inverse Filtration Problem of a Bidisperse Suspension

Liudmila I. Kuzmina[1] and Yuri V. Osipov[2(✉)]

[1] HSE University, Moscow, Russia
[2] Moscow State University of Civil Engineering, Moscow, Russia
yuri-osipov@mail.ru

Abstract. Filtration problems of suspensions and colloids in porous media are considered when designing tunnels and underground structures. To strengthen weak soil, a liquid solution is injected into the rock, the particles of which are filtered in the pores and distributed far from the well. A deep bed filtration model of 2-particle suspension in a porous material is considered. The purpose of the work is to determine the model parameters from the measured outlet concentration of suspended particles. Using an explicit solution to the direct filtration problem on the concentration front, the inverse problem is reduced to a system of nonlinear algebraic equations, which is a special case of the moment problem. The system is solved by passing to a canonical basis in the space of symmetric polynomials. Conditions for the existence of a solution are obtained. An explicit solution is constructed. The inverse filtration problem of a suspension with particles of two types is solved, determining the initial partial concentrations and filtration coefficients.

Keywords: Deep Bed Filtration · Porous Medium · Inverse Problem · Analytical Solution

1 Introduction

The transport and deposition of particles in porous media must be taken into consideration when designing tunnels and underground structures, preparing foundations and consolidating loose soils [1–5]. Fine particles transported by groundwater are usually heterogeneous and differ in shape and size. Knowledge of the composition of a suspension filtered in a porous sample allows us to calculate the distribution of sediment and its composition.

At deep bed filtration, suspended particles are transferred and retained throughout the entire porous medium [6–8]. A porous medium contains millions of pores of different sizes. The retention mechanisms are associated with hydrodynamic, electrical and gravitational forces. Sedimentation depends on the shape, mass, size and charge of suspension particles. Size-exclusion mechanism of particle retention means that transport and blocking of particles depends on particle to pore size ratio [9–11]. An analytical solution to the one-dimensional filtration problem with identical particles was obtained in [12].

© The Author(s) 2024
G. Feng (Ed.): ICCE 2023, LNCE 526, pp. 585–593, 2024.
https://doi.org/10.1007/978-981-97-4355-1_56

When filtering a bidisperse suspension, the retention profiles of large and small particles are different [13]. A deep bed filtration scheme of a bidisperse suspension with particles of two sizes is presented in Fig. 1.

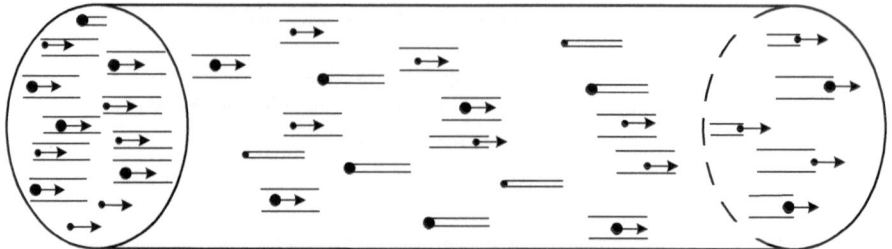

Fig. 1. Transport and retention of 2-particle suspension in porous medium

The mathematical filtration model of a bidisperse suspension in a porous medium includes balance equations for the concentrations of suspended and deposited particles of each type. The growth of the sediment is proportional to the concentration of particles and depends on the formed sediment of particles of both types and their initial concentrations. The model under consideration was constructed and solved numerically in [14], the analytical solution was obtained in [15, 16]. 4x4 PDE-system with given initial concentrations and filtration coefficients determines the unknown concentrations of suspended and retained particles.

Laboratory equipment makes it possible to measure the total concentration of suspended particles of a suspension at the inlet and outlet of a porous medium. The inverse filtration problem consists of finding the initial partial concentrations and filtration coefficients from the measured total concentration. To obtain partial concentrations, the total concentration at the outlet of the porous medium at the moment of breakthrough is measured and compared with the analytical solution to the direct filtration problem. By collecting the suspension at the outlet and passing it through a new sample of the porous medium, we obtain a new value of the total concentration at the moment of breakthrough. Based on several experiments, a closed system of algebraic equations for finding unknown concentrations of particles of different types and their filtration coefficients has been compiled. This problem is close to the discrete moment problem, studied in probability theory and mathematical statistics.

Inverse filtration problems, as a rule, do not have an analytical solution and are solved numerically [17]. The instability of such problems leads to a significant change in the solution with small errors in the initial data. The presence of an exact solution allows one to determine the parameters of the direct problem with high accuracy.

The inverse filtration problem is reduced to a nonlinear algebraic system of equations. Conditions for the existence of a solution are obtained, an exact solution to the inverse problem is given explicitly.

2 Mathematical Model of Direct and Inverse Filtration Problem

Consider 1-D deep bed filtration model in dimensionless form. Filtration of a bidisperse suspension in a homogeneous porous medium is determined by the system

$$\frac{\partial c_i}{\partial t} + \frac{\partial c_i}{\partial x} + \frac{\partial s_i}{\partial t} = 0, \tag{1}$$

$$\frac{\partial s_i}{\partial t} = (1 - b)\lambda_i c_i, \quad b = c_1^0 s_1 + c_2^0 s_2, \quad i = 1, 2 \tag{2}$$

with conditions

$$x = 0: \; c_i = c_i^0, \; c_i^0 > 0, \quad t = 0: \; c_i = 0, \; s_i = 0, \; i = 1, 2. \tag{3}$$

Here λ_i, c_i^0 are positive constants and $\lambda_1 > \lambda_2 > 0$, $c_1^0 + c_2^0 = 1$.

Before the concentration front at $0 < t < x$ the solution is zero: $c = 0$, $s = 0$, $b = 0$; after the front at $t > x$ the solution is positive. At the concentration front $s_i = 0$, $i = 1, 2$.

Substitute (2) into Eq. (4):

$$\frac{\partial c_i}{\partial t} + \frac{\partial c_i}{\partial x} + \lambda_i c_i = 0 \tag{4}$$

Solution to Eq. (4) with the first condition (3) sets partial suspended concentrations at the front $c_i^f(x) = c_i^0 e^{-\lambda_i x}$, $i = 1, 2$.

The breakthrough concentration at the porous medium outlet $x = 1$

$$c_i^1 = c_i^0 e^{-\lambda_i}, \quad i = 1, 2 \tag{5}$$

To find the initial partial concentrations c_1^1, c_2^1 and filtration coefficients λ_1, λ_2, Formula (5) is used several times. If you collect a suspension at the outlet of a porous medium at the breakthrough moment, measure its total concentration and filter it again through a porous sample, then at the outlet the breakthrough partial concentrations are $c_i^2 = c_i^0 e^{-2\lambda_i}$, $i = 1, 2$.

After the third filtration the breakthrough partial suspended concentrations at the outlet $c_i^3 = c_i^0 e^{-3\lambda_i}$, $i = 1, 2$.

Using 3 measurements of the total suspended concentration at the outlet, we obtain a non-linear system of algebraic equations

$$\begin{aligned}
c_1^0 + c_2^0 &= 1, \\
c_1^0 e^{-\lambda_1} + c_2^0 e^{-\lambda_2} &= m_1, \\
c_1^0 e^{-2\lambda_1} + c_2^0 e^{-2\lambda_2} &= m_2, \\
c_1^0 e^{-3\lambda_1} + c_2^0 e^{-3\lambda_2} &= m_3.
\end{aligned} \tag{6}$$

System (6) is a discrete finite moment problem complicated by additional conditions $\lambda_1 > \lambda_2 > 0$, $c_1^0 > 0$, $c_2^0 > 0$.

3 Analytical Solution to the Inverse Problem

Denote

$$k_1 = e^{-\lambda_1}, \quad k_2 = e^{-\lambda_2}, \quad 0 < k_1 < k_2 < 1. \tag{7}$$

The system (6) takes the form

$$c_1^0 + c_2^0 = 1, \tag{8}$$

$$c_1^0 k_1 + c_2^0 k_2 = m_1, \tag{9}$$

$$c_1^0 k_1^2 + c_2^0 k_2^2 = m_2, \tag{10}$$

$$c_1^0 k_1^3 + c_2^0 k_2^3 = m_3. \tag{11}$$

Using (8), (9), express unknowns c_1^1, c_2^1 through k_1, k_2:

$$c_1^0 = \frac{m_1 - k_2}{k_1 - k_2}, \quad c_2^0 = \frac{m_1 - k_1}{k_2 - k_1}. \tag{12}$$

Substitute (12) into Eqs. (10) and (11) and transform the equations

$$\begin{aligned} m_1(k_1 + k_2) - k_1 k_2 &= m_2, \\ m_1(k_2^2 + k_1 k_2 + k_1^2) - k_1 k_2(k_2 + k_1) &= m_3. \end{aligned} \tag{13}$$

Denote

$$u = k_1 + k_2, \quad v = k_1 k_2, \tag{14}$$

System (13) takes the form

$$\begin{aligned} m_1 u - v &= m_2, \\ m_1(u^2 - v) - uv &= m_3. \end{aligned} \tag{15}$$

Solution to system (15)

$$u = \frac{m_3 - m_1 m_2}{m_2 - m_1^2}, \quad v = \frac{m_1 m_3 - m_2^2}{m_2 - m_1^2}. \tag{16}$$

By Vieta's theorem and Formula (14) the unknowns k_1, k_2 are determined by quadratic equation $k^2 - uk + v = 0$ with two roots

$$k_{1,2} = \frac{u \pm \sqrt{u^2 - 4v}}{2} = \frac{(m_3 - m_1 m_2) \pm \sqrt{(m_3 - m_1 m_2)^2 - 4(m_1 m_3 - m_2^2)(m_2 - m_1^2)}}{2(m_2 - m_1^2)}. \tag{17}$$

Let us formulate the conditions on the measured total concentrations m_i under which the solution to system (8)–(11) satisfy the inequalities

$$c_1^0 > 0, \quad c_2^0 > 0 \text{ and } 0 < k_1 < k_2 < 1. \tag{18}$$

Necessary conditions follow from positive-semi definiteness of Hankel matrices:

$$0 < m_3 < m_2 < m_1 < 1, \quad m_2 > m_1^2, \quad m_3 > m_1 m_2, \quad m_1 m_3 > m_2^2. \tag{19}$$

These conditions can be derived directly from the inequalities (18):

$$1 = c_1^0 + c_2^0 > c_1^0 k_1 + c_2^0 k_2 = m_1 > c_1^0 k_1^2 + c_2^0 k_2^2 = m_2 > c_1^0 k_1^3 + c_2^0 k_2^3 = m_3 > 0,$$
$$m_2 - m_1^2 = c_1^0 c_2^0 (k_2 - k_1)^2 > 0,$$
$$m_3 - m_1 m_2 = c_1^0 c_2^0 (k_1 + k_2)(k_2 - k_1)^2 > 0,$$
$$m_1 m_3 - m_2^2 = c_1^0 c_2^0 k_1 k_2 (k_2 - k_1)^2 > 0. \tag{20}$$

From the necessary conditions follows an inequality

$$k_1 < m_1 < k_2. \tag{21}$$

Formulae (21) and (12) ensure the inequalities $c_1^0 > 0, \quad c_2^0 > 0$.

Sufficient conditions specify restrictions on measurements m_i under which the inequalities $0 < k_1 < k_2 < 1$ are satisfied. According to the properties of the parabola $y = k^2 - uk + v$, its roots satisfy the relation $0 < k_1 < k_2 < 1$, if

$$y(0) > 0, \quad y(1) > 0, \quad 0 < \frac{k_1 + k_2}{2} = \frac{u}{2} < 1, \quad y\left(\frac{u}{2}\right) < 0 \Leftrightarrow D > 0. \tag{22}$$

Figure 2 demonstrates the sufficiency of the inequalities (22) for finding two roots in the interval $(0;1)$.

In the variables u, v, inequalities (22) take the form

$$0 < u < 2, \quad v > 0, \quad v > u - 1, \quad u^2 - 4v > 0. \tag{23}$$

The inequalities $u > 0, \quad v > 0$ follow from (20). Other sufficient conditions are also necessary:

$$u - 2 = k_1 + k_2 - 2 < 0,$$
$$v - u + 1 = (k_1 - 1)(k_2 - 1) > 0, \tag{24}$$
$$u^2 - 4v = (k_2 - k_1)^2 > 0.$$

Sufficient conditions (24) do not follow from the necessary conditions (19). Let $m_1 = 0.5, \quad m_2 = 0.275, \quad m_3 = 0.25$.

All the necessary conditions are fulfilled:

$$0 < 0.25 < 0.275 < 0.5 < 1, \quad 0.275 > 0.5^2 = 0.25,$$
$$0.25 > 0.5 \cdot 0.275 = 0.1375, \quad 0.5 \cdot 0.25 = 0.125 > 0.275^2 = 0.075625,$$

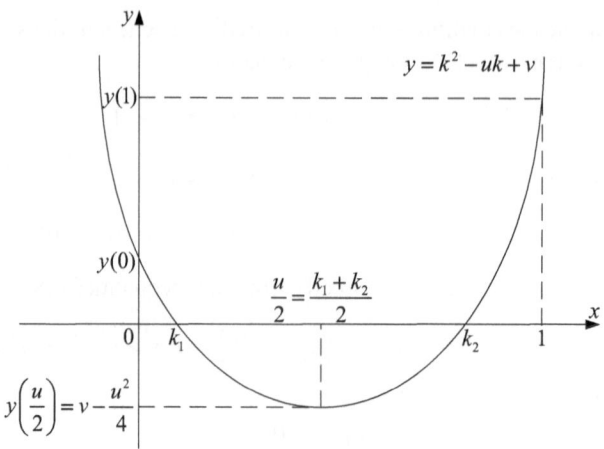

Fig. 2. Parabola graph.

but sufficient condition $u < 2$ is not valid:

$$u = \frac{m_3 - m_1 m_2}{m_2 - m_1^2} = 4.5 > 2. \tag{25}$$

In this case $0 < k_1 < 1 < k_2$ and conditions (18) are not true.

Thus, the inequalities (19) and (23) do not follow from each other, these conditions are necessary and sufficient to solve the inverse problem.

If all inequalities (19) and (23) are true, then c_1^0, c_2^0 are expressed by (12) and λ_1, λ_2 are obtained from (7).

4 Numerical Results

For example, let the measurements of the total breakthrough suspended concentration at the porous medium outlet are $m_1 = 0.5$, $m_2 = 0.33$, $m_3 = 0.25$.

All the necessary conditions (19) are valid:

$$0 < 0.25 < 0.33 < 0.5 < 1, \qquad 0.33 > 0.5^2 = 0.25,$$
$$0.25 > 0.5 \cdot 0.33 = 0.165, \qquad 0.5 \cdot 0.25 = 0.125 > 0.33^2 = 0.1089. \tag{26}$$

The variables $u = 1.0625$, $v = 0.20125$ and sufficient conditions (23) are fulfilled also:

$$0 < 1.0625 < 2, \quad 0.20125 > 0, \quad 0.20125 > 1.0625 - 1,$$
$$1.0625^2 - 4 \cdot 0.20125 = 0.324 > 0. \tag{27}$$

The roots k_1, k_2 are obtained from (17):

$$k_1 = 0.479, \quad k_2 = 0.584. \tag{28}$$

The required parameters are determined by Formulae (7) and (12):

$$\lambda_1 = 0.736, \quad \lambda_2 = 0.538, \quad c_1^0 = 0.8, \quad c_2^0 = 0.2. \tag{29}$$

Figure 3 shows the partial and total suspended concentrations at the front.

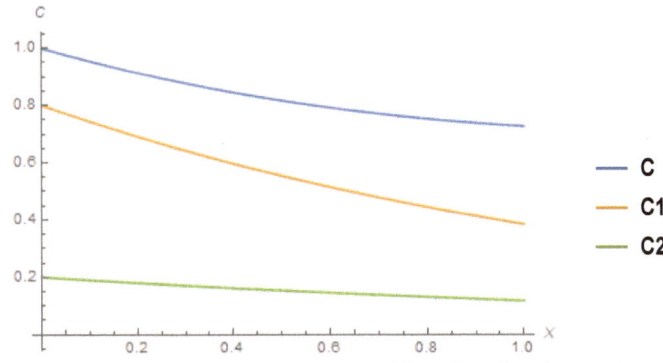

Fig. 3. Suspended concentrations $c^f = c_1^f + c_2^f$, c_1^f, c_2^f at the front.

5 Discussion

Inverse problems, as a rule, are unstable, that is, small measurement errors lead to large deviations in the solution. Such problems are solved numerically; they require a large number of calculations. The use of exact solutions makes the inverse problem correct and significantly simplifies the calculations [18, 19].

Filtration of a bidisperse suspension leads to uneven distribution of partial sediment: large particles are mainly deposited closer to the inlet, and small ones - at the outlet of the porous medium. Solving the inverse filtration problem allows to calculate the distribution of large and small suspension particles. This makes it theoretically possible to non-chemically separate different types of particles.

The refined filtration model of a bidisperse suspension contains additional parameters that determine the size of the places occupied by particles of different types on the porous medium frame. These parameters are not included in system (6) and can be determined separately using the asymptotics of the solution at the exit of the porous medium.

The generalization of the inverse filtration problem to the case of a polydisperse suspension is important for theory and practice [20, 21]. For n types of particles, only the necessary conditions for solving the inverse problem are known [22–24]. Finding sufficient conditions for solving the inverse problem of filtration of a polydisperse suspension and its solution is a separate complex problem.

6 Conclusion

The study of the inverse filtration problem of a 2-particle suspension in a porous medium lead to the following conclusions:

- A model for solving the inverse problem was constructed based on measurements of the concentration at the outlet of a porous medium.
- Necessary and sufficient conditions for the existence of a solution to the inverse problem in the form of a system of inequalities are derived.
- An exact solution to the inverse problem has been obtained in explicit form.
- The exact solutions regularize the inverse problem and make it resistant to measurement and calculation errors.

References

1. Zhou, Z., Zang, H., Wang, S., Du, X., Ma, D., Zhang, J.: Filtration behavior of cement-based grout in porous media. Transp. Porous Media **125**, 435–463 (2018)
2. Xie, B., et al.: Theoretical research on diffusion radius of cement-based materials considering the pore characteristics of porous media. Materials **15**, 7763 (2022)
3. Zhu, G., Zhang, Q., Liu, R., Bai, J., Li, W., Feng, X.: Experimental and Numerical study on the permeation grouting diffusion mechanism considering filtration effects. Geofluids **2021**, 1–11 (2021). https://doi.org/10.1155/2021/6613990
4. Wang, X., Cheng, H., Yao, Z., Rong, C., Huang, X., Liu, X.: Theoretical research on sand penetration grouting based on cylindrical diffusion model of tortuous tubes. Water **14**(7), 1028 (2022)
5. Christodoulou, D., Lokkas, P., Droudakis, A., Spiliotis, X., Kasiteropoulou, D., Alamanis, N.: The development of practice in permeation grouting by using fine-grained cement suspensions. Asian J. Eng. Technol. **9**(6), 92–101 (2021)
6. Herzig, J.P., Leclerc, D.M., Le Goff, P.: Flow of suspensions through porous media – application to deep filtration. J. Ind. Eng. Chem. **62**(8), 8–35 (1970)
7. Galaguz, Y., Kuzmina, L.I., Osipov, Y.: Problem of deep bed filtration in a porous medium with the initial deposit. Fluid Dyn. **54**(1), 85–97 (2019)
8. Kuzmina, L.I., Nazaikinskii, V.E., Osipov, Y.V.: On a deep bed filtration problem with finite blocking time. Russ. J. Math. Phys. **26**(1), 130–134 (2019)
9. Rabinovich, A., Bedrikovetsky, P., Tartakovsky, D.: Analytical model for gravity segregation of horizontal multiphase flow in porous media. Phys. Fluids **32**(4), 1–15 (2020)
10. Tartakovsky, D.M., Dentz, M.: Diffusion in porous media: phenomena and mechanisms. Transp. Porous Media **130**, 105–127 (2019)
11. Santos, A., Bedrikovetsky, P.: Size exclusion during particle suspension transport in porous media: stochastic and averaged equations. Comput. Appl. Math. **23**(2–3), 259–284 (2004)
12. Vyazmina, E.A., Bedrikovetskii, P.G., Polyanin, A.D.: New classes of exact solutions to nonlinear sets of equations in the theory of filtration and convective mass transfer. Theor. Found. Chem. Eng. **41**(5), 556–564 (2007)
13. Safina, G.: Calculation of retention profiles in porous medium. Lect. Notes Civil Eng. **170**, 21–28 (2021)
14. Malgaresi, G., Khazali, N., Bedrikovetsky, P.: Non-monotonic retention profiles during axisymmetric colloidal flows. J. Hydrol. **580**, 124235 (2020)
15. Kuzmina, L.I., Osipov, Y., Astakhov, M.D.: Filtration of 2-particles suspension in a porous medium. J. Phys. Conf. Ser. **1926**, 012001 (2021)
16. Kuzmina, L.I., Osipov, Y., Astakhov, M.D.: Bidisperse fltration problem with non-monotonic retention profiles. Annali di Matematica **201**, 2943–2964 (2022)
17. Vabishchevich, P., Vasil'ev, V., Vasilyeva, M., Nikiforov, D.: Numerical solution of an inverse filtration problem. Lobachevskii J. Math. **37**, 777–786 (2016)

18. Alvarez, A.C., Hime, G., Marchesin, D., Bedrikovetsky, P.: The inverse problem of determining the filtration function and permeability reduction in flow of water with particles in porous media. Transp. Porous Media **70**, 43–62 (2007)
19. Vaz, A., Bedrikovetsky, P., Fernandes, P.D., Badalyan, A., Carageorgos, T.: Determining model parameters for non-linear deep-bed filtration using laboratory pressure measurements. J. Petrol. Sci. Eng. **151**, 421–433 (2017)
20. Kuzmina, L., Osipov, Y.: Calculation of filtration of polydisperse suspension in a porous medium. MATEC Web Conf. **86**, 01005 (2016)
21. Kuzmina, L.I., Osipov, Y.V., Gorbunova, T.N.: Asymptotics for filtration of polydisperse suspension with small impurities. Appl. Math. Mech. (Engl. Ed.) **42**(1), 109–126 (2021)
22. Shohat, J.A., Tamarkin, J.D.: The Problem of Moments. American mathematical society, New York (1943)
23. Akhiezer, N.I.: The Classical Moment Problem And Some Related Questions in Analysis. Hafner Publishing Co, New York (1965)
24. Dio, P.J., Schmüdgen, K.: The multidimensional truncated moment problem: carathéodory numbers. J. Math. Anal. Appl. **461**, 1606–1638 (2018)

Lumped Plasticity Model's Accuracy in Estimating the Cyclic Response of Low and High-Ductile Reinforced Concrete Frames

Ooi Chin Ee[1], Mohammadreza Vafaei[1(✉)], and Sophia C. Alih[2]

[1] Faculty of Civil Engineering, University Teknologi Malaysia, 81310 Johor Bahru, Malaysia
vafaei@utm.my
[2] Institute of Noise and Vibration, Faculty of Civil Engineering, University Teknologi Malaysia, 81310 Johor Bahru, Malaysia

Abstract. The lumped plasticity model has been widely used for the nonlinear analysis of structures due to its simplicity, established guideline in seismic design codes, and less computational effort. However, it has been shown that the accuracy of the lumped plasticity model depends significantly on the plastic hinges' properties. In this study, the nonlinear behaviors of two similar RC frames but with two different reinforcing detailing were simulated using the lumped plasticity model and the results were compared with those obtained from the conducted experiments on the frames. The effects of using two different plastic hinge models (i.e., ASCE/SEI 41–17 and moment-curvature) and effective stiffness values (i.e. ASCE/SEI 41–17 and Eurocode 8) were investigated. It was observed that the employed plastic hinge models had insignificant effect on the frames' ultimate loads and both were able to estimate the ultimate loads accurately. However, the models underestimated the displacements corresponding to the ultimate loads of the frames. The failure modes of the frames were also estimated accurately.

Keywords: Lumped Plasticity Model · Plastic Hinge · RC Frame · Cyclic Response · Inadequate Lap Splice Length

1 Introduction

Although an elastic method is suitable for the seismic analysis of many regular structures, it cannot predict the failure mechanisms and the redistribution of forces during a progressive yielding. Therefore, for complex structures, an advanced seismic analysis that employs a nonlinear method should be used. The continuum finite element model, distributed plasticity model, and lumped plasticity model are among techniques used for the nonlinear analysis of structures. However, due to its simplicity, computational efficiency, and supports from seismic design codes, the lumped plasticity model has been broadly used by practice engineers and researchers [1–4]. Despite its wide application for seismic analysis of different structures, the accuracy of the lumped plasticity model requires further investigations [5]. It has been shown that the plastic hinge length, stiffness reduction factors, and the location of plastic hinges can significantly affect the

© The Author(s) 2024
G. Feng (Ed.): ICCE 2023, LNCE 526, pp. 594–603, 2024.
https://doi.org/10.1007/978-981-97-4355-1_57

results of a nonlinear analysis that use the lumped plasticity model [5–8]. Besides, the plastic hinges' moment-rotation relationships and their damage levels' acceptance criteria can later the results of a nonlinear analysis that make use of the lumped plasticity model. In this study, the results of an experiment conducted on two reinforced concrete frames (RC) are used to examine the efficiency of the lumped plasticity model in estimating their force-displacement relationships. The first frame represents a special moment resisting RC frame and the second one is a low-ductile RC frame with inadequate lap splice length.

2   Details of the Selected Frames and Conducted Experiment

The selected frames for this study are show in Fig. 1. As the Fig. 1 shows, the RC frames have a similar geometry, reinforcement ratio, and material properties. However, their seismic detailing is quite different. While the first frame (referred to as Frame 1) satisfies the requirements of ACI 318 [9] for a special moment frame, the second frame (referred to as Frame 2) suffers from many deficiencies. Inadequate lap splice length in the reinforcing bars of columns, the use of a 90-degree hook in stirrups, large distance

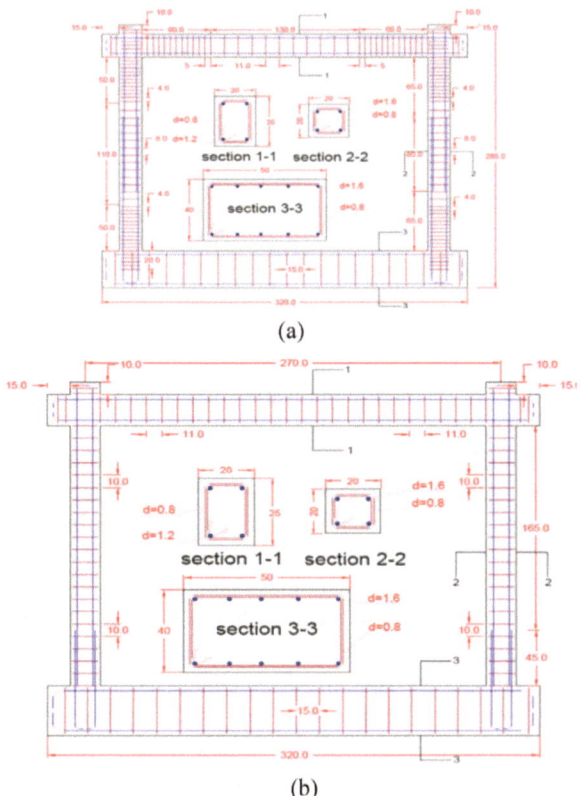

(a)

(b)

Fig. 1. Details of the tested frames (a) Frame 1 (b) Frame 2.

between stirrups at critical locations, and joints without any shear link are among the main deficiencies of the second frame. The 28-days compressive strength of the employed concrete measured on the standard cylinder was 30 MPa. Table 1 displays the mechanical properties of the employed reinforcing bars in frames. The frames were subjected to a similar quasi-static loading and their force-displacement responses were recorded at their beam level. Figure 2 depicts the obtained force-displacement relationships for the frames. As can be seen, Frame 1 has a slightly larger ultimate load than Frame 2 and its displacement at the ultimate load is also significantly larger than that of the Frame 2. Both frames have shown a linear response up to the lateral displacement of 3.2 mm. Frame 1 has reached the ultimate load of 61.48 kN at the lateral displacement of 90.3 mm. On the other hand, Frame 2 has reached an ultimate load of 54.5 kN at the lateral displacement of 66.8 mm. More detail about these frames can be found in [10].

Table 1. Mechanical properties of the frames' reinforcing bars

Diameter (mm)	Yield Stress (Mpa)	Ultimate Stress (Mpa)	Yield Strain (mm/mm)	Ultimate Strain (mm/mm)
8	532	693	0.0026	0.086
12	444	565	0.0022	0.112
16	537	681	0.0027	0.092

3 Finite Element Simulation

As mentioned earlier, the lumped plasticity model was employed to simulate the inelastic response of the frames under a gradually increasing lateral load (i.e., Pushover analysis). Figure 3 shows the considered locations for the plastic hinges in both frames. As can be seen from Fig. 3a, two plastic hinges were assigned to the both ends of beams and columns. The moment-rotation relationships of plastic hinges followed the shown multiline representation in Fig. 3b. In this figure, line AB shows the elastic stiffness of the element, with point B as the yield point. Besides, line BC displays the post-yield stiffness, with point C as the ultimate capacity of the plastic hinge. Line CDE represents the strength degradation of the plastic hinge after passing its ultimate capacity. In order to determine the required parameters of Fig. 3b (i.e., a, b and c) two different methods were employed. In the first approach, these parameters were determined based on the recommendation of ASCE/SEI 41 [11] as shown in Tables 2 and 3. The yield and ultimate capacities of the plastic hinges were calculated based on the elements' cross-sectional and material properties using the given equations in ACI 318 [9]. As shown in the tables, the beams were classified based on their stirrup spacing into conforming (i.e., stirrup spacing was less than 1/3 of beams' effective depth) and non-conforming (i.e., stirrup spacing was larger than 1/2 of beams' effective depth). In the second approach, the required parameters in Fig. 3b were determined based on the moment-curvature

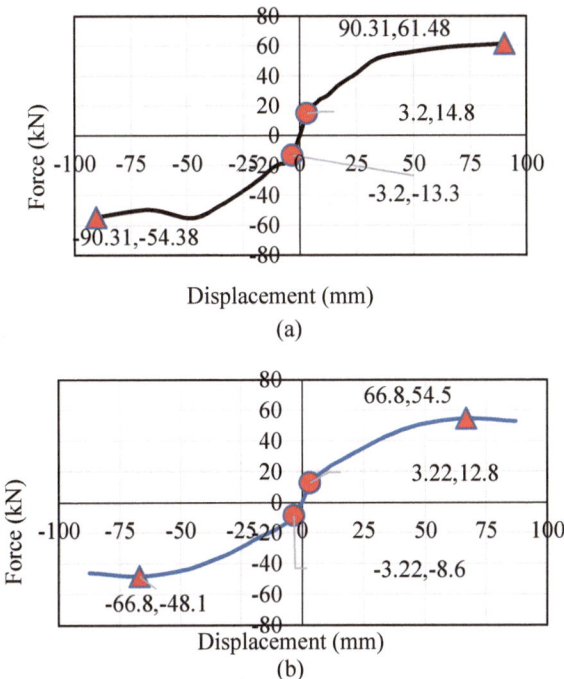

Fig. 2. The backbone curves of tested frames (a) Frame 1 (b) Frame 2.

analysis of elements' cross-section. For this purpose, at first, the cross-section of elements was divided into small fibres of steel and concrete materials. Then, nonlinear material properties were assigned to these fibres. For steel fibres, the values shown in Table 1 were used to determine the stress-strain relationships. However, the concrete's stress-strain relationships were developed based on Mender's equations [12] for confined and unconfined concrete elements. The plastic hinges' length was estimated using the equations suggested by Paulay and Priestley [13]. Table 3 displays the obtained results from the moment-curvature analysis. It should be motioned that the effect of inadequate lap splice length in the columns of Frame 2 was taken into account using the proposed method in ASCE/SEI 41 [11]. In this method, the yield stress of the reinforcing bars with inadequate lap splice length is reduced proportional to the ratio of the provided lap length to the required lap length. Based on this method, the yield stress of longitudinal reinforcing bar in the columns of Frame 2 (only at the location of lap length) was reduced from 537 MPa to 475 MPa. The effective stiffness of frames were taken into account using two different methods. First, the recommended values by ASCE/SEI 41 [11] were used (i.e., 0.3EI and 0.7EI for the flexural action of columns and beams, respectively), and then the proposed value by Eurocode 8 [14] (i.e., 0.5EI for beams and columns) was examined.

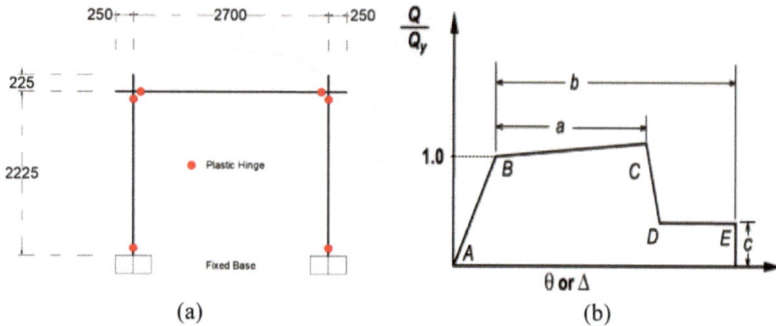

Fig. 3. (a) Location of plastic hinges in the FE models (b) Employed forced-displacement relationship for the plastic hinges.

Table 2. Moment-rotation parameters used for nonlinear analysis of beams

Method	Plastic rotation angle (rad)			Moment capacity (kN.m)	
	a	b	c	Yield	Ultimate
ASCE 41–17 (conforming)	0.025	0.050	0.200	21.51	23.66
ASCE 41–17 (non-conforming)	0.020	0.030	0.200	21.51	23.66
Moment-curvature (confined)	0.064	0.115	0.200	21.90	24.10
Moment-curvature (unconfined)	0.023	0.115	0.200	21.90	24.10

Table 3. Moment-rotation parameters used for nonlinear analysis of columns

Method	Plastic rotation angle (rad)			Moment capacity (kN.m)	
	a	b	c	Yield	Ultimate
ASCE 41–17 (adequate lap)	0.045	0.111	0.205	32.57	35.83
ASCE 41–17 (inadequate lap)	0.025	0.060	0.400	29.28	29.28
Moment-curvature (confined)	0.071	0.146	0.203	41.08	45.19
Moment-curvature (unconfined)	0.013	0.146	0.205	37.43	37.43

4   Results and Discussions

Figures 4 and 5 compare the experimentally obtained force-displacements relationships of Frame 1 and Frame 2 with those predicted by the finite element models. As shown in the figures, the employed methods for the calculations of the plastic hinge parameters (i.e., ASCE 41–17 and moment-curvature approaches) have had an insignificant effect on the obtained force-displacement relationships from the finite element models. However, the selected effective stiffness for structural elements has had a significant effect on the obtained force-displacement relationships. In both frames, regardless of the employed method for the calculation of plastic hinges parameters and the selected values for the effective stiffness of structural members, the ultimate loads have been estimated with good accuracy. The maximum difference between the obtained ultimate loads for Frame 1 and Frame 2 and that of the experiment are, respectively, 13.38% and 6.37%. The obtained results for the ultimate loads also show that the proposed method by ASCE/SEI 41–17 for the consideration of inadequate lap splice length has been efficient

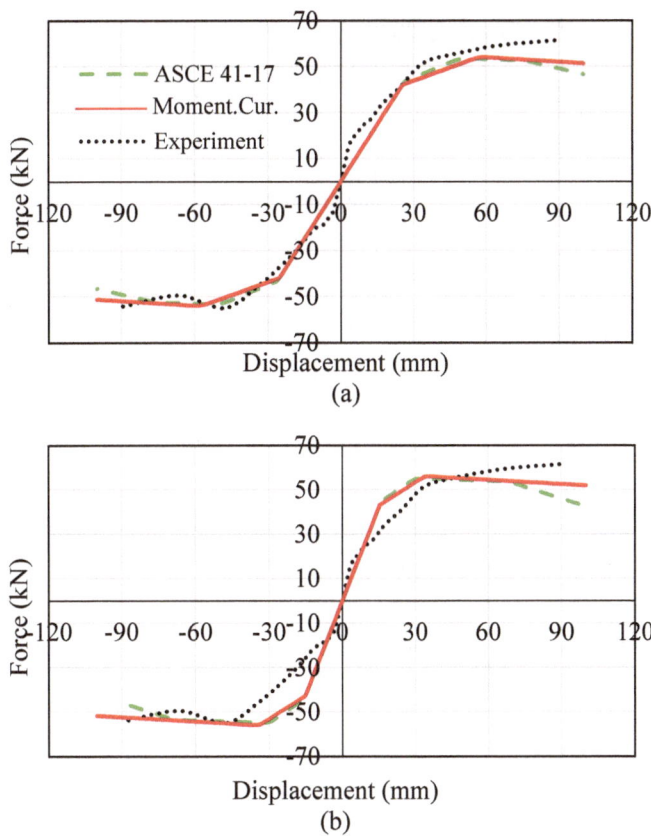

Fig. 4. Comparison between force-displacement relationships of finite element models and experiment for Frame 1 (a) using ASCE/SEI 41 effective stiffness values (b) using Eurocode 8 effective stiffness values.

as the ultimate load of Frame 2 has been estimated accurately. On the other hand, the displacement corresponding to the ultimate loads have been underestimated in all models. Although the use of ASCE/SEI 41–17's effective stiffness has reduced the difference between the displacements at ultimate loads, the differences in the results are around 33.9% for Frame 1 and 22.9% for Frame 2. It is also noteworthy that, although ASCE/SEI 41–17's effective stiffness values estimate the displacement at the ultimate load better than that of Eurocode 8's effective stiffness values, it has predicted the initial stiffness of frames with a lower accuracy. Figure 6 displays the predicted damage level of the plastic hinges at the ultimate load of frames. As can be seen, all models predict severe damage to the base of columns (i.e., CP hinges) while they expect a minor damage to the beam (i.e., IO hinges). This prediction correlates well with the observed damage to the beams and columns of both frames. In order words, the finite element models have been able to predict the damaged zone and the intensity of damage with a good precision.

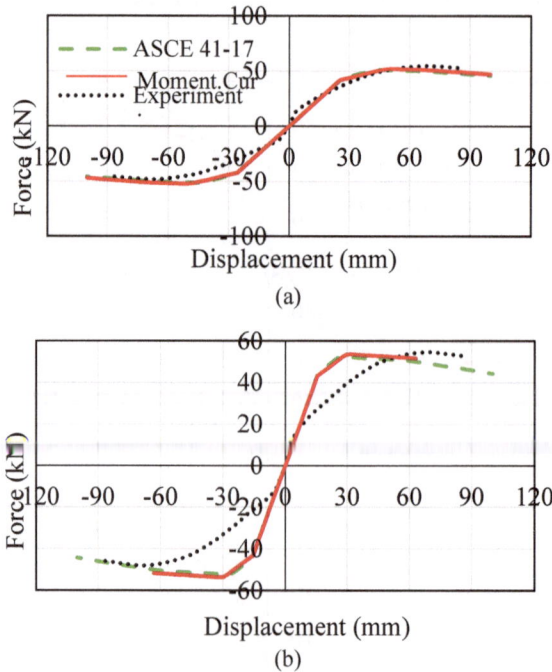

Fig. 5. Comparison between force-displacement relationships of finite element models and experiment for Frame 2 (a) using ASCE/SEI 41 effective stiffness values (b) using Eurocode 8 effective stiffness values.

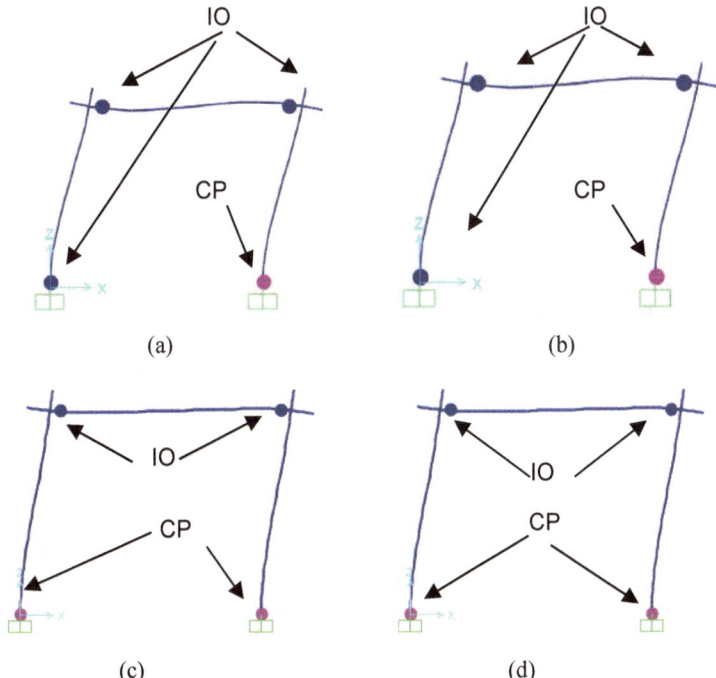

Fig. 6. Damage level of the plastic hinges (a) Frame 1 using ASCE/SEI 41 effective stiffness values (b) Frame 1 using Eurocode 8 effective stiffness values (c) Frame 2 using ASCE/SEI 41 effective stiffness values (2) Frame 2 using Eurocode 8 effective stiffness values.

5   Conclusion

This study investigated the accuracy of two different lumped plasticity models for estimating the force-displacement relationships of two similar RC frames but with different seismic detailing. Comparisons with the experimental results indicated that both ASCE/SEI 41–17 and moment-curvature lumped plasticity models estimated the ultimate loads of both frames with less than 15% error. However, both models underestimated the displacements corresponding to the ultimate loads by 33.9%. Therefore, the lumped plasticity model should be used in the performance-based seismic design methods like the capacity spectrum approach with precaution as the target displacement may be estimated inaccurately. The finite element models were able to predict the damage level and failure mode similar to the obtained results from the experiment.

Acknowledgments. The authors would like to thank the Ministry of Higher Education of Malaysia and Universiti Teknologi Malaysia for Fundamental Research Grant Scheme registered under FRGS/1/2022/TK06/UTM/02/33 (Cost Center : R.J130000.7851.5F545).

O. C. Ee et al.

References

1. Vafaei, M., Alih, S.C., Abdul Rahman, Q.: Drift demands of low-ductile moment resistance frames (MRF) under far field earthquake excitations. J. Teknol. **78**, 82–92 (2016). https://doi.org/10.11113/jt.v78.5076
2. Inel, M., Ozmen, H.B.: Effects of plastic hinge properties in nonlinear analysis of reinforced concrete buildings. Eng. Struct.Struct. **28**, 1494–1502 (2006). https://doi.org/10.1016/j.engstruct.2006.01.017
3. Kremmyda, G.D., Fahjan, Y.M., Tsoukantas, S.G.: Nonlinear FE analysis of precast RC pinned beam-to-column connections under monotonic and cyclic shear loading. Bull. Earthq. Eng.Earthq. Eng. **12**, 1615–1638 (2014). https://doi.org/10.1007/s10518-013-9560-2
4. Dadi, V.V.S.S.K., Agarwal, P.: Effect of types of reinforcement on plastic hinge rotation parameters of RC beams under pushover and cyclic loading. Earthq. Eng. Eng. Vib. **14**, 503–516 (2015). https://doi.org/10.1007/s11803-015-0040-3
5. Vafaei, M., Alih, S.C., Fallah, A.: The accuracy of the lumped plasticity model for estimating nonlinear behavior of reinforced concrete frames under gradually increasing vertical loads. Struct. Concr.. Concr. **21**, 65–80 (2019). https://doi.org/10.1002/suco.201800357
6. López-López, A., Tomás, A., Sánchez-Olivares, G.: Influence of adjusted models of plastic hinges in nonlinear behaviour of reinforced concrete buildings. Eng. Struct.Struct. **124**, 245–257 (2016). https://doi.org/10.1016/j.engstruct.2016.06.021
7. Shatarat, N., Shehadeh, M., Naser, M.: Impact of plastic hinge properties on capacity curve of reinforced concrete bridges. Adv. Mater. Sci. Eng. **2017**, 1–13 (2017). https://doi.org/10.1155/2017/6310321
8. Berry, M.P., Lehman, D.E., Lowes, L.N.: Lumped-plasticity models for performance simulation of bridge columns. ACI Struct. J.Struct. J. **105**, 270 (2008)
9. ACI 318, Building Code Requirements for Structural Concrete (ACI 318-14) and Commentary on Building Code Requirements for Structural Concrete (ACI 318R-14). American Concrete Institute, Farmington Hills, MI (2014)
10. Vafaei, M., Baniahmadi, M., Alih, S.C.: The relative importance of strong column-weak beam design concept in the single-story RC frames. Eng. Struct.Struct. **185**, 159–170 (2019). https://doi.org/10.1016/j.engstruct.2019.01.126
11. ASCE/SEI 41–17, Seismic Evaluation and Retrofit of Existing Buildings. American Society of Civil Engineering, Virginia, USA (2013). https://doi.org/10.1061/9780784414859
12. Mander, J.B., Priestley, M.J.N., Park, R.: Theoretical stress-strain model for confined concrete. J. Struct. Eng.Struct. Eng. **114**, 1804–1826 (1988). https://doi.org/10.1061/(ASCE)0733-9445(1988)114:8(1804)
13. Paulay, T., Priestly, M.J.N.: Seismic Design of Reinforced Concrete and Masonry Buildings. Wiley (1992). https://doi.org/10.1002/9780470172841
14. Eurocode 8: Design of structures for earthquake resistance-part 1: general rules, seismic actions and rules for buildings, European Committee for Standardization. EN 1998–1, Brussels (2004)

Comprehensive Zeta Potential Analysis of Moringa oleifera-Based Coagulants for Heavy Metal Removal

Ravikumar Karunakaran[✉]

Kerala Water Authority, WASH PMU, Jalabhavan, Thiruvananthapuram, India
ravikumarkwa@gmail.com

Abstract. This study investigates the efficacy of Moringa oleifera-derived coagulants for removing Cd, Cr, and Pb from water through zeta potential analysis. The surface charge characteristics of Moringa oleifera seed, gum powder, bentonite clay, and clay-polymer composites are explored, emphasizing their role in coagulation-flocculation processes. The isoelectric point (IEP) is identified as a crucial parameter, underlining its significance in the colloidal system. Results highlight promising heavy metal removal by Moringa oleifera seed, gum, and their composite coagulants. The research offers unique insights into the zeta potential characteristics of these coagulants, emphasizing pH's importance in heavy metal removal. This holistic examination of Moringa oleifera-derived coagulants presents a promising avenue for sustainable water purification practices.

Keywords: Zeta Potential · Moringa Oleifera Seed Composite · Moringa Oleifera Gum Composite · Coagulo-Adsorption · Heavy Metal Removal

1 Introduction

Zeta potential measurement plays a crucial role in determining the stability of colloidal suspensions. The stability of dispersions relies on the magnitude of zeta potential, influencing particle aggregation. These studies are closely linked to the coagulation-flocculation process, wherein the isoelectric point (IEP) becomes a critical indicator of colloidal system stability [1]. Effective coagulation processes hinge on an understanding of IEP, where the potential energy barrier opposing coagulation disappears. Previous studies have indicated the suitability of Moringa oleifera seed/gum coagulant and composites for removing turbidity, fluoride, and heavy metals from water and wastewater [2–8].

Despite these individual studies, there exists a noticeable research gap in comprehensively understanding the zeta potential characteristics of Moringa oleifera-derived coagulants, their interactions with other coagulants, and their practical applications in heavy metal removal. This gap highlights the necessity for a holistic investigation that integrates these key elements to address the complexities of water purification.

G. Feng (Ed.): ICCE 2023, LNCE 526, pp. 604–612, 2024.
https://doi.org/10.1007/978-981-97-4355-1_58

This research significantly contributes to bridging the existing gap in the literature. The study not only explores the zeta potential characteristics of Moringa oleifera seed, gum powder, bentonite clay, and clay-polymer composites but also delves into their interactions within the colloidal system. Furthermore, the systematic examination of the application of these coagulants in heavy metal removal enhances our understanding. This holistic approach provides valuable insights that can inform more effective and sustainable water purification practices.

2 Methodologies

The Zeta potential analyzer, Malvern Zetasizer Ver. 7.01, analyzed Zeta potentials of coagulants at various pH conditions for heavy metal removal. pH significantly influences Zeta potential, with a positive curve at low pH and a lower or negative curve at high pH. The point where the curve crosses zero is the isoelectric point, crucial for stability. Zeta potential measurements, conducted on Moringa oleifera seed (MOS) cake coagulant, Moringa oleifera gum (MOG) powder coagulant, and bentonite clay (BC) and its composite Moringa oleifera seed composite (MOSC), Moringa oleifera gum composite (MOGC) in water at room temperature, help optimize coagulant dosage in water treatment.

Electrophoresis, measuring electrophoretic mobility (EM) using Henry's Equation, determined Zeta potential for each coagulant. Electrophoresis involved injecting 25 ml of aqueous dispersions into the Zeta potential instrument's cell at room temperature. The electrophoresis cell, designed for microscopy, comprises two electrode chambers connected by an optically polished tube. Zeta potential was determined by directly measuring electrophoretic mobility using Henry's Equation [9].

$$\mu = \frac{2\varepsilon_0 \varepsilon_r \zeta f(\kappa r)}{3\eta} \tag{1}$$

where, μ = electrophoretic mobility, η = viscosity of medium, $\square$ = permittivity of a vacuum, $f(\kappa r)$ = Henry's function, $\varepsilon 0$ = permittivity of a vacuum, εr = medium dielectric constant (or permittivity), ζ = zeta potential, κ = Debye-Hückle parameter and r = hydrodynamic radius of particle.

Previous studies in 2020, Ravikumar and Udayakumar investigate a green clay-polymer nanocomposite for the removal of heavy metals [7] and in 2021, Ravikumar and Udayakumar introduce Moringa oleifera gum composite as a novel material for heavy metals removal [8]. This research delves into the preparation and characterization of the nanocomposite, showcasing its potential in addressing heavy metal pollution. Figures 1 and 2 show the isoelectric pH and apparent pH of MOS, MOG and BC.

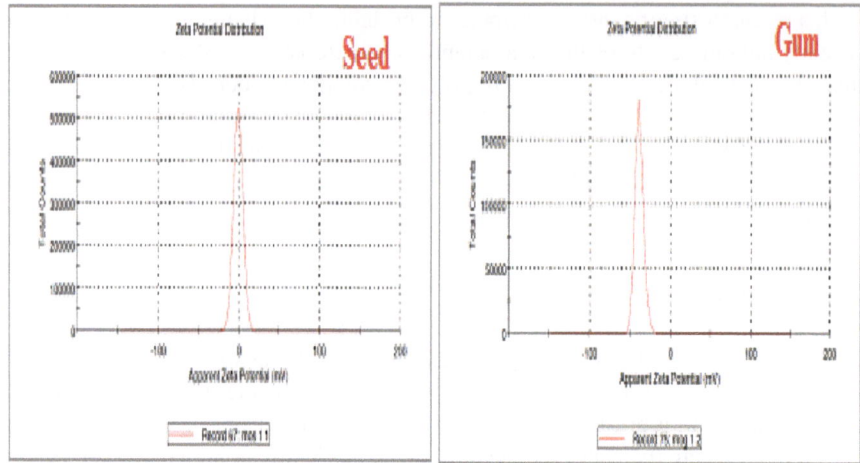

Fig. 1. Apparent zeta potential values of *Moringa oleifera* seed and gum

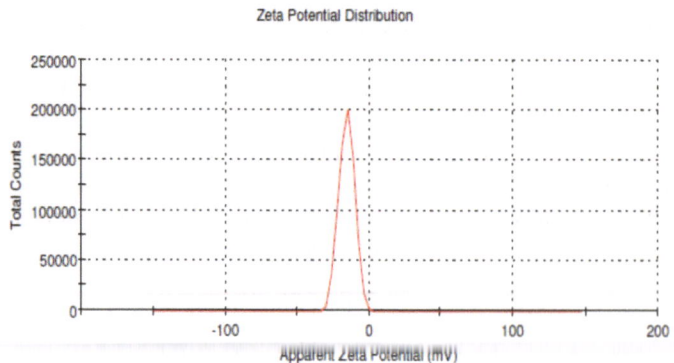

Fig. 2. Apparent zeta potential of bentonite

The modified coagulants, MOSC and MOGC, demonstrated varying zeta potential values at pH 5.5. The clay-polymer composite media exhibited enhanced heavy metal removal efficiency compared to natural clay, emphasizing the significance of the composite in coagulo-adsorption processes.

The measurement of the zeta potential is also a method that provides us with insight into the character of the particle surface itself and the processes occurring on this surface (e.g., adsorption, ion exchange, modification).

Figure 3 show that the apparent zeta potential of MOSC (3.78 mV) was less positive than that of MOGC (-1.21 mV) at a pH of 5.5.

Table 1 shows the isoelectric pH and apparent pH of BC, MOS and MOG, MOSC, and MOGC. The isoelectric point indicates that at the experimental pH > pHpzc, heavy metal species are attracted to the surface sites of the coagulant, which are negative. To avoid the possible precipitation of metals, pH studies were not performed at pH > 8.

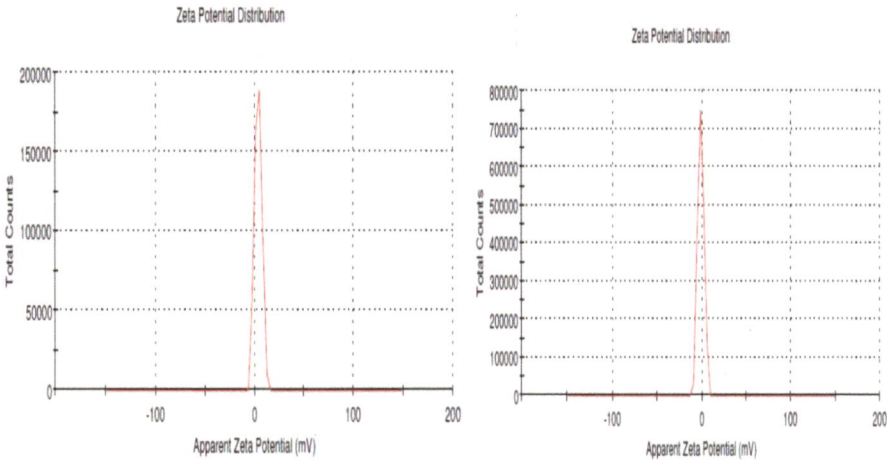

Fig. 3. Apparent zeta potential of MOSC and MOGC

The comprehensive influences of all functional groups determine pHpzc (point of zero charge) of a coagulant and the pH at which the charge on the adsorbent surface is zero.

Table 1. Isoelectric pH and apparent pH of MOSC and MOGC

Sample	Isoelectric pH	Apparent zeta potential (pH 5.5)
BC	–	−15.5mV
MOS	6 .2	−0.516mV
MOG	1.5	−38.4mV
MOSC	5.7	3.78mV
MOGC	5.3	1.21mV

3 Result and Discussion

The pH influenced the adsorption process for metal ions. This observation was attributed to the surface charge of MOSC/MOGC, which could be modified by changing the pH of the solution.

It could be mentioned that both MOSC and MOGC particles have a higher adsorption affinity to adsorb metal cations (cadmium and lead) at high pH values. As the pH increases and the balance between H_3O^+ and OH^- becomes equal, more of the positively charged metal ions in the solution are adsorbed on the negative nano clay-polymer composite surface and thus, the removal percentage of the metal cations (Cd (II) and Pb (II)) increases.

When the pH decreased toward acidic conditions, the zeta potentials of MOSC and MOGC decreased and converted to positive values caused by the protonation of the carboxylic and hydroxide ions on the surface. The results of zeta potential measurements indicated that the isoelectric point (pHzpc) of the MOSC and MOGC nanocomposite was 5.7 and 5.3 and the surface charge of the composite was positive at pH < 5. This positively charged surface at pH < 5 favours the retention of anionic contaminants. With the increase of pH, the zeta potential value decreased, which will in turn reduce the Cr (VI) removal efficiency.

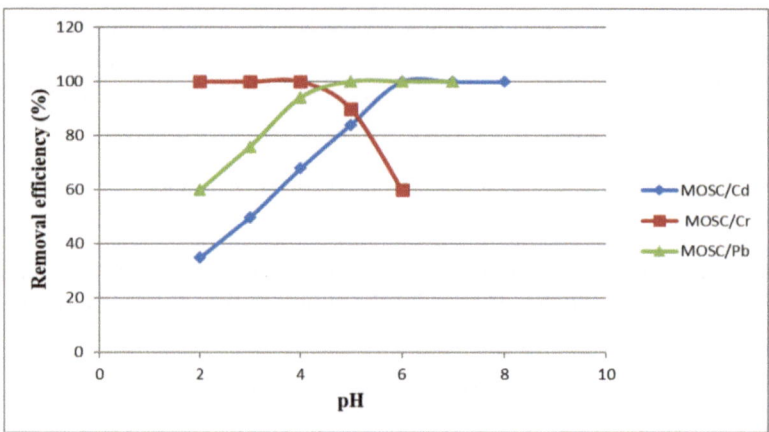

Fig. 4. Influence of pH in the removal process of cadmium, chromium and lead by MOSC coagulo-adsorbent

Figures 4 and 5 show the influence of pH in the heavy metal's removal efficiency by coagulation/flocculation with 5 g/l of MOSC (at 40 rpm of rapid mixing and 15 rpm of slow mixing) and MOGC (at 45 rpm of rapid mixing and 20 rpm of slow mixing) composite coagulo-adsorbent with an initial heavy metal concentration of 6 mg/l.

The clay-polymer coagulo-adsorption studies with MOSC confirmed that optimum condition for metal ions removal were pH 6–8 for cadmium, pH 2–4 for chromium and pH 5–7 for lead. At very low pH values metal uptake has been found very less in the case of cadmium and lead. But chromium has more removal efficiency at very low pH values.

The adsorption studies with MOGC confirmed that optimum condition for metal ions removal were pH 7–8 for cadmium, pH 1–4 for chromium and pH 6–7 for lead.

MOSC and MOGC have a tendency to chelate with metal ions like Cd, Cr, and Pb, etc. The functional groups present in the clay-polymer chain are strongly active with metal ions, as follows:

$$M^{n+} + RNH_2 \rightarrow M(RNH_2)^{n+} \tag{2}$$

The amino group forms coordinate bonds with the metal ions by donating free electrons present on nitrogen and oxygen in the amino groups and hydroxyl groups,

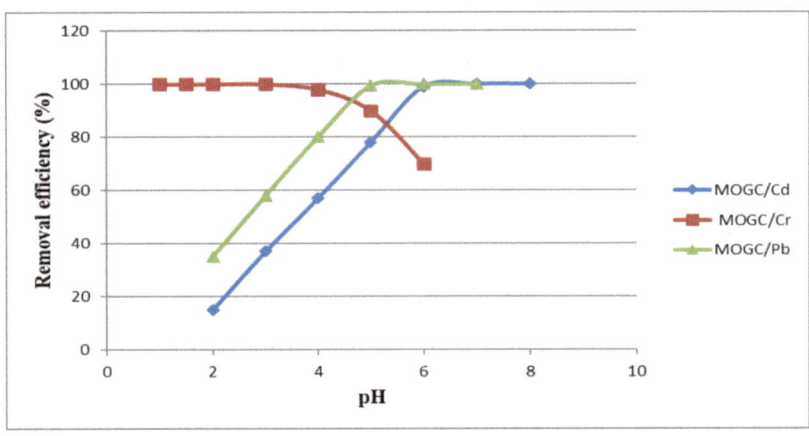

Fig. 5. Influence of pH in the removal process of cadmium, chromium and lead by MOGC coagulo-adsorbent

respectively, to the vacant orbitals of the metals. The metal binding efficiency of the clay-polymer depends on the availability of the amino groups for interaction with metal ions, chain length, and the extent of inter/intra-molecular hydrogen bonding, etc.

Solution pH determines the level of electrostatic or molecular interaction between the adsorption surface and adsorbate owing to charge distribution on the materials. The zero point charge or isoelectric point (pHzpc) of the MOS was about 6.2 and then shifted to 5.7 after being modified by bentonite clay. Similarly, in the case of MOG, it shifted from 1.5 to 5.3. The pHzpc of MOSC and MOGC was obtained at pH = 5.7 and 5.3. At pH below 5.3, more amino groups in the MOSC and MOGC were protonated (i.e., from $-NH_2$ to $-NH_3$ +). From pH 5.7 to 10, the negative zeta potential of MOSC and MOGC, the amino group in the composite media was not deprotonated under this pH condition (i.e., from $-NH_2$ to $-NH^-$). Moreover, the adsorption capacity of composite media highly depended on the positive charge at pH < 5.3.

Clay-polymer composites were found to be much more effective than natural clay for the removal of heavy metals. If a bare particle may have a high zeta potential value, a polymer is adsorbed, giving an increase in the adsorbed layer thickness; the zeta potential would be reduced to some extent. The adsorbed clay particles with Moringa oleifera shift the net surface charge of clay from slightly positive to negative in the case of MOSC; the point of zero charge of clay shifted from 6.2 to 5.7. The adsorbed clay particles with Moringa oleifera shift the net surface charge of clay from negative to positive from 1.5 and 5.3 in the case of MOGC. The heavy metal ion adsorption on composite media was found to be affected by the ionic attraction between the protonated surface groups on Moringa oleifera and the heavy metal ions. But, the heavy metal ions adsorption on natural clay is governed by the positively charged clay particle edges formed by broken bonds of Al–O and Si–O.

Moringa oleifera seed/gum composite presents interesting adsorption capacities because it contains deprotonated hydroxyl groups and is able to give electrostatic inter-action with heavy metals, which mainly include complexing with non-binding doublets on nitrogen and oxygen atoms. The intercalation part of Moringa chains in the clay minerals and the porous network of the composite ensured a good transport of the heavy metal solution to the remaining reactive sites of the bentonite.

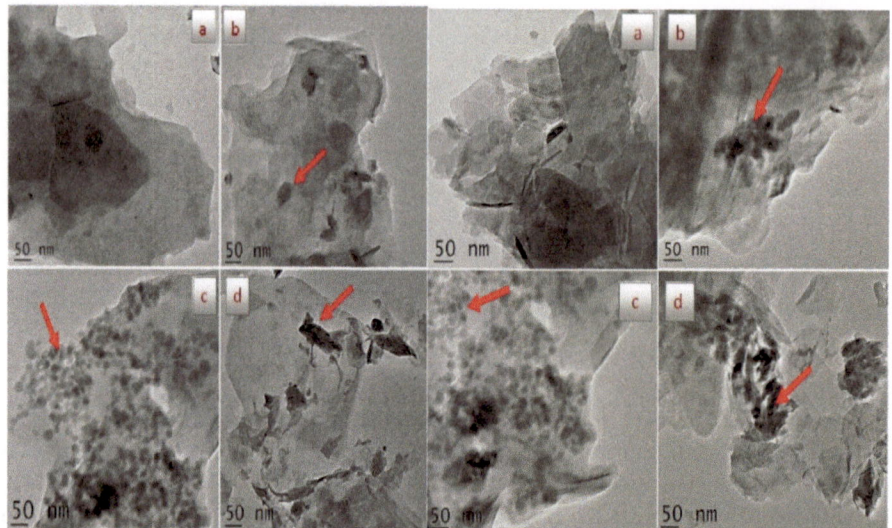

Fig. 6. TEM image in 50nm scale (a) MOSC and MOGC before heavy metal removal (b) Cd loaded MOSC and MOGC (c) Cr loaded MOSC and MOGC (d) Pb loaded MOSC and MOGC

TEM images of MOSC/ MOGC clay-polymer metal complexes indicate the formation of nano structure. TEM images of MOSC/MOGC before and after coagulo-adsorption (Cd, Cr and Pb) was undertaken in order to locate the active adsorptive sites of the composite coagulants to form its metal complexes. In each figure (a) is pure MOSC/MOGC, (b) MOSC/MOGC-Cd complex, (c) MOSC/MOGC-Cr complex and (d) is MOSC/MOGC-Pb complex. TEM images with 50 nm magnification the adsorbed heavy metal ions assume more clarity in their shapes in the form of dark dots as shown in Fig. 6.

Another study [10] contributes to this research by investigating a composite of kaolinite clay and moringa seedcake, which effectively removes methylene blue and acid orange-7 dyes. This investigation includes a comprehensive examination involving batch and column tests, chemical modifications, and characterization techniques. The isoelectric pH values of kaolinite clay, Moringa seedcake, and their composite are presented in Table 2.

The study achieved optimum removal rates of 86% and 94% at pH 2 and 10, respectively, for acid orange-7 (AO-7) and methylene blue (MB). These results were obtained using a 1 g/L adsorbent dose and 50 mg/L initial dye concentration. This study underscores the potential of the moringa-based clay-polymer composite for dye removal.

Table 2. Isoelectric pH of kaolinite clay, Moringa seedcake, and its composite

Sample	Isoelectric pH
Kaolinite clay	5.50
Moringa seedcake	6.00
Composite	7.50

4 Conclusion

In conclusion, the zeta potential analysis revealed distinctive characteristics of Moringa oleifera seed (MOS) and gum (MOG) coagulants at pH 6.5, with MOS exhibiting a zeta potential of -0.516 mV and MOG showing -38.4 mV. Notably, MOS demonstrated superior coagulant activity at pH 7.0, making it more effective for cadmium removal within the pH range of 7–8. Bentonite, with an apparent zeta potential of -15.5 mV at pH 6.5, demonstrated a pH-sensitive zeta potential, underscoring its role in system stability. The study revealed that heavy metal ion adsorption on the clay surface altered bentonite's zeta potential, highlighting its significance in the removal process.

Furthermore, the modified coagulants, MOSC and MOGC, exhibited varying zeta potential values at pH 5.5. The clay-polymer composite media, particularly MOSC/MOGC, demonstrated enhanced heavy metal removal efficiency compared to natural clay, emphasizing the composite's pivotal role in adsorption processes. Transmission electron microscopy (TEM) images of MOSC/MOGC composite coagulant confirmed the mono-dispersed and spherical nature of nanoparticles, highlighting their uniform distribution and absence of agglomeration. The visualization of adsorbed heavy metal ions in the clay-polymer nanocomposites at 50 nm elucidated their shape and size.

In essence, this comprehensive analysis of zeta potential characteristics in the context of heavy metal removal using Moringa oleifera-based coagulants provides valuable insights into the underlying mechanisms. The findings not only advance our understanding of coagulant behavior but also carry practical implications for the development of effective and sustainable water purification methods. Exploring the scalability and long-term efficacy of these coagulants in real-world scenarios would be instrumental in realizing their potential for widespread application in water treatment processes.

References

1. Endut, Z., Yoon, Y.M., Jusoh, A., Ali, N., Wan Ngah, W.S.: Zeta potential in the coagulation process: Its role and measurement techniques. J. Ind. Eng. Chem. **33**, 1–10 (2016)
2. Ravikumar, K., Sheeja, A.K.: Water clarification using Moringa oleifera seed coagulant. In: IEEE 2012 International Conference on Green Technologies (ICGT), pp. 064–070 (2012)
3. Ravikumar, K., Sheeja, A.K.: Heavy metal removal from water using Moringa oleifera seed coagulant and double filtration. Int. J. Sci. Eng. Res. **4**(5), 09–12 (2013)
4. Ravikumar, K., Sheeja, A.K.: Removal of fluoride from aqueous system using Moringa oleifera seed cake and its composite coagulants. Int. J. Current Eng. Technol. **4**(3), 1356–1360 (2014)

5. Ravikumar, K., Udayakumar, J.: Moringa oleifera biopolymer coagulation and bentonite clay adsorption for hazardous heavy metals removal from aqueous systems. Geosyst. Eng. **23**, 265–275 (2020)
6. Ravikumar, K., Udayakumar, J.: Moringa oleifera seed composite: a novel material for hazardous heavy metals (Cd, Cr and Pb) removal from aqueous systems. J. Mater. Environ. Sci. **11**(1), 123–138 (2020)
7. Ravikumar, K., Udayakumar, J.: Preparation and characterisation of green clay-polymer nanocomposite for heavy metals removal. Chem. Ecol. **36**(3), 270–291 (2020)
8. Ravikumar, K., Udayakumar, J.: Moringa oleifera gum composite: a novel material for heavy metals removal. Int. J. Environ. Anal. Chem. **11**, 123–138 (2021)
9. Cantuaria, M.L.: Double layer and critical coagulation concentration in colloids with variable charge. Colloids Surf. A A **469**, 213–221 (2015)
10. Rawat, S., Ahammed, M.M.: Clay-moringa seedcake composite for removal of cationic and anionic dyes. Chemosphere **350**, 141083 (2024)

Analysis of Dynamic Soil Properties by a Systematic Approach

Ripon Hore[1](✉) and Shoma Hore[2]

[1] Local Government Engineering Department, Dhaka, Bangladesh
riponhore@gmail.com
[2] Bangladesh University of Engineering and Technology, Dhaka, Bangladesh

Abstract. The aim of the paper is to review the different research manuscripts on dynamic soil properties analysis in Bangladesh. In this case, Wrap faced embankment is significantly better result to build earthquake resilient infrastructure. Wrap faced embankment is basically a sand embankment wrapped around by geotextile layers as a steep embankment. This wrap-faced embankment is filled with different characteristics of sand. In this research, the wrap faced embankment is model by reduce scale in BUET Geotechnical Lab. The shaking table test machine was used to induce earthquake and wave action. The fifteen sensors have been placed in the wrap faced embankment and clay soil layer which was discussion on Hore's research from 2020 to 2023. The different soil characteristics like accelerations, displacement, pore water pressure and strain has been analyzed in this research. This embankment showed better response to the earthquakes and wave action means stable to seismic and wave action. The results of the research have also been verified by numerical analysis by PLAXIS 3D software. Without knowing the characteristics of the wrap faced embankment during the earthquake, it is impossible to build this type earthquake/wave resisting embankment. The dynamic behavior of wrap faced embankment on soft soil in Bangladesh has been invented by the Hore's research. Now, it is possible to design and construction of earthquake/wave resistance wrap faced embankment on soft Bangladeshi soil.

Keywords: Soil Structure · Embankment · Wrap-Faced · Geotechnical Properties

1 Introduction

The shake table test on the different dynamic loading based on the slopy area developed a calibrated numerical model and analyzed the input ground motions [1, 2] at the base of the rigid-faced reinforced soil-retaining wall. The seismic behavior under dynamic loads, these series of tests was performed using two different slope angles, and reinforcements [3–10]. The research on seismic response of slurry wall and sandy soil was presented by [11]. The effectively performed more tests to investigate the behavior of excess pore water pressure in different soft soil-foundation [12–15]. Latha and Manju [5] described the performance of geo-cell retaining walls inside a laminar box which were under seismic

G. Feng (Ed.): ICCE 2023, LNCE 526, pp. 613–618, 2024.
https://doi.org/10.1007/978-981-97-4355-1_59

shaking conditions. A recent study by Gidday and Mittal [16] on reinforced soil retaining wall on soil which is facilitated by the shake table test. The embankment analysis of the soft clay soil in Bangladesh is the vital effect on the soil structure interaction. The experimental and numerical analysis (PLAXIS 3D) which were performed by the shaking table platform with laminar box on soft clay soil. The dynamic soil analysis platform is very significant for the analysis of the soil behavior as per seismic response. Moreover, cyclic loading is the vital role on the analysis of the Bangladeshi soft soil [17–34].

In this research, the response of earthquake and wave action on the soft soil in Bangladesh has been analyzed where the different soil type of wrap faced embankment (local and Sylhet sand) on the shaking table test machine used in the experiment Subjected to cyclic loading. This paper is reviewed the different research paper from 2020 to 2023 on wrap Faced embankment at lab of Bangladesh University of Engineering and Technology (BUET).

2   Shaking Table, Laminar Box and Testing

A computer-controlled servo-hydraulic single degree of freedom shaking table facility was used in this experiment, where the platform used for testing was made of steel, the measurement is of 2 m by 2 m size, and with a payload capacity of 1500 kg. as shown in Fig. 1 [30]. The acceleration range is 0.05 g to 2 g. A frequency range is 0.05 Hz to 50 Hz. The maximum amplitude was ± 200 mm. The maximum velocity was 30 cm/sec. Twenty-four (24) hollow aluminum layers of a large-sized shear box is used for this experiment. The friction between the layers is minimum, as shown in Fig. 1. In the present study, the height of clay soil foundation is 300 mm. On the other hand, the thickness of the sand blanket is 50 mm as shown in Fig. 2 (Chakraborty et al. 2022). The prototype to model scale being N = 10 and scale factor 1/N. Figure 3 [26] shows the Wrap-faced soil retaining wall.

Fig. 1. Seismic analysis of the model embankment

3 Results and Discussion

Displacement profile with respect to surcharge and base acceleration has been analyzed in the results and discussion section. Horizontal face displacement along the height of the wall was scrutinized based on different sinusoidal motion where different frequency and acceleration level are fixed presented in Figs. 4 to 5 [26]. Here elevation is denoted as z and horizontal displacements as δh are presented in non-dimensional from after normalizing.

The height of the wall is denoted as H. Figure 4 depicts the normalized displacement profile for different base accelerations of 0.05 g, 0.10 g, 0.15 g and 0.20 g. The tests are ST72, ST80, ST88, and ST96 respectively. From the Fig. a maximum horizontal displacement of 2.26% of the total wall height (H), for 0.20 g, was observed compared with 2.16% for 0.05 g base accelerations. The maximum displacement is 9.06 mm at a acceleration of 0.2 g, whereas it is decreased to 8.67 mm at a acceleration of 0.05 g. The numerical analysis results are 3.97% and 3.85% higher than the experimental results respectively. The effect of different surcharge loadings of 1.72 kPa, 1.12 kPa and 0.7 kPa as shown in Fig. 5. The displacement response against surcharge variation was inversely proportional at all elevations. The maximum displacement of the wall was (δh/H = 2.19%) at a surcharge pressure of 0.7 kPa, whereas it was decreased to (δh/H = 2.10%) at a surcharge pressure of 1.72 kPa. The numerical analysis results are 3.57% and 3.67% higher than the experimental results respectively.

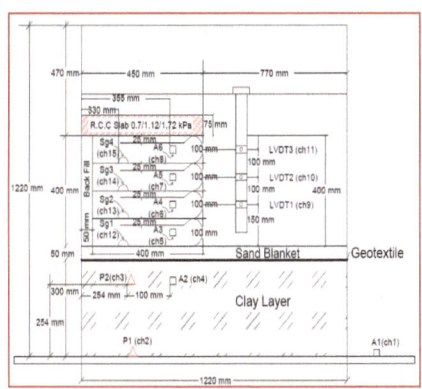

Fig. 2. Experimental setup reinforced

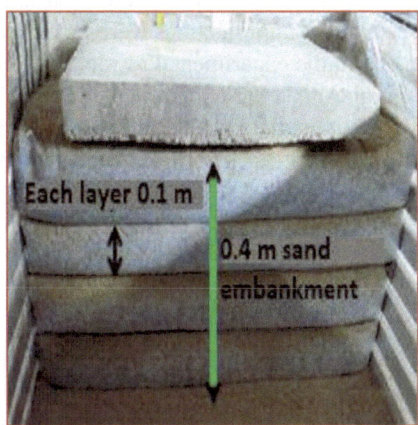

Fig. 3. Wrap-faced geotextile

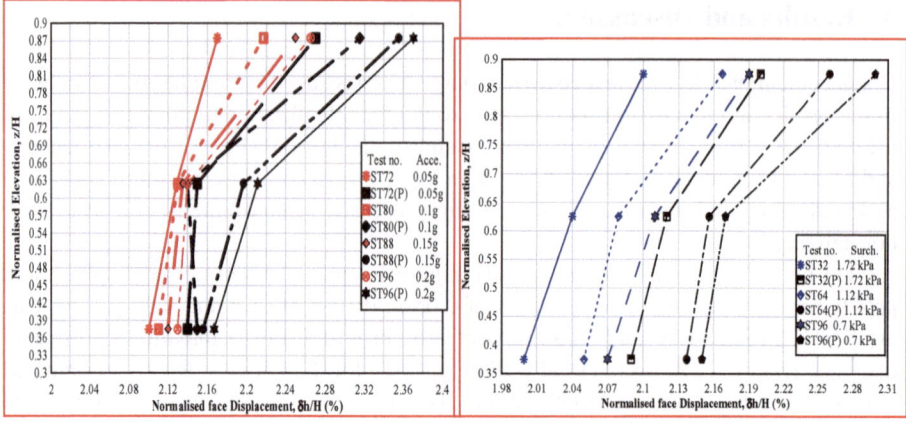

Fig. 4. Displacement profile (base acceleration)

Fig. 5. Displacement profile

4 Conclusions

The behavior of wrapped-face retaining wall has a significant effect on the soft clay soil in Bangladesh which was presented in Hore's research from 2020 to 2023. Accelerations at the top of the wall were inversely proportional to the surcharge pressures and acceleration response against frequency variation is not directly proportional. On the other hand, acceleration amplifications were increased with increased base accelerations which are proved in the experimental results. Moreover, in all cases, numerically obtained values are higher than the experimental results. The experimental result is found to be lower than the numerical result that is used in PLAXIS 3D for all parameters. These research outcomes are very helpful to analysis of the future research on dynamic behavior of soft soil and forecast a future scenario of the soil profile specially the middle and southern part of the country.

References

1. Murali Krishna, A., Bhattacharjee, A.: Seismic analysis of reinforced soil retaining walls. In: Ilamparuthi, K., Robinson, R.G. (eds.) Geotechnical Design and Practice. DGE, pp. 159–171. Springer, Singapore (2019). https://doi.org/10.1007/978-981-13-0505-4_14
2. Krishna, A.M., Latha, G.M.: Seismic response of wrap-faced reinforced soil-retaining wall models using shaking table tests. Geosynth. Int. **14**(6), 355–364 (2007). https://doi.org/10.1680/gein.2007.14.6.355
3. Latha, G.M., Krishna, A.M.: Shaking table studies on reinforced soil retaining walls. Indian Geotech. J. **36**(4), 321–333 (2006)
4. Latha, G.M., Krishna, A.M.: Seismic response of reinforced soil retaining wall models: influence of backfill relative density. Geotext. Geomembr. **26**(4), 335–349 (2008). https://doi.org/10.1016/j.geotexmem.2007.11.001
5. Madhavi Latha, G., Manju, G.S.: Seismic response of geocell retaining walls through shaking table tests. Int. J. Geosynthetics Ground Eng. **2**(1), 1–15 (2016). https://doi.org/10.1007/s40891-016-0048-4

6. Latha, G.M., Nandhi Varman, A.M.: Shaking table studies on geosynthetic reinforced soil slopes. Int. J. Geotech. Eng. **8**(3), 299–306 (2014). https://doi.org/10.1179/1939787914Y.000 0000043

7. Krishna, A.M., Bhattacharjee, A.: Behavior of rigid-faced reinforced soil-retaining walls subjected to different earthquake ground motions. Int. J. Geomech. **17**(1), 06016007 (2017). https://doi.org/10.1061/(ASCE)GM.1943-5622.0000668

8. Sabermahani, M., Ghalandarzadeh, A., Fakher, A.: Experimental study on seismic deformationmodes of reinforced-soil walls. Geotext. Geomembr. **27**(2), 121–136 (2009). https://doi.org/10.1016/j.geotexmem.2008.09.009

9. Sakaguchi, M., Muramatsu, M., Nagura, K.: A discussion on reinforced embankment structures having high earthquake resistance. In: Proceedings of the International Symposium on Earth Reinforcement Practice, IS-Kyushu 1992, Fukuoka, Japan, vol. 1, pp. 287–292 (1992)

10. Sakaguchi, M.: A study of the seismic behaviour of geosynthetic reinforced walls in Japan. Geosynth. Int. **3**(1), 13–30 (1996). https://doi.org/10.1680/gein.3.0051

11. Xiao, M., Ledezma, M., Hartman, C.: Shake table test to investigate seismic response of a slurry wall. In: Geo-Congress 2014: Geo-characterization and Modeling for Sustainability, pp. 1234–1243 (2014). https://doi.org/10.1061/9780784413272.120

12. Yazdandoust, M.: Investigation on the seismic performance of steel-strip reinforced-soil retaining walls using shaking table test. Soil Dyn. Earthq. Eng. **97**, 216–232 (2017). https://doi.org/10.1016/j.soildyn.2017.03.011

13. Siavoshnia, M., Kalantari, F., Shakiba, A.: Assessment of geotextile reinforced embankment on soft clay soil. In: Proceedings of The 1st International Applied Geological Congress, Department of Geology, Islamic Azad University, pp. 1779–1784 (2010)

14. Sahoo, S., Manna, B., Sharma, K.G.: Seismic response of a steep nailed soil slope: shaking table test and numerical studies. In: Sundaram, R., Shahu, J., Havanagi, V. (eds.) Geotechnics for Transportation Infrastructure, vol. 28, pp. 611–623. Springer, Singapore (2019). https://doi.org/10.1007/978-981-13-6701-4_39

15. Gidday, B.G., Mittal, S.: Dynamic response of wrap-faced cement treated reinforced clayey soil retaining walls. Innov. Infrastruct. Solut. **5**(42), 1–9 (2020). https://doi.org/10.1007/s41 062-020-00295-x

16. Guan, Y., Zhou, X., Yao, X., Shi, Y.: Seismic performance of prefabricated sheathed cold-formed thin-walled steel buildings: shake table test and numerical analyses. J. Constr. Steel Res. **167**, 105837 (2019). https://doi.org/10.1016/j.jcsr.2019.105837

17. Hore, R., Ansary M.A.: SPT-CPT correlations for reclaimed areas of Dhaka. J. Eng. Sci. JES, KUET (2018)

18. Hore, R., Arefin, M.R., Ansary, M.A.: Development of zonation map based on soft clay for Bangladesh. J. Eng. Sci. **10**(1), 13–18 (2019)

19. Hore, R., Chakraborty, S., Ansary, M.A.: A field investigation to improve soft soils using prefabricated vertical drain. Transp. Infrastruct. Geotechnol. **7**, 127–155 (2020). https://doi.org/10.1007/s40515-019-00093-8

20. Hore, R., Chakraborty, S., Shuvon, A.M., Ansary, M.A.: Effect of acceleration on wrap faced reinforced soil retaining wall on soft clay by performing shaking table test. In: Proceedings of Engineering and Technology Innovation (2020)

21. Hore, R., Chakraborty, S., Bari, M.F., Shuvon, A.M., Ansary, M.A.: Soil zonation and the shaking table test of the embankment on clayey soil, Geosfera Indonesia (2020). https://doi.org/10.19184/geosi.v5i2.17873

22. Hore, R., Chakraborty, S., Ansary, M.A.: Experimental investigation of embankment on soft soil under cyclic loading: effect of input surcharges. J. Earth Eng. (JEE) **5**(1), 1–8 (2020)

23. Hore, R., Ansary, M.A.: Different soft soil improvement techniques of dhaka mass rapid transit project. J. Eng. Sci. **11**(2), 37–44 (2020). https://doi.org/10.3329/jes.v11i2.50896

24. Hore, R., Al-Mamun. S.: Climate change and its diverse impact on the rural infrastructures in Bangladesh. Disaster Adv. **13**(9) (2020)
25. Hore, R., Chakraborty, S., Ansary, M.A.: Liquefaction Potential Analysis based on CPT and SPT. Geotech. Eng. J. SEAGS AGSSEA (2020)
26. Hore, R., Chakraborty, S., Ansary, M.A.: Seismic response of embankment on soft clay based on shaking table test. Int. J. Geosynthetics Ground Eng. **7**, 1–18 (2021). https://doi.org/10.1007/s40891-020-00246-7
27. Hore, R., Chakraborty, S., Shuvon, A.M., Ansary, M.A.: Dynamic response of reinforced soil retaining wall resting on soft clay. Transp. Infrastruct. Geotechnol. **8**, 607–628 (2021). https://doi.org/10.1007/s40515-021-00156-9
28. Hore, R., Chakraborty, S., Kamrul, K., Shuvon, A.M., Ansary, M.A.: Numerical verification for seismic response of reinforce soil embankment on soft clay foundation. Geotech. Eng. J. SEAGS AGSSEA **53**(2), 18–28 (2022)
29. Hore, R., Chakraborty, S., Kamrul, K., et al.: Earthquake response of wrap faced embankment on soft clay soil in Bangladesh. Earthq. Eng. Eng. Vib. **22**, 703–718 (2023). https://doi.org/10.1007/s11803-023-2194-8
30. Chakraborty, S., Hore, R., Shuvon, A.M., et al.: Effect of surcharge pressure on model geotextile wrapped-face wall under seismic condition. Iran J. Sci. Technol. Trans. Civ. Eng. **46**, 4409–4423 (2022). https://doi.org/10.1007/s40996-022-00900-2
31. Chakraborty, S., Hore, R., Shuvon, A.M., Ansary, M.A.: Physical and numerical analysis of reinforced soil wall on clayey foundation under repetitive loading: effect of fineness modulus of backfill material. Arab. J. Geosci. **14**, 1108 (2021). https://doi.org/10.1007/s12517-021-07317-7
32. Chakraborty, S., Hore, R., Shuvon, A.M., Ansary, M.A.: Dynamic responses of reinforced soil model wall on soft clay foundation. Geotech. Geol. Eng. **39**, 2883–2901 (2021). https://doi.org/10.1007/s10706-020-01665-z
33. Chakraborty, S., Hore, R., Ahmed, F., Ansary, M.A.: Soft ground improvement at the Rampal coal based power plant connecting road project in Bangladesh. Geotech. Eng. J. SEAGS AGSSEA, AIT **48**, 69–75 (2017)
34. Talukder, A.H., Hore, S., et al.: A review of soil chemical properties for Bangladesh perspective. West. Eur. J. Modern Exp. Sci. Methods **1**(1), 52–65 (2023)

Technology for Performing Emergency Dismantling Works at an Industrial Facility Destroyed Due to Military Actions

V. Naumov[1], A. Bilokon[2], I. Sokolov[1], Ye. Plakhtii[3(✉)], and P. Nesevrya[2]

[1] Department of Organization and Management in Construction, Prydniprovska State Academy of Civil Engineering and Architecture, 24-a Architect Oleg Petrov St., Dnipro, Ukraine
[2] Department of Construction Technologies, Prydniprovska State Academy of Civil Engineering and Architecture, 24-a Architect Oleg Petrov St., Dnipro, Ukraine
[3] Department of Computer Science, Information Technologies and Applied Mathematics, Prydniprovska State Academy of Civil Engineering and Architecture, 24-a Architect Oleg Petrov St., Dnipro, Ukraine
plakhtii.ev@gmail.com

Abstract. This investigation develops an innovative dismantling methodology for industrial facilities impacted by a 2022 missile strike, aiming to restore operations with minimal interruption. The purpose of the research is to establish a rapid and secure dismantling process that integrates seamlessly with ongoing industrial activities, ensuring safety and efficiency. It focuses on the damaged workshop facilities, analyzing the affected structures and utilities to inform emergency dismantling and recovery efforts. Advanced technologies, including specialized excavator attachments and carts, are introduced for precise dismantling, maintaining the integrity of adjacent structures. Our findings illustrate the efficacy of integrating advanced dismantling technologies within active industrial settings, significantly enhancing operational safety and efficiency. The successful application of these methodologies not only aids in the rapid recovery of damaged facilities but also sets a new benchmark for emergency industrial operations. **Object of Research:** The primary focus is on the damaged industrial workshop facilities, specifically examining the structures, utilities, and operational frameworks affected by the missile strike. This includes the physical site, the technological layout, and the existing industrial processes within the context of emergency dismantling and restoration efforts.

Keywords: Dismantling Process · Disassembling Buildings · Prefabricated Elements · Demolition · Technology of Work

1 Introduction

One of the main tasks of construction is the rational use of resources. This issue becomes especially important during war or hostilities.

The industrial building suffered threatening destruction as a result of a missile attack on the workshop facilities at the end of 2022. In order to restore the functioning of the

G. Feng (Ed.): ICCE 2023, LNCE 526, pp. 619–629, 2024.
https://doi.org/10.1007/978-981-97-4355-1_60

workshop, it became necessary to carry out dismantling and restoration works of the load-bearing and enclosing building structures.

The affected site is systematically divided into four work sections, emphasizing areas most impacted by the rocket attack. Following the completion of emergency measures, specialists conduct a comprehensive diagnosis of structures beyond the hazardous zone, providing an approximate assessment of the scope of emergency demolition works.

The consequences of the damage are presented in (Fig. 1). The characteristics of the constructive and object-planning solution of the facility are provided in Table 1.The existing roof of the building is made of ribbed precast concrete slabs measuring 3×6 m. On top of the slabs, there is 80 mm insulation, 20 mm cement screed, and four layers of roofing felt on bitumen mastic. The roof slabs are welded to the trusses at three points. The dismantled building is conditionally divided into 6 sections (Fig. 2).

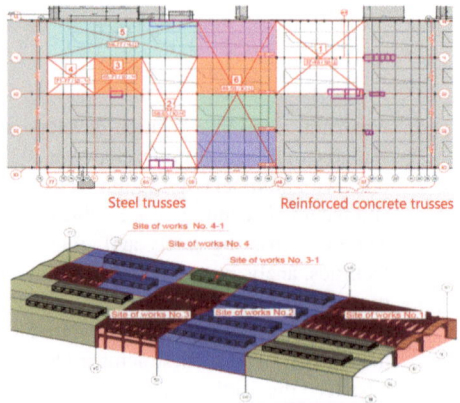

Fig. 1. Nature of object destruction.   **Fig. 2.** The diagram and 3^d scheme of the building being restored (divided into sections).

2   Analysis of Recent Research

In this study, we created methodology for the emergency dismantling of industrial facilities impacted by military activities, a subject that remains largely unexplored within the current corpus of research.

Recent literary sources on the research topic demonstrate a consistent trend towards so-called "green dismantling" and recycling of materials generated as a result of dismantling. For instance, the article [1] describes a methodology for developing a dismantling plan and minimizing CO_2 emissions during the dismantling process.Authors in article [2] describe new technologies for dismantling concrete structures that utilize sound-absorbing chemical agents to minimize the environmental and health impacts of dismantling activities. The article outlines the main components of these agents and their influence on the dismantling process, as well as presents an environmental impact assessment system.

Article [3] investigates and explains the dismantling practices in cities in the United States and Germany, while article [4] presents solutions based on completed projects in Ukraine. While article [3] focuses more on economic aspects and typology, article [5] highlights the need for selective dismantling, which involves disassembling buildings into components and materials for further reuse. The dismantling practices issue is also discussed using examples from China, the Czech Republic and other countries [6–8]. Another important topic is the technical assessment of residual resources of dismantled buildings [9, 10]. Timely conducted technical assessment can serve as a legal basis for. can identify reserves for their further utilization [9]. Accurate information on the composition (quantity and quality) of materials is necessary for designing the subsequent use of construction materials obtained from building and structure dismantling [10]. In the research conducted [11], quantitative data on the activities of reuse, recycling, and demolition of construction materials have been analyzed.

Table 1. Characteristics of the destroyed object

Parameter Type	Value
Building type:	Industrial complex
The number of spans of the building:	4
The length of each span of the building:	30 m
Total length of the building:	120 m
Capacity of bridge cranes in the building	12 t, 20 t
Length of reconstructed area	236 m
Column spacing (main)	12 m
Column spacing (specific areas with different spacing):	6 m, 18 m, 24 m, 36 m
Year of commissioning:	1961
Load-bearing structures:	Reinforced concrete
Roof structure:	Trusses with metal in axes 78-48, reinforced concrete in axes 48-37

To create structures and buildings with high transformability and dismantling potential, more effort needs to be focused on the development of prefabricated elements and building systems to support the potential for transformation during possible dismantling stages [12]. A successful example of addressing waste disposal and overall waste management can be seen in the case of AZS company (Czech Republic) [13]. An attempt to summarize current trends in the field of construction waste recycling is being conducted in the work [14].

Considering demolition as a stage preceding the main construction process, several authors in their works focus on the issues of designing the technology and making key decisions regarding the dismantling and destruction of building elements, with a special emphasis on controllability, safety, and process efficiency.

In the work [15], the corresponding requirements are outlined to ensure the specified conditions when choosing the method of building demolition, many factors to be considered. The article [16] presents the process of mechanical demolition through step-by-step investigation and structural collapse. Initially, the structure is weakened in a cross-section using a cutter, and then it is brought down by applying lateral (horizontal) loading. Demolishing structures in this manner allows for the maximum amount of material to be recycled.

The work [17] introduces a Java application with a Building Information Modelling-based Deconstruction Assessment Score (BIM-DAS) to assess the deconstructability of structures during the design stage. Additionally, careful image pre-processing can remove noise and smooth the original information for later analysis [18–20].

Recent studies mainly address standard dismantling and construction waste recycling, overlooking emergency scenarios. This article aims to show how conventional dismantling technologies and digital tools can be adapted for use in sites damaged by the missile strike. The identified research gap in emergency dismantling underlines the importance of exploring advanced technologies for managing industrial crises.

3   Results and Discussion

In the first hours following the occurrence of an emergency situation, ensuring the safety of personnel and the environment becomes the primary concern. To achieve this, the following actions needed to be taken as soon as possible: 1) disconnect all utilities and power systems in the building; 2) extinguish the fire resulting from the explosion of the missile; 3) assess the extent of damage to the building; 4) identify key issues and develop a plan for the dismantling of destroyed and damaged structures; 5) disconnect all utilities and power systems in the building.

Since an emergency situation had arisen, it was necessary to expedite the process of dismantling the affected structures to minimize potential consequences such as environmental pollution and the risk of damaging other buildings and systems on the premises.

The first step was to dismantle the damaged structures and equipment that could potentially pose a danger, such as falling structures or damaged power systems.

Subsequently, the dismantling process was carried out step by step, taking into account the specific characteristics of the building and potential risks.

The building was conventionally divided into 4 general work sections, with the areas most affected by the rocket attack highlighted in purple. The schematics of the most affected site are shown in Fig. 3a, 3b. After completing all the emergency measures, specialists conducted a diagnosis of the structures located outside the hazardous zone of the collapsed emergency structures. After the inspection and examination, an approximate assessment of the scope of emergency works was carried out (Table 2).

At the next stage, it was necessary to remove the already collapsed structures and those that were partially destroyed but still secured and posed a risk of collapse (Fig. 4a, 4b). This was done in order to subsequently install cranes and safely proceed with the demolition of less affected structures.

The next task was to dismantle the remaining structures that were attached to the frame, such as wall panels and beams, which were in a critical condition and at risk of collapse at any time. In general, the dismantling technology for panels at considerable height (7 m and above) is not significantly different from installation, except that it is performed in reverse order and involves the following steps: 1) establishing a safety perimeter around the hazardous zone, with a distance of 7 m from the building; 2) installing a crane and a hydraulic lift; 3) assessing the condition of the panel and identifying its attachment points to the building; 4) rigging the structure to be dismantled, which involves creating openings in the structure or welding lifting lugs to metal parts of the panel, and ensuring proper tensioning of the rigging.

However, in the current conditions, it is not feasible to use this technology because the panels are severely damaged and lack structural rigidity. Even if they could be rigged, when cutting the embedded parts, they may collapse in any direction, posing a risk to construction machinery and workers involved in the dismantling operations. In such cases, it is advisable to use mechanical demolition by excavators. However, even a 30–40-ton excavator with a standard boom would not be able to handle the task, as the upper panels are situated at a significant height (over 17 m), and the excavator would need to be positioned outside the hazardous zone, i.e., at a radius of more than 7 m from the hazardous structures.

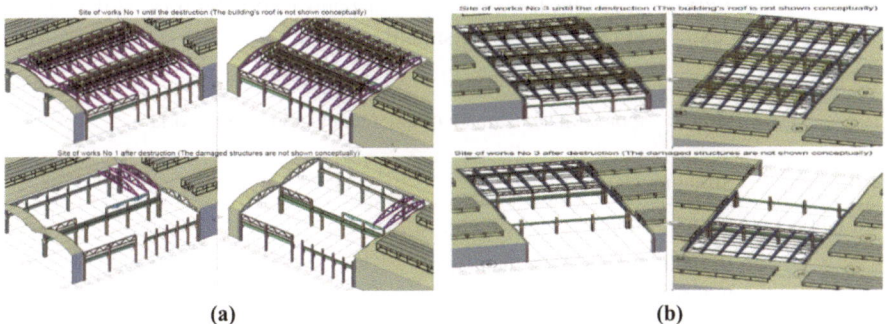

(a)   (b)

Fig. 3. a. The 1st section (diagrams of before and after destruction), **b.** The 3rd section (diagrams of structures structures before and after destruction)

(a)   (b)

Fig. 4. a. Clearing the site from the demolished structures (photo before), **b.** Photo of the site after clearing.

One of the presented solutions is the use of an excavator with an extended boom for demolition works. The diagram for demolition using an excavator with an extended boom is shown below (Fig. 5). The technology for carrying out work during the dismantling of panels using an excavator with a boom extension involves the following steps: 1) recting a stable enclosure around the work area; 2) attaching the boom extension instead of the bucket; 3) driving the excavator into the demolition zone;4) securing the damaged wall panels using the boom extension and using the excavator's hydraulic system to lower them to ground level; 5) after completing the dismantling, moving to another parking area.

Using another excavator equipped with a hydraulic hammer positioned outside the dangerous zone of panel collapse to further process the panels into a convenient size for loading and transportation.

After the dismantling of the hanging damaged panels, the next step is to gradually dismantle the damaged fencing structures. During the inspection, it was found that the roofing panels have also been damaged and are in a critical condition, making it impossible for workers to be on the roof. Therefore, the standard method of dismantling a part of the roofing panels cannot be used when workers are present on or under the roof covering. To dismantle the structures, high-capacity crawler cranes of 100 and 250 tons with movable hooks were selected and utilized. The main feature of these cranes is that they have two hook suspensions on one hook (Fig. 6), each with its own mechanism (brake, winch, safety device). One hook is used to suspend a work platform with workers, while the other can be used to attach the dismantled panel.

Table 2. Approximate estimate of emergency demolition works

No	Structure Description	Unit	Quantity
1. Light-aeration lantern structures			
1.1	Demolition of metal structures of industrial light-aeration lanterns	tons	100
1.2	Demolition of metal sheet coverings for lantern structures	tons	195
1.3	Demolition of assembled ribbed plate coverings for lantern structures	m^3/ton	220/514
1.4	Demolition of profiled sheet fencing for lantern structures	m^2/ton	2270/72
1.5	Demolition of metal sheet fencing for lantern structures	m^2/ton	1290/25
1.6	Demolition of side panels of lantern structures	m^3/ton	40/85
2. Framework structures			
2.1	Labor safety of metal truss structures (length: 30 m)	tons	510

(*continued*)

Table 2. (*continued*)

No	Structure Description	Unit	Quantity
2.2	Demolition of reinforced concrete truss structures (length: 30 m)	m^3/ton	144/336
2.3	Demolition of metal sub-purl in truss structures (length: 12–36 m)	tons	135
2.4	Demolition of metal crane beam structures (length: 6–36 m)	tons	728
2.5	Demolition of outer row reinforced concrete columns	m^3/ton	225/510
2.6	Demolition of middle row reinforced concrete columns	m^3/ton	432/993
2.7	Demolition of metal ties on reinforced concrete framework	tons	165
2.8	Demolition of metal platforms and fencing	tons	150
3. Roofing on trusses			
3.1	Demolition of metal sheet coverings on truss structures	tons	515
3.2	Demolition of assembled ribbed plate coverings on truss structures	m^3/ton	325/811
4. Enclosure structures			
4.1	Demolition of metal framework structures for glazing	m^2/ton	1512/75
4.2	Demolition of assembled reinforced concrete facade panels (length: 12 m)	m^3/ton	140/460
4.3	Demolition of assembled reinforced concrete facade panels (length: 6 m)	m^3/ton	60/155
4.4	Demolition of profiled sheet fencing for facades	m^2/ton	667/21

This technology was adopted for the dismantling of the damaged panels at this site. The work execution scheme is shown (Fig. 7).

When we have completed the dismantling of the parts of the building that were in an emergency state, it became possible to move on to the dismantling of the surviving parts of the coating, which are set aside for dismantling. The following work method can partially overcome this problem. The task was to remove the coating/metal sheets in undamaged areas and out of reach of the crane. To solve this task, the following technological scheme was adopted for performing working a specialized cart.

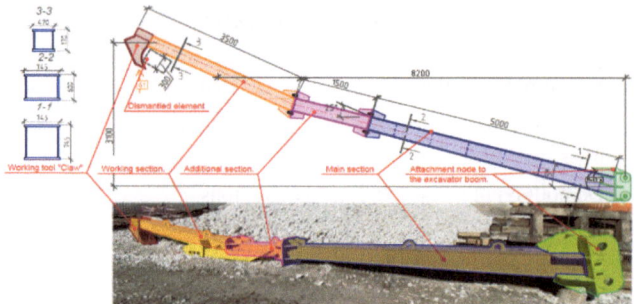

Fig. 5. Photo of the boom extension

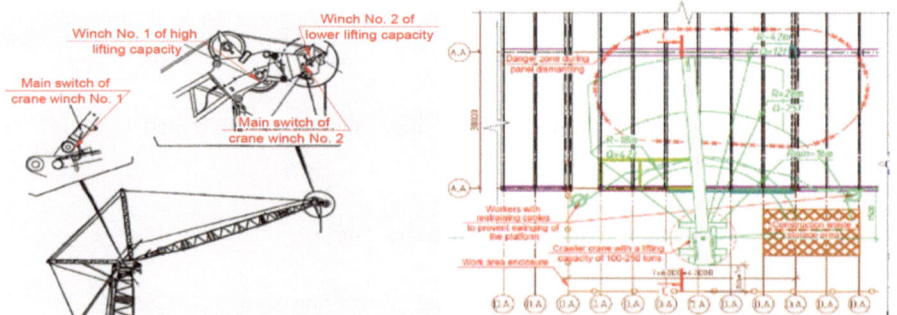

Fig. 6. Winch arrangement diagram.

Fig. 7. Scheme of dismantling the roof panels.

Preparatory work: 1) Inspect the roof, if it is in satisfactory condition and can accommodate workers; 2) Workers climb onto the roof using existing ladders. 3) Secure safety ropes to existing structures and place wooden planks on the roof (Fig. 8a). Workers must be constantly secured with a safety harness attached to the structure or the installed safety rope. 4) Attach specialized safety systems (Fig. 8b) to existing structures.

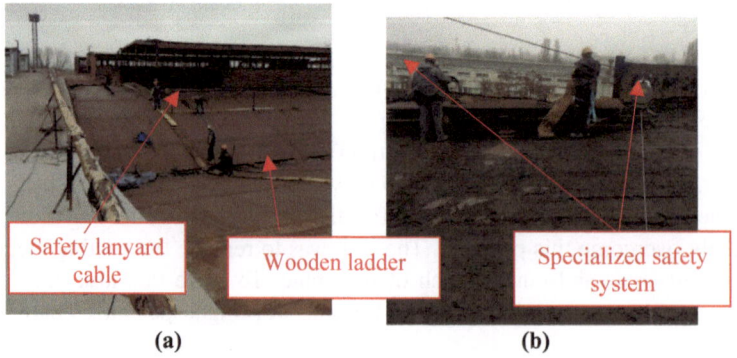

(a)   (b)

Fig. 8. **a**. Scheme for installing ladder, **b**. Schemes of specialized safety systems.

The principle of the safety system is that it allows free movement for the worker, stretches under normal movement speed, but if the speed exceeds 1 m/s (falling speed), it stops (brakes are activated) and keeps the harness taut. 5) Install guide rails for the workers (Fig. 9a). 6) Position the trolley on the guide rails in the working position (Fig. 9).

Main technological process: 1) Position the carriage above the plate/sheet to be dismantled. 2) Make holes for slinging in the reinforced concrete plate and weld attachment points on the metal sheet. 3) Sling the plate/sheet using the hooks attached to the carriage. 4) Separate the reinforced concrete plate using manual tools (break the joints and trim embedded parts). For metal sheets, detach their embedded parts. 5) Lift the detached sheet 200–300 mm from the roof using the hoists. 6) Transport the detached sheet along the rail tracks using the carriage to the crane work area. 7) Lower the plate/sheet onto the roof and release it from the slings in the crane work area. 8) Remove the plate/sheet using the crane in storage area.

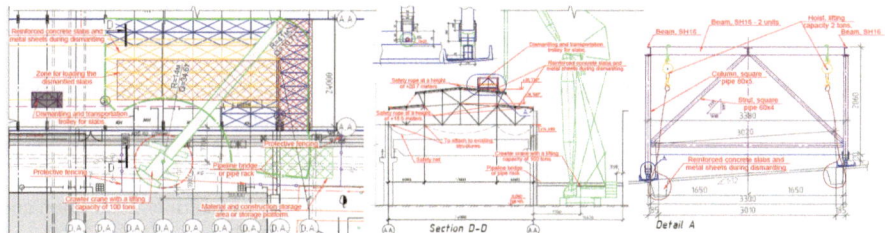

Fig. 9. Dismantling scheme of reinforced concrete panels/metal sheets using a specialized carriage.

4 Conclusions

Summing up the highlights of the project efficiency, with the dismantling of 2470 tons of metal structures, coverings, beams, and crane equipment, along with nearly 650 m^3 of prefabricated and monolithic concrete structures, and also replaced nearly 7,800 m^2 of damaged covering concrete slabs on metal sheets completed within a four-month period in complex industrial conditions. This underscores the method's effectiveness and potential for broader application.

The study provides a detailed exploration of the methodologies necessary for addressing the consequences of incidents at industrial complexes, specifically focusing on the dismantling of structures within a facility compromised by a missile strike. By introducing advanced technological strategies, the research significantly reduces the need for manual labor, maximizes the effectiveness of construction machinery, and notably improves on-site safety conditions. These technological advancements are versatile, making them suitable for a wide range of industrial settings, thus underscoring the universal applicability of the proposed solutions. The applicability of these technological schemes extends across diverse industrial environments, highlighting the broad relevance of the study. Future research will aim to develop a systematic framework and typology for the dismantling methods applicable to both industrial and civilian infrastructures, thereby contributing to the standardization and efficiency of dismantling practices globally. Moreover, the integration of digital tools and real-time data analysis is anticipated to further streamline the dismantling process, enabling more precise decision-making and enhancing operational agility in response to unforeseen challenges.

References

1. Jiang, Z., Zheng, W., Wang, Y., Li, S., Sun, L.: Green demolition technology of reinforced concrete slab: application of soundless chemical demolition agents and its evaluation system. Struct. Concr. **24**(4), 4549–4564 (2023). https://doi.org/10.1002/suco.202201162
2. O'Grady, T., Minunno, R., Chong, H.Y., Morrison, G.M.: Design for disassembly, deconstruction and resilience: a circular economy index for the built environment. Resour. Conserv. Recycl. **175**, 105847 (2021)
3. Gordon, M., Batallé, A., De Wolf, C., Sollazzo, A., Dubor, A., Wang, T.: Automating building element detection for deconstruction planning and material reuse: a case study. Autom. Constr. **146**, 104697 (2023). https://doi.org/10.1016/j.autcon.2022.104697
4. Bilokon, O.I., Nesevrya, P.I., Naumov, V.O.: Analiz osnovnykh tekhnichnykh rishen' u proiektakh znesennia budivel ta sporud [Analysis of main technical solutions in building and construction demolition projects] Ukrainian Journal of Construction and Architecture 3, 15–26 (2022). [in Ukrainian]. https://doi.org/10.30838/J.BPSACEA.2312.050722.15.860
5. Hoang, N.H., Ishigaki, T., Watari, T., Yamada, M.: Current state of building demolition and potential for selective dismantling in Vietnam. Waste Manage. **149**, 218–227 (2022)
6. Wang, Q., Jiang, T., Liu, L., Zhang, S., Kildunne, A., Miao, Z.: Building a whole process policy framework promoting construction and demolition waste utilization in China. Waste Manage. Res. **41**(4), 914–923 (2023). https://doi.org/10.1177/0734242X221126393
7. Mesa, J.A., Fúquene-Retamoso, C.: Life cycle assessment on construction and demolition waste: a systematic literature review. Sustainability **13**(14), 1–22 (2021)
8. Ilić, M., Nikolić, M.: Waste management benchmarking: a case study of Serbia. Habitat Int. **53**, 453–460 (2016). https://doi.org/10.1016/j.habitatint.2015.12.022
9. Honic, M., Kovacic, I., Aschenbrenner, P.: Material passports for the end-of-life stage of buildings: challenges and potentials. J. Clean. Prod. **319**, 128702 (2021)
10. Casprini, E., Passoni, C., Belleri, A., Marini, A.: Demolition-and-reconstruction or renovation? Towards a protocol for the assessment of the residual life of existing RC buildings. In: IOP Conference Series: Earth and Environmental Science, p. 012010 (2019)
11. Riosa, F.C., Chong, W.K., Grau, D.: Design for disassembly and deconstruction-challenges and opportunities. Procedia Eng. **118**, 1296–1304 (2015)
12. Salama, W.: Design of concrete buildings for disassembly: an explorative review. Int. J. Sustain. Built Environ. **6**(2), 617–635 (2017)
13. Vondráčková, T., Podolka, L., Voštová, V.: Handling construction waste of building demolition. In: MATEC Web of Conferences, vol. 146, pp. 03012 (2018)
14. Quéheille, E., Ventura, A., Saiyouri, N., Taillandier, F.: A life cycle assessment model of end-of-life scenarios for building deconstruction and waste management. J. Clean. Prod. **339**, 130694 (2022). https://doi.org/10.1016/j.jclepro.2022.130694
15. Ivanica, R., Risse, M., Weber-Blaschke, G., Richter, K.: Development of a life cycle inventory database and life cycle impact assessment of the building demolition stage: a case study in Germany. J. Clean. Prod. **338**, 130631 (2022)

16. Walls, R.: Demolition of steel structures: structural engineering solutions for a more sustainable construction industry. In: Bahei-El-Din, Y., Hassan, M. (eds.) Advanced Technologies for Sustainable Systems. LNNS, vol. 4, pp. 3–9. Springer, Cham (2016). https://doi.org/10.1007/978-3-319-48725-0_1

17. Janani, S.E., Renuka, S.M., Umarani, C.: Quantification of the deconstruction potential of buildings with innovative connections using BIM based DAS (deconstructability assessment score) tool. Mater. Today: Proceedings **65**, 1964–1975 (2022)

18. Kovalenko, A.V., Vovk, S.M., Plakhtii, Y.G.: Smoothing photoluminescence spectra and their derivatives for identification of individual bands. Funct. Mater. **27**(2), 424–433 (2020). https://doi.org/10.15407/fm27.02.424

19. Kovalenko, A.V., Vovk, S.M., Plakhtii, Y.G.: Removal of narrow spectral lines from experimental photoluminescence spectra of ZnS: Mn nanocrystals. J. Appl. Spectrosc. **87**(6), 995–999 (2021). https://doi.org/10.1007/s10812-021-01099-2

20. Kovalenko, A.V., Vovk, S.M., Plakhtii, Y.G.: Sum decomposition method for Gaussian functions comprising an experimental photoluminescence spectrum. J. Appl. Spectrosc. **88**(2), 357–362 (2021). https://doi.org/10.1007/s10812-021-01182-8

Review on Impacts of Geometric Imperfections on Behavior of Cold-Formed Structural Members

Quoc Anh Vu and Ngoc Hieu Pham[✉]

Faculty of Civil Engineering, Hanoi Architectural University, Hanoi, Vietnam
hieupn@hau.edu.vn

Abstract. Cold-formed structural members have been demonstrated to be highly sensitive to buckling modes due to their small thickness. Cold-formed sections have been identified to exhibit significant geometric imperfections which have been illustrated to significantly affect the stability and should be considered in structural analyses of such cold-formed structural members. Their effects were presented in a variety of previous investigations, but these obtained results were still discrete without general evaluations. Therefore, this paper will provide an overview on the impacts of various imperfection components on the behavior of cold-formed structural members. Subsequently, several previous investigations on structural behaviors of cold-rolled structural members were summarized and thoroughly analyzed to enhance understanding of the influence of each geometric imperfection component. It was found that sectional imperfection components should be considered in the analyses for short and intermediate structural members whereas global imperfection components should be included in the analysis of long structural members. Also, bucking behaviors of a structural member were governed by combinations of various imperfection components instead of any single component. These findings will be the base for future investigations of such members under the effects of geometric imperfections.

Keywords: Impacts · Geometric Imperfections · Behaviour · Cold-formed Structural Members

1 Introduction

Geometric imperfections can be observed in cold-formed structural members due to unavoidable factors in the manufacturing, transportation, and assembly processes. Considered as thin-walled structures, cold-formed structural members are highly sensitive to various forms of instability influenced significantly by geometric imperfections. Therefore, addressing geometric imperfections in the analysis of cold-formed structural components is essential. These geometric imperfections are categorized into global and sectional imperfections corresponding to different buckling modes. Global imperfections include initial twist (G_3) and flexural components (G_1 and G_2), while sectional imperfections involve deformations of flat sections, including local imperfections (d_1) and distortional imperfections (d_2) (refer to Fig. 1). The influence of geometric imperfections

© The Author(s) 2024
G. Feng (Ed.): ICCE 2023, LNCE 526, pp. 630–639, 2024.
https://doi.org/10.1007/978-981-97-4355-1_61

leads to a gradual buckling occurrence from pre-buckling, buckling to post-buckling, which makes the unclear buckling point.

These geometric imperfections are required to measure on the actual specimens for investigations. A variety of methods have been employed for this measurement, including the use of displacement gauges [1], optical observation [2], strain measurement devices [3], two-dimensional and three-dimensional laser scanning devices [4–6], and imaging devices [7]. The geometric imperfections are processed to incorporate them into structural analysis models. This processing procedure is detailed in the report by Pham et al. [8]. The results of integrating geometric imperfections into the analysis model are illustrated in Fig. 2.

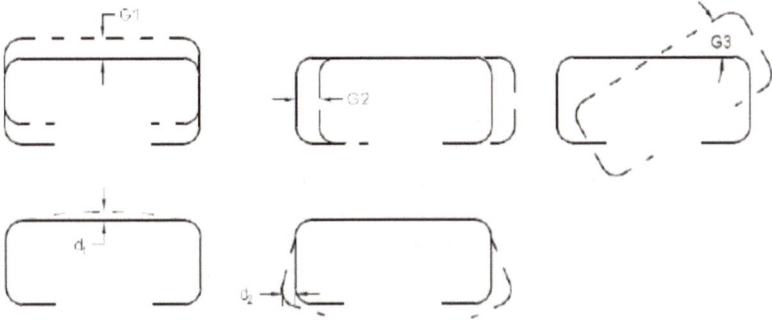

Fig. 1. The representatives of global and sectional geometric imperfections

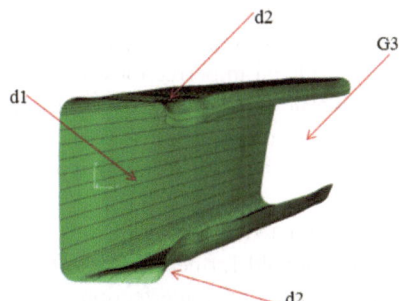

Fig. 2. Actual geometric imperfections of a specimen

Figure 2 illustrates that the geometric imperfections of cold-formed structural members are significant, impacting both their behavior and strength. Therefore, these parameters need to be carefully examined and incorporated into structural analyses in studies related to this type of structure.

The paper focuses on providing a comprehensive overview regarding the influence of geometric imperfections on the behavior of cold-formed steel or aluminum structural members. The aim is to enable readers to gain a deeper understanding of the impact of these parameters. Subsequently, the paper thoroughly presents the behavioral analysis of cold-formed aluminum columns influenced by geometric imperfections, based on the investigated results conducted by Pham [9–11]. Based on these results from the previous investigations, recommendations can be given for the consideration of geometric imperfection components in the buckling analyses of cold-formed structural members.

2 Overview on the Impact of Geometric Imperfections on the Behavior of Cold-Formed Structural Members

Numerous studies have examined geometric imperfections and their impact on the capacity and behavior of cold-formed steel structural members. Pi and Trahair's have investigated the influence of torsional geometric imperfections on cold-formed steel channel or Zed beams [5, 12–15]. These studies have highlighted the effect of the twist direction on the strength of investigated beams, whereas this twist direction was determined by the initial twist geometric imperfections (G_3). Dubina and Ungureanu's study [16] also explored the influence of the initial twist (G_3) and flexural imperfections (G_1); and it was found that these two components significantly affect the bending capacity of the examined structural members, while the sectional geometric imperfections have a negligible impact and can be disregarded. Subsequently, Dinis [17] investigated the separate impact of two sectional geometric imperfection components (d_1) and (d_2) on the post-buckling behavior of cold-formed steel channel columns. The study revealed that the imperfection component (d_2) significantly affects the capacity of the investigated sections. Schafer and Zeinoddini [18] examined the influence of geometric imperfection (d_1) on the strength of columns and provided design recommendations. Katarzyna and Andrzej [19] conducted a stability analysis of cold-formed steel sigma section columns influenced by both overall and local geometric imperfection components. The research results indicated a 20% reduction in the column capacities due to overall imperfections and only a 10% reduction due to local imperfections. Dominik et al. [20] developed a probabilistic approach to geometric imperfections to study the instability behavior of eccentrically loaded I-section columns. Andrei et al. [21] conducted sensitivity analyses to identify the most critical geometric imperfection shapes affecting the compressive strength of perforated steel columns.

Bassem and Hanna [22] conducted a study to examine how geometric imperfections affect the ultimate moment of cold-formed sigma section beams. The investigation indicated that compression flange sectional imperfections significantly influence the performance of short and medium-length beams, while longer beams are more critically affected by global imperfections Chao and Yong-Lin [23] examined the impact of imperfections (G_1) on the capacity of box-shaped columns. Random values were generated based on collected data, and these values were then input into the analytical model to obtain the limit load values. Unfavorable geometric imperfections were proposed to obtain detrimental capacities. Dinis et al. [24] studied the behavior of cold-formed steel channel columns influenced by global imperfections and sectional imperfections (d_2). The investigated results served as a basis for identifying the most unfavorable imperfections, which were subsequently used in numerical models. Additionally, studies also pointed out that the influence of geometric imperfections varies between beams and columns. Sectional imperfections have negligible effects and can be ignored in beam models [12, 16, 25, 26], whereas they should be considered in column models [27–32].

Regarding the geometric imperfections in aluminum cold-formed structures, research on the influence of geometric imperfections is still limited, as this type of structure is relatively new worldwide. Pham [33] investigated the impact of geometric imperfections on the strength of aluminum cold-formed members under compression or bending subject to global buckling. Detrimental modes of geometric imperfections were suggested for further extensive studies. Pham [9, 10] also examined the influence of various geometric imperfections on the behavior and capacity of short and intermediate lengths of aluminum cold-formed columns.

A review of research studies on geometric imperfections in cold-formed steel and aluminum members has been reported. This allows the readers to have an overview of the influence of geometric imperfections on the behaviors of cold-formed structural members. To better understand of these influences, the paper also analyses several research results from Pham [9–11], which explored the influence of various geometric imperfections on the behavior and capacity of cold-formed structural columns regarding their lengths as presented in Sect. 3.

3   Summary and Analysis of the Influence of Geometric Imperfections on the Behavior and Capacity of Cold-Formed Structural Columns

The paper summarizes studies conducted by Pham [9–11] on the impact of geometric imperfections. Geometric imperfection data were collected from the Cold-formed Aluminium Structure Project with the reference number ARC LP140100863, carried out at the University of Sydney, Australia. Based on the results of studies on the influence of geometric imperfections [27–32] as presented in Sect. 2, various sectional imperfection modes were examined for short and intermediate-length columns corresponding to local and distortional bucklings, while global imperfections were considered for long columns in the case of global buckling.

3.1   Short and Intermediate-Length Columns

The geometric imperfections considered in the study include local and distortional imperfection modes. These two types are combined to create various model shapes, as shown in Fig. 3 for short columns and Fig. 4 for intermediate columns, where L and D stand for local and distortional imperfection components respectively. The obtained results depicting the behavior of the members and average limit load values are illustrated in Figs. 5 and 6.

For short columns, the obtained results are reported as following:

- The results of models 1.2 and 1.3 are higher than the other two models, corresponding to cases where the signs of local and distortional imperfections are opposite. Models 1.1 and 1.4 provide detrimental results with both local and distortional imperfections having the same signs, exhibiting lower strength by up to 8% compared to the strength

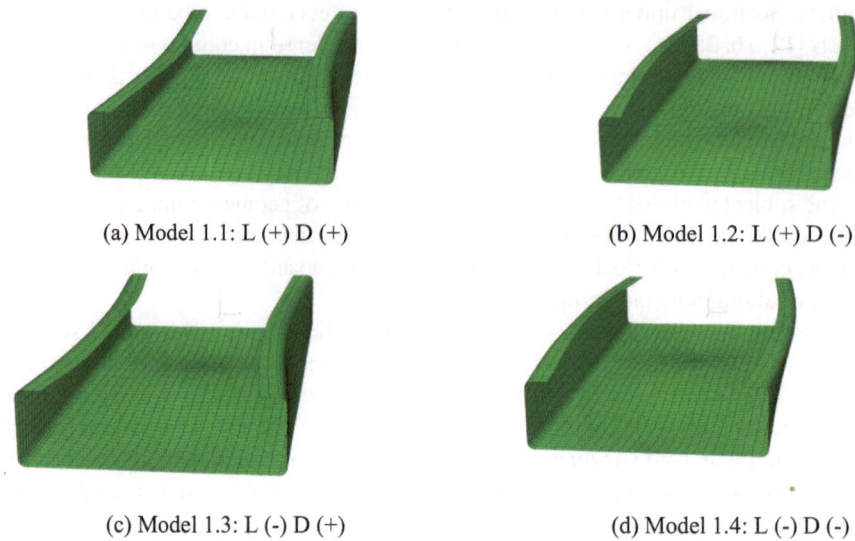

(a) Model 1.1: L (+) D (+)   (b) Model 1.2: L (+) D (-)

(c) Model 1.3: L (-) D (+)   (d) Model 1.4: L (-) D (-)

Fig. 3. Sign conventions of short column models. Note: L and D stand for local and distortional imperfection components respectively.

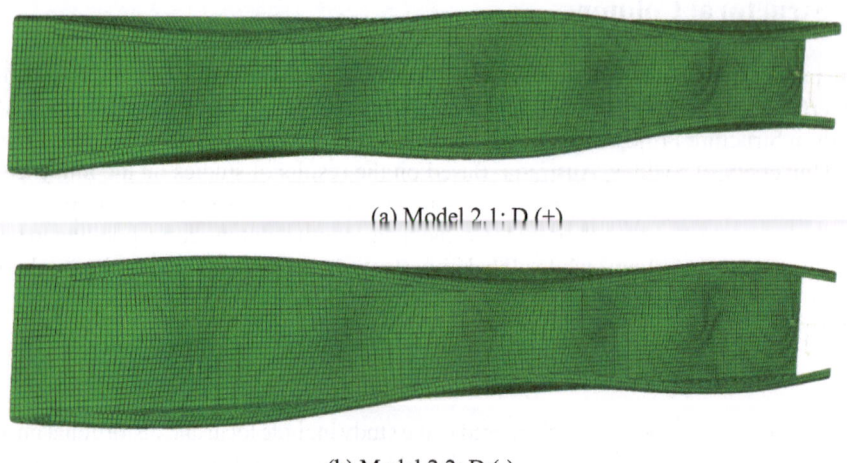

(a) Model 2.1: D (+)

(b) Model 2.2: D (-)

Fig. 4. Sign conventions of intermediate columns. Note: D stands for the distortional imperfection component.

of the former models. Therefore, the later models are considered in the development of numerical investigations to propose design recommendations.

- Fig. 5 illustrates the sectional behavior depending on the direction of the local imperfection. These models exhibit the same deformation behaviors when they have the same direction of local imperfection.

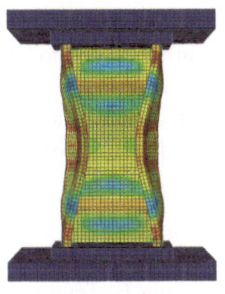

(a) Model 1.1: L (+) D (+) (Capacity: 129 kN)

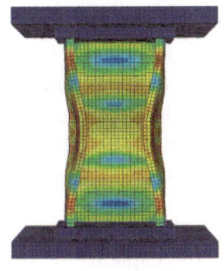

(b) Model 1.2: L (+) D (-) (Capacity: 137 kN)

(c) Model 1.3: L (-) D (+) (Capacity: 132 kN)

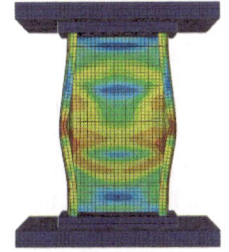

(d) Model 1.4: L (-) D (-) (Capacity: 126 kN)

Fig. 5. Behavior and average strength of short columns.

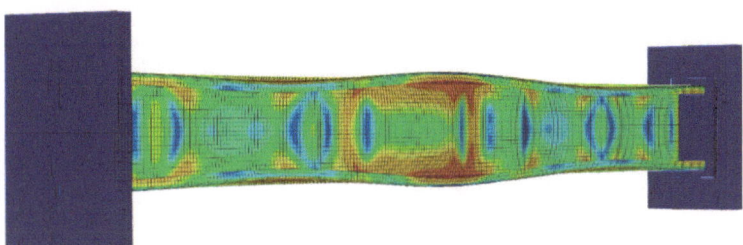

(a) Model 2.1: D (+) (Capacity: 104 kN)

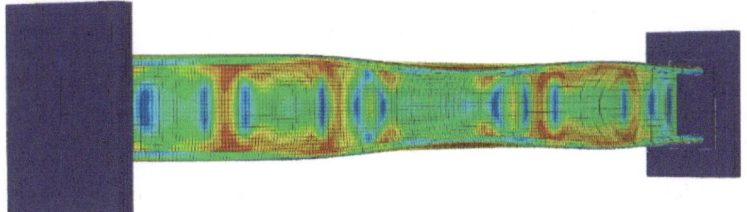

(b) Model 2.2: D (-) (Capacity: 102 kN)

Fig. 6. Behavior and average strength of intermediate columns.

– For intermediate columns, the obtained results are reported as following:
– Although the results of model 2.2 are slightly lower than those of model 2.1, the difference is not significant, being approximately 1%.
– Fig. 6 depicts the sectional behavior depending on the direction of the distortional imperfection. The model exhibits different behaviors as the directions of distortional imperfections changes.

The research results from Pham [10] also indicate that the influence of imperfection values on the capacity of short and medium-length columns is negligible and can be disregarded in the design recommendations. Further details can be found in Pham's work [10].

3.2 Long Columns

The configuration model of the long aluminum column is illustrated in Fig. 7 under boundary conditions allowing the column to rotate around the y-y axis. The geometric imperfection (G_1) attributed to this component is significant in the research model, as discussed in Pham [11]. For long columns, a nominal eccentricity value is determined as presented in Pham [11]. Due to the asymmetry of the cross-section about the y-y axis, two cases of eccentricity (E) and geometric imperfection (G_1) are defined as shown in Fig. 8.

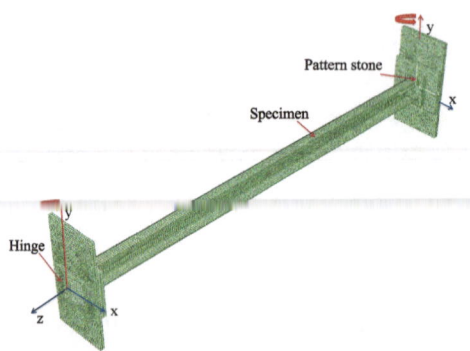

Fig. 7. Configuration model for long columns.

Based on the investigated results presented by Pham [11], several remarks are made as follows:

– The most detrimental loading condition is specified as the eccentricity E has a positive value E(+) and the flexural imperfection has a negative value G_1(-).
– The impact of imperfection should be considered in the proposed design according to the regulations of American Specification [34] with a coefficient of variation (CoV) equal to 0.2, as analysed in Pham's work [11].

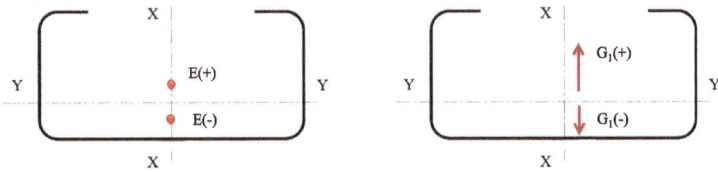

Fig. 8. Sign conventions of eccentricity and flexural imperfections.

4 Conclusions

The paper provides an overview of investigations and related studies on the influence of geometric imperfections on the behavior of cold-formed steel and aluminum structures. Also, several studies regarding the impact of various modes of geometric imperfections on the behavior and capacity of cold-formed aluminium columns are summarized and analyzed with a variety of column lengths. Remark conclusions can be given as follows:

Sectional imperfection components should be incorporated in the buckling analyses for short and intermediate structural members whereas global imperfection components can be included in the simulation models for long structural members.

A combination of geometric imperfection components should be considered in the buckling analyses to get the detrimental output instead of only using the single imperfection component.

These remarks are recommended for further investigations of cold-formed structural members with the consideration of geometric imperfections.

References

1. Mulligan, G.P.: The influence of local buckling on the structural behavior of singly-symmetric cold-formed steel columns. Cornell University Ithaca (1983)
2. Dat, D.T., Pekoz, T.P.: The strength of cold-formed steel columns. New York (1980)
3. Young, B.: The Behaviour and Design of Cold-Formed Channel Columns. The University of Sydney, Sydney (1997)
4. Becque, J.: The Interaction of Local and Overall Buckling of Cold-Formed Stainless Steel Columns. University of Sydney, Sydney (2008)
5. Niu, S.: Interaction Buckling of Cold-Formed Stainless Steel Beams. The University of Sydney, Sydney (2014)
6. Zhao, X., Schafer, B.W.: Measured geometric imperfections for Cee, Zee, and Built-up cold-formed steel members. In: Proceeding of Wei-Wen Yu International Specialty Conference on Cold-Formed Steel Structures, Baltimore, Maryland, pp. 73–78 (2016)
7. MCAnallen, L.E., Padilla-Llano, D.A., Zhao, X., Moen, C.D., Schafer, B.W., Eathertion, M.R.: Initial geometric imperfection measurement and characterization of cold-formed steel C-section structural members with 3D non-contact measurement techniques. In: The Annual Stability Conference Toronto, Toronto (2014)
8. Pham, N.H., Pham, C.H., Rasmussen, K.J.R.: Incorporation of measured geometric imperfections into finite element models for cold-rolled aluminium sections. In: Proceeding of 4th Congres International de Geotechnique-Ouvrages-Structures, Ho Chi Minh City, pp. 161–171 (2017)

9. Pham, N.H.: Influence of sectional imperfections on strength and behavior of cold-rolled aluminium alloy channel stub columns. In: Akimov, P., Vatin, N., Tusnin, A., Doroshenko, A. (eds.) FORM 2022. LNCE, vol. 282, pp. 189–200. Springer, Cham (2022). https://doi.org/10.1007/978-3-031-10853-2_18

10. Pham, N.H.: Numerical investigation of sectional buckling behaviors of cold-rolled aluminium alloy channel columns. Key Eng. Mater. **942**, 173–180 (2023)

11. Pham, N.H.: Influence of geometric imperfections on global buckling strengths of cold-rolled aluminium alloy channel columns. In: Proceedings of the 8th International Conference on Mechanical, Automotive and Materials Engineering, pp. 171–180 (2023)

12. Pi, Y.L., Put, B.M., Trahair, N.S.: Lateral buckling strengths of cold-formed channel section beams. J. Struct. Eng. **10**, 1182–1191 (1998)

13. Pi, Y.L., Put, B.M., Trahair, N.S.: Lateral buckling strengths of cold-formed Z-section beams. Thin-Walled Struct. **34**, 65–93 (1999)

14. Put, B.M., Pi, Y.L., Trahair, N.S.: Lateral buckling tests on cold-formed channel beams. J. Struct. Eng. **125**, 532–539 (1999)

15. Put, B.M., Pi, Y.L., Trahair, N.S.: Lateral buckling tests on cold-formed Z-beams. J. Struct. Div. **125**, 1277–1283 (1999)

16. Dubina, D., Ungureanu, V.: Effect of imperfections on numerical simulation of instability behaviour of cold-formed steel members.Thin-Walled Struct. **40**, 239–262 (2002)

17. Borges Dinis, P., Camotim, D., Silvestre, N.: FEM-based analysis of the local-plate/distortional mode interaction in cold-formed steel lipped channel columns. Comput. Struct. **85**, 1461–1474 (2007)

18. Schafer, B.W., Zeinoddini, V.M.: Impact of global flexural imperfections on the cold-formed steel column curve. In: The 19th International Specialty Conference on Recent Research and Developments in Cold-Formed Steel Design and Construction, pp. 81–95 (2008)

19. Rzeszut, K., Garstecki, A.: Modeling of initial geometrical imperfections in stability analysis of thin-walled structures. J. Theor. Appl. Mech. **47**, 667–684 (2009)

20. Schillinger, D., Papadopoulos, V., Bischoff, M., Papadrakakis, M.: Buckling analysis of imperfect I-section beam-columns with stochastic shell finite elements. Comput. Mech. **46**, 495–510 (2010)

21. Crisan, A., Ungureanu, V., Dubina, D.: Behaviour of cold-formed steel perforated sections in compression: Part 2 - numerical investigations and design considerations. Thin-Walled Struct. **61**, 97–105 (2012)

22. Gendy, B.L., Hanna, M.T.: Effect of geometric imperfections on the ultimate moment capacity of cold-formed sigma-shape sections. HBRC J. **13**, 163–170 (2017)

23. Dou, C., Pi, Y.L.: Effects of geometric imperfections on flexural buckling resistance of laterally braced columns. J. Struct. Eng. **142** (2016)

24. Dinis, P.B., Young, B., Camotim, D.: Local-distortional-global interaction in cold-formed steel lipped channel columns: behavior, strength and DSM design. In: Structural Stability Research Council Annual Stability Conference, pp. 654–687 (2016)

25. Pit, Y.L., Trahair, N.S.: Lateral-distortional buckling of hollow flange beams. J. Struct. Eng. **123**, 695–702 (1997)

26. Wilkinson, T., Hancock, G.J.: Predicting the rotation capacity of cold-formed RHS beams using finite element analysis. J. Constr. Steel Res. **58**, 1455–1471 (2002)

27. Kaitila, O.: Imperfection sensitivity analysis of lipped channel columns at high temperatures. J. Constr. Steel Res. **58**, 333–351 (2002)

28. Young, B., Yan, J.: Finite element analysis and design of fixed-ended plain channel columns. Finite Elem. Anal. Des. **38**, 549–566 (2002)

29. Young, B., Yan, J.: Channel columns undergoing local, distortional, and overall buckling. J. Struct. Eng. **128** (2002)

30. Demao, Y., Hancock, G.J., Rasmussen, K.J.R.: Compression tests of cold-reduced high strength steel sections. II: long columns. J. Struct. Eng. **130**, 1782–1789 (2004)
31. Young, B., Yan, J.: Numerical investigation of channel columns with complex stiffeners-part I: test verification. Thin-Walled Struct. **42**, 883–893 (2004)
32. Narayanan, S., Mahendran, M.: Ultimate capacity of innovative cold-formed steel columns. J. Constr. Steel Res. **59**, 489–508 (2003)
33. Pham, N.H.: Strength and Behaviour of Cold-rolled Aluminium Members. The University of Sydney, Sydney (2019)
34. Aluminum Association: Aluminum Design Manual. Washing DC (2015)

Redesign of a Non-electrified Urban Railway Line with Hydrogen-Fuelled Trains

Giuseppe Fabri[1], Antonio Ometto[1], Haitao Li[2], and Gino D'Ovidio[1(✉)]

[1] University of L'Aquila, L'Aquila, Italy
gino.dovidio@univaq.it
[2] Southwest Jiaotong University, Chengdu, China

Abstract. The passenger rail transportation system is of strategic importance to the decarbonization of the transportation sector. The use of green hydrogen is an environmentally sustainable option where highly polluting diesel trains currently operate on non-electrified rail lines. This paper proposes a novel adaptive power flow management strategy for urban railway trains powered by a hydrogen fuel cell stack and electrochemical batteries. The fuel cell stack is not dynamically controlled as usual, but it operates in on-off conditions to improve its overall efficiency. It always operates at the maximum efficiency operating point when it provides electrical power (on state) without following the load power variations. Furthermore, the state of the fuel cell stack depends on the state of charge of the electrochemical batteries which is the controlled quantity. As a case study, it is proposed a simulation of a suitably redesigned hydrogen-fuelled railway train operating, over an existing non-electrified line, for the L'Aquila (Italy) urban transportation service. The main components of the railway line and vehicle powertrain are designed, and the hydrogen consumption for railway operation is estimated.

Keywords: Railway Transport · Hydrogen Fuelled-Rail Train · Fuel Cell · Decarbonization

1 Introduction

Transport systems currently are responsible for a quarter of the greenhouse gas (GHG) emissions in Europe.

The European Commission has adopted a set of proposals to make EU climate, energy, and transport policies aligned with the community's purposes of reducing net greenhouse gas emissions by at least 55% by 2030, compared to 1990 levels, and then achieving climate neutrality in 2050 [1].

The European rail industry is implementing and financing research activities to achieve the sustainable performance target set up by the European Commission. The goal is to perform the widest research activities in the rail sector, to get the greatest enhancements, able to introduce operative and technological changes in the railway system, which enable it to meet the Sustainable Development Goals (SDG).

G. Feng (Ed.): ICCE 2023, LNCE 526, pp. 640–648, 2024.
https://doi.org/10.1007/978-981-97-4355-1_62

The use of alternative fuels, such as hydrogen, rather than fossil ones offers further potential in reducing emissions in railway transport [2]. However, nowadays the production and refuelling hydrogen chain has not been fully developed. To substitute diesel traction vehicles running in secondary railways there would be necessary high infrastructural investments, which now are economically justified only in the so-called "primary" lines and high speed/high-capacity service that have significant operative frequencies and flows (for passengers and goods).

Therefore, highly polluting diesel trains operate on secondary lines, which range from 30% to 70% of the extension of European national railway networks [3], due to low traffic density and high electrification costs.

Undoubtedly, the transition to electric mobility is a fundamental step towards cutting direct environmental emissions but will be not fully effective if the energy used is from fossil combustibles. At the same time, change in the rail sector requires the full use of energy carriers produced from renewable sources.

In addition, the development of innovative technologies that can reduce the motion resistance of vehicles makes it possible to increase their operational efficiency and, consequently, reduce energy consumption. [4–6].

For railway applications on non-electrified lines, one of the most environmentally friendly technology options involves the use of hydrogen as a fuel in vehicle traction. This has zero greenhouse gas emissions, can be produced from renewable energy sources, and overcomes the need for infrastructure electrification [7].

Operational examples on this topic have been realized, in the international context, both urban and extra-urban environments [8–10].

Most of the above applications use technological solutions in which the hydrogen FC is hybridized with energy storage systems (ESSs), that usually are electrochemical batteries [11]. As an alternative for electrochemical batteries' usage, such as ESS, to support FCs, some authors have investigated more environmentally sustainable technological options based on the Flywheel Energy Storage System (FESS) [12–18].

The work aims to simulate the dynamic behaviour of a hydrogen-powered railway train traveling along an existing non-electrified line, suitably redesigned for urban transportation service in L'Aquila City (Italy).

2   Vehicle Design Method

The scheme of the proposed railway train is shown in the Fig. 1. The traction motors are placed in the two end side trolleys; FCs and the hydrogen vessels are distributed on the roof of the rail train. The proposed powertrain uses electric traction motors (EM), fed by the hybrid power unit (HPU), for each rail car. The HPU consists of FC stacks operating in on-off conditions and ESS electrochemical battery-based. The traction motor and the FC are connected to the DC power bus (continuous red line) by converters, respectively (CM and CFC), to manage the power flows, as required by the master control system (CS) via a communication bus (green line).

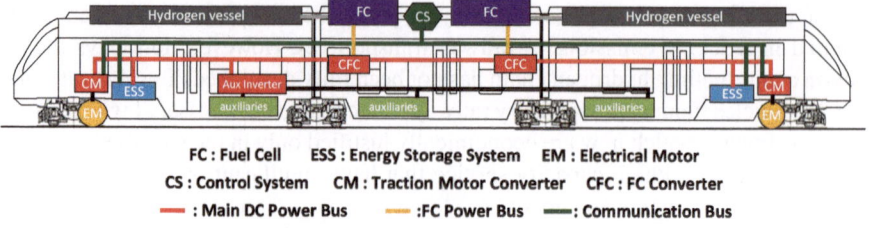

FC : Fuel Cell   ESS : Energy Storage System   EM : Electrical Motor
CS : Control System   CM : Traction Motor Converter   CFC : FC Converter
—— : Main DC Power Bus   —— :FC Power Bus   ——: Communication Bus

Fig. 1. Sketch of the vehicle system architecture

The control strategy of the HPU is based on the ESS state of charge. Due to the FC's slow dynamic response and to improve the FC's overall efficiency, the master controller imposes a constant operating point for the FC stack. In this way, the ESS handles the load variations with the purpose of:

a) providing power when the load power is higher than the FC power
b) recovering power when the FC power is higher than the load power and during the regenerative electrical braking.

2.1 Model-Based Approach

The vehicle's powertrain is described by using a model-based approach [19]. A parametric dynamic simulator has been developed, by the authors, in a MATLAB-Simulink environment. It consists of interconnected analytical-numerical sub-models and allows to simulate the travel of a given vehicle along a path and evaluate the performance of each component. The submodule interconnection is illustrated in the block diagram in Fig. 2.

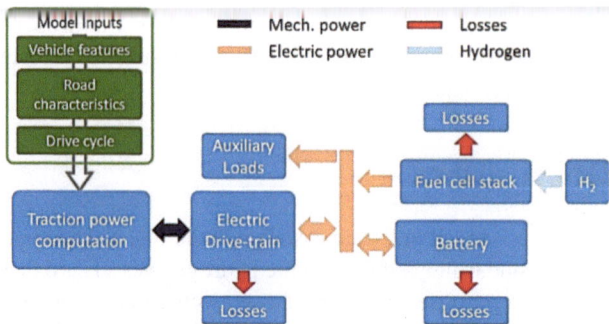

Fig. 2. Block diagram of the system simulator

The implemented software calculates the power flows and energy consumption starting from the definition of the inputs concerning the following three blocks:

- "Vehicle", addressing the vehicle (mass, dimensions, mechanics, efficiency, payload, etc.);
- "Railway path", concerning the topography characteristics;
- "Drive cycle", describing the vehicle mission's driving profile by specifying the speed vs. time;
- "Traction power" block computes the mechanical power needed to drive over the input path at the specified speed profile.

The vehicle simulation software, developed by the authors, was used for the investigations presented in this paper. The simulator solves the following vehicle equations of motion utilizing numeric integration:

$$T(v(t)) - m \cdot a \cdot \frac{dv}{dt} = \sum R(v(t)) = R_W(v(t)) \mp R_S + R_W(v(t)) + R_A(v(t)) \quad (1)$$

where R_W is the rolling resistance, R_S is the slope resistance, R_A is the air resistance, m_{ax} is the average mass vehicle per axle, m is the gross mass, v is the speed of the train, respectively, g is the acceleration of gravity, β is the angle of the track slope, C_A is the drag coefficient, S is the vehicle frontal area, n_c is the number of coaches, and α is the rotational mass inertial coefficient. Through the traction thrust $T(v(t))$ the mechanical power of the traction motors P_M can be evaluated as:

$$P_M(t) = \frac{1}{\eta_t} T(v(t)) \cdot v(t) = \frac{1}{\eta_t} \sum R(v(t)) \cdot v(t) \quad (2)$$

where η_t is the transmission efficiency. In regenerative braking $1/\eta_t$ becomes η_t. Considering regenerative electrical braking, the electrical traction power P_U is:

$$P_U(t) = \begin{cases} \frac{P_M(t)}{\eta_M} & \text{if } P_M(t) > 0 \\ \eta_M \cdot P_M(t) & \text{if } P_M(t) < 0 \end{cases} \quad (3)$$

where η_M is the electromechanical conversion efficiency of the EM and CM. The energy recovered during braking is stored according to the ESS stack's State of Charge (SOC). Moreover, the use of on-board auxiliary devices is considered, whose operation requires energy $E_{aux}(t)$ that is calculated taking in consideration the absorbed power $P_{aux}(t)$. The electric traction motor is simply modeled using its torque and power capability, i.e., the torque/speed and power/speed limit curves. The FC is assumed to be a non-linear voltage generator modeled by its voltage-current static characteristic, neglecting its dynamic behavior and any time delay. Since the FC works at constant power, the maximum output power is chosen as the operating point to minimize the rating of the whole FC stack. This choice has been possible because the efficiency value does not differ too much from the maximum one. Therefore, an FC constant efficiency (η_{fc}) is assumed in the model. The fuel consumption is calculated using relation (4):

$$m_{Fuel} = \frac{E_o}{\eta_{fc} \cdot H_i} \quad (4)$$

where E_0 is the output energy, m_{Fuel} (kg) is the mass of fuel and H_i (MJ/kg) is the fuel lower heating coefficient.

3   Case Study

To validate the proposed vehicle performance, a real urban path with four stations was identified. L'Aquila City's (Italy) railway has been selected for application study. This is an existing non-electrified single-track line where diesel trains currently operate. The railway line has been redesigned, introducing further 8 stations, to increase the line accessibility. Figure 3 shows the map of the selected railway line on which the round-trip route (47.2 km long) is simulated.

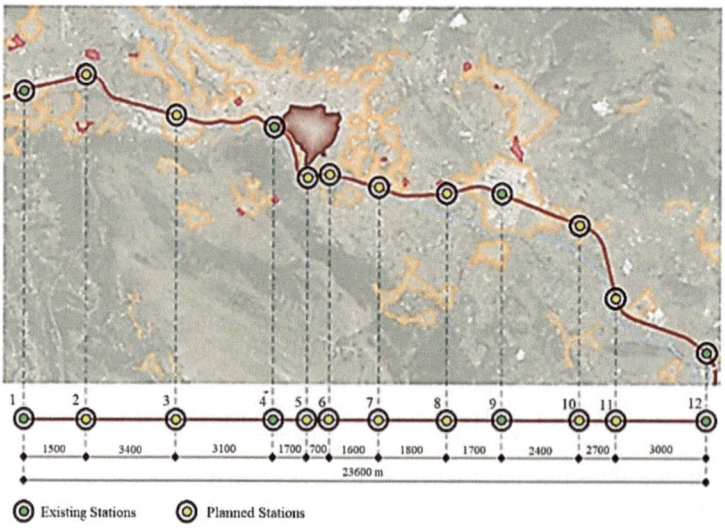

Fig. 3. Case study railway line section.

The calculated driving cycle is illustrated in Fig. 4.

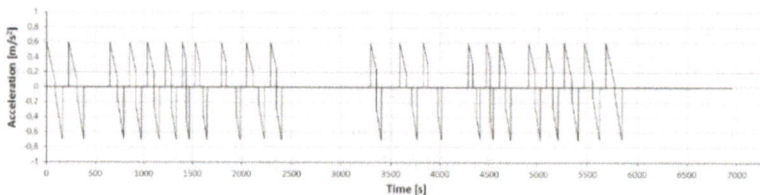

Fig. 4. Driving cycle

Vehicle mission requires an overall time of 115 min, dwell time included, to cover a 47.2 km distance. A maximum speed of 80 km/h is reached during the mission with corresponding peaks, of positive and negative acceleration, of 0.6 m/s^2. Logged peaks and grade average are, respectively, of 13‰ and 3.38‰. The cycle time is calculated by including an average dwell time per station of 60 s and railhead turn-around times of, respectively, 25 min and 15 min. The overall roundtrip time is about 115 min, including dwell times on terminals. An urban rail train topology, carrying a full load capacity of 300 passengers, was considered to run along the selected railway line shown above. The main characteristics of the baseline vehicle are listed in Table 1.

Table 1. Main rail vehicle data.

Parameters	Symbol	Unit	Value
Number of coaches per train	n_c	-	3
Carrying capacity (passengers)	n_p	-	300
Gross mass	m	t	104
Auxiliaries Power (peak)	P_{aux}	kW	114
Fuel Cells Power	P_{fc}	kW	250
Fuel Cells Efficiency	η_{fc}	-	0.55
DC/DC Efficiency	$\eta_{DC/DC}$	-	0.91
Fuel Cell Stacks	k	-	1
Vehicle Front Area	S	m2	9.8
Drag coefficient	C_A	-	0.45
Motors power for traction (total, peak)	P_M	kW	1200
ESS Energy	E_{ESS}	kWh	100
ESS Power (peak)	P_{ESS}	kW	1400
ESS Efficiency	η_{ESS}	-	0.9
Maximum efficiency of traction motor/generator	η_M	-	0.94
Transmission efficiency	η_t	-	0.92
Inertia coefficient of rotating masses	α	-	1.18
Radius of wheel	r_W	m	0.425

3.1 Simulation and Results

A simulation of the rail train based on the adaptive control logic have been carried out on the selected rail path whose features are described above.

Each single block of the whole system's power flow has been evaluated by using the developed dynamic model to simulate the vehicle mission in the application case.

The electrical power is provided by the FC and ESS, managed by the control strategy, and is based on the ESS SOC reference value. The FC behaviour, in terms of output power and of the ESS SOC over the vehicle mission, is reported in Fig. 5.

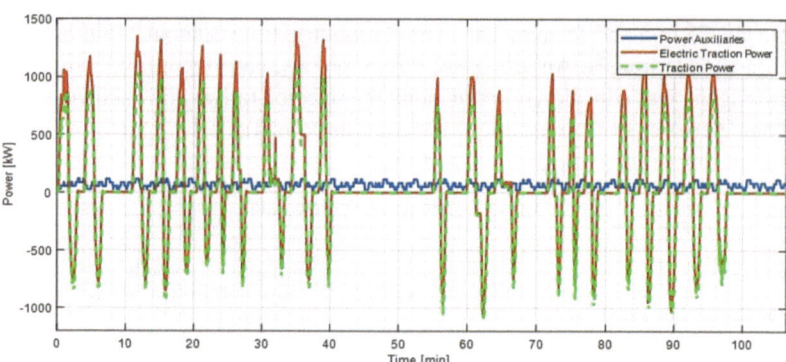

Fig. 5. Power profiles of traction, electric traction and Auxiliaries.

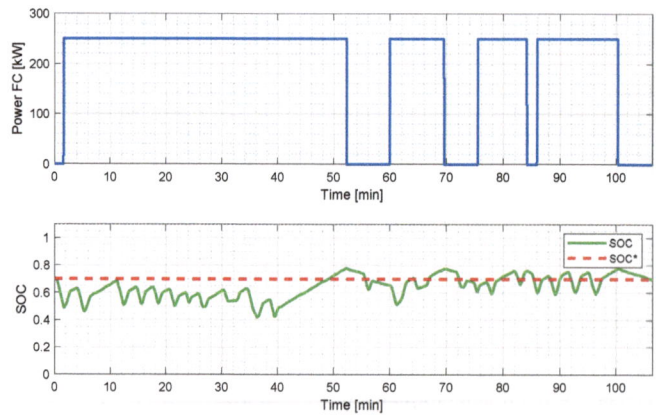

Fig. 6. FC output power (top), reference and actual FESS SOC (bottom).

Figure 6 shows the most significant energy values during the mission cycle. Given that the ESS SOC value at the end of the cycle is equal to the initial one and it can be neglected, the following energy balance considerations can be drawn. The electric recovery braking energy is almost 310 kWh, the useful energy recovered is about 170 kWh and the amount of energy generated by FC is about 350 kWh. Those results demonstrate the ESS bank's significant braking energy recovery capabilities, while the FC control appears to be well suited to respond to the energy output variations that occur during driving.

The effect of the auxiliary loads is significant in terms of energy consumption; nevertheless, their energy absorption during braking operation or driving downhill helps to keep the ESS SOC within its boundaries. The results show a railway train fuel consumption of 0.4 kg H2/km (Fig. 7).

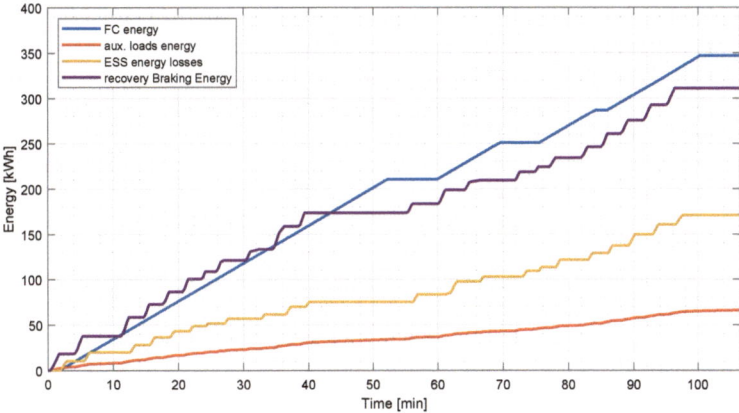

Fig. 7. Vehicle cycle: energy count of FC stack, electric drivetrain (losses included), auxiliary loads and the energy losses in the ESS.

4 Conclusion

This paper provided a novel adaptive energy flow management strategy for an urban railway electric train powered by a hydrogen fuel cell stack and electrochemical batteries.

An existing non-electrified single-track railway for the urban transport service in the L'Aquila city (Italy) has been properly redesigned and considered as case study.

The main components of the railway and the vehicle drive train were designed and the hydrogen consumption for railway operation was estimated. The results show a rail train fuel consumption of 0.4 kg H_2/km. Moreover, the results prove that the new control strategy of the fuel cell stack is suitable for urban applications. This is relevant result because the proposed control strategy increases the system efficiency while reduces the energy consumption and traction costs.

Future research will be focused on the development of a fully predictive control strategy aiming to reduce railway train fuel consumption, by knowing the actual passenger's load, and to minimize the power unit.

References

1. European Council. European Council Meeting-Conclusions, 11 December 2020; EUCO 22/20; European Council: Brussels, Belgium (2020)
2. European Council, Directive 2014/94/EU (2014 Directive 2014/94/EU of the European Parliament and of the Council of 22 October 2014 on the deployment of alternative fuels infrastructure (Latest Consolidated Release in 2021)
3. European Commission, Electrification of the Transport System: studies and reports (2017)
4. D'Ovidio, G., Crisi, F., Navarra, A., Lanzara, G.: Comparison of maglev behavior of three inductors with static and dynamic field interacting with a HTC superconductor: Test and evaluation, Physica C: Superconductivity and its Applications, pp. 15–20 (2006)
5. D'Ovidio, G., Crisi, F., Lanzara, G.: On the magnetic resistance of YBaCuO bulk superconductor dynamically interacting with perturbed flux of iron-homopolar magnetic track. J. Optoelectron. Adv. Mater. **10**(5), 1011–1016 (2008)

6. Fabri, G., Ometto, A., Villani, M., D'Ovidio, G.: A battery-free sustainable powertrain solution for hydrogen fuel cell city transit bus application, Sustainability, **14**(9), 5401 (2022)
7. Marin, G.D., Naterer, G.F., Gabriel, K.: Rail transportation by hydrogen vs. electrification - case study for Ontario, Canada, II: energy supply and distribution. Int. J. Hydrogen Energy **35**(12), 6097–6107 (2010)
8. Bloomberg, China's Hydrogen-Powered Future Starts in Trams, Not Cars. https://www.bloomberg.com. Accessed 17 Dec 17
9. Elsevier, Japanese fuel cell rail vehicle in running tests. Fuel Cells Bulletin, Dec. 2006 **12** (12), 2–3 (2006)
10. Alstom, Alstom unveils its zero-emission train Coradia iLint at InnoTrans, http://www.alstom.com, last accessed 17/12/2023
11. Xin D, Li J, Liu C.: Research on the Application and Control Strategy of Energy Storage in Rail Transportation. *World Electric Vehicle Journal*. 2023; 14(1):3. (2023)
12. Spiryagin, M., Wolfs, P., Szanto, F., Sun, Y.Q., Cole, C., Nielsen D.: Application of flywheel energy storage for heavy haul locomotives. Appl. Energy, 607–618 (2016)
13. Thelen, R.F., Herbst, J.D., Caprio, M.T.: A 2MW flywheel for hybrid locomotive power. In: IEEE Vehicular technologic conference, Orlando, Florida, USA, pp. 3231-3235 (2003). 4-9
14. D'Ovidio, G., Carpenito, A., Masciovecchio, C., Ometto, A.: Preliminary analysis on advanced technologies for hydrogen light-rail train application in sub-urban non electrified routes. Ingegneria Ferroviaria, **11**, 868–878 (2017)
15. Ciancetta, F., Ometto, A., D'Ovidio, G., Masciovecchio, C.: Modeling, analysis and implementation of an urban light-rail train hydrogen powerd. Int. Rev. Electr. Eng. (IREE), 14(4), 237–245 (2019)
16. D'Ovidio, G., Ometto, A., Valentini, O.: A novel predictive power flow control strategy for hydrogen city rail train. Int. J. Hydrogen Energy **45**(2020), 4922–4931 (2020)
17. D'Ovidio, G., et al: Hydrogen fuel cell-powered rail trains for passenger transport applications on non-electrified secondary lines. In: ICTD 2023. International Conference on Transportation and Development 2023, pp 276–290 (2023)
18. Pielecha, I., Dimitrov, R., Mihaylov, V.: Energy flow analysis based on a simulated drive of a hybrid locomotive powered by fuel cells. Rail Vehicles/Pojazdy Szynowe. **1–2**, 68–76 (2022)
19. Teng, J., Li, L., Jiang, Y., Shi, R.: A review of clean energy exploitation for railway transportation systems and its enlightenment to China. Sustainability **14**, 10740 (2022)

Expansive Soil Subgrade: Soil Treatment Using Waste Ceramic Dust and Cement

Bernard Oruabena[1]([✉]) and Okoh Elechi[2]

[1] Department of Civil Engineering, Federal Polytechnic Ekowe, Ilaro, Nigeria
boruabena2@gmail.com
[2] Department of Chemical Engineering, Federal Polytechnic Ekowe, Ilaro, Nigeria

Abstract. This study examines the potential application of leftover ceramic dust for stabilising expansive soil in infrastructure projects. We know that expansive soils have low strength and bearing capacity, which makes them troublesome in the natural world. Therefore, it's crucial to level off weak or powerless soil to increase the sublevel's bearing limit and support a suitable, long-lasting wearing course. To stabilise the soil, the poor soil was taken from Yenagoa, Bayelsa state in the South-South region of Nigeria, and mixed with varying amounts of waste ceramic dust. Standard Proctor compaction, soaked/unsoaked CBR testing, unconfined compressive strength tests, and consistency limit tests were used to evaluate the applicability of stabilised soil. The OMC appreciated at 7.5% of CD, which was utilized as an additive at 17.9 KN/m^2, according to the results. From 1.72 to 1.74 KN/m^2, MDD increased. From 9.3% at 0% to 16.77% (unsoaked) at 10% admixture (WCD + PLC) and 2.60% at 0%, according to the CBR test results, there was a rise. The highest (CBR) value was obtained when the mixture was soaked to 11.52%, as opposed to 16.77% when it was unsoaked. It was revealed that stabilisation in UCS improved with replacement ratios of 2.5%, 5%, 7.5%, and 10%, in that order. In conclusion, it was found that expanding soil stabilisation can be achieved without failure by using the ideal mix design.

Keywords: Stabilization · WCD · Expansive Soil · CBR · UCS

1 Introduction

Increasing environmental challenges and tighter budgets are facing engineers in both the public and commercial sectors. Costs associated with traditional road construction techniques are rising and building and maintaining roads is more expensive in developing countries. There are fewer soil aggregates, such as gravel, available for building road bases, and the expense of transporting these resources for road construction has increased dramatically. In locations where traditional aesthetics are desired, the use of existing soil in road construction not only lowers costs but also contributes to preserving the natural beauty of unpaved roads. A significant decrease in construction costs is maintained by using stabilisation-treated soils for the construction of road pavement. Clay minerals found in expansive soils can absorb water, increasing its volume. But during the dry season, it contracts and develops fissures that let water seep through

© The Author(s) 2024
G. Feng (Ed.): ICCE 2023, LNCE 526, pp. 649–655, 2024.
https://doi.org/10.1007/978-981-97-4355-1_63

deeply when the weather is moist [1]. Expansive soil problems lead to cracks and crumbling in the pavement, embankments, building foundations, and other structures [2–5]. Pavements deteriorate due to the properties that soils cohabit. Some researchers have examined the level of damage brought about by this alternative swell behavior of soil, and various strategies for enhancing this condition for appropriate construction have also been considered. Although the application of beach sand and waste ceramic dust to improve mangrove soil conditions suitable for pavement design has not been explored, the project's success will not only close the gap in the knowledge base by providing a new source of construction materials for road design and projects, which will lower construction costs, but it will also address waste management issues about waste ceramics. According to [6], the tile industry produces over thirty percent of its waste each day, which is disposed of, polluting the air, water, and soil. The goal of this study is to determine whether the properties of expansive soil—such as its index properties, compaction properties, MDD, UCS, soaked and unsoaked CBR, shear strength parameters, and swelling pressure—can be improved by using and waste ceramic dust for stabilizing expansive soil [7].

2 Methodologies

2.1 Materials

Expansive Soil. The soil employed in this study is called expansive soil, and it was collected 1.5 m below the surface of the land at Yenagoa, Yenagoa L G A, Bayelsa State. The BS1377 (1990) code of practice is used to determine the index and engineering parameters of Expansive soil. The soil around mangroves has a comparatively high moisture content. Therefore, to provide the appropriate moisture and improved cohesiveness to create a cementing action that functions as a waterproofing material, soil stabilising agents are used.

Waste Ceramics Dust (Tiles Waste).Ceramic tile is recognized as a crystalline, inorganic, non-metallic substance. Kermos, which translates to "potter's clay," is where the word "ceramic" originated. History demonstrates that in the beginning, ceramics were created by people. Clay-based things were either created entirely of clay or by combining it with other substances like silica, porcelain, and brick. To produce a smooth, long-lasting, and corrosion-free product, later ceramics were hardened at temperatures between 1,600 and 1,800 degrees Celsius. Clay minerals, including feldspar, that are extracted from the earth's crust make up ceramics. When tile debris is utilised to stabilise soil, it might be a significant issue to dispose of. It is crushed by hand until it passes through a 90-micron filter and is then mixed with soil. The majority of the tile wastes are made up of 59.12% silica and 1.60% CaO [8]. Additionally, waste generated during the ceramics production process is waste ceramic dust. An estimate indicates that every day, thirty percent of useless tiles are created. These wastes pollute our water, air, and land when they are disposed of in the environment.

Portland Cement. Portland limestone cement under the Dangote brand was bought from a Yenagoa building materials vendor, which meets. [9, 10]. Portland limestone cement is a hydraulic cement that forms a waterproof composite by solidifying in water.

Water.The use of potable water in construction is frequently permitted. Presumably, the water utilised in this study's studies is safe to drink and devoid of any dangerous impurities.

2.2 Methods

Physicochemical Assets of Cement, Waste Ceramic Dust, and Expansive Soil.The physicochemical characteristics of the soil from black cotton, WCD, and cement were examined. To evaluate the average particle sizes and particle absorbance of the soil, tests were conducted to determine its important constituents in conjunction with cement and WCD. The following test (BS1377–1, 1990) was conducted using the UV/VIS Spectrophotometer instrument; the results are shown in Table 1 (Tables 2 and 3).

Table 1. Physicochemical Asset of Soil and Treatment Agents

Properties	Portland cement (42.5R)	Waste Ceramic Dust	Expansive soil
SG	3.02	2.61	2.44
pH	12	7.16	11
SiO_2 (%)	18.92	51.99	62.96
Fe_2O_3(%)	3.04	6.85	3.57
Al_2O_3 (%)	6.11	10.10	17.18
CaO (%)	65.02	12.51	0.16
MgO (%)	1.33	2.10	1.05
SO_3(%)	1.93	2.74	0.76
K_2O (%)	1.12	3.39	2.09
P_2O_3(%)	0.19	-	-
TiO_2 (%)	0.29	0.20	-

Table 2. Comparison of OMC and MDD against Percentage Replacement

Mix proportion %	Soaked CBR (%)	Unsoaked CBR %	MDD (%)	OMC (gm/cc)
100	2.50	8.26	-	-
97.5	6.08	9.26	1.67	16.1
95	7.68	10.54	1.71	18.1
92.5	9.24	13.32	1.69	17.2
90	11.48	15.72	1.62	15.8

Mix Proportion. The expansive soil was combined with varying proportions of WCD (0, 2.5, 5, 7.5, and 10%), and it was then put through a series of tests, including a wet and

Table 3. Effect of WCD and PLC on Compressibility Characteristics of Expansive Soil

Stress KN/m^2 90%	100%Natural Soil + 0%WCD + 0%PLC	97,5% Natural Soil + 1.25%WCD + 1.25%PLC	95% Natural Soil + 2.5%WCD + 2.5%PLC	92.5% Natural Soil + 3.75%WCD + 3.75%PLC	Natural Soil + 5%WCD + 5%PLC
22	0.984	0.813	0.743	0.52	0.312
44	0.961	0.80	0.72	0.492	0.283
66.6	0.876	0.78	0.70	0.490	0.25

dry grain size analysis test, specific gravity, liquid limit, plastic limit, Proctor compaction, CBR, UCS, and indirect tensile strength tests. After mixing, the Atterberg limit tests were conducted both instantly and 24 h later. Furthermore, the Un-soaked conditioning test for the CBR test was conducted right after mixing, and the Soaked condition was tested three days later.

Index Properties. Every laboratory examination and method, such as [11–14], was carried out in compliance with the standard operating procedures specified in the applicable Codes. Furthermore, a comprehensive examination of the components employed was carried out and documented.

CBR Tests. To make different CBR samples, the collected coastal soil was dried, homogenized, sieved, and combined with WCD. The CBR measurements at various WCD percentages (2.5, 5, 7.5, and 10%) were obtained in both wet and dry situations.

Compaction Tests. The MDD and optimal moisture levels of the expansive marine clay were assessed using the standard heavy compaction test, which varied the amount of WCD added. OMC and MDD were evaluated for every test.

3   Result and Discussion

3.1   The Soaked/Unsoaked

The CBR test findings (Fig. 1) for the soil show that, as a result of the higher percentages of both cement and WCD, the soaked CBR value is lower than the unsoaked CBR values. At 0% replacement, CBR values drop from 9.3% (unsoaked) to 2.60% (soaked) to 10.75% (unsoaked) to 6.12%, 11.55% to 7.75%, 13.82% to 9.55%, and (soaked) 16.77% to 11.52% (unsoaked) at 2.5, 5, 7.5, and 10% replacements, in that order. The CBR value for 0% replacement (100% expansive soil) is 9.3%, as shown in the graph (Fig. 2) which compares the CBR values of wet and dry soil. Nonetheless, a steady rise was observed between 2.5 and 10% in contrast to 0% replacement.

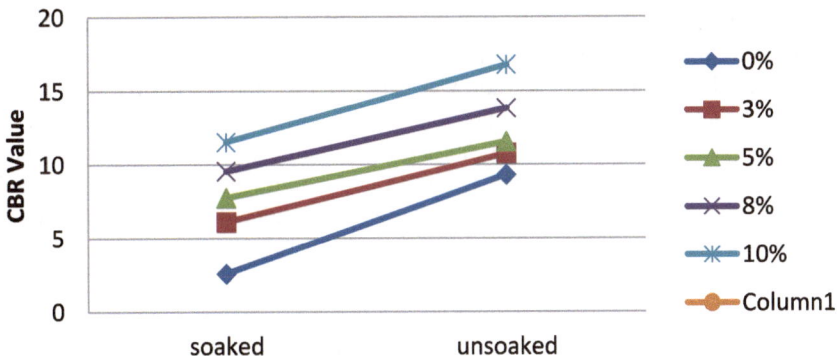

Fig. 1. CBR Test Result for WCD Stabilized Expansive Soil

3.2 Compaction Test

The OMC and MDD variations are depicted in the above Fig. The greatest OMC of 17.9 was obtained at a matching MDD of 1.62 gm/ml at 0% to 10% substitution of cement plus CD. Such behavior results from the substitution of low-specific-gravity soil particles for WCD (2.68).

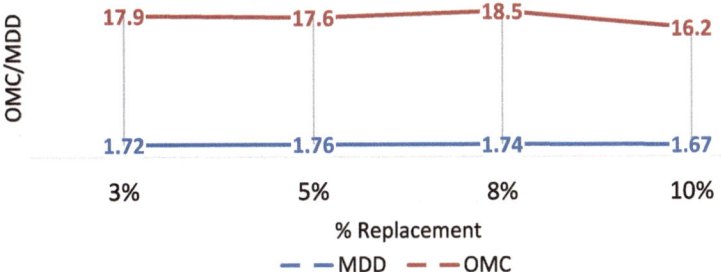

Fig. 2. OMC/MDD Curve for WCD Stabilized Expansive Soil

3.3 One-Dimensional Consolidation Test

The Oedometer test is intended to replicate the drainage conditions and one-dimensional deformation that soils encounter in real-world scenarios. Loads are transferred from the beam to the column and down to the foundations when constructions are built on the subgrade. Load impacts on the soil often reach a depth of two to three times the foundation's breadth. The forces placed on the soil at this distance cause it to become compacted. The reduction in volume of the mass caused by the compaction of the soil mass results in the settling of the structure.

By adding up the movements of individual mass components brought on by strains arising from alterations in the stress system, it is possible to rationally ascertain the movements that emerge at any given border of the soil mass. The time-dependent or virtually immediate firmness of the soil mass resulting from induced pressures can be determined by the permeability characteristics of the soil. At 22.2, 44.20, and 66.61 KN/m² vertical pressure, the void ratio was 1.187, 1.142, and 0.951, in that order. The void ratio of the stabilized marine clay sol falls for all percentages as WCD and PLC percentages rise, creating a high parking structure that strengthens the soil. Figure 3. 4.24 above depicts the impact of WCD and PLC on stabilized marine clay. The soil's response to a change in effective stress in the field is predicted using the data from the One-Dimensional Consolidation experiments.

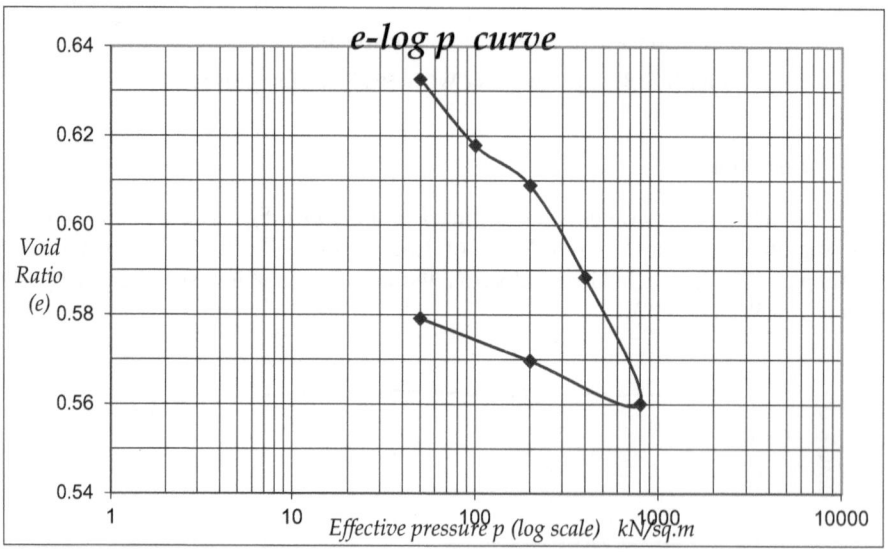

Fig. 3. Consolidation Curve

4   Conclusion

The results of the laboratory experiments demonstrate that the volume variations that arise with changes in the moisture content of the soil were lessened by adding cement and WCD as stabilizing agents to the expansive soil. The CBR is higher in the expanding soil that was treated with cement and WCD. Layer thickness and fatigue performance will therefore be impacted by the use of cement and WCD as stabilizers in the design of the flexible pavement. The results of this study also suggest that WCD may be utilized as stabilizing materials for new roads and as a potential remedy for problems with the disposal of solid waste, both of which will reduce the degradation of the environment.

References

1. Saygili, A.: Use of waste marble dust for stabilization of clayey soil. Mater. Sci. (Medziago-tyra) **21**(4), 601–606 (2015)

2. Gromko, J.: Expansive soils. J. Geotech. Eng. Div. **100**, 666–687 (1974)
3. Wayne, M., El-faith.: Construction on expansive soils in sudan. J. Constr. Eng. Manage. **110**, 359–374 (1984)
4. Mowafy, Y.B.G., Sakeb, F.: Treatment of expansive of soils. Transp. Res. Rec. (1985)
5. Kehew, E.: Geology Foe Engineering and Environmental Scientists 2nd Ed," Prentice Englewood cliff, New Jersey, pp. 295–302 (1995)
6. Binici, H.: Effect of crushed ceramic and basaltic pumice as fine aggregates on concrete mortars properties. Constr. Build. Mater. **21**, 1191–1197 (2007)
7. Akshaya.: Stabilization of expansive soil using waste ceramic dust. Electron. J. Geotech. Eng. **17**, 3915–3926 (2012)
8. Sabat.: Stabilization of expansive soil using waste ceramic dust. Electron. J. Geotech. Eng. 3915–3926 (2012)
9. C. ASTM, "Standard Practice for Ordinary Portland Cement Specification," West Conshohocken (2013)
10. Onyelowe, K.C.: Nanosized palm bunch ash as stabilization of lateritic soil for construction purposes. Int. J. Geotech. Eng. 1–9 (2017)
11. BS1377–1, "Method of Test for Soil for Civil Engineering Purposes," General Requirements and Sample Preparation (1990)
12. ASTMD6913–04, "Standard Test Methods for Particle Size Distribution (Gradation of Soils using Sieve Analysis." (2009) www.astm.org
13. ASTMD2488–09a, "Standard Practice for Description and Identification of Soils (Visual Manual Procedure)" (2015). www.astm.org
14. BS5930, "Code of Practice for Site Investigation" (2015). www.bsigroup.com
15. Onyelowe, K.: Nanostructured waste paper ash treated lateritic soil & its california bearing ratio optimization. Global J. Tech. Optim. **8**(2), 1–6 (2017)
16. Gopal, R., Rao, A.: Basic Applied Mechanics, New Delhi: New Age Int'l (2011)
17. Otoko, R., Fubara-Manuel, I., Chinweike, S., Oyebode, J.: Soft soil stabilization using palm oil fibre ash. J. Multidisc. Eng. Sci. Technol. **3**(5), 4954–4958 (2016)

The Influence of Controlled Vibration Effects on Fluid Flow in Technological and Engineering Processes

Alexey Fedyushkin[✉]

Ishlinsky Institute for Problems in Mechanics of Russian Academy of Sciences Moscow,
Moscow, Russia
fai@ipmnet.ru

Abstract. This article presents the results of studies demonstrating the influence of nonlinear effects of laminar flow under vibrational harmonic effects on fluid flow and heat transfer. The paper summarizes the results of research on the influence of vibrations in various fluid flow problems. The effect of periodic oscillations on the symmetrization of an asymmetric flow in a diffuser, on Rayleigh-Bernard convection and on the wide of boundary layers in various single crystal growth processes are shown.

Keywords: Vibrations · Fluid Flow · Symmetrization Flow · Heat Transfer · Boundary Layers · Crystal Growth · Rayleigh-Bernard Convection · Numerical Simulation

1 Introduction

During vibrational action on continuous media, their anomalous nonlinear peculiarities and resonant properties may manifest themselves [1–3]. Nonlinear peculiarities of the moving under vibration action are manifested not only in liquids, but also in the movement of bulk granular media [2]. The study of the effects of vibrations on liquid media has been carried out since the works of M. Faraday (1831) and L. Rayleigh (1883). Vibrational control of the heat exchange in the melt is more energy-efficient and simpler than controlling the melt flow by changing the gravitational or magnetic field. Therefore, the study of vibration effects on the hydrodynamics of the melt is an actual task. Reviews of works on vibrational convective flow can be found in [3–6]. Many theoretical papers [1–6] and experimental papers [7–10] have been devoted to the study of vibrations.

This paper presents and summarizes the results of mathematical modeling of the following problems: on flow symmetrization in a flat diffuser, on Rayleigh-Benard convection, and on the hydrodynamics of melt and heat and mass transfer in the processes Bridgman and Czochralski of crystal growth [10–18]. The results of numerical modeling have shown also that vibrations can reduce the thickness of dynamic and temperature boundary layers and increase the temperature gradient at the crystallization front, which can intensify heat and mass transfer and the rate of crystal growth [10–16]. The fact

G. Feng (Ed.): ICCE 2023, LNCE 526, pp. 656–664, 2024.
https://doi.org/10.1007/978-981-97-4355-1_64

of increasing the crystal growth rate up to four times under vibrational action on the crystal was discovered experimentally in [7], which is an experimental confirmation of an increase in the temperature gradient at the crystallization front. The paper [17] shows the change in the beginning time and in structure of Rayleigh-Benard convection under vertical vibrations in a long-confined layer heated from below (the Rayleigh-Benard problem with vibration of the heated wall).

The study of the problem symmetrization of asymmetric fluid flows by means of vibration action on the flow is also important in a lot off applications, for example, in mechanical engineering for fuel injection in engines, as well as in biomedicine when creating new technologies and methods for the precise targeted delivery of drugs to the necessary areas of organs in human treatment. This paper presents the results on the symmetrization of the flow in a flat diffuser (for (Jeffery-Hamel problem [19–23]) using two technique of vibration action.

2 Mathematical Model

The mathematical model is based on the numerical solution of a system of non-stationary planar 2D Navier-Stokes equations for natural convection of an incompressible liquid in the Boussinesq approximation (1–3):

$$\nabla \cdot \mathbf{u} = 0 \tag{1}$$

$$\rho_0 d\mathbf{u}/dt + \nabla p = \nabla \cdot (\rho_0 \nu \nabla \mathbf{u}) - \rho_0 g \beta (T - T_0) \mathbf{e}_z \tag{2}$$

$$\rho_0 c_V dT/dt = \nabla \cdot (k_T \nabla T) \tag{3}$$

where traditional notation is used. The problems were considered for flat cases or for conditions of axial symmetry with or without rotation. Therefore, for a cylindrical coordinate system r, θ, z, then u, v, w are radial, circumferential and axial velocity projections, ν, k_T are kinematic viscosity, heat conduction coefficients, β is the buoyancy coefficient, T_0 is a reference temperature, ρ_0 is a reference density, g – acceleration of gravity opposite directed to the vertical coordinate axis (z). The boundary conditions were as follows: for velocity - no friction on a free surface, no slip condition on solid surfaces and setting the velocity of the vibrator or moving at the vibrating wall (on law y(orz) $= A\sin(2\pi ft)$ with a frequency f and an amplitude A, $Re_{vibr} = A^2 2\pi f/\nu$ – is vibration Reynolds number); for temperature - were conditions of the first kind or thermal insulation conditions and at the crystallization interface, either the crystallization temperature or the Stefan condition with latent heat release was set.

The results presented in this paper were obtained using different numerical methods: the finite-difference scalar method, the fully implicit matrix finite-difference and the finite element methods [24, 25]. The good accuracy of numerical results was confirmed by comparison with experimental data and comparison of numerical results obtained by various numerical models.

3   Result and Discussion

3.1   Symmetrization of Fluid Flow in a Flat Diffuser by Vibrational Effect

The problem of the flow of a viscous incompressible liquid in a flat diffuser in the approximation of flow symmetry was solved by the authors of [19, 20]. It is known that when the Reynolds number increases above the critical Re* number, the flow loses symmetry, staying steady state and laminar. [21–23]. This article shows two methods of symmetrization of the asymmetric flow of a viscous incompressible liquid in a flat diffuser using periodic vibration action: 1 - from the side of the input stream, 2 - from the side of the walls of the diffuser. The research was carried out based on solving the complete two-dimensional Navier-Stokes equations for an incompressible fluid (1, 2) for case g = 0. The harmonic effects of vibration effects (in the form of $A\sin(2\pi ft)$, where A and f are the amplitude and the frequency of the changing velocity) on velocity are considered.

The Problem Statements. The laminar flow of a viscous incompressible fluid driven.

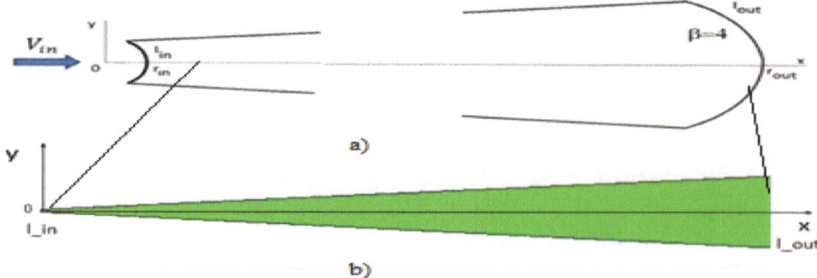

Fig. 1. Scheme of the computational domain for a flat diffuser: a) it is details of domain near the inlet and outlet of the diffuser; b) the numerical region with mesh ($\beta = 4°$,$L = 0.495$ m).

Through a channel bounded by two flat walls inclined towards each other at a small angle β is considered. In this paper we consider flat diffuser bounded by two arcs ("input" and "output" boundary) with the one center (Fig. 1a). The geometry of the mathematical model was chosen to be able to compare our results with the results of well-known works [19–22]. Geometric model of the diffuser is as follows: opening angle is$\beta = 4°$, of an arc the input boundary has the form l_{in} (r_{in}=0.005 m) where r is calculated by formula $r^2 = x^2 + y^2$ (Fig. 1). The initial conditions are$t = t_0 = 0$, $V(t_0) = 0$, $P = 0$. The velocity scale is chosen by the velocity V_{in} and the Reynolds numbers are defined as $Re = Re_{in} = V_{in}l_{in}/\nu$, $Re_{vibr} = Afl_{in}/\nu$, $y_{dimless} = y/r \sin(\beta/2)$, $V_{x\_dimless} = V_x/V_{x\_in}$, $V_{y\_dimless} = V_y/V_{x\_in}$.

The Fluid Flows in the Diffuser Without Vibrations. The results for the case asymmetric fluid flows (Re = 279) in the diffuser without vibration effects are presented in Fig. 2 [23]. The results coincide with results of paper [22].

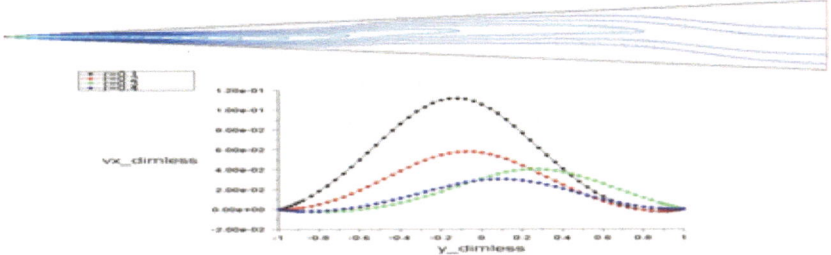

Fig. 2. The isolines and the profiles in vertical cross- sections of horizontal component V_x of velocity vector for the case of asymmetrical steady state fluid flows (Re = 279).

Symmetrization of the Fluid Flow in the Diffuser Due to the Effect of Vibration on the Inlet Velocity. The effect of a periodic vibrational disturbance $V = V_{in} + A \sin(2\pi f)$ (f = 10 Hz, A = 0.1 m/s Re_{vibr} = 2.4) on the basic flow with Re = 279 are presented in Fig. 3.

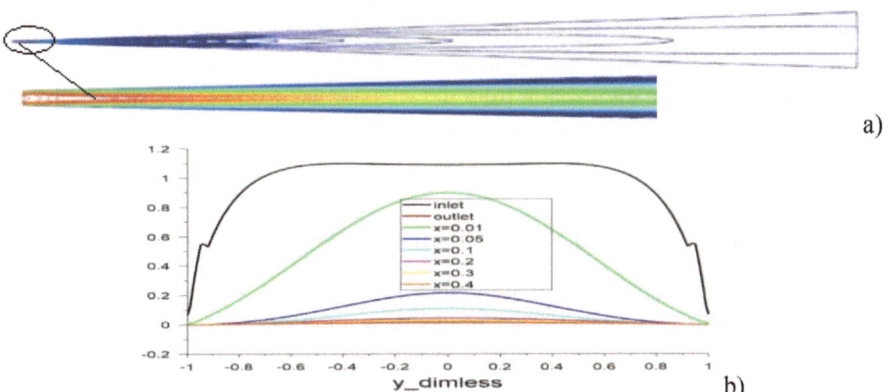

Fig. 3. The isolines of the averaged longitudinal component of the velocity meanV_x, (the isolines of the meanV_x velocity near the entrance to the diffuser are shown below) (a), the profiles of the longitudinal component of the mean_V_x velocity (b) for case V_{in}= 11.7m/s, A = 0.1m/s, f = 10Hz (Re = 279, Re_{vibr} = 2.4)

Comparison of the results in Fig. 2 and Fig. 3 shows that the effect of vibrations (Re_{vibr} = 2.4), even at amplitudes less than 1% of the velocity V_{in} (Re = 279) can lead to symmetrization of the fluid flow in the diffuser.

Symmetrization of the Fluid Flow in the Diffuser Due to the Effect of Vibration From the Walls. An example second approach of symmetrization of the fluid flow velocity in a flat diffuser by vibration action along normal to the walls of the diffuser according to the harmonic law $V_n = A \sin(2\pi f)$ with a small amplitude A and a frequency f is shown in Fig. 5. In Fig. 4 mean_V_x – is the time-average velocity profiles for Re = 279, A = 0.001 m/s, f = 10 Hz (Re_{vibr} = 0.02) are shown.

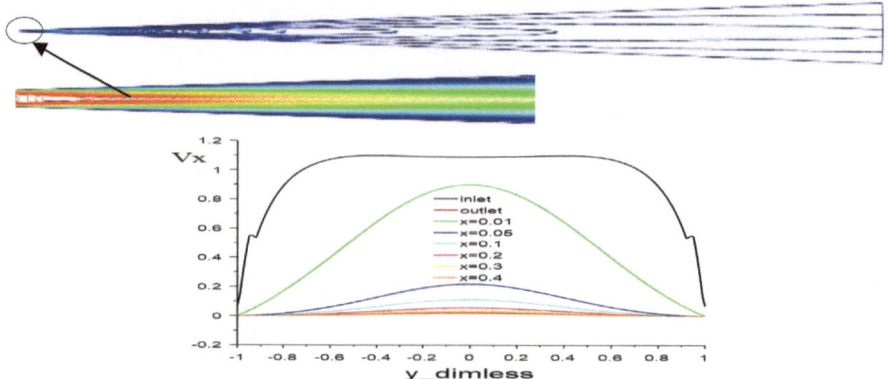

Fig. 4. The isolines and the profiles of time average velocity (mean_V_x) for fluid flow in a flat diffuser with vibration action from the walls of the diffuser for Re = 279, A = 0.001 m/s, f = 10Hz (Re$_{vibr}$ = 0.02).

The results of numerical simulation have shown two ways of symmetrization of asymmetric laminar flows of viscous incompressible fluid in a flat diffuser: the first - due to a weak periodic effect on the flow velocity at the entrance to the diffuser and the second – due to vibration action from the walls of the diffuser. It is shown that the impact of vibration, even at amplitudes less than 1% of the velocity V_{in}, can lead to the symmetrization of the fluid flow in the diffuser.

3.2   The Effect of Controlled Vibrations on Rayleigh-Benard Convection

The problem of convective flow in a horizontal layer heated from below is called the Rayleigh-Benard (R-B) problem. This problem has a threshold character of the occurrence of natural convection, which is determined by the critical Rayleigh number. R-B problem was considered for a horizontal layer with free top boundary with an aspect ratio of 1:10 and the Prandtl number Pr = 1 in a gravity field with specified temperatures on horizontal walls and with thermally insulated vertical walls. The results of numerical simulation presented in Fig. 5 show the influence of the lower horizontal wall oscillations on the structure of the convective flow in the Rayleigh-Benard problem. The number of Rayleigh-Benard rollers decreases from 10 to 9 during vertical harmonic vibrations of the lower wall (on law y = Asin(2πft) with a frequency f = 10 Hz and an amplitude A = 10^{-4} m, Re$_{vibr}$ = $A^2 2\pi f/\nu$ = 0.007), which indicates a decrease in the wave number of the periodic convective structure.

The simulation results also showed the possibility of a significant decrease in the critical Rayleigh number for the occurrence of R-B convection under vibration action. The time of occurrence and establishment of the quasi-stationary regime of convective flow is also significantly reduced. A decrease in the critical Rayleigh number and the time of occurrence of Rayleigh-Benard convection due to vertical vibrations of the lower wall was also shown in paper [17] (Fig. 6). This is important for boiling processes [18].

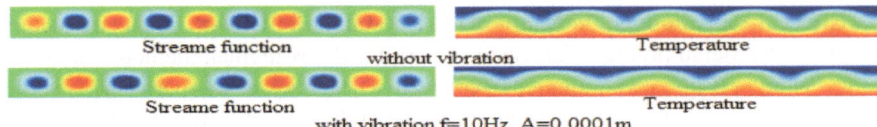

Fig. 5. Pictures of isolines of the stream function and isotherms with and without vibrations of the lower wall with $Re_{vibr} = 0.007$, $Ra = 4 \cdot 10^3$, $Pr = 1$.

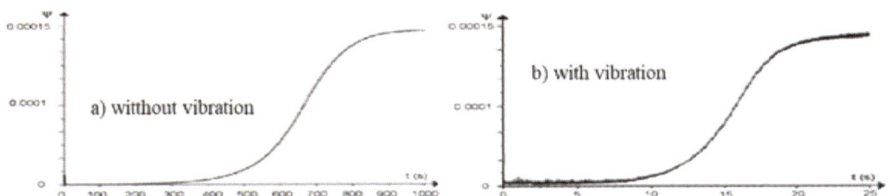

Fig. 6. The dependences of the maximum values of the stream function on time ($Ra = 4 \cdot 10^3$, $Pr = 1$): a) – without vibrations; b) – with vertical vibrations of the bottom wall with $Re_{vibr} = 0.007$[17].

3.3   The Effect of Vibrations in Crystal Growth Processes

Bridgman model. The calculation results were carried out for the following geometric configurations of crucibles for Bridgman method with submersible vibrators for a fixed flat and variable calculated shape of the crystallization front shown in Fig. 7. The area under consideration has the following dimensions: $R = 1.6 \cdot 10^{-2}$; $H = 3.2 \cdot 10^{-2}$; $r_{vibr} = 4 \cdot 10^{-3}$; $h_1 = 8 \cdot 10^{-3}$; $h_2 = 8 \cdot 10^{-3}$; $\delta = 10^{-3}$ (m) where R is the radius of the ampoule, H is the height of the ampoule, h_1 is the distance between the vibrator and the solid-liquid interface, h_2 is the thickness of the vibrator (the distance between its lower and upper surfaces), the gap δ (the distance between the vibrator and the side wall of the crucible. The following variants with size values $A = 5 \cdot 10^{-4}$ and 10^{-4} m, f = 0–100 (Hz) are calculated. The effect of vibrations on temperature boundary layers are shown in Fig. 8.

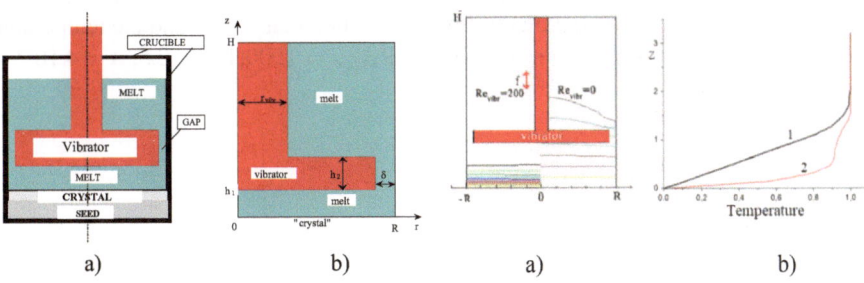

Fig. 7. The geometrical schemes for Bridgman crystal growth model with submerged vibrator, a) for Stefan problem, b) model with fixed flat shape of the melt-crystal interface

Fig. 8. a) - Isotherms in the melt ($Pr = 5.43$) (on the right – without vibrations, on the left – with vibrations $Re_{vibr} = 200$), b) - temperature profiles (r = 0.75) (line 1 – without vibrations, 2 – with vibrations)

The Effect of Vibrations on the Shape of the Crystallization Front. Using the method of solving the Stefan problem described in [25], for the Bridgman method with a sub-merged vibrator (Fig. 9), a simulation of convective heat transfer was performed in order to determine the effect of vibrations on the shape of the crystallization front.

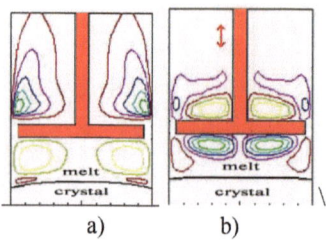 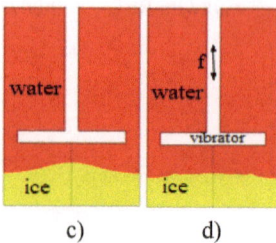

a)   b)   c)   d)

Fig. 9. The effect of vibrations on the shape of the front crystallization: a), b) – stream function in the melt of NaNO3, (a) without vibrations - f = 0; b) - with vibrations A = 10^{-4} m, f = 50 Hz), c), d) – water-ice interface, c) – without vibrations, f = 0, d) – with vibrations, A = 10^{-4} m, f = 30 Hz)

Czochralski Model with Submerged Vibrator. The scheme of the computational domain is shown in Fig. 10. The computational domain is a square with sides L = H = 3 cm crystal with a diameter of d = 1cm and immersed into the melt to a depth of 1mm, the vibrator has a diameter of 0.8 cm and thickness 1mm. It is assumed that the immersed vibrator is located under the crystal at a distance h. Irregular grid with refinement near the solid walls and the corners of the vibrator and the crystal were used in the calculations. The vibrator makes translational oscillatory movements along the vertical axis of the crystal according to the law: y = y_0+Asin(2πft), with frequency f and small amplitude A = 10^{-4} m, y_0 is initial location of vibrator.

The isotherm and structure of the averaged flow is presented in Fig. 11(a, b). It is showing how the vibrating immersed activator leads to the mixing of the entire volume of the melt. In Fig. 11c presents temperature profiles on vertical cross section (on axis) that show the effect of vibration on the temperature boundary layer and the temperature gradient near the crystal-melt interface (Pr = 7; Re_{vibr} = 1500; h/d = 0.5, A = 4 10^{-4} m, f = 20 Hz).

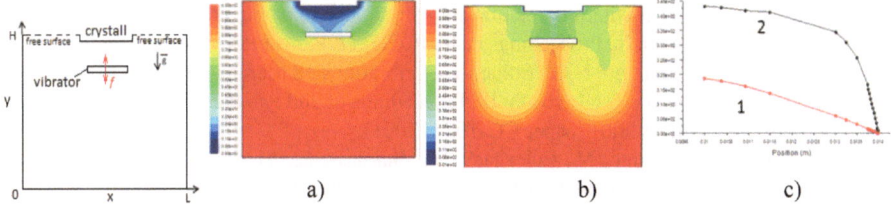

Fig. 10. Scheme of the computational domain

Fig. 11. Isotherms temperature: a) – without vibration, b) with vibrations, c) temperature profiles on the axis section: curve 1- is without vibration, curve 2 - is with vibration. (Pr = 7, $\mathrm{Re}_{vibr} = 1500$, Ra = 0).

4   Conclusion

It is possible to symmetrize the flow of viscous liquid in a flat diffuser using the effects of weak harmonic vibration from the inlet side or from the walls of the diffuser. It is also shown that the vibration effect can change the structure and time of occurrence of Rayleigh –Benard convection. This is important for boiling processes. By controlling the vibration effect on the convective fluid flow, the thickness of the boundary layers can be reduced. For the Bridgman model, it is shown that the surface of the crystallization front can be made flatter by means of vibration action. This is of fundamental importance in crystal growth and for controlling temperature gradients to control the kinetics and rate of crystal growth through vibration.

Acknowledgements. This work was supported by the Russian Science Foundation grant 24–29-00101.

References

1. Chelomei V.N. Mechanical paradoxes caused by vibrations. Dokl. Akad. Nauk SSSR, vol. 270, no.1, pp.62–67 (1983). (In Russian)
2. Blekhman, I.I., Blechman, L.I., Weisberg L.A., et al.: "Abnormal" phenomena in a liquid under the action of vibration. Dokl. Akad. Nauk SSSR, vol. 422, no. 4, pp. 470–474 (2008). (In Russian)
3. Ganiev, R.F., Ukrainsky, Y.E.: Dynamics of particles under the influence of vibration Kiev: Naukova dumka, p. 168 (1975). (In Russian)
4. Capper, P, Zharikov, E.: Oscillatory-Driven Fluid Flow Control during Crystal Growth from the Melt in Handbook of Crystal Growth 2 Edn. (vol. 2, pp. 950–993). Elsevier Inc (2015)
5. Gershuni, Z., Lubimov, D.V.: Termal Vibrational Convection, p. 357. Willey, Hoboken (1998)
6. Gershuni G.Z., Zhukhovitsky E.M., Nepomnyashchy A. A. Stability of convective flows. Moscow, Nauka publ., 1989, 320c. (In Russian)
7. Vitovsky, B.V.: Increasing the crystal growth rate by applying sound frequency vibrations to it. In: Proceedings of the ICANN of the USSR. 11. 1 (1955). (In Russian)
8. Kozlov, V.G.: Vibrational thermal convection in an enclosure performing high frequency rotational swingings. Fluid Dyn. (USSR) **3**, 138 (1988)
9. Zharikov, E.V., Prihod'ko, L.V., Storozhev, N.R.: Vibrational convection during the growth of crystals. Growth of Crystals. 19, 71–81 (1993)

10. Fedyushkin, A.: The gravitation, rotation and vibration - controlling factors of the convection and heat- mass transfer. In: Proceedings of 4th ICCHMT, Paris, FRANCE, pp.948–951 (2005)
11. Fedyushkin, A.I., Bourago, N.G.: Influence of vibrations on Marangoni convection and melt mixing in Czochralski crystal growth. In: Proceedings of 2nd Pan Pacific Basin Workshop on Microgravity Sciences 2001, paper CG-1072 (2001)
12. Fedyushkin, A., Bourago, N., Polezhaev, V., Zharikov, E.: The influence of vibration on hydrodynamics and heat-mass transfer during crystal growth. J. Crystal Growth **275**, e1557–e1563 (2005)
13. Bourago, N.G., Fedyushkin, A.I., Polezhaev, V.I.: Dopant distribution in crystals growth by the submerged heater method under steady and oscillatory rotation. Adv. Space Res. **24**(10), 1245–1250 (1999)
14. Fedyushkin, A.I., Burago, N.G., Puntus, A.A.: Convective heat and mass transfer modeling under crystal growth by vertical Bridgman method. J. Phys. Conf. Ser. **1479**, 012029 (2020)
15. Fedyushkin, A.I., Burago, N.G., Puntus, A.A.: Effect of rotation on impurity distribution in crystal growth by Bridgman method. J. Phys. Conf. Ser., Inst. of Phys. (UK), **1359**, 012045 (2019)
16. Fedyushkin, A.I.: Heat and mass transfer during crystal growing by the Czochralski method with a submerged vibrator. J. Phys. Conf. Ser., Inst. Phys. (UK), **1359**, 012054 (2019)
17. Fedyushkin, A.I.: The effect of controlled vibrations on Rayleigh-Benard convection. J. Phys. Conf. Ser. **2057**(1), 012012 (2021). https://doi.org/10.1088/1742-6596/2057/1/012012
18. Fedyushkin, A.I.: Numerical simulation of gas-liquid flows and boiling under effect of vibrations and gravity. J. Phys. Conf. Ser. **1479** (2020)
19. Jeffery, G.B.: The two-dimensional steady motion of a viscous fluid. Phil. Mag. Ser6 **29**(172), 455–465 (1915)
20. Hamel G Spiralformige Bewegu ngen zaher Flussigkeiten. Jahres her. Deutsch Math Ver Bd 25: 34–60. (1917)
21. Pukhnachev, VV.: Simmetrii v uravnenijah Nav'e–Stoksa. Uspehi Mekh. N 6: 3–76 (2006)
22. Akulenko, L.D., Georgievskii, D.V., Kumakshev, S.A.: Solutions of the Jeffery-Hamel problem regularly extendable in the Reynolds number. Fluid Dyn. **39**(1), 12–28 (2004)
23. Fedyushkin, A.I., Puntus, A.A., Volkov, E.V.: Symmetry of the flows of Newtonian and non-Newtonian fluids in the diverging and converging plane channels. In: AIP Conference Proceedings, Vol.2181, no. 1, pp. 020016-1–020016-8 (2019)
24. Polezhaev, V.I., Bello, M.S., Verezub, N.A., et al.: Convective processes in weightlessness, p. 240 (1991). (in Russian)
25. Burago N.G., Fedyushkin A.I.: Numerical solution of the Stefan problem. J. Phys. Conf. Ser. 1809(1), 012002 (2021). https://doi.org/10.1088/1742-6596/1809/1/012002

Analysis of Hugging Flow Through the Powerful Technique of Homotopy Asymptotic Method (HAM)

Qayyum Shah[✉]

Department of Basic Sciences, University of Engineering and Technology Peshawar, Peshawar 25000, KP, Pakistan
qayyumshah@uetpeshawar.edu.pk

Abstract. A hugging flow occurs when a fluid is hugged between two undistinguishable plates fronting one another. To this end, the hugging flow between two disks in drive of unstable nature is of immense importance for its technical and scientific uses such as, molding of fluids, the study of fluid machinery such as pumps, fans, blowers, windmills, air compressors, heat exchangers, jet & rocket engines, gas turbines, power plants, pollution control equipments, air-conditioning equipments, heating & ventilation systems, breathing aids, heat-lungs machines, among others makes fluid mechanics of massive importance to Mechanical Engineers. Similarly, expansion and contraction in blood flow, piston motion, brakes, and in cooling towers among others. The aim of this research is to investigate Newtonian fluid between two porous time-varying plates in hugging flow. The impact variable, magnetic field is taken into account. The treatment of obtained system of equations is done by HAM (Homotopy Asymptotic Method). After a comparative analysis between The results obtained through HAM and Numerical methods showed a great agreement of harmony. Variation in the flow fields is presented with the help of figures. The residual errors for fluid flow fields are calculated and shown with the help of table. All the computational work has been done with the help of computer software Mathematica Software.

Keywords: Non-Newtonian Fluid · Two Plates · HAM Solution

1 Introduction

Non-Newtonian fluid flow has established the researchers' interest in modern age as they are taking interest due to its excessive applications like pollution control equipments, air-conditioning equipments, heating & ventilation systems, breathing aids, heat-lungs machines and other nano-fluids are examples. Numerous scientists and engineers have gone through and scrutinized this concept from innumerable rheological viewpoints. A fluid is a non-Newtonian liquid with exceptional features. As for example Mohmand et al. [24–27] have thrown light on the graphical solution of fluid flow. Geniuses have been successfully applied to a number of nonlinear problems arising in the science and engineering by various researchers [3–5, 10–16, 18, 20, 33–35]. Shah et al. [17–19] have

G. Feng (Ed.): ICCE 2023, LNCE 526, pp. 665–673, 2024.
https://doi.org/10.1007/978-981-97-4355-1_65

investigated the behavior through graphical representation. Khan et al. [1, 2] discussed flow between rotating stretchable disks. What's more, Khan et al. [1] found that when both the discs rotate in the same sense then the fluid in the disks rotates with an angular velocity. Khan et al. [3] discussed Dynamics with Cattaneo–Christov heat and mass flux theory of bioconvection Oldroyd-B nanofluid. Notwithstanding, Khan et al. [1–3] further explored that the their study provides the best solutions and it has been proved that its solution is close to exact solution. Khan et al. [4] discussed Rotating flow assessment of magnetized mixture fluid suspended with hybrid nanoparticles and chemical reactions of species. Rasheed et al. [5] discussed Computational analysis of hydromagnetic boundary layer stagnation point flow of nano liquid by a stretched heated surface with convective conditions and radiation effect. Mohmand et al. [24–27] scrutinized oscillating and porous, and flow with heat transfer effect as well as vibratory flow. Usman et al. [6] discussed Computational optimization for the deposition of bioconvection thin Oldroyd-B nanofluid with entropy generation. Khan et al. [7] explored Lorentz forces effects on the interactions of nanoparticles in emerging mechanisms with innovative approach. Khan et al. [8] analyzed Solution of magnetohydrodynamic flow and heat transfer of radiative viscoelastic fluid with temperature dependent viscosity in wire coating analysis. Khan et al. [9] investigated A Framework for the Magnetic Dipole Effect on the Thixotropic Nanofluid Flow Past a Continuous Curved Stretched Surface. Khan et al. [10] studied Analytical solution of UCM viscoelastic liquid with slip condition and heat flux over stretching sheet: Galerkin Approach. Shah et al. [17–19] have explored the transient flow, with unsteady stretching surface and accompanied by Soret and Dufour effects. Khan et al. [11] discussed Mechanical aspects of Maxwell nanofluid in dynamic system with irreversible analysis. Khan et al. [12] studied Numerical simulation of double-layer optical fiber coating using Oldroyd 8-constant fluid as a coating material. Khan et al. [8, 10, 12] have also discussed heat and heat transfer. Shah et al. [13, 18] analyzed Gravity Driven Flow of an Unsteady Second Order Fluid as well as Heat transfer rate of the fluid at the belt is also calculated. Khan et al. [14] discussed Investigation of wire coating using hydromagnetic third-grade liquid for coating along with Hall current and porous medium. Khan et al. [15] studiedAnalytical Solution of the MHD Viscous Flow over a Stretching Sheet by Multistep Optimal Homotopy Asymptotic Method. Fiza et al. [16] explored Modifications of the multistep optimal homotopy asymptotic method to some nonlinear KdV-equations. Shah et al. [17]discussed. The ADM solution of MHD non-Newtonian fluid. Khan et al. [1–4] have discussed solution through tables. Shah et al. [18] studied the Heat transfer and hydromagnetic effects on the unsteady thin film flow of Oldroyd-B fluid over an oscillating moving vertical plate. Shah et al. [19] explored for Soret and Dufour effect on the thin film flow over an unsteady stretching surface. Khan et al. [14, 20, 21] discussed for Mechanical aspects of Maxwell nanofluid in dynamic system with irreversible analysis as well as the impact of emerging parameters involved in the solutions are discussed through graphs on the velocity and temperature profiles in detail. Khan et al. [14] researched on the investigation of Wire Coating using Hydromagnetic Third-Grade Liquid for Coating along with Hall Current and Porous Medium. Khan et al. [21] discussed the Analytical Solution of UCM Viscoelastic Liquid with Slip Condition and Heat Flux over Stretching Sheet: The Galerkin Approach. Mohmand et al. [22] discussed the Engineering Investigations of Dufour and Soret effect on MHD heat and mass transfer with radiative heat flux in a liquid over a rotating dick. Mohmand et al. [23] explored the Engineering applications and analysis of vibratory motion fourth order fluid film over the time dependent heated flat plate. Mohmand et al. [24] analyzed for Time dependent Oldroyd-B liquid film flow over an oscillating and porous vertical plate with the effect of thermal radiation. Mohmand et al. [25] studied Time dependent second grade fluid between two vertical oscillating plates with heat transfer effect. Mohmand et al. [26] investigated the Vibratory motion of fourth order fluid film over a unsteady heated flat. Mohmand et al. [27] discussed Engineering applications and analysis of vibratory motion fourth order fluid film over the time dependent heated flat plate. Mohmand et al. [28] explored for Heat transfer and hydromagnetic effects on the unsteady thin film flow of Oldroyd-B fluid over an oscillating moving vertical plate. Shah et al. [29] discussed Soret and dufour effect on the thin film flow over an unsteady stretching surface. Likewise Khan et al. [20] have also discussed about the Brownian motion and thermophoresis with thermal radiation and buoyancy effects are encountered in the governing equations. The oscillating parallel plates were discussed by Shah et al.[13]. Shah et al. [17–19] have also explored some more properties of fluids. Similarly, Fiza et al. and Khan et al. [14, 16] have discussed flow through numerical results (Runge-Kutta method). Rasheed et al. [30] and Shah et al. [29] have discussed fluid flow through shooting technique and numerical approach. Moreover, flow of Oldroyd-B fluid over an oscillating was discussed by Shah et al. [28]. Likewise Mohmand et al.[27] have discussed time-dependent heated plate. Moreover, Rasheed et al. [31, 32] have discussed the fluid flow. To this end, Khan et al. [2] have investigated pressure distribution and entropy generation rate and then their solution through HAM approach. Moreover, Khan et al. [3–5] Darcy–Forchheimer law is used to study heat and mass transfer flow and microorganisms motion in porous media as well as flow of Maxwell nanofluid induced by two parallel rotating disks were analyzed. Furthermore Rasheed et al.[30–32] have discussed fluid motion with thermal analysis. Shah et al. [33] have also discussed MHD flow. Khan et al. [1–4] have analyzed the fluid motion through graphs. Similarly, Shah et al. [17–19] have also investigated the MHD fluid motion through graphs. Likewise, Khan et al. [7–12] have also discussed fluid flow through graphs and found excellent harmony with the already published works. Shah et al. [33, 34] have scrutinized a mathematical and computational analysis of MHD fluid with heat source effect and the chemically reactive Casson fluid to add knowledge to the existing one. OHAM technique was also applied by Shah [35] and as concern of PDE or ODE was discussed by Khan et al. [36].

The aim of the author in this study is to examine the non-Newtonian fluid model, which includes two dimensional fluid flow of a viscous, incompressible, and electrically conducting fluid in two parallel plates, and to represent the obtained results in graphical as well as in a tabulated form.

2 Problem Formulation and Solution

Consider the two dimensional fluid flow of a viscous, incompressible, and electrically conducting fluid in two parallel plates. The temperature of the lower and upper plates are T0 and T1. The lower plate is hugging in the direction of lower plate with velocity $(\frac{dh(t)}{dt})$. The two-dimensional coordinate system is taken as that the x-axis is along the lower plate while y-axis is normal to it. The velocity is represented by $\vec{u} = \vec{u}(u(x, y, t))$, temperature is shown by $T = T(x, y, t)$, and magnetic field is considered by $T = T(x, y, t)$. The magnetic field is considered in both directions. In view of these assumptions, the leading equations are in Fig. 1:

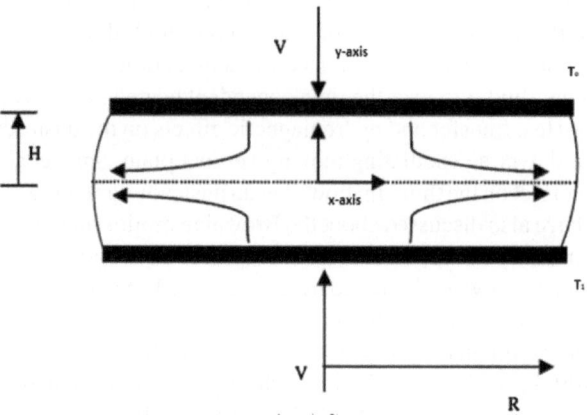

Fig. 1. Geometry

$$\frac{\partial u}{\partial y} + \frac{\partial u}{\partial x} = 0, \tag{1}$$

$$u\frac{\partial u}{\partial x} + v\frac{\partial u}{\partial y} + \frac{\partial u}{\partial t} = -\frac{1}{\rho}\frac{\partial P}{\partial x}v\left(\frac{\partial^2 u}{\partial y^2} + \frac{\partial^2 u}{\partial x^2}\right) - \frac{\sigma B_0}{\rho}(uB_0 + vb), \tag{2}$$

$$v\frac{\partial u}{\partial y} + u\frac{\partial u}{\partial x} + \frac{\partial u}{\partial t} = -\frac{1}{\rho}\frac{\partial P}{\partial x}v\left(\frac{\partial^2 u}{\partial y^2} + \frac{\partial^2 u}{\partial x^2}\right) - \frac{\sigma B_0}{\rho}(uB_0 - vb), \tag{3}$$

$$v\frac{\partial T}{\partial y} + u\frac{\partial T}{\partial x} + \frac{\partial T}{\partial t} = \sigma\left(\frac{\partial^2 T}{\partial y^2} + \frac{\partial^2 T}{\partial x^2}\right), \tag{4}$$

$$\frac{\partial b}{\partial t} - \delta B_0\frac{\partial u}{\partial x} - \delta u\frac{\partial B_0}{\partial y} + \delta b\frac{\partial v}{\partial y} + \delta v\frac{\partial b}{\partial y} - \frac{1}{\delta\mu_c}\frac{\partial^2 b}{\partial y^2} = 0 \tag{5}$$

$$\frac{\partial B_0}{\partial t} - \delta B_0\frac{\partial u}{\partial x} - \delta b\frac{\partial v}{\partial x} + \delta b\frac{\partial v}{\partial y} = -\frac{1}{\delta\mu_c}\frac{\partial^2 B_0}{\partial y^2} - \delta V\frac{\partial b}{\partial y} \tag{6}$$

with

$$\tilde{u} = 0, \tilde{v} = -v_w = -\frac{a\cdot}{c}, T = 0, at \, \tilde{y} = h(t), \tag{7}$$

$$\tilde{u} = 0, \tilde{v} = 0, T = T_0\left(1 - e^{-\eta t}\right)\ldots AT \ldots y = 0, \tag{8}$$

By these transformations, the leading equations are reduced to,

$$S\left(f''' + 3f'' + f'f' - ff'''\right) = A\left(2f'GH + 2FG^2\right)$$
$$+f'''' - A\left(2f'HH + H^2f' + f'HG + fHG' + fH'G\right),$$

(9)

$$2\theta'' - f\theta l^2 - \theta' \eta l^2 = 0$$

(10)

$$G'\eta - 2G - \delta Hf'' - \delta f'H' - \delta f'H' - \delta Gf' - fG' - G''M = 0,$$

(11)

$$H''Q + H'\eta + H\left(1 - f'\right) = 0,$$

(12)

3   Results and Discussions

Figure 2 portrays the complete harmony of MAM and numerical method for f(η). Notwithstanding, table shows a very small error which lead us to believe on the authenticity of our current research.

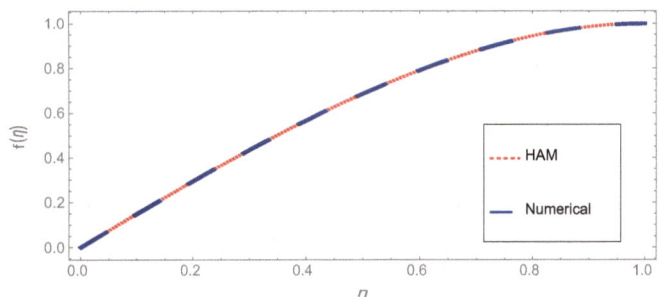

Fig. 2. AS and NS comparison for $f(\eta)$

Table 1 shows the individual Residual Error of $f'(\eta)$, $\theta(\eta)$, and $H(\eta)$.

Table 1. Errors among different quantities

No	$f(\eta)$	$\theta(\eta)$	$G(\eta)$	$H(\eta)$
2	3.45099×10^{-4}	2.79452×10^{-11}	0.00675467	2.7712924
4	1.00488×10^{-3}	1.77377×10^{-13}	0.000406648	2.43464×10^{-3}
6	1.11916×10^{-4}	1.64768×10^{-14}	9.33238×10^{-6}	2.3744×10^{-4}
8	1.11381×10^{-5}	1.61982×10^{-15}	2.40518×10^{-6}	1.44139×10^{-4}
10	1.2227×10^{-6}	1.88767×10^{-15}	1.00453×10^{-7}	1.76503×10^{-6}
12	1.12164×10^{-7}	1.6313×10^{-16}	2.11491×10^{-8}	1.47125×10^{-6}
14	1.23155×10^{-8}	1.0007×10^{-18}	1.23349×10^{-9}	1.07647×10^{-7}
16	1.95402×10^{-9}	1.17776×10^{-18}	1.01639×10^{-10}	1.0624×10^{-8}
18	1.10497×10^{-10}	1.23983×10^{-21}	1.11093×10^{-11}	1.20031×10^{-9}
20	1.23668×10^{-11}	1.19905×10^{-20}	1.0113×10^{-12}	6.65142×10^{-11}
22	1.10174×10^{-12}	1.25915×10^{-22}	1.00199×10^{-13}	4.16438×10^{-11}
24	1.56872×10^{-13}	1.12765×10^{-22}	1.0015×10^{-14}	4.00878×10^{-13}
26	1.22234×10^{-13}	1.27271×10^{-23}	1.20496×10^{-15}	4.00156×10^{-14}
28	1.00903×10^{-15}	1.13626×10^{-24}	1.01007×10^{-17}	4.03467×10^{-15}
30	1.00566×10^{-15}	1.27023×10^{-25}	1.00222×10^{-17}	4.01952×10^{-16}

4   Conclusion

Concluding remarks of the current study are listed as under: a. The current study has found that both techniques are agreed to treat the modeled system of equations in a best possible way. b. It is logical that the larger values of S show falling influence on $f'(\eta)$. c. It is noticed that the upper values of P show declining impact on $f'(\eta)$. d. It is witnessed that the higher values of R show the heightening impact on $\theta(\eta)$. e. The current analysis suggests that the growing values of R enhances $G(\eta)$ and vice versa.

5   Future Work

For the upcoming researchers, one can reformulate this mathematical model for compressible flow. One can calculate the results by finite element method among others to open more doors of intriguing knowledge in this area of research for other scientists in the field.

References

1. Khan, N.S., Shah, Q., Bhaumik, A., Kumam, P., Thounthong, P., Amiri, I.: Entropy generation in bioconvection nanofluid flow between two stretchable rotating disks. NAT.-Sci. Rep. **30**(01), 52–88 (2020)

2. Khan, N.S., Zuhra, S., Shah, Q.: Entropy generation in two phase model for simulating flow and heat transfer of carbon nanotubes between rotating stretchable disks with cubic autocatalysis chemical reaction. Appl. Nanosci.Nanosci. **9**(8), 1797–1822 (2019)
3. Saeed Khan, N., Shah, Q., Sohail, A.: Dynamics with Cattaneo-Christov heat and mass flux theory of bioconvection Oldroyd-B nanofluid. Adv. Mech. Eng. **12**(8), 1687814020930464 (2020)
4. Khan, N.S., Shah, Q., Sohail, A., Ullah, Z., Kaewkhao, A., Kumam, P., Thounthong, P.: Rotating flow assessment of magnetized mixture fluid suspended with hybrid nanoparticles and chemical reactions of species. Sci. Rep. **11**(1), 11277 (2021)
5. Rasheed, H.U., Islam, S., Khan, Z., Khan, J., Mashwani, W.K., Abbas, T., Shah, Q.: Computational analysis of hydromagnetic boundary layer stagnation point flow of nano liquid by a stretched heated surface with convective conditions and radiation effect. Adv. Mech. Eng. **13**(10), 16878140211053142 (2021)
6. Usman, A.H., Khan, N.S., Humphries, U.W., Ullah, Z., Shah, Q., Kumam, P.: Computational optimization for the deposition of bioconvection thin Oldroyd-B nanofluid with entropy generation. Sci. Rep. **11**(1), 1–23 (2021)
7. Khan, N.S., et al.: Lorentz forces effects on the interactions of nanoparticles in emerging mechanisms with innovative approach. Symmetry **12**(10), 1700 (2020)
8. Khan, Z., Khan, M.A., Siddiqui, N., Ullah, M., Shah, Q.: Solution of magnetohydrodynamic flow and heat transfer of radiative viscoelastic fluid with temperature dependent viscosity in wire coating analysis. PLoS ONE **13**(3), e0194196 (2018)
9. Khan, N.S., et al.: A framework for the magnetic dipole effect on the thixotropic nanofluid flow past a continuous curved stretched surface. Crystals **11**(6), 645 (2021)
10. Khan, Z., et al.: Analytical solution of UCM viscoelastic liquid with slip condition and heat flux over stretching sheet: the Galerkin approach. Math. Prob. Eng. **1**(1), 7563693 (2020)
11. Khan, N.S., Shah, Q., Sohail, A., Kumam, P., Thounthong, P., Muhammad, T.: Mechanical aspects of Maxwell nanofluid in dynamic system with irreversible analysis. ZAMM-J. Appl. Math. Mech./Zeitschrift für Angewandte Mathematik und Mechanik **101**(12), e202000212 (2021)
12. Khan, Z., Rasheed, H.U., Shah, Q., Abbas, T., Ullah, M.: Numerical simulation of double-layer optical fiber coating using Oldroyd 8-constant fluid as a coating material. Opt. Eng. **57**(7), 076104 (2018)
13. Shah, Q., Gul, T., Shah, R.A., Ullah, W., Idrees, M., Mohmand, M.I.: Gravity driven flow of an unsteady second order fluid between two parallel and vertical oscillating plates. J. Appl. Environ. Biol. Sci **5**(3), 213–220 (2015)
14. Khan, Z., et al.: Investigation of wire coating using hydromagnetic third-grade liquid for coating along with Hall current and porous medium. Math. Prob. Eng. **2020**(1), 4218717 (2020)
15. Fizaa, M., Ullaha, H., Islama, S., Chohanb, F.: Analytical solution of the viscous flow over a stretching sheet by multi-step optimal homotopy asymptotic method. Int. J. Fluid Mech. Res. **45**(4), 369–375 (2018)
16. Fiza, M., Ullah, H., Islam, S., Shah, Q., Chohan, F.I., Mamat, M.B.: Modifications of the multistep optimal homotopy asymptotic method to some nonlinear KdV-equations. Eur. J. Pure Appl. Math. **11**(2), 537–552 (2018)
17. Shah, Q., Mamat, M.B., Gul, T., Tofany, N.: The ADM solution of MHD non-Newtonian fluid with transient flow and heat transfer. AIP Conference Proceedings, vol. 1775, no. 1, p. 030087 (2016)
18. Shah, Q., Mamat, M.B., Gul, T., Tofany, N.: Heat transfer and hydromagnetic effects on the unsteady thin film flow of Oldroyd-B fluid over an oscillating moving vertical plate. AIP Conference Proceedings, vol. 1775, no. 1) (2016)

19. Shah, Q., Gul, T, Mamat, M.B., Khan, W., Tofany, N.: Soret and dufour effect on the thin film flow over an unsteady stretching surface. AIP Conference Proceedings, vol. 1775, no. 1, p. 0300881 (2016)

20. Noor Saeed Khan, T.M., Shah, Q., Sohail, A., Kumam, P., Zamm, P.T.: Mechanical aspects of Maxwell nanofluid in dynamic system with irreversible analysis. J. Appl. Math. Mech. (2021) I.FACTOR-1.603 8, **03**(21) (2021)

21. Khan, Z., et al.: Analytical Solution of UCM viscoelastic liquid with slip condition and heat flux over stretching sheet: the Galerkin approach. Math. Prob. Eng. **2020**(1), 7563693 (2020)

22. Mohmand, M.I., Mamat, M.B., Shah, Q.: Engineering investigations of Dufour and Soret effect on MHD heat and mass transfer with radiative heat flux in a liquid over a rotating dick. Int. J. Eng. Technol. **7**(4.5), 439–441 (2018)

23. Mohmand, M.I., Mamat, M.B., Shah, Q.: Engineering applications and analysis of vibratory motion fourth order fluid film over the time dependent heated flat plate. In: AIP Conference Proceedings, vol. 1859, no. (1), p. 020073 (2017)

24. Mohmand, M.I., Shah, Q., Mamat, M.B., Shah, Z., Khan A.S.: Time dependent Oldroyd-B liquid film flow over an oscillating and porous vertical plate with the effect of thermal radiation. In: AIP Conference Proceedings, vol. 1846, no. (1), p. 050003 (2017)

25. Mohmand, M.I., Mamat, M.B., Shah, Q.: Time dependent second grade fluid between two vertical oscillating plates with heat transfer effect. In: AIP Conference Proceedings, vol. 1823, no. (1), p. 020093 (2017)

26. Mohmand, M.I., Mamat, M.B., Shah, Q., Gul, T.: Vibratory motion of fourth order fluid film over a unsteady heated flat. In: AIP Conference Proceedings, vol. 1823, no. (1), p. 020066 (2017)

27. Mohmand, M.I., Shah, Q., Mustafa, M.: Engineering applications and analysis of vibratory motion fourth order fluid film over the time dependent heated flat plate. In: AIP Conference proceeding

28. Shah, Q., Gul, T., Mamat, M.B., Tofany, N.: Heat transfer and hydromagnetic effects on the unsteady thin film flow of Oldroyd-B fluid over an oscillating moving vertical plate. In: AIP Conference Proceeding (2016)

29. Shah, Q., Gul, T., Mamat, M.B., Khan, W., Tofany, N.: Soret and dufour effect on the thin film flow over an unsteady stretching surface. In: AIP Conference proceeding

30. Rasheed, H.U., Zeeshan, Islam, S., Ali, B., Shah, Q., Ali, R.: Implementation of shooting technique for Buongiorno nanofluid model driven by a continuous permeable surface

31. Rasheed, H., Shah, Q., Khan, J., Abbas, T., Khan, W., Mohmand, M.I.: Analysis of 2D stagnation point micropolar fluid with MHD effect and modified heat and mass flux towards a permeable channel: a computational study. Published in Heat Transfer (2023)

32. Rasheed, H., Shah, Q., Khan, J., Abbas, T., Khan, W., Muhammad, M.I.: Physical insight into thermal analysis of MHD stagnation point flow of second grade fluid across a flexible surface equipped with porous medium and Fourier and Fick's law Heat Transfer(2023)

33. Shah, Q., Rasheed, H., Khan, M.I.: A mathematical and computational framework for MHD and heat source effect on nanofluid flow by a nonlinearly stretching sheet. Kongzhi yu Juece/Control and Decision (2023)

34. Shah, Q., Rasheed, H., Khan, M.I.: (2023) Analysis of MHD flow chemically reactive Casson liquid by an elongated permeable sheet with Lorentz force and heat reservoir effects. Kongzhi yu Juece/Control and Decision

35. Shah, Q.: Application of Optimal Homotopy Asymptotic Method (OHAM) to thirteenth order Boundary Value Problem. Springer Conference Series (2024)

36. Khan, S., et al.: A well-conditioned and efficient implementation of dual reciprocity method for Poisson equation. AIMS Mathematics (2021)

Applications of Optimal Homotopy Asymptotic Method (OHAM) to Tenth Order Boundary Value Problem

Qayyum Shah[✉]

Department of Basic Sciences, University of Engineering and Technology Peshawar, Peshawar 25000, KP, Pakistan
qayyumshah@uetpeshawar.edu.pk

Abstract. The aim of this paper is to apply the Optimal Homotopy Asymptotic Method (OHAM), a semi-numerical and semi-analytic technique for solving linear and nonlinear Tenth order boundary value problems. The approximate solution of the problem is calculated in terms of a rapidly convergent series. Two bench mark examples have been considered to illustrate the efficiency and implementation of the method and the results are compared with the Variational Iteration Method (VIM). An interesting result of the analysis is that, the OHAM solution is more accurate than the VIM. Moreover, OHAM provides us with a convenient way to control the convergence of approximate solutions. The obtained solutions have shown that OHAM is effective, simpler, easier and explicit.

Keywords: OHAM · Tenth Order Boundary Value Problems

1 Introduction

In this paper, it is observed that the OHAM is a powerful approximate analytical tool like HAM (Homotopy Asymptotic Method) that is simple and straight forward and does not require the existence of any small or large parameter as does traditional perturbation method. OHAM has been successfully applied to a number of nonlinear problems arising in the science and engineering by various researchers [3–5, 10–16, 18, 20, 34–36]. Shah et al. [17–19] have investigated the behavior through graphical representation. Whilst Khan et al. [1, 2] discussed flow between rotating stretchable disks. What's more, Khan et al. [1] found that when both the discs rotate in the same sense then the fluid in the disks rotates with an angular velocity. Khan et al. [3] discussed Dynamics with Cattaneo–Christov heat and mass flux theory of bioconvection Oldroyd-B nanofluid. Notwithstanding, Khan et al. [1–3] further explored that the their study provides the best solutions and it has been proved that its solution is close to exact solution. Khan et al. [4] discussed Rotating flow assessment of magnetized mixture fluid suspended with hybrid nanoparticles and chemical reactions of species. Rasheed et al. [5] discussed Computational analysis of hydromagnetic boundary layer stagnation point flow of nano liquid by a stretched heated surface with convective conditions and radiation effect. Mohmand et al. [25–28]

G. Feng (Ed.): ICCE 2023, LNCE 526, pp. 674–682, 2024.
https://doi.org/10.1007/978-981-97-4355-1_66

scrutinized oscillating and porous, and flow with heat transfer effect as well as vibratory flow. Usman et al. [6] discussed Computational optimization for the deposition of bio-convection thin Oldroyd-B nanofluid with entropy generation. Khan et al. [7] explored Lorentz forces effects on the interactions of nanoparticles in emerging mechanisms with innovative approach. Khan et al. [8] analyzed Solution of magnetohydrodynamic flow and heat transfer of radiative viscoelastic fluid with temperature dependent viscosity in wire coating analysis. Khan et al. [9] investigated A Framework for the Magnetic Dipole Effect on the Thixotropic Nanofluid Flow Past a Continuous Curved Stretched Surface. Khan et al. [10] studied Analytical solution of UCM viscoelastic liquid with slip condition and heat flux over stretching sheet: Galerkin Approach. Shah et al. [17–19] have explored the transient flow, with unsteady stretching surface and accompanied by Soret and Dufour effects. Khan et al. [1–4] have discussed solution through tables. Khan et al. [11] discussed Mechanical aspects of Maxwell nanofluid in dynamic system with irreversible analysis. Khan et al. [12] studied Numerical simulation of double-layer optical fiber coating using Oldroyd 8-constant fluid as a coating material. Khan et al. [8, 10, 12] have also discussed heat and heat transfer. Shah et al. [13, 18] analyzed Gravity Driven Flow of an Unsteady Second Order Fluid as well as Heat transfer rate of the fluid at the belt is also calculated. Khan et al. [14] discussed Investigation of wire coating using hydromagnetic third-grade liquid for coating along with Hall current and porous medium. Khan et al. [15] studied Analytical Solution of the MHD Viscous Flow over a Stretching Sheet by Multistep Optimal Homotopy Asymptotic Method. Fiza et al. [16] explored Modifications of the multistep optimal homotopy asymptotic method to some nonlinear KdV-equations. Shah et al. [17] discussed The ADM solution of MHD non-Newtonian fluid with transient flow and heat transfer. Shah et al. [18] studied the Heat transfer and hydromagnetic effects on the unsteady thin film flow of Oldroyd-B fluid over an oscillating moving vertical plate. Shah et al. [19] explored forSoret and Dufour effect on the thin film flow over an unsteady stretching surface. Khan et al. [20–22] discussed for Mechanical aspects of Maxwell nanofluid in dynamic system with irreversible analysis as well as the impact of emerging parameters involved in the solutions are discussed through graphs on the velocity and temperature profiles in detail. Khan et al. [21] researched on the investigation of Wire Coating using Hydromagnetic Third-Grade Liquid for Coating along with Hall Current and Porous Medium. Khan et al. [22] discussed the Analytical Solution of UCM Viscoelastic Liquid with Slip Condition and Heat Flux over Stretching Sheet: The Galerkin Approach. Mohmand et al. [23] discussed the Engineering Investigations of Dufour and Soret effect on MHD heat and mass transfer with radiative heat flux in a liquid over a rotating dick. Mohmand et al. [24] explored the Engineering applications and analysis of vibratory motion fourth order fluid film over the time dependent heated flat plate. Mohmand et al. [25] ana-lyzed for Time dependent Oldroyd-B liquid film flow over an oscillating and porous vertical plate with the effect of thermal radiation. Mohmand et al. [26] studied Time dependent second grade fluid between two vertical oscillating plates with heat transfer effect. Mohmand et al. [27] investigated the Vibratory motion of fourth order fluid film over a unsteady heated flat. Mohmand et al. [28] discussed Engineering applications and analysis of vibratory motion fourth order fluid film over the time dependent heated flat plate. Mohmand et al. [29] explored for Heat transfer and hydromagnetic effects on the unsteady thin film flow of Oldroyd-B fluid over an oscillating moving vertical plate. Shah et al. [30] discussed Soret and dufour effect on the thin film flow over an unsteady stretching surface. Likewise Khan et al. [20] have also discussed about the Brownian motion and thermophoresis with thermal radiation and buoyancy effects are encountered in the governing equations. The oscillating parallel plates were discussed by Shah et al. [13]. Shah et al. [17–19] have also explored some more properties of fluids. Similarly, Fiza et al. and Khan et al. [16, 21] have discussed flow through numerical results (Runge-Kutta method). Rasheed et al. [31] and Shah et al. [30] have discussed fluid flow through shooting technique and numerical approach. Moreover, flow of Oldroyd-B fluid over an oscillating was discussed by Shah et al. [29]. Likewise Mohmand et al. [28] have discussed time-dependent heated plate. Moreover, Rasheed et al. [32, 33] have discussed the fluid flow. To this end, Khan et al. [2] have investigated pressure distribution and entropy generation rate and then their solution through HAM approach. Moreover, Khan et al. [3–5] Darcy–Forchheimer law is used to study heat and mass transfer flow and microorganisms motion in porous media as well as flow of Maxwell nanofluid induced by two parallel rotating disks were analyzed. Furthermore Rasheed et al. [31–33] have discussed fluid motion with thermal analysis. Shah et al. [34] have also discussed MHD flow. Khan et al. [1–4] have analyzed the fluid motion through graphs. Similarly, Shah et al. [17–19] have also investigated the MHD fluid motion through graphs. Likewise, Khan et al. [7–12] have also discussed fluid flow through graphs and found excellent harmony with the already published works. Shah et al. [34] & [35] have scrutinized a mathematical and computational analysis of MHD fluid with heat source effect and the chemically reactive Casson fluid to add knowledge to the existing one. As regard to the validity solution of PDEs one can consult the research done by Khan et al. [36]. Similarly, Shah [37] has discussed the fluid motion through the treatment of OHAM. This paper is organized as follows. First, we formulate the problem. Then, we present basic principles of OHAM. The OHAM solution is also given. After that, we analyze the comparison of the solution using OHAM with existing solution of VIM and DTM. Last but not the least, drew the conclusion.

2   Formulation of the Problem

In the present paper, thirteen-order boundary value problems are solved using OHAM. The following thirteen-order boundary value problems are considered

$$\left. \begin{array}{c} u^{(13)}(x) = f(x, u(x)), a \le x \le b \\ u^{(i)}(a) = A_i, \\ u^{(j)}(b) = B_j \end{array} \right\} \tag{1}$$

where for i = 0,1, 2, 3, ..., 6 and j = 0,1, ..., 5 A$'_i$s and B$'_j$s are finite real constants. Also f(x, u(x)) is a continuous function on [a, b].

3   Fundamental Mathematical Theory of OHAM

Consider the differential equation of the following form:

$$\mathcal{A}(u(x)) + f(x) = 0, x \in \Omega \tag{2}$$

$$\mathcal{B}(u, \partial u/\partial x) = 0, x \in \Gamma \tag{3}$$

where $\mathcal{A}$ is a differential operator, $u(x)$ is an unknown function, and x and t denote spatial independent variable, Γ is the boundary of Ω and $f(x)$ is a known analytic function. $\mathcal{A}$ can be divided into two parts: $\mathcal{L}$ and $\mathcal{N}$ such that:

$$\mathcal{A} = \mathcal{L} + \mathcal{N} \tag{4}$$

$\mathcal{L}$ is the linear part of the differential equation which is easier to solve, and $\mathcal{N}$ contains the nonlinear part of $\mathcal{A}$.

According to OHAM, one can construct an optimal homotopy $\phi(x, p) : \Omega \times [0,1] \rightarrow \mathfrak{R}$ which satisfies:

$$H(\phi(x, p), p) = (1 - p)\{\mathcal{L}(\phi(x, p)) + f(x)\} - H(p)\{\mathcal{A}(\phi(x, p)) + f(x)\} = 0, \tag{5}$$

where the auxiliary function H(p) is nonzero for $p \neq 0$ and $H(0) = 0$. Equation (5) is called optimal homotopy equation. Clearly, we have:

$$p = 0 \Rightarrow H(\phi(x, 0), 0) = \mathcal{L}(\phi(x, 0)) + f(x) = 0, \tag{6}$$

$$p = 1 \Rightarrow H(\phi(x, 1), 1) = H(1)\{\mathcal{A}(\phi(x, p)) + f(x)\} = 0, \tag{7}$$

Obviously, when $p = 0$ and $p = 1$ we obtain $\phi(x, 0) = u_0(x)$ and $\phi(x, 1) = u(x)$ respectively. Thus, as p varies from 0 to 1, the solution $\phi(x, p)$ approaches from $u_0(x)$ to $u(x)$, where $u_0(x)$ is obtained from Eq. (5) for $p = 0$:

$$\mathcal{L}(u_0(x)) + f(x) = 0, \mathcal{B}(u_0, \partial u_0/\partial x) = 0. \tag{8}$$

Next, we choose auxiliary function H(p) in the form

$$H(p) = p^1 C_1 + p^2 C_2 + \dots \tag{9}$$

To get an approximate solution, we expand $\phi(x, p, C_i)$ by Taylor's series about p in the following manner,

$$\phi(x, p, C_i) = u_0(x) + \sum_{k=1}^{\infty} u_k(x, C_i)p^k, i = 1, 2, \dots \tag{10}$$

Substituting Eq. (10) into Eq. (5) and equating the coefficient of like powers of p, we obtain Zeroth order problem, given by Eq. (8), the first and second order problems are given by Equations. (11–12) respectively and the general governing equations for $u_k(x)$ are given by Eq. (13):

$$\mathcal{L}(u_1(x)) = C_1 \mathcal{N}_0(u_0(x)), \mathcal{B}(u_1, \partial u_1/\partial x) = 0 \tag{11}$$

$$\mathcal{L}(u_2(x)) - \mathcal{L}(u_1(x)) = C_2 \mathcal{N}_0(u_0(x)) + C_1[\mathcal{L}(u_1(x)) + \mathcal{N}_1(u_0(x), u_1(x))], \mathcal{B}(u_2, \partial u_2/\partial x) = 0 \tag{12}$$

$$\mathcal{L}(u_k(x)) - \mathcal{L}(u_{k-1}(x)) = C_k\mathcal{N}_0(u_0(x)) + \sum_{i=1}^{k-1} C_i\big[\mathcal{L}(u_{k-i}(x))$$
$$+\mathcal{N}_{k-i}(u_0(x), u_1(x), \dots, u_{k-i}(x))\big], k = 2, 3, \dots B(u_k, \partial u_k/\partial x) = 0 \,\text{,} \tag{13}$$

where $\mathcal{N}_{k-i}(u_0(x), u_1(x), \dots, u_{k-i}(x))$ are the coefficient of p^{k-i} in the expansion of $\mathcal{N}(\phi(x, p))$ about the embedding parameter p.

$$\mathcal{N}(\phi(x, p, C_i)) = \mathcal{N}_0(u_0(x)) + \sum_{k\geq 1} \mathcal{N}_k(u_0, u_1, u_2, \dots, u_k)p^k \tag{14}$$

It should be underscored that the u_k for $k \geq 0$ are governed by the linear equations with linear boundary conditions that come from the original problem, which can be easily solved.

It has been observed that the convergence of the series Eq. (10) depends upon the auxiliary constants $C_1, C_2, \dots$. If it is convergent at $p = 1$, one has:

$$\tilde{u}(x, C_i) = u_0(x) + \sum_{k\geq 1} u_k(x, C_i) \tag{15}$$

Substituting Eq. (15) into Eq. (1), it results the following expression for residual:

$$R(x, C_i) = \mathcal{L}(\tilde{u}(x, C_i)) + f(x) + \mathcal{N}(\tilde{u}(x, C_i)) \tag{16}$$

In actual computation $k = 1, 2, 3, \dots, m$.

If $R(x, C_i) = 0$ then $\tilde{u}(x, C_i)$ is the exact solution of the problem. Generally it doesn't happen, especially in nonlinear problems.

For the determinations of auxiliary constants, $C_i = 1, 2, \dots, m$, there are different methods like Galerkin's Method, Ritz Method, Least Squares Method and Collocation Method. One can apply the Method of Least Squares as under:

$$J(C_i) = \int_a^b R^2(x, C_i)dx \tag{17}$$

$$\frac{\partial J}{\partial C_1} = \frac{\partial J}{\partial C_2} = \dots = \frac{\partial J}{\partial C_m} = 0 \tag{18}$$

The mth order approximate solution can be obtained by these constants so obtained. The constants C_i can also be determined by another method as under:

$$R(h_1; C_i) = R(h_2; C_i) = \dots = R(h_m; C_i) = 0, i = 1, 2, \dots, m. \tag{19}$$

at any time t, where $h_i \in \Omega$.

The more general auxiliary function H(p) is useful for convergence, which depends upon constants $C_1, C_2, \dots$ can be optimally identified by Eq. (18) and is useful in error minimization.

4   Solution of the Problem via OHAM

In this section we will apply OHAM to a linear boundary value problem and non-linear boundary value problem. Example 1: The linear thirteen-order BVP, is considered as

$$u^{(13)}(x) = cosx - sinx, 0 \leq x \leq 1, \tag{20}$$

Subject to the boundary conditions,

$$\left.\begin{array}{l} u(0) = 1, u(1) = \cos(1) + \sin(1), \\ u^{(1)}(0) = 1, u^{(1)}(1) = \cos(1) - \sin(1), \\ u^{(2)}(0) = -1, u^{(2)}(1) = -\sin(1) - \cos(1), \\ u^{(3)}(0) = -1, u^{(3)}(1) = -\cos(1) + \sin(1), \\ u^{(4)}(0) = 1, u^{(4)}(1) = \cos(1) + \sin(1), \\ u^{(5)}(0) = 1, u^{(5)}(1) = \cos(1) - \sin(1), \\ u^{(6)}(0) = -1, \end{array}\right\} \qquad (21)$$

The exact solution of the problem is,

$$u(x) = cosx + sinx. \qquad (22)$$

According to above equation, we have

$$\mathcal{L}(u(x)) = u^{(13)}(x); \mathcal{N}(u(x)) = 0; \text{ and } f(x) = -cosx + sinx; \qquad (23)$$

5 Comparison of OHAM Solution with VIM Solution

Comparison of absolute error of OHAM with VIM in Table 1.

Table 1. Shows the comparison of absolute error of OHAM with VIM [10]

x	Exact	OHAM	Abs: Error in present method	Abs: Error in VIM
0	1.00000	1.00000	0.00000	0.00000
0.1	1.105170918	1.10517	4.32987×10^{-14}	4.17444×10^{-14}
0.2	1.221402758	1.2214	2.44249×10^{-14}	2.64144×10^{-12}
0.3	1.349858808	1.34986	5.99520×10^{-15}	2.99314×10^{-11}
0.4	1.491824698	1.49182	1.44329×10^{-14}	1.67101×10^{-10}
0.5	1.648721271	1.64872	2.64677×10^{-13}	6.30955×10^{-10}
0.6	1.8221188	1.82212	1.02807×10^{-12}	1.84757×10^{-9}
0.7	2.013752707	2.01375	8.95284×10^{-13}	4.47866×10^{-9}
0.8	2.225540928	2.22554	6.89315×10^{-12}	9.21592×10^{-9}
0.9	2.459603111	2.4596	2.60871×10^{-11}	1.58906×10^{-8}
1	2.718281828	2.71828	3.41771×10^{-12}	2.09057×10^{-8}

6 Conclusion of the Current Analysis

In the above table, it is clearly observed that OHAM is better than VIM and give wonderful results of Tenth order boundary value problems for both linear and nonlinear. Therefore, we conclude that OHAM is reasonably good method for any type of T order boundary value problem. Below Fig. 1: represents the exact solution of the Tenth order nonlinear boundary value problem, while Fig. 2: shows the OHAM solution of the Tenth order nonlinear boundary value problem.

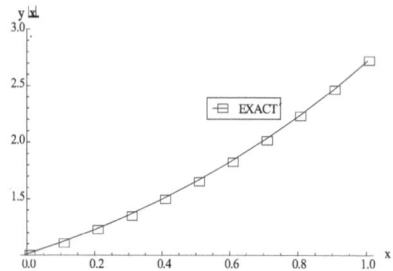

Fig. 1. Exact Solution   **Fig. 2.** OHAM Solution

References

1. Khan, N.S., Shah, Q., Bhaumik, A., Komam, P.: Entropy generation in bioconvection nanofluid flow between two stretchable rotating disks Nat.-Sci. Rep. **30**(01), 52–88 (2020)
2. Khan, N.S., Zuhra, S., Shah, Q.: Entropy generation in two phase model for simulating flow and heat transfer of carbon nanotubes between rotating stretchable disks with cubic autocatalysis chemical reaction. Appl. Nanosci. **9**(8), 1797–1822 (2019)
3. Khan, N.S., Shah, Q., Sohail, A.: Dynamics with Cattaneo–Christov heat and mass flux theory of bioconvection Oldroyd-B nanofluid. Adv. Mech. Eng. 12(8), 1687814020903046 (2020)
4. Khan, N.S., et al.: Rotating flow assessment of magnetized mixture fluid suspended with hybrid nanoparticles and chemical reactions of species. Sci. Rep. **11**(1), 11277 (2021)
5. Rasheed, H.U., et al.: Computational analysis of hydromagnetic boundary layer stagnation point flow of nano liquid by a stretched heated surface with convective conditions and radiation effect. Adv. Mech. Eng. **13**(10), 168781402110531 (2021)
6. Usman, A.H., et al.: Computational optimization for the deposition of bioconvection thin Oldroyd-B nanofluid with entropy generation. Sci. Rep. **11**(1), 1–23 (2021)
7. Khan, N.S., et al.: Lorentz forces effects on the interactions of nanoparticles in emerging mechanisms with innovative approach. Symmetry **12**(10), 1700 (2020)
8. Khan, Z., Khan, M.A., Siddiqui, N., Ullah, M., Shah, Q.: Solution of magnetohydrodynamic flow and heat transfer of radiative viscoelastic fluid with temperature dependent viscosity in wire coating analysis. PLoS ONE **13**(3), e0194196 (2018)
9. Khan, N.S., et al.: A framework for the magnetic dipole effect on the thixotropic nanofluid flow past a continuous curved stretched surface. Crystals **11**(6), 645 (2021)
10. Khan, Z., et al.: Analytical solution of UCM viscoelastic liquid with slip condition and heat flux over stretching sheet: the galerkin approach. Math. Prob. Eng. **2020**, 1–7 (2020)
11. Khan, N.S., Shah, Q., Sohail, A., Kumam, P., Thounthong, P., Muhammad, T.: Mechanical aspects of Maxwell nanofluid in dynamic system with irreversible analysis. ZAMM-J. Appl. Math. Mech./Zeitschrift für Angewandte (2021)

12. Khan, Z., Rasheed, H.U., Shah, Q., Abbas, T., Ullah, M.: Numerical simulation of double-layer optical fiber coating using Oldroyd 8-constant fluid as a coating material. Opt. Eng. **57**(07), 076104 (2018)

13. Shah, Q., Gul, T., Shah, R.A., Ullah, W., Idrees, M., Mohmand, M.I.: Gravity driven flow of an unsteady second order fluid between two parallel and vertical oscillating plates. J. Appl. Environ. Biol. Sci **5**(3), 213–220 (2015)

14. Khan, Z., et al.: Investigation of wire coating using hydromagnetic third-grade liquid for coating along with Hall current and porous medium. Math. Prob. Eng. **2020**, 1–8 (2020)

15. Fiza, M., Islam, S., Ullah, H., Chohan, F., Shah, Q.: Analytical solution of mhd viscous flow over a stretching sheet by multistage optimal homotopy asymptotic method. Int. J. Fluid Mech. Res. **45**(4), 369–375 (2018)

16. Fiza, M., Ullah, H., Islam, S., Shah, Q., Chohan, F.I., Mamat, M.B.: Modifications of the multistep optimal homotopy asymptotic method to some nonlinear KdV-equations. Eur. J. Pure Appl. Math. **11**(2), 537–552 (2018)

17. Shah, Q., Mamat, M.B., Gul, T., Tofany, N.: The ADM solution of MHD non-Newtonian fluid with transient flow and heat transfer. In: AIP Conference Proceedings, vol. 1775, no. 1, p. 030087 (2016)

18. Shah, Q, Mamat, M.B., Gul, T., Tofany, N: Heat transfer and hydromagnetic effects on the unsteady thin film flow of Oldroyd-B fluid over an oscillating moving vertical plate. In: AIP Conference Proceedings, vol. 1775, no. 1, p. 030085 (2016)

19. Shah, Q., Gul, T., Mamat, M.B., Khan, W., Tofany, N.: Soret and Dufour effect on the thin film flow over an unsteady stretching surface. In: AIP Conference Proceedings, vol. 1775, no. 1, p. 0300881 (2016)

20. Khan, N.S., Shah, Q., Sohail, A., Kumam, P., Thounthong, P.: Mechanical aspects of Maxwell nanofluid in dynamic system with irreversible analysis. Zamm J. Appl. Math. Mech. I.FACTOR **101**(12) (2021)

21. Khan, Z., et al.: Investigation of wire coating using hydromagnetic third-grade liquid for coating along with hall current and porous medium. Math. Probl. Eng. **2020**, 1–8 (2020). https://doi.org/10.1155/2020/4218717

22. Khan, Z., et al.: Analytical solution of UCM viscoelastic liquid with slip condition and heat flux over stretching sheet: the galerkin approach. Math. Prob. Eng. **2020**, 1–7 (2020). https://doi.org/10.1155/2020/7563693

23. Mohmand, M.I., Mamat, M.B., Shah, Q.: Engineering Investigations of Dufour and Soret effect on MHD heat and mass transfer with radiative heat flux in a liquid over a rotating dick (2018)

24. Mohmand, M.I., Mamat, M.B., Shah, Q.: Engineering applications and analysis of vibratory motion fourth order fluid film over the time dependent heated flat plate. In: AIP Conference Proceedings, vol. 1859. no. 1, p. 020073 (2017)

25. Mohmand, M.I., Shah, Q., Mamat, M.B., Shah, Z., Khan, A.S: Time dependent Oldroyd-B liquid film flow over an oscillating and porous vertical plate with the effect of thermal radiation. In: AIP Conference Proceedings, vol. 1846, no. 1, p. 050003 (2017)

26. Mohmand, M.I., Mamat, M.B., Shah, Q.: Time dependent second grade fluid between two vertical oscillating plates with heat transfer effect. In: AIP Conference Proceedings, vol. 1823, no. 1, p. 020093 (2017)

27. Mohmand, M.I., Mamat, M.B., Shah, Q., Gul, T.: Vibratory motion of fourth order fluid film over a unsteady heated flat. In: AIP Conference Proceedings, vol. 1823, no. 1, p. 020066 (2017)

28. Mohmand, M.I., Shah, Q., Mustafa, M: Engineering applications and analysis of vibratory motion fourth order fluid film over the time dependent heated flat plate. In: AIP Conference Proceeding (2017)

29. Shah, Q., Gul, T., Mamat, M.B., Tofany, N.: Heat transfer and hydromagnetic effects on the unsteady thin film flow of Oldroyd-B fluid over an oscillating moving vertical plate. In: AIP Conference Proceeding (2016)
30. Shah, Q., Gul, T., Mamat, M.B., Khan, W., Tofany, N.: Soret and dufour effect on the thin film flow over an unsteady stretching surface. In: AIP Conference Proceeding (2016)
31. Rasheed, H.U., Zeeshan, S.I., Ali, B., Shah, Q., Ali, R.: Implementation of shooting technique for Buongiorno nanofluid model driven by a continuous permeable surface. Heat Transf. 52(4), 3119–3134 (2023)
32. Rasheed, H., Shah, Q., Khan, J., Abbas, T., Khan, W., Mohmand, M.I.: Analysis of 2D stagnation point micropolar fluid with MHD effect and modified heat and mass flux towards a permeable channel: a computational study. Heat Transf. (2023)
33. Rasheed, H.U., Shah, Q., Khan, J., Abbas, T., Khan, W., Mohmand, M.I.: Physical insight into thermal analysis of magnetohydrodynamic stagnation point flow of micropolar nanofluid across a flexible surface equipped with porous medium and Fourier and Fick's law. Heat Transf. 53(2), 512–532 (2024)
34. Shah, Q., Rasheed, H., Khan, M.I: A mathematical and computational framework for MHD and heat source effect on nanofluid flow by a nonlinearly stretching sheet. Kongzhi yu Juece/Control and Decision (2023)
35. Shah, Q., Rasheed, H., Khan, M.I.: Analysis of MHD flow chemically reactive Casson liquid by an elongated permeable sheet with Lorentz force and heat reservoir effects. Kongzhi yu Juece/Control Decis. (2023)
36. Shah, Q.: Application of Optimal Homotopy Asymptotic Method (OHAM) to Tenth order Boundary Value Problem. Springer Conference Series (2024)
37. Khan, S., et al.: A well-conditioned and efficient implementation of dual reciprocity method for Poisson equation. AIMS Math. 6, 12560–12582 (2021)

Study on the Control Effect of Pretreatment Measures on the Deformation of Subway Caused by the Excavation of Foundation Pit

Yahui Jia[1,2], Yanliang Shang[1,2], Jianhua Du[1,2], and Hongqian Dang[1(✉)]

[1] Department of Railway Engineering, Shijiazhuang Institute of Railway Technology, Shijiazhuang 050041, China
xrzhsh@sohu.com
[2] Application Technology R&D Center of Bridge and Tunnel Intelligent Construction of Hebei Colleges, Shijiazhuang, China

Abstract. At present, there are many control measures for the deformation of subway caused by the excavation of foundation pit. It is worthy of attention that pretreatment measures can effectively reduce the influence of deformation on the subway tunnel under different working conditions. Therefore, the finite element model is established by Midas/GTS to explore the reinforcement effect of two measures of pit bottom reinforcement and isolation pile reinforcement. The results show that: in the excavation of foundation pit, the two schemes have different effects on controlling the deformation of tunnel structure, and the inclination of diameter line is generally smaller. The deformation characteristics of each component structure of the diameter line after the pit bottom reinforcement are similar to those of the unreinforced condition, while the displacement changes of each structure after the isolation pile reinforcement become uniform and reasonable. Isolation pile reinforcement is better than pit bottom reinforcement, especially in reducing the vertical displacement of the structure. The reduction degree of the maximum horizontal displacement after the isolation pile reinforcement is about 24%–31%. Therefore, when selecting the pretreatment method, priority is given to the isolation pile reinforcement method.

Keywords: Foundation Pit Excavation · Displacement · Pit Bottom Reinforcement · Isolation Pile Reinforcement

1 Introduction

The excavation of foundation pits in cities inevitably causes disturbance to nearby subways. It is necessary to take corresponding reinforcement measures when encountering additional deformation caused by excavation of foundation pits in subway tunnels [1, 2]. Previous researchers have done sufficient work on the deformation control of corresponding tunnel structures. Isolation pile reinforcement is currently one of the most widely used reinforcement methods, playing an effective role in controlling building deformation [3]. Although there has been extensive research on reinforcement methods

G. Feng (Ed.): ICCE 2023, LNCE 526, pp. 683–690, 2024.
https://doi.org/10.1007/978-981-97-4355-1_67

by predecessors, there are few reports on the application of full pit bottom reinforcement and isolation pile reinforcement methods in the same project, and comparing their deformation control effects. This article will rely on the adjacent underground diameter line project of Tianjin Jiahai Deep Foundation Pit, adopting pre-treatment schemes of bottom grouting reinforcement and isolation pile reinforcement respectively. Based on the established finite element model, the control effect of reinforcement measures on diameter line deformation caused by foundation pit excavation will be analyzed. The analysis results can provide reference and reference for similar projects.

2   Project Overview

The south side of the Tianjin Jiahai Foundation Pit Project is the underground diameter line from Tianjin West Station to Tianjin Station. As this foundation pit belongs to a super large deep foundation pit, it is divided into three parts: the north, middle, and south. The excavation depth of the southern foundation pit is about 10.8 m, with a distance of 16–20 m from the diameter line and a parallel distance of about 250 m. Multiple construction methods are used for the underground diameter line. The tunnel is buried at a smaller depth and is constructed using the open excavation method. The section with a deeper burial depth in the middle is constructed using the shield tunneling method. The side of the foundation pit close to the diameter line adopts a support form of drilled cast-in-place piles + concrete internal support, while the west side of the foundation pit adopts double row cast-in-place piles + counter pressure soil. The enclosure structure of the open cut tunnel adopts the form of drilled pile + water stop curtain, and underground connecting wall structure (the deeper section of the foundation pit adopts underground connecting wall, and the shallower section adopts drilled pile + water stop curtain). Two pre-treatment methods are considered for reinforcement, and the pre-treatment scheme adopted for foundation pit construction is shown in Table 1.

Table 1. Foundation Pit Excavation pretreatment scheme

Scheme	Concrete content
Scheme 1	Strengthen the soil of 3 m below the pit bottom. The reinforcement area is the southern foundation pit with full reinforcement
Scheme 2	A separation wall (pile) is used to reinforce the soil between the southern foundation pit and the shield shaft. The reinforcement range is: the length range of the shield tunnel is extended by 20 m towards the open excavation section and the shield tunnel section, with a reinforcement width of 3 m and a depth of 30 m below the surface

3 Model Establishment

The dimensions of the 3D model are 250 m in length, 135 m in width, and 50 m in depth. In order to consider the above two working conditions in the same model, the geometric entities will be cut accordingly during modeling, and the entities in the grouting reinforcement area at the bottom of the pit and the reinforcement area of the separation wall (pile) will be cut in advance, so that the transformation of the two models can be easily realized by modifying the material properties. See Fig. 1 and Fig. 2 for the specific segmentation. The constitutive relationship between soil and structure is determined using the D-P model and the linear elastic constitutive model, respectively. The soil within a depth range of 50 m in the model is approximately divided into four layers. Simulation of the excavation process of foundation pits. CS1: Apply gravity and initial water head to form an initial stress field, and then clear the displacement to zero; CS2: The first step of excavation is completed, and the support and waist beam are completed; CS3: Excavate to the bottom of the pit.

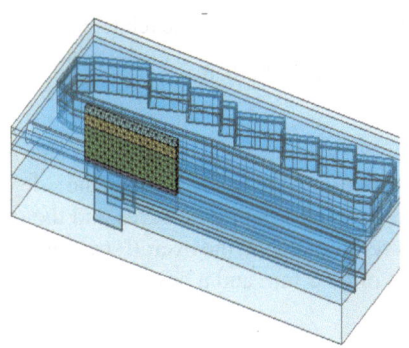   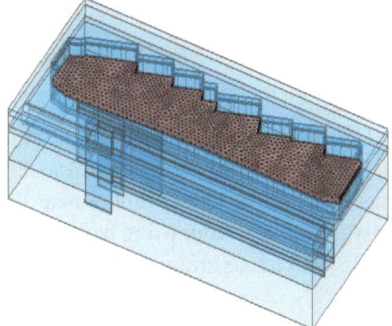

Fig. 1. Pit bottom reinforcement   **Fig. 2.** Isolation pile reinforcement

4 Analysis of Reinforcement Measures

Select the CS3 (excavation to the bottom of the pit) process for analysis, and the model specifies that the horizontal direction is towards the side of the pit as the positive direction, and the vertical direction is upwards as the positive direction.

4.1 Displacement of Shield Shaft

The horizontal displacement of the shield shaft for two preprocessing schemes is shown in Fig. 3. From the figure, it can be concluded that both schemes are effective in reducing the absolute displacement of the shield shaft bottom plate. However, after using bottom grouting reinforcement, a reverse displacement of 1.18 mm occurred at the top of the shield shaft (i.e. moving away from the foundation pit), resulting in a relative displacement of 8.72 mm. Compared to the relative displacement before reinforcement, the value increased, resulting in a more pronounced inclination of the shield shaft. After adopting the isolation pile reinforcement scheme, the horizontal displacement of the shield shaft varies uniformly along the height, resulting in a maximum horizontal displacement of 6.26 mm, which occurs in the middle of the side wall near the foundation pit. The minimum displacement at the top is 5.73 mm, resulting in a relative displacement value of only 0.53 mm for the shield shaft. This reinforcement method reduces relative displacement, which means that the inclination of the shield shaft is also reduced.

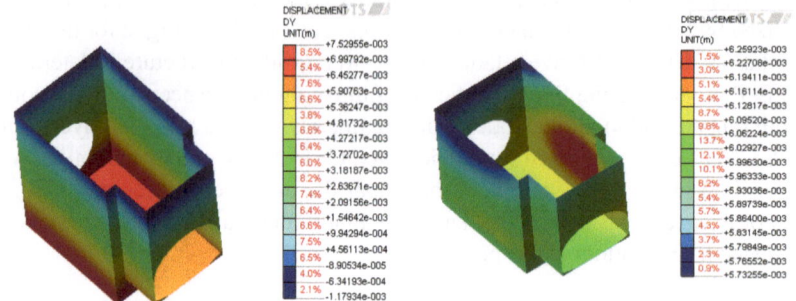

（a）Grouting reinforcement plan at the bottom of the pit   （b）Isolation pile reinforcement plan

Fig. 3. Horizontal displacement of shield shaft

The vertical displacement of the shield shaft in Scheme 1 and Scheme 2 is shown in Fig. 4. The grouting reinforcement scheme at the bottom of the pit resulted in the side wall (inner side) of the shield tunnel near the foundation pit floating upwards and the outer side wall sinking. After using isolation piles for reinforcement, the overall structure tends to sink, with a maximum vertical displacement of 0.32 mm and a downward direction.

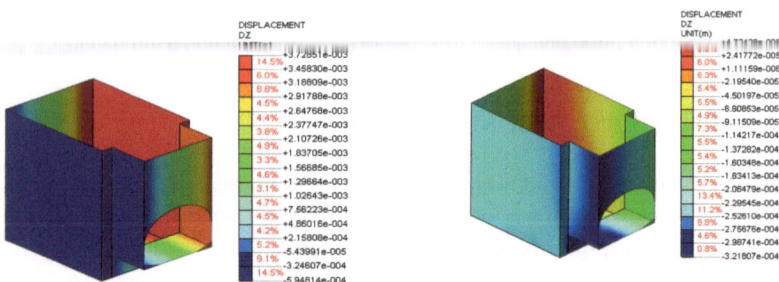

（a）Grouting reinforcement plan at the bottom of the pit   （b）Isolation pile reinforcement plan

Fig. 4. Vertical displacement of shield shaft

4.2 Displacement of Shield Tunnel Segments

The horizontal displacement of the reinforced shield tunnel segment is shown in Fig. 5. The horizontal displacement variation of the shield tunnel segment after grouting reinforcement at the bottom of the pit is similar to that of the unreinforced scheme. The maximum horizontal displacement occurs at the bottom of the segment near the shield shaft, and the minimum displacement occurs at the top of the segment far from the shield shaft end. After the reinforcement of the isolation pile, the horizontal displacement trend of the shield tunnel segment changes, and the displacement of the segment is relatively uniform along the cross-sectional direction. The horizontal displacement of the segment near the shield shaft is the largest, and gradually decreases along the longitudinal direction of the segment.

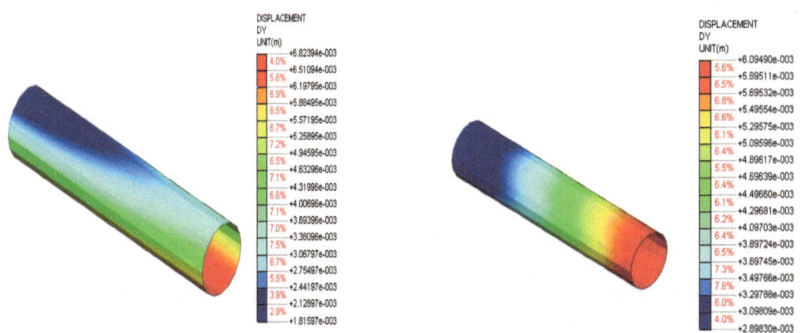

（a）Grouting reinforcement plan at the bottom of the pit　（b）Isolation pile reinforcement plan

Fig. 5. Horizontal displacement of shield segment

The vertical displacement of the reinforced shield tunnel segment is shown in Fig. 6. After grouting reinforcement at the bottom of the pit, the maximum vertical displacement occurs at the "arch waist" position near the shield shaft and closer to the foundation pit, with the displacement direction being upward. The minimum displacement occurs at the "arch waist" position on the side far from the shield shaft and the foundation pit, with a downward displacement direction. After using isolation piles for reinforcement, the vertical displacement change law generated by the shield tunnel segment is similar to the horizontal displacement change law, that is, the displacement is evenly distributed along the cross-sectional direction, which also indicates that the existence of isolation piles makes the segment more evenly stressed during the excavation process of the foundation pit.

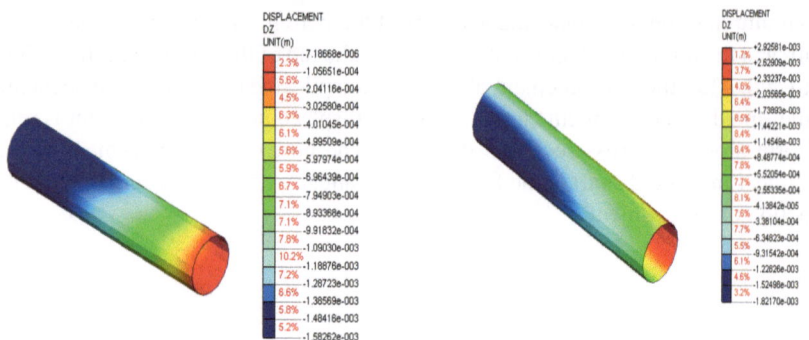

(a) Grouting reinforcement plan at the bottom of the pit   (b)Isolation pile reinforcement plan

Fig. 6. Vertical displacement of shield segment

4.3   Displacement of the Main Structure of Open Excavation

The horizontal displacement of the reinforced open cut main structure is shown in Fig. 7. After using bottom grouting reinforcement, the maximum horizontal displacement of the main structure of the open excavation occurs at the bottom plate near the shield shaft, and the minimum horizontal displacement occurs at the arch far from the shield shaft end, similar to the pattern in the unreinforced situation. After using isolation piles for reinforcement, the displacement variation pattern of the open excavation section is consistent with that of the shield tunnel section using the same reinforcement method. The displacement near the shield shaft is the largest, and there is a clear uniform decreasing trend along the longitudinal direction of the structure.

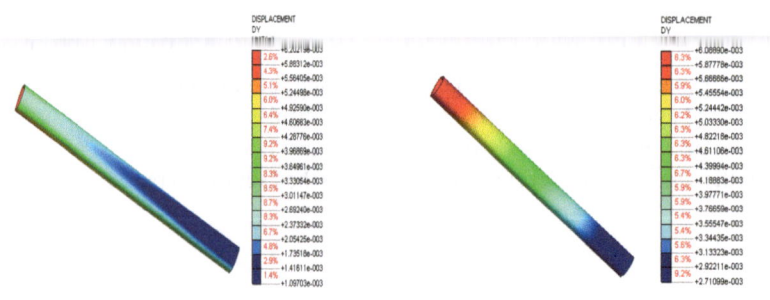

（a）Grouting reinforcement plan at the bottom of the pit   （b）Isolation pile reinforcement plan

Fig. 7. Horizontal displacement of open-cut main structure

The vertical displacement of the reinforced open cut main structure is shown in Fig. 8. After adopting Scheme 1 for reinforcement, the maximum and minimum vertical displacement of the main structure of the open excavation is similar to that of the shield tunnel segment. After adopting Scheme 2 reinforcement, the displacement of the main structure of the open excavation is similar to that of the shield tunnel segment, with the maximum vertical displacement occurring near the shield shaft.

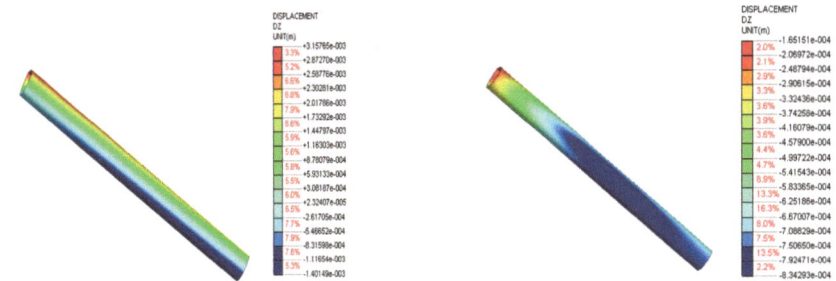

（a）Grouting reinforcement plan at the bottom of the pit（b）Isolation pile reinforcement plan

Fig. 8. Vertical displacement of open-cut main structure

5 Conclusion

(1) Both pre-treatment schemes have played an effective role in controlling the deformation of the diameter line, and the overall inclination of the diameter line structure is reduced.

(2) The control effect of isolation pile reinforcement on deformation is better than that of grouting reinforcement at the bottom of the pit. After the reinforcement of the isolation pile, the deformation of each structure shows a uniform trend, and the stress situation becomes reasonable compared to the unreinforced one; The decrease in absolute and relative displacement of the structure after adopting the isolation pile reinforcement scheme is greater than that of the pit bottom grouting reinforcement scheme.

(3) When selecting the pre-treatment method, priority can be given to the reinforcement method of isolation piles.

Acknowledgments. This study was founded by Science Research Project of Hebei Education Department (Grant No. ZD2022027), the Hebei Province Construction Science and Technology Research Guidance Plan Project (Grant No. 2022-2021), and the Science and Technology Project of Hebei Province (Grant No. 16215408D). These financial supports are gratefully acknowledged.

References

1. Wang Weidong, W., Jiangbin, W.Q.: Numerical simulation of the impact of excavation and unloading of foundation pits on subway section tunnels. Geotech. Mech. **S2**, 251–255 (2004)
2. Hongwei, H., Huang, X., Helmut, S.F.: Numerical simulation study on the influence of foundation pit excavation on the operation of shield tunnels under the ground. J. Civ. Eng. **45**(03), 182–189 (2012)

3. Lei, Y.: Research on protection measures for the underpass City Wall and bell tower of Xi'an metro line 2. Geotech. Mech. **31**(01), 223–228+236 (2010)
4. Gang, W., Jing, L., Haili, X., Dong Lizheng, X., Yongyong, Z.S.: Experimental analysis of the impact of large-scale deep foundation pit excavation on nearby subway shield tunnels. J. Railw. Sci. Eng. **15**(03), 718–726 (2018)
5. Hongwei, Y., Tao, L., Yongwen, Y., Xinyu, X.: Analysis of the effectiveness and engineering application of deep foundation pit partition walls for protecting adjacent buildings. J. Geotech. Eng. **33**(07), 1123–1128 (2011)

Research on the Influence of Aggregate Content on the Abrasion Resistance of High Performance Concrete

Jingjing He[✉], Yong Zhang, Haodan Lu, Peng Huang, and Lihao Fan

Powerchina Xibei Engineering Corporation Limited, Power China, Xi'an 710065, Shanxi, China
hejingji@nwh.cn

Abstract. Performance tests were conducted on high abrasion resistance concrete with different coarse aggregate contents and sand replacement rates. The influence of aggregate content on the slump, compressive strength, and impact and wear resistance of concrete was analyzed. The results show that when the ratio of slurry to aggregate is 5:5, the slump and abrasion resistance strength of high abrasion resistance concrete decrease with the increase of sand replacement rate, while the compressive strength increases with the increase of sand replacement rate. At a ratio of 5:5–5, the abrasion resistance of hydraulic concrete reaches its optimal performance.

Keywords: Aggregate Content · Abrasion Resistance · Hydraulic Concrete

1 Introduction

In hydraulic and hydroelectric projects, high speed mud, sand and stone flows wash hydraulic structures for a long time. This leads to the aggravation of abrasion and cavitation on the overflow surface of concrete buildings, resulting in large-area denudation of surface concrete, which greatly shortens the service life of buildings. The research and application of using high performance concrete (HPC) instead of traditional concrete to improve the impact and wear resistance of hydraulic structures has gradually become a hot spot. The studies show that [1–3], when adding coarse aggregate into HPC, it can reduce the amount of cementitious materials and cut the cost. This can not only reduce the plastic shrinkage of HPC, but also improve the overall stability and strength of concrete because of the bite action between aggregates. The study [4] pointed out that the low water binder ratio of HPC makes the internal structure of concrete more compact. In particular, the interfacial bonding ability between coarse aggregate and matrix is much higher than that of ordinary concrete [5]. This conclusion provides a theoretical basis for adding coarse aggregate to HPC. At present, the research on HPC containing coarse aggregate mostly focuses on its mechanical properties and durability, but the research on abrasion resistance is rarely reported.

© The Author(s) 2024
G. Feng (Ed.): ICCE 2023, LNCE 526, pp. 691–698, 2024.
https://doi.org/10.1007/978-981-97-4355-1_68

Based on the above analysis, high performance grouting material is used as the base material, and different proportions of coarse aggregates are added. Some of the coarse aggregates are then replaced with sand to prepare hydraulic high performance abrasion resistant concrete, and performance tests are conducted. Analyze the influence of aggregate content on the abrasion resistance strength of concrete, and obtain the optimal aggregate content. This can provide a basis for the application of hydraulic abrasion resistant concrete engineering, thereby improving the economic efficiency of abrasion resistant concrete.

2   Test Overview

2.1   Raw Materials

2.1.1   Grouting Material

A high performance grouting material is used as the base material, including cement, sand, admixture and other materials. The design strength of high performance grouting material is C80. Table 1 shows the test results of physical and mechanical properties of grouting materials.

Table 1. Physical and mechanical property index of grouting material

Fluidity (mm)		Bleeding rate (%)	Compressive strength (MPa)			Vertical expansion rate (%)*	
Initial fluidity	Fluidity at 30 min		1 d	3 d	28 d	3 h	24 h–3 h
300	278	0	25	61	82	0.14	0.07

*3 h express the vertical expansion rate of grouting material in 3h, 24 h–3 h express the difference between 24 h and 3 h expansion rates.

2.1.2   Aggregate

The test aggregates are all artificial aggregates, in which the fineness modulus of sand is 2.6, the particle size of small stone is 5 mm–20 mm, and the particle size of medium stone is 20 mm–40 mm. The physical performance test results of aggregates are shown in Table 2.

Table 2. Physical performance indicators of aggregates

Project	Sand	Small stone	Medium stone
Saturated surface dry density (kg/m^3)	2600	2620	2630
Saturated surface dry water absorption (%)	1.38	0.90	0.62
Needle and flake content (%)	—	6.8	2.0
Crushing index (%)	—	8.9	—
Robustness	2.5	3.9	0.1

2.2 Experimental Design

High abrasion resistance concrete is prepared by adding different proportions of fine aggregate and coarse aggregate into high performance grouting material. The ratio of grouting material to coarse aggregate is 5:5 and 4:6. The ratio of small stone to medium stone is 5:5. The replacement rate of fine aggregate (sand) is 0%, 5%, 10% and 15%. The water consumption is 10% of the quality of grouting material. The workability of concrete can be improved by adding a certain amount of fine aggregate. Each concrete is calculated as 100 kg and its mix proportion is shown in Table 3.

Table 3. Test mix proportion

test number	sand substitution rate	sand/kg	medium stone/kg	small stone/kg	grouting material/kg	water/kg
5:5–0	0%	0.0	25.00	25.00	50	5
5:5–5	5%	2.5	23.75	23.75	50	5
5:5–10	10%	5.0	22.50	22.50	50	5
5:5–15	15%	7.5	21.25	21.25	50	5
4:6–0	0%	0.0	30.00	30.00	40	4
4:6–5	5%	3.0	28.50	28.50	40	4
4:6–10	10%	6.0	27.00	27.00	40	4
4:6–15	15%	9.0	25.50	25.50	40	4

3   Test Method and Analysis of Test Results

3.1 Slump Test and Result Analysis

The slump test of concrete mixture shall be carried out according to Chinese specification DL/T 5150-2017. And the workability of the mixture is analyzed. The slump test process of high abrasion resistance concrete mixture at a ratio of 5:5 is shown in Fig. 1. The mixture state and slump test process of high abrasion resistance concrete with a ratio of 4:6 are shown in Fig. 2 and Fig. 3, respectively.

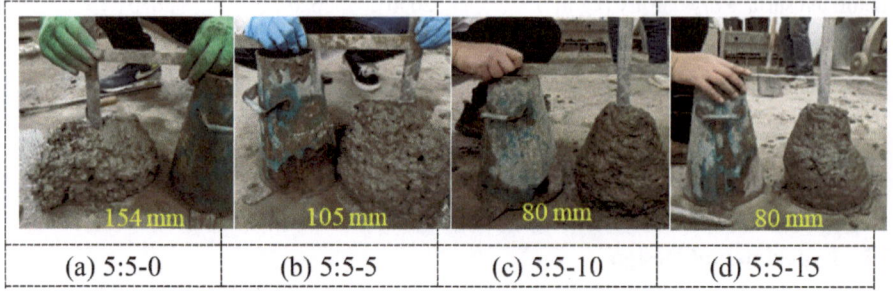

| (a) 5:5-0 | (b) 5:5-5 | (c) 5:5-10 | (d) 5:5-15 |

Fig. 1. Slump test process of concrete (5:5)

Fig. 2. State of concrete mixture (4:6)   **Fig. 3.** Slump test process of concrete (4:6)

It can be seen from Fig. 1 that when the ratio of slurry to stone is 5:5, the cohesion of the mixture is good. The aggregate shall be evenly wrapped and there shall be no water precipitation on the surface. From Figs. 2 and 3, it can be seen that when the ratio of slurry to stone is 4:6, the concrete mixture basically has no slump. Due to the lack of sufficient water for mixing, the proportion of cement slurry and inclusion in aggregate can not be completely reduced.

It can be seen from Fig. 1 that the slump of concrete mixture is 80 mm–154 mm, and the slump decreases with the increase of sand substitution rate. When the sand substitution rate is 5% and 10%, the concrete slump is 32% and 48% lower than that of 5:5–0% group respectively. This is because the sand has a large specific surface area and needs more cement paste to wrap, which reduces the free water in the concrete mixture and reduces the fluidity of the concrete. With the increase of sand replacement rate, the wrapping of cement mortar to aggregate and the workability of mixture are improved. However, when the sand replacement rate exceeds 15%, the specific surface area of aggregate increases, resulting in the relative decrease of water binder ratio of concrete mixture. The workability is reduced to a certain extent.

3.2 Compressive Strength

The compressive strength test of concrete shall be carried out according to the test method in standard DL/T 5150-2017. The compressive test piece is a standard cube with a side length of 150 mm. The 28 d compressive strength comparison of high-performance impact resistant and wear-resistant concrete is shown in Fig. 4 and Fig. 5.

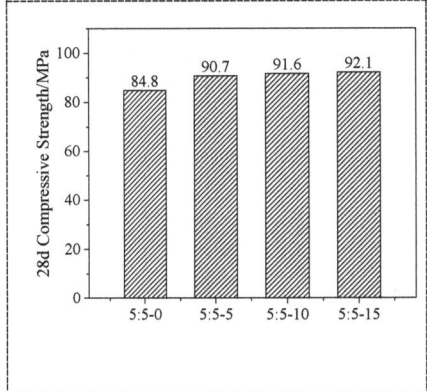 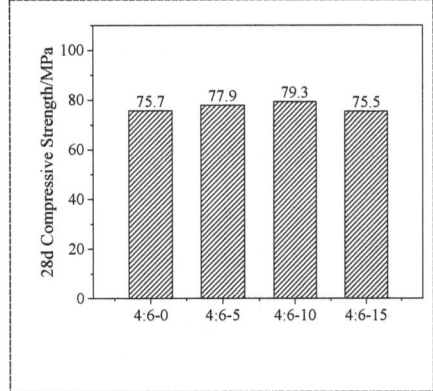

Fig. 4. Compressive strength of concrete (5:5)   **Fig. 5.** Compressive strength of concrete (4:6)

As shown in the Fig., when the ratio of slurry to stone is 5:5, the 28 d compressive strength of high abrasion resistance concrete is between (84–92) MPa. And the compressive strength slightly increases with the increase of sand replacement rate. This is because a certain proportion of sand can fill the gaps in the coarse aggregate, making the aggregate grading more uniform. While improving the workability of concrete mixtures, it also makes the interior of the concrete denser, thereby improving the compressive strength of the concrete.

When the ratio of slurry to stone is 4:6, the 28 d compressive strength of high abrasion resistance concrete is between (75–80) MPa. The sand replacement rate has little effect on the compressive strength of concrete. On the one hand, due to the inherent high proportion of coarse aggregate, the workability of concrete decreases sharply after sand replaces a portion of coarse aggregate, leading to an increase in internal defects in the concrete. On the other hand, when the amount of aggregate added increases too much, the cement slurry cannot fully wrap around the aggregate interface or the wrapping layer is relatively thin, which will cause insufficient adhesion between the aggregate and the matrix, resulting in a decrease in the compressive strength of the concrete.

3.3 Abrasion Resistance

The high abrasion resistance concrete containing coarse aggregate shall be tested for abrasion resistance according to the underwater steel ball method in chinese specification DL/T 5150-2017. This method can be used to measure the relative resistance of concrete to underwater flow medium wear and evaluate the relative wear resistance of concrete surface.The abrasion index of concrete is expressed by the abrasion strength. The abrasion strength according to Eq. (1). Figure 6 shows the abrasion strength of high abrasion resistant concrete at 28 days under two kinds of coarse aggregate content.

$$fa = \frac{tA}{\Delta m} \tag{1}$$

where fa represents the impact wear strength, which is the time required per unit area to be worn per unit mass, in $h/(kg/m^2)$;t represents the cumulative test time, in h; A represents the area of the specimen subjected to impact and wear, in m^2; Δ represents the cumulative mass loss of the specimen after grinding during the t period, in kg.

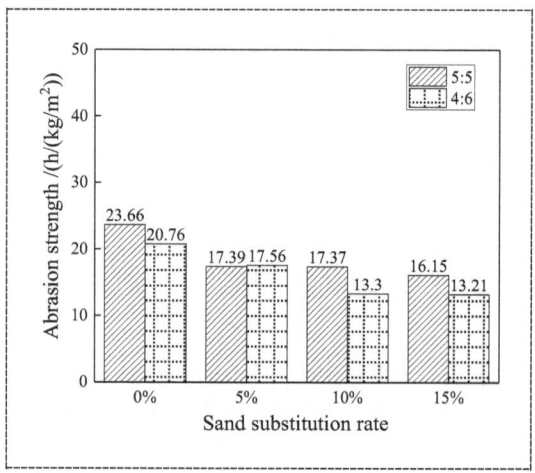

Fig. 6. Abrasion strength of concrete

It can be seen from Fig. 4 that after adding 50% and 60% coarse aggregate into the grouting material, the abrasion strength of high abrasion resistant concrete at 28 days reaches 23.66 $h/(kg/m^2)$, 20.76 $h/(kg/m^2)$ respectively. With the increase of aggregate proportion, the slump of concrete mixture decreases obviously, and the workability becomes worse, which increases the internal defects of concrete and leads to the reduction of impact and wear strength. After sand replaces part of coarse aggregate, the impact and abrasion strength of concrete in 28 days is reduced to (13–18) $h/(kg/m^2)$. The abrasion strength of concrete decreases slightly with the increase of sand replacement rate. This is because the impact of fine aggregate on the impact and abrasion resistance of concrete is more significant than that of coarse aggregate. The artificial aggregate used in the study has high crushing index and relatively low aggregate strength, which will change the first failure trend of the interface transition zone. Therefore, the abrasion strength of concrete will be greatly reduced after adding fine aggregate.

Figure 7 shows the impact abrasion test failure diagram of high abrasion resistance concrete when the coarse aggregate content is 50%. It can be seen from the Fig. 7 that the form of abrasion resistance is surface aggregate exposure, indicating that the interfacial bonding strength between aggregate and rubber material is high. And the exposure degree of aggregate increases with the increase of sand substitution rate.

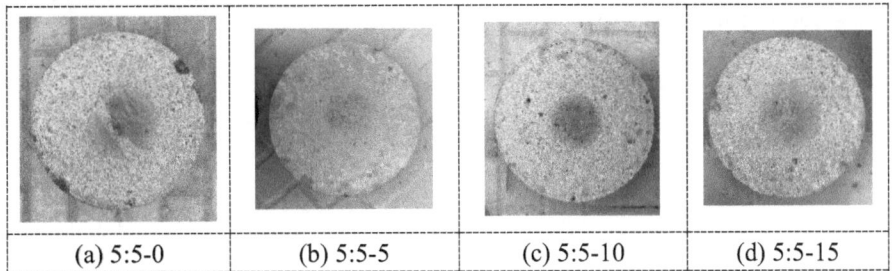

| (a) 5:5-0 | (b) 5:5-5 | (c) 5:5-10 | (d) 5:5-15 |

Fig. 7. Damage diagram of abrasion sample

Based on the analysis of the above test results, when the content of coarse aggregate is 50% and the sand replacement rate is 5%, the workability and abrasion resistance of concrete are better. Under this mix proportion, the slump of concrete mixture is 105 mm. The compressive strength is 90.7 MPa. And the abrasion strength is 17.39 h/(kg/m^2).

4 Conclusion

(1) The ratio of slurry to aggregate has a certain impact on the workability, compressive strength, and impact and wear resistance of high impact and wear resistance concrete.
(2) When the ratio of slurry to aggregate is 5:5, the slump and impact wear strength of high abrasion resistance concrete decrease with the increase of sand replacement rate, while the compressive strength increases with the increase of sand replacement rate.
(3) When the ratio of slurry to aggregate is 5:5 and the sand replacement rate is 5% (5:5–5), the workability of concrete is good, with a compressive strength of 90.7 MPa and an abrasion resistance of 17.39 (h/(kg/m^2). This ratio not only ensures certain mechanical properties and impact wear resistance, but also reduces engineering costs.

Acknowledgments. The authors gratefully acknowledge the support of the Natural Science Basic Research Program of Shaanxi China (Program No. 2021JQ-983).

References

1. Peng, G.F., Teng, Y., Huang, Y.Z.: Fire resistance of high performance concrete with coarse aggregate: a research review. J. North China Univ. Water Res. Electric Power **34**(1), 1–6 (2013)
2. Yu, Z.R., Wang, B.H., An, M.Z.: Experimental study on bending performance of ultra high performance fiber-reinforced concrete slab. J. Build. Struct. **40**(9), 131–139 (2019)

3. Cheng, P., Wu, F.H., Zeng, Y.Q., Shi, Y.C.: Experimental research on flexural properties of high performance concrete with coarse aggregate. J. Water Res. Arch. Eng. **17**(2), 41–45 (2019)
4. Aictin, P.C.: High Performance Concrete, pp. 199–203. Routledge, New York (2004)
5. Richard, P., Cheyrezy, M.: Composition of reactive powder concretes. Cem. Concr. Res.. Concr. Res. **25**(7), 1501–1511 (1995)

Design Optimization of Jacket Substructure of Offshore Wind Turbine

Ben He, Na Lv$^{(\boxtimes)}$, and Ruilong Shi

Power China Huadong Engineering Corporation Limited, Hangzhou 311121, Zhejiang, China
lv_n@hdec.com

Abstract. The design optimization of the jacket substructure could reduce the cost of energy and the correlated carbon cost. The mass of jacket is optimized with respect to maximum stress and displacement responses. Wind and dead loads were applied at the top of the jacket referenced point. Wave load and foundation boundary were applied using Dload and UEL function in the Abaqus subroutine. Six different parameters were evaluated, including the pile height above mudline, the base width, brace thickness and diameter, and the main leg thickness and diameter. The parametrically optimized model has mass minimized by 16%. The maximum stress at the top of the optimal jacket reduced to 274.91 MPa from 288.70 MPa. The maximum displacement at the top increased to 87.37 mm from 83.88 mm, albeit, lower than the code limit of 175 mm. The absolute value of the maximum rotation at the top reduced from 0.309° to 0.304°, which is lower than the code limit of 0.382°. The pile rotation at the mudline increased by 14%, from 0.120° to 0.137°, nonetheless, lower than the DNV code limit of 0.250°. Initial numerical basis has been developed for the parametric design optimization of the jacket substructure. From the parameter studies conducted mathematical models could be developed for efficient numerical optimization.

Keywords: Offshore Wind Turbine · Jacket Substructure · Abaqus-Python Scripting · Fortran Subroutine · Geometric Design Optimization

1 Introduction

Research and investment in offshore wind turbines have been in progress since the last 40 years. Wind energy life cycle production process has low carbon effect and is sustainable [1]. For water depth of above 25 m the jacket substruction configuration comes as the best option [2]. The X-brace jacket configuration was chosen because previous studies confirmed it has good performance compared with other bracings configuration [3, 4].

The wind, wave and dead loads were the main ultimate limit state loads for the offshore wind turbine jacket substructure. The wind loads are the design driving force whiles the wave loads have secondary effect on the response of the structure under ultimate load condition [5]. The key variables for the design are the wind speed, wind intensity, wave height and wave period. These data are normally taking from actual site records for a period.

© The Author(s) 2024
G. Feng (Ed.): ICCE 2023, LNCE 526, pp. 699–706, 2024.
https://doi.org/10.1007/978-981-97-4355-1_69

Engineering optimization is generally classified into topology, shape, and size optimization [6]. Element types are specific to the different types of the optimization. Sizing optimization is a common process used to improve the performance of engineering designs [7]. The common geometry variables in sizing optimizations are diameter, thickness, length, width, and height. Motlagh optimized the jacket structure by using the diameter and the thickness of the jacket pipes as the design variables. The mass was minimized by 15% under ultimate loading [8]. In a similar research, Sandal et al., optimized the jacket using the thickness and diameter as optimization variables. The mass reduced by 40% and with reduced performance of the maximum horizontal displacement, increasing by 80 mm. In general, for the jacket substructure, the variables influencing the optimization process are the base width, the jacket height, the member diameter and thickness [9]. Therefore, for this research the base width, the diameter and thickness were the main variables. After the optimization process the responses were checked against code limitations and were within code limits whiles minimizing the mass. Therefore, the performance of the jacket with respect strength have been maintained, whiles reducing the mass.

2   The Jacket Boundary Conditions and Finite Element Modeling

2.1   The Boundary Conditions

To model the support as realistic as possible, the soil structure interaction was considered in the foundation behaviour, therefore spring support was used. The ultimate load on top of the jacket structure applied at the reference point (RP) is made of estimated wind load, rotor nacelle assembly (RNA), and the tower. The unfactored loads from wind are $Fx = Fy = 5071$ kN, $Mx = My = 423875$ kNm, $Mz = -33324$ kNm, and RNA, and tower load $Fz = -14972$ kNm. The wave load calculation was base on the Morisons equations. The velocities and acceleration of the wave were estimated by Stokes 5th order calculation model. The wave loads were based on the significant wave height (Hs) of 13.98 m, peak period (Tp) of 18.06 s, and wave length (L) of 308.07. The ultimate load used for the design was based on load combination in Table 1, which is based on Chinese code GB 50009-2012 [10]. The jacket configuration is shown in Fig. 1.

2.2   Finite Element Model

From Fig. 1, a total of 196 components were modeled and members having similar geometric properties classified into 26 groups. In the absence of the tranistion piece, the top was model with 50 mm thick 2-dimensional plate element and 4-top pipe elements of 1550 mm diameter and 220 mm thickness to give enougth stiffness for transmission of applied load. The high water level of (HWL) 66.39 m was used in the model. The coordinates of the jacket substruture were modeled with python-Abaqus-scripting. The yield strength of the steel is 355 MPa, and the tensile strength used is 560 MPa. The B31 CPS3 linear pipe element type was used.

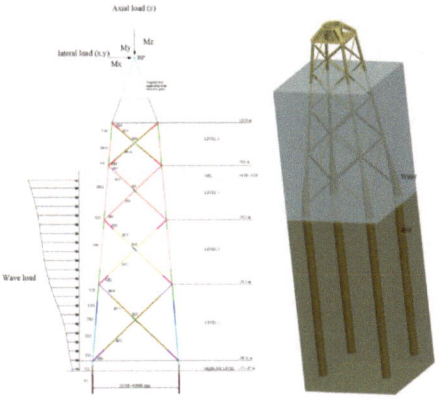

Fig. 1. The jacket model with spring support *(MSL- mean water level)*

Table 1. Load combination for design

Load case	wind	Wave	Deadload
LC	1.5*(wind1)	1.35*0.7*(wave1)	1.0*DL

3 Parameter Studies and Optimization

3.1 Parameter Studies

Six parameters were studied for the optimization; 1) pile length (PL) above the mudline, 2) the base width (BW), 3) brace diameter (BD), 4) brace thickness (BT), 5) leg diameter (LD), and 6) Leg thickness (LT). The responses with respect to maximum displacement, maximum stress, maximum rotation, mudline displacement, mudline rotation and mudline stress were analysed.

Increasing the PL (1400 mm to 5900 mm), has little effect on the maximum stress (+0.8 MPa), the maximum displacement increased by 9.4% (+7.69 mm), the mudline displacement increased by 10.3% (+1.74 mm), the maximum top rotation increased by 2.3% (+0.007°), the maximum pile rotation at the mudline increased by 34.3% (+0.037°).

Increasing the BW (22500 mm to 40000 mm), has significant effect on the maximum stress (−42.98 MPa), the maximum displacement reduced by 49.3% (−70.29 mm), the mudline displacement decreased by 5.7% (−1.06 mm), the maximum top rotation decreased by 8.4% (−0.028°), the maximum pile rotation at the mudline increased by 36.2% (+0.034°).

Increasing the BD (by + 100 mm @ + 20 mm for each model), has little effect on the maximum stress (−1.68 MPa), the maximum displacement reduced by 2.7% (−2.28 mm), the mudline displacement decreased by 4.4% (−0.81 mm), the maximum top rotation decreased by 4.7% (−0.015°), the maximum pile rotation at the mudline decreased by 17.1% (−0.024°).

Increasing the BT (by + 25 mm @ + 5 mm for each model), has significant effect on the maximum stress (−85 MPa), the maximum displacement reduced by 10.4%

(−9.61 mm), the mudline displacement decreased by 2.3% (−0.41 mm), the maximum top rotation decreased by 22.2% (−0.086°), the maximum pile rotation at the mudline decreased by 24.7% (−0.038°).

Increasing the LD (by + 100 mm @ + 20 mm for each model), has moderate effect on the maximum stress (−6.84 MPa), the maximum displacement reduced by 5.4% (−4.7 mm), the mudline displacement decreased by 3.4% (−0.62 mm), the maximum top rotation decreased by 2.8% (−0.009°), the maximum pile rotation at the mudline decreased by 0.8% (−0.001°).

Increasing the LT (by + 25 mm @ + 5 mm for each model), has significant effect on the maximum stress (−86.69 MPa), the maximum displacement reduced by 24.1% (−25.52 mm), the mudline displacement decreased by 4.8% (−0.87 mm), the maximum top rotation decreased by 4.1% (−0.013°), the maximum pile rotation at the mudline increased by 3.4% (+0.004°).

Based on the trends of the results achieved, the jacket was parametrically optimized. The PL and BW of the original model were maintained, so the response of the jacket with respect to the BW and PL has been retain to the original model. Increasing the BD has little effect on the maximum stress, and displacement, the some of the BD features were increased. to compensate for the increment in BD, BT parameters were decreased. Decreasing the BT, is expected to have significant effect on the maximum stress increment, moderate effect on the increment of the maximum displacement, significant effect on both the maximum top rotation and the rotation at the mudline level. To counteract the effect of the BT decrement, the LD parameter was moderately increased level 2 to level 4 members (original features maintained for level 1). Increment of LD, has moderate to little impact on all the responses studied. Finally, the LT was decreased to balance the reduction in stress by the increment of the LD.

3.2 Parametric Optimization

Based on the optimization philosophy developed from the above, the jacket model was parametrically optimized. The detail responses of the optimized solution are shown in Table 2, Figs. 2, 3, 4 and 5. Compared to the original design the optimized option (OPT1) mass reduced by 16%. The maximum stress reduced from 288.7 MPa to 274.91 MPa (Fig. 2). The Maximum displacement increased from 83.88 mm to 87.37 mm (Fig. 3), the maximum rotation reduced from 0.309° to 0.304° (Fig. 4) and rotation at the mudline increased from 0.1203° to 0.1384° (Fig. 5). From Fig. 2, the stress contour of the optimized model is well distributed compare to the stress contour of the original model.

3.3 Comparing Results with Design Code Limits

The maximum stress response of the optimized structure is below the elastic limit of 355 MPa. The jacket has been parametrically optimized. The limiting value of the current design horizontal displacement based on the ASCE-7-16 is 175 mm. The maximum displacement from the parameter optimization is 87.37 mm. From Eurocode 7 [11], the relative maximum rotation acceptable for ultimate limit state is 1/150 rad (0.38°), the optimized solution satisfied the rotation condition under ultimate state. The optimized model has maximum top rotations of 0.3044°, and being lower than the original jacket

Table 2. Results of optimized solution

Model	Mass (ton)	Maximum top stress (MPa)	Base width (mm)	Maximum displacement at top (mm)	Maximum rotation at top (deg)	Maximum displacement at mudline (mm)	Maximum rotation at mudline (deg)
OPT1	1490	274.91	35000	87.37	0.3044	18.96	0.1384
ORIGH	1781	288.7	35000	83.88	0.3088	17.56	0.1203

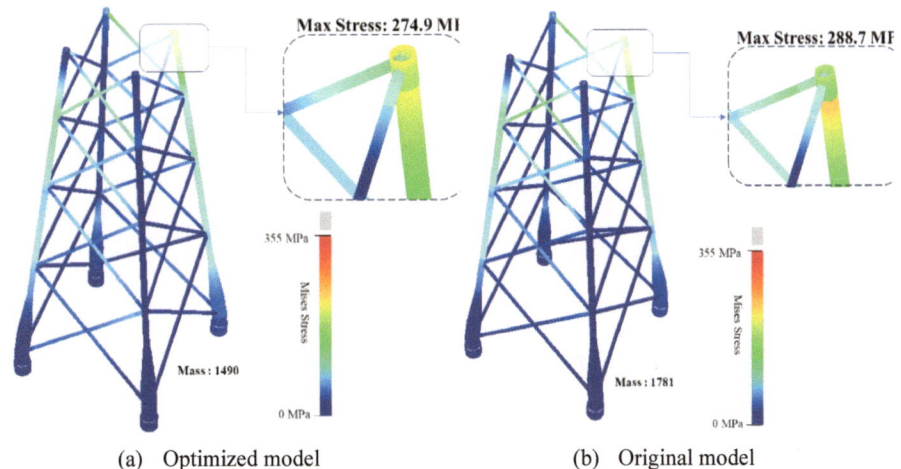

(a) Optimized model   (b) Original model

Fig. 2. Stress contours of optimized and original model

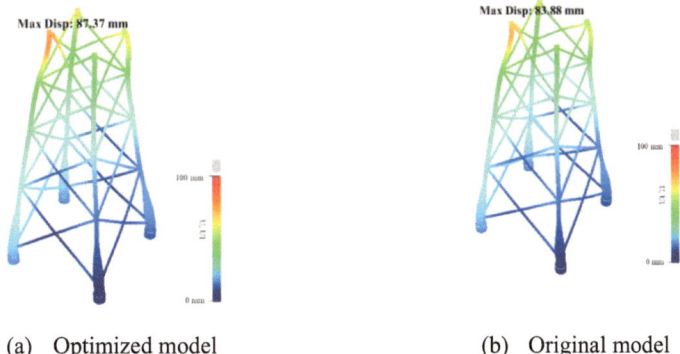

(a) Optimized model   (b) Original model

Fig. 3. Displacement contours of optimized and original model

maximum rotation of 0.3088°. The DNV code J101, limit the mudline rotation to 0.2500° under serviceability loading. The mudline rotations for the optimized solution under

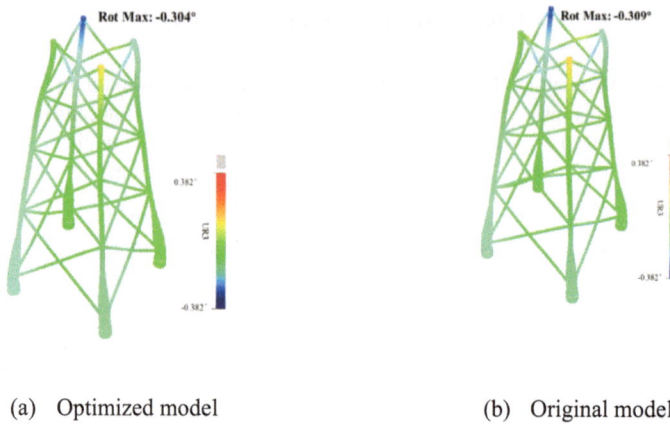

(a)   Optimized model   (b)   Original model

Fig. 4. Rotation contours of optimized and original model

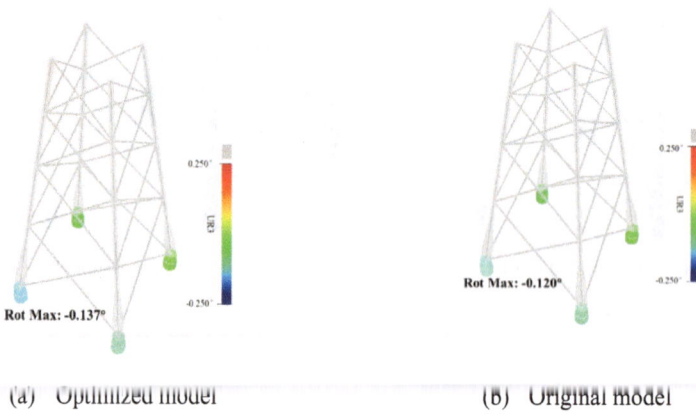

(a)   Optimized model   (b)   Original model

Fig. 5. Mudline rotation contours of optimized and original models

ultimate loading is 0.1384°, which is lower than the DNV codes limits but greater than the original model mudline rotation of 0.1203°.

4   Conclusions

The FEA models of the jacket substructure were built using GMSH module in python-scripting environment. The models were then converted to Abaqus-readable inp files. The foundation boundary conditions and the wave load were applied through Fortran subroutine using the UEL and Dload functions respectively. Parameter studies was conducted on the models of jacket substructure under extreme ultimate loading. Base on the parameter studies the jacket substructure was quickly optimized. The following are the summaries of the investigation:

Six different parameters of the jacket were studied for the parametric optimization.

From the results of the parameter studies, the maximum stress was mainly affected by the variation of BW, BT, and LT. The maximum displacement was mainly affected by the variation of BW and LT. The maximum rotation at the top was affected by the movement in BT only. The pile mudline rotation was affected by the movement in BW, PL, BD, and BT.

The parametrically optimized mass reduced 16%. The maximum stress reduced from 288.7 MPa to 274.9 MPa. The absolute value of the maximum rotation at top reduce from 0.309° to 0.304°. The maximum displacement increased to 87.37 mm from 83.88 mm. The mudline rotation increased by 14.2% (from 0.120° to 0.137°).

A quick optimized solution has been developed through the trend developed from the parameter studies. The works shows an efficient modelling of the jacket substructure for optimization. The work can be enhanced by developing mathematical models from the parameter studies for an efficient numerical optimization using a good optimization algorithm.

Acknowledgments. The authors gratefully acknowledge financial support from National Natural Science Foundation of China (No. 52271294).

References

1. Wang, X., Zeng, X., Li, J., Yang, X., Wang, H.: A review on recent advancements of substructures for offshore wind turbines. Energy Convers Manag. **158**, 103–119 (2018)
2. Shi, W., Park, H., Chung, C., Baek, J., Kim, Y., Kim, C.: Load analysis and comparison of different jacket foundations Renew. Energy **54**, 201–210 (2013)
3. Zhang, P., et al.: Bearing capacity and load transfer of brace topological in offshore wind turbine jacket structure. Ocean Eng. **199**, 107037 (2020)
4. Tran, T.-T., Kim, E., Lee, D.: Development of a 3-legged jacket substructure for installation in the southwest offshore wind farm in South Korea. Ocean Eng. **246**, 110643 (2022). https://doi.org/10.1016/j.oceaneng.2022.110643
5. Shittu, A.A., Mehmanparast, A., Amirafshari, P., Hart, P., Kolios, A.: Sensitivity analysis of design parameters for reliability assessment of offshore wind turbine jacket support structures. Int. J. Naval Arch. Ocean Eng. **14**, 100441 (2022). https://doi.org/10.1016/j.ijnaoe.2022.100441
6. Assimi, H., Jamali, A.: A hybrid algorithm coupling genetic programming and Nelder-Mead for topology and size optimization of trusses with static and dynamic constraints. Expert Syst. Appl. **95**, 127–141 (2018)
7. Gerzen, N., Clausen, P.M., Pedersen, C.B.W.: Sizing optimization for industrial applications and best practice design process. In: High Performance and Optimum Design of Structures and Materials II, pp 41–49. WIT Press (2016)
8. Motlagh, A.A., Shabakhty, N., Kaveh, A.: Design optimization of jacket offshore platform considering fatigue damage using Genetic Algorithm. Ocean Eng. **227**, 108869 (2021)
9. Sandal, K., Latini, C., Zania, V., Stolpe, M.: Integrated optimal design of jackets and foundations. Mar. Struct.Struct. **61**, 398–418 (2018)
10. Chinese Standard 2012 Chinese code GB 50009-2012 (2012)
11. EN 1997-1: Eurocode 7: Geotechnical design - Part 1: General rules (2004)

Integration of Structure and Enclosure System (Large Span Curtain Wall Structure) Monitoring Technology and Linkage Analysis

Wentao Gao[✉] and Jian Liu

Shenzhen Branch of China Academy of Building Research, Greater Bay Area Research Institute
National Engineering Research Center of Building Technology, Shenzhen, China
306253795@qq.com

Abstract. The integrated monitoring technology and linkage analysis of the structure and enclosure system (large span curtain wall structure) are studied by studying the impact of deformation of the main structure on the stress and deformation performance of the curtain wall structure, as well as considering the impact of the load transmitted by the curtain wall structure to the main structure on the main structure. Conduct research on three scenarios: pure large-span curtain wall structural model, pure main body structural model, and collaborative work between large-span curtain wall structural model and main body structural model. Compare and study indicators such as period, deformation, and interlayer displacement angle. The analysis shows that the load transmitted by the large-span curtain wall structure to the main structure has very little impact on the deformation of the main structure; The deformation of the main structure has little impact on the relative deformation of the large-span curtain wall structure.

Keywords: Main Structure · Large Span Curtain Wall Structure · Monitoring · Overall Analysis

1 Project Overview

Due to the difficulty of monitoring the main structure, such as high-altitude operations, structural components being covered by decorative layers, etc. In order to achieve the monitoring task of the enclosure structure and the main structure, data collection is carried out by arranging monitoring points on the inner side of the building enclosure structure room or the outer surface of the enclosure structure; Using mature calculation software, establish a calculation model for the enclosure structure and the main structure. By comparing and analyzing their respective periods, displacements, stresses, etc., determine the mutual influence and loading form of the enclosure structure and the main structure; Calculate the enclosure structure and main structure by applying loads obtained from monitoring data; And compare with the corresponding monitoring results to identify the reasons for the differences in results, and modify the calculation model and assumptions to achieve the goal of matching the final calculation results with the monitoring results [1–3].

© The Author(s) 2024
G. Feng (Ed.): ICCE 2023, LNCE 526, pp. 707–713, 2024.
https://doi.org/10.1007/978-981-97-4355-1_70

As a part of the building, the large-span curtain wall structure needs to be considered in collaboration with the main structure (overall analysis or simplified analysis). The load, deformation, and stress of the curtain wall structure are linearly distributed. It is necessary to study the impact of the deformation of the main structure on the stress and deformation performance of the curtain wall structure, and also consider the impact of the load transmitted by the curtain wall structure to the main structure on the main structure.

This article conducts research on three scenarios: pure main structure model, pure large span curtain wall structure model, and collaborative work between large span curtain wall structure model and main structure model. It mainly compares and studies indicators such as period, deformation, and interlayer displacement angle, analyzes the mutual influence between the main structure and large span curtain wall, determines the loading method between them, and explores the feasibility of separate calculation between large span curtain wall structure and main structure, By simplifying the model for analysis, the purpose of monitoring is achieved.

Project situation: A rectangular building with a podium length of 92 m and a width of 42 m; The tower is 63 m long and 23 m wide; The span of the large-span curtain wall structure is 10.4 m, with one frame every 2 floors (i.e. 7.8 m); The large-span curtain wall structure is a steel structure, with the main structure being a steel frame structure. The analysis software is Midas Gen.

In order to simplify the calculation, the long-span curtain wall structure only considers dead load and wind load, DL (dead load): 7.8 KN/M, W + (wind load): 13.3 KN/M, W - (wind load): 13.3 KN/M [4–6].

2   Comparative Analysis of Models

2.1   Analysis Model

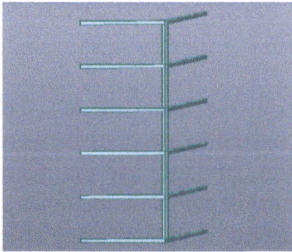

Fig. 1. A Structural Model of Pure Long Span Curtain Wall

2.2   Analysis Ideas

Step 1: The initial model support of the pure large-span curtain wall structure is a rigid joint support, and the support reaction forces F_{1i} and M_{1i} (i taken as x, y, z) are calculated and analyzed in Fig. 1, Fig. 2 and Fig. 3;

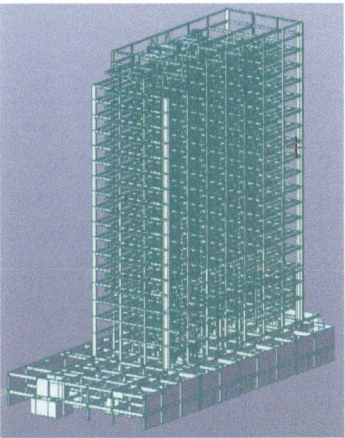

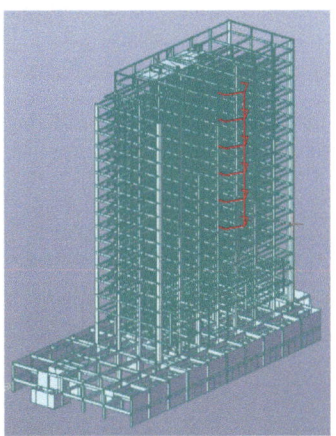

Fig. 2. Pure subject structure model   **Fig. 3.** Final assembly model

Step 2: Apply the calculated support reaction force of the large-span curtain wall structure to the pure main structure, obtain the displacement values Di and Ri (i takes x, y, z) of each support, and calculate the stiffness of each support through Ki = F1i/Di and Ki = M1i/Ri (i takes x, y, z). This stiffness value is the stiffness value after considering the deformation of the main structure;

Step 3: Add the stiffness value Ki after considering the deformation of the main body to the pure main body structural model for calculation and analysis, and obtain the support reaction forces F2i and M2i. This analysis result is the result after considering the influence of the main body;

Step 4: Re add the support reaction forces F2i and M2i from the previous step to the pure main body model for calculation and analysis;

Step 5: Collaborate between the large-span curtain wall structure model and the main structure model, and compare the calculation and analysis results with the pure main structure model and the pure large-span curtain wall structure model analysis results;

2.3 Cycle Comparison

Through modal comparison analysis, the pure main structure model T1 = 4.7805 corresponds to the final assembly model T1 = 4.766, with a very similar period. The periods of other corresponding modes are also very similar, and the modal trend is consistent. The pure curtain wall structure has minimal impact on the main structure (Table 1).

2.4 Comparison of Deflection Between Pure Main Structure and Final Assembly Model

When the stiffness of the main structure is considered in the large-span curtain wall structure, the generated reaction force is applied to the pure main structure. At this time, the pure main structure model (considering the influence of the load transmitted by the curtain wall) is analyzed, and compared with the final assembly model analysis.

Table 1. Period comparison

Model	Vibration mode	Cycle
Pure main structure model (considering the influence of curtain wall structure)	1	4.7805
	2	3.5343
	3	3.3238
Mold assembly model	1	4.766
	2	3.5418
	3	3.336

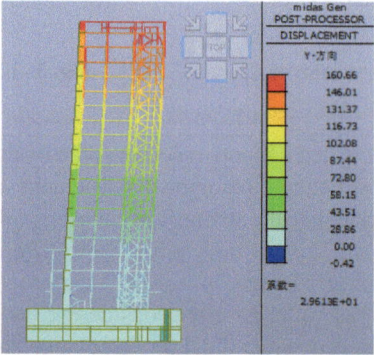

Fig. 4. Pure main body (considering the influence of curtain wall) Y-direction deflection

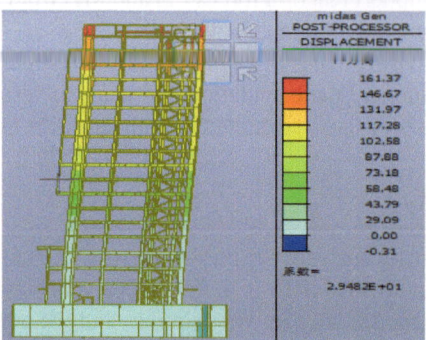

Fig. 5. Y-axis deflection of the final assembly model

Under 1.0 constant + 1.0 wind conditions, the deflections of the pure main structure model (considering the influence of the load transmitted by the curtain wall) are DY = 160.66 mm and DX = 119.24 mm, respectively; The deflections of the final assembly model are DY = 161.37 mm and DX = 118.14 mm, as shown in Table 2 for comparative analysis. From this, it can be seen that the deflection difference in the Y direction is 0.44%,

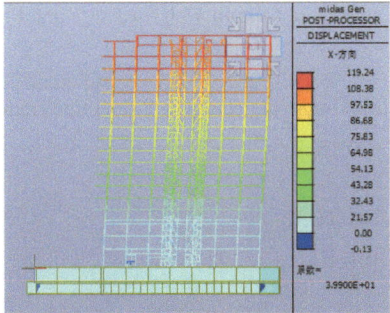

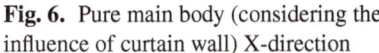

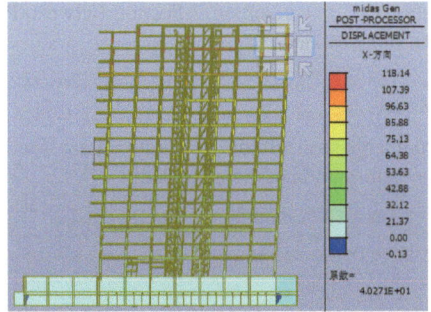

Fig. 6. Pure main body (considering the influence of curtain wall) X-direction

Fig. 7. Final assembly model X-direction deflection

the deflection difference in the Y direction is 0.93%, and the deflection differences in the X and Y directions are very small. According to this simplified method, the deflection of the pure main structure analysis and the deflection of the final assembly mold are basically consistent in Fig. 5, Fig. 6 and Fig. 7 (Fig. 4).

Table 2. Comparison of Deflection Between Pure Main Structure and Final Assembly Model

Model	Maximum deflection in Y direction (mm)	Maximum deflection in X direction (mm)
Pure main body (considering the influence of curtain walls)	160.66	119.24
Final assembly model	161.37	118.14

2.5 Comparison of Deflection Between Pure Large-Span Curtain Wall Structure and Final Assembly Model

When analyzing the pure large-span curtain wall structure, the influence of the main stiffness was considered, and compared with the final assembly model analysis, only the Y-direction deflection was studied, as shown in Table 3. According to the analysis in the previous section, the deformation of the pure main structure in the Y-direction is 160.66 mm. When considering the stiffness of the main structure, the Y-direction deformation of the large-span curtain wall structure is 0.16 mm, and the relative deformation of the pure main structure in the Y-direction is $161.37 - 0.16 = 161.21$ mm, which is 0.34% different from the deformation of 160.66 mm when analyzing the pure main structure alone, and the difference is very small. According to the analysis, the main structure has a linear relationship with the deformation of the large-span curtain wall structure, and the deformation is coordinated from top to bottom. Therefore, the main structure has little impact on the large-span curtain wall structure, and this simplified model method is feasible.

Table 3. Comparison of Deflection Between Pure Large-Span Curtain Wall Structure and Final Assembly Model

Model	Maximum deflection in Y direction (mm)
Pure large span curtain wall (without considering the deformation of the main body)	0.16
Final assembly model	161.37
relative deformation	161.21

2.6 Comparison of Interlayer Displacement Angles Between Pure Main Structure and Final Assembly Mode

Taking 16F as an example, when considering the influence of curtain wall load transfer, the interlayer displacement angle of the pure main structure is 1/3075 < 1/550; The inter story displacement angle of the main structure of the final assembly model is 1/3153 < 1/550, with a difference of 2.5%, which is relatively small. Analogy analysis of other layers shows that the differences are relatively small. From the analysis results, it can be seen that the main structure considers the influence of load transfer from the large-span curtain wall structure, and the interlayer displacement angle is very close to the calculation results of the final assembly model, indicating that this simplified model method is feasible.

3 Conclusion

By comparing the pure large-span curtain wall model (considering the main stiffness), pure main structure model (considering the influence of large-span curtain wall), and final assembly model on the period, interlayer displacement angle, and deflection of large-span curtain wall structure and main structure, it can be seen from the period that the modal trends of the sub model and final assembly model are consistent, and the interlayer displacement angle and deflection are also very close. The main conclusions are as follows:

1) The most accurate way to monitor various indicator data of the main structure is to use the final assembly model of the main body and large-span curtain wall structure for analysis and monitoring. However, in actual projects, many models are relatively large in size, and if we want to use the final assembly model for calculation and analysis, both configuration and efficiency requirements are relatively high. So it is necessary to simplify the model for analysis and processing to achieve the goal of improving efficiency.

2) The pure large-span curtain wall structural model is a simplified method of applying a large-span curtain wall structural reaction force to the main model, calculating and analyzing the lateral stiffness of the main structure, and then adding it back to the large-span curtain wall structural model for analysis and calculation. This simplified method greatly improves the efficiency of calculation and analysis. By comparing and

analyzing the results of the combined model calculation, it is found that the cycle and modal trends are consistent, and the differences in interlayer displacement angles and deflections are also very small. Therefore, the pure large-span curtain wall structure model can be analyzed separately in this simplified model method. However, in actual projects, comprehensive consideration needs to be given to the actual situation and engineering requirements.

3) The pure subject model is a simplified method of directly applying the reaction force obtained from the separate analysis of the pure large-span curtain wall structure to the subject, which takes into account the stiffness of the main structure. This simplified method also greatly improves the efficiency of calculation and analysis. By comparing and analyzing the results of the combined model calculation, it is found that the cycle and modal trends are consistent, and the deformation difference is also very small. Therefore, the pure main structure model can be analyzed separately in this simplified model method. However, in actual projects, comprehensive consideration needs to be given based on the actual situation and engineering requirements.

References

1. Building structure load specification: GB 50009-2012. China Construction Industry Press, Beijing (2012)
2. Steel structure design standard: GB 50017-2017. China Construction Industry Press, Beijing (2018)
3. Seismic specification for building structures: GB 50011-2010. China Construction Industry Press, Beijing (2010)
4. Technical procedures for space grid structure:JGJ 7—2010. China Construction Industry Press, Beijing (2010)
5. Concrete Structure Design Specification:GB 50010—2010. China Construction Industry Press, Beijing (2011)
6. Technical regulations for steel structures in high-rise civil buildings: JGJ 99-2015. China Construction Industry Press, Beijing (2015)

Hamiltonian-Jacobi Equation for Torsional Problems of Corrugated Steel Web Box Girders

Xin Shu[✉], Guohua Li, and Jun Dong

Daxing Campus of Beijing University of Civil Engineering and Architecture, Beijing, China
sx17783531385@163.com

Abstract. Due to the use of corrugated steel plate in the web, the torsional performance of the beam body will be weakened to a certain extent. In this paper, the Hamiltonian-Jacobian equation is used to analyze the torsion of the corrugated steel web composite box girder. The Lagrangian function of the system is obtained by taking the warping function and the relative torsion angle of the cross-section as the two generalized coordinates, and the Hamiltonian-Jacobian equation for the torsion problem is established by using the regular transformation. The accuracy of the proposed method is verified by comparing the obtained results with the literature values, and the fact that the torsional performance of the beam is weakened by using corrugated steel plate is verified by using the results of this paper.

Keywords: Corrugated Steel Web · Torsional Properties · Warpage Function; Relative Torsion Angle · Hamiltonian-Jacobian Equations

1 The Hamiltonian-Jacobian Equation for the Torsion Problem

The warping function of the cross-section and the relative torsion angle are selected as the two generalized coordinates of the torsion problem [5], they are both functions of the coordinates z, $\beta = \beta(z)$ and $\varphi = \varphi(z)$, and the time variable t in Hamiltonian mechanics is replaced by the coordinate z, and the Lagrangian function of the torsion problem is [3, 4]:

$$L = \frac{1}{2}\left(\sum_{i=1}^{4} E_i I_{\omega i}\right)\dot{\beta}^2 + \frac{1}{2}\left(\sum_{i=1}^{4} G_i J_i\right)\dot{\varphi}^2 + \frac{1}{2}\left(\sum_{i=1}^{4} G_i(I_{Pi} - J_{Bi})\right)(\dot{\varphi} - \beta)^2 - m\varphi \tag{1}$$

where $I_{\omega i}$ is the main sector moment of inertia of the cross-section, J_i is the torsional moment of inertia of the cross-section, I_{Pi} is the polar moment of inertia of the cross-section, and J_{Bi} is the Brett torsional moment of inertia of the cross-section [6], which are all cross-section constants that can be found and calculated by the engineering manual. m is the moment of the external force acting on the beam per unit length. Let $A = \frac{1}{2}\left(\sum_{i=1}^{4} E_i I_{\omega i}\right)$ $B = \frac{1}{2}\left(\sum_{i=1}^{4} G_i J_i\right)$, $C = \frac{1}{2}\left(\sum_{i=1}^{4} G_i(I_{Pi} - J_{Bi})\right)$

G. Feng (Ed.): ICCE 2023, LNCE 526, pp. 714–719, 2024.
https://doi.org/10.1007/978-981-97-4355-1_71

Then the generalized momentum corresponding to the generalized coordinates is [1]:

$$P_\beta = \frac{\partial L}{\partial \dot\beta} = 2A\dot\beta$$
$$P_\varphi = \frac{\partial L}{\partial \dot\varphi} = 2B\dot\varphi + 2C(\dot\varphi - \beta) \qquad (2)$$

Substituting (2) into the Lagrangian function yields the Hamiltonian function of the system:

$$H = 0.25A^{-1}P_\beta^2 + 0.25(B+C)^{-1}P_\varphi^2 - BC(B+C)^{-1}\beta^2 + C(B+C)^{-1}\beta P_\varphi + m\varphi \qquad (3)$$

The Hamiltonian function H does not contain coordinates z, so there is a generalized energy integral h (h is a constant), and the parent function of the canonical transformation can be selected as follows:

$$S(\beta, \varphi; z) = -hz + W(\beta, \varphi) + K \qquad (4)$$

where K is the integral constant of the addition, and the regular transformation relation is as follows:

$$P_\beta = \frac{\partial W}{\partial \beta}$$
$$P_\varphi = \frac{\partial W}{\partial \varphi} \qquad (5)$$

Substituting (2) and (5) into (3), we get the Hamiltonian-Jacobian equation for the torsion problem as:

$$0.25A^{-1}\left(\frac{\partial W}{\partial \beta}\right)^2 + 0.25(B+C)^{-1}\left(\frac{\partial W}{\partial \varphi}\right)^2 - BC(B+C)^{-1}\beta^2 + C(B+C)^{-1}\beta\frac{\partial W}{\partial \varphi} + m\varphi = h \quad (6)$$

2 The Solution to the Torsion Problem

2.1 Simplification of Equations

(6) is a first-order nonlinear partial differential equation, which has been shown to be an important factor affecting the torsional characteristics of the corrugated steel plate, while the coefficient of the nonlinear term is a dimensionless coefficient, which no longer includes the stiffness of the corrugated steel, so the influence of the nonlinear term is negligible, and the equation can be simplified as:

$$\left[0.25A^{-1}\left(\frac{\partial W}{\partial \beta}\right)^2 - BC(B+C)^{-1}\beta^2\right] + \left[0.25(B+C)^{-1}\left(\frac{\partial W}{\partial \varphi}\right)^2 + m\varphi\right] = h$$
$$\qquad (7)$$

2.2 Separation of Variables

(7) It can be solved by using the separation variable method [2], let $W(\beta, \varphi) = W_1(\beta) + W_2(\varphi)$ Substituting (7) gives us two ordinary differential equations:

$$0.25A^{-1}\left(\frac{dW_1}{d\beta}\right)^2 - BC(B+C)^{-1}\beta^2 = \eta$$

$$0.25(B+C)^{-1}\left(\frac{dW_2}{d\varphi}\right)^2 + m\varphi = h - \eta$$

(8)

In this case, the parent function of the regular transformation is:

$$S(\beta, \varphi; z) = -hz + W(\beta, \varphi) + K = \eta\left(A\tfrac{B+C}{BC}\right)^{\frac{1}{2}} \ln\left[\left(\tfrac{ABC}{B+C}\right)^{\frac{1}{2}}\beta + \left(A\eta + \tfrac{ABC}{B+C}\beta^2\right)^{\frac{1}{2}}\right]$$

$$+\beta\left(A\eta + \tfrac{ABC}{B+C}\beta^2\right)^{\frac{1}{2}} - \frac{4(B+C)^{\frac{1}{2}}(h-\eta-m\varphi)^{\frac{3}{2}}}{3m} - hz + K$$

(9)

where η, h, K are the three integration constants, which are determined by the boundary conditions.

2.3 Analytic Expressions for $\varphi(z)$ and $\beta(z)$

The regular transformation relation expressed in terms of the parent function is as follows:

$$\frac{\partial S}{\partial \eta} = \gamma_1$$

$$\frac{\partial S}{\partial h} = \gamma_2$$

(10)

Substituting the expression (10) of S into (11) gives the expression of $\varphi(z)$ as follows:

$$\varphi(z) = \frac{h - \eta}{m} - \frac{m}{4(B+C)}(z + \gamma_2)^2$$

(11)

The expression for $\beta(z)$ is as follows:

$$\left(\tfrac{A}{B+C}\right)^{\frac{1}{2}} \ln\left[(BC)^{\frac{1}{2}}\beta + \left(B\eta + C\eta + BC\beta^2\right)^{\frac{1}{2}}\right] + \tfrac{\beta}{2}\left(B\eta + C\eta + BC\beta^2\right)^{-\frac{1}{2}}$$

$$+\eta(B+C)\left(\tfrac{A}{BC}\right)^{\frac{1}{2}}\left[(BC)^{\frac{1}{2}}\beta + \left(B\eta + C\eta + BC\beta^2\right)^{\frac{1}{2}}\right]^{-1}\left(B\eta + C\eta + BC\beta^2\right)^{-\frac{1}{2}} = \gamma_1 - \frac{z+\gamma_2}{(B+C)^{\frac{1}{2}}}$$

(12)

where $h, \eta, \gamma_1, \gamma_2$ are the 4 integration constants, which are determined by the boundary conditions.

3 Analysis and Discussion

3.1 Case Analysis

In order to verify the effectiveness of the proposed algorithm, a specific example Ref. [9] is selected for calculation.

The beam bears the mid-span torque, which is solved by the algorithm in this paper. The results of the warpage function are shown in Fig. 1.

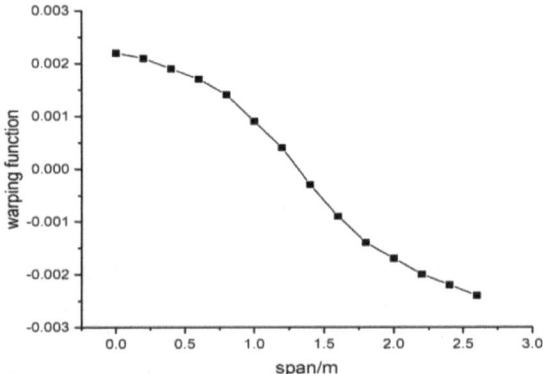

Fig. 1. Calculation of the warpage function

Then, according to the warping function [7], the values of the normal stresses at each point on the cross-section can be calculated, and the expression of the normal stresses is as follows:

$$\sigma_z = -E_i \dot{\beta} \omega \tag{13}$$

where ω is called the main sector coordinate of the box section, and the specific calculation method can be referred to Ref. [10], which will not be repeated in this article. The following Table 1 shows the normal stress values of some measurement points at a span of 0.975 m, where point A is the edge point of the roof flange, point B is the intersection point of the roof and the web, and point C is the edge of the bottom plate. It can be seen that the error between the normal stress value of the obtained measurement point and the literature value is not more than 3%, and the correctness and accuracy of the proposed algorithm are verified by the comparison of stress values.

Table 1. Comparison of normal stress values of cross-sections(MPa)

Measuring points	Measuring points A	Measuring points B	Measuring points C
Calculated values in this article	0.973	−0.206	−10.39
Calculated values from Ref. [9]	0.950	−0.200	−10.13

3.2 Torsional Performance Analysis

The torsional performance of a beam can be measured using the torsional angle per unit length. Deriving the expression (11) yields the absolute value of the torsion angle per unit length as follows:

$$|\dot{\varphi}(z)| = (B+C)^{-1}\left|0.25(z+\gamma_2)^2\dot{m}(z) + 0.5(z+\gamma_2)m(z)\right| \tag{14}$$

When the boundary conditions and the external force couple moment are constant, the magnitude of the torsion angle per unit length is only related to the magnitude of (B + C). It can be seen from the above that when the corrugated steel plate is used instead of the concrete slab, the shear stiffness of the cross-section will decrease [8], and the value of (B + C) will increase, so the torsional angle per unit length will also be larger, so the torsional performance of the beam will be weakened, which is consistent with the existing research conclusions.

4 Conclusion

In this paper, the Hamiltonian-Jacobian equation for the torsion problem of the corrugated steel web is given, and an approximate solution to the torsion problem is obtained. The accuracy of the proposed algorithm is verified by comparing with the calculated results of the existing literature. The results of this paper can well meet the needs of the project, and its engineering significance is as follows:

Although the use of corrugated steel plate has the advantages of light weight and good crack resistance, the torsional performance of the beam will be weakened to a certain extent, so special attention should be paid to the torsion-related strength and stiffness calculation in the engineering design to ensure the safe operation of the beam.

References

1. Jin, S.: Theoretical Mechanics, 3rd edn. Higher Education Press, Beijing (2020)
2. Zhou, Y.: Tutorial of Theoretical Mechanics, 4th edn. Higher Education Press, Beijing (2018)
3. Li, L.: Deflection Mechanical Properties of Corrugated Steel Web Composite Box Girder Considering the Influence of Shear Deformation and its Experimental Study. Lanzhou Jiaotong University, Lanzhou (2019)

4. Sun, C., Zhang, Y.: Confined torsion analysis of corrugated steel web composite box girder. Chin. J. Comput. Mech. **37**(06), 709–714 (2020)
5. Deng, W., Mao, Z., Liu, D., et al.: Analysis and experimental study on torsion and distortion of a single box three chamber corrugated steel web cantilever beam. J. Build. Struct. **41**(02), 173–181 (2020)
6. Zhang, Y., Liu, Y., Li, Y.: Experimental and theoretical research on the torsional performance of corrugated steel web composite box beams]. Shijiazhuang J. Rail. Univ. (Nat. Sci. Ed.) **34**(02), 1–9 (2021)
7. Jiang, K., Ding, Y., Yang, J., et al.: Experimental study on torsional bearing capacity of PC composite box beams with corrugated steel webs under pure torsion. Eng. Mech. **30**(06), 175–182 (2013)
8. Wei, N., Zhang, Y., Yao, X.: Analysis of restrained torsional shear stress in box girders with corrugated steel webs. Appl. Math. Mech. (2023)
9. Mao, Y.: Research on the torsional mechanical characteristics of corrugated web steel bottom plate composite box beams. Lanzhou Jiaotong Univ. **41**(04), 386–395 (2020)
10. Chen, Z.: Bending and Torsion Analysis of Corrugated Steel Web Composite Box Beams Under Hamiltonian Mechanics. Hebei University of Engineering, Handan (2023)

Experimental Study on the Influence of Superplasticizer on the Performance of Ecotype Ultra-High Performance Concrete (E-UHPC)

Gao Xu[1], Chungang Deng[2], and Weiyu Xu[1(✉)]

[1] Dachengkechuang Foundation Construction Co., Ltd., Wuhan, Hubei, China
Xuwy@126.com
[2] Wuhan Port and Waterway Construction Group Co., Ltd., Wuhan, Hubei, China

Abstract. A kind of Ecotype Ultra High Performance Concrete (E-UHPC) with silicate high strength cement and a variety of industrial tailings as cementing system, green sand as aggregate and ecotype organic fiber as reinforcing material is proposed. The fluidity, flexural strength and compressive strength of E-UHPC under different factors were studied by different superplasticizer content and sand particle size. The results show that the variation trend of fluidity with particle size can be expressed as an exponential function. The peaceability of E-UHPC with Superplasticizer content of 8% is better, and the flexural strength and compressive strength are better than that of E-UHPC with superplasticizer content of 15%. The research results can provide reasonable suggestions for the content of superplasticizer and the optimization of sand particle size of E-UHPC.

Keywords: E-UHPC · Superplasticizer · Fluidity · Flexural Strength · Compressive Strength

1 Introduction

With the rapid development of domestic infrastructure construction, and the demand for concrete materials is becoming more urgent. Ultra High Performance Concrete (UHPC) is a kind of concrete with large amount of cementing material, low water-binder ratio and large shrinkage characteristics, which usually requires the use of high-quality raw materials and the incorporation of appropriate admixtures to ensure its workability and mechanical properties [1].

There are many researches on UHPC at present, such as Men et al. conducted an experiment based on the effects of slag powder, silica fume and steel fiber on the mechanical properties of UHPC, which results showed that the steel fiber content had the greatest impact on the mechanical properties of UHPC [2]. Wang et al. studied the content of superplasticizer in UHPC based on the closest packing theory, his research showed that when the content of superplasticizer was 3%, the obtained UHPC had the highest strength, and the UHPC prepared with the content of superplasticizer was 4% had the

G. Feng (Ed.): ICCE 2023, LNCE 526, pp. 720–727, 2024.
https://doi.org/10.1007/978-981-97-4355-1_72

best working performance [3]. Although UHPC has excellent mechanical properties, but its low water-binder ratio, large amount of cementing material and lack of coarse aggregate increase the viscosity and decrease the fluidity of UHPC [4], resulting in increased difficulty in the pumping and construction process of ultra-high performance concrete mixes [5].

Traditional UHPC preparation processes often require particularly high-quality raw materials such as high-quality aggregate, special cement, copper-plated fine steel fibers, etc. [6]. To achieve high strength, special curing and protection measures such as high-temperature steam curing are often required [7].

This topic aims to prepare low-energy-consuming ecological E-UHPC by optimizing the proportion and gradation of aggregate, cementing material, and admixture, using ordinary sand as aggregate, industrial tailings as admixture, and organic dispersible fiber instead of steel fiber, under conventional technology (i.e., without special mixing equipment, steam curing, etc.). A set of basic mix design methods for ecological ultra-high performance concrete (E-UHPC) and the proposed E-UHPC basic material ratio are proposed.

2 Methods and Material

2.1 Test Material and Scheme Design

The test materials were made up by P.O 52.5 cement, river sand, and A and B polycarboxylic acid superplasticizers with content of 15% and 8%, respectively, and the water-binder ratio is 0.19. The particle size of river sand is 2.5–4.75 mm, 0.45–0.9 mm, 0.22–0.45 mm, 0.125–0.22 mm, respectively. The test scheme of different superplasticizer content and sand particle size is shown in Table 1.

Table 1. Test scheme

Test number	Dosage of superplasticizer/%	cement/g	Water/g	Sand/g	Sand size/mm
1-A	15	1100	209	900	2.5–4.75
2-A					0.45–0.9
3-A					0.22–0.45
4-A					0.125–0.22
1-B	8				2.5–4.75
2-B					0.45–0.9
3-B					0.22–0.45
4-B					0.125–0.22

2.2 Test Methods

Fluidity Test. The fluidity test was carried out in accordance with GB/T 50448-2015 "Technical code for application of cementitious grout". The tests is shown in Fig. 1 and Fig. 2.

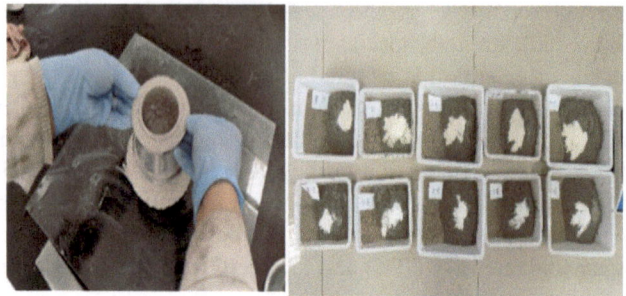

Fig. 1. Variation of fluidity with sand particle size

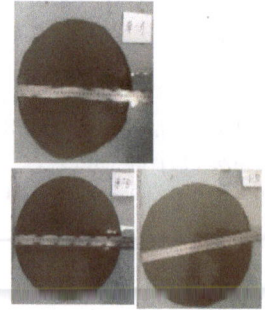

Fig. 2. Flow state diagram raw materials and fluidity for water reducer test

Flexural Strength Test. The test of mechanical properties of E-UHPC mainly includes flexural strength and compressive strength, which is in accordance with GB/T 17671-2021 "Test method of cement mortar strength (ISO method)". The flexural strength places one side of the specimen on the support column of the testing machine, and the axis of the specimen is perpendicular to the supporting cylinder, and the load is uniformly applied vertically on the opposite side of the prism until it is broken through the loading rate of 50 N/s. The flexural strength is calculated according to Eq. (1)

$$R_f = \frac{1.5 F_f L}{b^3} \tag{1}$$

In the Eq. (1), Rf represents the flexural strength, MPa; and Ff represents the load applied to the middle of the prism when its broken, N; and L represents the distance between the supporting columns, mm; and b is the side length of a prismatic square section, mm.

Compressive Strength Test. After the flexural strength test, the two halves were taken out and the compressive strength test was carried out. In the whole loading process, the rate of 2400 N/s is uniformly loaded until it is destroyed. The compressive strength is calculated according to Eq. (2)

$$R_c = \frac{F_c}{A} \tag{2}$$

In the Eq. (2), Rc represents the compressive strength, MPa; and Fc represents the maximum load when the material is failure, N; and A is compression area, mm^2.

3 Results

3.1 Analysis on the Fluidity of E-UHPC

The fluidity of E-UHPC under different particle sizes is shown in Fig. 3. Figure 3(a) shows the fluidity of E-UHPC with 15% of superplasticizer. It can be seen from the Fig. 3a that the fluidity value shows an exponential growth trend with the particle size. When the particle size is less than 1 mm, the fluidity increases linearly with the increase of the particle size. And when the particle size is larger than 1mm, the increasing trend of fluidity is slow. Figure 3(b) shows the fluidity of E-UHPC with 8% of superplasticizer. The development trend of the fluidity value of E-UHPC with 8% of superplasticizer is consistent with that of with 15%. Which can be expressed by Eq. (3).

$$f = ae^{k/b} + c \tag{3}$$

In the Eq. (3), f represents the fluidity value, mm, k represents the sand particle size, and a, b, and c are the fitting parameters, as shown in Table 2.

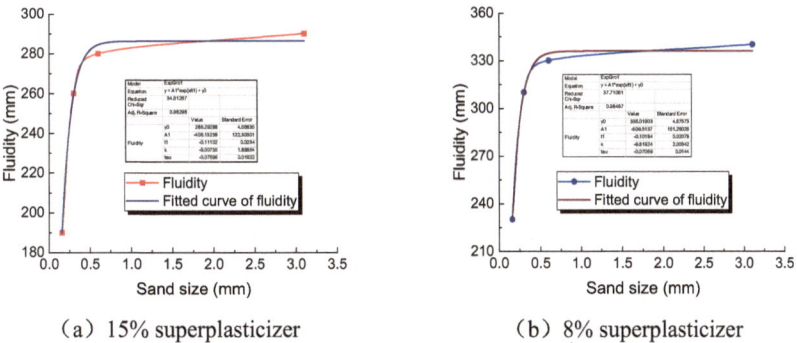

(a) 15% superplasticizer   (b) 8% superplasticizer

Fig. 3. Variation of fluidity with sand particle size

Table 2. Parameter of the exponential function

superplasticizer dosage/%	a	b	c	R^2
15	−406.13	−0.110	−9.0	0.983
8	−509.50	−0.102	−9.8	0.985

The fluidity of E-UHPC with two types of superplasticizer content at different grading is shown in Fig. 4. As can be seen from the Fig. that the greater dosage of superplasticizer, the smaller the fluidity. The fluidity of cement mortar with different grading is the same, which will decreases with the decrease of sand particle size. The main reason is that the smaller the particle size of sand particles, the larger the specific surface area, and the amount of cement paste required to wrap the surface of sand particles increases. Therefore, under the same content of water and superplasticizer, the fluidity of cement mortar decreases significantly. The fluidity of cement mortar with 8% superplasticizer is 17%–21% higher than that with 15%, hence the effect of 8% superplasticizer content on the flow performance of cement mortar is more significant.

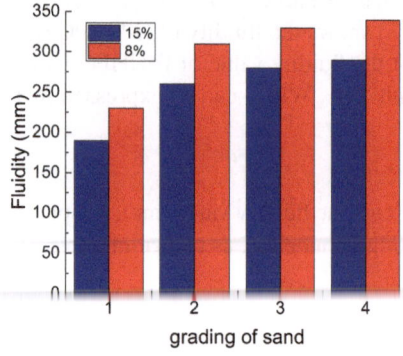

Fig. 4. Comparison of fluidity of E-UHPC with 15% and 8% superplasticizer

3.2 Analysis on the Flexural Strength of E-UHPC

The 28-day flexural strength E-UHPC under different particle sizes is shown in Fig. 5. The flexural strength of E-UHPC corresponding to 8% and 15% water-reducing agent is basically the same, which is increased first and then decreased with the increase of particle size. When the particle size is at range of 0.125–0.9 mm, the flexural strength increases linearly with the change of particle size.

When the sand particle size is at range of 0.125–0.22 mm, the flexural strength of E-UHPC corresponding to two types of superplasticizers is at a small value. The main reason is that the smaller the sand particle size is, the larger the surface area is, and the bonding property between sand particles is weakened, which affects the mechanical properties of E-UHPC to a certain extent. When the sand particle size is at range of 0.45–0.9 mm, the flexural strength corresponding to the two types of superplasticizers reaches the maximum, and the flexural strength corresponding to the 8% superplasticizers is about 22% of the 15% superplasticizers Dosage. With the particle size increasing to 2.5 mm–4.75 mm, the flexural strength corresponding to the two content superplasticizers showed a linear decrease trend, mainly because the sand surface area decreased due to the doubling of the sand particle size.

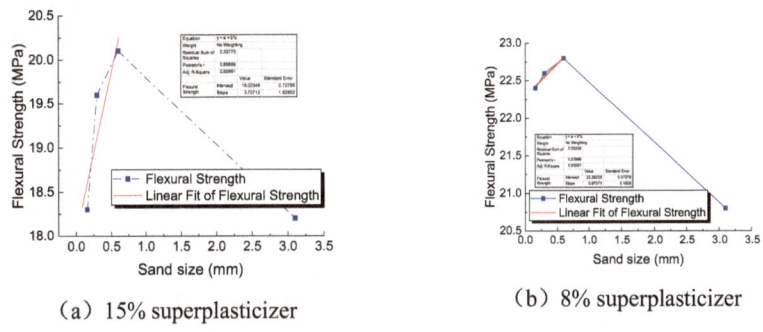

(a) 15% superplasticizer   (b) 8% superplasticizer

Fig. 5. Variation of flexural strength with sand particle size

Comparison of 28-day flexural strength of E-UHPC with 15% and 8% superplasticizer is shown in Fig. 6. As can be seen from the Fig. That the flexural strength of E-UHPC corresponding to the 8% superplasticizers is larger than that to 15% superplasticizers at every partical size, which is about 14–22%.

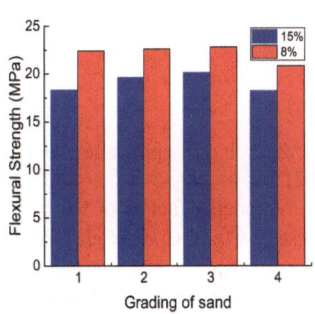   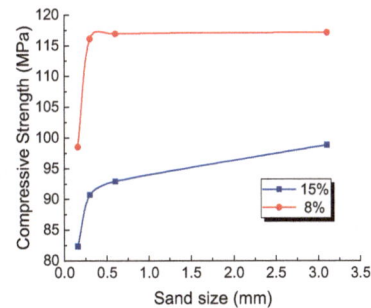

Fig. 6. Comparison of 28-day flexural strength of E-UHPC with 15% and 8% superplasticizer

Fig. 7. Variation of compressive strength with sand particle size

3.3 Analysis on the Compressive Strength of E-UHPC

After the flexural strength test, the two halves were taken out and the compressive strength test was carried out. The 28-day compressive strength of E-UHPC under different particle sizes is shown in Fig. 7. The results in the Fig. show that the compressive strength of E-UHPC corresponding to 15% superplasticizer increases rapidly at first and then steadily with the increase of particle size, and reaches the maximum when the particle size is at range of 2.5 mm–4.75 mm. The compressive strength of E-UHPC corresponding to 8% superplasticizer increases rapidly with the increase of particle size at first and then remains basically unchanged, and reaches the maximum when the particle size is at range of 0.22–0.45 mm.

By comparing the 28-day compressive strength of E-UHPC with two types of super-plasticizer content, it can be found that the 28-day compressive strength of E-UHPC with 8% superplasticizer content under the same particle size is about 14% to 28% larger than that with 15%.

4 Conclusions

By comparing the fluidity, flexural strength and compressive strength of E-UHPC with two different contents of water reducing agent and four kinds of sand particle sizes, The fluidity of E-UHPC with the superplasticizer content of 8% is 17%–21% higher than that with the superplasticizer content of 15%, which shows a better peaceability. The 28-day flexural strength of E-UHPC with two types of superplasticizer content increases first and then decreases with the increase of particle size, while the 28-day compressive strength increases rapidly first and then steadily. The 28-day flexural strength and 28-day compressive strength of flexural strength with superplasticizer content of 8% are obviously higher than that of 15%. Therefore, E-UHPC with superplasticizer content of 8% has better working and mechanical properties.

Acknowledgment. This work was financially supported by the Wuhan Science and Technology Plan Project 2022022202015074.

References

1. Qian, Z., Shihua, L., Limin, L., et al.: Overview of research and application of ultra high performance concrete. Value Eng. **A41**, 186–190 (2022)
2. Guangyu, M., Xiaolong, J., Wenbo, Z.: Study on the influencing factors of workability and mechanical properties of ultrahigh performance concrete, China Conc. Cem. Prod. 44–49 (2023)
3. Zhe, W., Xiongguang, X., Zhonghe, S., et al.: Rational selection of the dosage of superplasticizer for UHPC based on the close packing theroy. Bull. Chin. Ceramic Soc. **38**, 1503–1509 (2019)
4. Qianqian, Z., Jianzhong, L., Huaxin, Z.: Rheological properties of ultra high performance concrete and its effect on the fiber dispersion within the material. Mater. Rep. **31**, 73–77 (2017)
5. Reddy, G., Kishore, G., Ramadoss, P.: Influence of alccofine incorporation on the mechanical behavior of ultra-high performance concrete (UHPC). Mater. Today **33**, 789–797 (2020)
6. Mishra, O., Singh, S.P.: An overview of microstructural and material properties of ultra-high-performance concrete. J. Sustain. Cem.-Based Mater. **8**, 97–143 (2019)
7. Jin, Y., Fazhou, W.: Influence of assumed absorption capacity of superabsorbent polymers on the microstructure and performance of cement mortars. Constr. Build. Mater. **204**, 468–478 (2019)

Analysis of the Hydraulic Characteristics of the Debris Flow Disaster Chain in Baozhuping Gully and Analysis of the Conditions for Debris Flow Initiation

Hai Yue[1,2,3](✉), Dingde Wu[1,2,3], Lifeng Hou[1,2,3], and Mingliang Zhang[1,2,3]

[1] Sichuan Geological Environment Survey and Research Center, Chengdu 610081, Sichuan, China
114721535@qq.com

[2] Sichuan Province Engineering Technology Research Center of Geohazard Prevention, Chengdu 610081, Sichuan, China

[3] Sichuan 915 Construction Engineering Co., Ltd., Meishan 620010, Sichuan, China

Abstract. Taking the Baozhuping gully landslide-debris flow disaster chain in Ya'an City as an example, through field investigation and on-site measurement, this study analyzed the characteristics of the collapse and breakthrough during the "7.16" debris flow outbreak, and provided the types of dam break and its hydraulic characteristics after the landslide slide. The results show that the instability of the dam body of the pond is mainly due to the damage caused by the overflow of the top, and the maximum flow rate of the dam break reaches 30.12 m^3/s, which triggers the initiation of debris flow downstream, providing reference for the prevention and control of similar debris flows.

Keyword: Debris Flow · Disaster Chain · Dam Break · Baozhuping Gully

1 Introduction

On July 16, 2022, heavy rainfall occurred in Yucheng District, Ya'an City. A landslide occurred on the upstream slope of Baozhuping gully, a right-bank tributary of Zhougong River, on the right bank of Qingyi River. The slope soil rapidly slid to the foot of the slope, and the water in the pond quickly rushed into Baozhuping gully, triggering a debris flow that lasted for about 15 min. It burst out of the channel at Nanba East Street, damaging three vehicles. The debris flow accumulated on the surface of Nanwai Ring Road and Nanba East Street, blocking the roads and causing damage to houses on both sides of Baozhuping gully. This resulted in road interruption, vehicle destruction, injuries to people, and property losses. People's Daily, China News Network, and other media outlets reported on this incident promptly, generating strong social response [1]. The dam break played an important role in triggering the debris flow in Baozhuping gully.

Currently, extensive research has been conducted both domestically and internationally on the formation of blockage and failure points in channel sedimentary dams.

G. Feng (Ed.): ICCE 2023, LNCE 526, pp. 728–736, 2024.
https://doi.org/10.1007/978-981-97-4355-1_73

Takahashi [2] identified three main causes of debris flow: 1) Infiltration of water into the deposited material in the channel bed, which forms surface flow and destabilizes the material, leading to debris flow. 2) Soil blocks that collapse from the slope break apart and mix with water during movement, transforming into debris flow. 3) Slope collapse material blocking the channel bed, which becomes unstable due to the accumulation of upstream water and subsequently forms debris flow as a large amount of water flows out. Wu [3] discussed the generation of flow from blockage and failure of landslides, debris flows, and glacial deposits in channels. David and Froehlich [4] analyzed a large amount of data on failed sedimentary dams and used multiple regression to derive a new empirical formula for predicting peak outflow during dam failure. Xiang [5] analyzed the hydraulic characteristics and initiation conditions of debris flows within the Qipanguo dam failure in Wenchuan County. Yucheng District in Ya'an City is known for its frequent occurrence of geological hazards due to rainfall [6]. This article takes the "7·16" debris flow in Baozhuping gully, Yucheng District, Ya'an City as an example to analyze the characteristics of hydraulic conditions and initiation conditions after the landslide sliding and pond failure. The results have certain representativeness and can serve as a reference for the prevention and control of similar geological hazard chains.

2 Basic Situation of Baozhuping Gully

Baizhuping Gully is located in Chengqing Village, Dongcheng Street, Yucheng District, Ya'an City. Its geographical coordinates are 103°01′28.13″ E and 29°59′12.13″ N. It is approximately 4 km away from the Yucheng District People's Government. There is a rural cement road leading to the exploration point, and the gully mouth is connected to the South Outer Ring Road, providing good transportation conditions.

Baizhuping Gully is a right-bank tributary of Zhougong River, a tributary of Qingyi River. It has a drainage area of 0.26 km^2 and flows from southeast to northwest. The upstream area has a funnel-shaped topography, while the middle and downstream areas have a strip-shaped topography. The main gully is 1.4 km long with a gradient of 313‰ and a relative elevation difference of 430 m.

2.1 Geological Conditions

The main geological structure in the area of Baizhuping Gully is folding structure, with no regional faults observed. The internal structure of the area exhibits a north-south orientation, forming part of the north-south tectonic belt of Sichuan and Yunnan. The basin's geological formations mainly consist of sandstone and mudstone of the Cretaceous Guankou Formation, with additional exposures of loose deposits from the Quaternary period. These loose deposits are primarily found in the sedimentary areas of the middle and lower reaches of the gully and on the surface of the gully slopes.

The main gully has a narrow valley, typically with a width of 10–20 m, and often exhibits a "U" shape. According to the Comprehensive Vulnerability Score based on the "Technical Code for Investigation of Debris Flow Disaster Prevention and Control Engineering (Trial)" (TCAGHP 006–2018), Baizhuping Gully has a score of 74, indicating a mild susceptibility to debris flows. The completeness coefficient of the Baizhuping

Gully basin is relatively small (0.133), indicating poor convergence conditions and a relatively gentle flood process. Therefore, the likelihood of debris flows occurring is low in the absence of rapid changes in hydraulic conditions [7–9].

Considering the surrounding environment, there are several tributaries with similar geological conditions to Baizhuping Gully (see Figs. 1 and 2). These tributaries share similarities in terms of basin area, main gully length, terrain slope, and lithology. No debris flows were reported in any of these tributaries on July 16, 2022. From the video footage taken by Chuan Guan News after the debris flow event in Baizhuping Gully, it can be seen that there was still a large amount of floodwater in the gully, while the left tributary remained clear (see Figs. 3 and 4). This indicates that the debris flow in Baizhuping Gully was likely caused by rapid sliding of slope soil and rock from the upstream slope into the foot pond, resulting in a significant change in hydraulic conditions due to the rapid influx of water [10–12].

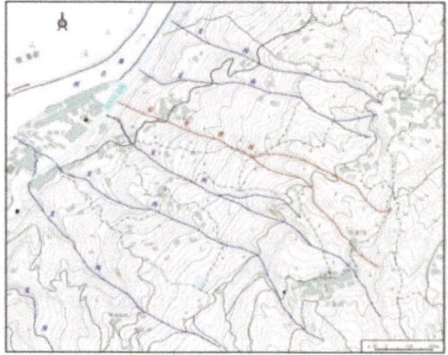

Fig.1. Distribution map of Baozhuping gully and surrounding branch gully

Fig. 2. Image of Baozhuping gully and surrounding branch gully

2.2 Hydrometeorological Conditions

The average annual rainfall in Yucheng District, Ya'an City is 1614 mm. The largest annual rainfall in 50 years was 2367.2 mm in 1966, and the largest daily rainfall in 50 years was 339.7 mm in 1959. The maximum hourly rainfall was 86 mm, and the maximum 10-min rainfall was 30.1 mm.

According to the measured data from Zhougongshan Meteorological Station (the nearest meteorological station to Baizhuping Gully), which is located upstream of the gully, there were a total of 18 times with daily maximum rainfall greater than 100 mm in the past five years, ranging from 103.1—179.2 mm. The maximum hourly rainfall during these 18 heavy rainfalls ranged from 15.7—85.6 mm. The debris flow in Baizhuping Gully occurred on July 16, 2022, with a rainfall of 159.1 mm and a maximum hourly rainfall of 48 mm. In mountainous areas of Sichuan Province, the rainfall that usually triggers debris flows is around 48—50 mm for a single rainfall or 8.0—12.2 mm for a 10-min rainfall, with a rainfall intensity of about 0.8—1.2 mm per minute (Wu et al., 1993). Before July 16, 2022, there were 13 heavy rainfalls, and after that, there were four heavy rainfalls, but none of them triggered debris flows. Therefore, it is unlikely that Baizhuping Gully was directly triggered by rainfall to cause large-scale debris flows. The main cause is the fast sliding of slope soil and rocks from the upstream slope into the foot pond, causing a significant change in hydraulic conditions due to the rapid influx of water into Baizhuping Gully in Table 1.

Table 1. Statistical table of heavy rainfall in the past 5 years at Zhougongshan Meteorological Station in Yucheng District, Ya'an City

Number	Data	Maximum daily rainfall (mm)	Maximum hourly rainfall (mm)	Note	Number	Data	Maximum daily rainfall (mm)	Maximum hourly rainfall (mm)	Note
1	2019.8.6	141	20.1		10	2021.8.22	150.1	80.4	
2	2019.8.22	112.1	22.9		11	2022.5.9	162.2	26.9	
3	2019.9.13	134.5	20.6		12	2022.5.13	121.2	27.5	
4	2020.8.11	179.2	85.6		13	2022.7.12	194	74.2	
5	2020.8.16	147.8	15.7		14	2022.7.16	159.1	48	Debris flow occurred in Baozhuping Gully
6	2020.8.18	155	47.2		15	2023.7.11	103.1	41.1	
7	2020.8.30	185	50.8		16	2023.7.16	137.5	24.2	
8	2021.8.18	139.8	24.1		17	2023.8.10	108.3	21.1	
9	2021.8.20	123.1	43.5		18	2023.8.12	106.0	46.2	

3 Hydrodynamic Characteristics After Dam Failure

3.1 Basic Characteristics of Ponds

The main cause of the mudslide in Baizhuping Gully on "7·16" was the significant change in hydraulic conditions due to the failure of the upstream reservoir. The reservoir is located on the right bank tributary of Baizhuping Gully, about 300 m from the watershed, with a drainage area of approximately 0.05 km^2.

According to the investigation and analysis of pre-and post-failure images (Fig. 3), the reservoir was excavated by local residents as a slope foot, with a front section built as a dam primarily for aquaculture. It has an elongated rectangular shape, with a length of about 85m and a width of 50—20 m. The depth of the reservoir is approximately 4m, with a storage capacity of about 11,900 m^3.

Based on the data from Zhougongshan Meteorological Station on the Rain City Natural Resources and Planning Bureau's meteorological information service platform, there was continuous rainfall from July 12th to July 16th, 2022, with a total accumulated rainfall of 402.3 mm. The reservoir was already at full capacity before the occurrence of the mudslide.

Due to the heavy rainfall over several days, the slope formed by excavating and constructing the fish pond at the front edge became saturated. The strength and stability of the soil and rock decreased continuously, eventually resulting in a rapid sliding along the mudstone layer towards the reservoir. The sliding distance was approximately 30 m, with a volume of 9,600 m^3. The soil and rock mass squeezed the water in the reservoir, causing it to rapidly overflow into Baizhuping Gully.

Fig. 3. Aerial images on August 11, 2022

3.2 Types of Reservoir Dam Breach

By using the slope stability calculation formula, the stability safety factor of the reservoir dam at half and full capacity was calculated. The overall stability factor of the most dangerous surface under both conditions was greater than 1, indicating that the reservoir dam would not experience overall movement instability. The damage to the dam was due to the erosion caused by a large amount of water overflowing to the outside of the dam after the reservoir was full. The failure mode was an overflow erosion and collapse type (Kuang, 1993). This calculation result is consistent with the on-site investigation.

3.3 Hydrodynamic Characteristics After Dam Failure

Regarding the overflow erosion and collapse type dam, according to the peak flow formula proposed by David C. Froehlich (1996), the peak flow during the collapse was calculated to be 30.12 m³/s (Table 2), causing significant damage downstream.

$$Q_p = 0.607 V_w^{0.295} H_w^{1.24} \tag{1}$$

Here, VW represents the capacity of the reservoir at the time of the incident (m³), and HW represents the depth from the bottom of the final breach to the surface of the reservoir when the dam fails (m).

Table 2. Calculation table for peak flow when the reservoir bursts

Dam location	The capacity of the pond at the time of the accident (m³)	The depth of the pond from the bottom of the final breach to the surface of the reservoir when the dam failed (m)	Peak discharge flow (m³/s)	Note
Baozhuping gully upper reaches reservoir dam	11900	2.5	30.12	

4   Conditions for Triggering Debris Flow After Reservoir Dam Failure

Based on the morphology of the on-site investigation, the Baizhuping Gully debris flow is a dilute debris flow (water and sediment flow) containing certain fine particles. According to the research results of Fei (2004), it is assumed that the fine particles are uniformly distributed along the vertical line.

In the Fig., θ is start critical inclination angle for channel, h is the deep flow; h' is the thickness of the channel coarse-grained layer, driving shear stress in fixed loose materials (τ) and starting resistance (τ_L) are:

$$\tau = \left[S'_{vm}(\gamma_s - \gamma_f)h' + \gamma_f h \right] \sin \theta \tag{2}$$

$$\tau_L = \left[S'_{vm} (\gamma_s - \gamma_f) h' \right] \cos \theta \tan \alpha' + \tau_f \tag{3}$$

In the equation γ_f is the bulk density of a suspension composed of fine particles and water, $\tan \alpha'$ is the friction coefficient in the presence of fine particles, γ_s is the solid bulk density, s_{vm}' is the channel activation layer concentration, the suspension shear stress τ_f is very small, can be ignored, In this way, the starting condition is

$$\tau = \tau_L \tag{4}$$

or

$$\tan \theta = \frac{S'_{vm}(\gamma_s - \gamma_f) \tan \alpha'}{S'_{vm}(\gamma_s - \gamma_f) + \gamma_f (h / h')} \tag{5}$$

When the channel slope is greater than or equal to θ, debris flow starts. Select the location of the pond and dam for starting condition analysis, and calculate the peak flow rate at this location in 50 years to be 1.99 m^3/s based on the inference formula. The peak flow rate after the collapse is 30.12 m^3/s (Table 3), Compared with normal conditions, the flow rate is enlarged by 15.1 times (table). Based on experience, it is judged that the possibility of debris flow under this flow rate is greater.

Table 3. Calculation table of cross-section flood peak flow at the location of the dam

Calculate position	catchment area km^2	Main groove length km	Main ditch longitudinal slope‰	peak flow (m^3/s)(p = 1%)	peak flow (m^3/s)(p = 2%)	peak flow (m^3/s)(p = 5%)	peak flow (m^3/s)(p = 10%)
Section at the location of the dam	0.05	0.4	240	3.30	1.99	1.69	1.16

Using Formula (5), the determination of whether a debris flow will be triggered after the Baizhuping Gully reservoir dam failure is conducted. The calculation parameters are based on field measurements and empirical data, and the calculation results are shown in Table 4. From Table 4, it can be observed that the slope of the channel is greater than the initiation slope of the debris flow. Therefore, when the reservoir overflows and erodes, it will trigger the scouring of sediment at the bottom of the channel, initiating a debris flow (Zhao et al., 2021).

Table 4. Summary of calculation parameters and results

$s_{vm}{}'$	γ_s	γ_f	h	h'	$\tan\alpha'$	θ	Channel slope
0.9	2.4	1.1	3	1	0.75	11.1	12.0

5 Conclusion

The fact that multiple tributaries with similar engineering geological conditions on both sides of Baizhuping Gully have never experienced debris flow indicates a lower integrity coefficient and poor convergence conditions in the Baizhuping Gully watershed. The likelihood of debris flow occurrence is lower in the absence of rapid changes in hydraulic conditions.

In the past five years, the region has experienced 18 heavy rainfall events (daily rainfall ranging from 103.1 to 179.2 mm, maximum hourly rainfall ranging from 15.7 to 85.6 mm). Except for the event on "7·16", no debris flow occurred, suggesting that rainfall is not the main triggering factor for debris flow in Baizhuping Gully.

Through stability calculations and on-site investigations, the failure mode of the upstream reservoir in Baizhuping Gully is identified as overflow erosion and collapse type.

The peak flow rate during the failure of the upstream reservoir in Baizhuping Gully is 30.12 m³/s, with an initiation slope of 11.1°. The downstream channel slope of the dam is 12.0°. The failure flood triggers scouring of sediment at the bottom of the downstream channel, initiating a debris flow and causing significant damage downstream.

References

1. Zhang, X., et al.: Fine investigation project report on geological hazards along the south outer ring of Yucheng district (2022)
2. Takahashi, T.: Debris flow. Annu. Rev. Fluid Mech. **13**, 57–77 (1981)
3. Wu, J., Cheng, Z., Geng, X.: Formation mechanism of debris flow blocking dam in Southeastern Tibet. J. Mt. Sci. **04**, 4399–4405 (2005)
4. David, C., Froehlich: Peak discharge of breached earth and rock dams. Dam and Safety **4**, 63–69 (1996)
5. Xiang, G., Zhang, D., Chang, M., Qu, Y.: Hydraulic characteristics and initiation conditions of debris flow in Qipanguo after the breach of barrier lake. Water Resour. Power **33**(04), 143–146 (2015)
6. Li, Y.: Preliminary study on the critical value of rainfall-induced landslides in Yucheng district, Ya'an city Sichuan province. Hydrogeol. Eng. Geol. **01**, 26–29 (2005)
7. Yue, H.: Geological Hazard Characteristics and Prevention Suggestions for Yucheng District, Ya'an City after the Lushan Earthquake on April 20th. Southwest Jiaotong University (2015)
8. Wu, J., Tian, L., Kang, Z., Zhang, Y., Liu, J.: Debris Flow and Its Comprehensive Control. Science Press, Beijing (1993)
9. Kuang, S.: Formation mechanism and mathematical model of debris flow from natural dam breach. J. Sed. Res. **04**, 42–57 (1993)
10. Fei, X., Shu, A.: Mechanism of Debris Flow Movement and Disaster Prevention. Tsinghua University Press, Beijing (2004)

11. Fei, X.: Sediment concentration and flow velocity of water-sediment flow. J. Sed. Res. **04**, 8–12 (2002)
12. Zhao, S., Zhao, Z., Yuan, G.: Study on characteristics of debris flow source in Jiuzhaigou seismic zone. J. Sichuan Geol. 41(01) (2021)

Research on the Application of Double-Pipe Split-Grouting Anchor in Deep Fill Slope Engineering

Dianjun Liang[1], Zhao Li[2(✉)], Xunhai Sun[2], Haitao Yang[2], Binbin Ma[2], and Hao Fan[2]

[1] PowerChina Real Estate Group Ltd., Beijing 100097, China
[2] Institute of Foundation Engineering, China Academy of Building Research, Beijing 100013, China
lz417130724@126.com

Abstract. In recent years, with the continuous development of China's construction industry, the site selection of construction projects has become more and more complex, and more and more construction projects have encountered deep and complex filling problems. This paper is based on a project in the north of China, which is a deep and complex filling slope and the support form is pile-anchor support. Considering the risk of deep and complex soil filling on slope stability, in order to ensure the construction quality and safety, the double-pipe split anchor test is studied and analyzed. Through the comparison of engineering test, numerical simulation and actual monitoring data, it is shown that the scheme has a good effect on improving the quality of anchor, and has certain guiding significance for similar projects.

Keywords: Complex Fill · Anchorage · Double-Pipe Spilt-Grouting · Numerical Simulation

1 Introduction

In recent years, with the continuous development of China's construction industry, more and more construction project site selection involves deep and complex fill areas. Due to the uneven nature of deep and complex fill, loose structure, large thickness difference, low bearing capacity, high porosity, strong permeability and other factors, there are great risks in slope support. At present, the most widely used slope support method is pile-anchor support, which has good effect, simple construction, strong adaptability and strong economy [1–4].

This paper takes a project in northern China as an example. The project has a deep and complicated filling up to 20 m, which has great risks. In order to ensure the construction quality and safety of the project, the construction of ordinary anchor and double-pipe splitter were carried out respectively, and the multi-stage overtension test was carried out respectively, and the effect of improving the quality of anchor was comprehensively analyzed through the test data and quantitative indicators. Subsequently, numerical simulation combined with actual monitoring data was used for comparative analysis, the overall effect is good.

© The Author(s) 2024
G. Feng (Ed.): ICCE 2023, LNCE 526, pp. 737–743, 2024.
https://doi.org/10.1007/978-981-97-4355-1_74

2 Engineering Project

2.1 Project Overview

The project is located in a city in northern China, the above-ground use function is a residence, the above-ground 4 floors, the underground 1–3 floors. The project is a permanent slope support project, including supporting piles, permanent anchor rods and a variety of retaining wall forms.

The proposed site is originally a mountain trench, the terrain is complex, the height difference is large, the site has been filled with soil treatment, the thickness of the site is about 20 m, the backfill time is about one year from the project construction, the site has been DDC construction treatment, after backfilling to form a permanent slope of 9–12 m.

In this project, the slope support system is adopted, and 3–5 anchors are set according to the slope height, the anchor aperture is 200 mm, the anchor length is 20–38 m, the steel strands are 4–5 1 * 7–17.8 (1860 MPa), the construction Angle is 35–40°, and the standard tension value is 290–570 kN. The locking value is 390–770 kN, the bolt adopts P.O 42.5 cement, the water-cement ratio is 0.5–0.55, and the construction technology adopts the casing follow-up bolt drill to make holes and the secondary split grouting.

2.2 Engineering Geological Condition

The special soil of this project is artificial fill and weathered bedrock.

The artificial fill is mainly mixed fill and gravel fill. The thickness of the layer is 0.5–20.0 m. According to the investigation, the artificial soil filling was formed when the site was leveled. The soil layer mainly includes mixed fill soil, gravel mixed fill soil, fully weathered sandstone, strongly weathered sandstone. The artificial filling time is short, about one year, the artificial filling composition is complex, the distribution is uneven, and the mechanical properties are poor. No groundwater was found within 20.0 m of the site.

2.3 Anchor Design Scheme

The design cement is P.O 42.5, and the water-cement ratio is 0.5–0.55. The first grouting pressure is 0.5 MPa, and the second high pressure grouting pressure is 2.0–2.5 MPa. When grouting, the grouting pipe should be inserted 250–500 mm away from the bottom of the hole, and the mixed cement slurry should be injected into the bottom of the hole and pumped outward from the bottom of the hole. Due to the shrinkage of the cement slurry, the grouting treatment should be done in time after grouting to make the hole section full of slurry.

In this test, the fifth anchor rod in the double-row pile area was selected. The anchor rod was composed of 4–17.8 (1860 MPa) steel strands. The anchoring section length was 15 m, the free section length was 5 m, the standard pulling force value was 290 kN, and the locking value was 390 kN.

According to the drawings, the maximum load of acceptance test is 1.5 Nak, and the maximum load of acceptance test of anchor rod on this road is 435 kN. In the process of tensioning lock acceptance, the multistage overtension verification test effect will be carried out according to the actual tensioning situation.

According to CECS 22 Technical Specification for Ground Anchors, the formula for calculating bolt displacement is shown as follows:

$$\Delta s = \frac{P * (Lf + \frac{La}{2})}{E * A} = \frac{435 * (0.8 * 5)}{1.95 * 4 * 193} \sim \frac{435 * (5 + 15/2)}{1.95 * 4 * 193} = 1.16 \sim 3.61 \, cm \quad (1)$$

According to Article 9.4.6, "The total displacement measured by tension type anchor rod under the maximum test load shall exceed 80% of the theoretical elastic elongation value of the length of the free section of the rod body under the load, and be less than the theoretical elastic elongation value of the sum of the length of the free section of the rod body and the length of the 1/2 anchoring section", In this project, the displacement of the bolt at the maximum load value of 435 kN is between 1.16 and 3.61 cm.

3 Double Tube Splitting Test

In view of the situation that the backfill soil in the field area is thick and the soil is loose and easy to collapse the hole, two ways of ordinary anchor rod and double pipe splitting are carried out respectively in this project. The double-pipe splitting technology is to add a second splitting pipe on the basis of the original one-time grouting and secondary splitting. Through the effect of double splitting, it can make up for the shortcomings that cement slurry is easy to lose in loose soil, so as to increase the solid holding force of anchor bolt and improve the overall anchoring strength.

In order to verify the test results, the maximum load value was increased to 1.95Nak during the tensioning process. The test records and curves are as follows Fig. 1 and Fig. 2.

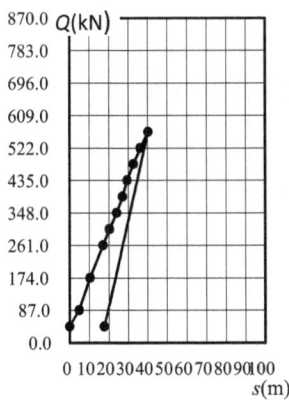 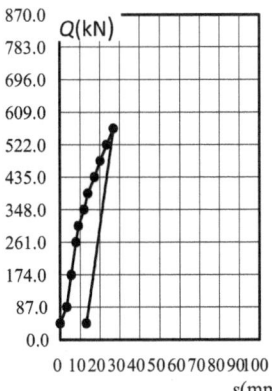

Fig. 1. Q-s curve of ordinary anchor   **Fig. 2.** Q-s curve of double pipe split bolt

Through the comparative analysis of the above two sets of curves and corresponding load-displacement data, the results show that the displacement of the double-pipe split grouting test bolt is significantly higher than that of the ordinary anchor, which indicates that the construction quality of the anchor can be improved by the double-pipe split grouting method for the loose soil and other easily collapsing holes.

In this project, the construction technology of double-pipe split-grouting is adopted in the deep and complicated fill areas. At present, all the anchor rods in the site have been constructed, and the construction quality is good. The basic test and acceptance test meet the requirements.

4   Numerical Simulation

Taking a section of pile and anchor area as an example, the numerical simulation of pile and anchor support is carried out, and the numerical simulation results are compared with the measured results.

In the numerical simulation area, the pile diameter is 1.2 m, the pile length is 19 m, and the pile spacing is 2 m. The top of the pile has a bolt, the bolt bore is 200 mm, composed of 4 steel strands, the anchoring section length is 25 m, the free section length is 7 m, the standard pulling force value is 400 kN, the locking value is 540 kN. The soil layer in the pile and anchor supporting area consists of fill soil layer, gravel layer and strong weathered sand rock layer from top to bottom. The thickness from the top elevation of the pile to the bottom is 15,5 and 20 m, and there is no groundwater in the depth range.

After the establishment of the slope support model, the software including gravity self-stability, pile construction, anchor cable construction and slope excavation will carry out corresponding calculations to solve the internal force and deformation increment of the support system, and realize the controllability of simulated excavation. The simulation and analysis results are as follows Fig. 3, 4, 5, 6 and 7.

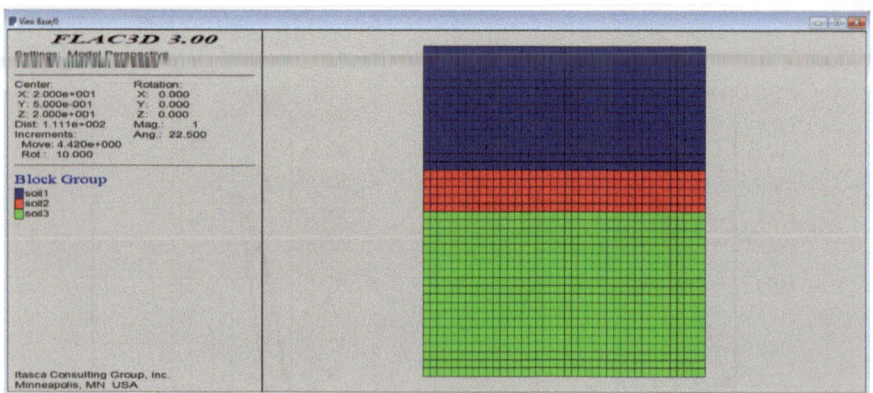

Fig. 3. Pile-anchor support model

The maximum horizontal displacement generated by the set monitoring record is shown in the Fig. above. It can be seen from the simulation results that the maximum horizontal displacement of the pit wall occurs in the middle and lower part of the foundation pit slope, and the horizontal displacement of the soil behind the slope gradually

Fig. 4. Displacement cloud map of advance support of slope protection pile

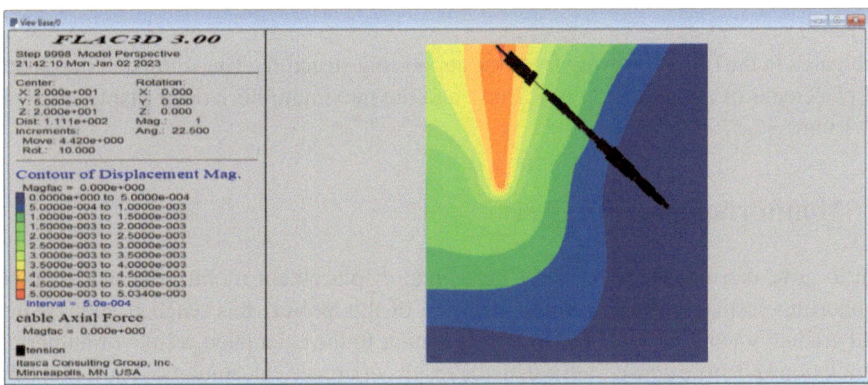

Fig. 5. Cloud map of advance support of anchor cable

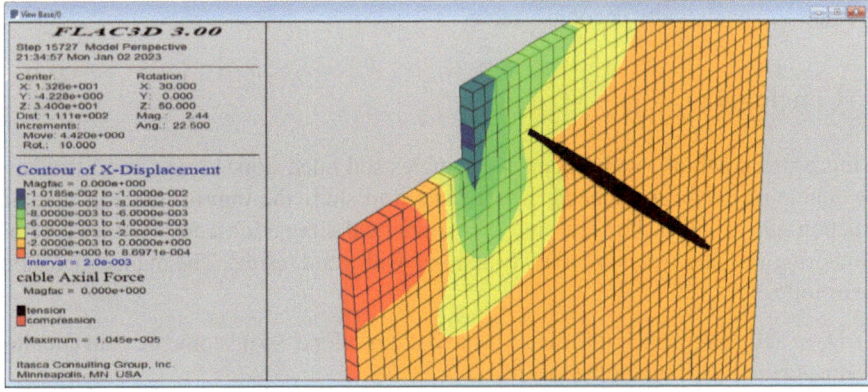

Fig. 6. Horizontal displacement contour map

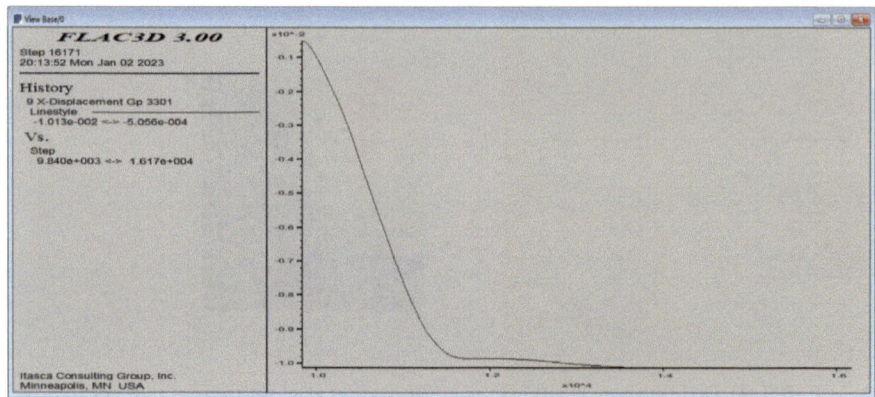

Fig. 7. Maximum horizontal displacement curve

decreases in the direction away from the supporting structure. The simulated horizontal displacement of the slope top is 8.7 mm and the maximum horizontal displacement is 10.1 mm.

5　Monitoring Data Analysis

Up to now, the maximum value of horizontal displacement monitoring of the actual supporting section slope top in the test area of this project has reached about 8 mm and gradually tends to be stable, which is similar to the calculated results of numerical simulation and more conservative than the results of numerical simulation, and generally meets the requirements of monitoring and alarm value. According to its changing trend, the horizontal displacement gradually increases with the construction of pile protection, anchor cable and slope excavation. With the completion of soil consolidation on the pile side, the curve gradually becomes stable.

6　Conclusion

In this paper, combined with a deep and complex soil filled slope project in north China, the double-pipe splitting scheme was first used to study the improvement of the construction quality of anchor rod, and then numerical simulation combined with actual monitoring was used for comparative analysis, with good results. The main conclusions are as follows.

(1) Deep and complicated fill is harmful to building slope, so it should be paid attention to and measures should be taken.
(2) For the pile and anchor support system of slope in deep and complex soil filling area, the displacement and deformation of double pipe split grouting are reduced compared with that of ordinary bolt. For loose soil and other strata prone to collapse holes, the construction quality of bolt can be improved by double pipe split grouting.

(3) FLAC3D software was used to numerically simulate the support situation of pile and anchor slope. After the advance support of slope protection pile and anchor cable, soil was excavated, and the excavation and displacement conditions were studied. Through the comparison between the simulated situation and the actual monitoring situation, it was found that the numerical simulation was basically consistent with the actual situation.

References

1. Chongqing Housing and Urban Rural Construction Commission 2013 Technical code for building slope engineering (GB 50330-2013). China Architecture & Building Press, Beijing
2. China Metallurgical Construction Association 2015 Technical code for engineering of ground anchorage and shotcrete support (GB 50086-2015). China Planning Press, Beijing
3. Central Research Institute of Building and Construction Co., Ltd. MCC Group 2005 Code for design of building foundation (CECS 22:2005). China Planning Press, Beijing
4. China Academy of Building Research 2008 Technical code for building pile foundations (JGJ94-2008). China Architecture & Building Press, Beijing

Analysis of the Inclined Impact Resistance Characteristics of Concrete Structures Against Projectiles

Xiangyu Xu, Yingxiang Wu[✉], Yifan Jia, and Songping Gan

Defense Engineering Institute, AMS, PLA, Beijing 100850, China
1224846321@139.com

Abstract. The research involved the use of commonly utilized concrete materials as the target plate, with a specially designed 76 mm projectile intended to impact the concrete structure at a specific angle. The goal was to conduct numerical simulation research on the deflection of the projectile under oblique impact conditions and analyze the concrete structure's ability to resist penetration under such conditions. Through numerical simulation calculations, various aspects of the oblique impact were studied, including ricochet patterns under different projectile velocities and angles of incidence, as well as the resulting damage to both the projectile and the target plate, penetration depth, and deflection angle of the projectile. The findings showed that, compared to vertical penetration, the concrete structure exhibited superior deflection and cover-plate effectiveness under oblique impact conditions. The ricochet patterns observed could potentially inform the design of concrete anti-ballistic structures.

Keywords: Concrete Structure · Projectile Ricochet · Oblique Impact

1 Introduction

With the continuous development of precise guidance technology, the threat of penetrating weapons is increasing. Therefore, the research and development of new camouflage technology is crucial to improve the protection capabilities of important engineering facilities on the battlefield, and the study of yaw structures has attracted widespread attention. For example, Reference [1, 2] developed a yaw layer composed of electrical ceramics and PRC spherical columns, and conducted a shooting test with a 57 mm semi-armor-piercing bullet. The results showed that the yaw layer had a significant effect on the deflection of the projectile, with a maximum deflection angle of 64°. References [3, 4] analyzed the impact resistance characteristics of ruby ball concrete. The study of oblique impact of bullet-target is the key to the study of yaw structure. Reference [5] conducted tests on the oblique penetration of C30 and C60 reinforced concrete by projectiles at speeds of 250–430 m/s. The results showed that the critical ricochet angle of C30 reinforced concrete was between 38°–44°, and that of C60 reinforced concrete was between 34°–42°. Reference [6] conducted oblique penetration tests on C40 concrete

© The Author(s) 2024
G. Feng (Ed.): ICCE 2023, LNCE 526, pp. 744–753, 2024.
https://doi.org/10.1007/978-981-97-4355-1_75

target plates using a 10 mm diameter projectile, and the results showed that when the inclination angle of the projectile increased to a certain degree, ricochet occurred. References [7–10] used cavity expansion theory and numerical simulation methods to study the ricochet problem when projectiles obliquely collided with concrete target plates. Currently, the yaw theory analysis of concrete spallation layer is difficult to achieve with high precision. Most research is based on projectile impact tests and numerical simulation results. Therefore, using simulation to analyze the ricochet situation of the projectile after oblique impact on the target, as well as the penetration depth and deflection angle of the projectile, is of great significance for studying the deflection effect of special structure concrete. It can also provide a reference for the design and optimization of spallation layer structures in protective engineering. Therefore, this paper carries out a numerical simulation study on the deflection of concrete structures under oblique impact conditions.

2 Simulation Conditions and Model Selection

2.1 Simulated Operating Conditions

The simulated projectile has a diameter of 76 mm, a length-to-diameter ratio of approximately 9, and a caliber radius head (CRH) of 7. The specific dimensions and weight parameters are shown in Table 1 below.

Table 1. Dimensions of the cartridges

Diameter /mm	L/D	CRH	Mass/Kg
76	9	7	15.24

The projectile velocity conditions range from 400 to 800 m/s, encompassing three different velocity conditions. The projectile impacts the C60 concrete target at two different angles. Using the dynamic finite element simulation software LS-DYNA, a numerical simulation of the oblique penetration of a 76 mm projectile into the concrete target is conducted. The pre-processing software HyperMesh is utilized to establish the finite element model of the oblique penetration of the 76 mm projectile into the concrete target. Through simulation calculations, the ricochet situation of the 76 mm projectile obliquely penetrating the C60 concrete target, as well as the damage situation of the projectile and the target plate, are obtained under different projectile velocities and different projectile incidence angle conditions. The deflection angle, penetration depth, projectile damage, and the diameter of the concrete target pit are statistically analyzed for each condition. The schematic diagram of the statistical results is shown in Fig. 1, where the projectile incidence angle is denoted as β, the projectile deflection angle as γ, the penetration depth as H, and the diameter of the concrete target pit as D.

A finite element model of the 76 mm projectile oblique impact on a C60 concrete target was established, with a total of 4 working conditions selected. Projectile velocities included 400 m/s, 500 m/s, and 600 m/s, while the incident angles were 45° and 60°,

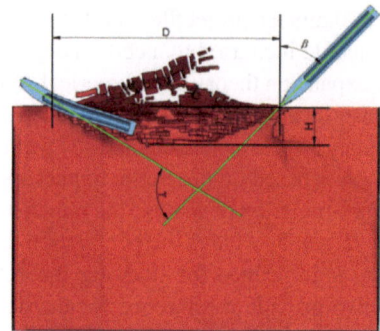

Fig. 1. Schematic diagram of intrusion results

as shown in Table 2. By calculation, the ricochet situation of the 76 mm projectile obliquely penetrating the C60 concrete target under different projectile velocities and incident angles, as well as the damage situation of the projectile and the target plate, was obtained.

Table 2. Numerical simulation calculation table

Operating Condition	projectile velocity/m/s	Angle of incidence/°	Analysis of results
1	500	45	Bouncing condition
2	400	60	Deflection angle
3	500	60	Crater depth Crater Diameter
4	600	60	

2.2 Concrete Target Board Instructions

Based on the numerical simulation scheme for the oblique penetration of a 76 mm projectile into a C60 concrete target, a 3D model was created using the SolidWorks software platform. The concrete target has a diameter of 2000 mm and a height of 1200 mm, surrounded by 5 mm steel plates on all sides. Finite element mesh division was conducted using HyperMesh to construct the corresponding finite element mesh model. The finite element mesh model of the concrete target is shown in Fig. 2, with mesh refinement in the projectile-target contact area. The finite element model of the projectile is shown in Fig. 3.

2.3 Unit and Algorithm Selection

The simulation software selected for this simulation calculation is ANSYS/LS-DYNA program, due to the need to establish a three-dimensional finite element model of the

Fig. 2. Finite element model of concrete target　**Fig. 3.** Finite element model of the projectile

76 mm bullet oblique impact C60 concrete target, so the choice of three-dimensional 8-node solid cell SOLID164 solid mesh division, the unit has eight nodes, each node has six degrees of freedom. The following Fig. 4 describes the geometric properties, node positions and coordinate system of SOLID164 cell.

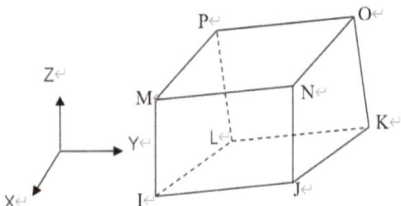

Fig. 4. SOLID164 solid cell geometry characteristics

2.4　Load and Constraint Setting

According to the actual working conditions of the simulation calculation of the 76 mm projectile hitting the C60 concrete target obliquely, a series of specific solution conditions are set up in the simulation calculation process, specifically in the following aspects:

(1) Because of the huge number of units and nodes in the calculation model, the solver selects LS-DYNA software and sets up multi-node parallel calculation;
(2) In order to ensure the energy conservation of the system as a whole, the hourglass control of the calculation system is carried out, and viscous damping is added to reduce the energy loss;
(3) The contact between the concrete target and the elastomer is set as erosion contact.

2.5　Selection of Material Constitutive Model

In the calculation process, the elastic body of the principal model selection of viscoplastic damage principal model (*MAT_JOHNSON_COOK), the elastic body in the process of invasion are embodied in the strain rate effect, the commonly used viscoplastic damage model that can take into account both the strain rate effect and the temperature effect is the JOHNSON COOK model, the model consists of two main parts. The first part only involves stress:

$$\sigma = \left[A + B\left(\bar{\varepsilon}^P\right)^n\right]\left[1 + C\ln\dot{\varepsilon}^*\right]\left[1 - \left(T^*\right)^m\right] \tag{1}$$

where: σ is the VON MISES flow stress, A, B, C, n, and m are the input constants associated with the material, $\bar{\varepsilon}^P$ is the equivalent plastic strain, and $\dot{\varepsilon}^*$ is the relative equivalent plastic strain rate. $T^* = (T - Tr)/(Tm - Tr)$ is the dimensionless temperature, and Tm and Tr are the melting point and room temperature of the material.

The first factor n of the above equation represents the flow stress as a function of equivalent plastic strain ($\dot{\varepsilon}^* = 1.0$, $T^* = 0$); the second factor C represents the strain rate effect; and the third factor m represents the temperature effect.

The second part deals with strain:

$$\varepsilon^f = \left[D_1 + D_2 expD_3\sigma^*\right]\left[1 + D_4\ln\dot{\varepsilon}^*\right]\left[1 + D_5T^*\right] \tag{2}$$

where ε^f is the fracture strain, σ^* is the ratio of the pressure to the equivalent force of VON MISES; D1, D2, D3, D4, and D5 are the damage coefficients. Fracture occurs when $D = \sum \frac{\Delta\varepsilon}{\varepsilon^f} = 1$ ($\Delta\varepsilon$ is the equivalent plastic strain increment during the integration cycle).

The JOHNSON COOK model is generally used in conjunction with the GRUNEISEN equation of state, which is expressed as follows:

$$p = \frac{\rho_0 C^2\mu[1 + (1 - \frac{\gamma_0}{2})\mu - \frac{a}{2}\mu^2]}{[1 - (S_1 - 1)\mu - S_2\frac{\mu^2}{\mu+1} - S_3\frac{\mu^3}{(\mu+1)^2}]^2} + (\gamma_0 + a\mu)E \tag{3}$$

where C is the intercept of the vs-vp curve; S1, S2, and S3 are the slope coefficients of the vs-vp curve; γ_0 is the GRUNEISEN constant, and a is the first-order volumetric correction for γ_0 and $\mu = \frac{\rho}{\rho_0} - 1$.

Concrete materials, available material models are more abundant. At present, the concrete material models applicable to impact conditions include HJC model, RHT model, TCK model, etc. From the numerical simulation of a large number of projectile penetration conditions and the comparison of test results, the HJC model (Holmquist-Johnson-Cook Concrete mode) is more advantageous in describing the dynamic mechanical behavior of concrete under high strain rate and high pressure conditions. The HJC model has been clearly defined in the LS-DYNA program and is divided into three aspects: equation of state, strength model, and damage model.

The equation of state is as follows:
Resilience phase:

$$p = k_e\mu \tag{4}$$

Plastic loading stage:

$$p = p_1 + \frac{(p_1 - p_c)(\mu_1 - \mu_c)}{\mu - \mu_1}p_c \leq p \leq p_1 \tag{5}$$

Plastic unloading stage:

$$p - p_{max} = \left[\left(1 - \frac{\mu_{max} - \mu_c}{\mu_1 - \mu_c}\right)K_e + \frac{\mu_{max} - \mu_c}{\mu_1 - \mu_c}K_1\right](\mu_1 - \mu_{max})p_c \leq p \leq p_1 \quad (6)$$

Fully compacted loaded section:

$$p = k_1 \frac{\mu - \mu_1}{1 + \mu_1} + k_2 \left(\frac{\mu - \mu_1}{1 + \mu_1}\right)^2 + k_3 \left(\frac{\mu - \mu_1}{1 + \mu_1}\right)^3 p > p_1 \quad (7)$$

Fully compacted unloading section.

$$p - p_{max} = k_1 \left(\frac{\mu - \mu_1}{1 + \mu_1} - \left(\frac{\mu - \mu_1}{1 + \mu_1}\right)_{max}\right) p > p_1 \quad (8)$$

At this stage the concrete is non-porous and dense and the material is completely destroyed.

Yield equation:

$$\sigma^* = \left[A(1 - D) + BP^{*N}\right](1 - C \ln \varepsilon^*) \quad (9)$$

where: σ^* is the normative equivalent stress, $\sigma^* = \sigma/f_c$, σ is really the equivalent stress, f_c is the quasi-static uniaxial compressive strength; P^* is the normative pressure, $P^* = P/f_c$; D is the damage parameter; ε^* is the dimensionless strain rate.

Injury Equation:

$$D = \sum \frac{\Delta \varepsilon_P + \Delta \mu_P}{D_1(P^* + T^*)^{D_2}} \quad (10)$$

where: $\Delta \varepsilon_P$ is the equivalent plastic strain; $\Delta \mu_P$ is the equivalent plastic bulk strain; D1 and D2 are material constants; T^* is the standardized maximum tensile hydrostatic pressure, $T^* = T/f_c$.

Destructive strength:

$$DS = f_c' min\left[SFMAX, A(1 - D) + BP^{*N}\right][1 + C \ln \varepsilon^*]P^* > 0 \quad (11)$$

$$DS = f_c' max\left[0, A(1 - D) + A\left(\frac{P^*}{T}\right)\right][1 + C \ln \varepsilon^*]P^* < 0 \quad (12)$$

3 Simulation Results

The simulation results for four different working conditions corresponding to the oblique impact on the C60 concrete target plate at projectile speeds of 400, 500, and 600 m/s and angles of 45° and 60° are shown in Fig. 5. The stress distribution of the projectile at different times is shown in Fig. 6.

The results show that in Case 1, the slanting impact of the projectile on the concrete target produces a ricochet, the deflection angle of the projectile is 123°; the depth of

OP1
OP2
OP3
OP4

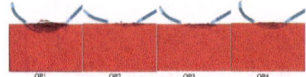

Fig. 5. Results of oblique penetration of projectile into concrete target

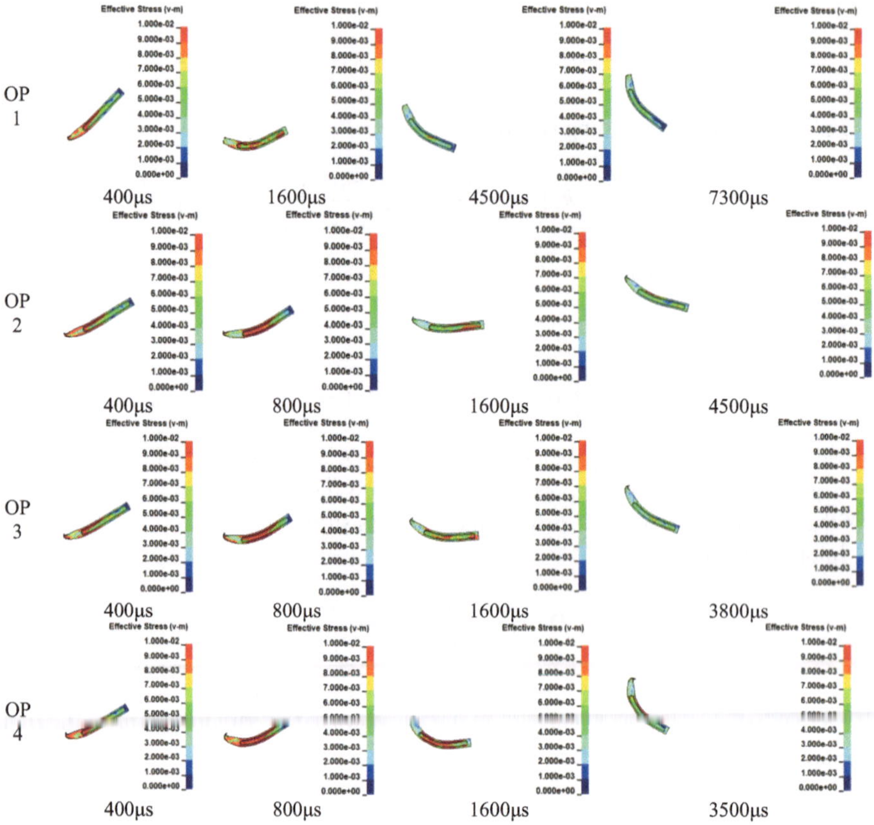
Effective Stress (v-m)
1.000e-02
9.000e-03
8.000e-03
7.000e-03
6.000e-03
5.000e-03
4.000e-03
3.000e-03
2.000e-03
1.000e-03
0.000e+00
OP 1
400μs
1600μs
4500μs
7300μs
OP 2
400μs
800μs
1600μs
4500μs
OP 3
400μs
800μs
1600μs
3800μs
OP 4
400μs
800μs
1600μs
3500μs

Fig. 6. Stress Distribution Cloud Map of Projectile Penetrating Concrete Target

penetration of the projectile is 198.8 mm; the diameter of the concrete target pit is 1098.2 mm; the head of the projectile has been eroded and the projectile body is deformed obviously. Case 2: oblique penetration of the projectile into the concrete target, resulting in ricochet, the deflection angle of the projectile is 71°; the depth of penetration of the projectile is 71.4 mm; the diameter of the concrete target pit is 928 mm; the head of the projectile has been eroded, and the body of the projectile has obvious deformation. Case 3: oblique penetration of the projectile into the concrete target, resulting in ricochet, the deflection angle of the projectile is 89°; the depth of penetration of the projectile is 98.6 mm; the diameter of the concrete target pit is 941.5 mm; the head of the projectile has been eroded, and the body of the projectile has been deformed significantly. In Case 4, the slanting penetration of the projectile into the concrete target produces ricochet, the deflection angle of the projectile is 117°; the penetration depth of the projectile is 139.2 mm; the diameter of the concrete target pit is 942.3 mm; the head of the projectile has been eroded, and the body of the projectile has obvious deformation.

4 Analysis of Simulation Results

4.1 Analysis of Projectile Deflection and Target Damage

According to the requirements of the working conditions, the finite element model of the 76 mm projectile impacting obliquely on the C60 concrete target was established, and a total of four working conditions were calculated, as shown in Table 3 below. Through the calculation of different projectile velocity and different projectile incidence angle conditions, the 76 mm projectile oblique impact C60 concrete target ricochet, as well as the damage of the projectile and the target plate.

Based on the simulation results, the following conclusions can be obtained:

(1) From the calculation results, it can be concluded that for the same projectile and under the same velocity conditions, a larger angle of incidence leads to a smaller projectile deflection angle, shallower penetration depth, and smaller crater diameter.
(2) From the calculation results, it can be concluded that for the same projectile and under the same angle of incidence, a higher velocity leads to a larger projectile deflection angle and greater penetration depth.
(3) Compared with the vertical penetration of concrete armor-piercing structures under the same conditions, the advantage of concrete structures against penetration is more pronounced under oblique impact conditions. The design of new concrete armor-piercing structures should fully utilize the advantage of concrete structures in causing projectile deviation under oblique impact conditions.

Table 3. Summary of numerical simulation results

Operating Condition	projectile velocity/m/s	angle of incidence/°	Angle of deflection/°	penetration depth /mm	Diameter of crater /mm	Condition of the projectile
1	500	45	123	198.8	1098.2	Warhead abrasion, body deformation
2	400	60	71	71.4	928	
3	500	60	89	98.6	941.5	
4	600	60	117	139.2	942.3	

4.2 Comparison Results with Vertical Penetration of Concrete Structures

Using the same numerical model to calculate the 76 mm projectile with 400 m/s vertical penetration of C60 concrete structure, the results are shown in Fig. 7, penetration depth of 0.68 m. Penetration according to the penetration formula proposed by Forrestal [11] can be calculated 76 mm projectile penetration of C60 concrete depth of about 0.7 m, and the numerical simulation results coincide with the 76 mm projectile in large angle In the four working conditions under the condition of oblique impact on C60 concrete structure, the effective depth of penetration is about 0.07 m, which is much smaller than the vertical penetration depth of the projectile, compared with the anti-invasion effect is very obvious, which proves that the large-angle oblique impact of the projectile target can be effective in deflecting the projectile to achieve the purpose of protection at the same time.

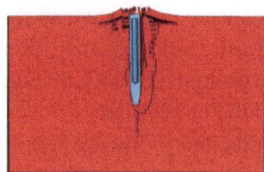

Fig. 7. Schematic diagram of vertical penetration results

5 Conclusion

This article focuses on the numerical simulation of the oblique impact of a 76 mm projectile on a C60 concrete target, and a total of 4 working conditions were selected for calculation. The simulation results were used to obtain the skipping behavior of the 76 mm projectile and the damage to the projectile and the target plate under different projectile velocities and oblique impact angles.

(1) The calculation results show that the oblique penetration of the projectile into the concrete target can be divided into the formation of a projectile pit, sliding zone, and tunnel. This is due to the loosening and shedding of the surface concrete when the projectile impacts the concrete target, resulting in minimal resistance to the projectile's movement. As the penetration continues, an asymmetrical resistance force causes the projectile to deviate laterally. With increasing penetration depth, the resistance force becomes approximately uniform, resulting in the formation of a tunnel. When the tunnel length reduces to zero, skipping of the projectile occurs.

(2) From the calculation results, it can be observed that for the same type of projectile and the same velocity, a larger oblique impact angle leads to a smaller deviation angle, shallower penetration, and smaller pit diameter.

(3) Similarly, for the same type of projectile and oblique impact angle, a higher velocity results in a larger deviation angle and deeper penetration.

(4) Compared to the vertical penetration of the projectile into the concrete, the concrete structure exhibits a more significant advantage in resisting penetration under oblique impact conditions. The design of new concrete shielding structures should fully exploit the advantage of deflecting projectiles under oblique impact conditions.

References

1. Chen, W., Guo, Z., Wu, H., et al.: Mechanism and experiment of inducing projectile deviation by surface irregular shape ablative layer. J. Ballistics **23**(04), 66–69, 74 (2011)
2. Mustfer, M., Hameed, A., Wood, D.: Ricochet quantification using a multiple sensor approach. Defense Technol. **17**(02), 305–314 (2020)
3. Zhou, B., Tang, D., Chen, X., et al.: Analysis of projectile impact resisting characteristics of concrete with single layer tightly arrayed corundum spheres. J. Vibr. Shock **21**(4), 87–89 (2002)
4. Zhou, B., Chen, X., Tang, D., et al.: An experimental study on anti-penetration characteristics of concretes shielded with single layer of tightly arrayed corundum spheres. Explosion Shock Waves **23**(2), 173–177 (2003)
5. Duan, J., Wang, K., Zhou, G., et al.: Critical skipping of projectiles penetrating concrete. Explosion Shock Waves **36**(06), 797–802 (2016)
6. Xue, J., Shen, P., Wang, X.: Experimental study and numerical simulation of oblique penetration of projectiles into concrete targets. Explosion Shock Waves **37**(03), 536–543 (2017)
7. Warren, T.L., Poormon, K.L.: Penetration of 6061–T6511 aluminum targets by ogive-nosed VAR 4340 steel projectiles at oblique angles: experiments and simulations. Int. J. Impact Eng **25**(10), 993–1022 (2001)
8. Yoo, Y.H., Kim, J.B., Lee, C.W.: Effects of the projectile geometries on normal and oblique penetration using the finite cavity pressure method. Appl. Sci. **9**(18), 3939 (2019)
9. Cho, H., Choi, M.K., Park, S., et al.: Determination of critical ricochet conditions for oblique impact of ogive-nosed projectiles on concrete targets using semi-empirical model. Int. J. Impact Eng **165**, 104214 (2022)
10. Choi, M.K., Han, J.H., Park, S., et al.: Efficient method to evaluate critical ricochet angle of projectile penetrating into a concrete target. Math. Probl. Eng. 3696473–3696479 (2018)
11. Forrestal, M., Altman, B.S., Cargile, J.D., et al.: Empirical equation for penetration depth of ogive-nose projectiles into concrete targets (1992)

Experimental Study on the Effect of Mixing Time on the Performance of High Strength Concrete

Xiangwei Ma[1], Xiuzhi Zhu[1], Junjie Zeng[2], Shangchuan Zhao[3], and Longlong Liu[3(✉)]

[1] Guangdong Provincial Highway Construction Co., Ltd., Guangzhou 511447, Guangdong, China
[2] Road & Bridge South China Engineering Co., Ltd., Zhongshan 528400, China
[3] Research Institute of Highway, Ministry of Transport, Beijing 100088, China
liulong4517@126.com

Abstract. To solve the problem of concrete quality caused by non-standard mixing time of high-strength concrete and improve the durability life of concrete structures, based on the working performance and mechanical properties of concrete, effect of mixing time on the workability and strength was studied. In this paper, for high-strength concrete with water-binder ratios of 0.31 and 0.38, the mixing time was 120 s, 160 s, 180 s and 240 s. The effects of different mixing times on the fluidity, cohesion and water retention of high-strength concrete and the compressive strength of 28 days were studied. The test results show that the appropriate mixing time benefits the working performance of concrete and the development of concrete strength. Based on the workability and mechanical properties of concrete, the mixing time of concrete with a water-binder ratio of 0.38 is 160 s, and the mixing time of concrete with a water-binder ratio of 0.31 is 180 s. To ensure the excellent uniformity, compactness, workability and mechanical properties of concrete, the mixing time should consider the slump and the water-binder ratio of concrete. For concrete with a water-binder ratio of 0.31 to 0.38, the mixing time can change from 180 to 160 s.

Keywords: Mixing Time · Water-Binder Ratio · Workability · Admixture

1 Introduction

Concrete mixing quality is mainly controlled by mixing method, feeding sequence and mixing time. To make the concrete mixture uniform and the mixing time appropriate, Germany first issued the national industry standard DIN1045. However, the concrete homogeneity and mixing time have not been precisely defined. At present, forced mixing is often used in on-site construction. 45–120 s is selected as the mixing time of concrete. The main reason is that too long a mixing time will lead to the bleeding of concrete, affecting its workability, strength, and durability. If the mixing time is longer, the composition of the mixture will be uneven, the homogeneity will be good, the honeycomb will appear inside the structure, and the phenomenon of pockmark will appear

G. Feng (Ed.): ICCE 2023, LNCE 526, pp. 754–762, 2024.
https://doi.org/10.1007/978-981-97-4355-1_76

on the surface [1, 2]. Therefore, a specific internal relationship exists between the uniformity of concrete mixing and bleeding. As shown in Fig. 1, the uniformity of concrete and the development process of segregation exceed the optimal mixing time, and the concrete will begin to segregate. After the concrete reaches a specific mixing time, there will always be a time point for the best mixing quality. On the graph, the position of this point is determined by the sum of mixing and segregation [3]. Zheng Donghao et al. [4] believed that the uniformity of the concrete mixture was the best when the mixing time was 90 s. Ren Changxi [5] believes that the mixing time has little effect on the air content of the concrete mixture. When the mixing time is between 90 s and 120 s, the slump and workability of the concrete are the best. Yang et al. [6] used a double spiral concrete mixer to test the influence of mixing time and mixing rate on the strength and uniformity of concrete. Through comparative analysis, it was found that better quality concrete could be obtained when the mixing speed of concrete was between 1.6~ and 1.8 s and the mixing time was between 85~ and 95 s. He et al. [7] used the discrete element analysis software EDEM to simulate the material mixing process. Rong Xin of Chang'an University established a mathematical model to characterize the workability, and the optimal mixing time of concrete was between 90 s and 110 s by comparing the established mathematical model with the traditional slump test [8]. Fang [9] modified the VC value and strength performance of concrete with different consistency, and determined that the concrete performance was the best when the mixing time was 120 s.

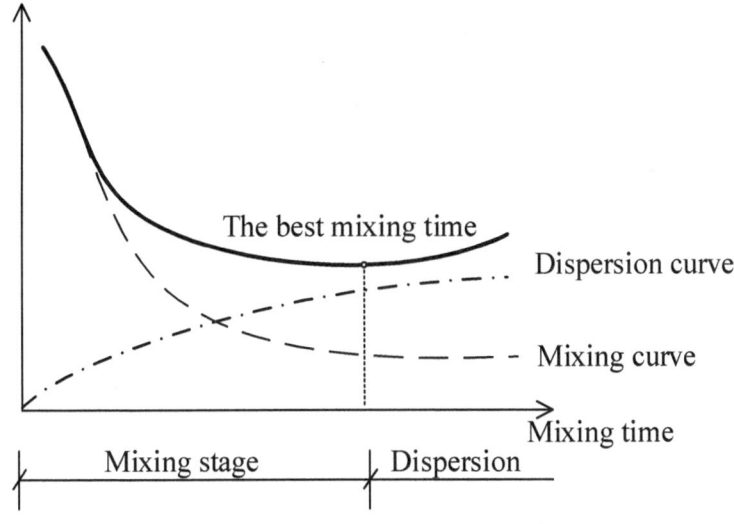

Fig. 1. Curve of mixing process

Ngo [10] designed a method to calculate concrete's shortest mixing time (stabilization time) based on the mixer's power and reduced the energy consumption by 17%. The 'Technical Specification for Concrete Durability Quality Control of Hong Kong-Zhuhai-Macao Bridge' stipulates the shortest continuous mixing time according to the mixer model and capacity, as shown on Table 1.

Table 1. The shortest continuous mixing time of concrete

Mixing machine type	Mixer capacity (L)	The shortest continuous mixing time (min)
Forced (non-vertical)	≤ 1500	2.5
	> 1500	3.0

The mixing time of ordinary concrete is in the range of 90 s–120 s, and the performance of concrete is the best. However, for concrete with high strength and a small water-binder ratio, it is easy to cause insufficient mixing, leading to uncompacted concrete pouring and insufficient strength. To solve the problem of concrete quality caused by the non-standard mixing time, based on the working performance of concrete such as slump and workability, the influence of mixing time on the working performance, mechanical properties and crack resistance of concrete was studied.

2 Test Scheme

2.1 Raw Materials

Raw materials used in the test process: Gravel, limestone, particle size 5–25 mm, cement grade PII42.5, the manufacturer is Yingde Conch Cement Co., LTD., water reducing agent 3301C-HM03, Sika (Jiangsu) Building Materials Co., LTD., first-class fly ash, provided by Zhuhai Yuezhu Environmental Protection Technology Development Co., LTD., slag powder for S95 type, Provided by Zhuhai Yueyufeng Steel Co., LTD.

2.2 Test Methods

The mixing equipment and methods used are different in the concrete mixing process, and the mixing time may also be different when the concrete has the best working performance. The optimum concrete mixing time with different mix ratio designs is also different. For concrete designed with different water-binder ratios, the same HJS-60 horizontal double-axis forced mixer was used to carry out the mixing test, as shown in Fig. 2, and the mixing time required to achieve the best working performance was studied.

Fig. 2. HJS-60 horizontal double-axis forced mixer.

Two mix ratios of w/b = 0.31 and 0.38 were selected, and the mix ratio is shown in Table 2. When the mixing time was 120 s, 160 s, 180 s and 240 s, the working performance, fluidity, cohesiveness and water retention of concrete were observed. At the same time, specimens with the size of 100 × 100 × 100 mm were poured and maintained for 28 days to test the strength of concrete.

Table 2. Mix proportion ratio of concrete

Number	Water-binder	Cement	ground slag	Fly ash	Sand	Stone	Water	Superplasticizer
1	0.31	350	130	0	702	1098	150	4.8
2	0.38	155	90	120	769	1106	140	3.65

3   Test Results

3.1   Influence of Mixing Time on Working Performance

According to the test data analysis in Figs. 3, 4 and Table 3, the concrete with a water-binder ratio of 0.31 and 0.38 has a segregation phenomenon and poor cohesion without water retention when mixing for 120 s. With the increasing mixing time, the cohesion in the range of 120 s–180 s gradually increases, and the cohesion of concrete worsens after mixing for 240 s. The cohesiveness and liquidity achieves the best when the mixing time is 180 s.

Fig. 3. Slump of concrete at different mixing time (Water-binder ratio 0.31)

Fig. 4. Slump of concrete at different mixing time (Water-binder ratio 0.38)

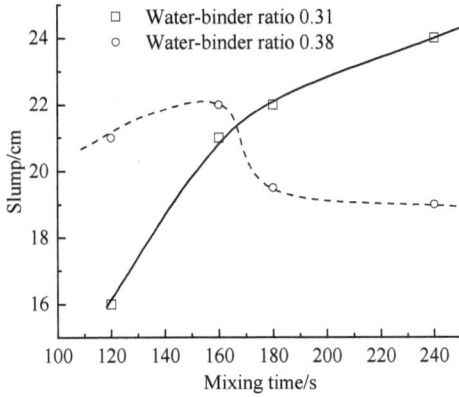

Fig. 5. The relationship between the slump and the mixing time

Table 3. Workability of concrete with different water-binder ratio at different mixing time

Water-binder ratio	Mixing time/s	Slump/mm	Stickiness	Water-retaining property	Cohesiveness
0.31	120	160	mid	no	poor
0.31	160	210	superior	no	range
0.31	180	220	superior	mickle	good
0.31	240	240	superior	no	fair
0.38	120	210	mid	no	poor
0.38	160	220	superior	no	fair
0.38	180	195	superior	no	good
0.38	240	190	superior	no	fair

As shown in Fig. 5, according to different mixing time (120 s, 160 s, 180 s, 240 s), the concrete resistance to segregation, bleeding performance and cohesion. Analysis of the influence of concrete quality, such as concrete strength and dispersion, shows that concrete with the water-binder ratio of 0.31 and 0.38 has better concrete workability in the range of 120 s to 180 s, and the concrete workability become worse after mixing 240 s. The cohesiveness of the two kinds of water-binder ratio concrete reaches its best when the mixing time is 180 s. The results show that the concrete with the water-binder ratio of 0.31 and 0.38 has the best workability, water retention and cohesiveness at 180 s.

3.2   Effect of Mixing Time on Mechanical Properties

The relationship between the compressive strength of concrete and the mixing time in Fig. 6 shows that the compressive strength increases with the incremental mixing time. In addition to the influence of mixing time, the changing trend of compressive strength of concrete relates to the mix ratio of concrete materials. The compressive strength increases from 54 MPa to 59 MPa (an increase of 9.3%) when the mixing time changes from 120 s to 180 s. The change in compressive strength of concrete with a water-binder ratio of 0.38 is not apparent, and the compressive strength increases from 46 MPa at 120 s mixing to 48 MPa at 160 s mixing (an increase of 4.3%).

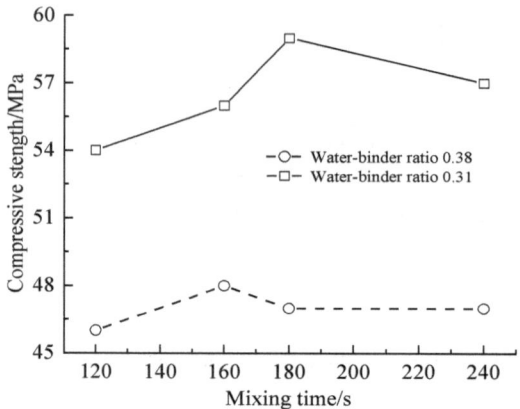

Fig. 6. The relationship between the compressive strength and the mixing time

3.3 Analysis

Because the performance of concrete is not only related to the material itself but also to the mixing time of the material, and the hydration reaction will occur during the mixing process, the change of mechanical properties of concrete will affect the workability of fresh concrete. Therefore, using the same mixer, the mixing time is an essential indicator of workability. The mixing process includes four stages: Stage 1 (about 0–20 s): Due to the lack of water, the particles between the aggregates are seriously staggered due to lack of lubrication. The second stage (about 20 s–70 s): After the contact of water with cement and mineral admixtures, the aggregate particles begin to form agglomerates, and the uniformity of the concrete mixture reaches its best. The water and cement have not fully reacted at this time, so the workability cannot reach the best. In the third stage (about 70–120 s): the lubrication effect of water weakens, and the fluidity and cohesion of concrete begin to increase. The fourth stage (after 120 s): As the hydration continues, the colloid formed by cement, mineral admixtures, and water gradually encapsulates the aggregate, and the workability gradually increases. When the workability of the concrete mixture reaches the optimal value, the colloid and aggregate form a whole of fluidity; if the stirring continues, some particles begin to peel off the concrete mixture, and the colloid will also flow out of the mixture in a liquid state.

Comparing the different mixing time on the working performance of different water-binder ratio concrete. The concrete mixing time of the water-binder ratio of 0.38 is 160 s, and the concrete mixing time of the water-binder ratio of 0.31 is 180 s. The working performance of concrete is the best. With the decrease in the water-binder ratio, the water content in the mixture decreases, and the stirring time should be extended appropriately. From the test process, it is found that not only the mixing time should be controlled according to the slump of concrete, but also the mixing time should be controlled according to the water-binder ratio of concrete. For concrete with a water-binder ratio of 0.31 to 0.38, the mixing time can be controlled between 160 s and 180 s. For concrete with a smaller water-binder ratio, the mixing time should be appropriately increased and verified by experiments.

4 Conclusions

In this paper, concrete with water-binder ratio w/b = 0.31 and 0.38 and mixing time of 120 s, 160 s, 180 s and 240 s were used to observe the working performance, fluidity, cohesiveness, and water retention of concrete. Effect of mixing time on the working performance and strength of concrete with a small water-binder ratio was analyzed. Specific conclusions are as follows:

(1) When mixing concrete with a small water-binder ratio and high strength, the mixing time is between 160 s and 180 s, and the concrete has good workability, water retention and cohesiveness.
(2) To obtain good uniformity, compactness, and good workability of concrete, not only the slump should be considered but also the water-binder ratio of concrete, the water-binder ratio of 0.38 concrete mixing time 160 s, the water-binder ratio of 0.31 concrete mixing time 180 s is more appropriate.
(3) If the mixing time exceeds 240 s, the concrete fluidity is good, but the phenomenon of bleeding begins, which will also affect the compressive strength of the concrete.
(4) The compressive strength of concrete is related to workability, and the concrete strength with better workability, water retention and cohesiveness are more significant.

References

1. Ding, Z., An, X.: A method for real-time moisture estimation based on self-compacting concrete workability detected during the mixing process. Constr. Build. Mater. **139**, 123–131 (2017)
2. Zeyad, A.M., Almalki, A.: Influence of mixing time and superplasticizer dosage on self-consolidating concrete properties. J. Market. Res.Market. Res. **9**(3), 6101–6115 (2020)
3. Wang, M.: Effect of mixing time on the quality of concrete. Concr. Reinf. Concr. (05), 28–31 (1984)
4. Zheng, D., Hou, Y., Si, B.: Effect of mixing time on working and strength of high flowing concrete. J. Beijing Univ. Civil Eng. Architect. **33**(04), 18–21+32 (2017)
5. Ren, C.: Effect of mixer mixing time on the uniformity of concrete. Sichuan Cem. (06), 299 (2015)
6. Yang, P., Weng, J., Lv, C., Zhang, H.: Research of mixing technology of double helix concrete mixer mixing high-strength concrete. Concrete (02), 155–160 (2013)

7. He, Z., Wang, T.: Optimal design of the key structure of planetary concrete mixers based on EDEM. Int. J. Mater. Prod. Technol. **62**(4), 295–310 (2021)
8. Rong, X., Liu, H., Li, C.: A proposed method and monitoring system for evaluating workability of Portland cement concrete during mixing. Heliyon **8**(11), e11355 (2022)
9. Fang, D., Zhou, M.K., Wang, H.G.: Influence of mixing time on workability and strength properties of concrete in different consistency. Adv. Mater. Res. **217–218**, 1224–1228 (2011)
10. Ngo, H.-T., Kaci, A., Kadri, E.-H., Ngo, T.-T., Trudel, A., Lecrux, S.: Energy consumption reduction in concrete mixing process by optimizing mixing time. Energy Procedia **139**, 810–816 (2017)

Research on Permanent Deformation and Maintenance Measures of Existing Asphalt Payment Based on Rut Damage

Jiangjiao Wang[1], Fei Fang[1], Jianjun Song[1], Futao Chen[2], Yang Xu[1], and Peng Zhu[2(✉)]

[1] Sichuan Weiru Construction Engineering Co., Ltd., Chengdu 610000, China
[2] Zhonglu Jiaojian (Beijing) Engineering Materials Technology Co., Ltd., Beijing 100088, China
277626264@qq.com

Abstract. In order to study permanent deformation of Expressway Asphalt Payment and the corresponding maintenance counter measures which have run away. In this paper, full thickness loading test of asphalt surface is used to analyze the cumulative run deformation capacity and residual bearing capacity of the existing pavement with different run depths and analyze the deformation layer of the core sample after loading and the test results after painting and curing. It is included that the peak rutting depth of existing pavement line to structural damage is 17.6 mm and the permanent rutting deformation depth is 27.3 mm. At the same time, the pavement maintenance counter measures and evaluation standard for permanent deformation of existing asphalt pavement are observed for the three rutting depth ranges.

Keywords: Existing Aspect Payment · Rutting Deformation · Residual Carrying Capacity · Maintenance Countermeasure

1 Introduction

Transportation is the foundation of national economic development and an important criterion for measuring a country's level of modernization [1, 2]. But with the increasing vehicle load, overloading, and complex and ever-changing environmental effects, diseases such as ruts on asphalt pavement on highways are also increasing. The appearance of rutting diseases will to some extent shorten the service life of the road surface, cause economic losses, and even cause local structural damage in a short period of time. This not only affects the service performance of the road surface, but also affects the comfort of road driving. In serious cases, it can pose a threat to safe driving and affect personal safety [3–5]. Based on this background, this article focuses on the rutting disease of existing highways. Through indoor asphalt surface layer full thickness accelerated loading tests, a set of evaluation methods for permanent deformation of existing asphalt pavement is studied. At the same time, reasonable maintenance strategies are specified

G. Feng (Ed.): ICCE 2023, LNCE 526, pp. 763–774, 2024.
https://doi.org/10.1007/978-981-97-4355-1_77

according to different rutting depths. The research results of this article are helpful for maintaining the good service function of existing pavement. The preservation, preservation, and even extension of the service life of existing pavement assets are of great significance.

2 Research Sample

This article focuses on on-site sampling of a highway in Jiangsu Province, and the pavement structure types are shown in Table 1.

Table 1. Types of Study Road Section Structures

Horizon	Thickness/cm	Material type
Upper layer	4	Modified asphalt AK-13
Middle surface layer	6	Modified asphalt AC-20
Lower layer	8	Ordinary asphalt AC-25
basic level	40	Cement stabilized crushed stone
Soil foundation	20	Lime stabilized soil

The highway is currently in the middle and later stages of its design life. With the increasing traffic volume and the increasing proportion of trucks, overloading and over limit phenomena are severe. The road surface has been operating under overload for a long time, and some sections have already experienced serious rutting diseases. The various performance of the road surface is still declining, and the depth of rutting on the road surface is increasing year by year. The material performance of the road surface is declining, and the overall performance of the structural layer is declining.

In order to study the development trend of rutting diseases on the expressway and formulate different maintenance plans to prevent further development of pavement diseases. On the basis of highway road condition detection data, this study first selected suitable sections for research, and then used a three meter straightedge (0.1 mm level) to measure on-site. Samples were taken in a gradient of 1 mm. Take 2 different rut depths at the same position on the right wheel track of the driving lane Φ 300 mm core samples were used for parallel testing. In order to facilitate subsequent research, core samples with the same rut depth were assigned the same number. The core sample numbers are shown in Table 2.

Table 2. Core sample number

Disease type	rut depth						
Rutting	8 mm	9 mm	10 mm	11 mm	12 mm	13 mm	14 mm
Sample number	X_8	X_9	X_{10}	X_{11}	X_{12}	X_{13}	X_{14}
Disease type	rut depth						
Rutting	15 mm	16 mm	17 mm	18 mm	19 mm	20 mm	21 mm
Sample number	X_{15}	X_{16}	X_{17}	X_{18}	X_{19}	X_{20}	X_{21}

Note: The rut depth measured on site is rounded off using the rounding method for the convenience of this study

3 Research on the Development Trend of Existing Road Rutting Diseases

3.1 Experimental Design

The rutting disease is one of the main forms of asphalt pavement diseases in China. The factors affecting pavement rutting are complex, and many influencing factors overlap with each other [6–9]. It is difficult to simulate the actual environment and stress situation of the pavement using simple indoor tests. Therefore, combined with relevant research results [10–12], this paper develops the full thickness accelerated loading test of asphalt pavement layer, which can effectively simulate the environment and stress situation of the pavement, It is also possible to obtain data on changes in road performance in a relatively short period of time.

The full thickness accelerated loading test of asphalt surface layer adopts specimens as Φ 300 mm core sample, core sample and test mold are shown in Fig. 1, and the equipment used for the test is the standard rut testing machine (RP-0719A). This experiment applies a load of 0.7 MPa to the core sample under 70 °C conditions and repeatedly applies it, while automatically recording the rutting depth value under loading. The experiment is conducted until the core sample structure is completely damaged, and the general loading can reach more than 200000 times.

The full thickness accelerated loading test of the asphalt surface layer adopts an iron mold with a length of 30 cm * width of 30 cm * height of 18 cm. The mold is detachable around it, and to prevent the deviation of the rut trajectory during the test, cement mortar [13] (by weight, cement: sand: water = 1:3:0.5) is used to fill the gaps around the core sample and the mold. The middle and lower layers are wrapped in tin paper to prevent mortar from infiltrating the gaps of the core sample opening. The test starts after 24 h of room temperature curing.

Fig. 1. Accelerated loading test with full thickness of asphalt surface

3.2 Analysis of Accumulated Rutting Deformation

The standard rut testing machine (RP-0719A) [14] was used for the test. Based on past experience, after 72 h of loading (180000 times), the core sample basically exhibits collapse type failure or stable rut depth. The unified plan for continuous loading of the pavement core sample in this design is 72 h. Under the premise of controlling the loading time to be consistent, the variation trend of the rut depth of the pavement core sample after continuous loading is analyzed. The typical curve of the comprehensive layer thickness asphalt mixture rutting test is shown in Fig. 2.

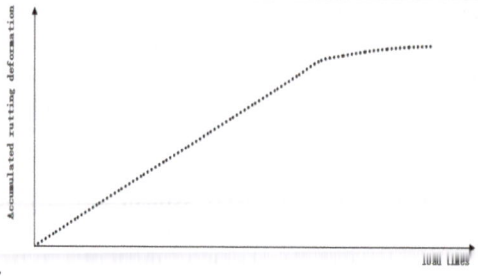

Fig. 2. Typical curve of full layer thickness aspect mixture rutting test

The full thickness accelerated loading test of asphalt surface layer can be idealized into two stages, as shown in Fig. 3. In the first stage (rapid development zone), with the increase of loading times, the cumulative rutting deformation increases rapidly, and develops to the second stage (stable/collapse period). The cumulative rutting deformation tends to stabilize, and the slope of the curve is significantly smaller than that of the first stage. The transition node from the first stage to the second stage is called the stable/collapse node.

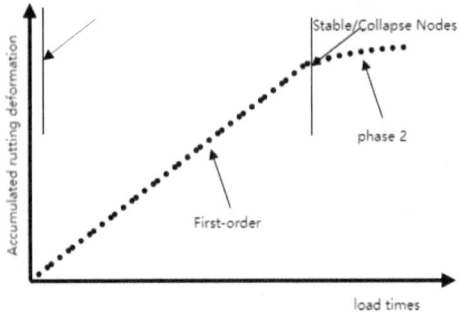

Fig. 3. The relationship between cumulative rutting deformation and loading time

After the on-site core sample underwent the full thickness accelerated loading test of the asphalt surface layer, the loading rut depth and stable collapse/collapse nodes are shown in Table 3. The relationship between the on-site rut depth and the loading rut depth is shown in Fig. 4, and the relationship between the on-site rut depth and stable/collapse nodes is shown in Fig. 5.

Table 3. Cumulative total rutting depth and stability/collapse node

number	Load rut depth (mm)	Stable/Collapse Nodes	number	Load rut depth (mm)	Stable/Collapse Nodes
X_8	five point two	64 h	X_{15}	fifteen point two	44 h
X_9	seven point nine	60 h	X_{16}	fifteen point three	32 h
X_{10}	nine point four	56 h	X_{17}	sixteen point two	36 h
X_{11}	eleven point eight	48 h	X_{18}	thirteen point eight	28 h
X_{12}	twelve point five	52 h	X_{19}	fourteen point one	24-h
X_{13}	fourteen point four	48 h	X_{20}	sixteen point eight	28 h
X_{14}	fourteen point nine	40 h	X_{21}	fifteen point two	20 h

The depth of on-site rutting and the depth of on-site core asphalt surface layer under full thickness accelerated loading test are regressed using polynomial regression, and the regression equation is shown in formula (1):

$$y = -0.1087x^2 + 3.8248x - 17.784 \quad \left(R^2 = 0.9231\right) \tag{1}$$

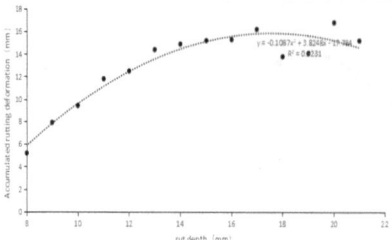

Fig. 4. Relationship between total rutting depth and total rutting depth

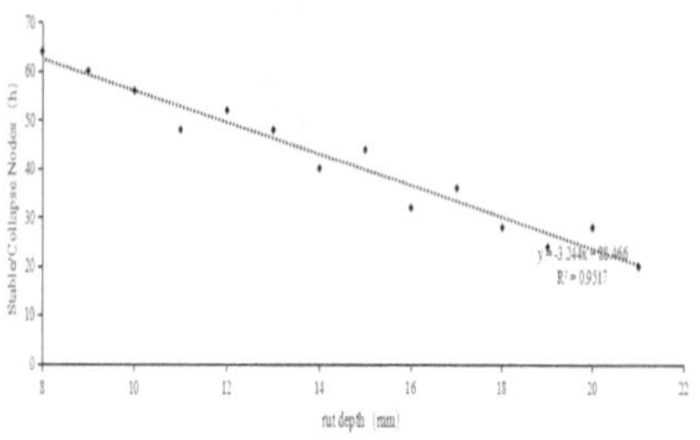

Fig. 5. Relationship between running depth and stability/collapse node

Linear regression is used to determine the stability/collapse node between the on-site rutting depth and the full thickness accelerated loading test of the asphalt surface layer of the on-site core sample. The regression equation is shown in formula (2):

$$Y = -3.244x + 88.466$$
$$\left(R^2 = 0.9517\right) \tag{2}$$

Regression formulas (1) and (2) both show high correlation.

From Fig. 6, it can be seen that the larger the on-site rut depth, the greater the loaded rut depth. From regression formula (1), it can be calculated that the on-site rut depth corresponding to the curve turning point is 17.6 mm. It can be inferred that when the on-site rut depth reaches 17.6 mm, the maximum vertical deformation borne by the road surface gradually stabilizes and obvious lateral deformation begins to appear. When loaded, the observation of the core sample change state indicates that bulges begin to appear on both sides of the rut.

From Fig. 4, it can be seen that as the depth of on-site rute increases, the stable/collapsed nodes show a gradually decreasing trend. From regression formula (2), it can be calculated that the on-site rut depth corresponding to the stable/collapsed node of 0 h is 27.3 mm, which corresponds to the permanent deformation of road structure ruts [15–17] studied in this paper being 27.3 mm. Therefore, when the rut depth reaches 27.3 mm, with the increase of vehicle load, the road surface no longer compresses and deforms, and structural damage begins to occur, which poses great risks to driving safety.

3.3   Research on Rut Deformation Rate

The full thickness accelerated loading test of asphalt surface layer was used to analyze the rutting deformation rate of existing asphalt pavement [18–20]. This article uses the rutting deformation rate, which is the amount of deformation caused by each loading, as the evaluation index for permanent deformation of asphalt pavement. Due to the fact that the rutting depth in the second stage (stable/collapse period) no longer increases with the increase of loading times, the permanent deformation of asphalt pavement is not closely related to it. Therefore, this section only studies the rutting deformation rate in the first stage (rapidly developing zone). Assuming that after the first stage (rapid development zone) of loading, the permanent deformation of the pavement core sample has been achieved, the experimental loading variation is shown in Table 4. The relationship between the rutting deformation rate of the loading test and the number of loading cycles is shown in Fig. 6.

Table 4. Rutting deformation rate of road core specifications after loading

number	Load rut depth/mm	Stable/Collapse Nodes	Loading times (10000 times)	Rut deformation rate (μ M/time)
X_8	five point two	sixty-four	sixteen point one	zero point zero three
X_9	seven point nine	sixty	fifteen point one	zero point zero five
X_{10}	nine point four	fifty-six	fourteen point one	zero point zero seven
X_{11}	eleven point eight	forty-eight	twelve point one	zero point one zero
X_{12}	twelve point five	fifty-two	thirteen point one	zero point one zero
X_{13}	fourteen point four	forty-eight	twelve point one	zero point one two
X_{14}	fourteen point nine	forty	ten point one	zero point one five
X_{15}	fifteen point two	forty-four	eleven point one	zero point one four
X_{16}	fifteen point three	thirty-two	eight point one	zero point one nine
X_{17}	sixteen point two	thirty-six	nine point one	zero point one eight
X_{18}	thirteen point eight	twenty-eight	seven point one	zero point two zero
X_{19}	fourteen point one	twenty-four	six	zero point two three
X_{20}	sixteen point eight	twenty-eight	seven point one	zero point two four
X_{21}	fifteen point two	twenty	five	zero point three zero

According to the linear regression equation between the change in each loading test and the number of loading tests, the R2 reaches 9.65, indicating a high correlation between the two. From Fig. 6, it can be seen that as the depth of on-site ruts increases, the rut deformation rate gradually increases, indicating that the larger the depth of on-site ruts, the greater the rut deformation rate.

In order to propose the evaluation criteria for permanent deformation of asphalt pavement with different rutting depths in the full thickness accelerated loading test of asphalt surface layer, the difference between the rutting deformation rate of each core sample and the average rutting deformation rate of all core samples is used as the interval node. The difference between the two is shown in Table 5.

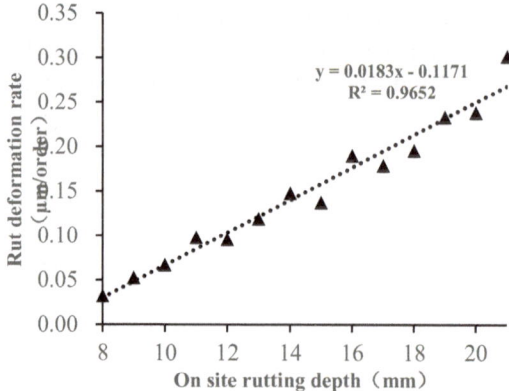

Fig. 6. Relationship between running deformation rate and loading times

Table 5. Difference between average rutting deformation rate and rutting deformation rate

number	Difference	number	Difference	number	Difference	number	Difference	number	Difference
X_8	-0.1	X_{11}	-0.1	X_{14}	zero	X_{17}	zero	X_{20}	zero point one
X_9	-0.1	X_{12}	-0.1	X_{15}	zero	X_{18}	zero	X_{21}	zero point two
X_{10}	-0.1	X_{13}	zero	X_{16}	zero	X_{19}	zero point one		

According to the difference in the table above, it can be roughly divided into three intervals, which are less than or equal to 12 mm, less than or equal to 18 mm if it is greater than 12 mm, and greater than 18 mm. The corresponding rutting rate is less than or equal to 0.1, respectively μ M/time, greater than 0.1 μ M/time less than or equal to 0.2 μ M/time, greater than 0.2 μ. Based on the previous conclusion, the evaluation criteria for permanent deformation of the pavement are proposed as follows:

(1) When the rut deformation rate is ≤0.1 μ At m/time, the old road surface has good resistance to rutting deformation;
(2) When the rut deformation rate is greater than 0.1 μ M/time, ≤0.2 μ At m/time, the resistance of the old road surface to rutting deformation weakens, posing a risk of accelerated damage;
(3) When the rut deformation rate RD is greater than 0.2 μ At m/time, the old road surface has poor resistance to rutting deformation, and timely measures should be taken to treat the existing road surface.

4   Research on Maintenance Strategies

The use of a new asphalt surface layer as an overlay on an existing road surface is a very typical reinforcement method, which has been widely used both domestically and internationally. However, before adding pavement, it is necessary to treat the existing pavement diseases reasonably to prevent unnecessary economic losses caused by the insufficient anti rutting deformation ability of the lower bearing layer in the short term. Therefore, this article focuses on all core samples after the full thickness accelerated loading test of the asphalt surface layer, and uses a concrete drilling and coring machine (HZ-20 type) along the center position of the rut for coring (Φ 100 mm), the loaded specimen and drilled core sample are shown in Fig. 7. After drilling the core sample, the deformation layer after loading can be determined to guide the treatment layer of road surfaces with different rut depths before paving. In order to study the pavement overlay schemes under different rutting deformation rates, this paper also conducted secondary on-site core sampling work.

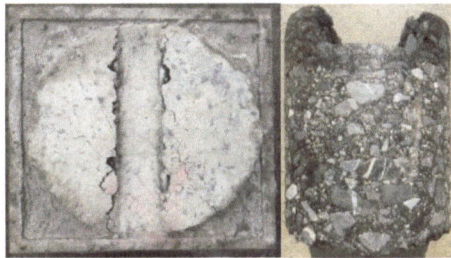

Fig. 7. Loaded specifications and drilled core sample

By manually observing the rutting deformation of the drilled core sample, combined with the evaluation criteria of the road surface's resistance to rutting deformation in Sect. 2.3, the results are as follows:

(1) When the rut deformation rate is ≤ 0.1 μ At m/time, the rutting deformation layer occurs in the middle and upper layers, without involving the lower layer;
(2) When the rut deformation rate is greater than 0.1 μ M/time, ≤ 0.2 μ At m/time, 66.7% of the rutting deformation layer occurs in the middle and upper layers, and 33.3% involves the lower layer;
(3) When the rut deformation rate is greater than 0.2 μ At m/time, the rutting deformation layer occurs in the upper, middle, and lower layers.

In order to determine the layers that need to be treated before adding pavement for different rut deformation rates, this article adopts Φ The 300 mm core sample cutting method is used to simulate milling at different layers on site, and the full thickness accelerated loading test of the asphalt surface layer is also used (the test conditions are consistent with the previous text), combined with the rut rate to determine the overlay scheme for different rut depth ranges on the road surface.

For different rutting deformation rates, this article proposes the following three overlay schemes:

- Option 1: Directly lay 4 cm modified SMA-13 mixture;
- Option 2: Milling and planning the 4 cm original upper layer, backfilling with 4 cm modified SMA-13 mixture, and then adding 4 cm modified SMA-13 mixture;
- Option 3: Milling and planning 10 cm of the original upper and middle surface layer, backfilling 6 cm of modified AC-20 mixture, and then adding 4 cm of modified SMA-13 mixture.

Three schemes involve fully coating modified emulsified asphalt with a solid content of 0.2 kg/m^2 between layers. The deformation rate of ruts in the indoor loading test of the core sample after laying is shown in Table 6.

Table 6. Rutting deformation rate after additional payment

Plan number	Range of rutting deformation rate	Load rut depth (mm)	Stable/Collapse Nodes	Rut deformation rate（μM/time)
Plan One	≤ 0.1	thirteen point three	sixty-eight	zero point zero seven eight
	(0.1, 0.2)	eighteen point five	sixty	zero point one two two
	> 0.2	twenty point five	fifty-six	zero point one four five
Option 2	≤ 0.1	ten point four	sixty	zero point zero six nine
	(0.1, 0.2)	twelve point eight	sixty-four	zero point zero seven nine
	> 0.2	sixteen point eight	fifty-two	zero point one two eight
Plan Three	≤ 0.1	nine point eight	sixty-eight	zero point zero five seven
	(0.1, 0.2)	twelve point seven	seventy-two	zero point zero seven zero
	> 0.2	fourteen point five	sixty-eight	zero point zero eight five

From Table 6, it can be seen that after adding the overlay, the core sample undergoes the full thickness accelerated loading test of the asphalt surface layer, and the rutting deformation rate decreases. This indicates that adding the overlay causes the existing layer to move downwards, increasing the overall deformation resistance of the pavement structure. According to the research conclusion in Sect. 2.3, it is believed that the rutting deformation rate is ≤0.1 μ At m/time (blue background in Table 6), the old road surface has good resistance to rutting deformation. According to the rutting deformation rate, it can be seen that the more layers of treatment before paving, the greater the tolerance for existing road ruts. Scheme 3 is applicable to all road rut sections studied in this article. Based on economic considerations, this article proposes the following maintenance strategies for different road rut sections:

(1) When the rut deformation rate is ≤ 0.1 μ. When m/time, adopt scheme one;
(2) When the rut deformation rate is greater than 0.1 μ M/time, ≤ 0.2 μ. When m/time, scheme 2 is adopted;
(3) When the rut deformation rate is greater than 0.2 μ. When m/time, scheme three is adopted.

5 Conclusion

The main conclusions drawn from this article are as follows:

(1) This article studies the full thickness accelerated loading test of asphalt pavement, which can be effectively used for permanent deformation of existing asphalt pavement.
(2) Through the analysis of permanent deformation of asphalt pavement, this article proposes the rutting deformation rate index, and proposes evaluation standards for permanent deformation of asphalt pavement for different rutting deformation rate ranges.
(3) Through maintenance strategy research, three different types of overlay maintenance strategies are proposed for the pavement structure studied in this article:
   a) When the rut deformation rate is ≤ 0.1 μ. When m/time, adopt scheme one;
   b) When the rut deformation rate is greater than 0.1 μ M/time, ≤ 0.2 μ. When m/time, scheme 2 is adopted;
   c) When the rut deformation rate is greater than 0.2 μ. When m/time, scheme three is adopted.
   d) The experiment studied in this article is time-consuming and the sample size is limited. Readers can further expand their research.

Funding. NingXia Construction Investment Group Co., Ltd. Research Project JTKY 2023-14.

References

1. Wang, Y.: Looking Back at the 30 Years of Reform and Opening Up and Looking Forward to the New Scale of Highway Construction - Press Conference on the Achievements of Highway Transportation Development in the 30 Years of Reform and Opening Up by the Ministry of Transport. Transportation World (2008)
2. Sha, Q.: The phenomenon, causes, and prevention of early damage to asphalt pavement on highways. Transportation World (U09) 17–18 (2004)
3. Huang, X., Fan, Y., Zhao, Y., et al.: Investigation and experimental analysis of high-temperature rutting on asphalt pavement of highways. Highw. Transp. Technol. **24**(5), 5 (2007)
4. Highway Incident Report: Motorcoach Medical Crossover and Collision with Sport Utility Vehicle, Hewitt, Texas, 14 February 2003 (2005)
5. Xu, S.: The rutting depth of asphalt pavement and driving safety. J. Beijing Inst. Arch. Eng. **10**(1), 5 (1994)
6. Shen, J., Li, F., Chen, J.: Analysis and prevention measures for early damage of asphalt pavement on highways. People's Transportation Press (2004)

7. Eisenmann, J., Hilmer, A.: Influence of wheel load and impact pressure on the rusting effect at asphalt pavement experiences and theoretical investments. In: Sixth International Conference, Structural Design of Asphalt Pavement, vol. i, Procedures, University of Michigan. Publication of Michigan University, Ann Arbor (1987)
8. Dawley, C.B., Hogewiede, B.L., Anderson, K.O.: Mitigation of facility rutting of asphalt concrete payments in letterbridge, Alberta, Canada (with discussion and closure). J. Assoc. Asphalt Payment Technol. 59 (1990)
9. Kim, J.R., Drescher, A., Newcomb, D.E.: Rational test methods for predicting permanent deformation in asphalt concrete pavement. Final report Compression (1991)
10. Zhang, Z., Zhang, Y., Li, D., et al.: Exploration of using rutting as a design indicator for asphalt pavement. J. Beijing Inst. Technol. **34**(8), 4 (2008)
11. Wu, J., Ye, F.: Analysis of rutting deformation on asphalt pavement based on MLS66 accelerated loading test. J. Build. Mater. **17**(3), 8 (2014)
12. Ji, X., Zheng, N., Liu, Y., et al.: Prediction of rutting under full scale accelerated loading on asphalt pavement. J. Beijing Inst. Technol. **39**(3), 5 (2013)
13. Wang, J.: The strength characteristics of "cement sand." Northwest Water Res. Water Eng. **8**(1), 6 (1997)
14. Liu, Z.: Rut testing machine: CN2357036Y (2000)
15. Zhang, J.: Numerical simulation of permanent deformation of asphalt pavement and prediction of rutting. J. Hebei Univ. Technol. **40**(4), 4 (2011)
16. Peng, S.: Research on rutting tests at different temperatures and analysis of permanent deformation parameters. Abstract Highw. Transp. (10), 39–41
17. Cao, L., Li, L., Sun, D.: Prediction of permanent deformation of asphalt layer based on rutting test. J. Build. Mater. **12**(5), 4 (2009)
18. Zhang W.: Research on the Ultimate Bearing Capacity and Service Life Prediction of Asphalt Pavement. Hunan University (2009)
19. Sun, Y.: Analysis of bearing capacity and residual life evaluation of pavement structure. Heilongjiang Transportation Technology **3**, 2 (2010)
20. Cao, D., Xu, B., Ding, R., et al.: An indoor evaluation method for the bearing capacity of drainage asphalt pavement: CN108896421A (2018)
21. Ni, L.: New Concept and Detection Technology Application in Maintenance Design of Asphalt Pavement on Freeways. Highway Transportation Technology: Applied Technology Edition (2006)
22. Wang, C.: Research on Asphalt Pavement Overlay Technology. Chang'an University (2008)
23. Lai is full Research on the Structure and Material Properties of Asphalt Overlay on Old Cement Concrete Pavement. Southeast University

The Impact of Lean Construction Tools on Environmental Sustainability in Morocco: A Structured Survey Analysis

Mohamed Saad Bajjou[1]([⊠]), Salma Arabi[2], and Anas Chafi[2]

[1] Laboratory of Applied Mathematics and Business Intelligence (LMAID), Higher National School of Mines Rabat (ENSMR), Rabat, Morocco
bajjou@enim.ac.ma
[2] Laboratory of Industrial Techniques, Sidi Mohamed Ben Abdellah University, Fez, Morocco

Abstract. This study explores the influence of Lean Construction (LC) techniques on environmental sustainability in the Moroccan construction industry. Known for minimizing waste and maximizing value in production systems, LC not only holds promise for improving efficiency and competitiveness but also has the potential to address environmental concerns. This study aims to fill this gap by assessing how LC can contribute to the environmental performance in Morocco based on a survey by structured questionnaire involving 330 Moroccan construction professionals. The methodology includes a comprehensive survey and analysis using SPSS V26.0. Key findings indicate a significant positive impact of LC on environmental performance, particularly in reducing material use, energy consumption, pollutant release, and non-product output. Notably, non-product output emerged as the most significantly influenced factor by LC practices. The study also reveals a differential impact of LC based on organizational characteristics, with distinct influences observed between contractors and consultants, especially in pollutant releases. These insights underscore the potential of LC in fostering environmental sustainability within the Moroccan construction sector. This survey is crucial for a more holistic understanding of LC's impact, aligning with global environmental goals and the specific needs of the Moroccan construction industry.

Keywords: Lean Construction · Survey study · Environmental Performance · Construction Industry

1 Introduction

The construction industry is a major contributor to economic growth. In Morocco, it employs almost one million people, or 9.3% of the active population, and contributes 6.6% of total value added. [1, 2]. The construction sector is important for reducing unemployment and strengthening the national economy. However, it faces several challenges, including project delays, cost overruns, and compromised quality. Additionally, it is a significant contributor to environmental pollution [3, 4].

© The Author(s) 2024
G. Feng (Ed.): ICCE 2023, LNCE 526, pp. 775–785, 2024.
https://doi.org/10.1007/978-981-97-4355-1_78

In Morocco, construction projects often struggle with inefficiency, largely attributed to significant delays and cost overruns [5]. This challenge is compounded by the environmental concerns arising from substantial waste generation by Moroccan construction firms. This scenario underscores the urgent need for innovative management systems that not only enhance project efficiency but also prioritize environmental sustainability by minimizing waste impact.

To address these dual objectives of performance improvement and environmental responsibility, the Moroccan construction sector is gradually adopting the LC philosophy. Originating from Lean Manufacturing principles, particularly the Toyota Production System, LC is still emerging in this context [6]. Its core goal is to execute construction projects that align with customer demands while emphasizing waste reduction and value maximization, thereby reducing environmental footprint. Additionally, LC fosters a culture of continuous improvement, actively involving individuals at all levels, from frontline employees to top management, in sustainable practices and decision-making.

In total, 330 experts in the construction field from various regions of Morocco participated in this survey. The study's focused goal is to examine how the application of LC can improve the environmental efficiency of construction projects within the Moroccan context. These insights underscore the potential of LC in fostering environmental sustainability within the Moroccan construction sector.

2   Literature Review

LC is a project delivery process that aims to maximize stakeholder value while minimizing waste and improving efficiency [7]. It emphasizes collaboration between various stakeholders in the construction process. The goal is to ensure a continuous, reliable workflow, leading to increased productivity, profitability, and innovation in the industry [8]. LC is a methodology based on the principles of lean manufacturing, which was popularized by the Toyota Production System [9]. It involves the application of various tools, methods, and systems to translate lean thinking into the construction industry. The benefits of LC include improved process efficiency, quality, timely project delivery, and cost savings. By adopting a mindset of continuous improvement, companies can achieve cumulative benefits over time. LC allows collective knowledge to solve industry problems and prepare for future challenges [10, 11].

LC, as a methodology, offers numerous benefits to the construction industry. It enhances collaboration, innovation, delivery, control, and quality [12]. By eliminating waste and improving process cycles, it reduces costs, improves quality, and increases efficiency [13]. This approach is particularly beneficial in the South African construction industry, where it has been found to reduce waste, improve material administration, and enhance the overall cost of construction projects [14]. Despite its potential, the practical application of LC in the industry is still limited, with its principles often implicitly embedded in other practices such as building information modelling, low or zero-carbon building, prefabrication, and modular construction [15–18]. A range of studies have highlighted the potential environmental benefits of LC practices. Babalola et al. [19] found that these practices can significantly reduce construction waste, while Yao et al. [20] identified prefabricated construction as particularly effective in this regard. Solaimani and Sedighi [21] emphasized the need for a more comprehensive understanding of how LC contributes to sustainability, particularly in terms of social and environmental values. Bajjou et al. [22] further underscored the importance of integrating sustainability dynamics into the LC system. These findings collectively suggest that LC practices have the potential to significantly reduce the environmental impact of construction activities.

The growing awareness of environmental issues has prompted construction organizations to concentrate not only on operational excellence but also on reassessing their operations and processes for greater environmental sustainability. This shift reflects an increasing recognition of the significance of ecological concerns in manufacturing, although research in this area remains relatively limited.

This survey evaluates four key environmental performance indicators: material usage, energy consumption, non-product output, and pollutant emissions. These indicators align with those employed in [23]'s research assessing lean management's environmental impact, which includes mitigating air emissions, managing wastewater, reducing solid waste, and curtailing the use of toxic, hazardous, or harmful materials.

3   Research Methodology

The research methodology involved conducting a survey to gather insights from professionals in Morocco's construction industry. The survey was distributed in two primary formats. Firstly, an online version was created using Google Forms and shared via LinkedIn with individuals working in both private and public sectors of the construction industry. Secondly, printed versions of the questionnaire were distributed to a range of contractors and consulting firms to ensure an acceptable response rate. The questionnaire consisted of two main sections. The initial section aimed to collect basic information about the respondents. Section 2 assessed the level of implementation of LC methods such as Just in time, Poka Yoke and Visual management in Moroccan constructions firms. While, Sect. 3 of the questionnaire examined respondents' perceptions to determine whether their have achieved any progress in terms of environmental performance measures studied (i.e., material use, non-product output, pollutant releases, energy consumption). In order to evaluate the content quality, to verify the suitability with the Moroccan context, to adjust the order of the questions and their comprehension by the respondents, semi-structured interviews with six construction professionals, with no less than ten years, were performed. Hence, the preliminary variables have been validated. Furthermore, additional items were recommended to be included, especially in the Sect. 2, to deeply explore the LC techniques such as: Prfebarication and The Failure Mode and Effect Criticality Analysis (FMECA).

In an extensive survey targeting the Moroccan construction sector, we reached out to 440 organizations involved in various construction projects, including building, road, and bridge construction. This survey was initiated in June 2023 in Morocco. Out of the contacted organizations, we received 330 responses, indicating a substantial participation rate of approximately 75%. The data from this survey was thoroughly analyzed using SPSS, specifically version 26.0 for Windows, to ensure accurate and reliable results. The research encompassed a broad spectrum of participants, reflecting various segments of the Moroccan construction sector. The diversity in the survey pool was intentional, aiming to capture a holistic view of the industry. Participants were segmented based on their affiliated organization's size. The distribution of respondents across these categories was telling; a significant 48.8% hailed from smaller firms, whereas 27.2% and 24.0% belonged to larger and medium-sized companies, respectively. This pattern indicates a prevalence of small-scale enterprises in Morocco's construction landscape. In terms of participant background, the study did not limit itself to a single sector or specialization. It drew from a pool that included various domains within both the private and public segments of the construction industry, aiming for a comprehensive understanding of stakeholder perspectives. A notable aspect of the participant demographic was the high level of education and experience. A majority, 66.1%, possessed a master's degree or higher. In terms of industry experience, 61.1% had been in the construction field for over five years. These facets of the participant demographic - the mix of company sizes, varied specializations, and high educational and experience levels - collectively enhance the credibility and depth of the study's findings.

4   Results

4.1   The Current Level of Awareness Oflean Construction Practices

The analysis of the study on the adoption of lean methods in Morocco revealed the use of 17 distinct lean techniques, at varying levels and forms. According to the data presented in Fig. 1, 39% of the respondents are not familiar with any LC methods, indicating that they have neither heard of nor implemented them. Conversely, 35% of respondents have knowledge of LC practices but have not yet incorporated them into their projects. Meanwhile, 26% of respondents are both familiar with and actively utilizing LC techniques. Interestingly, Fig. 1 also highlights that certain lean techniques are more widely adopted, with over half of the survey participants employing methods such as prefabrication and continuous improvement, noted at 62% and 51%, respectively. In contrast, many lean techniques are not widely used in the Moroccan construction sector. For example, a majority of professionals are unfamiliar with techniques such as the kanban system, Value Stream Mapping (VSM), and Poka-Yoke, with unfamiliarity rates at 63%, 58%, and 68%, respectively. This trend can be attributed to the relatively recent introduction of these methods in the Moroccan context and a lack of extensive technical training available for these specific techniques.

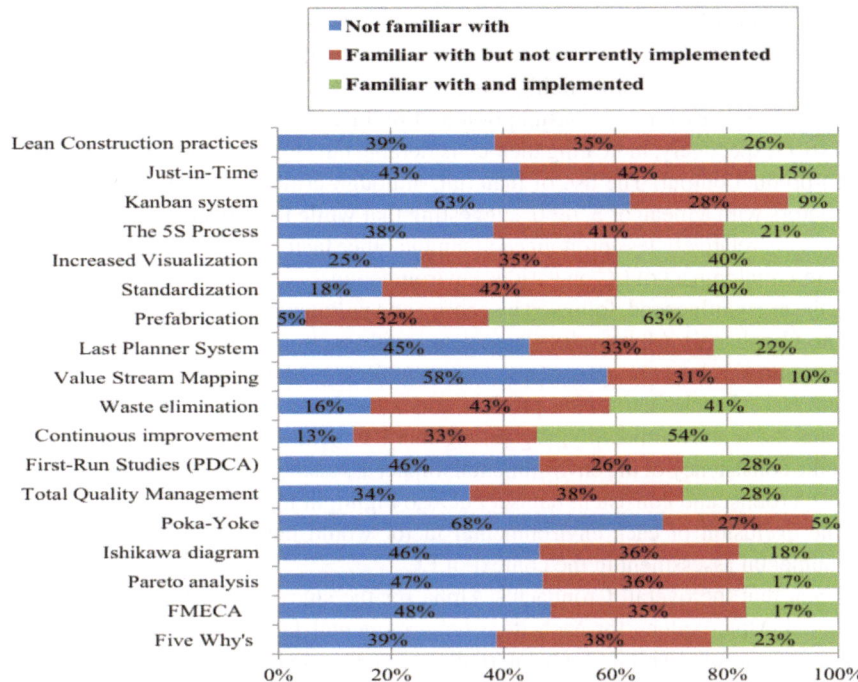

Fig. 1. The current level of awareness of lean construction practices among Moroccan construction professionals

4.2 Benefits of LC Implementation on the Environmental Performance Projects

Table 1 illustrates the prioritization of the environmental advantages associated with LC application, as deduced from the survey's average scores. The reliability of the survey's five-point scale is substantiated by a Cronbach's alpha coefficient of 0.872, which exceeds the 0.8 threshold, confirming the robustness of the measurement approach in this context.

Table 1. Ranked environmental benefits of LC implementation (Overall Cronbach's Alpha = 0.872)

	Mean	Standard deviation	Rank	Cronbach's Alpha if Item Deleted
Energy consumption	4.244	1.051	2	0.843
Pollutant releases	4.019	1.016	3	0.864
The use of material	3.787	1.036	4	0.829
Non-product output	4.417	0.931	1	0.808

The ranking of the means of various environmental factors impacted by LC practices reveals insightful trends. Non-product output, with the highest mean (4.417), is ranked as the most significant factor, highlighting its substantial influence in LC environments. This is followed by energy consumption and pollutant releases, with means of 4.244 and 4.019 respectively, indicating their considerable but slightly lesser impact compared to non-product output. The use of material, although critical, ranks lowest among the four factors with a mean of 3.7860, suggesting that while it is affected by LC practices, its impact is relatively less pronounced than the other factors. This ranking underscores the differential impact of LC on various environmental aspects, with non-product output being the most influenced, followed by energy and pollutants, and finally material usage.

The Cronbach's Alpha values associated with each environmental factor provide insights into the internal consistency and reliability of the dataset when considering the removal of each factor. Non-product output, with a Cronbach's Alpha of 0.8808, indicates the highest reliability, suggesting that its exclusion would most significantly affect the consistency of the environmental impact assessment. These values collectively suggest a robust and consistent dataset but also highlight the varying degrees of impact that the exclusion of each environmental factor would have on the reliability of the environmental assessment in the context of LC.

Given that the overall Cronbach's Alpha for the study is 0.872 is higher than the individual Alpha values for each item (environmental factor), we can infer that the dataset as a whole exhibits high internal consistency and reliability in assessing the environmental impacts of LC. This overall Alpha value is a measure of how well the set of items (in this case, environmental factors) are interrelated, indicating a strong degree of coherence and reliability in the measurement of the study's constructs.

4.3 Spearman Rank Correlation

To evaluate the relationship and its strength and direction between two groups of clusters, the non-parametric Spearman rank correlation was employed. This method calculates the Spearman rank correlation coefficient (rs) to quantify the correlation. The coefficient can be computed as follows:

$$r_s = 1 - \frac{6 \sum d^2}{N(N^2 - 1)} \tag{1}$$

where:

The Spearman's rank correlation coefficient, denoted as rs, is a statistical measure. It is calculated based on the differences in rankings (d) assigned by two respondents for a specific item, and N represents the total number of rank pairs. The value of rs can vary from $+1$ to -1, where $+1$ signifies a perfect positive relationship, -1 indicates a perfect negative relationship, and a value of 0 implies no correlation between the rankings.

Table 2. Spearman's rank inter-correlation test

			Energy consumption	Pollutant releases	The use of material	Non-product output
Spearman's coefficient	Energy consumption	Correlation coefficient	1.000	0.504**	0.547**	0.433**
		Sig. (bilateral)	.	0.000	0.000	0.000
		N	330	330	330	330
	Pollutant releases	Correlation coefficient	0.504**	1.000	0.736**	0.455**
		Sig. (bilateral)	0.000	.	0.000	0.000
		N	330	330	330	330
	The use of material	Correlation coefficient	0.547**	0.736**	1.000	0.403**
		Sig. (bilateral)	0.000	0.000	.	0.000
		N	330	330	330	330
	Non-product output	Correlation coefficient	0.433**	0.455**	0.403**	1,000
		Sig. (bilateral)	0.000	0.000	0.000	.
		N	330	330	330	330

** Correlation is significant at the 0.01 level (two-tailed).

To study the level of inter-correlation between the different items, a Spearman rho correlation matrix was drawn up. In Table 2, cells highlighted in black in the matrix indicate a high correlation (Spearman correlation coefficient > 0.5); between Energy consumption and (Pollutant releases and The use of material); between Pollutant releases and The use of material. Cells highlighted in grey indicate moderate correlation (0.3 < pearman correlation coefficient ≤ 0.5), between Non-product output and (Energy consumption, Pollutant releases, and The use of material). Most correlations were found to be significant at a two-sided confidence level of 99% (**). Consequently, the data set could be analyzed as a whole for one way test Anova.

4.4 One Way Anova Analysis

The sample was divided into four distinct subgroups, each based on specific organizational characteristics, allowing for a comprehensive analysis of the diverse aspects of the construction industry, as illustrated in Table 3. Subgroup 1 was categorized according to the type of organization, distinguishing between contractors and consultants. Subgroup 2 was classified based on specialization fields, including bridge and road projects, and building projects. Subgroup 3 was delineated based on the activity area, distinguishing between public and private sectors. Finally, subgroup 4 was delineated by the size of the organization, with distinctions made for small, medium, or large organizations.

Table 3. One way Anova test (items: environmental benefits)

Identifier	According to the type of organization	According to the field of specialization	According to the area of activity	According to the size of the organization
	P-value	P-value	P-value	P-value
energy consumption	0.069	0.944	0.798	0.693
pollutant releases	0.012*	0.272	0.869	0.591
The use of material	0.129	0.578	0.996	0.602
non-product output	0.160	0.236	0.982	0.673

Based on an Anova test, all p-values are above 0.05 for sub-groups 2, 3, and 4 indicating that field of specialization, activity area and type of organization do not significantly impact the scores given to the four items listed in the current survey, as shown in Table 3. However, it's essential to highlight that the only statistically significant result in our analysis is for pollutant releases (p = 0.012), a level which is lower than the typical alpha threshold of 0.05. This strongly suggests that the type of organization indeed has a significant impact on pollutant releases. To further elaborate on this finding, we can examine the mean scores for pollutant releases among different types of organizations. On average, contractors have a mean score of 4.250, whereas consultants have a lower mean score of 3.914. The reason for this observed difference in pollutant releases between contractors and consultants could be attributed to several factors. One possible explanation might be that contractors, due to their direct involvement in various construction and industrial activities, could be subject to more stringent environmental regulations and monitoring, which motivates them to better manage and mitigate pollutant releases. Consultants, on the other hand, may have less direct control over on-site operations and, therefore, may not be as focused on pollution reduction measures.

Additionally, contractors may have more resources and expertise dedicated to environmental management, given their hands-on role in projects, whereas consultants might prioritize other aspects of their work. It is also possible that contractors have adopted more advanced pollution control technologies and practices due to their greater exposure to environmental compliance requirements.

Further research and a detailed analysis of specific operational practices and policies within these types of organizations could provide deeper insights into the reasons behind this significant difference in pollutant releases. The other factors (energy consumption, use of material, and non-product output) are not significantly affected by the type of organization.

5 Conclusion

The survey aimed to understand how LC can contribute to improving environmental performance in Morocco, specifically by reducing negative impacts such as material use, energy consumption, pollutant release, and non-product output. The study's results demonstrate that LC philosophy has significant potential to enhance environmental sustainability in the Moroccan construction industry. The mean values for all evaluated environmental factors indicate that LC practices can have a significant positive impact. The findings show that LC has a differential impact on various environmental aspects. Non-product output had the highest mean value and was identified as the most significantly influenced factor, followed by energy consumption and pollutant releases. Material use, while important, had a relatively less pronounced impact compared to the other factors. The study also analyzed the role of organizational characteristics in LC implementation and their impact on environmental performance. The analysis showed that the type of organization, such as contractors versus consultants, could significantly influence certain environmental factors, particularly pollutant releases. This indicates that different types of organizations may adopt and benefit from LC practices in distinct ways. In conclusion, the paper argues that LC methodologies can substantially contribute to environmental sustainability in the Moroccan construction industry. The study's comprehensive statistical analysis supports the conclusion that LC can significantly reduce the environmental impact of construction activities. The findings can inform policymakers and industry leaders in Morocco and other developing countries about the potential environmental benefits of LC. This could lead to the development of targeted strategies and policies to promote LC methodologies in the construction sector. In addition, the study provides a benchmark for environmental performance in the Moroccan construction industry. Organizations can use these findings to monitor their performance and strive for continuous improvement in environmental sustainability.

References

1. Habchi, H., Taoufiq, C., Aziz, S.: Last planner® system: implementation in a moroccan construction project. In: IGLC 2016 - 24th Annual Conference of the International Group for Lean Construction, pp. 193–202 (2016)
2. Bajjou, M.S., Chafi, A.: Empirical study of schedule delay in Moroccan construction projects. Int. J. Constr. Manag. **20**, 783–800 (2020)
3. Bajjou, M.S., Chafi, A.: Identifying and managing critical waste factors for lean construction projects. EMJ Eng. Manag. J. **32**, 2–13 (2020)
4. Bajjou, M.S., Chafi, A.: Assessing the critical sources of wastes in the Moroccan construction industry. In: The 3rd International Conference on Smart City Applications, pp. 1–5 (2018)

5. Bajjou, M.S., Chafi, A.: Exploring causes of wastes in the Moroccan construction industry. In: Ben Ahmed, M., Boudhir, A.A., Younes, A. (eds.) SCA 2018. LNITI, pp. 57–64. Springer, Cham (2019). https://doi.org/10.1007/978-3-030-11196-0_6

6. Bajjou, M.S., Chafi, A.: Lean construction implementation in the Moroccan construction industry: awareness, benefits and barriers. J. Eng. Des. Technol. 16, 533–556 (2018)

7. Sarhan, J.G.: Implementation of lean construction management practices in the Saudi Arabian construction industry. Constr. Econ. Build. 17, 46–69 (2017)

8. Ballard, G., Howell, G.: Implementing lean construction: stabilizing work flow. Lean Constr. 105–114 (1997)

9. Arabi, S., Bajjou, M.S., Chafi, A., El Hammoumi, M.: Evaluation of critical success factors (CSFs) to lean implementation in Moroccan SMEs: a survey study. In: 2022 2nd International Conference on Innovative Research in Applied Science, Engineering and Technology (IRASET), pp. 1–10. IEEE, March 2022

10. Gao, S., Low, S.P.: Lean Construction Management. Springer, Singapore (2014). https://doi. org/10.1007/978-981-287-014-8

11. Bajjou, M.S., Chafi, A., Ennadi, A.: A conceptual model of lean construction: a theoretical framework. Malay. Const. Res. J. 26, 67–86 (2018)

12. Bajjou, M.S., Chafi, A.: Exploring the critical waste factors affecting construction projects. Eng. Constr. Archit. Manag. 29, 2268–2299 (2022)

13. Ben Ruben, R., Vinodh, S., Asokan, P.: Development of structural equation model for Lean Six Sigma system incorporated with sustainability considerations. Int. J. Lean Six Sigma 11, 687–710 (2020)

14. Akinradewo, O., Oke, A., Aigbavboa, C., Ndalamba, M.: Benefits of adopting lean construction technique in the South African construction industry. In: International Conference on Industrial Engineering and Operations Management, pp. 1271–1277, November 2018

15. Bos, F., Wolfs, R., Ahmed, Z., Salet, T.: Additive manufacturing of concrete in construction: potentials and challenges of 3D concrete printing. Virtual Phys. Prototyp. 11, 209–225 (2016)

16. Tay, Y.W.D., Panda, B., Paul, S.C., Noor Mohamed, N.A., Tan, M.J., Leong, K.F.: 3D printing trends in building and construction industry: a review. Virtual Phys. Prototyp. 12, 261–276 (2017)

17. Ramadany, M., Bajjou, M.S.: Applicability and integration of concrete additive manufacturing in construction industry: a case study. Proc. Inst. Mech. Eng. Part B J. Eng. Manuf. 8, 1338–1348 (2021)

18. Arabi, S., Chafi, A., Bajjou, M.S., El Hammoumi, M.: Exploring lean production system adoption in the moroccan manufacturing and non-manufacturing industries: awareness, benefits and barriers. Int. J. Autom. Mech. Eng. 18, 9312–9332 (2021)

19. Babalola, O., Ibem, E.O., Ezema, I.C.: Implementation of lean practices in the construction industry: a systematic review. Build. Environ. 148, 34–43 (2019)

20. Yao, F., et al.: Evaluating the environmental impact of construction within the industrialized building process: a monetization and building information modelling approach. Int. J. Environ. Res. Public Health 17, 1–22 (2020)

21. Solaimani, S., Sedighi, M.: Toward a holistic view on lean sustainable construction: a literature review. J. Clean. Prod. 248, 119213 (2020)

22. Bajjou, M.S., Chafi, A., Ennadi, A., El Hammoumi, M.: The practical relationships between lean construction tools and sustainable development: a literature review. J. Eng. Sci. Tech. Rev. 10, 170–177 (2017)

Analysis and Evaluation of the Slope Stability of K49 + 800 Section of S207 Road in Shizong County

Fengfeng Ding[1] and Jinyong Jiao[2(✉)]

[1] College of Resources and Environment, Yunnan Vocational College of Land and Resources, Kunming 652501, Yunnan, China
[2] China Power Construction Group Kunming Survey Design and Research Institute Co., Ltd., Kunming 650000, Yunnan, China
121560297@qq.com

Abstract. With the development of urban construction, slope engineering has become more and more important and difficult projects, and the hidden dangers of slope can not be ignored, which requires judgment and protection of the stability and safety of slope, so as to promote the economic development and construction of the city under the premise of ensuring the safety of people and property. Based on the analysis and study of engineering geological conditions and factors affecting slope stability of K49 + 800 section of S207 road in Shizong County, this paper analyzes and evaluates its stability and puts forward corresponding treatment measures.

Keywords: Highway Slope · Stability Evaluation · Treatment Measures

1 Overview of the Study Area

1.1 Project Overview

The research area is located at the south end of Gaolang Township government residence in Shizong County, on the left side of milepost No. K49 + 800 section of S207 Highway, which generally runs north-south. The inner part of the highway is an artificial steep ridge caused by the construction of the highway, with a height of about 25 m and good vegetation development on the upper part of the steep ridge. The slope of the road to be studied is located on the outside of the road. The slope is a soil slope, distributed under the road in an inverted U-shape, with a total length of about 38 m and a width of about 35 m. It is a small slope. The overall slope of the mountain is 32–36°, the slope direction is 67°, the foot of the slope is the Fengwei River, the rear wall of the slope is about 4.5 m high steep ridge, the slope height difference is about 20 m. The highway is sTable and open to traffic. The slope slope is about 34°, located on the bank of the mountain river, and vegetation is relatively developed around the slope body and on the slope body. When the sliding body slides, vegetation will slip synchronously with the sliding body. Bedrock exposure can be seen on both upper slopes (See Fig. 1).

G. Feng (Ed.): ICCE 2023, LNCE 526, pp. 786–797, 2024.
https://doi.org/10.1007/978-981-97-4355-1_79

Fig. 1. Slope body

1.2 Meteorology and Hydrology

Shizong County is located in the middle of Yunnan-Guizhou Plateau, affected by the elevation of the terrain and the atmospheric circulation, belongs to the north subtropical monsoon climate area, which is characterized by mild climate all year round, no cold and dry winter, warm and dry spring, cool and humid autumn, dry and wet seasons, small annual temperature difference, large daily temperature difference, concentrated rainfall, sufficient light, there are four seasons like spring, rain into winter. The average annual temperature is 14.7 °C, the average temperature in the coldest January is 7 °C the average temperature in the hottest July is 20.1 °C and the annual frost-free period is 246 days. The average annual relative humidity is 74%, and the average annual rainfall is 976.5 mm. The seasonal distribution of rainfall is extremely uneven.

2 Engineering Geological Condition

2.1 Topography and Landform

It is intended to study that the bedrock near the slope is well exposed, and the bedrock occurrence is 146°∠31°. The slope is generally high in the west and low in the east, descending from west to east. Above the slope is a rocky scarp of silty mudstone with a height difference of about 25 m, a slope of about 75° and a scarp dip of 67°. There are shrubs above the steep ridge, and the vegetation is well developed. At present, there are no protective measures for the steep ridge, and the rubble falls under the action of running water in the rainy season, posing a threat to vehicles and pedestrians. To the west of the site is a river, named Fengwei River, Hanoi perennial flow, flow with seasonal changes, the bank is covered by the fourth system alluvial alluvial layer. On the other side of the river is farmland and a large number of houses. Along the highway, there are civil houses on both sides of the highway, and the civil houses are discontinuous and dislocated (See Fig. 2, Fig. 3, Fig. 4 for details).

Fig. 2. Topography

Fig. 3. Inner rock slope of highway

Fig. 4. The status of both sides of the highway

2.2 Geological Structure

The study area is located in the Yangtze platform (I) and the eastern Yunnan platform wrinkle belt (I). The regional geological structure is dominated by folds, and the faults are not developed. The regional tectonic lines are distributed in the northeast direction, and are composed of four major structural systems: the northeast tectonic system, the North-South tectonic system, the northwest tectonic system and the joint peak arc tectonic system. In terms of a large area, the study area is located in the north of Luoping Shizong fold bundle in the southeast Yunnan fold belt of the South China fold system, and the structure is dominated by compression-shear faults extending to the north east, followed by near-east-west tensile faults and northeast-trending folds. The west of the study area is the Xiaojiang strong seismic fault, which runs through the middle of Yunnan Province from north to south along Huize, Dongchuan, Xundian and Tonghai. Xundian-xuanwei fault crosses Yiliang and Xiaojiang fault (east branch) through Xuanwei and Xundian fault; Qujing fault crosses Qujing, Luliang, Shilin and Maitrea-Shizong fault in Maitreya on the east side of the study area; the study area is located in Gaoliang Township, Shizong

County, far away from each fault zone, and no fault passes through the field, so it is a sTable block.

2.3 Formation Lithology

Through the investigation of the slope body [1], the borehole revealed that the formation is composed of quaternary mixed fill (Q_4^{ml}) and grey gravel silty clay (Q_4^{dl+el}) and the strong to medium weathering silty mudstone (T_2bc) of the first member of Banna Formation in the Middle Geold Formation of the Triassic. According to the characteristics of geotechnical layer in the field, in-situ testing and laboratory rock and soil test data, the geotechnical layer in the site is divided into four layers, which are as follows:

① Miscellaneous fill (Q_4^{ml}):

Variegated, light grey, mainly composed of crushed stone, containing a small amount of bricks and silt, smelly, surface containing a small amount of plant roots. Slightly wet, loosely structured. This layer is formed by the accumulation of fill for highway construction. The slope is a temporary dump, with a large amount of domestic garbage piled up in the upper part. In the center of the site there are north-south concrete and rubble retaining walls. The retaining wall slides down with the slope, and the structural integrity is not damaged. The layer is thicker in the center of the site. Due to the large changes in the layer, uneven distribution, containing concrete and other domestic construction waste. The thickness of the layer is 0.90–4.00 m, with an average thickness of 2.82 m.

② Pebbly silty clay (Q_4^{dl+el}):

Light gray, light grayish yellow, slightly wet, hard plastic, slightly dense structure, section is not smooth, rough fracture, high dry strength, medium toughness, local mudstone gravel. It is distributed in the site, the buried depth of the top layer is 0.9–4.0 m (816.50–831.77) meters, the thickness of the layer is 2.60–5.80 m, and the average thickness is 4.78 m.

③ Silty mudstone (Strong weatheringT_2bc):

Gray, light gray, silty, argillaceous clastic structure, strongly weathered like earth, containing about 25% gravel. It is distributed in the field, the layer thickness is 1.20–2.80 m, and the average thickness is 1.82 m.

④ Silty mudstone (moderate weatheringT_2bc):

Gray, dark gray, silty argillaceous structure, meso-like structure, is a soft rock, moderately weathered, joint fissure development, the gaps are mostly filled with mud, the core is broken, mostly fragmented, some short columnar, there are exposed silty mudstone on the upper and both sides of the slope in the site, the occurrence of exposed silty mudstone is 146°∠31°. The saturated uniaxial compressive strength of the drilling rock is 17.65 MPa, the hardness grade of the rock is relatively soft rock, the quality of the rock mass is poor, the integrity degree of the rock mass is broken, the basic quality grade of the rock mass is IV, and the rock quality index RQD is about 50%. The survey did not expose this layer, and no karst cave was found within the range of drilling control depth. The maximum exposed thickness was 9.40 m, the exposed layer thickness was 6.20–9.40 m, the average thickness was 7.75 m, and the top buried depth of the exposed layer was 6.80–12.10 m.

3   Slope Stability Analysis and Evaluation

3.1   Analysis of Influencing Factors of Slope Stability

Soil condition and influence of rainfall. According to the investigation, no tension crack was found at the top of the slope, and the exploration drilling revealed the existence of weak surface between the soils, and the soil of the slope was in an unsTable state. With the coming of the rainy season, the surface water permeates, the shear strength of the soil decreases, and the stability of the slope becomes worse [2].

The impact of human engineering activities. The human engineering activities in the study area are mainly manifested as the building load on the top of the slope and cutting the slope to build the road.

With the construction of Gaoliang Township, limited by geographical conditions, houses had to be built on the top of steep slopes. The formerly fragile geological environment is worsening. At the same time, in order to meet the traffic needs, the provincial road is built on the top of the slope, which forms a lot of excavation and filling, and the rock mass of the slope is broken, which affects the stability of the slope even more (See Fig. 5).

Fig. 5. Slope body

3.2   Evaluation of Slope Stability

Qualitative Evaluation. The slope area is a steep terrain with an average topographic slope of 34°. According to the survey data, the surface of the survey area is composed of quaternary mixed fill and residual slope deposits, and the rock and soil properties are mainly clay, gravel, broken strong to moderately weathered silty mudstone, with loose structure and poor physical and mechanical properties. In the rainy season, groundwater infiltration softenes the contact surface, which is easy to form a weak structural plane. Field investigation shows that the slope is in an unsTable state.

Quantitative Evaluation. The upper part of the slope is composed of quaternary mixed fill (Q_4^{ml}), grey gravel silty clay (Q_4^{dl+el}) and strong to medium weathering silty mudstone (T_2bc) of the first member of Banna Formation in the Middle Geold Formation of the Triassic. The sliding surface of the slope is in the quaternary loose soil. The sliding surface is regarded as a broken line, and the section parallel to the sliding direction is used to calculate the stability and thrust of the slope.

Calculation formula of slope stability. According to the analysis of the soft structural plane between rock and soil bodies of slope, the soft structural plane between rock and soil bodies of slope is regarded as the sliding surface, and the sliding surface is approximately folded. According to the relevant requirements of the "Code for Exploration of Landslide Prevention and Control Engineering" [3], the transfer coefficient method is adopted, the sliding surface is regarded as the broken line, the unit width of landslide is taken as 1m, and the stability and thrust of landslide are simplified into a two-dimensional problem.

(1) Landslide stability calculation formula:

$$F_s = \frac{\sum_{i=1}^{n-1} (R_i \prod_{j=1}^{n-1} \psi_j) + R_n}{\sum_{i=1}^{n-1} (T_i \prod_{j=1}^{n-1} \psi_j) + T_n} \tag{1}$$

$$\psi_j = \cos(\theta_i - \theta_{i+1}) - \sin(\theta_i - \theta_{i+1})\tan\varphi_i + 1 \tag{2}$$

$$\prod \psi_j = \psi_i \cdot \psi_{i+1} \cdot \psi_{i+2}\ldots\ldots\psi_{n-1} \tag{3}$$

$$T_j = W_i \sin\theta_i + PW_i \cos(a_i - \theta_i) \tag{4}$$

$$R_i = N_i \tan\varphi_i + c_i l_i \tag{5}$$

$$N_i = W_i \cos\theta_i + P_{Wi} \sin(\alpha_i - \theta_i) \tag{6}$$

$$W_i = V_{iu}\gamma + V_{id}\gamma' + F_i \tag{7}$$

$$P_{wi} = \gamma_w i V_{id} \tag{8}$$

$$i = \sin|\alpha_i| \tag{9}$$

$$\gamma' = \gamma_{sat} - \gamma_w \tag{10}$$

Calculation formula of hydrodynamic pressure:

$$P_{Wi} = \gamma_{Wi} V_{id} \qquad i = \sin|\alpha_i| \tag{11}$$

Of the form:

F_S—Coefficient of landslide stability;

Ψj—i calculates the transfer coefficient when the remaining sliding force of the bar is transferred to the $i + 1$ bar ($j = i$);

Ri—The anti-sliding force of the sliding body of the strip is calculated on the i(kN/m);

Ti—When the sliding component force (kN/m) on the sliding surface of the strip is applied to the i, and the sliding component force is opposite to the sliding direction, Ti should be negative;

Ni—i calculates the reaction force of the strip sliding body on the normal of the sliding surface (kN/m);

Ci—The standard value of the bond strength of the rock and soil mass on the sliding surface of the strip is calculated (kPa);

φi—i calculate the standard value of the internal friction Angle of strip sliding soil (°);

li—i Calculating strip sliding surface length (m);

αi—i calculate the average dip Angle of the strip groundwater flow line. In general, the average value (°) of the dip Angle of the infiltration line and the dip Angle of the slip surface is taken, and the negative value is taken when the dip is reversed;

Wi—i calculate the sum of strip weight and building and other ground loads (kN/m);

θi—i calculate the dip Angle (°) of the bottom surface of the strip, and take a negative value when reversing;

P_{Wi}—The i section calculates the osmotic pressure per unit width of the strip, and the inclination Angle of the action direction is αi (kN/m);

I—Seepage gradient of groundwater;

γ_W—The bulk density of water (kN/m^3);

Viu—Section i calculates the volume above the infiltration line of the rock mass per unit width of the strip (m^3/m);

Vid—Section i calculates the volume below the infiltration line of the rock mass per unit width of the strip (m^3/m);

γ—Natural bulk density of rock and soil (kN/m^3);

γ'—Floating bulk density of rock and soil mass (kN/m^3);

γ_{sat}—Saturated bulk density of rock and soil mass (kN/m^3);

Fi—Section i calculates the ground load on the strip (kN).

(2) Calculation formula of residual slide thrust of landslide

$$P_i = P_{i-1} \cdot \psi + F_{st} \cdot T_i - R_i \tag{12}$$

Of the form:

Pi, Pi-1—They are the residual sliding force of the sliding body of the i block and the I-1 block respectively (kN/m);

Fst—The safety factor of landslide thrust is calculated according to the harm degree of landslide;

Ti—Sliding component acting on the sliding surface of segment i (kN/m);

Ri—The skid resistance acting on the i of all segment (kN/m).

Notes:

(1) The groundwater is not revealed during the investigation of the slope area, and the groundwater pressure is not taken into account in the calculation;

(2) The seismic fortification intensity of the study area is 7°, the design earthquake is divided into the third group, and the basic design earthquake acceleration is 0.10 g. The comprehensive seismic influence coefficient is 0.20.

2. Calculate the load combination

There is no building load on the slope surface. The value of the trailing edge of the landslide is 0 kN/m.

Condition I: Sliding body soil natural weight load + Groundwater;
Condition II: Sliding body soil natural weight load + Groundwater + Heavy rain;
Condition III: Sliding body soil natural weight load + Groundwater + Earthquake [4];

3. Calculate the value of the parameter

The values of soil mass weight and shear strength (cohesion C, internal friction angle φ) of slope: according to the results of indoor geotechnical tests and combined with field observations, the average value and experience value of indoor soil test results are used for heavy soil mass, and the direct shear value and experience value are used for shear strength of soil mass. See Table 1 and Table 2 below.

Table 1. Table of weights

Soil layer number	Name of soil layer	Heavy (γ) KN/m^3
①	Fill	18.0*
②	Gravelly silty clay	18.9
③	Strongly weathered silty mudstone	21.3*
④	Medium weathered silty mudstone	26.7

Note: The above values are the average of the geotechnical test, with * for the empirical value

Table 2. Shear strength of slope soil

Soil layer number	Name of soil layer	Straight cut (q)	
		Cohesion (C)	Angle of internal friction (φ)
②	Gravelly silty clay	30.25	9.25
③	Strongly weathered silty mudstone	11.3*	20.8*
④	Medium weathered silty mudstone	18.9*	26.6*

Note: The above values are all the minor values in the geotechnical test, and the values with * are the empirical values

Safety factor of control engineering design: the grade of S207 slope regulation project in Gaoliang Township of Shizong County is II. According to the Technical Specifications for Design and Construction of Landslide Control Engineering (DZ/T0219-2006) [5], the following data is the anti-sliding safety factor of the slope under various working conditions:

ConditionsII: Ks = 1.02~1.15 Value 1.10;
ConditionsIII: Ks = 1.02~1.15 Value 1.05.

The infiltration depth of rainstorm is 3 m [6].

Evaluation of slope stability. According to the "Code for Exploration of Landslide Prevention and Control Engineering" (DZ/T0218-2006), the sTable state of slope should be determined according to the stability coefficient of slope according to Table 3. After calculation, the stability evaluation of S207 highway slope in Gaoliang Township is shown in Table 4.

Table 3. Slope stability Table

Stability FactorK_f	$K_f < 1.00$	$1.00 \leq K_f < 1.05$	$1.05 \leq K_f < 1.15$	$K_f \geq 1.15$
Stability	UnsTable	Less sTable	Basically sTable	STable

Table 4. Table of evaluation results of slope stability

Section	Conditions					
	ConditionsI		ConditionsII		ConditionsIII	
	Stability Factor	Stability	Stability Factor	Stability	Stability Factor	Stability
1–1' Section	0.751	UnsTable	0.349	UnsTable	0.546	UnsTable
2–2' Section	0.636	UnsTable	0.320	UnsTable	0.583	UnsTable
3–3' Section	0.435	UnsTable	0.154	UnsTable	0.433	UnsTable

The calculation results show that the slope stability coefficient k = 0.435–0.751 under condition I, that is, the slope is unsTable under natural condition, and K = 0.154–0.349 under condition II, the stability coefficient k = 0.433–0.583 is less than the stability coefficient 1.00, and the slope is in the unsTable state.

4   Slope Hazard and Treatment

4.1   Slope Hazards

The slope has a large height and is a soil slope with weak structural planes between soils, low shear strength of soils, easy to form weak structural planes, and poor stability of the slope body. In external forces or heavy rainfall seasons, the slope body may lose stability and continue to slide, causing the slide body to slide into the river, which may cause a small range of river blockage.

4.2 Governance Suggestions

According to the geotechnical geological characteristics, distribution and regional experience in the site, the slope is high and the slope is large. After field investigation, the houses outside the nearby highway were built with pile foundation leveling, and there was no obvious strain deformation of the surrounding houses. The rock of the back wall of the slope is exposed, the geological condition of the site foundation is good, and the slope body has the risk of instability. In line with the economical and effective management measures, combined with the local building construction experience, it is recommended to adopt the following schemes for support [7]:

Gravity retaining wall. Gravity retaining wall is used for continuous support, and corresponding supporting facilities such as drainage are built on the upper part. The retaining wall has the advantages of simple structure, materials and convenient construction, large section size and heavy wall body, and the lateral earth pressure on the back of the wall is mainly balanced by the gravity of the wall body, so it occupies a lot of land, can not give full play to the strength performance of building materials, and is not easy to implement construction mechanization.

Cantilever retaining wall. The cantilever retaining wall is used for supporting, the wall section is small, and the stability of the wall is not or not completely maintained by its own gravity, so the structure is lighter, occupies less land, and increases the anti-overturning stability. Conducive to mechanized construction. However, the retaining wall is only suiTable for retaining walls less than 6 m high.

Buttress retaining wall. The use of arm-type retaining wall for supporting, the wall body section is small, the stability of the wall is not or not completely dependent on its own gravity to maintain, so the structure is lighter, occupies less land, and increases the anti-overturning stability, is conducive to mechanized construction. The retaining wall is suiTable for a wall height not higher than 15 m.

5 Conclusions and Recommendations

5.1 Conclusion

1. The seismic fortification intensity of Shizong County is the third group of $7°$, and the designed basic seismic acceleration value is 0.10 g. The type of site soil is medium site soil, and the type of site is Class II building site, which is an unfavorable seismic area.
2. There is Fengwei River under the slope with low terrain, which has a great influence on the stability of the slope.
3. Unsaturated sand and silt within the proposed study site are not considered in the liquefaction of seismic sand (silt).
4. The slope body is affected by its own geotechnical geological characteristics, rainfall and human engineering activities, and there is a risk of sliding.

5.2 Suggestions

1. Remove the floating soil on the side slope to reduce the upper load of landslide;
2. The base of the slope is supported by gravity rubble or cantilever or armrest reinforced concrete retaining wall. The retaining layer of the retaining wall foundation can be the third layer of strongly weathered silty mudstone and the fourth layer of moderately weathered silty mudstone.
3. Make drainage measures such as truncation and drainage ditch for extension of the hilltop highway, and do a good job of drainage system to avoid rainwater scouring the slope and water softening the foundation; The foundation pit should be closed in time after excavation.
4. Landslide damage will have an impact on highway construction and social development, should be carried out in a timely manner before the landslide slide, it is suggested to improve the preliminary work in a timely manner, and pay close attention to the implementation of prevention and control projects.
5. Corresponding engineering measures should be taken according to the deformation mechanism of landslide and the special engineering geological conditions.
6. Before the landslide prevention and control project is completed, the monitoring, early warning and forecasting measures and publicity work of all landslides should be strengthened, and the disaster prevention awareness of site construction personnel and surrounding residents should be strengthened, especially during the rainy season, the relevant departments should do a good job of landslide emergency measures.
7. The proposed projects or engineering activities around the landslide should be supervised to prevent them from aggravating the deformation of the slope.

Funding Resources. Scientific Research Fund Project of Yunnan Education Department.

References

1. Code for geotechnical engineering investigation (GB50021-2001)
2. Ali, A., Huang, J., Lyamin, A., et al.: Boundary effects of rainfall-induced landslides. Comput. Geotech. **61**, 341–354 (2014)
3. Code for investigation of landslide prevention and Control Engineering (DZ/T0218-2006)
4. Code for seismic design of buildings (GB50011-2010)
5. Technical code for design and construction of Landslide Control Engineering (DZ/T0219-2006)
6. Vennari, C., Gariano, S., Antronico, L., et al.: Rainfall thresholds for shallow landslide occurrence in Calabria, southern Italy. Nat. Hazard. **14**(2), 317–330 (2014)
7. Chao, Q.H.: Study on analysis and treatment measures of a high slope landslide of an expressway. Shandong University, Jinan (2018)

Inspection of Old Brick-Concrete Structure Buildings for Remodeling: A Case Study on an Office Building in Yunnan

Shaohong Pan and Yanfeng Zhao[✉]

Yunnan Land and Resources Vocational College, Kunming, Yunnan, China
315753649@qq.com

Abstract. Aiming at the problems of old brick-concrete structure houses built for a long time, the safety cannot meet the requirements of normal use, and the overall seismic performance of the house is poor, we use the comparison of the testing and appraisal data before and after the reinforcement and transformation, to analyze the feasibility and reasonableness of the reinforcement design scheme. Taking an office building as an example, the theoretical and empirical results show that: by choosing the appropriate 'method, it can effectively improve the comprehensive seismic capacity and use function of the house and guarantee the normal and safe use of the house.

Keywords: Brick-Concrete Structure · Testing And Appraisal · Reinforcement and Retrofitting · Seismic Capacity

1 Introduction

With the rapid development of China's economy, social productivity and people's living standards have been greatly improved. Nowadays, many old buildings, due to the technical and economic of construction restrictions, there are already hidden safety hazards or cannot continue to use the old building reinforcement design, functional transformation and upgrading, to enhance the comprehensive seismic capacity of housing and buildings is the current urgent need to solve the problem [1]. Old buildings in the use of function, load, structural form and other changes, the original housing structure of the test and identification is essential to the development of reinforcement design should be collected before the design program, consult, analyze the relevant information on housing, through the testing and identification to determine the weak points of the house and the bearing capacity of the components, through the modeling and calculation of the selection of safe and suitable, economically rational, technologically advanced reinforcement and transformation of the design program. Reinforcement construction is completed again after the completion of testing and identification of base modeling, according to the actual test results of the reinforcement construction quality assessment, verification of the rationality of the reinforcement design [2].

2 Project Overview

An office building houses built in the 1990s, the original as equipment room and office use, now the use of the building functions and room area has been unable to meet the owner's use of the needs of the owner, and thus the owner intends to remodel all the office building and new housing one floor area to continue to use the house all the original information is missing, the house basic information questionnaire see Table 1 the original house one floor plan and reinforced after the transformation The building plan of the first floor of the original house and the reinforcement and remodeling after the removal of the wall, new columns and beams and plate design drawings are shown in Fig. 1.

Table 1. Basic Housing Questionnaire

Architectural overview	Era of construction	1990s	Building area (m^2)	Original floor area of approximately 620.0
	Building height (m)	9.6	Number of layers	3 floors
	Usage	Formerly equipment room and office building, now all changed to office building use	Plane shape	L-shaped
	Floor height (m)	First floor: 3.6; second and third floors: 3.0	Intensity of fortification	7° (0.10 g)
	body structure	brick hybrid structure	(Math.) basic form	burrstone foundation (geology)
Structural overview	Ring girder setting	Setting on each floor	Structural column setup	yet unsettled
	Load-bearing wall materials	240 mm mixed mortar sintered solid clay bricks	Floor and roof panels	100 mm thick cast-in-place concrete slab
Operating environment	vibratory	not have	corrosive medium	not have
	extend a building	Proposed first floor extension	lamination	not have
	Surroundings	Perimeter is normal		

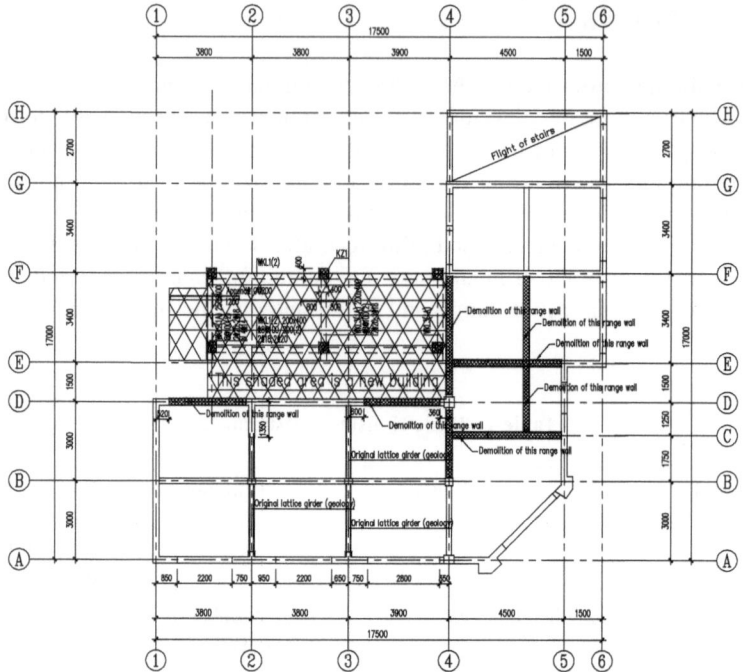

Fig. 1. Floor plan of the first floor of the original house and the design of wall removal and new columns and beams after reinforcement and remodeling

3 Structural Inspection and Appraisal

In November 2021, an office building house was inspected and appraised [3].

3.1 Appearance Quality Inspection

Appearance quality inspection of the appraised house's foundation foundation, super-structure, enclosure system and other areas with inspection conditions: according to the site field survey, the appraised house site has no bad geology and is suitable for construction. There is no corrosion, crispy alkali, loose and obvious subsidence defects in the part of foundation foundation, the house is slightly disconnected around the connection with the ground, the measured maximum tilt value of the house is 13 mm, the tilt rate is 1.37‰, the lateral displacement in the plane of the structure has not exceeded the requirements of the limits of the "Reliability Appraisal Standard for Civil Buildings" [4](limit: ≤ H/300) and the "Code for the Design of Building Ground Foundations" (limit: ≤ 4‰); the house is partially Water seepage occurs in the roof panel, the stucco layer falls off, the wall surface of the balustrade is seriously skinned by moisture, and the floor tiles are seriously hollowed out.

3.2   Testing of Component Material Properties

With reference to the provisions of Technical Procedures for Testing Concrete Strength by Rebound Method, 4 concrete members were sampled on the spot, and the presumed value of concrete strength of 2 members did not meet the specification requirements [5]; with reference to the provisions of Technical Standards for On-site Inspection of Masonry Works, 8 members were sampled on the spot, and the strength of the sampled masonry bricks were MU10 The strength grade of masonry bricks sampled is MU10, which meets the specification requirements; with reference to the Technical Procedures for Masonry Mortar Detection by Penetration Method, 8 members are sampled on site, and the strength conversion values of masonry mortar of all the sampled members do not meet the specification requirements. Specific test results are shown in Table 2.

Table 2. Material performance test results of components

serial number	Component Name Location	Concrete presumptive value (MPa)	Concrete Specification Limits	Component Name Location	Strength class of bricks at current age	Brick Strength Specification Limits	Component Name Location	Conversion value of mortar strength at present age (MPa)	Mortar Strength Specification Limits
1	1F 1/B-D Ring girder	15.8	C15	1F 1/B-D Walls	MU10	MU7.5	1F 1/B-D Walls	1.8	M2.5
2	1F 2/B-D Ring girder	13.5	C15	1F 1–2/D Walls	MU10	MU7.5	1F 1–2/D Walls	1.5	M2.5
3	2F 1–2/D Ring girder	18.3	C15	1F 1–3/B Walls	MU10	MU7.5	1F 1–3/B Walls	1.9	M2.5
4	2F 2/B-DRing girder	13.6	C15	2F 1–2/D Walls	MU10	MU7.5	2F 1–2/D Walls	2.1	M2.5
5				2F 2/B-D Walls	MU10	MU7.5	2F 2/B-D Walls	1.8	M2.5
6				3F 3–4/B Walls	MU10	MU7.5	3F 3–4/B Walls	2.2	M2.5
7				3F 2/B-D Walls	MU10	MU7.5	3F 2/B-D Walls	2.0	M2.5
8				3F 2–3/D Walls	MU10	MU7.5	3F 2–3/D Walls	2.4	M2.5

3.3   Detection of Reinforcement in Members

Referring to the provisions of Code for Quality Acceptance of Construction of Concrete Structural Engineering (GB 50204-2015), on-site sampling of 8 concrete members (4 ring beams + 4 floor slabs) of the reinforcement: member batch of reinforcement protective layer qualified point rate of 78.8%, does not meet the specification requirements of the batch of the qualified rate of 90%; member batch of reinforcement spacing to meet the specification requirements. With the picking method to check the diameter of 3 bars, the diameter of the measured bars meets the minimum requirements of the specification.

3.4 Appraisal Unit Safety Appraisal Ratings

Based on the actual inspection of the house, PKPM software JDJG module is used for modeling and calculation. Referring to the "Reliability Appraisal Standard for Civil Buildings", the structural safety rating is made. Structural safety rating is divided into components, sub-units (foundation, upper load-bearing structure, load-bearing part of the enclosure system), and appraisal unit at three s layer by layer: (1) safety appraisal of masonry structural components, according to the four inspection items of load-bearing capacity, structure, displacement and cracks or other damages which are not suitable to continue to carry: empirically calculated, the walls of the house are under the action of normal load, and the ratio of resistance to effects of the limbs of the wall of the first floor and second floor is less than 1, and the compressive load-bearing capacity is not suitable to continue to carry. The ratio of resistance to effect is less than 1, and the compressive bearing capacity does not meet the specification requirements. The statistical results are shown in Table 3, and the results of the first-floor wall limb compressive bearing capacity checking are shown in left of Fig. 2; the structural items of the structure's main load-bearing member construction are comprehensively evaluated as c_u; the items of the deformation that is not suitable for continued load-bearing are comprehensively evaluated as b_u; and the items of the cracks and other items are comprehensively evaluated as b_u; (2) The subunit safety rating of the foundation is comprehensively evaluated as B_u, the safety appraisal rating of the upper load-bearing structure subunit is assessed as D_u according to its structural bearing function, structural integrity and structural lateral

Table 3. Statistical table for wall limbs with insufficient compressive bearing capacity

Resistance to load effect ratio$(\frac{R}{\gamma_0 S})$	Story	Percentage of wall limb with insufficient compressive bearing capacity
$1.0 > (\frac{R}{\gamma_0 S}) \geq 0.95$ Level: b_u	Sheet	2.8%
$0.95 > (\frac{R}{\gamma_0 S}) \geq 0.90$ Level: c_u		11.1%
$0.90 > (\frac{R}{\gamma_0 S})$ Level: d_u		27.8%
$1.0 > (\frac{R}{\gamma_0 S}) \geq 0.95$ Level: b_u	Second floor	0%
$0.95 > (\frac{R}{\gamma_0 S}) \geq 0.90$ Level: c_u		2.9%
$0.90 > (\frac{R}{\gamma_0 S})$ Level: d_u		11.8%
$1.0 > (\frac{R}{\gamma_0 S}) \geq 0.95$ Level: b_u	Triple	0%
$0.95 > (\frac{R}{\gamma_0 S}) \geq 0.90$ Level: c_u		0%
$0.90 > (\frac{R}{\gamma_0 S})$ Level: d_u		0%

displacement, and the load-bearing part of the enclosure system is assessed as D_u; (3) In conclusion, the safety rating of the housing appraisal unit is assessed as D_{su}, and the safety of the housing seriously fails to comply with the standard regulations for A_{su}, which severely affects the overall bearing, and the safety of the housing structure does not meet the subsequent normal use under the action of static loading. Under static load, the structural safety of the house does not meet the subsequent normal use requirements [6].

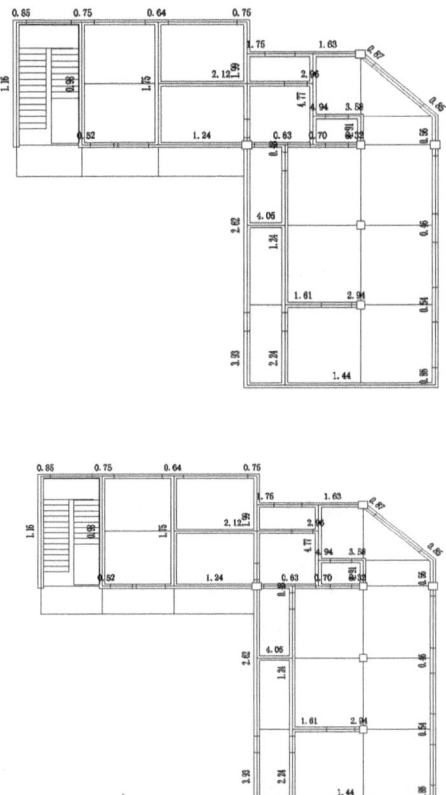

Fig. 2. Results of inspection for compressive bearing capacity and Seismic Calculation of first floor wall limb

3.5   Seismic Identification

With reference to the relevant provisions of the Building Seismic Identification Standard, the house is located in Yunnan Province, built in the 1990s, the seismic intensity of 7°, the design of the basic seismic acceleration value of 0.10 g, the design of the seismic grouping of the third group, according to the seismic appraisal of class B masonry houses: seismic measures identification of structural columns set up, the house's local dimensions, the strength of the mortar, the wall tension reinforcement does not meet the specification requirements. In the seismic load capacity calculation, the ratio of the resistance of the wall limb on the first floor of the house under seismic load (the percentage of the seismic load capacity of the wall is 8.4%) to the load effect is less than 1.0, and the seismic load capacity does not meet the requirements, and the results of the seismic calculation of the wall limb on the first floor are shown in the right of Fig. 2. In conclusion, the comprehensive seismic capacity of the house structure does not meet the standard requirements.

4   Reinforcement and Renovation Program Design

4.1   Engineering Characteristics

This project has a large remodeling surface and complex remodeling content, which is a typical comprehensive remodeling project. In addition to a variety of different reinforcing technology reinforcing trades, but also involves demolition, civil engineering, decoration, water, electricity and other specialties. Various types of work procedures are complex, more cross operations, how to coordinate between different floors, different components, different technologies, different processes, is the project's major difficulties. Another tight schedule, spring and summer remodeling construction efficiency is low, the construction area adjacent to the residential construction noise, to a certain extent, increased the difficulties of strengthening and remodeling.

4.2   Reinforcement and Retrofit Program Design

Parameter value. The design adopts the standard value of uniform live load: 2.0 KN/m^2 for functional rooms, 2.5 KN/m^2 for corridors and stairwells, 2.5 KN/m^2 for bathrooms, 0.5 KN/m^2 for uninhabited roofs, and 0.3 KN/m^2 for basic wind pressure. Floor constant load = plate weight + 1.5 KN/m^2 decoration load, roof constant load = plate weight + 3.0 KN/m^2 decoration load. Strength of materials: strength of brick and mortar is taken according to actual test value, strength of new concrete column is taken as design value C40, strength of high ductility concrete surface is taken as design value C30. Structural calculation parameters are taken as follows: seismic intensity 7°, design basic seismic acceleration 0.10 g, design seismic grouping group III, characteristic period 0.45 s, building site soil category II, ground roughness class B, class B ground floor. Frame structure building (subsequent service life 40 years), seismic defense category C, structural safety class II, structural importance coefficient 1.0.

Reinforcement program design. According to the owner's demand for use and the results of the inspection and appraisal report, the structural form of the house is transformed from a brick-concrete structure to a ground-floor frame brick house structure, with the first floor using columns, beams, and walls for joint load-bearing. The original structure of the beam bottom reinforcement insufficient position using increased cross-section and sticky carbon fiber cloth way to reinforcement. The wall limbs with insufficient seismic load bearing capacity are reinforced by adding high ductility reinforced concrete slab walls. The demolition position of the wall is reinforced with additional reinforced concrete columns and new conversion beams. The position of wall end without structural column is reinforced by adding four sides of reinforced concrete surface layer to form structural column [7]. This project is the existing building reinforcement design, mainly based on the "Building Seismic Reinforcement Technical Regulations", according to the above design ideas modeling calculation analysis, the results meet the expected assumptions. The sketch of the reinforcement design model is shown in Fig. 3.

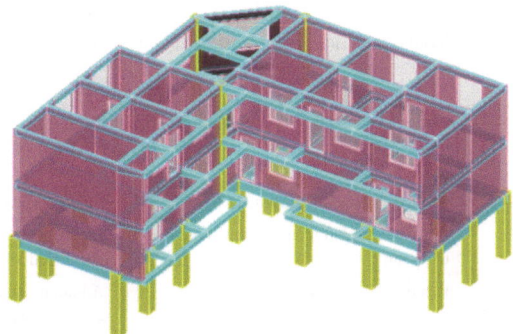

Fig. 3. Sketch of house structural reinforcement design model

(3) Reinforcement and remodeling content and method.

Remove part of the wall of 1–5/A-F axis on the first floor of the house, remove the wall and add new concrete columns with dimensions of 400 mm × 400 mm, 600 mm × 600 mm and other four models, and add new conversion beams with dimensions of 400 mm × 500 mm, 400 mm × 550 mm and other three models. The new columns and beams in the area of 1–5/A-F axes on the first floor, the location and reinforcement diagram are shown in Fig. 1. The new columns are added by adding new independent foundations, and the new columns are added by removing the walls and using the method of enlarging the original foundation cross section. The retained wall of the first floor is reinforced with additional high ductility reinforced concrete slab wall, with the thickness of 50 mm on one side; the wall of the second and third floors is reinforced with high ductility concrete slab wall, with the thickness of 20 mm on one side; the bottom of the beams and plates of the second and third floors, as well as all the staircases, are reinforced with sticky carbon fiber cloth. The contents and methods of house reinforcement and renovation are shown in Table 4, and the construction of the reinforcement site is shown in Fig. 4.

Table 4. House reinforcement and remodeling content and methods

	Reinforcement of the main contents of the renovation	Main Methods of Reinforcement and Retrofitting
Tectonic plates	Second floor and roof	Carbon Fiber Cloth Reinforcement
	No structural columns at the end of the wall	Addition of a four-sided envelope of reinforced concrete facing to form structural columns
	staircases	Carbon Fiber Cloth Reinforcement
Carrying capacity component	Insufficient seismic bearing capacity of walls	Reinforcement of single and double sided high ductility reinforced concrete slab walls
	Inadequate combined floor seismic capacity index	Reinforcement of single and double sided high ductility reinforced concrete slab walls
	Beam cross section too small	Enlargement of beam section
	Insufficient reinforcement at the bottom of the beam	Enlargement of cross-section, attachment of carbon fiber cloth
	Removal of wall locations	Additional reinforced concrete columns and new transition beams

Fig. 4. Construction plan of the reinforcement site

5 Results of House Inspection and Appraisal After Reinforcement and Remodeling

In April 2023, after the completion of strengthening and remodeling, the house was re-inspected and appraised, and no obvious quality defects were found in the exterior quality. The strength presumption value of the new concrete columns meets the design requirements, and the strength presumption value of the high ductility concrete surface layer meets the design requirements, and the test results are shown in Table 5. Modeling calculations under static load, checking the reinforcement of the first floor of the concrete members of the house, the stress ratio of the steel members, and the stress sketch of the lower flange stability check, the frame columns and beams did not show any over-reinforcement; checking the calculation of the axial compression ratio of the first floor of the concrete columns, the axial compression ratio was between 0.05 and 0.17, which was lower than the limit value of 0.05. 0.05–0.17, which is lower than the limit value of 0.9 and meets the specification requirements; the compressive bearing capacity of the wall limb and the height-thickness ratio of the second and third floors meet the specification limits.

Table 5. Concrete strength test result table

Location of components	1F 2/A spine	1F 4/B spine	1F 5/D spine	1F 6/G spine	1F 2–3/Cvertical wall	2F 3/B-Chorizontal wall	2F 4–5/Fvertical wall	3F 1–2/Bvertical wall	3F 6/E-Fhorizontal wall
design strength	C40	C40	C40	C40	C30	C30	C30	C30	C30
(statistics) standard deviation	1.9	1.5	1.8	1.4	1.0	1.1	0.8	1.1	0.8
Presumptive strength value (MPa)	43.5	42.1	42.8	42.7	31.1	31.5	31.9	31.8	31.7

Comprehensive testing indicators and results, with reference to the "Civil Building Reliability Appraisal Standards" to make the structural safety rating: normal use of housing appraisal unit safety rating for B_{su} level, that is, the safety of the house is slightly lower than the standard provisions of the A_{su} level, is not yet a significant impact on the overall load bearing.

Referring to the relevant provisions of "Building Seismic Identification Standard", the relevant parameters of reinforcement design and measured data are taken for modeling and calculation, and the seismic identification is carried out according to Class B ground floor frame brick house: the maximum height of the house, the number of floors, the maximum spacing of seismic transverse wall, the strength of the material, and the overall connecting construction of the house satisfy the requirements of verification of seismic measures of the specification. Seismic load capacity calculation, the house of the first floor of concrete structural members reinforcement to meet the requirements of the concrete column axial compression ratio is less than the normative limit, the second floor, the third-floor wall limb seismic calculation results are shown in Table 6. In summary, the house structure of the comprehensive seismic load capacity to meet the standard requirements.

Table 6. Seismic calculations of wall limbs at the second and third floors

Second floor wall limb	a double-check	β_{i0}	β_{si0}	ξ_{oi0}	η_{i0}	β_{i90}	β_{si90}	ξ_{oi90}	η_{i90}
		2.85	1.99	0.0478	2.27	2.61	1.83	0.0545	2.31
	a double-check	β_{i44}	β_{si44}	ξ_{oi44}	η_{i44}	λ			
		2.79	1.95	0.0508	2.30	1.00			
Triple wall limb	a double-check	β_{i0}	β_{si0}	ξ_{oi0}	η_{i0}	β_{i90}	β_{si90}	ξ_{oi90}	η_{i90}
		5.05	3.53	0.0277	2.33	4.14	2.90	0.0301	2.50
	a double-check	β_{i44}	β_{si44}	ξ_{oi44}	η_{i44}	λ			
		4.70	3.29	0.0286	2.41	1.00			

Notes: β_i-Average floor seismic capacity index; β_{si}-Combined seismic capacity index of strengthened floors; ξ_{oi}-Base area ratio for seismic walls; η_i--Enhancement factor for seismic capacity; λ --Intensity impact factor.

6   Conclusion

This paper comprehensively considers the old brick-concrete structure of the house itself problems, reinforcement goals, subsequent use of the life and the owner's needs and other factors, an office building as an example, choose a reasonable reinforcement design program targeted to the house seismic structural measures do not meet the standard requirements, the comprehensive seismic bearing capacity is insufficient to enhance the use of functionality and other issues to deal with the house reinforced before the assessment of the safety level of the Dsu level, after the reinforced transformation to increase the safety level to Bsu level, effectively guaranteeing the normal safe use of the house to enhance the comprehensive seismic capacity and the use of functionality. The safety grade of the house was Dsu grade before reinforcement, and the safety grade was upgraded to Bsu grade after reinforcement and remodeling, which effectively guarantees the normal safe use of the house and improves the comprehensive seismic capacity of the house and the use function [8]. This project has a lot of transformation contents, complex reinforcement process, and many kinds of cross work, through the comparison of testing and appraisal data before and after reinforcement and transformation, it verifies that the reinforcement design scheme is reasonable and feasible, and it will provide reference for other similar projects in the future.

Funding Resources. 1. Scientific Research Fund project of Yunnan Education Department in 2019 (2019J0485); 2. Scientific Research Fund project of Yunnan Education Department in 2022 (2022J1362);3. The second batch of vocational education teacher teaching innovation team in Yunnan Province - Civil engineering inspection technology professional team of Yunnan Land and Resources Vocational College.

References

1. Huanqiang, W.: Reinforcement design method for remodeling of brick-concrete houses–an example of a brick-concrete house remodeling project. Xiamen Sci. Technol. **04**, 23–30 (2022)
2. Liu, C., Lu, Y., Li, J.: Example of structural testing identification and reinforcement of office building. Constr. Technol. **47**(11), 1036–1039 (2016).https://doi.org/10.13731/j.issn.1000-4726.2016.11.023
3. Zhang, M.Q., Liu, B.D., Li, Y.Z., et al.: Safety and seismic testing and identification of typical rural masonry structure houses. Eng. Seismic Reinforcement Retrofitting **38**(03), 124–129 (2016). https://doi.org/10.16226/j.issn.1002-8412.2016.03.020
4. Xingyuan, L., Chengjiu, F., Yang, L.: Discussion on the application of reliability appraisal standard for civil buildings GB50292-2015. Chongqing Constr. **16**(05), 54–57 (2017)
5. Kovler, K., Wang, F., Muravin, B.: Testing of concrete by rebound method: Leeb versus Schmidt hammers. Mater. Struct. **51**(5), 138 (2018)
6. Li, L., Lu, J., Zhang, H., et al.: Structural inspection and safety appraisal of multi-story brick-concrete houses before overall remodeling. Build. Struct. **37**(S1), 243–244+247 (2007).https://doi.org/10.19701/j.jzjg.2007.s1.079
7. Zheng, L.: Design of remodeling and reinforcement of brick-concrete office buildings. Constr. Technol. **40**(01), 50–52 (2009)
8. Repapis, C., Zeris, C., Vintzileou, E.: Evaluation of the seismic performance of existing RC buildings II: a case study for regular and irregular buildings. J. Earthq. Eng. **10**(03), 429–452 (2006)

Modelling and Analysis of the Seattle Space Needle Based on ANSYS

Shu Chen[1](✉), Hanwen Chen[1], Xinyi Xie[2], and Linqing Xu[3]

[1] Department of Civil and Environmental Engineering, Zhejiang University-University of Illinois Urbana-Champaign Institute, Haining 314400, China
shu.20@intl.zju.edu.cn

[2] Department of Mechanical Engineering, Rice University, Houston 77005, USA

[3] International Department, Chongqing BI Academy, Chongqing 401120, China

Abstract. Modal analysis is an effective method that helps engineers understand how a building responds to different types of dynamic loads, such as wind, earthquakes, or vibrations caused by human activities. It provides valuable insights into a building's natural frequencies, mode shapes, and damping characteristics. In this paper, a simplified model of the Space Needle Tower is established using ANSYS SPACE CLAIM, followed by modal analysis using ANSYS WORKBENCH. For the purpose of simplification, the equivalent density of the model is set to the actual mass of the Space Needle tower divided by the volume of the model. The space needle tower model consisting of 11 components is then divided into 126144 nodes and 66287 units. The first 6 modes of deformation and the corresponding natural frequencies are obtained. In the end, the paper points out the tower's potential weaknesses or areas that require reinforcement and provides potential solutions to improve the stability of the tower under dynamic loads.

Keywords: ANSYS · Modal Analysis · Natural Frequency · Finite Element Method

1 Introduction

The Space Needle in Seattle, Washington, is one of the most popular tourist spots around the world. Approximately 1.3 million guests visit the Space Needle per year, and nearly 60 million visitors have visited the tower since it opened in 1962. The tower was designed by John Graham & Company.

The structure of Space Needle comprises a steel tripod, with each of the three legs pinched just above the middle of their height and topped by a multi-level tophouse reminiscent of a flying saucer. This tophouse consists of five stacked layers: a revolving restaurant, a mezzanine level, an observation deck, a mechanical equipment level, and at the tower's pinnacle, an elevator penthouse. The structure was also originally crowned by a 15 m natural gas torch. On the basis of its structural features, the tower is 42 m wide, weighs 8660 metric tons, and is built to withstand winds of up to 32 0 km/h and earthquakes of up to 9.0 magnitude, as strong as the 1700 Cascadia earthquake [1].

© The Author(s) 2024
G. Feng (Ed.): ICCE 2023, LNCE 526, pp. 810–823, 2024.
https://doi.org/10.1007/978-981-97-4355-1_81

Earlier studies on the Space Needle tower mainly focus on the analysis of its acoustical design and commercial values. However, the natural frequencies and deformation type of the tower are merely studied. Thus, this study focus on calculation of natural frequencies and analysis of the tower's weakness [2].

2 Method

2.1 Model Establishment

For the purpose of simplification, the model constructed in the ANSYS WORKBENCH is composed with 11 components [3].

Component 1 is a triangle base with a total height of 154 m, which can be further divided into three platforms. The length of the six support bars on platform one (the lower one) is 11.3 m, and the thickness is 2.3 m. The length of the support bars on platform two (the middle one) is 7.1 m, and the thickness is 1.2 m. The highest platform is composed with two cylinders whose edges are linearly connected. The diameter of the upper cylinder is 33 m, and the diameter of the lower cylinder is 21 m (Fig. 1).

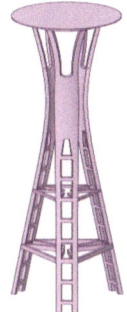

Fig. 1. Component 1

As shown in Fig. 2, the curvature of the upper part of the tripod changes from 0.011 m^{-1} to 0.008 m^{-1} from top to bottom.

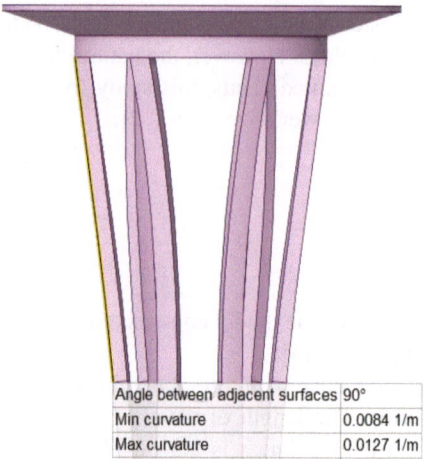

Angle between adjacent surfaces	90°
Min curvature	0.0084 1/m
Max curvature	0.0127 1/m

Fig. 2. Component 1 Upper Part Dimension

The curvature of the middle part ranges from 0.0002 m^{-1} to 0.0078 m^{-1}, see Fig. 3.

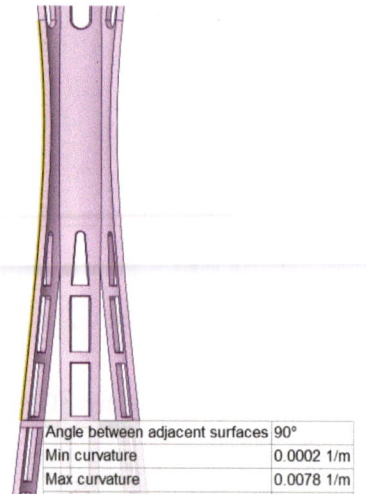

Angle between adjacent surfaces	90°
Min curvature	0.0002 1/m
Max curvature	0.0078 1/m

Fig. 3. Component 1 Middle Part Dimension

As shown in Fig. 4, the curvature of the lower part is constant, which is 0.0003 m^{-1}. And each hollow cuboid has a length of 9 m, a width of 3.3 m, and a thickness of 1.2 m.

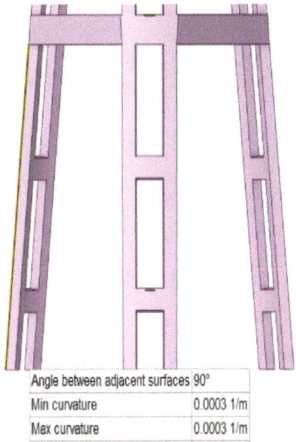

Angle between adjacent surfaces	90°
Min curvature	0.0003 1/m
Max curvature	0.0003 1/m

Fig. 4. Component 1 Lower Part Dimension

In addition to Component 1, component 2 is another support for the Space Needle Tower (See Fig. 5). It was constructed by creating a circle and then using the circular pattern button to make it a hexagon with an edge length of 3.4 m and finally extruded it to a height of 151.5 m.

Fig. 5. Component 2

Component 3 is composed with two cylinders (See Fig. 6). A big circle with a diameter of 21.1 m is first created and then it is stretched to a thickness of 0.6 m. The smaller one is created by repeating the same process, its diameter is 15.1 m and its thickness is 1.6 m.

Fig. 6. Component 3

As shown in Fig. 7, component 4, 5, 6 have the same shape but are different in size. Three types of edges are named Edge # 1, 2, and 3 based on their length, from shortest to longest. The dimensions of each component are recorded in Table 1.

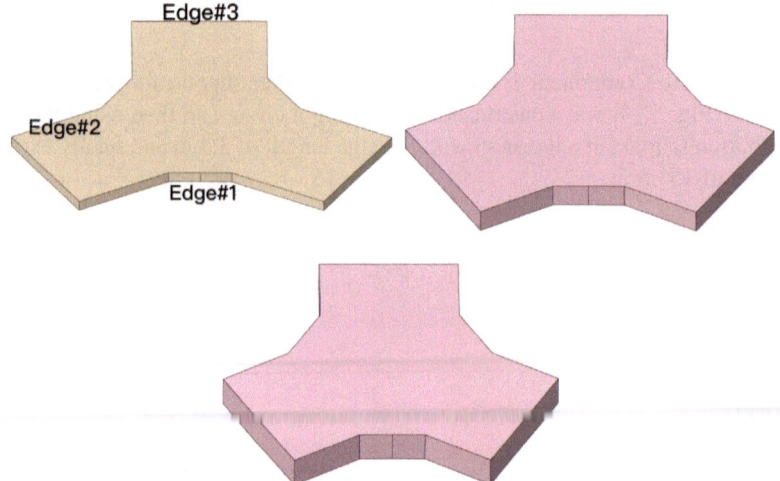

Fig. 7. Component 4, 5, 6

Table 1. Dimensions of Component 4, 5, 6

Component	Edge #1	Edge #2	Edge #3	Height	Angle between Edge #1&2
#4	7.3 m	12 m	15.7 m	1.2 m	150 °
#5	7.3 m	7.4 m	15.7 m	2.4 m	150 °
#6	7.3 m	7.4 m	15.7 m	3.5 m	150 °

Figure 8 shows component 7, the top of the tower, which was composed with four cylinders. The cylinders are named Cylinder #1, 2, 3, 4 from top to bottom. The curvature range of the connecting surface between cylinders 3 and 4 is 0.27 m^{-1} to 0.47 m^{-1}. The dimensions of each cylinder are recorded in Table 2.

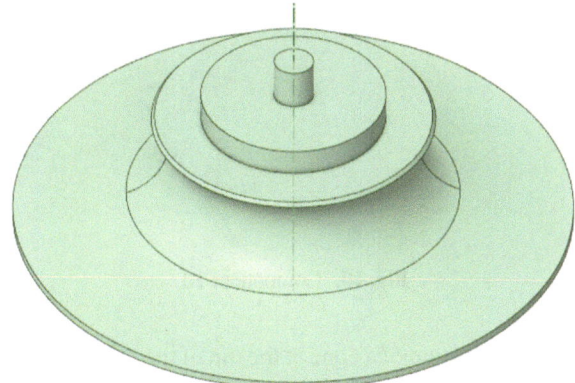

Fig. 8. Component 7

Table 2. Dimensions of the Cylinders in Component 7

Cylinder	Diameter	Height
#1	2.1 m	2.2 m
#2	10 m	1.4 m
#3	15 m	0.3 m
#4	30.4 m	0.4 m

Figure 9 shows component 8&9, which are two cylinders with the same diameter of 30.2 m, the height of component 8 is 3.1 m and the height of component 9 is 2.8 m.

Fig. 9. Component 8, 9

As shown in Fig. 10, the longer edges of component 10 have a length of 11.7 m and the shorter edges are 4.5 m. The angle between two adjacent short edges is 120°.

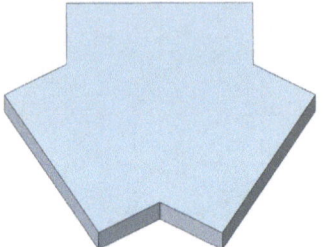

Fig. 10. Component 10

Component 11 is the frustum of a cone at the top of the model, which has a diameter of 0.6 m at the upper bottom, 2.1 m at the lower bottom, and a height of 11.7 m.

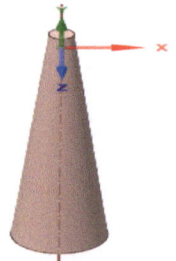

Fig. 11. Component 11

All components are combined into a whole after being established by using the combine and joint button. The final modeling result is shown in Fig. 12.

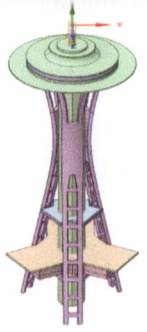

Fig. 12. Final Model of the Space Needle

2.2 Material Settings

The main material of the Space Needle Tower is structural steel. The actual total weight is 9550 tons (excluding the foundation), and the volume of the model is 25958.26 m³. We chose structural steel as the only material, and as a simplification, we set the equivalent density of the model to the actual mass of the building divided by the model volume [4]:

$$\rho = \frac{M}{V} = \frac{9550000kg}{25958m^3} = 367.9kg/m^3 \tag{1}$$

The corresponding material settings are shown in Table 3.

Table 3. Material Settings

Item	Value
Material	Structural Steel
Volume	26000 m³
Mass	9550 ton
Equivalent Density	367.9 kg/m3
Young's Modulus	200 GPa
Poisson's Ratio	0.3

2.3 Boundary Conditions

The bottom of the space needle tower is actually connected to a foundation, as our model didn't consider the foundation, the seven faces at the bottom of the model are considered fixed and supported (see Fig. 13) [5].

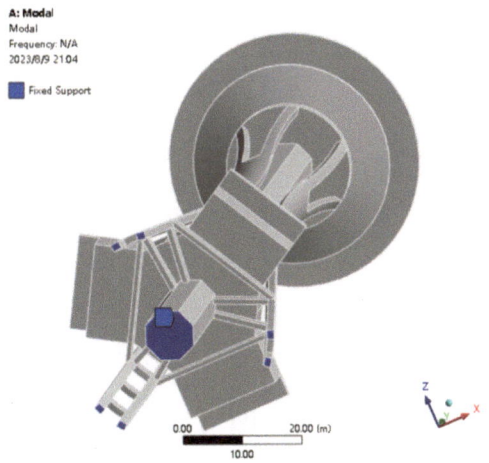

Fig. 13. Boundary Conditions

2.4 Mesh Generation

Appropriate mesh generation is the key to accurate model analysis. To determine the appropriate grid size, an independence study was conducted to test the dependence of the result on mesh density. By changing the mesh size, the relationship between the number of nodes and the fundamental frequency is shown in Fig. 14 [6, 7].

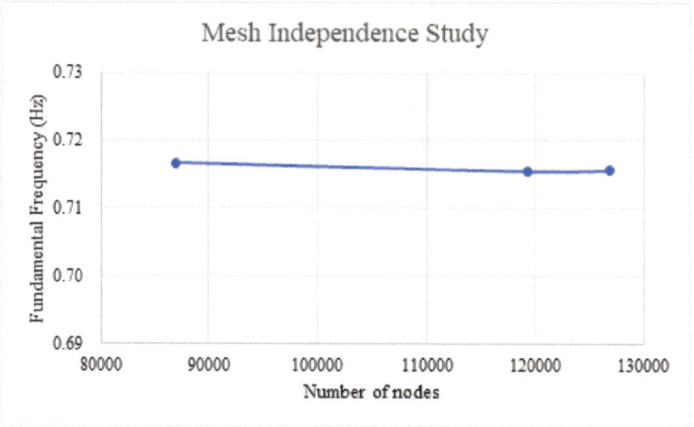

Fig. 14. Mesh Independence Study

According to Fig. 11, the fundamental frequency hardly varies with the number of nodes in the range of 80000 to 130000 nodes, so the weak dependence can be ignored. In order to obtain the most accurate results, a grid size of 0.767 m was used. The model was then divided into 126144 nodes and 66287 units (see Fig. 15).

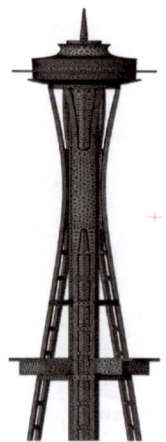

Fig. 15. Mesh Generation Result

3   Results

The modal analysis was carried out by using ANSYS Student Version. Six modes and the corresponding natural frequencies were obtained (see Fig. 16).

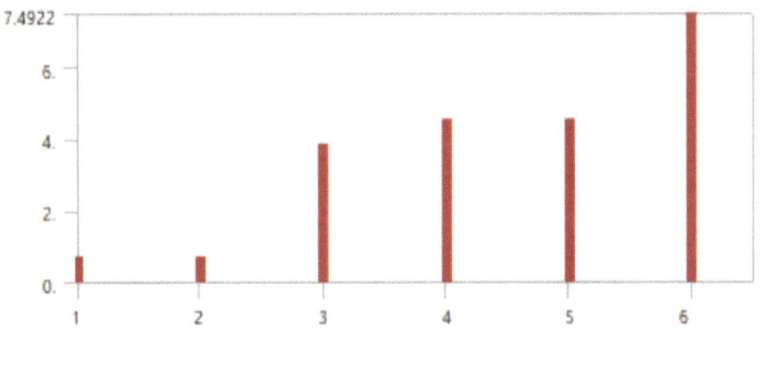

Fig. 16. Frequencies of Different Modes

For six modes with different frequencies, deformation analysis was carried out separately, and the result was shown in Fig. 17, 18, 19, 20, 21 and 22.

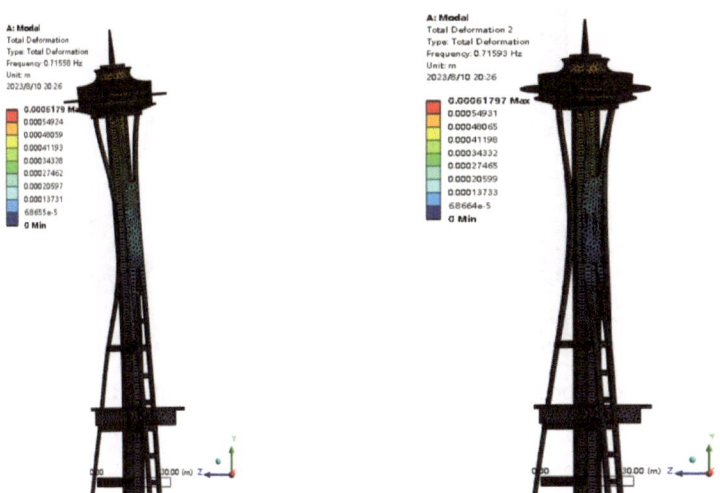

Fig. 17. Deformation Diagram Mode 1   **Fig. 18.** Deformation in Diagram in Mode 2

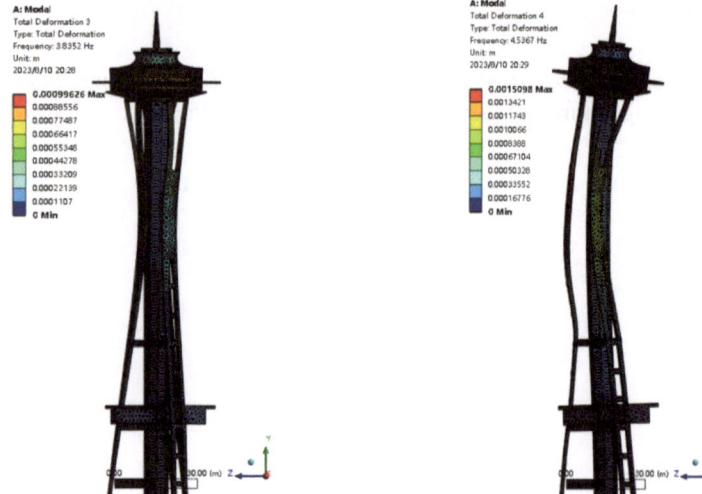

Fig. 19. Deformation Diagram in Mode 3   **Fig. 20.** Deformation in Diagram in Mode 4

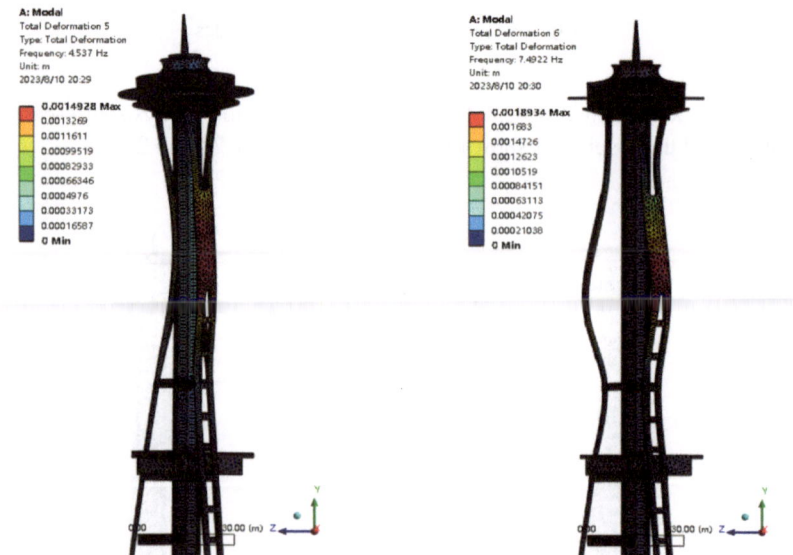

Fig. 21. Deformation Diagram Mode 5   **Fig. 22.** Deformation Diagram in in Mode 6

Figure 17, 18, 19, 20, 21 and 22 shows the deformation of six modes with different frequencies. According to the Figs, the area where the maximum normalized deformation occurs and the tower's weakest points can be clearly identified (See Table 4).

Table 4. Deformation at Six Different Frequencies

Mode	Frequency (Hz)	Maximum Normalized Deformation (cm)	Type of Deformation
1	0.71558	0.06179	Z-axis Bending
2	0.71593	0.061797	X-axis Bending
3	3.8352	0.099626	Y-axis Torsion
4	4.5367	0.15098	Z-axis Sway
5	4.537	0.14928	X-axis Sway
6	7.4922	0.18934	Y-axis Sway

4 Conclusions

4.1 Summary

Modal analysis is performed on the simplified model of the Space Needle tower. Proper meshing and boundary conditions are applied during the simulation process. Six mode types with corresponding natural frequencies have been obtained. The natural frequencies range from 0.72 Hz to 7.49 Hz, the corresponding maximum normalized deformation is from 0.062 cm to 0.189 cm. Component 1, the triangle base has the greatest degree of deformation in all six modes. Component 4, 5, 6 (which together form the first platform) have the smallest degree of deformation. In mode 1&2, bending is the most obvious and primary form of deformation [7]. Within a component, the part at a higher altitude tends to show a larger degree of deformation. In mode 3, torsion centered on Y-axis appears and greatest deformation occurs at the highest platform (UFO shaped) throughout the entire tower. In mode 4, 5, 6, sway occurs and the location where the maximum deformation occurs is in the middle of the two connection points (component 1& platform 1, component 1 and platform 3).

4.2 Discussions

According to an empirical study, the natural frequency of tall buildings can be simply given by:

$$f = \frac{46}{L} \text{ (Hz, L is the height of the building in m} \tag{2}$$

Thus, the empirical natural frequency should be around 0.3 Hz and the difference between the value of mode 1 and the empirical value is relatively small. As shown in the deformation diagrams, the triangular structure and the complex curves of the steel bars (component 1) effectively make the overall structure safer and more stable. It should be pointed out that for the purpose of simplification, the complicated surface structure and the foundation of the tower is not considered during the modeling process, and the weight of the tower is distributed uniformly throughout the whole tower. Actually, the designers took into account the characteristics of seismic waves and specially used a heavy and solid foundation (5850 ton, including 250 tons of refined steel) to frame the entire tower. Therefore, the real tower must exhibit a much more stable state than the model does since its center of gravity is much lower. These steps of simplification are the major causes of inaccuracies in the simulation results. As for future maintenance, one practical method is to install a dynamic monitoring system at the top of the Space Needle tower. Continuously recorded signals can automatically identify the natural frequencies of key vibration modes of the structure and provide engineers with latest information as references.

Acknowledgement. This research project is guided by Professor Ronaldo I. Borja from Stanford University and his teaching assistants. We are grateful for their endless willingness to support our study.

References

1. Fernandez, S., Caruso, N., Paige, N.: Seattle Space Needle: How Did They Build That? (2021)
2. Bruck, D.C., Penzell, J.: The Seattle space needle renovation: acoustical design considerations and challenges at 600 ft above ground level. J. Acoust. Soc. Am. **150**(4), A248 (2021)
3. Meng, S., Kang, W., Jiang, P., Liu, W.: Research on modular modeling technology of transmission tower based on ANSYS. In: 2012 Asia-Pacific Power and Energy Engineering Conference, pp. 1–4. IEEE (2012)
4. Wei, L.: ANSYS Civil Engineering Application Examples. China Water Power Press, Beijing (2007)
5. Hu, G., Zhu, S., Gao, R., Xiao, L.: Modeling and analysis of Shanghai tower based on ANSYS Workbench. In: 2021 4th International Symposium on Traffic Transportation and Civil Architecture (ISTTCA), pp. 312–319. IEEE (2021)
6. Jalammanavar, K., Pujar, N., Raj, R.V.: Finite element study on mesh discretization error estimation for Ansys workbench. In: 2018 International Conference on Computational Techniques, Electronics and Mechanical Systems (CTEMS), pp. 344–350. IEEE (2018)
7. Yan-Zhong, J., Guan-nan, Z., Jun-Feng, B., Xun-Jiang, Z.: Finite element analysis and test study of dynamic behavior of single-tower and double-face cable-stayed bridge. In: 2010 International Conference on Mechanic Automation and Control Engineering, pp. 2814–2817. IEEE. Connor JJ (2003). Introduction to Structural Motion Control (2010)
8. Dym, C.L., Rossmann, J.S.: Introduction to engineering mechanics: a continuum approach. In: Gentile, C., Guidobaldi, M., Saisi, A., (eds.) (2015, July). Structural Health Monitoring of a Historic Masonry Tower. In 2015 IEEE Workshop on Environmental, Energy, and Structural Monitoring Systems (EESMS) Proceedings, pp. 168–173. IEEE (2008)

Reimagining Household Cooking: A Critical Assessment of Improved Cookstoves Implementation for Sustainable Development

Yejashva Kaneriya[1], Devendra Dohare[2], and Milad Khatib[3,4](✉) 🆔

[1] Civil Engineering and Applied Mechanics Department, Shri G.S., Indore, India
[2] Institute of Technology and Science, Indore, Madhya Pradesh, India
[3] School of Engineering, Lebanese International University, Beirut, Lebanon
milad.khatib@liu.edu.lb
[4] Civil and Environmental Engineering Department, University of Balamand,
Souk El Ghareb, Lebanon

Abstract. Food is one of the most important components for the survival of humans. Women suffer the most impacts of the cooking harmful gases and particulate matters that are 2.5 microns (PM2.5) or less such as PM10. Improved cookstoves are used to reduce firewood consumption, lower the smoke released from the kitchen, help women improve their health, and conserve native trees in the area. Sustainably managed biomass, unlike the burning of fossil fuels, is considered carbon neutral since it does not contribute to the overall carbon emissions in the atmosphere. In 2015, the United Nation mentioned 17 Sustainable Development Goals (SDGs) to be achieved by 2030. This research aims to assess an improvement in the cookstove implementation level within the clean development, by reducing the carbon emissions following the Indian perspective for the program of better cookstoves.

Keywords: Sustainability · Carbon Offsets · Improved Cookstoves · Clean Development · Emission Reductions · India

1 Introduction

Clean cooking provides a highly effective and affordable approach to tackle the urgent problems of pollution, climate change, and biodiversity loss. Around 2.4 billion individuals, which accounts for approximately one-third of the global population, depend on inefficient stoves or open fires fuelled by coal, biomass (like wood, animal dung, and crop waste), and kerosene for their cooking needs. In 2020 alone, this practice causing household air pollution (HAP), which was responsible for approximately 3.2 million deaths, including over 237,000 deaths of children under 5 years [1].

Inefficient and frequently lacking in appropriate ventilation, traditional cooking stoves consume the majority of solid fuels. One of the primary advantages of biomass is to be a renewable energy source. Properly managed biomass is considered carbon-neutral since it does not contribute to an overall increase in carbon emissions in the atmosphere, unlike burning fossil fuels [2].

G. Feng (Ed.): ICCE 2023, LNCE 526, pp. 824–833, 2024.
https://doi.org/10.1007/978-981-97-4355-1_82

The production and consumption of fuelwood and charcoal contribute to the release of 1–2.4 billion metric tons of greenhouse gases (GHGs) in the form of carbon dioxide equivalents (CO2e) annually. Improper use of fuelwood is a significant human activity that degrades the environment and harms forests. The primary contributors to greenhouse gas (GHG) emissions were the burning of wood and diesel. Moreover, the incomplete combustion of firewood releases substantial amounts of black carbon (soot) and carbon-based greenhouse gases. In rural areas of India, approximately 90% of the population lacks access to modern fuels, leading to a significant annual usage of 150 million tonnes of biomass for cooking purposes [3].

Eliminating the use of solid fuels and kerosene for cooking will help to address environmental degradation and mitigate global warming that reduce smoke exposure impacts (3rd Sustainable Development Goal, SDG3). Furthermore, this clean cooking can be used to increase productivity impacts SDG8 on economic growth by saving time and money, in addition to many of the other SDGs, which established by the United Nations on having access to clean, modern, sustainable, and affordable energy [4].

2 Methodologies

The "improved" cooking technology consume less fuel and smokeless from incomplete combustion. The co-benefits include enhanced fuel savings, enhanced health outcomes, reduced indoor air pollution, and positive environmental effects. Solid-fuel stoves, or Improved Cookstoves (ICS), are produced to be more effective than conventional biomass technology, which saves fuel, and reduces particle' emissions. Recent studies, such as Project Surya, have demonstrated that improved forced draft stoves can reduce Black Carbon (BC) concentrations nearly twice as much as natural draft stoves in controlled kitchen environments, and contribute to climate mitigation efforts [5]. By reducing the amount of biomass required for cooking, these stoves alleviate pressure on forests and other vegetation sources promote sustainable resource management techniques, and help to conserve natural resources. Furthermore, it offer significant health advantages by reducing indoor air pollution (IAP) that associated with conventional cooking methods. The inclusion of grating in ICS is a vital feature that enhances combustion and efficiency. Typically, a 20 × 20 cm cast iron grating is attached at the bottom of the ICS, with a channel generated beneath it to facilitate airflow into the cookstove. This design promotes complete combustion of firewood, and reduces smoke emissions. Furthermore, the unimpeded airflow within the channel improves combustion efficiency. Cookstoves are categorized in the literature based on various factors, including the extent of modifications, performance requirements in terms of efficiency and emissions, and the type of fuel used [6].

Cookstove projects can lead to the generation of two types of carbon credits: Certified Emissions Reduction (CER) credits and Voluntary Emissions Reduction (VER) credits. These credits are recognized under the Kyoto Protocol, which is overseen by the United Nations Framework Convention on Climate Change (UNFCCC), which is responsible for issuing CER credits. Both types of projects undergo a rigorous testing and validation process to demonstrate their ability to offset a certain number of emissions [7]. Under the Clean Development Mechanism (CDM), two different approaches are available for cookstove projects AMS II.G and AMS I.E. The AMS II.G is applicable when a more efficient cookstove introduced to reduce the consumption of non-renewable biomass. Furthermore, the AMS I.E approach applies when renewable technologies, such as biogas or solar cookers, are implemented to replace the use of non-renewable biomass [8]. A comparison between those two approaches is presented in Table 1.

Table 1. CDM-AMS II.G versus CDM-AMS I.E.

Program	CDM-AMS II.G	CDM-AMS I.E
Measure of biomass fuel consumption	KPT, WBT, or CCT (Controlled Cooking Test)	
Baseline Scenario	Consider using fossil fuels to meet cooking and heating demands	
Covered Project types	Installing enhanced thermal devices can reduce the consumption of fossil fuels biomass	Renewable energy technology can replace fossil fuels biomass as a thermal energy source

This methodology measures the reduction of emissions by calculating the amount of nonrenewable biomass or fossil fuels that are consumed less. For non-renewable biomass, the methodology is the same as that outlined in AMS-II.G. Such reduction in a certain year (ER_y) is related to the tons weight saved for the biomass ($By,vings,i,j$) per each project i and batch j. The percentage of woody biomass ($fNRB,y$) with some unpredictability reduction (ud), which considered as not renewable, should be determined, in addition to the CO_2 ($EFwf,CO2$) and the non-CO_2 ($EFwf,non CO2$) emission factors (t CO_2/TJ). It can be calculated through the following equation:

$$ER_y = \sum_i \sum_j B_{y,savings,i,j} \times N_{0,i,j} \times n_{y,i,j} \times \mu_y \times f_{NRB,y} \times NCV_{biomass} \times (EF_{wf,co2} + EF_{wf,nonco2}) \times Adj_{LE} \times (1 - u_d)$$

(1)

where:

$NCVbiomass$ is the net calorific value of the non-renewable woody biomass (replaced or reduced). $N0$, is the project devices (i) and batch (j) assigned. However, ny,i,j is the assigned remain operating ratio of the same project devices and batch within the same year, using certain adjustment factor (μy) during the same year, and another adjustment factor ($AdjLE$) that is related to the leakage for the non-renewable woody biomass saved.

The Chetak cookstove is specifically designed to burn dung cakes, firewood, and agricultural waste as fuel. To accommodate small utensils, the Chetak cookstove has undergone certain modifications, with a thermal efficiency of 21% (Fig. 1). Udairaj cookstove, an enhanced biomass cookstove has two pots, with the first having a diameter of 17 cm, and 15 cm for the second pot. The wood burns efficiently and safely in this cookstove. It offers a thermal efficiency of 25%, ensuring effective utilization of the fuel's heat [9].

The Patsari cookstove incorporates several design elements to enhance its efficiency and reduce smoke emissions [10]. The design of the Patsari cookstove includes tunnels that lead to secondary chambers with smaller combustion areas. These secondary chambers are used for low-power cooking tasks. The combustion gases are expelled through these tunnels, contributing to improved combustion efficiency and reducing fugitive smoke emissions. This design ensure to make it more efficient and environmentally friendly [11].

The Lakech and the Merchaye stoves offer options for efficient charcoal cooking, but with some differences in their specifications and characteristics. The Lakech Cookstoves have been estimated to have a CO2e (carbon dioxide equivalent) emission mitigation potential of 0.14 tCO2e (tons carbon dioxide equivalent) per stove per year. This indicates the potential reduction in greenhouse gas emissions compared to traditional cooking methods. However, the Merchaye stove has a higher estimated CO2e emission mitigation potential compared to the Lakech stove, with a value of 0.33 tCO2e per stove per year [12].

By fostering the processes of gasification (the conversion of solid biomass into combustible gases) and pyrolysis (the decomposition of organic material by heat in the absence of oxygen), TLUD and IDD gasifier stoves are made to maximize the burning of biomass fuel, such as wood or agricultural waste (Fig. 2, 3, 4 and 5). These stoves' Map Predicted Fire (MPF) enables the production of combustible gases to produce cooking heat. Top-lit up-draft gasifier stoves are highly regarded for their ability to effectively burn the volatile gases produced by biomass fuels when heated in the absence of oxygen. These stoves are designed with reduced primary air inlets to allow for a natural draft, which pulls the gases up the chimney. This combustion process ensures that the gas fuels would be released into the atmosphere with lower emissions. In addition, it improves indoor air quality, benefiting the health and well-being of users. TLUDs, with their small fuel batch sizes and moderate burn rates, offer exceptional fuel efficiency and minimal emissions during the combustion process. Additionally, TLUD stoves contribute to lower greenhouse gas emissions, aligning with environmental sustainability goals. Furthermore, their fuel efficiency lead to cost savings and better fuel economy in cooking methods [13].

Fig. 1. Modified Single Pot Chetak Cookstove [9].

Fig.2. Patsari wood-burning cookstove [10].

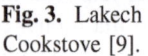

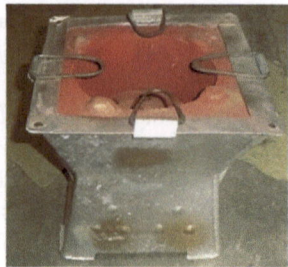

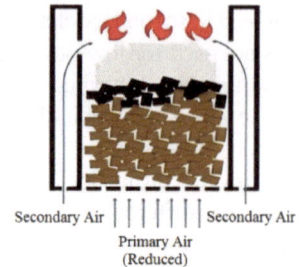

Secondary Air Secondary Air
Primary Air
(Reduced)

Fig. 3. Lakech Cookstove [9].

Fig. 4. Merchaye Cookstove [9].

Fig. 5. Top-Lit Up-Draft Technology [13]

In performance studies of biomass cookstoves, various tests are commonly used to assess their effectiveness and efficiency in real-world cooking scenarios. Two commonly used tests are the Water Boiling Test (WBT) and the Kitchen Performance Test (KPT).

(a) The Water Boiling Test (WBT) proposed to measure the emissions during cooking and to assess the stove uses fuel to heat water in a cooking pot. The main goal of the WBT is to evaluate the performance of various cookstoves by comparing them according to parameters like thermal efficiency (ηth) and emissions mass, furthermore, to make sure that these stoves adhere to the standards set by various governmental and non-governmental organizations. It is divided into three distinct phases: the cold-start high-power in which water is heated to room temperature, then brought to a boil in the second phase (the hot-start high-power), and simmering is the last phase (Fig. 6), which entails holding the boiling water at a temperature below 45 min (Fig. 7) [14].

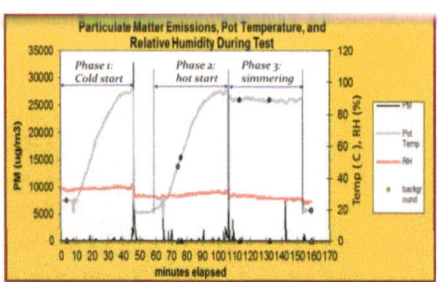

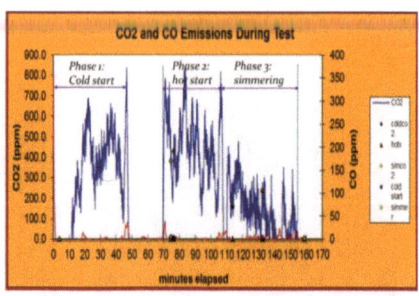

Fig. 6. Particulate Matter (PM), Pot temperature, and relative humidity during WBT [14]

Fig. 7. CO_2 and CO emissions during WBT [14]

To assess the emissions performance of biomass cookstoves, during WBT, a Laboratory Emissions Monitoring System (LEMS) is often used (Fig. 8) that includes sensors for measuring carbon monoxide (CO), carbon dioxide (CO_2), and Particulate Matter (PM). These sensors play a crucial role in evaluating the emissions generated by the stove under test. The CO sensor in the LEMS utilizes an electrochemical cell with two electrodes to measure the amount of CO produced by the stove. The CO_2 sensor in the LEMS uses non-dispersive infrared technology to measure the amount of CO_2 produced by the stove. This sensor generates a voltage that corresponds to the CO_2 concentration, enabling the quantification of CO_2 emissions. The gravimetric system, which provides a direct measurement of PM concentration, measures total PM utilizing filter-based sampling in Fig. 8.

Fig. 8. Laboratory Emissions Monitoring System with the Nozzle type Energy Efficient Cookstove mounted inside the metal hood [14]

The performance parameters of biomass cookstoves can be usefully revealed by emissions testing utilizing LEMS during water boiling tests. These emissions are significant considerations for assessing the environmental impact and possibility of decreasing Indoor Air Pollution (IAP) [14].

(b) Kitchen Performance Test (KPT) is considered the major field-based technique for evaluating the stove improvements that affect household fuel use. The aim of this test is to study the impact of the improved stove(s) on fuel consumption in actual kitchen contexts of families and to assess qualitative elements of stove functionality through household surveys. It includes qualitative analyses of stove performance and user satisfaction in addition to quantitative measurements of fuel usage in order to meet these objectives [15].

2.1   Firepower (FP)

In biomass cookstoves, it is a thermal performance parameter that measures the amount of thermal energy generated in a given time period (kW). Firepower is calculated by dividing the amount of thermal energy produced (kJ) by the time taken to generate that energy (s). The firepower of a cookstove is an important characteristic as it represents its heating capacity (cooking efficiency and performance). A higher firepower indicates a larger heat output, which can result in faster cooking times. It is important to consider other thermal performance parameters such as specific fuel consumption, turndown ratio, as well as emission performance parameters such as emission factors of pollutants, to evaluate the overall performance of biomass cookstoves and their environmental impact [16].

2.2 Thermal Efficiency (H)

It is a measure of how effectively it converts the energy generated during the combustion of biomass fuel into usable thermal energy for cooking. It is typically expressed as a percentage. It is evaluated by dividing the actual energy used by the pot and its contents for cooking by the firepower available due to the combustion of the fuel. This parameter is important since it reflects the efficiently of the cookstove in utilizing the energy from the biomass fuel for cooking. Higher thermal efficiency indicates a more efficient cookstove that is able to utilize more of the available energy for cooking, resulting in lower fuel consumption and reduced environmental impact [16].

2.3 Specific Fuel Consumption (SFC)

It measures the amount of dry fuel, typically expressed in grams, required to produce a specific unit of output. It is commonly reported as grams of fuel per kilogram of unit output (g/kg) by dividing the mass of the dry fuel used during the test by the mass of the unit output. The specific fuel consumption provides insights into the efficiency of a cookstove by quantifying the amount of fuel required to produce a given output. Lower values of SFC indicate better fuel efficiency, as less fuel is needed to achieve the desired cooking result [16].

2.4 Emission Factor (EF)

It evaluates the mass of a specific pollutant emitted per unit of fuel burned per kilojoule (kJ), or per megajoule (MJ) of energy released during cooking. It quantifies the pollutant emissions associated with the combustion of the fuel. The indoor concentration of a pollutant refers to the amount of exposure experienced by the user per cubic meter of air in the room or cooking area (ng/m3), which impact indoor air quality and the health of individuals exposed to them. Improved biomass cookstoves are known for their fuel efficiency. They can consume 20–50% less fuel compared to conventional biomass cookstoves, reduce lower emissions, and potentially improved indoor air quality. By using less fuel, improved cookstoves can also provide economic benefits by reducing fuel expenses for households and communities [16].

3 Result and Discussion

Despite variations in thermal efficiency, the Patsari stove exhibited several advantages over conventional stoves. During the low-power period, the Patsari stove demonstrated lower specific fuel usage compared to the high-power cold start phase. Furthermore, households that solely relied on fuelwood experienced an important decrease in energy consumption of when using the Patsari stove. It proves to be a more efficient and effective option, resulting in substantial energy savings for households relying on fuelwood.

Efficiency can be defined in various ways depending on different aspects of stove operation: Combustion efficiency, Heat transfer efficiency, Efficacy of the control, Pot effectiveness, and Efficiency of the cooking process.

3.1 Housing Improvements

Household exposure to air pollution can be considerably reduced by making improvements and adjustments to the housing. This can be accomplished by taking steps like building new or larger kitchen windows, adding flues and smoke hoods, widening roof spaces, elevating cooking surfaces to waist height, and dividing cooking rooms from other living areas. The promotion of home renovations for better health outcomes has typically been done through education and information dissemination. The success of housing renovation programs for bettering health in poor nations depends on the strict enforcement of building codes to ensure that changes are made to reduce exposure to indoor air pollution. Developing nations may achieve major advancements in housing conditions and reduce exposure to household air pollution by giving priority to the implementation of building standards, which will improve the health of their populations.

3.2 Behavioral Change

Recent analyses of behavioral modification programs have demonstrated that these tactics can dramatically lower the exposure of young children to household air pollution. Cooking outside, spending less time in the kitchen, keeping the door open while cooking, avoiding leaning over the fire while attending to cooking, not carrying children while cooking, and keeping kids away from the kitchen area are some of the advised behavioral changes to reduce exposure to household air pollution. To include household air pollution and clean cooking education in the training programs for frontline health workers, the Global Alliance for Clean Cookstoves (GACC) and other country alliances should contact the national health authorities. This may help raise public awareness of healthy cooking methods and encourage their adoption, which may minimize household exposure to air pollution and improve health outcomes in developing nations.

3.3 Carbon Offsets

Three billion people use traditional cookstoves or open flames to cook their food, which is a serious environmental and public health concern and a barrier to sustainable economic growth. Cleaner and more effective cooking technologies are available, but many households in underdeveloped nations cannot afford them or cannot get them in their local marketplaces. Businesses are now using carbon offsets or credits to help low-carbon development in developing nations, notably in the clean cooking industry, to address this problem. Cookstoves and fuels that are cleaner and more effective have the potential to achieve important annual reducing carbon dioxide (CO_2) emissions, in addition to enhancing livelihoods, quality of life, and health. Several SDGs could be aided by projects including efficient cookstoves. A more complete combustion is achieved by efficient cookstoves, which results in fewer emissions of methane and other pollutants, as well as using less fuel and/or switching to fuel that is less GHG-intensive, since few types of stoves meet the World Health Organization's (WHO) definition of "clean" for carbon monoxide and particulate matter emissions.

4 Conclusion

The benefits of Improved Cookstoves (ICS) are two-fold, with positive impacts reported at both the household and regional levels. At the household level, users of ICS will have less smoke, reduced risk of burns, improved taste of food, and decreased expenditures on fuelwood. At the regional level, widespread adoption of ICS can contribute to mitigating forest degradation and result in significant savings of tons of CO_2 emissions per year, making ICS an effective and efficient means of securing carbon storage in forests. The effectiveness of cookstove has been evaluated into reducing greenhouse gas (GHG) emissions, measuring emissions reductions, reducing exposures to harmful pollutants, and improving health and well-being as highlighted in the Intergovernmental Panel on Climate Changes (IPCC) Sixth Assessment Report, which underscores the need to protect the environment and combat climate change.

References

1. WHO, Household Air Pollution and Health (2022)
2. Ezzati, M., Kammen, D.: Indoor air pollution from biomass combustion and acute respiratory infections in Kenya: an exposure–response study. Lancet **358**, 619–624 (2001)
3. Kulkarni, J., et al.: Reduction in firewood consumption due to implementation of improved cookstoves in Melghat tiger reserve. India Asia-Pac. J. Rural Dev. **32**(1) (2022)
4. Nerini, F.F., et al.: Mapping synergies and trade-offs between energy and the sustainable development Goals. Nat. Energy **3**, 10–15 (2018)
5. Patange, O.S., et al.: Reductions in indoor black carbon concentrations from improved biomass stoves in rural India. Environ. Sci. Technol. **49**, 4749–4756 (2015)
6. Clean V. Cookstoves: Impact and Determinants of Adoption and Market Success (2021)
7. Cox, P.: Analysis of Cookstove Change-Out Projects Seeking Carbon Credits Environmental Sustainability Clinic of the University of Minnesota Law School (2011)
8. Loo, C.M., et al.: Assessing the climate impacts of cookstove projects: issues in emissions accounting. Challenges Sustain. **1**(2), 53–71 (2013)
9. Panwar, N.L., et al.: Mitigation of greenhouse gases by adoption of improved biomass cookstoves. Mitig. Adapt. Strat. Glob. Change **14**, 569–578 (2009)
10. Perez Maldonado, I.N., et al.: Indoor air pollution in Mexico. InTech (2011)
11. Johnson, M., et al.: Quantification of carbon savings from improved biomass cookstove projects. Environ. Sci. Technol. **43**(7), 2456–2462 (2009)
12. Mamuye, F., et al.: Emissions and fuel use performance of two improved stoves and determinants of their adoption in Dodola, southeastern Ethiopia. Sustain. Environ. Res. **28**(1), 32–38 (2017)
13. Berry, J., et al.: Design for Mass Production and Dissemination of Clean Cookstoves in Developing Countries (2019)
14. Okafor, I.F.: Energy efficient biomass cookstoves: performance evaluation, quality assurance and certification science. J. Energy Eng. **7**(4), 54–62 (2019)
15. Bailis, R., et al.: Kitchen Performance Test (KPT) version 3.0 Household Energy and Health Programme, Shell Foundation (2007)
16. Sutar, K.B., et al.: Energy Efficiency, emissions and adoption of biomass cookstoves renewable and sustainable energy. Reviews **41**, 1128–1166 (2015)

Design and Application of Multidimensional Monitoring and Early Warning System for Tunnel Safety

Hongliang Jiang[1,2,3], Chaobo Lu[1,2,3(✉)], Tingwei Yang[1,2,3], and Chunfa Xiong[1,2,3]

[1] Guangxi Transportation Science and Technology Group Co., Ltd, Nanning 530007, Guangxi, China
64059089@qq.com

[2] Guangxi Highway Tunnel Safety Warning Engineering Research Center, Nanning 530007, Guangxi, China

[3] Guangxi Key Lab of Road Structure and Materials, Nanning 530007, Guangxi, China

Abstract. The tunnel project has a large one-time investment, a long construction period, and potential safety hazards. During the operation period, structural cracks, voids, deformation, dislocation, water leakage and other diseases may occur, which greatly reduces the use function of the tunnel and threatens its safety in construction and operation. Therefore, it is necessary to carry out the development and research of tunnel safety monitoring and early warning system to pre-perceive and predict the working state and disease characteristics of the tunnel structure and timely warn. Aiming at the working state and disease characteristics of the tunnel, this paper discusses how to obtain the dynamic changes of the working state and disease characteristics of the key parts of the tunnel in real time, and optimizes the design of the tunnel safety monitoring and early warning system. A multi-dimensional monitoring and early warning system for tunnel safety is recommended, which can control the working state and disease changes of the tunnel in real time, and improve the service quality and service life of the tunnel.

Keywords: Tunnel Safety · Multi-Dimensional Monitoring · 3-D Laser · Oblique Photography

1 Introduction

At present, the scale and mileage of tunnel construction in China have become the world's largest, but compared with other non-underground projects, the probability of tunnel safety accidents and quality accidents is higher, and the consequences are serious [1]. Although the construction technology and level of tunnel structure have been improved year by year, due to the influence and restriction of various factors such as geological, topographic and hydrological characteristics, tunnel engineering may suffer from water leakage, deformation, structural crack and other diseases during operation and service [2], which greatly reduces the quality and service life of the tunnel, has a great impact on society and traffic, and even threatens the safety of life and property.

G. Feng (Ed.): ICCE 2023, LNCE 526, pp. 834–843, 2024.
https://doi.org/10.1007/978-981-97-4355-1_83

Many scholars at home and abroad have studied the early warning and prediction of tunnel engineering. Such as Zhang Junru [3] developed the tunnel health monitoring and intelligent information management evaluation system, and put forward the new development trend of tunnel intelligent monitoring technology and information management system. Based on Java language and Access system, Fang Yu [4] developed a tunnel construction monitoring data analysis and management system. Frank et al. [5] proposed the problems and solutions of wireless monitoring in tunnel and other fields. TSUNO et al. [6] introduced a tunnel monitoring method based on wireless sensor network. ALBAM et al. [7] introduced a tunnel section analysis method based on rapid measurement of digital image displacement. Frank et al. [5] proposed the problems and solutions of wireless monitoring in tunnel and other fields. Bennett et al. [8] carried out continuous deformation monitoring research on tunnel structure based on wireless sensor network.

In view of the current situation of tunnel structure and operation and maintenance requirements, considering the problems of tunnel structure cost input, technical status and service life, this paper proposes a multi-dimensional monitoring and early warning system for tunnel operation safety based on the working status and disease characteristics of the tunnel during operation. The dynamic changes of the working status and disease characteristics of the key parts of the tunnel are monitored and analyzed, and the technical status of the tunnel structure is dynamically controlled in real time. The potential tunnel operation diseases are identified and judged accurately and in advance, so as to avoid the structural damage of the tunnel caused by "small danger and big risk "and" small disease and big disease. "In order to improve the durability and prolong the service life of the tunnel structure, the development trend of the tunnel structure disease is perceived in advance and the early warning is carried out in time, so as to serve the early warning and scientific decision-making of the tunnel danger, provide scientific data support for the tunnel design and preventive maintenance, and promote the progress and development of the tunnel structure material design technology.

2 Multidimensional Monitoring and Early Warning Technology

Traditional monitoring methods rely on a large number of manual participation, and the degree of automation participation is not enough. The safety monitoring and evaluation technology of tunnel structure has the problems of time-consuming and low accuracy, and the treatment methods of tunnel diseases mostly stay on the surface treatment of tunnel structure, ignoring the pre-perception and prediction of tunnel structure diseases. Multi-dimensional monitoring and early warning technology is based on a series of monitoring methods such as multiple monitoring sensors + three-dimensional laser scanning + tilt photography combined with BIM model, which is a comprehensive and full-life cycle comprehensive intelligent monitoring and early warning technology integrating structural monitoring, condition assessment and safety early warning. Compared with the traditional monitoring methods, multi-dimensional monitoring and early warning technology can dynamically control the technical status of tunnel structure in real time, accurately identify and judge the potential tunnel operation diseases in advance, pre-perceive and predict the development trend of tunnel structure diseases and give early warning in time. It can significantly improve the durability of tunnel structure and prolong its service life, and has a wide application prospect.

3 Research on Design of Monitoring and Early Warning System

3.1 General System Design

Based on the current situation of tunnel structure and operation and maintenance requirements, a multi-dimensional monitoring and early warning system for tunnel safety is designed by comprehensively considering the cost input, technical status and service life of tunnel structure. It can control the working state and disease changes of tunnel in real time, improve the durability of tunnel structure and prolong the service life, and provide scientific data support for tunnel risk early warning and scientific decision-making, as well as tunnel design and preventive maintenance. The multi-dimensional monitoring and early warning system of tunnel safety is composed of multi-dimensional monitoring unit, monitoring data acquisition and transmission unit and monitoring data platform. The multi-dimensional monitoring unit is composed of multi-dimensional monitoring sensor + three-dimensional laser scanning + oblique photography + BIM model. The multi-dimensional monitoring unit is arranged in the key monitoring section of the tunnel, and the monitoring data acquisition and transmission unit is connected with the monitoring data platform respectively to intelligently monitor the dynamic changes of the working status and disease characteristics of the key parts of the tunnel in real time and give early warning in time.

3.2 System Unit Design

The multi-dimensional monitoring unit is composed of multiple monitoring sensors + 3D laser scanning + oblique photography + BIM model. According to the engineering characteristics of the tunnel structure, the GNSS displacement monitoring system, crack meter, strain gauge and other multi-dimensional monitoring sensors are arranged in the uphill slope of the tunnel portal and the end wall of the tunnel portal. Based on the three-dimensional point cloud data obtained by 3D laser scanning and the panoramic aerial photography data of the tunnel structure obtained by oblique photography, combined with the BIM model, the multi-dimensional monitoring model of the tunnel structure is constructed to intelligently monitor the dynamic changes of the working state and disease characteristics of the key parts of the tunnel. The monitoring data acquisition and transmission unit is a data acquisition control box that transmits the monitoring data obtained by the safety monitoring unit to the safety monitoring data platform through the communication transmission module. The data acquisition control box and the communication transmission module are jointly arranged with the multi-monitoring unit. The monitoring data platform through the data interface server, data application server and data analysis server, the tunnel structure monitoring data storage, data screening, display management, emergency plan processing and intelligent early warning analysis.

4 Application Research of System

4.1 Project Overview

A tunnel is a separated tunnel. The right line of the tunnel is 812 m long and the left line is 760 m long. The stratum in the tunnel area is mainly the third stratum of the Indosinian period. The main structures are the Nading fault and the Nalang-Datong fault, which are non-new active faults and have little influence on the engineering geological stability of the tunnel area. The surface water in the tunnel area is mainly gully water, and there are gullies at the entrance and exit of the tunnel. The gullies have a great influence on the tunnel. The groundwater in the tunnel area is mainly pore water and bedrock fissure water in the Quaternary overburden. The tunnel layout is shown in Fig. 1.

Fig. 1. Site layout of a tunnel

4.2 Monitoring Site Arrangement

The design concept of a tunnel is "people-oriented, safety first, and overall safety in the whole life cycle." Considering the functions of highway tunnels, driving safety, natural environment and other factors, taking into account the requirements of construction and later tunnel structure operation, based on the multi-element monitoring sensor, the whole life cycle health monitoring is mainly carried out for the parameters such as secondary lining concrete stress, secondary lining crack, vault subsidence and peripheral convergence. Based on the three-dimensional point cloud data obtained by three-dimensional laser scanning and the panoramic aerial photography data of tunnel structure obtained by oblique photography. Combined with BIM model, a multi-dimensional monitoring model of tunnel structure is constructed to intelligently monitor the dynamic changes of working status and disease characteristics of key parts of tunnel. The layout diagram of the monitoring points of the tunnel monitoring section, the overall design diagram of the tunnel health monitoring, and the on-site installation diagram of the tunnel multi-sensor are shown in Fig. 2, 3 and 4.

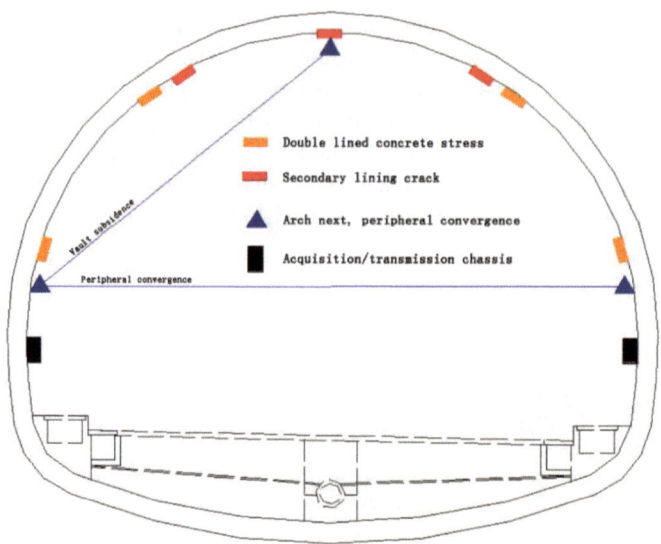

Fig. 2. The schematic diagram of measuring point layout of tunnel monitoring section

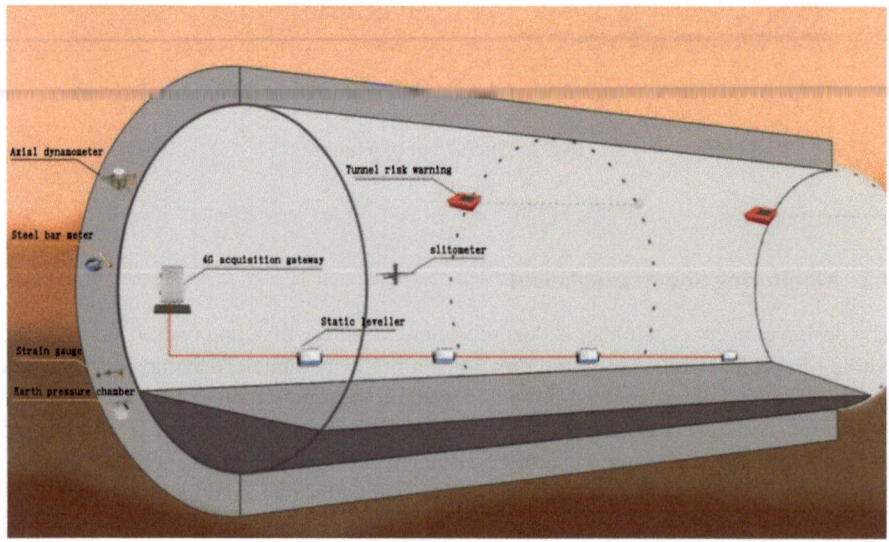

Fig. 3. Overall design diagram of tunnel health monitoring

Fig. 4. Tunnel multi-sensor on-site installation diagram

4.3 Monitoring and Early Warning Platform

The multi-dimensional monitoring and early warning platform for tunnel safety can query, browse, manage and analyze the monitoring data obtained by all multi-monitoring sensors and the three-dimensional point cloud data obtained by three-dimensional laser scanning in real time. Based on the tunnel BIM model, the monitoring layout of multiple monitoring equipment can be queried and displayed in real time. Based on the three-dimensional point cloud data obtained by three-dimensional laser scanning, all tunnel deformation monitoring data can be consulted and displayed in real time, and the analysis and research of tunnel deformation monitoring data can be carried out. Based on a series of monitoring methods such as multiple monitoring sensors + three-dimensional laser scanning + BIM model, the tunnel safety multi-dimensional monitoring and early warning system is an all-round and full-life cycle comprehensive intelligent monitoring and early warning system integrating structural monitoring, condition assessment and safety early warning.The BIM model diagram of the tunnel outside, the sensor of the tunnel monitoring section and the BIM model of the acquisition chassis are shown in Fig. 5, 6. The three-dimensional point cloud data map of the tunnel and the tunnel deformation monitoring data map based on the tunnel safety multi-dimensional monitoring and early warning system are shown in Fig. 7, 8.

Fig. 5. BIM model diagram outside the tunnel

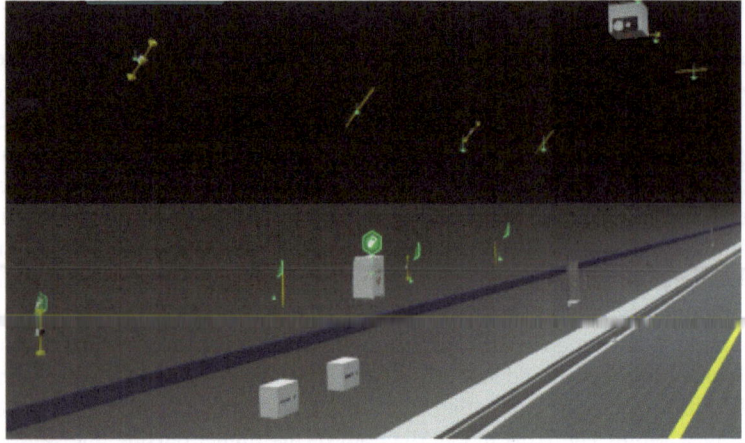

Fig. 6. Tunnel monitoring section sensor and acquisition chassis BIM model diagram

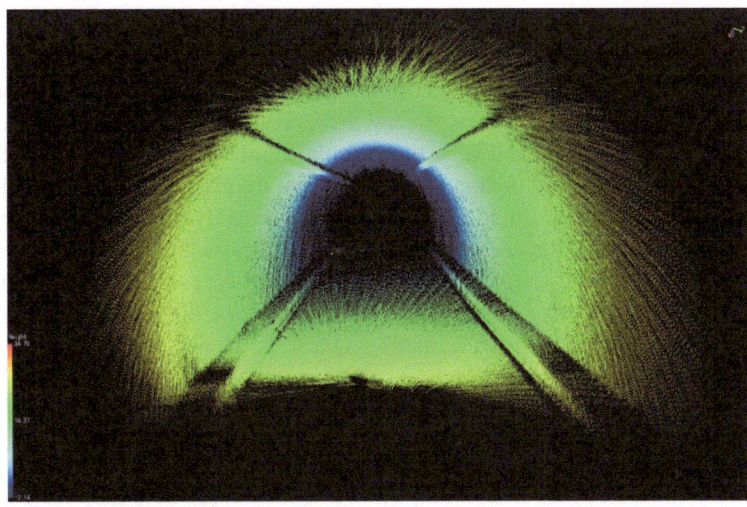

Fig. 7. Tunnel three-dimensional point cloud data map

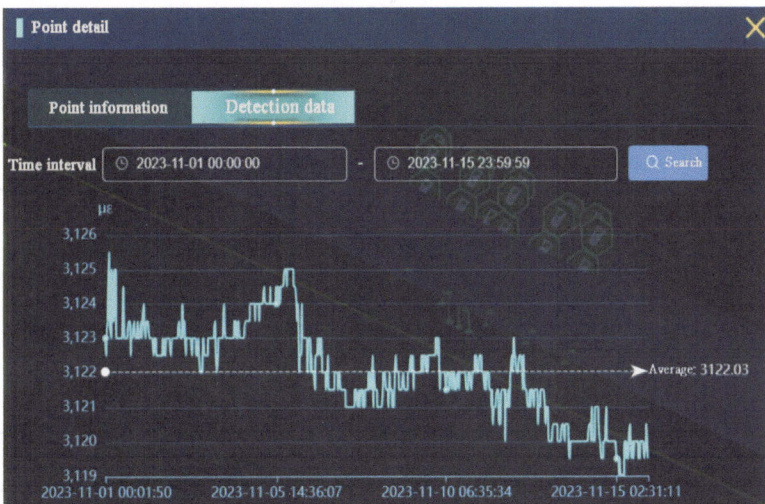

Fig. 8. Tunnel deformation monitoring data map based on tunnel safety multidimensional monitoring and early warning system

5 Conclusion and Suggestion

In this paper, the tunnel engineering project has a large one-time investment, long construction period, potential safety hazards, and structural cracks and deformation may occur during the operation period. The multi-dimensional monitoring and early warning system for tunnel safety is designed and applied. Compared with the traditional monitoring technology, the multi-dimensional monitoring and early warning technology is a comprehensive intelligent monitoring and early warning system for the full life cycle of the structure monitoring, condition assessment and safety warning. The system can query and display the monitoring and layout of multiple monitoring equipment in real time, query and manage all monitoring data in real time, control the technical status of tunnel structure in real time, identify and judge the potential tunnel operation diseases accurately and in advance, perceive and predict the development trend of tunnel structure diseases in advance and give early warning in time. It can significantly improve the durability of tunnel structure, prolong the service life, provide scientific decision-making services for tunnel danger warning and scientific decision-making, provide scientific data support for tunnel design and preventive maintenance, and promote the progress and development of tunnel structure material design technologyQ2.

Acknowledgements. The research of this article is supported by Guangxi's key research and development plan (Guike 22080027).

Declaration of Competing Interest. The authors declare that they have no known competing financial interests or personal relationships that could have appeared to influence the work reported in this paper.

References

1. Li, T.: Research and platform development of multi-objective fine management mode of tunnel based on data-driven [D].Southwest Jiaotong University (2019)
2. Qihu, Q.I.A.N.: State, issues and relevant recommendations for security risk management of China's underground engineering. Tunnel Constr. **37**(3), 251–263 (2017)
3. Zhang, J., Yan, B., Gong, Y., et al.: Research status and prospects of intelligent monitoring technology and information management system for tunnel engineering. Chinese J. Undergr. Space Eng. 17(2), 567–579 (2021)
4. Fang, Y., Liu, K.: Design and development of analysis and management system of tunnel monitoring data. Tunnel Constr. **30**(03), 231–234+245 (2010)
5. Stajano, F., Hoult, N., Wassell, I., et al.: Smart bridges, smart tunnels: transforming wireless sensor networks from research prototypes into robust engineering infrastructure. Ad Hoc Netw. **8**(8), 872–888 (2010)
6. Tan, G.H., Chua, K.G.: Developing an operational automated real time tunnel monitoring system
7. Kristof, M., Wim, S., Gerrit, F., et al.: Anomaly detection in long-term tunnel deformation monitoring. Eng. Struct.
8. Bennett, P.J., Kobayashi, Y., Soga, K., et al.: Wireless sensor network for monitoring transport tunnels. Proc. Inst. Civ. Eng. Geotech. Eng.Geotech. Eng. **163**(3), 147–156 (2010)

Author Index

© The Editor(s) (if applicable) and The Author(s) 2024
G. Feng (Ed.): ICCE 2023, LNCE 526, pp. 845–848, 2024.
https://doi.org/10.1007/978-981-97-4355-1